Cytochrome *c*

A MULTIDISCIPLINARY APPROACH

LIST OF CONTRIBUTORS

Numbers in parentheses indicate the pages on which the authors' contributions begin.

Paul A. Adams (635), MRC Biomembrane Research Unit, Department of Chemical Pathology, University of Cape Town Medical School, Cape Town, South Africa

Douglas S. Auld (203), Department of Chemistry, CB3290 Venable and Kenan Laboratories, University of North Carolina, Chapel Hill, North Carolina 27599-3290

David A. Baldwin (635), Materials Science Division, Council for Scientific and Industrial Research, Pretoria, South Africa

Stephen F. Betz (203), Department of Chemistry, CB3290 Venable and Kenan Laboratories, University of North Carolina, Chapel Hill, North Carolina 27599-3290

Hans Rudolf Bosshard (373), Biochemisches Institut der Universität Zürich, Winterthurerstrasse 190, CH-8057, Zürich, Switzerland

Gary D. Brayer (103), Department of Biochemistry and Molecular Biology, University of British Columbia, 2146 Health Sciences Mall, Vancouver, British Columbia V6T 1Z3, Canada

Michael A. Cusanovich (489), Department of Biochemistry, Biosciences West, University of Arizona, Tucson, Arizona 85721

Ben De Kruijff (449), Centre for Biomembranes and Lipid Enzymology, University of Utrecht, Padualaan 8, 3584 CH Utrecht,The Netherlands

Bill Durham (573), Department of Chemistry and Biochemistry, University of Arkansas, Fayetteville, Arkansas 72701

L. Liliana Garcia (203), Department of Chemistry, CB3290 Venable and Kenan Laboratories, University of North Carolina, Chapel Hill, North Carolina, 27599-3290

Colin Greenwood (611), School of Biological Sciences, University of East Anglia, Norwich NR4 7TJ, United Kingdom

Marilyn R. Gunner (347), Department of Physics, City College of New York, Convent Avenue and 138 Street, New York, New York 10031

L. H. Guo (317), Department of Chemistry, University of Rochester, Rochester, New York 14627

Peter Hildebrandt (285), Max Planck Institut für Strahlenchemie, Stiftstrasse 34–36, D-45470 Mülheim, Germany

Sharon E. Hilgen–Willis (203), Department of Chemistry, CB3290 Venable and Kenan Laboratories, University of North Carolina, Chapel Hill, North Carolina 27599–3290

H. Allen O. Hill (317), Inorganic Chemistry Laboratory, Oxford University, South Parks Road, Oxford OX1 3QR, United Kingdom

Barry Honig (347), Department of Biochemistry and Molecular Physics, Columbia University, 630 West 168 Street, New York, New York 10032

Wilco Jordi (449), AB–DLO, Bornsesteeg 65, 6708 PD Wageningen, The Netherlands

Nenad M. Kostić (593), Department of Chemistry, Iowa State University, Ames, Iowa 50011

Emanuel Margoliash (3), Laboratory for Molecular Biology, Department of Biological Sciences, University of Illinois at Chicago, 840 West Taylor, Chicago, Illinois 60607-7020

Helder M. Marques (635), Centre for Molecular Design, Department of Chemistry, University of Witwatersrand, Johannesburg, South Africa

George McLendon (317),Department of Chemistry, University of Rochester, Rochester, New York 14627

Terry E. Meyer (33), Department of Biochemistry, University of Arizona, Tucson, Arizona 85721

Francis S. Millett (475), (573), Department of Chemistry and Biochemistry, University of Arkansas, Fayetteville, Arkansas 72701

Michael E. P. Murphy (103), Department of Biochemistry and Molecular Biology, University of British Columbia, 2146 Health Sciences Mall, Vancouver, British Columbia V6T 1Z3, Canada

Barry T. Nall (167), Department of Biochemistry, University of Texas Health Science Center, 7703 Floyd Curl Drive, San Antonio, Texas 78284

Scott H. Northrup (543), Department of Chemistry, Tennessee Technological University, Cookeville, Tennessee 38505

Yvonne Paterson (397), Department of Microbiology, University of Pennsylvania, 209 Johnson Pavilion, Philadelphia, Pennsylvania 19104-6076

Gary J. Pielak (203), Department of Chemistry, CB3290 Venable and Kenan Laboratories, University of North Carolina, Chapel Hill, North Carolina 27599-3290

Abel Schejter (3), (335), Sackler Institute of Molecular Medicine, Sackler Medical School, Tel-Aviv University, Tel-Aviv 69978, Israel

Robert A. Scott (515), Center for Metalloenzyme Studies, University of Georgia, Athens, Georgia 30602-2556

Gordon Tollin (489), Department of Biochemistry, Biosciences West, University of Arizona, Tucson, Arizona 85721

Carmichael J. A. Wallace (693), Department of Biochemistry, Dalhousie University, Sir Charles Tupper Medical Building, Halifax, Nova Scotia B3H 4H7, Canada

Michael T. Wilson (611), Department of Chemistry and Biological Chemistry, University of Essex, Wivenhoe Park, Colchester C04 3SQ, United Kingdom

Cytochrome c

A MULTIDISCIPLINARY APPROACH

EDITED BY

ROBERT A. SCOTT
University of Georgia

A. GRANT MAUK
University of British Columbia

UNIVERSITY SCIENCE BOOKS
Sausalito, California

About the cover:
Horse heart cytochrome *c* viewed in the standard orientation with the exposed heme edge (red space-filling) facing the viewer. In this orientation, the methionine-80 ligand is on the left and the histidine-18 ligand is on the right of the heme (ligands are black space-filling).

University Science Books
55 D Gate Five Road
Sausalito, CA 94965
Fax (415) 332-5393

Production manager: *Steve Peters*
Manuscript editor: *Kristin Landon*
Designer: *Robert Ishi*
Compositor: *ASCO Trade Typesetting Ltd.*
Printer and Binder: *Maple-Vail Book Manufacturing Group*

This book is printed on acid-free paper.

Library of Congress Cataloging-in-Publication Data

Scott, Robert A., 1953–
      Cytochrome *c* : a multidisciplinary approach / Robert A. Scott and A. Grant Mauk.
          p.   cm.
      Includes bibliographical references and index.
      ISBN 0–935702–33–4
        1. Cytochrome *c*.   I. Mauk, A. Grant (Arthur Grant), 1947–
II. Title.
QP671.C8S36   1995
574.1'2454 -- dc20   94–40359
                                                    CIP

Printed in the United States of America
10 9 8 7 6 5 4 3 2 1

Contents

Preface

In every scientific field, certain systems become such foci of investigation that they evolve into defacto paradigms that provide an unending reservoir of inquiry and insight as new plateaus of experimental and theoretical achievement emerge. Whatever pragmatic or historical factors were responsible for cytochrome c attaining paradigm status, it is clear that the process was autocatalytic: The more studies of cytochrome c that appeared, the more reason to use it as a benchmark in development of new experimental and theoretical strategies.

This book was conceived in an effort to collect in one place expert surveys of contemporary work concerning the continuing evolution of this metalloprotein paradigm. It showcases the wide variety of techniques that have been and are being used to understand its properties. Thus, this collection not only provides insights into structure/function relationships exhibited by cytochrome c, but it also illustrates how a combination of seemingly divergent physical, biochemical, and genetic techniques is required to establish an understanding of these relationships. As such, we expect this book to be useful not only to experts in the study of cytochrome c, but also to be generally useful to students and practitioners of inorganic biochemistry. We believe that this book complements other books published on cytochrome c by providing a broad perspective from many authors with diverse backgrounds.

The original idea for an edited collection of reviews of cytochrome c was provided by Harry B. Gray and Ivano Bertini, and we acknowledge their encouragement to pursue this project. We also acknowledge the people who aided in the final stages of editing and publishing this book: Bruce Armbruster, Jane Ellis, and Stephen Peters. Most of all, we acknowledge the work, dedication, and patience of the authors who contributed to what turned out to be a major editorial effort.

Robert A. Scott
A. Grant Mauk

Cytochrome *c*

A MULTIDISCIPLINARY APPROACH

PART

I

Introduction

1

How Does a Small Protein Become So Popular?: A Succinct Account of the Development of Our Understanding of Cytochrome c

EMANUEL MARGOLIASH

Laboratory for Molecular Biology
Department of Biological Sciences
University of Illinois at Chicago

ABEL SCHEJTER

Sackler Institute of Molecular Medicine
Sackler Medical School
Tel-Aviv University

I. Introduction

When one follows the development of our knowledge of cytochrome c, a small electron transport heme protein, from its rediscovery by David Keilin in 1924 to the present, one can only be struck by the wide array of fields of biochemistry, molecular biology, and cell biology, often quite unrelated to each other, in which it holds a significant place. Cytochrome c has become an extremely popular protein, and its popularity gives no sign of stabilizing, but rather of continuing to increase. The objective signs of this situation are quite obvious: *Chemical Abstracts* lists 10,145 articles in which cytochrome c is mentioned in title or abstract, from 1967 to October 1992, as compared to 5,148 for ribonuclease, a small basic protein of very similar size to cytochrome c and which could probably be used equally well in the protein model studies for which cytochrome c has been regularly employed. It would appear that such popularity feeds on itself, and the more that is known about a protein, the more likely it is to be chosen for the next study with novel equipment or novel conceptual approaches. This is not unreasonable, since the better an object is understood, the less likely it is that errors will be made when it is examined in previously untested ways.

One can adduce many reasons for such widespread popularity. Possibly among the most important are the following: (i) Cytochrome c is a small, stable protein that can reversibly withstand rather extreme conditions, and is thus particularly suitable for testing the mechanisms by which such conditions affect protein structure. (ii) Following the early introduction of cation exchange chromatography for its purification (Paléus and Neilands, 1950; Margoliash, 1952, 1954a,b; Boardman and Partridge, 1953, 1954, 1955), this basic globular protein became very easy to obtain in large quantities in pure form. (iii) The protein carries a large reporter group in it, the heme, which is in contact with many amino acid residues (Dickerson et al., 1971), and so serves admirably to allow changes in the structure of the protein to be detected and followed by a wide range of spectroscopic methods, including optical rotatory dispersion, circular dichroism, and fluorescence. (iv) Conversely, because it can readily be obtained in pure form, in large amounts, and dissolved at high concentrations, cytochrome c became the favorite subject of every new type of spectroscopic technique developed in the last five decades that can be applied to the study of heme and iron: electron paramagnetic resonance, nuclear magnetic resonance, EXAFS, Mössbauer, and resonance Raman. Moreover, cytochrome c has become the prototypical representative of the low-spin class of heme iron, and as such it has served to identify d–d (metal) and π–d (porphyrin to metal) electronic transitions underlying near infrared bands, and corroborate theoretical predictions of their existence. (v) Following the early determination of the amino acid sequence of horse cytochrome c (Margoliash et al., 1961), the third protein to have its primary structure determined, it became relatively easy to obtain other sequences, and this information accumulated rapidly for the cytochrome c of many species, well in advance of primary structures for other proteins. Thus, there were seven by 1963 (Margoliash, 1963) and 20 by 1967 (Fitch and Margoliash, 1967). (vi) This led to cytochrome c becoming the prime object for studies of protein evolution and the relation between phylogeny and protein primary structure (Fitch and Margoliash, 1967, 1970; Margoliash and Fitch, 1970; Margoliash et al., 1971), further increasing the attraction of examining even more species. (vii) Mitochondrial cytochromes c, being universally distributed in eukaryotes, provide a large reservoir of homologous proteins having different amino acid sequences, making it possible to relate these differences to structural and functional attributes (Margoliash, 1972). (viii) In turn, cytochrome c became a favorite protein for studying the effect of chemical modifications (see Margoliash and Bosshard, 1983). The ease with which it can be chromatographed on cation exchangers and the extreme sensitivity of the chromatographic behavior of the protein to changes in the charge distribution on the molecular surface make it possible to separate individual derivatives in pure form from large sets of isomeric chemically modified products (see, for example, Brautigan et al., 1978a;

Osheroff *et al.*, 1980). This has yielded many of the 19 lysines in horse cytochrome *c* in singly modified chemical derivatives (see Margoliash and Bosshard, 1983; Ferguson-Miller *et al.*, 1979). Such a *tour de force* of protein chemistry allowed extensive mapping of the roles of different domains on the protein surface in enzymic and other interactions (see, for example, Koppenol and Margoliash, 1982). (ix) Most recently, the severe limitations of the chemical approaches to modifying proteins have been overcome by recombinant DNA procedures made possible through the development of molecular genetics (see Smith, 1985), allowing the expression in yeast cultures of mutant cytochromes *c* bearing any desired amino acid change which nevertheless maintains a modicum of the protein's mitochondrial function. Once more a major spurt in cytochrome *c* structure–function studies is taking place. (x) A more difficult procedure, termed semisynthesis, is also providing useful artificially modified cytochromes *c*. Most of this work is too recent to be included here. However, among the earliest successes with the semisynthesis of modified cytochromes *c*, one may cite Wallace and Offord (1979), Boon *et al.* (1979), Westerhuis *et al.* (1982), and Wallace and Rose (1983).

This introductory chapter is intended as a succinct overview of the historical process by which our understanding of cytochrome *c* developed, up to the start of the molecular biology era, roughly in the late 1970s to mid-1980s. The first products of this new era have only recently been introduced into this field, and the work, still in its early stages, is mentioned in the appropriate chapters of this volume. Only in those relatively few cases in which this is unavoidable will reference also be made to more recent work. This account will also strictly limit itself to the classical mitochondrial cytochromes *c*, classified in the so-called Class I. To cover the many varieties and classes of prokaryotic and photosynthetic cytochromes *c* would require far too extensive a treatment.

Whenever appropriate, reference will be made to chapters in this volume that detail the subject under discussion. It is also useful at this point to mention a few of the many reviews written on cytochrome *c* that over the years have marked either its history or the advancing front of knowledge about the protein, as follows:

(i) Keilin, D. (1966) *The History of Cell Respiration and Cytochrome*, University Press, Cambridge.
(ii) Margoliash, E., and Schejter, A. (1966) *Adv. Prot. Chem.* **21**, 113–280.
(iii) Dickerson, R. E., and Timkovich, R. (1975) in *The Enzymes, Vol. XI, Oxidation–Reduction* (Boyer, P. D., ed.), Academic Press, New York.
(iv) Salemme, F. R. (1977) *Annu. Rev. Biochem.* **46**, 299–326.
(v) Ferguson-Miller, S., Brautigan, D. L., and Margoliash, E. (1979) in *The Porphyrins, Vol. VII* (Dolphin, D., ed.). pp. 149–240, Academic Press, New York.
(vi) Pettigrew, G. W., and Moore, G. R. (1987) *Cytochromes c. Biological Aspects*, Springer-Verlag, Berlin.
(vii) Moore, G. R., and Pettigrew, G. W. (1990) *Cytochromes c. Evolutionary, Structural and Physicochemical Aspects*, Springer-Verlag, Berlin.

II. The Early Period

In the early 1920s, David Keilin, then a classical parasitologist, became intrigued by the changing distribution of color in the tissues of the larval form of a fly, *Gastrophilus*, that spends most of that stage of its life cycle hanging on the inside of

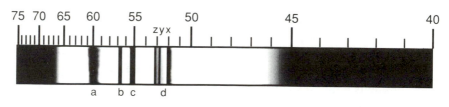

Figure 1.1

Spectrum of cytochrome as seen through the low-dispersion spectroscope in the thoracic muscles of the bee. According to Keilin (1925).

the stomach of the horse and carries a considerable amount of a hemoglobin-like pigment. This disappeared during metamorphosis outside the host, and a red color reappeared in the flight muscles of the adult fly. When in the summer of 1923 he turned a low-dispersion spectroscope on these flight muscles, he observed a typical four-banded cytochrome spectrum (Fig. 1.1). The same spectrum was found in a large variety of species, covering the entire taxonomic scale of eukaryotes, including both animals and plants. By analogy to the spectra of other already known hemochromes, the three bands at 604, 566, and 550 nm were ascribed to the α bands of components defined as *a*, *b*, and *c*, while the band at 520 nm was the combined β bands of all three (Keilin, 1925).

The postulated three components were termed "cytochrome" for "cellular pigment" and were shown to undergo oxidation and reduction *in vivo*, as in the classical study of the flight muscles of a live wax moth, *Galleria mellonella*, or only slightly less directly in yeast cells. When the moth was at rest in an atmosphere containing oxygen, the sharp spectral bands of the ferrous hemochromes disappeared, while under nitrogen or during active twitching of wings, they promptly reappeared. This simple but incisive experiment bypassed the argument between Thunberg (1927) and Wieland (1932) on one hand, who maintained that substrate-specific dehydrogenations were the essential characteristic of tissue-catalyzed oxidations, and Warburg (1924, 1925, 1949), on the other hand, who argued that the oxidation of physiological substrates was nothing but oxygen activation by a "respiratory enzyme" containing iron. Keilin's experiment merely indicated that the reactions at both ends of the oxido-reduction scale had to occur, and it laid the groundwork for the eventual realization that the cytochromes mediated the intermediate steps, as a final common electron transport pathway from substrates to oxygen.

An interesting aside was the result of an extensive literature search undertaken by Keilin at that time, without the help of computers. It turned up articles dating from the 1880s by MacMunn (1886, 1887, 1889, 1890), a practicing physician whose hobby was spectroscopy. He had seen the same four-banded spectrum in numerous tissues and organisms, oxidized and reduced it chemically, and had postulated that the pigments he called "histohematin" or "myohematin" had a respiratory function. This work was vigorously attacked by Hoppe-Seyler (1890a,b) when a student of his demonstrated that the spectroscopic phenomena observed by MacMunn (1887) when he attempted to extract the pigment from tissues could be obtained with mixtures of hemoglobin derivatives (Levy, 1889). However, to extend this justified criticism to the totality of MacMunn's work was obviously erroneous, since the four-banded spectrum had been seen in invertebrates that make no hemoglobin. This episode represents not only the common resistance to new ideas in what appeared to be a well-founded field, but also general ignorance of some basic chemistry. Thus, "hemochromogens" could not transport oxygen, in the sense that hemoglobin did, and hemoglobin and its derivatives were the only natural heme compounds known at the time, so that it could not be appreciated that the movement of oxidation equivalents and the movement of oxygen by oxygenation of a carrier were merely two aspects of the same fundamental biological function. MacMunn's discovery was largely forgotten until Keilin rediscovered the four-banded spectrum as well as MacMunn's publications, some four decades later. This brought to a close a classic instance of how a fundamentally new finding is unlikely to be accepted at a time when the field to which it applies is too little developed to be able to confidently encompass its ramifications.

Another interesting point in connection with these early developments is that had MacMunn and Keilin worked with spectrophotometers instead of low-dispersion ocular spectroscopes, then the only instruments available for such experiments, the discovery of cytochrome would have been much more difficult if not impossible. The differences between one instrument and the other are illustrated in Fig. 1.2. The emphasis of shoulders on spectra as clear lines, the relative insignificance of partial opacity or turbidity of samples, and the extreme ease of its operation gave crucial advantages to the ocular spectroscope.

The early history of the discovery of the cytochromes has been described several times (see, for example, Keilin and Slater, 1953; Summer and Somers, 1953; Nicholls, 1963), perhaps most completely and lucidly by Keilin, in his *History of Respiration* (1966), in which the conflicting conceptual backgrounds are thoroughly assessed.

By the late 1920s, the stage had been set for the identification of component *a* of cytochrome with the system that oxidized ferrous cytochrome *c*. Warburg and co-workers (Warburg and Negelein, 1929a,b,c; Warburg, 1932; Kubowitz and Haas, 1932) had obtained a photochemical action spectrum of the respiratory enzyme by measuring the reversal of the inhibition of oxygen uptake by carbon monoxide, using light of defined wavelengths across the entire spectrum. The spectrum obtained was clearly that of a heme protein. Furthermore, Keilin (1927) had found that indophenol oxidase—the enzyme discovered in 1885 by Ehrlich (1885) that, in the presence of oxygen, condensed a mixture of dimethyl-*p*-phenylenediamine and α-naphthol into indophenol—was, just like Warburg's "*Atmungsferment*," light-reversibly inhibited by carbon monoxide. By 1930, Keilin (1930) had obtained the first relatively crude preparation of cytochrome *c*, from Delft yeast. This made it possible to show that the indophenol oxidase activity of heart muscle particle preparations, much later realized to consist essentially of mitochondrial inner membrane vesicles and used to this day to study the interactions of cytochrome *c* with the mitochondrial respiratory chain, depended not only on the enzyme, which could be inhibited by cyanide or carbon monoxide (Keilin, 1930), but also on the cyto-

8

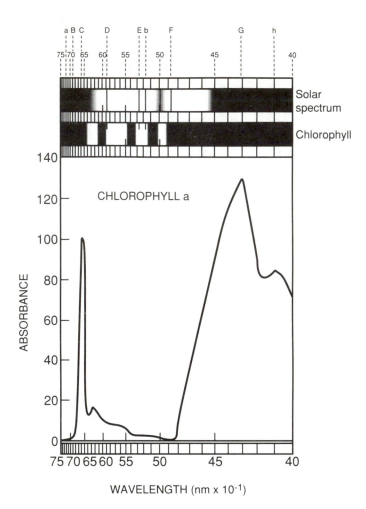

Figure 1.2
Comparison of the spectrum of chlorophyll as recorded photographically by MacMunn (1914) through his low-dispersion spectroscope and as recorded by a contemporary spectrophotometer, but redrawn to the wavelength scale of the spectroscope. The letters at the top refer to the positions of the Fraunhofer lines and the numbers to the wavelength in tens of nanometers. The top band is the solar spectrum, the one under it the spectrum of chlorophyll. The precise quantitation of absorbances by the spectrophotometer as referred to a colorless standard is accompanied by a flattening of low-intensity features, while the low-dispersion microspectroscope sharpens the bands, so that shoulders recorded by the spectrophotometer become distinct bands as observed through the low-dispersion microspectroscope. This, and the possibility of shining a strong light through the sample, are crucial advantages presented by the earlier instrument in the study of biological materials which are only partially transparent. From Schejter and Margoliash (1984), with permission.

chrome *c* present in the preparation. The cytochrome *c* was alternately reduced to the ferrous form by the substrate, dimethyl-*p*-phenylenediamine, and oxidized to the ferric form by the enzyme. Thus, the indophenol oxidase could now be termed cytochrome *c* oxidase (Keilin and Hartree, 1938a). By the late 1930s, cytochrome oxidase, the *a* component of the original spectrum, had been shown to consist of two hemochromes, cytochrome *a* and a newly named cytochrome a_3, the latter being responsible for the reactions with oxygen, carbon monoxide, and cyanide (Keilin and Hartree, 1938b, 1939).

Since that time, cytochrome oxidase (Complex IV) has been the object of extensive attention, which is still continuing (see reviews by Powell and Marsh, 1992; Babcock and Wikström, 1992; Cooper *et al.*, 1991; Malatesta *et al.*, 1990; Kadenbach *et al.*, 1991; Mueller and Azzi, 1991; Malmström, 1990; Capaldi, 1990). In mammals it is a very complex protein with four redox centers, two hemes, and two coppers, and some 13 different protein subunits. To discuss it further here is beyond the present scope. However, from the point of view of cytochrome *c* it is interesting to note that its steady-state kinetics of reaction with cytochrome oxidase are very clearly biphasic (Ferguson-Miller *et al.*, 1976), and this phenomenon must be part of any credible proposed mechanism (see Fig. 1.3). To date, three such general types of explanations, and several variants of each have been proposed (see Garber and Margoliash, 1990, and references therein). One simply assumes the existence of two sites of electron transfer on the oxidase, with different binding affinities and rates of turnover. Another considers that cytochrome *c*, in addition to binding to the electron-accepting site on the enzyme, binds to a second site or sites, at much lower affinity. This second binding results in increasing the rate of dissociation of the first molecule from the electron transfer site, the proposed rate-limiting step, electrostatically or allosterically. The third states that since electron transfer by

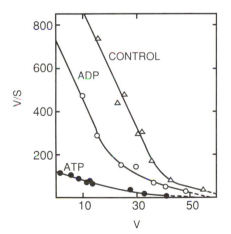

Figure 1.3

Effect of ATP and ADP (both 3 mM) on the kinetics of reaction of horse cytochrome *c* with beef heart Keilin–Hartree particle preparation cytochrome *c* oxidase, as determined polarographically using tetramethylphenylenediamine as reductant at pH 7.8 (25 mM cacodylate–Tris). From Ferguson-Miller *et al.* (1976), with permission.

cytochrome oxidase is coupled in intact mitochondria with the translocation of protons out of the organelle matrix, any titration of a parameter of one reaction must lead to nonlinear steady-state kinetics, if that parameter enters into the two rate expressions differently. Whether the same can be expected of the totally uncoupled systems (pure enzyme, or mitochondrial particle preparations) commonly used to study these kinetics is not obvious. The present authors are partial to the second type of model, having been involved in its development (Speck *et al.*, 1984), but clearly no consensus has been achieved on this subject (See Chapter 13).

Though the observation of cytochrome oxidase function had played a crucial role in uncovering the significance of electron transport in the respiratory chain, it gradually became clear that cytochrome *c* was much more versatile than originally thought in its redox reactions. Indeed, from early on it was clear that in the respiratory chain the electron was transferred to cytochrome *c* from the *b* component, cytochrome reductase (Complex III) (for some more recent reviews see Weiss, 1987; Weiss *et al.*, 1990; von Jagow *et al.*, 1987). Over the next several decades, it was found that ferric cytochrome *c* could react with a number of other redox proteins located in the intermembrane space of mitochondria. Thus, it could be reduced by cytochrome b_5 (see von Jagow and Sebald, 1980), and by sulfite oxidase (Heinberg *et al.*, 1953; Fridovich and Handler, 1956a,b; Oshino and Chance, 1975; Rajagopalan, 1980). Further, ferrocytochrome *c* could be oxidized by yeast cytochrome *c* peroxidase (Abrams *et al.*, 1942; George, 1953a,b; Yonetani, 1965, 1971; Erman and Yonetani, 1975), using hydrogen peroxide as donor of oxidizing equivalents, a reaction that was shown directly to occur *in vivo*. Moreover, the protein could also react with a large variety of nonphysiological oxidants and reductants—both macromolecules, such as plastocyanin, stellacyanin and azurin (see, for example, Zhou *et al.*, 1992; Augustin *et al.*, 1983; Armstrong *et al.*, 1986), and numerous small molecules, including a variety of inorganic oxidants and reductants, as first observed by MacMunn. Of the latter, a 1979 review (Ferguson-Miller *et al.*, 1979) lists studies with 19 reductants and eight oxidants. This large number of electron transfer partners became one of the many attractive aspects of the protein, making it possible to study its redox reactions, their kinetics, and their mechanisms under a wide range of conditions, and with a range of electron donors and acceptors having widely different properties. What these studies have revealed is discussed in general terms and very succinctly in this chapter's Section VII, "Chemical Reactivity of Heme and Reduction Potentials." The outstanding fact about these oxidation and reduction reactions is that, notwithstanding their enormous variety, all seem to involve approximately the same area of the surface of cytochrome *c*, at the so-called "front" of the molecule, including the solvent-accessible edge of the heme-containing pyrrole rings II and III (see, for example, Margoliash and Bosshard, 1983; Augustin *et al.*, 1983; Butler *et al.*, 1983; Armstrong *et al.*, 1986; Roberts *et al.*, 1991). Examples of such interaction domains are illustrated in Fig. 1.4. Whether this definitively proves that all electron transfer reactions are *via* the exposed heme edge or a closely approximated structure remains under discussion.

With regard to the reactivity of cytochrome *c* with its multitude of electron exchange partners, it may be interesting to note that the kinetics of reactions that are completed with a single electron exchange, even in a complex system, have generally not presented any difficulty, as for example with mitochondrial cytochrome *c*

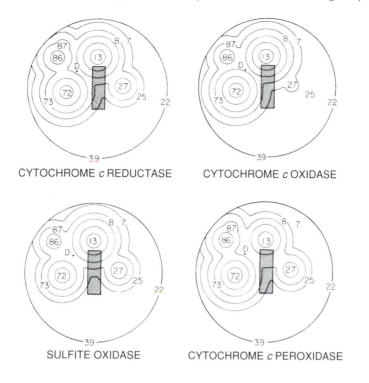

Figure 1.4

The domains on the surface of cytochrome c involved in the interaction with beef cytochrome c reductase and oxidase, yeast cytochrome c peroxidase, and beef sulfite oxidase. The shaded rectangle represents the solvent-accessible heme edge. The numbers indicate the relative positions of the α-carbons of lysyl residues, and D is the point near the β-carbon of phenylalanine 82, at which the dipole axis of native cytochrome c crosses the front surface of the protein. The number of circles around a given lysyl residue is proportional to the percentage of the inhibition of the carboxydinitrophenyl-cytochrome c modified at that lysine. The radii of these circles are multiples of an arbitrary value of 2.5 Å. From Koppenol and Margoliash (1982), with permission.

reductase (Speck and Margoliash, 1984). However, this does not appear to be the case when two electrons are required from cytochrome c, as with mitochondrial cytochrome oxidase, or yeast cytochrome c peroxidase (see, for example, Ferguson-Miller *et al.*, 1976; Kang *et al.*, 1977). In both these cases, after many years of investigation, there is still only partial agreement on the interpretation of the observed kinetics. Whether the complexities result only from the presence of more than one electron-accepting redox center, as is clearly the case for the peroxidase (Stemp and Hoffman, 1993; Yu and Margoliash, 1995), or can have other definite physicochemical bases, remains to be decided.

Finally, with the determination of its amino acid sequence and spatial structure and the development of protein chemistry in general, cytochrome c has in recent years become one of the favorite objects for studies of intramolecular protein electron movement between an artificially introduced metal center and ordinary heme iron (see Chapter 15).

III. The Preparation and Crystalization of Cytochrome *c*

The first preparation of cytochrome *c* was obtained by Keilin (1930) in 1930 from Delft yeast. From its hematin content (Hill and Keilin, 1933) it can be calculated to have been about 25% pure. What was remarkable about it was that though its reduced spectrum was that of a typical hemochrome, unlike the usual hemochromes from heme and nitrogenous organic bases such as pyridine, it did not autooxidize or combine with carbon monoxide over the pH range of 4 to 12, only outside it. Dixon, Hill, and Keilin (1931) measured its spectrum quantitatively, and this yeast cytochrome *c* added to the horse heart muscle particle preparation behaved like that in the particle preparation: It was oxidized in air, this oxidation being inhibited by cyanide, and light-reversibly by carbon monoxide, and it was reduced by sodium succinate. This was probably the first indication that cytochrome *c*, over the taxonomic range from yeast to mammals, is an evolutionarily highly conserved protein, not only in terms of its spectrum, as evident from Keilin's first observations (1925), but also in terms of its typical respiratory chain functional activities.

By 1936, Theorell (1936) developed a large-scale procedure for preparing heart muscle cytochrome *c* that was improved and much simplified by Keilin and Hartree (1937, 1940). These cytochrome *c* preparations contained 0.34% iron, which happens to be the iron content of hemoglobin, thus giving the impression that such a preparation was probably pure. This was, however, disproved when Theorell and Åkesson (1939, 1941) used prolonged electrophoresis at a pH near the alkaline isoelectric point of cytochrome *c* to give preparations having 0.43% iron. Similar preparations were obtained by Keilin and Hartee (1945) employing fractionation with ammonium sulfate at pH 10; by Roche *et al.* (1946) by boiling in 10% ammonium sulfate; and by Tsou (1951) by boiling in the presence of chloroform. These preparations were in turn superseded with the introduction of cation exchange resins for chromatographic purification (Paléus and Neilands, 1950; Neilands, 1952; Boardman and Partridge, 1953, 1954, 1955; Margoliash, 1952, 1954a,b), leading to products that contained 0.45% iron, corresponding to a molecular weight of about 12,400 and a polypeptide chain a little larger than 100 residues. After the problems involved in recovering an enzymically active protein from such chromatographies had been resolved (Margoliash, 1952, 1954a), it became clear that most preparations, in addition to the native monomeric form, contained deamidated and polymeric forms of the protein. These stubborn impurities require careful cation exchange chromatography to be separated out (Margoliash and Walasek, 1967; Brautigan *et al.*, 1978c) and some experiments can lead to erroneous conclusions because of their uncontrolled presence, particularly when commercial preparations not further purified are employed.

Finally, in 1955, for the first time a cytochrome *c* was prepared in the crystalline form (Bodo, 1955). It was the protein from the muscles of the King penguin and was crystallized from a solution of near-saturated ammonium sulfate. This was quickly followed by similarly obtained crystals from the cytochromes *c* from many species (Kuby *et al.*, 1956; see Okunuki, 1960). None appeared to be suitable for x-ray analysis (Bodo, 1957; Blow *et al.*, 1964). However, when in the 1960s, crystals of a size and quality that allowed x-ray crystallographic studies were obtained, initially by one of the present authors (EM), spatial structures were successively determined for the proteins from horse (Dickerson *et al.*, 1968, 1971; Bushnell *et al.*, 1990), tuna (Takano and Dickerson, 1981a,b), rice (Ochi *et al.*, 1983), and yeast (Louie *et al.*, 1988; Louie and Brayer, 1990). It should be noted that in all these cases the crystals were grown in near-saturated ammonium sulfate solutions, namely at ionic strengths of 8 to 9 M, and it is only recently that a crystallization at an ionic strength of about 40 mM, employing polyethylene glycol, has been reported (Walter *et al.*, 1990; Sanishvili *et al.*, 1994). To what extent the structures differ significantly at very high and low ionic strength is being examined. However, it would not be surprising if some such differences existed, in view of the very high positive charge on cytochrome *c*, at a net value of about +8, as estimated from the amino acid sequence and the heme.

IV. Covalent Structures

An extremely important aspect of the covalent structure of cytochrome *c* is the binding of the heme prosthetic group by thioether bonds to cysteine residues 14 and 17 of the polypeptide chain, formally resulting from the addition of the amino acid sulfhydryl groups across the double bonds of the vinyl groups of pyrrole rings I and II, respectively, so that the sulfur atom is bound to the α-carbon of the vinyl side chains. This prevents the heme from separating off under conditions that denature the protein, allowing an experimental flexibility available with very few heme proteins. Porphyrin *c*, the water-soluble product obtained by hydrochloric acid hydrolysis under reducing conditions, was first isolated by Hill and Keilin (1930) and studied by Theorell (1937, 1938, 1939), who postulated the correct structure for it. However, this structure remained in doubt because the harsh acid conditions employed also hydrolyzed the protein and could cause the secondary condensation of cysteine with free porphyrin vinyls. These doubts were finally removed by Paul (1950, 1951b), who used a much milder method for separating the apoprotein from the prosthetic group; he obtained an optically active hematoporphyrin, confirming Theorell's (1939) structure and showing that this hematoporphyrin had the properties expected for its postulated structure. Similar conclusions were reached by Neilands and Tuppy (1960), who prepared a crystalline synthetic porphyrin *c*, and by Sano and Granick (1961), who found that sulfhydryl compounds that do not react with porphyrin or heme, presumably because the vinyl double bonds are part of a stabilizing conjugated system, react readily with the porphyrinogen, the partial reduction product with six extra atoms of hydrogen, in which the conjugated system is broken. This made possible the condensation of appropriate peptides with porphyrin (Sano *et al.*, 1964) or even of the whole apoprotein (Sano and Tanaka, 1964) and apparently opened the way to total chemical synthesis of cytochrome *c*. This was tried some time later by Sano (1972) without success, possibly because the essential tryptophan residue could not be handled and had to be replaced by a phenylalanine. Finally, the relative orientations of the amino acid sequence and heme pyrroles, noted earlier, were definitively established by the first crystallographic structure study of the protein, that of horse cytochrome *c* (Dickerson *et al.*, 1967b).

In connection with this prolonged line of work on the binding of heme to polypeptide chain, it should be noted that Paul's (1950, 1951a,b) development of a procedure for breaking the thiother bonds by heating (60 °C for several hours) in the presence of metal ions, such as those derived from silver sulfate, in acid solution (0.1 M acetic acid), though it served excellently to yield an unmodified heme, was not as mild with respect to the protein. This method and its later modifications (Ambler, 1963) nevertheless yielded the first preparation of cytochrome c apoprotein that could readily be used for chemical work, including the determination of amino acid sequences. However, in experiments more sensitive to whether certain residues were deamidated, such as uptake of the apoprotein into intact mitochondria, this type of apoprotein preparation performed much more poorly than material obtained by *in vitro* transcription/translation of an appropriate gene construct (Miller, D., and Margoliash, E., unpublished). A more recent procedure by Fontana *et al.* (1973) is milder for the protein. However, it involves reaction with 2-nitrophenyl-sulfenylchloride, which leaves the tryptophan residue derivatized.

With Sanger's work on insulin (Sanger and Tuppy, 1951; Sanger and Thompson, 1953) and the development of the automatic amino acid analyzer (Spackman *et al.*, 1958), by the late 1950s, the determination of the amino acid sequence of a rather small protein, such as cytochrome c, became readily feasible. The first cytochrome c to have its primary structure established was the horse protein (Margoliash *et al.*, 1961, 1962; Margoliash and Smith, 1962; Margoliash, 1962). It became the third protein of known amino acid sequence, following insulin and shortly after ribonuclease. The effort this work entailed was considerably less than for the preceding protein, largely because the use of a volatile pyridine acetate buffer in the ion-exchange chromatographic separation of peptide fragments bypassed the difficult task of isolating small amounts of peptide from large amounts of buffer salts. As a result, with the involvement of several groups of investigators, amino acid sequences of cytochromes c from different species accumulated much more rapidly than for any other protein, with rather striking consequences. As long expected, but never yet clearly demonstrated, there was an obvious correlation between the number of amino acid sequence differences and the phylogenetic relatedness of the species making these proteins. Thus, the first comparison (Margoliash, 1963), which included the cytochromes c of only seven species, listed the number of residue differences from the horse protein as 3 for pig, 8 for rabbit, 12 for man, 12 for chicken, 19 for tuna, and 44 for yeast iso-1 cytochrome c. This collection of primary structures became more and more useful in attempting to describe the relationship between molecular evolutionary changes and the phylogeny of species. It led to the first statistical phylogenetic tree, derived from amino acid sequences with strictly no other biological input, employing 20 different cytochromes c (Fitch and Margoliash, 1967). This remained a major paradigm in the field, posed several of the problems underlying such correlations, and was the basis for a number of significant further advances (see Fitch and Margoliash, 1970; Margoliash and Fitch, 1970; Margoliash *et al.*, 1971; Margoliash, 1972). Among these was the distinction between convergent and divergent evolution, the identification of invariant and hypervariable codons, and the concept of concomitantly variable codons or covarions. Indeed, it could be shown that although two cytochromes c could vary by more than 50% of their amino acid sequences, the number of codons that could accept a mutation that would be fixed in the course of evolution was, for any one cytochrome *c* at any one time in its descent, only about 10. This set of covarions itself changed as they were mutated, eventually causing the accumulation of many differences between cytochromes *c* that had diverged a long time ago.

One of the earliest effective uses of the ability to determine the amino acid sequence of cytochrome *c* was the demonstration that cytochrome *c* in yeast, even though it was a mitochondrial protein, was coded for by a Mendelian nuclear gene (Sherman *et al.*, 1966). That this conclusion applied even to mammals was shown much later by determining the primary structures of horse and donkey cytochromes *c*, which differ by a single residue, and of the proteins of their crosses, mules and hinnies (Walasek and Margoliash, 1977).

By now there are more than 100 known amino acid sequences of eukaryotic cytochromes *c* (Moore and Pettigrew, 1990). This provides a large range of variants of a protein that throughout has essentially the same function, and that can be used effectively to study numerous aspects of the relation between the amino acid sequence and the physical chemistry or the functional activities of the protein. The only drawback of such natural variants is that all have passed the stringent evolutionary test of selection for function, each in its own particular environment. Thus, in general, one finds differences in activity only by examining function with electron transfer partners other than the phylogenetically related ones, such as yeast cytochrome *c* with mammalian cytochrome oxidase and mammalian cytochromes *c* with yeast cytochrome oxidase (Dethmers *et al.*, 1979), or higher primate cytochromes *c* with nonprimate mammalian cytochrome oxidase and vice versa (Osheroff *et al.*, 1983).

Another way of changing the covalent structure of cytochrome *c* is by modifying reactive amino acid side chains with appropriate reagents and purifying the resulting mixture extensively to eliminate by-products commonly formed in such procedures. Numerous such chemically modified cytochromes *c* have been obtained with the proteins from several species, including derivatives at the single internal tryptophan, at each of the several tyrosines, at each of the histidines, at one or both of the methionines, at the arginines, at the non-heme bonded cysteine of the yeast protein, and at the lysines. These, and many others, are listed and discussed in a review (Ferguson-Miller *et al.*, 1979) that covers the material up to the late 1970s. A good example of how useful these derivatives could be is given by those at lysine residues, which in the 1970s led to major developments of our understanding of the interaction between cytochrome *c* and its physiological electron exchange partners. For example, from the mixture of 4-carboxy-2,6-dinitrophenyl derivatives of horse cytochrome *c* at the 19 lysines in the protein, 12 were isolated in pure singly modified form (Brautigan *et al.*, 1978a,b; Osheroff *et al.*, 1980). This made it possible to map the binding domains on the surface of the protein for several of its mitochondrial reaction partners, showing that they were all essentially the same, occupying a large segment of the "front" surface of the molecule, including the solvent-accessible edge of the heme prosthetic group at pyrroles II and III (Ferguson-Miller *et al.*, 1978; Kang *et al.*, 1978; Speck *et al.*, 1979; König *et al.*, 1980; Speck *et al.*, 1981) (see Fig. 1.4). Other derivatives gave similar results and extended them to other reaction partners (see Margoliash and Bosshard, 1983). These results also served to uncover and gradually define the enormous influence that the strongly asymmetric electrostatic charge of the protein has on its reactivity with physiological electron exchange partners (see, for example, Koppenol and Margoliash, 1982). There is no doubt that chemical modifications are likely to continue to be useful experimental approaches in some cases, even though site-directed mutants are now providing a much larger variety of possible modifications.

A third method for obtaining changes in the covalent structure of cytochrome c is by so-called semisynthesis (Offord, 1980). In this approach, fragments of the natural protein are ligated into the original covalent structure of the protein with one fragment that has been modified. Such modification is most readily accomplished by total organic synthesis of the fragment in question, although it is not restricted to that approach, appropriate partial synthesis or chemical modifications being often possible. Related, though much more restricted, approaches involve either sequential degradation and resynthesis at the amino or carboxyl terminus of the complete protein chain, or spontaneous reassociation of fragments of the protein into a noncovalent complex that has a spatial structure similar to that of the original native protein and that shows some of its properties (Juillerat et al., 1980).

The major advantage of the semisynthetic approach is that amino acids with any desired side chains can be introduced, not just one of the 20 natural protein amino acids represented in the genetic code. This includes amino acids modified for special purposes—for example, by isotopic labeling for NMR studies. The major disadvantages are the experimental difficulty of obtaining the semisynthetic correctly ligated protein that has not undergone any covalent degradation in the often harsh process, and the derived necessity of having to employ large amounts of protein. In turn, this leads to a requirement, often overlooked, of having to characterize as fully as possible the final product, not only in terms of its primary structure, but also in terms of its spatial conformation, a much more demanding process. Indeed, without such a demonstration, observed changes in properties are likely to be poorly interpretable.

The early observation that opened the way to semisynthetic modifications of cytochrome c was the demonstration of Corradin and Harbury (1974) that the two fragments of horse cytochrome c cleaved by reaction with cyanogen bromide at methionine 65 could recombine covalently into the complete chain. This occurred by the reformation of the peptide bond between the activated carboxyl of homoserine lactone 65, the product of cyanogen bromide cleavage, and the amino group of glutamate 66, presumably because the two had come in close approximation in the noncovalent complex of the two segments. This unexpected phenomenon made possible the production of hybrid cytochromes c from the proteins of natural species, and, for example, was effectively used in a detailed study of the immunological properties of pigeon cytochrome c, which developed into a favorite model antigen (Hannum and Margoliash, 1985; Hannum et al., 1985). Other early uses of semisynthetic cytochromes c included the development of the technique (Wallace and Offord, 1979), the study of the effects of modifications on the activity with cytochrome oxidase (Boon et al., 1979), the effect of deletion of a segment on the ability to form a functional complex (Westerhuis et al., 1982), the influence of modification of the two invariant arginine residues (Wallace and Rose, 1983), etc. Numerous other studies, most of which are too recent to be included in this historical survey, have by now made cytochrome *c* a favorite object of semisynthetic modification.

Finally, in more recent years, the cloning of cytochrome *c* genes (Montgomery *et al.*, 1978; Smith *et al.*, 1979; Scarpulla *et al.*, 1981; Limbach and Wu, 1983, 1985; Swanson *et al.*, 1985; Miller, 1988) and the development of efficient procedures for site-directed mutagenesis and for the expression of mutant genes in yeast (see Smith, 1985; Pielak *et al.*, 1985) have made available cytochromes *c* of any desired amino acid sequence, with the proviso that the sequence be capable of yielding a holo-cytochrome *c* in the host and that this holocytochrome *c* be able to complement the function of the host respiratory chain enough to permit its growth. This readily provides a much wider field of variants than available from natural cytochromes *c*, among them mutants that are only poorly adapted to their physiological electron-exchange partners. Such variants are never seen in the natural proteins, all of which are necessarily the result of successful experiments in evolutionary adaptation. Here again, as for the semisynthetic proteins, it must be emphasized that any conclusions about the properties of a mutant protein, however plausible, are at best likely to remain tentative until confirmed and expanded by precise information about its spatial conformation. This essential requirement appears to have been taken up actively by investigators employing recombinant cytochromes *c* (see Louie and Brayer, 1989, 1990; Louie *et al.*, 1988; Pielak *et al.*, 1988a,b), even though, as already noted, such studies have the inherent relative disadvantage that recombinant mutants can so far be obtained only if they are capable of at least partly reconstituting the mito-chondrial respiratory chain of a live host cell, the protein presumably retaining the minimal features for the appropriate degree of function.

This area of research is currently extremely active and likely to remain so for a long time. Since this historical review is intended to stop at the start of the molecu-lar genetic era of the study of proteins, no more will be said here about the several interesting site-directed mutants of cytochrome *c* that have been examined to date.

V. Spatial Structure

The first x-ray structural study of cytochrome *c* resulted from the inadvertent crys-tallization of horse cytochrome *c* in 1962 by one of the authors of this review (EM) in a near-saturated ammonium sulfate solution containing 0.5 M sodium chloride. These crystals rapidly grew to x-ray size and quality and readily gave isomorphous heavy metal replacements with mersalyl (a mercurial), platinum chloride ($PtCl_4^{-2}$) and both. This led to a collaboration that resulted in the first solved spatial struc-ture of the cytochrome *c* molecule, that of the horse protein, at a resolution of 4 Å (Dickerson *et al.*, 1967a,b, 1968). The molecule had the shape of a prolate spheroid with axial dimensions of $25 \times 25 \times 37$ Å. By 1971 (Dickerson *et al.*, 1971), struc-tures with 2.8 Å resolution of the ferric protein from horse and from bonito were already available, and the 2.0 Å maps of the tuna oxidized (Swanson *et al.*, 1977) and reduced cytochrome *c* (Takano *et al.*, 1977) were finally refined to 1.8 Å reso-lution in 1981 (Takano and Dickerson, 1981a,b). Several additional structures appeared in later years, as listed earlier, and these will be discussed in the corresponding article (Chapter 3).

While the more recent work has begun to show subtle differences between species, the early conclusions about the high degree of conservation of the cytochrome *c* fold (Margoliash, 1972) have maintained their original validity. The most important details of this structure that were already known in the mid-1970s include (i) the heme, embedded in the protein, with one of its edges abutting the solvent and within some 30° from the positive end of the molecular dipole (Koppenol and Margoliash, 1982); (ii) the covalent attachment of the heme to the protein through its "top" by way of two thioether bonds to cysteines 14 and 17; (iii) the "closed crevice" (George and Lyster, 1958) in which the heme iron is held, with the histidine 18 imidazole on the right, and the methionine 80 sulfur on the left; (iv) the presence of five α-helical segments, notably the *N*-terminal (residues 2 to 14), the *C*-terminal (residues 87 to 103), and the helical segments 49–55, 60–70, and 70–75. Certain features that appeared suggestive of structure–functional significance, such as the channel formed on the left side by the side chains of tryptophan 59, tyrosines 67 and 74, and methionine 80, have not yet been explained; others have become prominent in recent times, notably the five interior water molecules, and most especially that hydrogen-bonded to the invariant residues asparagine 52, tyrosine 67, and threonine 78, which has been implicated in the regulation of the redox potential (Takano and Dickerson, 1981b) and lately in the stabilization of the protein conformation (Das *et al.*, 1989; Luntz *et al.*, 1989; Hickey *et al.*, 1991; Schejter *et al.*, 1992) (Chapter 3). For a recent NMR study of structural water in cytochrome *c*, see Qi *et al.* (1994).

VI. Optical Spectroscopy and Magnetic Properties

The major characteristics of the optical absorption spectra of reduced and oxidized cytochrome *c* had already been described in 1931 (Dixon *et al.*, 1931), and even earlier Keilin had classified the ferric and ferrous molecules as a "parahematin" and a "hemochromogen," respectively (Keilin, 1925; Keilin, 1926) on the basis of his observations with the low-dispersion spectroscope. The extensive application of spectroscopy to the study of cytochrome *c* began, however, in 1941, with the publication of Theorell's four classic papers (Theorell and Åkesson, 1941; Theorell, 1941); a detailed spectrum of the purified protein was published by Margoliash in 1959 (Margoliash and Frohwirth, 1959), and the theoretical analysis currently accepted was based on the work of Gouterman in the 1960s (Gouterman, 1961; Zerner *et al.*, 1966). In Theorell's seminal studies, the optical and magnetic properties of the ferric molecule were investigated at different pH values, leading to the definition of five different states.

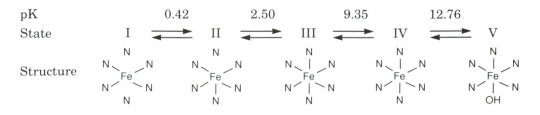

To each of these states was assigned a particular coordination of the iron atom by porphyrin and imidazole nitrogens, as shown in the preceding scheme. Two imidazoles were thought to be the iron ligands in the neutral pH State III. In the acid range, the pKs 2.5 and 0.5 were attributed to the successive cleavage of these metal–ligand bonds by protonation of the latter (Theorell and Åkesson, 1941), but later findings of specific anion effects were interpreted as due to iron binding by the anions (Boeri et al., 1953), to effects of the anions on the stabilities of States I and II (Fung and Vinogradow, 1968), and later to chaotropic effects on the protein structure (Aviram, 1973). That the effect was not due to direct binding of the anions to the iron was demonstrated by the similar spectroscopic changes caused by the nonbinding anion perchlorate (Aviram, 1973) and by the lack of effect of the paramagnetic iron center on the NMR spectrum of the anion methanesulfonate (Lanir and Aviram, 1975). Later work and present views of the acid pH ionizations were more recently reviewed (Dyson and Beattie, 1982).

Another much-investigated problem arose from the observation (Theorell and Åkesson, 1941) of the disappearance of the 695-nm band in the transition from State III to State IV. This characteristic band of cytochrome c (Theorell and Åkesson, 1939) was also lost when the heme iron was ligated by azide, cyanide, or imidazole (Horecker and Kornberg, 1946; Horecker and Stannard, 1948; Schejter and Aviram, 1969), heated or denatured (Schejter and George, 1964), or polymerized (Schejter et al., 1963). It was later shown that the band could be recovered when the cytochrome c heme peptide was ligated by the thioether sulfur of N-acetylmethionine (Schechter and Saludjian, 1967), and that it represented a porphyrin (π)-to-iron(d_{z^2}) electronic transition, polarized perpendicularly to the heme plane (Schejter and Aviram, 1969). Kinetic studies indicated that the disappearance of the band was associated with a rapid deprotonation and a subsequent conformation change involving displacement of the methionine 80 sulfur from iron ligation (Davis et al., 1974). While the displacement of the native ligand was demonstrated experimentally (Wüthrich, 1969), the nature of the replacing group remained moot and became the center of many investigations, because it is directly related to the question of which side chain replaces the methionine 80 sulfur when this atom is alkylated (Ando et al., 1966; Schejter and Aviram, 1970). Beginning with the finding on the crystallographic model that swinging the side chain of lysine 79 could bring its ε-amino group close to the iron (Margoliash et al., 1972), a variety of experimental observations dealing with this possibility were presented. On the one hand, EPR spectroscopy (Brautigan et al., 1977) tended to support the hypothesis of a lysine as ferric iron ligand, while kinetic (Brunori et al., 1972) and resonance Raman (Kitagawa et al., 1975) experiments argued for a similar ligation in the ferrous state; on the other hand, NMR investigations indicated pentacoordination of the ferrous iron (Keller et al., 1972), while chemical modification studies (Stellwagen et al., 1975; Pettigrew et al., 1976) argued against a ligand role for lysine. The latter results led to other possibilities, including a proposal that the sixth coordination remained empty (Aviram and Krauss, 1974). A more recent NMR study (Ferrer et al., 1993) again indicates that lysine 79 is the sixth ligand in one form of alkaline yeast ferricytochrome c. However, it is still questionable whether this issue has been satisfactorily answered (Moore and Pettigrew, 1990) (Chapter 19). Furthermore, the question as to whether the ferric low-spin spectrum at extremely high pH is due to a species with the metal ligated to an OH^- group, as Theorell had originally proposed (Theorell, 1941), also remains unanswered.

The optical spectrum of ferrocytochrome c is much less changeable with pH. The characteristic α and β bands were shown to increase in extinction at pH 14 (Butt and Keilin, 1962), but a pH-dependent transition at less alkaline pHs was never observed. The most interesting feature of the reduced spectrum is that of the sharpening and splitting of its bands at liquid-nitrogen temperatures (Keilin and Hartree, 1949; Estabrook, 1957; Estabrook and Sacktor, 1958). This is a property of all cytochromes c (Elliot and Margoliash, 1971; Goldin and Farnsworth, 1966). The observation of the low-temperature splitting, which can be greatly improved using second- and fourth-derivative spectra (Butler and Hopkins, 1970), is clearly linked to the conformational integrity of the protein or to its native ligation state, since it is lost (Estabrook, 1961) in dimers or higher polymeric forms, as well as in the heme peptides in which both structural constraints have been altered.

The absence of a pH-induced transition in the ferrous state was also implicit in Theorell's demonstration that cytochrome c released two more protons in the ferric than in the ferrous state when acidimetrically titrated between pHs 8.0 and 10.6 (Theorell and Åkesson, 1941). Strong though indirect evidence of a change in iron ligation accompanying the transition from State III to State IV was provided by an experiment in which reduction of the iron at alkaline pH by dithionite, which is a very fast process, was followed by a slow spectroscopic change of the reduced cytochrome c (Greenwood and Palmer, 1965). Two different species of ferricytochrome c appeared to exist at alkaline pH, one of which was not reducible by ascorbate (Lambeth and Palmer, 1973). Furthermore, an apparent paradox was found when temperature jump measurements (Brandt et al., 1966) indicated that the ratio of the kinetic constants in the equilibrium between ferri- and ferrocytochrome c, and ferri- and ferrocyanide, was independent of the pH, while the statically measured equilibrium constant, which should consequently also be pH-independent, increased instead by a factor of 4, in the pH range 7 to 9.4. All these results led to the kinetic study of the pH dependence of the transition from State III to State IV, from which the presently accepted mechanism was deduced: It is a deprotonation followed by a slow ligation change (Davis et al., 1974). The still unidentified ionizing group has a pK of 11.0, and the observed equilibrium pK of 9.0 results from the superimposed favorable ligation change, attended by an equilibrium constant $K = 10^2$ (Davis et al., 1974). The number of protons released per molecule of protein immediately after oxidation by ferricyanide was measured directly as 1.0 (Davis et al., 1974). Still, a discrepancy remains between this measurement and Theorell's measurement of a two-proton difference between the ferric and ferrous forms of cytochrome c (Theorell and Åkesson, 1941). Two protons were also reported in another kinetic study (Czerlinski and Dar, 1971).

The optical spectroscopy of the ferrous state was expanded, when new weak bands in the near infrared were reported; one, located at 613 nm, was assigned to the porphyrin-to-iron transition that is the counterpart of the ferric state 695-nm band; two others, at 720 nm and 840 nm were assigned to d–d transitions, the first such assignment in a heme protein (Eaton and Charney, 1969).

Theorell's study of the magnetic properties of ferricytochrome c (Theorell, 1941), was extended in the 1960s and 1970s in two different dimensions. The theoretical interpretation of magnetometric measurements in terms of the valence-bond theory (Pauling, 1948), which had been the basis for classifying heme proteins into paramagnetic and diamagnetic, was superseded by the ligand-field theory and its equivalent concepts of high-spin and low-spin compounds (Griffith, 1961). Moreover, on the experimental side, electron paramagnetic resonance became a major research tool.

The first EPR measurement of ferricytochrome *c* (Salmeen and Palmer, 1968) showed the magnetic asymmetry of the low-spin crystal field, and permitted the calculation of magnetic moments at various temperatures that were in keeping with experimental measurements (Tasaki *et al.*, 1967). In a new type of classification of low-spin heme proteins (Blumberg and Peisach, 1971)—based on the rhombic and tetragonal distortions of the crystal field that surrounds the iron, which could be calculated from the *g* values measured by EPR spectroscopy—cytochrome *c* appeared as the typical example of large rhombicity and low tetragonality. Another interesting outcome of the EPR spectroscopy of cytochrome *c* was the finding that the iron *d*-electrons were partly delocalized into ligand orbitals, as shown by calculations of the spin–orbit coupling constant of the metal, which was estimated as 20% lower than the value for free ferric iron (Salmeen and Palmer, 1968). This type of "orbital reduction" is usually interpreted as due to metal-to-ligand electron donation (Owen, 1965).

Ferrous low-spin iron, which has no unpaired electrons, is silent in EPR measurements, but the *d*-shell can be explored through its influence on the 4*s*-electrons, which are observed in Mössbauer spectroscopy. An interesting result derived from the Mössbauer spectrum of ferrocytochrome *c*, in which no quadrupole splitting should be expected because of the spherical symmetry of the ferrous low-spin d^6 iron, was that the experimental observations nevertheless indicated values of about 1.2 mm s^{-1} (Cooke and Debrunner, 1968; Lang *et al.*, 1968). This can again be explained as the consequence of the delocalization of the metal *d*-electrons that destroys the spherical symmetry of the charge distribution around the iron nucleus, the observed quadrupole splittings being compatible with a charge displacement of 15% from each of the d_{xz} and d_{yz} orbitals (Cooke and Debrunner, 1968; Lang *et al.*, 1968). Furthermore, considerations regarding the unusually small difference between the isomer shifts observed in the Mössbauer spectra of ferric and ferrous cytochrome *c* lead to essentially similar conclusions (Cooke and Debrunner, 1968).

VII. Chemical Reactivity of Heme and Reduction Potentials

A major characteristic of cytochrome *c* is the sluggishness of its reactions with ligands in the ferric state, and the total lack of reactivity in the ferrous state (Margoliash and Schejter, 1966). The reactions of ferricytochrome *c* with cyanide were found to be very slow and pH-dependent (George and Tsou, 1952; George *et al.*, 1967). An analysis of the pH dependence based on the "crevice" equations formulated by George (George and Lyster, 1958) led to the postulation that a group with a pK of 5.5 was displaced by cyanide (George and Tsou, 1952; George *et al.*, 1967) or azide (George *et al.*, 1967). This was in keeping with Theorell's hypothesis of iron coordination by two protein imidazoles (Theorell, 1941), but when later it was found that the displaced ligand is a methionine sulfur, this explanation of the pH effect had to be abandoned. No other alternative has been yet proposed.

The rates of these binding reactions depend on the ligand concentrations, and although they can become very fast, they are eventually saturated. The limiting rate constant, 30 to 60 s^{-1}, has been interpreted as representing the S_{N_1} opening of the crevice, namely the cleavage of the sulfur–iron bond (Sutin and Yandell, 1972).

The reaction of ferrocytochrome c with CO can be observed only at very high pH (Theorell and Åkesson, 1941; Butt and Keilin, 1962), unless the protein is denatured (Margoliash et al., 1959; Kaminsky et al., 1972) or polymerized (Margoliash et al., 1959; Dupré et al., 1974). At pH 14, two CO molecules are bound, although the second one is spectrally silent (Moore et al., 1975). That the lack of reaction at neutral pH is due to thermodynamic constraints was demonstrated by jumping the pH of a solution of CO bound ferrocytochrome c back to neutrality, resulting in the reappearance of the native spectrum (George and Schejter, 1973) upon the reformation of the sulfur–iron bond (Davis et al., 1974).

The oxidation–reduction equilibrium of the ferric/ferrous cytochrome c couple had been thoroughly studied by the mid-1960s (see Margoliash and Schejter, 1966), and the value of the potential of the purified protein at pH 7 and 25° was established as $+255$ mV (Henderson and Rawlinson, 1956). A decade later, it was found that at neutral pH, in a nonbinding tris-cacodylate buffer, the observed potentials behaved in typical Debye–Hückel fashion at low ionic strengths, indicating a net change of charge of $+7$ to $+6$ upon reduction. At zero ionic strength, the thermodynamic E_0 was found to be $+261$ mV (Margalit and Schejter, 1973a). The thermodynamic parameters for the oxidized/reduced cytochrome c couple were also determined by direct calorimetry (George et al., 1968; Wad and Sturtevant, 1969) and through the use of the van't Hoff equation (Margalit and Schejter, 1973a); at pH 7.0 and $I = 0.1$, ΔH is -18 kcal mol^{-1}, and ΔS is -30 cal mol^{-1} K^{-1}. Beyond pH 8, the values of E_0 were found to diminish by 60 mV per pH unit, this being consistent with the ionization of a group with pK 8.6 in the ferric form, without a counterpart in the reduced protein up to pH 11 (Margalit and Schejter, 1973a). From direct calorimetric studies, the pK of this group was estimated as 9.3 (Wad and Sturtevant, 1969). Clearly, this is the same group as that involved in the transition from State III to State IV.

Electrophoretic experiments (Margoliash et al., 1970) demonstrated that, while the ferric form has a specific affinity for anions, such as phosphate, ADP, and ATP, ferrocytochrome c has a specific affinity for cations. This was reflected in the ionic strength dependence of E_0' in binding buffers (Margalit and Schejter, 1973a) and confirmed in chromatographic studies (Margalit and Schejter, 1973b; Margalit and Schejter, 1974).

Thus, although the thermodynamics of the redox reaction and the effects of the environment on its derived parameters have been thoroughly investigated, there is still no consensus on the extent to which the structure of the porphyrin (Falk and Perrin, 1961), the iron ligands (Williams, 1961), the heme environment (Kassner, 1973), and the electrostatic charge of the whole molecule (George, 1961; George et al., 1966) contribute to determine the reduction potential of cytochrome c.

The reduction potential of cytochrome c indicates that the oxidation of its ferrous state by oxygen is thermodynamically favorable; nevertheless, it was found that this reaction is very slow at neutral pH, and becomes very fast only at extreme pH values (Boeri and Tossi, 1955) or in the presence of salts (Davison *et al.*, 1971). Hydrogen peroxide is also an oxidant of ferrocytochrome c, a reaction that is catalyzed by the product, ferricytochrome c (Mochan and Degn, 1969) as well as by the (bis)histidine complex of copper (Davison and Hulett, 1971). A product of the reaction, most probably the superoxide anion, which is known to be a cytochrome c reductant (Ballou *et al.*, 1969; Koppenol *et al.*, 1976), slowly reduces the ferric cytochrome c formed back to the ferrous state, resulting in a cyclical mechanism (Davison and Hulett, 1971).

The oxidation of ferrocytochrome c by metal complexes was exemplified by a study of the reaction of the reduced molecule with tris(1,10-phenanthroline)cobalt (III) (McArdle *et al.*, 1974) that showed that the experimental results agreed with the predictions of the Marcus theory (Marcus, 1963), as estimated from the kinetics of electron self-exchange between ferric and ferrous cytochrome c (Redfield and Gupta, 1971; Gupta *et al.*, 1972). This was interpreted as favoring the hypothesis that electrons are transferred from the ferrocytochrome c iron to oxidants through the heme group edge (MacArdle *et al.*, 1974). A consensus was reached regarding a similar hypothesis for the mechanism of electron transfer to the iron of ferricytochrome c by ferrocyanide (Sutin, 1972), dithionite (Creutz and Sutin, 1973; Ewall and Bennett, 1974), hexaammineruthenium(II) (Kurihara and Sano, 1970), and ferrous EDTA (Hodges *et al.*, 1974; Stein and Swallow, 1954), and for the self-exchange reaction (Redfield and Gupta, 1971; Gupta *et al.*, 1972). For the reduction with dithionite, two mechanisms have been proposed: one is "outer-sphere," where the reductant is the dithionite ion, and the other could be "inner-sphere," also with dithionite, or by electron transfer from the SO_2^- anion radical (Lambeth and Palmer, 1973; Creutz and Sutin, 1973). Reduction of ferricytochrome c by free radicals or by aquated electrons generated *in situ* by pulse radiolysis have also been extensively investigated (Mee and Stein, 1956; Rabani and Stein, 1962; Shafferman and Stein, 1974; Land and Swallow, 1971; Wilting *et al.*, 1972, 1975; Pecht and Faraggi, 1971, 1972). It was found that the first product of the radiolytic reaction was spectroscopically and enzymically identical to the native protein, but large doses of radiation resulted in a modified protein, with loss of enzymic activity and a new band at 606 nm (Mee and Stein, 1956; Van Buuren *et al.*, 1976) that was attributed to the reduction of a pyrrole ring double bond (Van Buuren *et al.*, 1976).

An intriguing phenomenon was described as the "autoreduction" of ferricytochrome c (Boeri *et al.*, 1955). In the absence of dissolved oxygen, the ferric molecule slowly becomes reduced, and the rate of this reaction increases in alkaline solutions. The possibility that the protein could be the origin of the reducing equivalents was raised (Brady and Flatmark, 1971), but it was later shown that the actual reductant was the glycine present in the buffering medium (Aviram, 1972).

VIII. Other Spectroscopies: Protein Conformation Studies

A substantial number of studies of natural and magnetic optical rotatory dispersion spectra of cytochrome c published in the 1960s (see Moore and Pettigrew, 1990) were soon thereafter superseded by the introduction of instrumentation capable of evaluating ellipticities of optical absorption bands from their circular dichroism. Natural and magnetic CD were used for studies of protein conformation in solution, including denaturation phenomena, and for the assignment of certain absorption bands. Examples of the latter were the assignments of $d-d$ transitions in the near infrared region of the reduced state (Eaton and Charney, 1969).

CD spectra in the far-ultraviolet region are commonly employed to determine the helical content of the protein. Contrary to results obtained from optical rotation measurements, it was found that change in oxidation state had no effect on the CD spectrum in that region (Myer, 1968a,b, 1970; Zand and Vinogradov, 1968). Helical contents of 30% were estimated, and the presence of β structures was suggested.

The conformation of the protein under diverse conditions was studied by a variety of classic as well as newly developed techniques. It was shown from viscosity measurements that the protein acquires an extended chain conformation at pH 2.0 (Babul and Stellwagen, 1972; Cohen et al., 1974), a result supported by the observed increase of the fluorescent emission of the single tryptophan residue (Cohen et al., 1974). The sensitivity of this fluorescence made it a useful tool in the study of cytochrome c conformation changes (Tsong, 1974, 1975) and led to the interesting finding that chloride ions could partly reverse the unfolding of the protein molecule caused by acidity (Cohen et al., 1974) or by denaturing agents (Tsong, 1975).

Magnetic circular dichroism of cytochrome c (Sutherland and Klein, 1972; Risler and Groudinsky, 1973) was employed to verify the assignment of optical absorption bands, especially those in the near infrared (Eaton and Hochstrasser, 1967; Eaton and Charney, 1969), and to measure the angular momentum of the Q_0 bands of the ferric and ferrous states, which were found close to the theoretical value of 9 Bohr magnetons (Gouterman, 1961). Unexpectedly, the MCD of the Q_0 band was found to be much larger in the ferrous than in the ferric state (Sutherland and Klein, 1972). This, however, was not considered to result from a change in the magnetic moment, but to reflect the well-known broadening of the visible absorption band that occurs when cytochrome c is oxidized. This broadening had been earlier attributed to the increase in mixing of iron and porphyrin orbitals upon oxidation (Hochstrasser, 1971). Interestingly, the low-temperature splitting of the visible bands in the reduced state (Estabrook, 1957) was also observed in the MCD spectrum (Davydov et al., 1978).

Resonance Raman spectroscopy was introduced into the field as a means of investigating the vibrations of the cytochrome c heme and their interactions with the different electronic states (Spiro and Strekas, 1972; Strekas and Spiro, 1972; Brunner, 1973; Friedman and Hochstrasser, 1973; Adar and Erecińska, 1974). Various lines of the Raman spectra were specifically assigned to oxidation states and spin states (Spiro and Strekas, 1974; Spiro, 1975), and Raman spectroscopy was used to investigate the heme-linked ionizations of cytochrome c and the changes in ligation of its iron atom (Kitagawa et al., 1977; Kihara et al., 1978), and the lifetimes of the heme excited states (Friedman et al., 1977).

Finally, NMR spectroscopy was and will continue to be used mainly for structural studies of cytochrome c in solution in different states and conditions, and of its covalently modified analogues. Since the first spectrum obtained in a 56.4 MHz spectrometer (Kowalsky, 1962), successive improvements in the resolution of the instruments have rendered many of the earlier results obsolete. Worthy of mention because of their relevance to particular aspects of cytochrome *c* chemistry are the studies of the kinetics of crevice closing (Wüthrich, 1969), the measurements of electron self-exchange (Redfield and Gupta, 1971; Gupta *et al.*, 1972), and, in the ferric protein, the demonstration of the asymmetric delocalization of the unpaired electron spin in the heme plane (Redfield and Gupta, 1971).

IX. References

Abrams, R., Altschul, A., & Hogness, T. R. (1942) *J. Biol. Chem.* **142**, 303.

Adar, F., & Erecińska, M. (1974) *Arch. Biochem. Biophys.* **165**, 570.

Ambler, R. P. (1963) *Biochem. J.* **89**, 349.

Ando, K, Matsubara, H., & Okunuki, K. (1966) *Biochim. Biophys. Acta* **118**, 240.

Armstrong, G. D., Chapman, S. K., Sisley, M. J., Sykes, A. G., Osheroff, N., & Margoliash, E. (1986) *Biochemistry* **25**, 6947.

Augustin, M. A., Chapman, S. K., Davies, D. M., Sykes, A. G., Speck, S. H., & Margoliash, E. (1983) *J. Biol. Chem.* **258**, 6405.

Aviram, I. (1972) *Biochem. Biophys. Res. Commun.* **47**, 1120.

Aviram, I. (1973) *J. Biol. Chem.* **248**, 1894.

Aviram, I., & Krauss, Y. (1974) *J. Biol. Chem.* **249**, 2575.

Babcock, G. T., & Wikström, M. (1992) *Nature* **356**, 301.

Babul, J., & Stellwagen, E. (1972) *Biochemistry* **11**, 1195.

Ballou, D., Palmer, G., & Massey, V. (1969) *Biochem. Biophys. Res. Commun.* **36**, 898.

Blow, D. M., Bodo, G., Rossmann, M. G., & Taylor, C. P. S. (1964) *J. Mol. Biol.* **8**, 606.

Blumberg, W. E., & Peisach, J. (1971) in *Probes of Function and Structure of Macromolecules and Membranes* (Chance, B., Yonetani, T., & Mildvan, A. S., eds.), vol. 2, p. 215, Academic Press, New York.

Boardman, N. K., & Partridge, S. M. (1953) *Nature* **171**, 208.

Boardman, N. K., & Partridge, S. M. (1954) *J. Polymer. Sci.* **12**, 281.

Boardman, N. K., & Partridge, S. M. (1955) *Biochem. J.* **59**, 543.

Bodo, G. (1955) *Nature* **176**, 829.

Bodo, G. (1957) *Biochim. Biophys. Acta* **25**, 428.

Boeri, E., & Tossi, L. (1955) *Ricerca Sci.* **25**, Suppl., 209.

Boeri, E., Ehrenberg, A., Paul, K. G., & Theorell, H. (1953) *Biochim. Biophys. Acta* **12**, 273.

Boeri, E., Scardi, V., & Tosi, L. (1955) *Ricerca Sci.* **25**, Suppl., 209.

Boon, P. J., van Raay, A. J. M., Tesser, G. I., & Nivard, R. J. F. (1979) *FEBS Lett.* **108**, 131.

Brady, R. S., & Flatmark, T. (1971) *J. Mol. Biol.* **57**, 529.

Brandt, K. G., Parks, P. C., Czerlinski, G. H., & Hess, G. P. (1966) *J. Biol. Chem.* **241**, 4180.

Brautigan, D. L., Feinberg, B. A., Hoffman, B. M., Margoliash, E., Peisach, J., & Blumberg, W. A. (1977) *J. Biol. Chem.* **252**, 574.

Brautigan, D. L., Ferguson-Miller, S., & Margoliash, E. (1978a) *J. Biol. Chem.* **253**, 130.

Brautigan, D. L., Ferguson-Miller, S., Tarr, G. E., & Margoliash, E. (1978b) *J. Biol. Chem.* **253**, 140.

Brautigan, D. L., Ferguson-Miller, S., & Margoliash, E. (1978c) *Methods Enzym.* **53**, 128.

Brunner, H. (1973) *Biochem. Biophys. Res. Commun.* **51**, 888.

Brunori, M., Wilson, M. T., & Antonini, E. (1972) *J. Biol. Chem.* **247**, 6076.

Bushnell, G. W., Louie, G. V., & Brayer, G. D. (1990) *J. Mol. Biol.* **214**, 585.

Butler, W. L., & Hopkins, D. W. (1970) *Photochem. Photobiol.* **12**, 439.

Butler, W. L., Chapman, S. K, Davies, D. M., Sykes, A. G., Speck, S. M., Osheroff, N., & Margoliash, E. (1983) *J. Biol. Chem.* **258**, 6400.

Butt, W. D., & Keilin, D. (1962) *Proc. Roy. Soc. London, Series B* **156**, 429.

Capaldi, R. A. (1990) *Ann. Rev. Biochem.* **59**, 569.

Cohen, J. S., Fisher, W. R., & Schechter, A. N. (1974) *J. Biol. Chem.* **249**, 1113.

Cooke, R., & Debrunner, P. (1968) *J. Chem. Phys.* **48**, 4532.

Cooper, C. E., Nicholls, P., & Freedman, J. A. (1991) *Biochem. Cell Biol.* **69**, 586.

Corradin, G., & Harbury, H. A. (1974) *Biochem. Biophys. Res. Commun.* **61**, 4100.

Creutz, C., & Sutin, N. (1973) *Proc. Natl. Acad. Sci. USA* **70**, 1701.

Czerlinski, G. H., & Dar, K. (1971) *Biochim. Biophys. Acta* **234**, 57.

Das, G., Hickey, D. R., McLendon, D., McLendon, G., & Sherman, F. (1989) *Proc. Natl. Acad. Sci. USA* **86**, 496.

Davis, L. A., Schejter, A., & Hess, G. P. (1974) *J. Biol. Chem.* **249**, 2624.

Davison, A. J., & Hulett, L. G. (1971) *Biochim. Biophys. Acta* **226**, 313.

Davison, A. J., Hamilton, R. T., & Kaminsky, L. S. (1971) *FEBS Lett.* **19**, 19.

Davydov, R. M., Magonov, S. N., Arutnyunyan, A. M., & Shronov, Yu. A. (1978) *Molekulyarnaya Biologiya* **12**, 1341.

Dethmers, J. K., Ferguson-Miller, S., & Margoliash, E. (1979) *J. Biol. Chem.* **254**, 11973.

Dickerson, R. E., Kopka, M. L., Borders, C. L., Varnum, J., Weinzierl, J. E., & Margoliash, E. (1967a) *J. Mol. Biol.* **29**, 77.

Dickerson, R. E., Kopka, M. L., Weinzierl, J. E., Varnum, J. C., Eisenberg, D., & Margoliash, E. (1967b) *J. Biol. Chem.* **242**, 3015.

Dickerson, R. E., Kopka, M. L., Weinzierl, J. E., Varnum, J. C., Eisenberg, D., & Margoliash, E. (1968) in *An Interpretation of a Two-Derivative 4Å Resolution Electron Density Map of Horse Heart Ferricytochrome c* (Okunuki, K., Kamen, M. D., & Sekuzu, eds.), p. 225, University of Tokyo Press, Tokyo.

Dickerson, R. E., Takano, T., Eisenberg, D., Kallai, O. B., Samson, L., Cooper, A., & Margoliash, E. (1971) *J. Biol. Chem.* **246**, 1511.

Dixon, M., Hill, R., & Keilin, D. (1931) *Proc. Roy. Soc. London, Series B* **109**, 29.

Dupré, S., Brunori, M., Wilson, M. T., & Greenwood, C. (1974) *Biochem. J.* **141**, 299.

Dyson, H. J., & Beattie, J. K. (1982) *J. Biol. Chem.* **257**, 2267.

Eaton, A. A., & Hochstrasser, R. (1967) *J. Chem. Phys.* **46**, 2533.

Eaton, W. A., & Charney, E. (1969) *J. Chem. Phys.* **48**, 2049.

Ehrlich, P. (1885) *Das Sauerstoff-Bedürfnis des Organismus*, Hirschwald, Berlin.

Elliot, W. B., & Margoliash, E. (1971) in *Developments in Applied Spectroscopy* (Grove and Perkins, eds.), vol. 9, p. 125, Plenum Press, New York.

Erman, J. E., & Yonetani, T. (1975) *Biochim. Biophys. Acta* **393**, 350.

Estabrook, R. W. (1957) *J. Biol. Chem.* **223**, 781.

Estabrook, R. W. (1961) in *Haematin Enzymes* (Falk, J. E., Lemberg, R., & Morton, R. K., eds.), vol. 2, p. 436, Pergamon Press, Oxford.

Estabrook, R. W., & Sacktor, B. (1958) *Arch. Biochem. Biophys.* **76**, 532.

Ewall, R. X., & Bennett, L. E. (1974) *J. Am. Chem. Soc.* **96**, 940.

Falk, J. E., & Perrin, D. D. (1961) in *Haematin Enzymes* (Falk, J. D., Lemberg, R., & Morton, R. K., eds.), Part I, p. 56, Pergamon Press, Oxford.

Ferguson-Miller, S., Brautigan, D. L., & Margoliash, E. (1976) *J. Biol. Chem.* **251**, 1104.

Ferguson-Miller, S., Brautigan, D. L., & Margoliash, E. (1978) *J. Biol. Chem.* **253**, 149.

Ferguson-Miller, S., Brautigan, D. L., & Margoliash, E. (1979) in *The Porplyrins*, Vol. III (Dolphin, D., ed.), p. 149, Academic Press, New York.

Ferrer, J. C., Guillemette, J. G., Bogumil, R., Inglis, S. C., Smith, M., & Mauk, A. G. (1993) *J. Am Chem. Soc.* **115**, 7507.

Fitch, W. M., & Margoliash, E. (1967) *Science* **155**, 279.

Fitch, W. M., & Margoliash, E. (1970) *Evolutionary Biology* **4**, 68.

Fontana, A., Veronese, F. M., & Boccu, E. (1973) *FEBS Lett.* **32**, 135.

Fridovich, I., & Handler, P. (1956a) *J. Biol. Chem.* **221**, 323.

Fridovich, I., & Handler, P. (1956b) *J. Biol. Chem.* **223**, 321.

Friedman, J. M., & Hochstrasser, R. M. (1973) *Chem. Phys.* **1**, 457.

Friedman, J. M., Rousseau, D. L., & Adar, F. (1977) *Proc. Natl. Acad. Sci. USA* **74**, 2607.

Fung, D., & Vinogradow, S. (1968) *Biochim. Biophys. Res. Commun.* **31**, 596.

Garber, E. A. E., & Margoliash, E. (1990) *Biochim. Biophys. Acta* **1015**, 279.

George, P. (1953a) *J. Biol. Chem.* **201**, 413.

George, P. (1959b) *Biochem. J.* **54**, 267.

George, P. (1961) in *Haematin Enzymes* (Falk, J. E., Lemberg, R., & Morton, R. K., eds.), Part I, p. 71, Pergamon Press, Oxford.

George, P., & Lyster, R. L. J. (1958) *Proc. Natl. Acad. Sci. USA* **44**, 1013.

George, P., & Schejter, A. (1973) *J. Biol. Chem.* **239**, 1504.

George, P., & Tsou, C. L. (1952) *Biochem. J.* **50**, 443.

George, P., Eaton, W. A., & Trachtman, M. (1968) *Fed. Proc.* **27**, 520.

George, P., Glauser, S. C., & Schejter, A. (1967) *J. Biol. Chem.* **242**, 1690.

George, P., Hanania, G. I. H., & Eaton, W. A. (1966) in *Hemes and Hemoproteins* (Chance, B., Estabrook, R. W., & Yonetani, T., eds.), p. 267, Academic Press, New York.

Goldin, H. H., & Farnsworth, M. W. (1966) *J. Biol. Chem.* **241**, 3590.

Gouterman, M. (1961) *J. Mol. Spect.* **6**, 138.

Greenwood, C., & Palmer, G. P. (1965) *J. Biol. Chem.* **240**, 3660.

Griffith, J. S. (1961) *The Theory of Transition Metal Ions*, Cambridge University Press, Cambridge, U.K.

Gupta, R. K., Koenig, S. H., & Redfield, A. G. (1972) *J. Magn. Resonance* **7**, 66.

Hannum, C. H., & Margoliash, E. (1986) *J. Immunol.* **135**, 3303.

Hannum, C. H., Matis, L. A., Schwartz, R. H., & Margoliash, E. (1985) *J. Immunol.* **135**, 3314.

Heinberg, M., Fridovich, I., & Handler, P. (1963) *J. Biol. Chem.* **204**, 913.

Henderson, R. W., & Rawlinson, W. E. (1956) *Biochem. J.* **62**, 21.

Hickey, D. R., Berghuis, A. M., Lafond, G., Jaeger, J. A., Cardillo, T. S., McLendon, D., Das, G., Sherman, F., Brayer, G. D., & McLendon, G. (1991) *J. Biol. Chem.* **266**, 11686.

Hill, R., & Keilin, D. (1930) *Proc. Roy. Soc.* **B107**, 286.

Hill, R., & Keilin, D. (1933) *Proc. Roy. Soc.* **B114**, 104.

Hochstrasser, R. M. (1971) in *Probes of Structure and Function of Macromolecules and Enzymes* (Chance, B., Lee, C. P., & Blasie, J. K., eds.), vol. 1, p. 57, Academic Press, New York.

Hodges, H. L., Holwerda, R. A., & Gray, H. B. (1974) *J. Am. Chem. Soc.* **96**, 3132.

Hoppe-Seyler, F. (1890a) *Z. Physiol. Chem.* **14**, 106.

Hoppe-Seyler, F. (1890b) *Z. Physiol. Chem.* **14**, 329.

Horecker, B. L., & Kornberg, A. (1946) *J. Biol. Chem.* **165**, 11.

Horecker, B. L., & Stannard, J. N. (1948) *J. Biol. Chem.* **172**, 589.

Juillerat, M., Parr, G. R., & Tanuichi, H. (1980) *J. Biol Chem.* **255**, 845.

Kadenbach, B., Stroh, A., Huether, F. J., Reimann, A., & Steverding, D. (1991) *J. Bioenerg. Biomembr.* **23**, 321.

Kaminsky, L. S., Burger, P. E., Davidson, A. J., & Helfet, D. (1972) *Biochemistry* **11**, 3702.

Kang, C. H., Ferguson-Miller, S., & Margoliash, E. (1977) *J. Biol. Chem.* **252**, 919.

Kang, C. H., Brautigan, D. L., Osheroff, N., & Margoliash, E. (1978) *J. Biol. Chem.* **253**, 6502.

Kassner, R. J. (1973) *J. Am. Chem. Soc.* **95**, 2674.

Keilin, D. (1925) *Proc. Roy. Soc. London, Series B* **98**, 312.

Keilin, D. (1926) *Proc. Roy. Soc. London, Series B* **100**, 129.

Keilin, D. (1927) *Nature* **119**, 670.

Keilin, D. (1930) *Proc. Roy. Soc.* **B106**, 418.

Keilin, D. (1966) *The History of Cell Respiration and Cytochrome*, University Press, Cambridge, U.K.

Keilin, D., & Hartree, E. F. (1937) *Proc. Roy. Soc.* **B122**, 298.

Keilin, D., & Hartree, E. F. (1938a) *Proc. Roy. Soc.* **B125**, 171.

Keilin, D., & Hartree, E. F. (1938b) *Nature* **141**, 870.

Keilin, D., & Hartree, E. F. (1939) *Proc. Roy. Soc.* **B127**, 167.

Keilin, D., & Hartree, E. F. (1940) *Proc. Roy. Soc.* **B129**, 277.

Keilin, D., & Hartree, E. F. (1945) *Biochem J.* **39**, 289.

Keilin, D., & Hartree, E. F. (1949) *Nature* **164**, 254.

Keilin, D., & Slater, E. C. (1953) *Brit. Med. Bull.* **9**, 89.

Keller, R. M., Aviram, I., Schejter, A., & Wüthrich, K. (1972) *FEBS Lett.* **20**, 90.

Kihara, H., Koyu, H., & Kitagawa, T. (1978) *Biochim. Biophys. Acta* **532**, 337.

Kitagawa, T., Kyogoku, Y., Iizuka, T., Ikeda-Saito, M., & Yamanaka, T. (1975) *J. Biol. Chem.* **78**, 719.

Kitagawa, T., Ozaki, Y., Teraoka, J., Kyogoku, Y., & Yamanaka, T. (1977) *Biochim. Biophys. Acta* **494**, 100.

König, B. W., Osheroff, N., Wilms, J., Muijsers, A. O., Dekker, H. L., & Margoliash, E. (1980) *FEBS Lett.* **111**, 395.

Koppenol, W. H., & Margoliash E. (1982) *J. Biol. Chem.* **257**, 4426.

Koppenol, W. H., Van Buren, K J. H., Butler, J., & Braams, R. (1976) *Biochim. Biophys. Acta* **449**, 157.

Kowalsky, A. (1962) *J. Biol. Chem.* **237**, 1807.

Kubowitz, F., & Haas, E. (1932) *Biochem. Z.* **225**, 247.

Kuby, S. A., Paléus, S., Paul, K.-G., & Theorell, H. (1956) *Acta Chem. Scand.* **10**, 148.

Kurihara, M., & Sano, S. (1970) *J. Biol. Chem.* **245**, 4804.

Lambeth, D. O., & Palmer G. (1973) *J. Biol. Chem.* **248**, 6095.

Land, E. J., & Swallow, A. J. (1971) *Arch. Biochem. Biophys.* **145**, 365.

Lang, G., Herbert, D., & Yonetani, T. (1968) *J. Chem. Phys.* **49**, 944.

Lanir, A., & Aviram, I. (1975) *Arch. Biochem. Biophys.* **166**, 439.

Levy, L. (1889) *Z. Physiol. Chem.* **13**, 309.

Limbach, K. J., & Wu, R. (1983) *Nucleic Acids Res.* **11**, 8931.

Limbach, K. L., & Wu, R. (1985) *Nucleic Acids Res.* **13**, 631.

Louie, G. V., & Brayer, G. D. (1989) *J. Mol. Biol.* **210**, 313.

Louie, G. V., & Brayer, G. D. (1990) *J. Mol. Biol.* **214**, 527.

Louie, G. V., Hutcheon, W. L. B., & Brayer, G. P. (1988a) *J. Mol. Biol.* **199**, 295.

Louie, G. V., Pielak, G. J., Smith, M., & Brayer, G. D. (1988b) *Biochemistry* **27**, 7870.

Luntz, T. L., Schejter, A., Garber, E. A. E., & Margoliash, E. (1989) *Proc. Natl. Acad. Sci. USA* **86**, 3524.

MacMunn, C. A. (1886) *Proc. Roy. Soc. London* **39**, 248.

MacMunn, C. A. (1887) *J. Physiol. London* **8**, 51.

MacMunn, C. A. (1889) *Z. Physiol. Chem.* **13**, 497.

MacMunn, C. A. (1890) *Z. Physiol. Chem.* **14**, 328.

MacMunn, C. A. (1914) *Spectrum Analysis Applied to Biology and Medicine*, Longmans, Green & Co., London.

Malatesta, F., Antonini, G., Sarti, P., & Brunori, M. (1990) *Ettore Majorana Int. Sci. Ser.: Phys. Sci.* **51**, 213.

Malmström, B. G. (1990) *Chem. Rev.* **90**, 1247.

Marcus, R. A. (1963) *J. Phys. Chem.* **67**, 853.

Margalit, R., & Schejter, A. (1973a) *Eur. J. Biochem.* **32**, 492.

Margalit, R., & Schejter, A. (1973b) *Eur. J. Biochem.* **32**, 500.

Margalit, R., & Schejter, A. (1974) *Eur. J. Biochem.* **46**, 387.

Margoliash, E. (1952) *Nature* **170**, 1014.

Margoliash, E. (1954a) *Biochem. J.* **56**, 529.

Margoliash, E. (1954b) *Biochem. J.* **56**, 535.

Margoliash, E. (1962) *J. Biol. Chem.* **237**, 2161.

Margoliash, E. (1963) *Proc. Natl. Acad. Sci. USA* **50**, 672.

Margoliash, E. (1972) *Harvey Lect.* **66**, 175.

Margoliash, E., & Bosshard, E. R. (1983) *TIBS* **8**, 316.

Margoliash, E., & Fitch, W. M. (1971) *Taxon* **20**, 51.

Margoliash, E., & Frohwirth, N. (1959) *Biochem. J.* **71**, 570.

Margoliash, E., & Schejter, A. (1966) *Adv. Prot. Chem.* **21**, 113.

Margoliash, E., & Smith, E. L. (1962) *J. Biol. Chem.* **237**, 2151.

Margoliash, E., & Walasek, O. F. (1967) *Methods Enzymol.* **10**, 339.

Margoliash, E., Frohwirth, N., & Wiener, E. (1959) *Biochem. J.* **71**, 559.

Margoliash, E., Smith, E. L., Kreil, G., & Tuppy, H. (1961) *Nature* **192**, 1121.

Margoliash, E., Kimmel, J. R., Hill, R. L., & Schmidt, W. R. (1962) *J. Biol. Chem.* **237**, 2148.

Margoliash, E., Barlow, G. H., & Byers, V. (1970) *Nature* **228**, 723.

Margoliash, E., Fitch, W. M., & Dickerson, R. W. (1971) in *Biochemical Evolution and the Origin of Life* (Schoffeniels, E., ed.), p. 52, North-Holland Publishing Company, Amsterdam.

Margoliash, E., Dickerson, R. E., & Adler, A. D. (1972) in *The Molecular Basis of Electron Transport* (Schultz, J., & Cameron, B. F., eds.), p. 153, Academic Press, New York.

McArdle, J. V., Gray, H. B., Creutz, C., & Sutin, N. (1974) *J. Am. Chem. Soc.* **96**, 5737.

Mee, L. K., & Stein, G. (1956) *Biochem. J.* **62**, 377.

Miller, D. D. (1988) Thesis, Northwestern University, Evanston, Illinois.

Mochan, E., & Degn, H. (1969) *Biochim. Biophys. Acta* **189**, 354.

Montgomery, D. L., Hall, B. D., Gillam, S., & Smith, M. (1978) *Cell* **14**, 673.

Moore, G. R., & Pettigrew, G. W. (1990) *Cytochromes c*. Springer-Verlag, Berlin.

Moore, T. A., Greenwood, C., & Wilson, M. T. (1975) *Biochem. J.* **147**, 335.

Mueller, M., & Azzi, A. (1991) *J. Bioenerg Biomembr.* **23**, 291.

Myer, Y. P. (1968a) *Biochemistry* **7**, 765.

Myer, Y. P. (1968b) *Biochim. Biophys. Acta* **154**, 84.

Myer, Y. P. (1970) *Biochim. Biophys. Acta* **214**, 94.

Neilands, J. B. (1952) *J. Biol. Chem.* **197**, 701.

Neilands, J. B., & Tuppy, H. (1960) *Biochim. Biophys. Acta* **38**, 351.

Nicholls, P. (1963) in *The Enzymes* (Boyer, P. D., Lardy, H., & Myrbäck, K., eds.), vol. 8, p. 3, Academic Press, New York.

Ochi, H., Hata, Y., Tanaka, N., & Kakude, M. (1983) *J. Mol. Biol.* **166**, 407.

Offord, R. E. (1980) *Semisynthetic Proteins*, J. Wiley & Sons, New York.

Okunuki, K. (1960) in *A Laboratory Manual of Analytical Methods of Protein Chemistry* (Alexander, P., & Block, R. J., eds.), vol. I, p. 32, Pergamon Press, Oxford.

Osheroff, N., Brautigan, D. L., & Margoliash, E. (1980) *J. Biol. Chem.* **255**, 8245.

Osheroff, N., Speck, S. H., Margoliash, E., Veerman, E. C. I., Wilms, J., Konig, B. W., & Muijsers, A. O. (1983) *J. Biol. Chem.* **258**, 5731.

Oshino, N., & Chance, B. (1975) *Arch. Biochem. Biophys.* **170**, 514.

Owen, J. (1955) *Proc. Roy. Soc. London, Series A* **227**, 183.

Paléus, S., & Neilands, J. B. (1950) *Acta Chem. Scand.* **4**, 1024.

Paul, K.-G. (1950) *Acta Chem. Scand.* **4**, 239.

Paul, K.-G. (1951a) *Acta Chem. Scand.* **5**, 379.

Paul, K.-G. (1951b) *Acta Chem. Scand.* **5**, 389.

Pauling, L. (1948) *The Nature of the Chemical Bond.* Cornell University Press, Ithaca, New York.

Pecht, I., & Faraggi, M. (1971) *FEBS Lett.* **13**, 221.

Pecht, I., & Faraggi, M. (1972) *Proc. Natl. Acad. Sci. USA* **69**, 902.

Pettigrew, G. W., Aviram, I., & Schejter, A. (1976) *Biochem. Biophys. Res. Commun.* **68**, 807.

Pielak, G. J., Mauk, A. G., & Smith, M. (1985) *Nature* **313**, 152.

Pielak, G. J., Boyd, J., Moore, G. R., & Williams, R. J. P. (1988a) *Eur. J. Biochem.* **177**, 167.

Pielak, G. J., Atkinson, R. A., Boyd, J., & Williams, R. J. P. (1988b) *Eur. J. Biochem.* **177**, 179.

Powell, G. L., & Marsh, D. (1992) in *Structure of Biological Membranes* (Yeagle, P. L., ed.), p. 1145, CRC Press, Boca Raton, Florida.

Qi, P. X., Urbauer, J. L., Fuentes, E. J., Leopold, M. F., & Wand, A. J. (1994) *Nature Struct. Biol.* **1**, 378.

Rabani, M., & Stein, G. (1962) *Radiat. Res.* **17**, 327.

Rajagopalan, K V. (1980) in *Molybdenum and Molybdenum Containing Enzymes* (Kaplan M., ed.), p. 243, Pergamon Press, New York.

Redfield, A. G., & Gupta, R. K. (1971) *Cold Spring Harbor Symp. Quant. Biol.* **36**, 405.

Risler, J. L., & Groudinsky, O. (1973) *Eur. J. Biochem.* **35**, 201.

Roberts, V. A., Freeman, H. C., Olson, A. J., Tainer, J. A., & Getzoff, E. D. (1991) *J. Biol. Chem.* **266**, 13431.

Roche, J., Derrien, Y., & Cahnmann, J. (1946) *Compt. Rend. Soc. Biol.* **140**, 146.

Salmeen, I., & Palmer, G. (1968) *J. Chem. Phys.* **48**, 2049.

Sanger, F., & Thompson, E. O. P. (1953) *Biochem. J.* **53**, 353.

Sanger, F., & Tuppy, H. (1951) *Biochem. J.* **49**, 463.

Sanishvili, R. G., Margoliash, E., Westbrook, M. L., Westbrook, E. M., & Volz, K. W. (1994) *Acta Cryst.* **D50**, 687.

Sano, S. (1972) in *Structure and Function of Oxidation Reduction Enzymes* (Åkesson, Å., & Ehrenberg, A., eds.), p. 69, Pergamon Press, Oxford.

Sano, S., & Granick, S. (1961) *J. Biol. Chem.* **236**, 1173.

Sano, S., & Tanaka, K. (1964) *J. Biol. Chem.* **239**, PC3109.

Sano, S., Ikeda, K., & Sakakibara, S. (1964) *Biochem. Biophys. Res. Commun.* **15**, 284.

Scarpulla, R. C., Agne, K. M., & Wu, R. (1981) *J. Biol. Chem.* **256**, 6480.

Schechter, E., & Saludjian, P. (1967) *Biopolymers* **5**, 788.

Schejter, A., & Aviram, I. (1969) *Biochemistry* **8**, 149.

Schejter, A., & Aviram, I. (1970) *J. Biol. Chem.* **245**, 1552.

Schejter, A., & George, P. (1964) *Biochemistry* **3**, 1045.

Schejter, A., & Margoliash, E. (1984) *TIBS* **9**, 364.

Schejter, A., Glauser, S. C., George P., & Margoliash, E. (1963) *Biochim. Biophys. Acta* **73**, 641.

Schejter, A., Luntz, T. L., Koshy, T. I., & Margoliash, E. (1992) *Biochemistry* **31**, 8336.

Shafferman, A., & Stein, G. (1974) *Science* **183**, 428.

Sherman, F., Stewart, J. W., Margoliash, E., Parker, J., & Campbell, W. (1966) *Proc. Natl. Acad. Sci. USA* **55**, 1498.

Smith, M. (1985) *Ann. Rev. Genet.* **19**, 423.

Smith, M., Leung, D. W., Gillam, S., Astell, C. R., Montgomery, D. L., & Hall, B. D. (1979) *Cell* **16**, 753.

Speck, S. H., & Margoliash, E. (1984) *J. Biol. Chem.* **259**, 1064.

Speck, S. H., Ferguson-Miller, S., Osheroff, N., & Margoliash, E. (1979) *J. Biol. Chem.* **253**, 8957.

Speck, S. H., Koppenol, W. H., Dethmers, J. K., Osheroff, N., Margoliash, E., & Rajagopalan, K. V. (1981) *J. Biol. Chem.* **256**, 7394.

Speck, S. H., Dye, D., & Margoliash, E. (1984) *Proc. Natl. Acad. Sci. USA* **81**, 347.

Spackman, D. H., Stein, W. H., & Moore, S. (1958) *Anal. Chem. 30*, 1190.

Spiro, T. G. (1975) *Biochim. Biophys. Acta 416*, 169.

Spiro, T. G., & Strekas, T. C. (1972) *Proc. Natl. Acad. Sci. USA 69*, 2622.

Spiro, T. G., & Strekas, T. C. (1974) *J. Am. Chem. Soc. 96*, 338.

Stein, G., & Swallow, A. J. (1954) *Nature 173*, 937.

Strekas, T. C., & Spiro, T. G. (1972) *Biochim. Biophys. Acta 278*, 188.

Stellwagen, E., Babul, J., & Wilgus, H. (1975) *Biochim. Biophys. Acta 405*, 115.

Stemp, E. D. A., & Hoffman, B. M. (1993) *Biochemistry 32*, 10848.

Summer, J. B., & Somers, G. F. (1953) in *Chemistry and Methods of Enzymes* (3rd ed.), p. 227, Academic Press, New York.

Sutherland, J. C., & Klein, M. P. (1972) *J. Chem. Phys. 57*, 76.

Sutin, N. (1972) *Chem. Brit. 8*, 148.

Sutin, N., & Yandell, J. K. (1972) *J. Biol. Chem. 247*, 6932.

Swanson, R., Trus, B. L., Mandel, N., Mandel, G., Kallai, O. B., & Dickerson, R. E. (1977), *J. Biol. Chem. 252*, 759.

Swanson, M. S., Zieminn, S. M., Miller, D. D., Garber, E. A. E., & Margoliash, E. (1985) *Proc. Natl. Acad. Sci. USA 82*, 1964.

Takano, T., & Dickerson, R. E. (1981a) *J. Mol. Biol. 153*, 79.

Takano, T., & Dickerson, R. E. (1981b) *J. Mol. Biol. 153*, 95.

Takano, T., Trus, B. L., Mandel, N., Mandel, G., Kallai, O. B., Swanson, R., & Dickerson, R. E. (1977) *J. Biol. Chem. 252*, 776.

Tasaki, A., Otsuka, J., & Kotani, M. (1967) *Biochim. Biophys. Acta 140*, 284.

Theorell, H. (1936) *Biochem. Z. 285*, 207.

Theorell, H. (1937) *Enzymologia 4*, 192.

Theorell, H. (1938) *Biochem. Z. 298*, 242.

Theorell, H. (1939) *Enzymologia 6*, 88.

Theorell, H. (1941) *J. Am. Chem. Soc. 63*, 1820.

Theorell, H., & Åkesson, Å. (1939) *Science 90*, 67.

Theorell, H., & Åkesson, Å. (1941) *J. Am. Chem. Soc. 63*, 1804, 1812, 1818.

Thunberg, T. (1927) *Skand. Arch. Physiol. 35*, 163.

Tsong, T. Y. (1974) *J. Biol. Chem. 249*, 1988.

Tsong, T. Y. (1975) *Biochemistry 14*, 1542.

Tsou, C.-L. (1951) *Biochem. J. 49*, 362.

von Jagow, G., & Sebald, W. (1990) *Ann. Rev. Biochem. 49*, 281.

von Jagow, G., Link, T. A., & Schaegger, H. (1987) in *Advances in Membrane Biochemistry Bioenergetics* (Kim, C. H., Tedeschi, H., Diwan, J. J., & Salerno, J. C., eds.), Plenum, New York.

Van Buuren, K. J. H., Wilting, J., & Braams, R. (1976) *Biochim. Biophys. Acta 423*, 586.

Wad, G. D., & Sturtevant, Y. M. (1969) *Biochemistry 8*, 4567.

Walasek, O. F., & Margoliash, E. (1977) *J. Biol. Chem. 252*, 830.

Wallace, C. J. A., & Offord, R. E. (1979) *Biochem. J. 179*, 169.

Wallace, C. J. A., & Rose, K. (1983) *Biochem J. 215*, 651.

Walter, M. H., Westbrook, E. M., Tykodi, S., Uhm, A. M. & Margoliash, E. (1990) *J. Biol. Chem. 265*, 4177.

Wang, Y., & Margoliash, E. (1995) *Biochemistry 34*, 1948.

Warburg, O. (1924) *Biochem. Z. 152*, 479.

Warburg, O. (1925) *Ber. Deut. Chem. Ges. 58*, 1001.

Warburg, O. (1932) *Z. Angew. Chem. 45*, 1.

Warburg, O. (1949) *Heavy Metal Prosthetic Groups and Enzyme Action*, Oxford University Press, Oxford.

Warburg, O., & Negelein, E. (1929a) *Biochem. Z. 204*, 495.

Warburg, O., & Negelein, E. (1929b) *Biochem. Z. 214*, 64.

Warburg, O., & Negelein, E. (1929c) *Biochem. Z. 214*, 101.

Weiland, H. (1932) *On the Mechanisms of Oxidation*, Yale University Press, New Haven, Connecticut.

Weiss, H. (1987) *Curr. Top. Bioenerg. 15*, 67.

Weiss, H., Leonard, K., & Neupert, W. (1990) *TIBS 15*, 178.

Westerhuis, L. W., Tesser, G. I., & Nivard, R. J. F. (1982) *Internatl. J. Peptide Protein Res. 19*, 290.

Williams, R. J. P. (1961) in *Haematin Enzymes* (Falk, J. D., Lemberg, R., & Morton, R. K, eds.), Part I, p. 41, Pergamon Press, Oxford.

Wilting, J., Braams, R., Nauta, H., & Van Buuren, K. J. H. (1972) *Biochim. Biophys. Acta* **283**, 543.

Wilting, J., Van Buuren, K. J. H., Braams, R., & Van Gelder, B. F. (1975) *Biochim. Biophys. Acta* **376**, 285.

Wüthrich, K (1969) *Proc. Natl. Acad. Sci. USA* **63**, 1071.

Yonetani, T. (1965) *J. Biol. Chem.* **240**, 4509.

Yonetani, T. (1971) *Ado. Enzymol.* **33**, 309.

Zand, R., & Vinogradov, S. (1968) *Arch. Biochem. Biophys.* **125**, 94.

Zerner, M., Gouterman, M., & Kobayashi, H. (1966) *Theoret. Chim. Acta* **6**, 363.

Zhou, J. S., Brothers, H. M., Neddersen, J. P., Peerey, L. M., Cotton, T. M., & Kostic, N. M. (1992) *Bioconjugate Chem.* **3**, 382.

2

Evolution and Classification of c-Type Cytochromes

T. E. Meyer

Department of Biochemistry
University of Arizona

I. Introduction

In an article on the evolution of c-type cytochromes, one expects to see phylogenetic trees. However, it is the author's opinion that such trees are only valid when applied to the higher eukaryotes, in which there are only two known kinds of c-type cytochromes. The quantification of the evolution of mitochondrial cytochrome c has been adequately dealt with in numerous other articles, and there is little more to add. Bacteria, on the other hand, are the center of abundance of cytochromes, and they are where one finds the greatest variation in cytochrome structure and function. For these proteins, the evolutionary changes have been too great to be treated in the same manner as have the mitochondrial cytochromes c. The description of their evolution is of necessity qualitative. Thus, there will be no conventional trees in this review. Nevertheless, bacterial cytochromes c can be classified based on their sequences and three-dimensional structures into at least four totally unrelated types and at least sixteen distinct subtypes. The list is still growing, particularly in the area of multiheme proteins. We are still at the descriptive level and have only begun to utilize genetic approaches and to assign function to these proteins. The outline of this article is similar to that of Meyer and Kamen (1982). It has been updated and expanded with as little repetition as possible. The emphasis is on evolutionary relationships to the extent that they can be discerned. One should refer to the previous review for sequence alignments.

The evolution of cytochromes has often been approached from a purely structural viewpoint without regard for function. However, structure and function are intimately connected and highly relevant to evolutionary studies. It is the function which provides direction to the evolutionary process and sets limits on the extent of variation. Gene duplication and gene transfer allow quantum leaps in structural variation without the normal restraints of function. As newly duplicated or transferred cytochromes adapt to new functions, there is a period of rapid change which eventually reaches a limit of variability necessitated by the new function. If there is no function to which a "new" cytochrome may adapt, it will not survive and will not be observed in more than a few anomalous species. Although the precise functional role of most cytochromes is unknown beyond generalized electron transfer, it can be assumed that a cytochrome found to have similar properties in a number of species has a useful role. What we are beginning to see is the emergence of distinct classes of cytochromes in bacteria which have readily distinguishable characteristics and metabolic pathways—e.g., purple phototrophs, green phototrophs, cyanobacteria and algae, sulfate-reducers, denitrifiers, methylotrophs, gram-negative aerobes, gram-positive aerobes, halophiles, fermenters, and methanogens. Eukaryotes are metabolically far more uniform than are bacteria, there are fewer kinds of eukaryotic cytochromes, and there is presumably no eukaryotic gene transfer. It is fair to say that bacterial systems are simpler to study (bacterial oxidases typically have three subunits instead of twelve). The bacterial cytochromes do not have the complication of N-terminal acetylation, nor are lysines trimethylated. Bacterial genes do not contain introns, and they are easier to express for overproduction of proteins in a variety of bacterial hosts.

Exciting new discoveries in the past decade include the three-dimensional structure of the photosynthetic reaction center with its associated tetraheme cytochrome, a cytochrome oxidase subunit 2 fused to a cytochrome c gene, and the genetic analysis of the role of cytochrome c_2. The sequences of cytochromes cd_1, flavocytochromes c, methanol and ethanol dehydrogenase–associated cytochromes c, and cytochromes c_1 and f all expand our understanding of the numbers of kinds of cytochromes. The $D.$ $vulgaris$ 16-heme cytochrome c_3, the tetraheme reaction center cytochrome c, the $Ps.$ $stutzeri$ tetraheme cytochrome c, and several diheme cytochromes such as $Ps.$ $stutzeri$ cytochrome c-552, $Pseudomonas$ cytochrome c_4, and $Ps.$ $aeruginosa$ cytochrome c peroxidase emphasize the role of tandem gene duplication and fusion. Finally, there are reports of additional multiheme cytochromes, which have not yet been sequenced, and therefore will not be reviewed, but which nevertheless promise to be interesting from the evolutionary standpoint.

A. Definition and Recognition

Cytochromes c may be defined as electron transfer proteins which have protoheme covalently attached via two thioether bridges between cysteine residues of the protein and the vinyl side chains of the heme. Although the original definition included all covalently bound heme, no other types of heme are known to be covalently bound, and cysteine is the only amino acid residue which is known to covalently bind heme. This also applies to the phycobiliproteins, which contain linear tetrapyrroles presumably derived from protoheme and bound to the protein via one or two cysteine thioether bridges. Thus, an evolutionary link between phycobiliproteins and c-type cytochromes may eventually be discovered, although the three-dimensional structure suggests a relationship to myoglobin (Schirmer *et al.*, 1985). There are, as with the biliproteins, some examples of c-type cytochromes which have a single covalent bond to heme; the mitochondrial cytochromes c and c_1 from *Euglena* and *Crithidia* (Pettigrew, 1972, 1973; Mukai *et al.*, 1989; Priest and Hajduk, 1992) are missing the first heme-binding cysteine. It is conceivable that there may be c-type cytochromes which have lost both cysteines through mutation and would therefore be indistinguishable from b-type cytochromes except in sequence and three-dimensional structure. An example is *E. coli* cytochrome b-562, which appears to be a type 2 cytochrome c which has lost both of its heme-binding cysteine residues (Weber *et al.*, 1981).

How might one recognize a c-type cytochrome from a gene sequence? In most cases, c-type cytochromes minimally have a CXXCH heme binding sequence, except as noted above. There are also unusual cytochromes c which have CXXXCH (Denariaz *et al.*, 1989) and CXXXXCH (Ambler, 1973) sequences (the histidine is required as one of the two extraplanar heme iron ligands). The sequence CXCH has not been found in any c-type cytochrome, but is present in the Rieske iron–sulfur proteins (Hauska *et al.*, 1988). On the other hand, there are proteins which are known to have typical heme-binding CXXCH sequences although they are not cytochromes c. Examples are provided by the cytochrome P-450 reductases known as adrenodoxin and putidaredoxin, which are 2Fe-2S ferredoxins (Tanaka *et al.*, 1973, 1974; Mittal *et al.*, 1988; Kagimoto *et al.*, 1988; Koga *et al.*, 1989). A 4Fe-4S ferredoxin subunit of *E. coli* DMSO reductase has a CXXCH sequence (Bilous *et al.*, 1988). Orf (open reading frame) 2 of the *E. coli* formate hydrogenlyase gene cluster contains a CXXCH sequence, which is part of a 4Fe-4S cluster binding site (Bohm *et al.*, 1990). Orf 132, associated with biogenesis of c-type cytochromes in *Bradyrhizobium*, appears to be homologous to thioredoxins and protein disulfide isomerases, but has a CXXCH site (Ramseier *et al.*, 1991). Two open reading frames adjacent to the fumarate reductase operon in *Proteus vulgaris* contain CXXCH sequences plus additional cysteines suggestive of iron–sulfur proteins (Cole, 1987). The large subunit of *Alcaligenes eutrophus* hydrogenase contains a CXXCH sequence, but is not known to bind heme (Kortluke *et al.*, 1992). One of the genes involved in PQQ cofactor biosynthesis in *Acetobacter* contains a CXXCH sequence as well as other cysteines, whereas the histidine in a homolog from *Klebsiella* is not conserved (Goosen *et al.*, 1989; Meulenberg *et al.*, 1992). Thus, it is not unusual for a protein to have CXXCH sequences and not bind heme. Ferredoxins often have CXXC sequences which are not followed by histidine and which in part bind iron–sulfur clusters, but there are other cysteines in these proteins which are necessary to complete the binding site (a minimum of four cysteines per iron–sulfur cluster). A copper-containing nitrous oxide reductase has both CH and CXXXCH sequences but does not bind heme (Viebrock and Zumft, 1988).

Small c-type cytochromes generally do not have more cysteines than required to bind the heme. Soluble cytochrome c_5 is an exception in that it has a cystine disulfide near the C-terminus (Ambler, 1973). The N-terminal extension of the membrane-bound form of cytochrome c_5 also has extra cysteines (Ambler, 1991). Yeast mitochondrial iso-1 cytochrome c has a free sulfhydryl at position 107 also near the C-terminus (Lederer *et al.*, 1972). Thus, to identify an open reading frame (ORF) in a DNA sequence as a c-type cytochrome gene, there would either have to be clear homology to a known cytochrome, or one would have to isolate the gene product and show that it had covalently bound heme.

Pseudogenes are ORFs which are related to genes for known proteins but which are presumably defective. The proteins encoded by pseudogenes are not expressed or are not folded properly and are quickly degraded after synthesis. Approximately 35 cytochrome c pseudogenes have been identified in mammals, although there is only a single functional cytochrome c in most (Scarpulla, 1984; Evans and Scarpulla, 1988). Eleven of the human cytochrome c pseudogenes have been sequenced and are obviously related to the functional gene, although they appear to be derived from two separate genes (Evans and Scarpulla, 1988). It is possible that there are even more divergent pseudogenes, which do not hybridize to the test probe, but which may have segments recognizable as derived from cytochrome c. Pseudogenes appear to be remarkably stable over time, because most of those from humans appear to be derived from an ancestral form of cytochrome c. This may provide an explanation for cytochromes c which have diverged more than expected—for example, rattlesnake (Ambler and Daniel, 1991), which may be a duplicated and rapidly mutated gene that was substituted for the single functional cytochrome c. In this case, we should look for pseudogenes having sequences closer to the expected sequence than that of the present functional gene.

B. Nomenclature

Assuming that we have isolated a new cytochrome c or that we are able to recognize the sequence of a cytochrome c gene, we are then faced with the dilemma of finding a name for the protein. Unfortunately, nomenclature has not kept up with developments in the cytochrome field ("Recommendations of the Nomenclature Committee of the International Union of Biochemistry," Palmer and Reedijk, 1991). Consequently, several approaches to nomenclature have been proposed, although none is widely accepted. Subscripts were originally used for newly discovered cytochromes (c_1, c_2, c_3, etc.), but this practice was discontinued when it appeared to be getting out of hand. The subscripts c_1 through c_6 remain useful because they now designate distinct subclasses of cytochromes rather than individual isolates. Even if confined to different classes or subclasses of cytochromes, numerical subscripts would eventually present the same problem as in the past because there are already more than sixteen kinds of type 1 cytochromes (defined below) which have been identified. Subscripts above c_6 are occasionally used in this more recent context to denote sub-classes of cytochromes, but these designations have not been used consistently and have not gained wide acceptance. For example, cytochrome c_8 has been used for cytochrome c' (Pettigrew and Moore, 1987) and for *Pseudomonas* cytochrome c-551 (Ambler, 1991). Cytochrome c_7 has been applied to the triheme cytochrome c_3 from *Desulfuromonas* (Ambler, 1971) and to *Pseudomonas* cytochromes c-551 (Pettigrew and Moore, 1987).

Individual low-spin cytochromes c are traditionally named according to the position of the reduced alpha peak maximum, e.g., cytochrome c-550 is a cytochrome c which has an alpha peak at 550 nm in the reduced state. The commonly used horse cytochrome c is a c-550. Because there can be many cytochromes c-550, the source species always has to be noted, e.g., *Rhodospirillum rubrum* cytochrome c-550. It is not often that more than one cytochrome in a single species will have the same reduced alpha band wavelength maximum. In those rare cases where it has occurred, authors have used roman numerals I and II to distinguish them, e.g., cytochrome c-550(I) and cytochrome c-550(II). This nomenclature, based on wavelength maxima, tells us nothing about the structural class to which a cytochrome may belong. Once the sequence and/or three-dimensional structure are known, then the name should be changed to reflect the class to which the cytochrome may be affiliated, if at all. Because the amino acid sequence shows that *Paracoccus denitrificans* cytochrome c-550 is a member of the cytochrome c_2 structural class (Ambler *et al.*, 1981a), then it can be designated either way, although the latter is preferred because it is more informative. However, most authors appear to be reluctant to change a name once it has become established.

In some instances, cytochrome c_2 has been used incorrectly in a broad sense to designate any high-redox-potential bacterial cytochrome. It has also been used to designate the electron donor to the photosynthetic reaction center. Cytochrome c_2 is actually a very well-defined class of cytochromes, which includes those bacterial cytochromes c most closely related to mitochondrial cytochrome c in structure. The currently accepted definition is based entirely on amino acid sequence and three-dimensional structure.

Some cytochromes have marked room-temperature splitting or asymmetry of the reduced alpha peak, which occasionally is reflected in the name by placing the wavelength of the shoulder in parentheses, e.g., *Chromatium vinosum* cytochrome c-553(550) (Cusanovich and Bartsch, 1969), which is now known to be a member of the cytochrome c_4 class (Van Beeumen, 1991). The alpha peak of most low-spin cytochromes is split into three well-resolved peaks at liquid nitrogen temperature, which may or may not be obvious at room temperature, but this has no diagnostic value. Nevertheless, split-alpha or split-Soret cytochromes sometimes appear in print. When there is considerable broadening or splitting of the alpha peak, the extinction coefficient is also lowered (e.g., 20–25 instead of 30 mM^{-1} cm^{-1}).

Occasionally, the bacterial genus or family in which a cytochrome was first described or in which it is most often observed is used to designate the structural class, such as "*Pseudomonas* cytochromes c-551" or "*Desulfovibrio* cytochromes c-553." This may result in some confusion because, for example, *Azotobacter vinelandii* is known to have a cytochrome homologous to *Pseudomonas* cytochrome *c*-551. In cases like this, a subscript, such as c_8 (Ambler, 1991), may be preferable.

The nomenclature of cytochromes c_H, c_L, and c_{LM} refers to isoelectric points of the proteins concerned and has been used for methylotrophic bacterial cytochromes (O'Keeffe and Anthony, 1980; Cross and Anthony, 1980). However, isoelectric points have no diagnostic value; very different proteins often have similar values, and members of the same class can have a wide range of values. For example, the sequence of a cytochrome c_H from one methylotrophic species was found to be related to cytochrome c_2, and that from another species was found to be related to *Pseudomonas* cytochrome *c*-551 (Ambler, 1991). Perhaps the "H" in cytochrome c_H should stand for "heterogeneous" instead of "high isoelectric point."

Use of redox potentials in nomenclature, such as in cytochrome b_{-245} (Nugent *et al.*, 1989), fortunately have not been applied to *c*-type cytochromes for the same reason as above. Not all members of a particular structural class necessarily have the same redox potential, and this property can vary with pH and ionic strength. The wavelength maximum of the carbon monoxide complex has been used for designation of hydroxylase enzymes such as the cytochromes P-450, but this practice has not been applied to *c*-type cytochromes, except in a general way as in cytochromes c_{co} (O'Keeffe and Anthony, 1980). The former is diagnostic because of the unique heme ligand cysteine and heme environment, but not the latter, because all known *c*-type cytochromes have a histidine fifth heme ligand and have a 416 nm wavelength maximum for the CO complex. Most high-spin *c*-type cytochromes rapidly bind carbon monoxide, but this property is randomly distributed among a handful of low-spin proteins from several different classes. Carbon monoxide binding is sometimes used as a test for denaturation of mitochondrial cytochromes *c* and is a property for which the kinetic or equilibrium values can change with conditions of measurement. The reactions are generally slow and incomplete. There is not one class of low-spin cytochromes which consistently binds carbon monoxide rapidly and stoichiometrically.

At one time, it was suggested that the presence of each heme in a multiheme cytochrome be reflected in the name, e.g., cytochrome cd_1, which is a cytochrome containing both a *c*-type and a d_1-type heme. However, the 4O-kDa tetraheme reaction center cytochrome from *Rhodopseudomonas viridis* (Deisenhofer *et al.*, 1985) would become cytochrome *c*-559 *c*-556 *c*-553 *c*-552 if this nomenclature were strictly applied. The wavelength maxima of the hemes in the *Chloroflexus* reaction center cytochrome have not even been resolved (Freeman and Blankenship, 1990). Thus, it is not likely that such an awkward name will ever be adopted. "Reaction center cytochrome *c*" is not sufficiently descriptive for the tetraheme protein because the monoheme green bacterial reaction center cytochrome *c* is not related to that from purple bacteria (Okkels *et al.*, 1992). However, "tetraheme reaction center cytochrome *c*" is sufficient to distinguish the purple bacterial protein. Occasionally, there are references to cytochromes *c*-556 and cytochromes *c*-552 in reaction centers, which incorrectly suggests that these hemes are attached to separate peptide chains, which they are not. It also suggests that only two of the four hemes have distinctive properties, which may also be untrue. Rather, "heme *c*-556 or heme *c*-552 in the tetraheme reaction center cytochrome" would be more appropriate. The four-heme and sixteen-heme cytochromes c_3 from *Desulfovibrio vulgaris* (Pollock *et al.*, 1991) are further instances where it would be impractical to name each heme.

The large molecular weight cytochromes c from *Desulfovibrio* species were labeled cytochromes cc_3 and cc_3' to reflect their multiheme nature, but are now thought to be isozymes of cytochrome c_3. These cytochromes have also been called octaheme cytochromes c_3 or 26-kDa cytochromes c_3 (DerVartanian and LeGall, 1974). The sequences of these proteins have not been determined, and it is unknown whether they contain subunits. If it turns out that these proteins are dimeric, then they should be referred to as iso-cytochromes c_3 rather than as octaheme cytochromes c_3. The gene sequence of a sixteen-heme cytochrome c_3 calls into question the very existence of eight-heme cytochromes, which in *D. vulgaris* has a similar N-terminal sequence and may be a proteolytic fragment of the larger protein (Pollock *et al.*, 1991).

The use of the name "cytochrome c" to designate the soluble cytochrome c from mitochondria is confusing because it is sometimes difficult to tell from the context whether the family as a whole or a specific class of protein in that family is intended. In our laboratory, we have been using "mitochondrial cytochrome c" to designate this class of proteins, although it is inappropriate for prokaryotic forms. For the latter, we have been using the name "cytochrome c_2" (a protein most often observed in purple phototrophic bacteria and in some cases structurally indistinguishable from mitochondrial cytochrome c when there are no insertions or deletions). To use "cytochrome c_2" for both prokaryotic and eukaryotic forms would make more sense, because cytochrome c_2 follows cytochrome c_1 in the electron transfer pathway of purple bacteria, and mitochondrial cytochrome c is situated in the same position after the cytochrome bc_1 complex. Although a broader definition of cytochrome c_2 was proposed several years ago to include mitochondrial cytochromes c (Pettigrew and Moore, 1987), it has not generally been adopted.

Cytochrome f is a c-type cytochrome which has the same heme bound in the same way as in the other c-type cytochromes. It is a component of the chloroplast cytochrome b_6f complex, which is analogous to the cytochrome bc_1 complex of mitochondria. Cytochromes c_1 and f are about the same size, but it is not obvious from their sequences that they are homologous. Using principal component analysis, Horimoto *et al.* (1991) reported a relationship among cytochromes c, c_1, and f. It would be desirable to change the name of cytochrome f to reflect its kinship to the c-type cytochromes (Pettigrew and Moore, 1987), but the lack of obvious homology to cytochrome c_1 suggests that it may be premature to change the name of cytochrome f to "chloroplast cytochrome c_1" and of the cytochrome b_6f complex to "chloroplast cytochrome bc_1 complex" until the three-dimensional structures of cytochromes c_1 and f are solved.

Mitochondrial cytochrome b is another example of the dual use of a generic designation for a specific protein. However, it is not often that cytochrome b is used in a specific context outside of the cytochrome bc_1 complex. Confusion arises when one considers that succinate dehydrogenase and fumarate reductase among others also contain a distinct type of membrane-bound cytochrome b subunit. Perhaps alphabetical subscripts, such as in cytochrome b_{QD}, b_{SD}, b_{FR}, and b_{NR} for the cytochrome b subunits of quinol dehydrogenase, succinate dehydrogenase, fumarate reductase, and nitrate reductase, respectively, would serve the purpose of designating membrane-bound b-type cytochromes which have been separated from their associated enzyme complexes as suggested in part by Sodergren and DeMoss (1988).

It was once thought that the dimeric cytochromes c' contained both a high-spin and a low-spin heme in a single peptide chain, which resulted in adoption of the name cytochrome cc'. It has since been shown that cytochrome c' contains only a single type of high-spin heme, and the nomenclature has been corrected. A cytochrome from *Methylophilus methylotrophus* is low-spin in the oxidized state and high-spin in the reduced state and has been temporarily designated cytochrome c'' supposedly by analogy to cytochrome c' (Santos and Turner, 1988; Berry *et al.*, 1990). A cytochrome which is high-spin in the oxidized state and low-spin in the reduced state has been found in *Chromatium vinosum* (Gaul *et al.*, 1983). Furthermore, one of the two hemes in *Pseudomonas* cytochrome c peroxidase changes spin-state with temperature (Ellfolk *et al.*, 1983; Villalain *et al.*, 1984). This points up a real problem in deciding how to describe new high-spin cytochromes not related to cytochrome c' because one cannot continue to add new superscripts indefinitely. Cytochrome c' is supposedly the designation for high-spin cytochromes c in general, but at the time the name was established, there was only a single type known. Another kind of high-spin cytochrome c has been described (Gaul *et al.*, 1983; Meyer and Cusanovich, 1985), but it does not seem appropriate to give it the name cytochrome c' without creating the same confusion as described above. We prefer to use the name cytochrome c' in a narrower context. We are now using the designation "type 2 cytochromes" for the family of proteins, both high- and low-spin, which are homologous to the original cytochrome c' (see below). The high-spin cytochrome c, which is not homologous to cytochrome c', we have tentatively called SHP (*Sphaeroides* heme protein) for lack of a better name and by analogy with an early designation for cytochrome c', which was originally known as RHP (*Rhodospirillum* heme protein).

A classification of cytochromes based on size was suggested by Dickerson (1980b). M-type or medium cytochromes are mitochondrial cytochromes c plus cytochromes c_2 which are the same size. L-type or large cytochromes c are the cytochromes c_2 which have internal insertions. The overall size was not considered (see the section on cytochrome c_2 for an expanded discussion). Obviously, L and M make sense only in reference to mitochondrial cytochromes c and cytochromes c_2. S-type or small cytochromes c are bacterial cytochromes having internal deletions relative to mitochondrial cytochrome c. S-type proteins include many distinct subclasses. There are M's which are larger than L's and S's which are larger than both because of N-terminal or C-terminal extensions. This classification has only limited significance and is confusing; it remains popular nevertheless.

Ambler (1982, 1991) presented a new system of classification based on amino acid sequences and three-dimensional structures. All proteins homologous to mitochondrial cytochrome c are called type 1 cytochromes c. Those homologous to cytochrome c' are called type 2, and those homologous to *Desulfovibrio* cytochrome c_3 are type 3. Ambler's type 4 (those cytochromes having additional chromophores, but unknown sequences) lacks the consistency of the other classes, and those "type 4" proteins, which have since been sequenced, actually belong in type 1.

Ambler's types 1, 2, and 3 provide the inspiration for the classification of cytochromes used in this review. Type 4 will be redefined as those proteins homologous to the tetraheme reaction center cytochrome. The subclasses adopted herein are based strictly on structural data (sequences and three-dimensional data). Some subclasses and their composition correspond to Ambler's, although not all. It is hoped that, in the future, authors will comply with the Recommendations of the Nomenclature Committee of The International Union of Biochemistry (Palmer & Reedijk, 1991) as closely as possible in assigning trivial names based on the reduced alpha peak wavelength maximum. When sequences and three-dimensional structures become available, a more permanent name may be assigned based on relationships to the well-characterized proteins. There are undoubtedly some proteins which cannot be assigned to existing categories, and new classes may eventually be necessary. Seven such proteins have been included in this review. Regrettably, many cytochromes have not been included because they have been incompletely characterized or because they have been found in only a single species. Authors are encouraged to determine at least N-terminal sequences as an aid to assignment of their proteins to one of the categories herein described.

II. Principal Types of Cytochromes c

There are currently four types of totally unrelated cytochromes c which are classified by sequences and three-dimensional structures. Sequence data suggest additional types, but three-dimensional structural data are needed for confirmation. The type 1 cytochromes c comprise the largest group and include mitochondrial cytochrome c and purple bacterial cytochrome c_2, among others. They are usually small (8–12 kDa), most of them are low-spin, they are generally water-soluble, and most have high redox potentials. The single heme is bound near the N-terminus, and the sixth ligand methionine is near the C-terminus.

Type 2 cytochromes are predominantly high-spin and include the 14-kDa subunit bacterial cytochromes c' and their low-spin counterparts (e.g., *Agrobacterium* cytochrome c-556). *E. coli* cytochrome b-562 is also homologous to members of this group. The basic structural motif is a four-alpha-helical bundle, and the single heme is bound near the C-terminus. The sixth ligand methionine, when present, is near the N-terminus.

The type 3 cytochromes c are also known as cytochromes c_3. They are usually small (14-kDa), tetraheme proteins with low redox potentials that show signs of at least one and perhaps two gene doublings. A degenerate triheme cytochrome c_3 has been described. There are also 66 kDa, 16-heme proteins which are made up of repeating tetraheme domains.

Type 4 cytochromes c are the large (40-kDa) tetraheme reaction center cytochrome c subunits for which there is also evidence for successive gene doublings. Cytochromes c_1 and f may represent one or more new types, although location of the single heme near the N-terminus (of the approximately 30-kDa peptide chain) suggests a relationship to the smaller type 1 cytochromes c. *R. sphaeroides* cytochrome c-551.5, which has two hemes in a 16-kDa peptide chain, *P. aeruginosa* cytochrome *c* peroxidase, which has two hemes in a 35-kDa peptide chain, *P. stutzeri* cytochrome *c*-552, which has two hemes in a 24-kDa peptide, and a *P. stutzeri* orf, which has four heme binding sites in a 20-kDa protein, may be representative of one or more additional types of unrelated cytochromes *c*, but this is uncertain. The three-dimensional structure is absolutely necessary to establish new types of unrelated cytochromes *c*, although subdivision of the four known types may be based on sequence data alone.

To establish a new subclass of type 1 cytochromes *c*, one first has to show sequence homology to type 1 cytochromes, then show that several species produce this new protein and that the sequences are more closely related to one another than they are to any of the existing subclasses. If the new proteins do not show greater differences from an existing class than members of that class show among themselves (ignoring insertions and deletions), then they belong to that class. Thus, bacterial cytochromes c_2 and mitochondrial cytochromes *c* are no more different from one another (on a percentage basis) than are members of either group alone, and therefore, they belong to the same subclass.

The presence of insertions and deletions alone should not be a criterion for subdivision. It is possible that some subclasses of type 1 cytochromes are more closely related to one another than to others, but to establish such relationships, we will need additional high-resolution three-dimensional structures to aid in sequence alignments. Low-resolution structures will only tell us that the cytochromes are homologous, something we should know from the sequences alone. We need to know the details of secondary structure and to identify the roles of particular amino acids which may be conserved. With this type of information, we may also be able to determine where insertions and deletions may actually have occurred and place these events on a stronger foundation than a mere exercise in sequence alignment. Having a computer alignment program assign penalties for creation of insertions and deletions is insufficient for the identification of actual events. Even more sophisticated software designed to simultaneously align several proteins at once (Murata *et al.*, 1985; Feng and Doolittle, 1987; Lipman *et al.*, 1989) is incomparable to three-dimensional structural data for establishing the locations of insertions and deletions. For cytochrome c_2, insertions and deletions are easy to assign without three-dimensional structural data, but for other proteins, it can be difficult to make assignments even with three-dimensional structures. We may never be able to precisely quantify all relationships among type 1 cytochromes *c*, but as additional sequences and high-resolution structures become available, we are gradually increasing our understanding.

A. Type 1 Cytochromes *c*

Type 1 cytochromes *c* are those which are homologous to mitochondrial cytochrome *c*. They can be subdivided into at least sixteen distinct subclasses. These are mostly small (8–20 kDa) proteins which have a single heme near the N-terminus. However, there are at least three cases of gene doubling (in cytochromes c_4, in the alcohol dehydrogenase cytochrome subunit, and in some flavocytochromes *c*) which are apparent from the sequences alone. There are also four complex proteins in this group, which contain additional chromophores. Flavocytochrome c from phototrophic bacteria and *Pseudomonas* has a small (10–20 kDa) type 1 cytochrome subunit firmly bound to a 45-kDa flavoprotein subunit. The cytochrome cd_1 type of nitrite reductase from *Pseudomonas aeruginosa* contains a small type 1 cytochrome domain fused to the N-terminus of a 60-kDa peptide. Another gene fusion, in *Thermus thermophilus* and *Bacillus* species, occurs between a type 1 cytochrome c and the C-terminus of subunit II of the a-type cytochrome oxidase (Buse *et al.*, 1989; Sone *et al.*, 1988; Ishizuka *et al.*, 1990). Likewise, a cytochrome c is fused to the C-terminus of the PQQ cofactor-containing alcohol dehydrogenase from *Acetobacter* (Tamaki *et al.*, 1991). The *Bradyrhizobium* bc_1 complex is synthesized as a large precursor, which is post-translationally cleaved into cytochrome b and c_1 peptides (Thony-Meyer *et al.*, 1989, 1991).

Not all type 1 cytochromes c are water-soluble, although none appears to span the membrane the way it is thought that mitochondrial cytochrome b does. *Bacillus subtilis* cytochrome c-550 is anchored to the membrane by an N-terminal segment similar to signal peptides (Von Wachenfeldt and Hederstedt, 1990a,b). The product of the *Bradyrhizobium* cycM gene is apparently a cytochrome c_2, also anchored to the membrane via its signal peptide (Bott *et al.*, 1991). Cytochrome c_5 has an N-terminal extension, which somehow seems to hold it to the membrane, although this extension does not appear to be related to signal peptides (Ambler, 1991). Likewise, it is not known how cytochrome c_4 is bound to the membrane because there is no large hydrophobic region of the sequence (Ambler *et al.*, 1984). There is at least one high-spin type 1 cytochrome (SHP) which has lost its sixth ligand through mutation (J. Van Beeumen *et al.*, unpublished). There is also an example of a type 1 cytochrome (algal cytochrome c-550), which has a low redox potential, presumably because of the substitution of a histidine for the usual sixth ligand methionine (Cohn *et al.*, 1989b).

Although all type 1 cytochromes are homologous by definition, some classes are apparently more closely related than others. High-resolution, refined, three-dimensional structures are known for *Pseudomonas aeruginosa* cytochrome c-551 (Matsuura *et al.*, 1982), for *Azotobacter vinelandii* cytochrome c_5 (Carter *et al.*, 1985), for *Desulfovibrio vulgaris* cytochrome c-553 (Nakagawa *et al.*, 1990), for several mitochondrial cytochromes c including yeast isozymes, tuna, horse, and rice (Takano and Dickerson, 1981; Ochi *et al.*, 1983; Louie *et al.*, 1988; Louie and Brayer, 1990; Bushnell *et al.*, 1990; Murphy *et al.*, 1992), and for bacterial cytochromes c_2, including *Rhodospirillum rubrum* and *Rhodobacter capsulatus* (Bhatia, 1981; Benning *et al.*, 1991). Low-resolution structures are also available for *Paracoccus denitrificans* cytochrome c_2 (Timkovich and Dickerson, 1976), *Chlorobium thiosulfatophilum* cytochrome c-555 (Korszun and Salemme, 1977), and *Pseudomonas putida* flavocytochrome c (Bellamy *et al.*, 1987; Mathews *et al.*, 1991). An alignment of the three-dimensional structures of cytochromes c was briefly reported (Johnson *et al.*, 1990).

Where the numbers of sequence identities are small, then analysis of insertions and deletions (gaps) can be more valuable for detecting relationships than if the matrix of sequence differences is used, but only when such gaps are demonstrated by three-dimensional structures. Through examination of structures, we can see that the cytochromes c_2 differ from the majority of bacterial type 1 cytochromes in that they have an approximately sixteen-residue insertion in the midsection which resulted in a rearrangement of adjacent residues such that it is impossible to tell exactly where the insertion was made or whether it was a single event.

Cytochrome c_5, *Pseudomonas* cytochrome c-551, and *Desulfovibrio* cytochrome c-553 are closer to one another in both size and folding pattern than they are to the cytochromes c_2. Nevertheless, the numbers of identical residues in these three proteins are so low that we may never be able to tell exactly how many insertions and deletions have taken place. From three-dimensional structures, we can see that cytochromes c_5 and c-551 share a helix that is not present in cytochromes c_2 or mitochondrial cytochromes c, which indicates that they may be more closely related to one another, although we cannot precisely quantify that relationship at present. Likewise, *Desulfovibrio* cytochrome c-553 has a helical segment not present in the other proteins that sets it apart. Evolutionary relationships based on three-dimensional structures were compared with those based on amino acid sequences (Johnson et al., 1990), although they were not sufficiently detailed to gain any further insights. The above discussion is intentionally vague because we have too few data to draw firm conclusions.

1. Mitochondrial cytochromes c and bacterial cytochromes c_2

Mitochondrial cytochrome c is the most thoroughly characterized c-type cytochrome in a structural sense, with approximately ninety known sequences and five high-resolution three-dimensional structures. Conserved features of the cytochrome c structure are as follows: the presence of interacting N-terminal and C-terminal helices, conserved glycine 1, aspartate 2, glycine 6, and aromatic residue 10 in the N-terminal helix (horse numbering), and five contiguous hydrophobic residues in the C-terminal helix at positions 94–98 (position 97 being aromatic). The heme-binding site is at cysteines 14 and 17. The heme fifth and sixth ligands are histidine 18 and methionine 80. Proline 30 H-bonds histidine 18, and tyrosine 67 H-bonds methionine 80. Arginine 38, tyrosine 48, and tryptophan 59 H-bond one heme propionate, and serine 49, asparagine 50, and threonine 78 H-bond the other propionate. Other highly conserved residues have less obvious roles, but still may be useful in comparison of divergent type 1 cytochromes. For example, a number of conserved basic residues have been shown to be important to function, namely lysines 8, 13, 25, 27, 72, 73, 79, 86, and 87. These nine basic residues surround the exposed heme edge where electron transfer usually takes place. The mitochondrial cytochrome c from *Tetrahymena* is unusual in that it possesses only two such basic residues (73 and 79) and two other sites (8 and 25) are substituted by acidic side chains. Otherwise, these features are highly conserved in mitochondrial cytochrome c. Some of these conserved residues are absent in other type 1 cytochromes c, as pointed out in later sections.

Sequence alignment is straightforward because only the nematode worm *Caenorhabditis elegans* (Vanfleteren et al., 1990), the flagellated protozoan *Euglena gracilis* (Pettigrew et al., 1975), and the ciliated protozoan *Tetrahymena pyriformis* (Tarr and Fitch, 1976) contain small internal deletions. An evolutionary tree was constructed by Baba *et al.*, (1981), which is more or less consistent with what we know of eukaryotic evolution. There are, however, some striking inconsistencies in the evolutionary tree such as rattlesnake (Ambler and Daniel, 1991) and *Arabidopsis* (Kemmerer *et al.*, 1991). Some of the discrepancies may be due to sequencing errors, while others may be explained in part by gene duplications with eventual replacement of the functional protein by rapidly mutated isozymes. Inconsistencies could also be due to rapid changes in the rate of amino acid substitution of a single gene for cytochrome *c* in response to changes in its reaction partners. Baba *et al.* (1981) attempted to reconcile the evolution of cytochromes *c* with that of other proteins from the same array of species to arrive at the best overall interpretation of eukaryotic evolution. Although this proposal provides general agreement with the fossil record and with other proteins, it still exhibits discrepancies which have not been adequately explained.

Evans and Scarpulla (1988) found that most human pseudogenes characterized to date were derived from the messenger RNA of an earlier form of cytochrome *c* rather than from the single extant functional cytochrome *c*. Either these pseudogenes have been carried for millions of years, or a cytochrome *c* isozyme has recently replaced the previously functional cytochrome *c*, which in turn would have been deleted. Thus, the presence of nonfunctional isozymes or pseudogenes with their associated rapid mutational rate may partially explain some of the discrepancies in the evolutionary tree.

Analyses of gene sequences for introns may be another way to follow the evolution of cytochrome *c* if it can be shown that there are differences along different lineages. Rat and human cytochrome *c* genes were found to contain the same two introns. On the other hand, the two yeast cytochrome *c* isozyme genes do not have introns (Smith *et al.*, 1979; Montgomery *et al.*, 1980; Russell and Hall, 1982). Isozymes of cytochrome *c* were found in *Saccharomyces* (Montgomery *et al.*, 1980) but not in *Schizosaccharomyces* (Russell and Hall, 1982). They are present in mouse (Hennig, 1975), rat (Virbasius and Scarpulla, 1988), bovine, and rabbit (Kim and Nolla, 1986). They are present in the housefly (Inoue *et al.*, 1984) and the fruit fly (Limbach and Wu, 1985a,b), but not in other flies (Inoue *et al.*, 1986) nor the honeybee (Inoue *et al.*, 1985). The evolutionary relationships of the isozymes were examined by Mills (1991). In conclusion, it seems that characterization of introns, isozymes, and pseudogenes may all be used to improve the evolutionary tree for cytochrome *c*.

Probably the most serious problem with evolutionary trees that has not been adequately addressed is that we do not have a good estimate of back and parallel mutations, collectively known as convergent mutations, for highly divergent proteins. Furthermore, it is not likely that reliable estimates can ever be obtained. It was shown by Meyer *et al.* (1986a) that most bacterial cytochromes c_2 have reached a limit of apparent change at which divergent and convergent mutations are balanced. This limit of change precludes construction of evolutionary trees based on matrices of sequence differences for highly divergent proteins. Some of the most divergent mitochondrial cytochromes, such as those from *Tetrahymena*, *Euglena*, and *Crithidia*, are very close to, if not at, this limit of change found for cytochromes c_2. Therefore, it is not surprising that Baba *et al.* (1981) could not distinguish *Tetra-*

hymena cytochrome *c* from bacterial cytochromes c_2. For the largest differences between proteins (approaching the limit of change), one must assign ever-increasing corrections and error limits to account for convergent mutations. Even if one applies a correction factor for convergence, as most authors claim to do, it is unlikely to be adequate (because too large a limit to change is assumed), so the uncertainty remains. Stated another way, the numbers of mutations should approach infinity asymptotically as sequence differences approach the limit of change for that protein. Evolutionary trees constructed from data near the limit of change will have large branch lengths and small differences between branch points. The errors associated with the correction for convergence will be larger than the distance between branch points; thus, the tree will have no validity. Unfortunately, it is as difficult to assign error limits as it is to correct for convergence. Consequently, such errors are seldom illustrated in evolutionary trees.

Sneath (1989) analyzed rRNA sequence–based evolutionary trees and assigned errors due to statistical sampling size, but he did not consider the errors due to convergence. A similar analysis should be applied to eukaryotic cytochromes. With the present information, it is impossible to determine the quantitative relationship of *Tetrahymena* to other protozoan, plant, animal, and fungal species. It is also impossible to quantify the relationships among most bacterial cytochromes c_2 or their relationship to mitochondrial cytochromes *c* based on percentage sequence differences. *Rhodopila globiformis* cytochrome c_2 is an example of a protein which appears to be closer to some mitochondrial cytochromes *c* than to other cytochromes c_2, but within the unavoidably large error limits due to convergence, we have to conclude that this relationship is anomalous (Ambler *et al.*, 1987b). Woolley and Athalye (1986) used principal coordinate analysis to clarify relationships. Although this procedure is superior to two-dimensional tree construction in its visual impact, errors due to convergence cloud any conclusions which might be drawn.

The cytochromes c_2 are the bacterial counterparts of mitochondrial cytochrome *c*. They have been most thoroughly characterized in the non-sulfur purple phototrophic bacteria, but they are present in non-phototrophs as well (Meyer and Kamen, 1982). The amino acid sequences of thirteen phototrophic bacterial cytochromes c_2 have been reported (Ambler *et al.*, 1979a). It has also been shown by sequence determination that the aerobic bacteria *Roseobacter denitrificans* (formerly *Erythrobacter*) Och 114 (Okamura *et al.*, 1987), *Aquaspirillum itersonii* (Woolley, 1987), *Thiobacillus novellus* (Yamanaka *et al.*, 1991), and *Nitrobacter agilis* (Tanaka *et al.*, 1982), in addition to the well-known *Paracoccus denitrificans* (Ambler *et al.*, 1981a) and *Agrobacterium tumefaciens* (Van Beeumen *et al.*, 1980), contain a cytochrome c_2. Gene sequences have been reported for *Rb. capsulatus* (Daldal *et al.*, 1986), *Rb. sphaeroides* (Donohue *et al.*, 1986), *R. rubrum* (Self *et al.*, 1990), *P. denitrificans* (Van Spanning *et al.*, 1990), and *Rps. viridis* (Grisshammer *et al.*, 1990) cytochromes c_2. All contain signal peptides which are cleaved after the cytochromes cross the membrane to the periplasmic space. The gene for an oxidase-associated membrane-bound 19-kDa protein from *Bradyrhizobium japonicum* is also homologous to the cytochromes c_2, but differs in that the signal peptide is retained (Bott *et al.*, 1991). A similar protein, called cytochrome c_y, is present in *Rb. capsulatus* and is interchangeable with cytochrome c_2 in photosynthetic electron transfer (Jenney and Daldal, 1993).

In the purple bacteria, cytochromes c_2 principally function to shuttle electrons from the cytochrome bc_1 complex to photosynthetic reaction centers. Cytochrome c_2 occasionally has been used in a broad functional context to represent any cytochrome which donates electrons to photosynthetic reaction centers (e.g., Brune, 1989). However, the narrower and more widely accepted definition of cytochrome c_2 based on structure is preferred. Cytochromes c_2 in the narrower structural sense are those bacterial cytochromes which are as close to mitochondrial cytochromes c as they are to one another in amino acid sequence. This can be illustrated by plots of the data in matrices of sequence differences (Meyer *et al.*, 1986a). The cytochromes c_2 average about 60% difference from one another and from the mitochondrial cytochromes c. In terms of the matrix of sequence differences, one cannot distinguish mitochondrial cytochromes c and bacterial cytochromes c_2. Some independent means must be used for this purpose.

The presence of several insertions and deletions in some cytochromes c_2 sets them apart from mitochondrial cytochrome c. *Rps. viridis*, *Rm. vannielii*, *Rps. acidophila*, *Nitrobacter agilis*, and *Rp. globiformis* cytochromes c_2 are most similar to the mitochondrial cytochromes c in that they require only a single-residue internal deletion for alignment. Recall that both *Euglena* and *Tetrahymena* mitochondrial cytochromes c contain one- and two-residue deletions, respectively. There are, in fact, two groups of cytochromes c_2, which can be distinguished by the pattern of insertions and deletions. "Large" cytochromes c_2 all contain three- and eight-residue insertions and a single-residue deletion which can be precisely placed based on three-dimensional structures (Bhatia, 1981; Timkovich and Dickerson, 1976; Benning *et al.*, 1991). There are cytochromes c_2 which do not have precisely the same insertions and deletions, but which have extended C-termini and a greater number of amino acid residues than do certain "large" cytochromes c_2 (Rott *et al.*, 1993). However, "large" cytochrome c_2 refers to the ten extra residues inserted internally. Dickerson (1980b) referred to the cytochromes c_2 as "large" and "medium" based on the above pattern of insertions and deletions. All other homologous bacterial cytochromes c, such as *Pseudomonas* cytochromes c-551, c_4, and c_5 amongst others, were called "short" cytochromes c in spite of the fact that some, like cytochrome c_4, are much larger than the average cytochrome c_2. The main problem with this categorization is that the "short" cytochromes c are a heterogeneous collection, whereas the "large" and "medium" cytochromes are indistinguishable from one another except by the above pattern of insertions and deletions. There is a distinct group of cytochromes c_2 which has the above three-residue insertion only [e.g., *Agrobacterium tumefaciens* (Van Beeumen *et al.*, 1980) and a *Rb. sphaeroides* isozyme (Rott *et al.*, 1993)]. It is reasonable to expect that cytochromes c with only the eight-residue insertion or the single-residue deletion will be found in the future. Such proteins need not be evolutionary intermediates, but could also be degenerate "large" cytochromes c_2.

One interesting observation is that in those species which generally express a "large" cytochrome c_2, there is direct interaction between the cytochrome c_2 and the photosynthetic reaction center, whereas in most of those species which have "medium" cytochrome c_2, there is a tetraheme cytochrome which is bound to the reaction center and which mediates electron transfer. One might guess from the foregoing that "medium" cytochrome c_2 would be a less efficient electron donor to reaction centers but horse mitochondrial cytochrome c was found to work better than the wild type or "large" cytochrome c_2 from *Rb. sphaeroides*. This suggests that the generalization concerning direct interaction of "large" cytochrome c_2 with reaction centers will not hold up when additional species are examined; in fact, it was reported that the *Roseobacter denitrificans* (formerly *Erythrobacter*) "large" cytochrome c_2 interacts with a tetraheme reaction center cytochrome c (Takamiya *et al.*, 1987).

The structural features characteristic of the mitochondrial cytochromes c are not as often conserved in the cytochromes c_2. For example, *A. itersonii*, *R. rubrum*, and *R. photometricum* lack the aromatic residue in the N-terminal helix. Tryptophan 59 (horse numbering) is absent in *R. molischianum* and *R. fulvum* iso-1 cytochromes c_2. Arginine 38 is absent in the *R. rubrum* and *Rm. vannielii* proteins. Tyrosine 67 is absent in *R. molischianum* and *R. fulvum* iso-2 c_2. The nine basic residues conserved in mitochondrial cytochrome c are all substituted to varying degrees in the cytochromes c_2, although not to an extent which would impair function. For example, all the cytochromes c_2 have a positively charged site of interaction with redox partners (Meyer *et al.*, 1984) including *R. rubrum*, *Rb. sphaeroides*, and *P. denitrificans*, which have an overall negative charge.

The recent reports of fused copper protein subunit and cytochrome c genes in *Bacillus* PS3, *Bacillus subtilis*, and *Thermus thermophilus* a-type oxidase operons (Buse *et al.*, 1989; Ishizuka *et al.*, 1990; Saraste *et al.*, 1991; Mather *et al.*, 1991) suggests that the copper protein is the site of interaction with soluble cytochromes c in all species. The fused cytochrome c appears to be most similar to mitochondrial cytochromes c and cytochromes c_2. If this is true, it would have experienced an eight-residue deletion between the heme binding site and the proline which H-bonds the histidine heme ligand. Only three residues would remain in this region. A single-residue insertion after the methionine heme ligand would also be necessary to complete the alignment. However, a three-dimensional structure is necessary for confirmation. If these assumptions prove to be correct, the *Bacillus* and *Thermus* cytochromes would be the most divergent of all, with only 24% similarity to the known cytochromes c and c_2. The postulated site of reaction with the copper protein would be slightly positive because of the presence of lysines "12 and 79" (horse numbering); aromatic residues in the N-terminal and C-terminal helices would be present, as would arginine "38" tryptophan "59" would be substituted by histidine and tyrosine "67" by tryptophan.

2. Pseudomonas Cytochromes c-551 or c_8

As the name implies, this protein is most often found in *Pseudomonas* species. It is also abundant in *Azotobacter*, which may be regarded as a nitrogen-fixing *Pseudomonas*. In these species, cytochrome c-551 is generally found in association with cytochromes c_4 and c_5. Protein sequences have been reported for the following *Pseudomonas* species: *P. aeruginosa*, *P. stutzeri*, *P. fluorescens*, *P. mendocina*, and *P. denitrificans* (Ambler and Wynn, 1973; Ambler, 1973, 1991). A sequence has been reported for the *P. aeruginosa* cytochrome c-551 gene (Nordling *et al.*, 1990; Arai *et al.*, 1990), which is adjacent to that for cytochrome cd_1 (nitrite reductase) (Silvestrini *et al.*, 1989). There is a leader sequence, which demonstrates that cytochrome *c*-551 is periplasmic. The nucleotide sequence of a 4.6-kb DNA fragment from *Pseudomonas stutzeri* encodes nitrite reductase and three other cytochromes in addition to cytochrome *c*-551 (Jungst *et al.*, 1991). However, a tetraheme and a diheme cytochrome *c* gene intervene between the cytochromes cd_1 and *c*-551 genes. *Pseudomonas* cytochrome *c*-551 type protein sequences have also been determined for *Azotobacter vinelandii* (Ambler, 1973), *Nitrosomonas europeae* (Miller and Nicholas, 1986), *Methylophilus methylotrophus* (Ambler, 1991), and *Hydrogenobacter thermophilus* (Sanbongi *et al.*, 1989, 1991). Perhaps more surprising is that *Pseudomonas* type cytochromes *c*-551 are found in a few purple phototrophic bacteria, namely *Rc. tenuis*, *Rc. purpureus*, and *Rc. gelatinosus* (Ambler *et al.*, 1979b; Ambler, 1991). These are species that apparently lack cytochrome c_2. Cytochromes c_4 and c_5 may be present, but proof in the form of sequences has not yet been obtained (see later sections).

Pseudomonas cytochrome *c*-551 is smaller than mitochondrial cytochrome *c* and bacterial cytochrome c_2, averaging about 82 amino acid residues. There is obvious homology in the location of the heme-binding site near the N-terminus and the sixth ligand near the C-terminus. The three-dimensional structure (Almassy and Dickerson, 1978; Matsuura *et al.*, 1982) shows that there are other regions of similarity, notably in the N- and C-terminal helices. However, the midsection of *Pseudomonas* cytochrome *c*-551 is about sixteen residues shorter; it has more helix, and this part cannot be superimposed on the mitochondrial cytochrome *c* structure. It is not surprising that the best guesses at alignment result in no more than about 20% similarity. The *Pseudomonas* cytochromes *c*-551 average about 60% similarity to one another, or about 40% if the purple bacterial examples are included.

The greater divergence of the purple phototrophic examples of cytochromes *c*-551 suggests that they have a different functional role and that they belong in a separate subclass, although for the present it is simpler to treat them here. They can be seen to be more closely related to *Pseudomonas* cytochromes *c*-551 than to any other group because of the pattern of conserved residues, which are known to have a structural role. For example, tryptophan 56 H-bonds the heme propionate in *Pseudomonas* cytochrome *c*-551 (although it is not homologous to horse tryptophan 59) and is expected to do so in the purple bacterial cytochromes *c*-551. *Pseudomonas* tryptophan 77 is more highly conserved than is tryptophan 56, and it corresponds to the aromatic residue present in the C-terminal helix of most type 1 cytochromes *c*. Undoubtedly there is a difference in the functional role of the purple bacterial proteins which permits more variation. In contrast, there is probably strong conservation of the functional roles of the *Pseudomonas*, *Azotobacter*, *Nitrosomonas*, *Hydrogenobacter*, and *Methylophilus* cytochromes *c*-551. It is interesting that the redox potentials of the purple bacterial proteins are more variable as well. *Rc. tenuis* has a redox potential of 405 mV, while that of *Rc. gelatinosus* is 28 mV and that of *Pseudomonas* is about 270 mV. This is somewhat similar to the relationship between redox potentials of the cytochromes c_2 relative to those of the mitochondrial cytochromes *c*. There is greater sequence variation in the cytochromes c_2, and there is also a greater range of redox potentials (250–450 mV).

Some of the other structural features of the mitochondrial cytochromes *c* are found in *Pseudomonas* cytochrome *c*-551, although not all (Matsuura *et al.*, 1982). For example, in the N- and C-terminal helices, there are conserved glycine and aromatic residues, and the proline that H-bonds the histidine heme ligand is present. However, there is a single-residue insertion between the N-terminal helix and the heme, a single residue deletion between the heme ligand methionine 61 and the C-terminal helix, and a three-residue deletion between the heme and proline 25. Differences from mitochondrial cytochrome c include asparagine 64, which H-bonds the methionine 61 heme ligand, which in turn is surrounded by four prolines. There is a fourteen-residue deletion in the midsection of cytochrome c-551 relative to mitochondrial cytochrome c, with a corresponding rearrangement of the peptide chain. Different residues H-bond the heme propionate than in mitochondrial cytochrome c (i.e., arginine 47, tryptophan 56, and perhaps tyrosine 34 and serine 52 in some species). The H-bonding arginine 47 is more often histidine or lysine in species other than *P. aeruginosa*. In *Rc. tenuis*, there is a five-residue insertion after the H-bonding tryptophan 56 (*Pseudomonas* numbering), and in *Rc. gelatinosus* there is a three-residue insertion after the heme ligand methionine 61 (Ambler *et al.*, 1979b). The basic residues found in mitochondrial cytochromes c and c_2 are not present in *Pseudomonas* cytochrome c-551, and the site of electron transfer is slightly negatively charged because of an overall negative charge on the protein (Meyer *et al.*, 1984). The phototrophic cytochromes c-551 have a slightly positive site of reaction due to an overall positive charge on the protein. There are relatively few charged residues at the exposed heme edge.

3. Diheme cytochromes c_4

Cytochrome c_4 occurs in many of the same bacterial species as cytochrome c_5 and *Pseudomonas* cytochrome c-551. The protein has been thoroughly characterized from *P. aeruginosa*, *P. stutzerii*, and *Azotobacter vinelandii* (Pettigrew and Brown, 1988). It is a peripheral membrane-bound protein, which along with cytochrome c_5 is solubilized by n-butanol/water (Tissieres, 1956; Swank and Burris, 1969). Furthermore, it is attached to the periplasmic face of the membrane (Hunter *et al.*, 1989). *Azotobacter vinelandii* cytochrome c_4 is the only complete sequence which has been reported (Ambler *et al.*, 1984). However, the N-terminal sequences of *P. aeruginosa*, *P. mendocina*, and *P. stutzeri* cytochromes c_4 demonstrate the general occurrence of this cytochrome in the pseudomonads (Ambler and Murray, 1973). The purple sulfur bacterium *Chromatium vinosum* definitely has a cytochrome c_4 based on its sequence (Van Beeumen, 1991). The physical–chemical properties of the *Chromatium* cytochrome, including solubilization with 50% acetone, were reported by Cusanovich and Bartsch (1969). There is spectral evidence for the occurrence of cytochrome c_4 in the purple bacteria *T. pfennigii* (Meyer *et al.*, 1973), *Rc. gelatinosus*, and *Rc. tenuis* (unpublished). A diheme, oxidase-associated cytochrome c gene from *Bradyrhizobium japonicum* shows some similarity to cytochromes c_4 (Preisig *et al.*, 1993).

Cytochrome c_4 has two hemes in a 20-kDa protein, which is the product of gene duplication of a smaller cytochrome similar to cytochromes c_5 and c-551. The two hemes have high redox potentials, which differ by 54–110 mV (Leitch *et al.*, 1985; Gadsby *et al.*, 1989). The wavelength maximum for the reduced alpha peak differs for the two hemes of *P. stutzeri*, but not for the other cytochromes c_4 (Leitch *et al.*, 1985). A low-resolution structure shows that the protein folds in two domains corresponding to the two halves of the sequence (Sawyer *et al.*, 1981). The segment connecting the two halves is about sixteen residues long. It is not obvious how cytochrome c_4 is held to the membrane because there are no particularly hydrophobic segments and no retained "leader" or "signal" peptide. Perhaps the interface between the two domains of the solubilized protein is exposed through a conformational change that allows it to bind to the membrane.

We cannot say how similar cytochrome c_4 is to cytochromes c_5 and c-551 until the three-dimensional structure of cytochrome c_4 is completed, but from the sequence it appears to be a little more divergent. In comparison of the two halves, there is only 24% identity and at least one five-residue insertion in the midsection of the first half. It is remarkable that the N-terminal seven residues of cytochrome c_4 (AGDAAAG...) are identical to those of *R. rubrum* cytochrome c_2. This N-terminal sequence is also found in a *Paracoccus* sp. (Ambler *et al.*, 1987a) and a *Thiobacillus neapolitanus* (Ambler *et al.*, 1985) cytochrome c, each of which has only one heme. There is a six-residue peptide in *Thiobacillus* which is also identical to that of *Azotobacter* cytochrome c_4 (FPKLAG), suggesting that these proteins may be closely related. The *Paracoccus* and *Thiobacillus* cytochromes are water-soluble and are more closely related to the first half of cytochrome c_4 than to any other cytochrome c (46% and 33%, respectively). *Thermus thermophilus* cytochrome c-552 is another monoheme cytochrome which appears to be distantly related to cytochrome c_4. This protein contains no insertions or deletions, although it possesses a 45-residue extension at the C-terminus (Titani *et al.*, 1985a,b).

Cytochrome c_4 appears to be slightly closer to mitochondrial cytochrome c than does *Pseudomonas* cytochrome c-551, in that it has a longer N-terminus containing the conserved glycines and aspartate. The aromatic residue typical of the N-terminal helix of type 1 cytochromes c is absent in cytochrome c_4, and there is a single-residue deletion. The C-terminal helix contains the normal aromatic and hydrophobic residues, but has a single-residue insertion. The proline 30 which H-bonds the histidine heme ligand is in the normal position and is followed by a conserved glycine 34 and arginine 38, similar to those of mitochondrial cytochrome c. There are the same numbers of residues in the midsection of the first half of *Azotobacter* cytochrome c_4 as in *Pseudomonas* cytochrome c-551, but none of the residues which might H-bond the heme propionates are present.

4. Cytochromes c_5

Cytochrome c_5 and cytochrome c_4 were originally discovered in the nitrogen-fixing bacterium *Azotobacter vinelandii* as peripheral membrane-bound proteins, which could be extracted by n-butanol/water (Tissieres, 1956; Swank and Burris, 1969). However, amino acid sequence studies show that they are both present in most if not all pseudomonads which have been examined, including *Pseudomonas mendocina*, *P. aeruginosa*, *P. stutzerii*, *P. denitrificans*, and *P. fluorescens* (Ambler, 1973; Ambler, 1991). There is a water-soluble form of cytochrome c_5, as well as a membrane-bound form, presumably related through proteolysis (Ambler, 1991). There is at least one purple phototroph, strain H1R, which has a water-soluble cytochrome c_5 (Ambler, 1991).

Two complete amino acid sequences of the water-soluble form of cytochrome c_5 have been published (Ambler and Taylor, 1973; Carter *et al.*, 1985). The N-terminal 33 residues of the membrane-bound form of the *A. vinelandii* cytochrome c_5 are apparently cleaved by endogenous proteolysis to give the 83-residue water-soluble form. The N-terminus of the bound form has five acidic and three basic residues uniformly spaced, which indicates that it is not a hydrophobic membrane anchor; thus, the mode of binding of the cytochrome to the membrane is still unknown (Ambler, 1991). The heme is attached at cysteines 48 and 51 in the membrane-bound form. Another unusual aspect of the water-soluble form of cytochrome c_5 is a disulfide preceding the C-terminal helix. There could also be a disulfide in the N-terminal part of the membrane-bound form of cytochrome c_5. The five-residue region around the sixth heme ligand methionine in a *Methylococcus capsulatus* cytochrome c-555 is the same as in cytochrome c_5 (NAMPP), although there is no other obvious similarity (Ambler *et al.*, 1986).

The three-dimensional structure of the water-soluble form of *Azotobacter* cytochrome c_5 (Carter *et al.*, 1985) shows obvious homology to *Pseudomonas* cytochrome c-551 although there are unique aspects of each protein. In fact, alignment of the structures of cytochrome c_5 and *Pseudomonas* cytochrome c-551 shows that they are similar at low resolution but differ in detail throughout their sequences. N- and C-terminal helices are present in both proteins, although the corresponding aromatic residues in these helices are absent in most cytochromes c_5. The phototrophic bacterial strain H1R cytochrome c_5 is water-soluble and contains the aromatic residues in the terminal helices. There is a three-residue insertion between the heme ligand methionine and the C-terminal helix in cytochrome c_5. Two of the four prolines near the methionine heme ligand in *Pseudomonas* cytochrome c-551 are present in cytochrome c_5. The proline which H-bonds the histidine heme ligand is present. The midsection of cytochrome c_5 contains two helices similar to those of cytochrome c-551. However, arginine 38 and glutamine 51 (*Azotobacter* water-soluble c_5 numbering) H-bond one of the heme propionates, a role filled by arginine 47 and tryptophan 56 in *Pseudomonas* cytochrome c-551. Glutamine 51 is replaced by lysine and serine in other cytochromes c_5. There is a six-residue insertion in *Pseudomonas* cytochrome c-551 relative to cytochrome c_5, which contains the tryptophan 56 (that H-bonds the heme propionate). Tryptophan 35 in cytochrome c_5 does not have a structural equivalent in *Pseudomonas* cytochrome c-551.

Like *Pseudomonas* cytochrome c-551, cytochrome c_5 has very little charge near the edge of the heme which is exposed to solvent and presumably where electron transfer most readily occurs. In fact, cytochrome c_5 crystallizes as a dimer, and the dimer interface is at the exposed heme edge. The hydrophobic heme edge region may also contribute to membrane binding. This is a pronounced difference from mitochondrial cytochromes c and bacterial cytochromes c_2, which have a concentration of positive charge near the heme edge that is thought to facilitate electron transfer through electrostatic attraction to oxidase and reductase or reaction center. An electrostatic map for cytochrome c_5 has been compared with that for *Pseudomonas* cytochrome c-551 and contrasted with that for cytochrome c_2 (Cheddar *et al.*, 1989). We have not been able to detect any functional difference between *Pseudomonas* cytochrome c-551 and cytochrome c_5, and it appears they could be interchangeable. If they function interchangeably, it is hard to understand why one of the two has not been naturally deleted in some species. The joint presence of these proteins in a wide range of species suggests that they have different roles.

5. *Chlorobium* cytochrome *c*-555

This protein is the green phototrophic bacterial equivalent of cytochrome c_5. It has been found in all species of green sulfur bacteria examined, including several strains of *Chlorobium thiosulfatophilum, Pelodictyon luteolum, Prosthecochloris aestuarii, C. limicola,* and *C. vibrioforme* (Meyer *et al.,* 1968; Olson and Shaw, 1969; Shioi *et al.,* 1972; Steinmetz and Fischer, 1981, 1982a,b; Steinmetz *et al.,* 1983; Fischer, 1989). Cytochrome *c*-555 is water-soluble and has a relatively low redox potential (80–160 mV), presumably reflecting the anaerobic environment in which these bacteria are found. There are only two distinct amino acid sequences known for cytochromes *c*-555, although several strains have been examined (Van Beeumen *et al.,* 1976). A preliminary low-resolution three-dimensional structure has been published for *C. thiosulfatophilum* strain Tassajara (Korszun and Salemme, 1977).

The amino acid sequences of cytochromes *c*-555 show that green bacterial classification is in need of revision. The definition of species primarily by cell morphology appears not to be valid. *Chlorobium* strains PM, L, and Tassajara have been found to have identical cytochrome *c*-555 sequences, whereas the *Prosthecococcus aestuarii* protein is 45% different. Nevertheless, the above three strains of *Chlorobium* have been placed in separate species (*C. limicola f. thiosulfatophilum* and *C. vibrioforme f. thiosulfatophilum*). We have been using the older name *C. thiosulfatophilum*, which includes all those strains having the following properties: freshwater growth, utilization of thiosulfate, and relatively high CG content (56–58%). The older definition of *C. limicola* includes primarily marine bacteria which are unable to use thiosulfate, and which have a lower CG content (52%). Those strains which utilize thiosulfate appear to be more uniform than those which do not. However, more sequences will have to be determined before it is known how many species of green sulfur bacteria there are and what their diagnostic features are.

The sequence of *Chlorobium* cytochrome *c*-555 has a two-residue insertion relative to that of cytochrome c_5 and is missing the C-terminal disulfide and the N-terminal extension. *Prosthecochloris* has a four-residue insertion relative to *C. thiosulfatophilum*. Cytochrome c_5 arginine 38 and glutamine 51, which H-bond the heme propionate, are present in *Prosthecochloris*. These positions are histidine and glycine in *C. thiosulfatophilum* cytochrome *c*-555. The aromatic residues in the N- and C-terminal helices in type I cytochromes are present in the green bacterial cytochromes *c*-555, although they are generally absent in cytochrome c_5. The proline which H-bonds the histidine heme ligand is also present. Cytochrome c_5 tryptophan 35 (role unknown) is retained in cytochrome *c*-555. There is no concentration of positive charge near the edge of the heme such as that found in mitochondrial cytochromes *c* and c_2, although the net charge of *Chlorobium* cytochrome *c*-555 is strongly cationic, which results in a small positive charge at the site of reaction with charged molecules (Meyer *et al.,* 1984).

6. Cytochromes c_6 or algal cytochromes *c*-553

Cytochromes c_6 are also known as algal cytochromes *c*-552 or *c*-553. They are found in both cyanobacteria and true algae, but not in plants. In the bacteria and algae, they are interchangeable with the copper protein plastocyanin, which in plants is the only electron donor to the photosystem I reaction center. Red algae and one-third of the cyanobacteria have the cytochrome only, whereas two-thirds of the cyanobacteria and three-quarters of the green algae have both the cytochrome and plastocyanin (Sandmann, 1986). One quarter of the green algae have plasto-cyanin only. *Anabaena variabilis* cytochrome c-553 is periplasmic (Serrano *et al.*, 1990).

The amino acid sequences of algal cytochromes c-553 are similar to the cyto-chromes c_5, averaging about 20% identity. On the other hand, they appear to be a little closer to the green bacterial cytochromes c-555. Unlike green bacterial cyto-chrome c-555, they generally have high redox potentials (about 350 mV). Fifteen sequences of algal cytochromes c-553 are now known (see Meyer and Kamen, 1982, for early reports, plus Sprinkle *et al.*, 1986; Okamoto *et al.*, 1987; Cohn *et al.*, 1989a). The gene sequence for the *Chlamydomonas reinhardtii* protein has also been determined (Merchant and Bogorad, 1987). There is apparently only a two-residue deletion in most algal cytochromes c-553 relative to the sequence of cytochrome c_5. This deletion occurs in the segment preceding the C-terminal helix. Aromatic resi-dues are present in the N-terminal and C-terminal helices like those of *Chlorobium* cytochromes c-555, but dissimilar to most cytochromes c_5. Algal cytochromes c-553 do not have the disulfide which is characteristic of cytochrome c_5, and they do not have the N-terminal extension which apparently holds cytochrome c_5 to the membrane. Most also appear to lack the proline which H-bonds the histidine heme ligand in most other type 1 cytochromes. Cytochrome c_5 glutamine 51, which H-bonds the heme propionate, is present in all the algal cytochromes c-553, but arginine 38 is absent. The algal cytochromes c-553 are generally negatively charged, although there are a few which are positively charged both overall and at the site of electron transfer (Meyer *et al.*, 1987). Cytochrome c_5 tryptophan 35 is absent in algal cytochrome c-553, but there is a conserved tryptophan of unknown function near the C-terminus.

There appears to be a closer than average relationship between cytochromes c-553 from the green algae *Bryopsis* (Okamoto *et al.*, 1987) and *Chlamydomonas* (Merchant and Bogorad, 1987) and that of *Euglena* (Pettigrew, 1974) based on a shared two-residue insertion in their sequences. The mitochondrial cytochrome c from *Euglena* (Pettigrew, 1973) is one of the most divergent known and is not very similar to that of *Chlamydomonas* (Amati *et al.*, 1988). On the other hand, *Euglena* appears to be related to the flagellated protozoa such as *Crithidia* (Pettigrew, 1972), as reflected in their mitochondrial cytochrome c sequences. Both proteins lack one of the heme-binding cysteines. These data suggest that *Euglena* is basically a flagellated protozoan which has incorporated a green algal chloroplast. This is the best evidence to date for separate origins of chloroplasts and mitochondria.

7. Cyanobacterial and algal cytochromes c-550

These cytochromes have a single heme, a molecular weight of 15,000, and very low redox potentials (about −150 mV). They were first described in the cyano-bacterium *Anacystis nidulans* (Holton and Myers, 1967) and have since been found in several other species (Krogmann, 1991), including a true alga, *Navicula pelliculosa* (Yamanaka *et al.*, 1967). The amino acid sequences of cytochromes c-550 from the cyanobacteria *Aphanizomenon flos-aquae* and *Microcystis aeruginosa* show that these proteins are type 1 cytochromes most similar to the algal cytochromes c-553 (Cohn *et al.*, 1989b). The low redox potentials appear to result from the substitution of a histidine for the sixth ligand methionine by analogy with the low-redox-potential cytochromes c_3.

Gene duplication of algal cytochromes c-553 to form cytochromes c-550 was not very recent, because there appear to be several insertions and deletions and because their alignment is not obvious. The algal cytochromes c-550 are only about 30% similar to the algal cytochromes c-553. There are about twenty-three extra residues at the N-terminus of algal cytochrome c-550, eleven extra residues in the midsection, and eight extra residues between the sixth heme ligand histidine and the presumed C-terminal helix. Like many of the algal cytochromes c-553, the cytochromes c-550 lack the proline which usually H-bonds the fifth heme ligand histidine. Also in common with the algal cytochromes c-553, there is a tryptophan of unknown function near the C-terminus.

8. *Ectothiorhodospira* cytochromes c-551

Ectothiorhodospiraceae is a family of marine and halophilic purple sulfur phototrophic bacteria distinct from the Chromatiaceae. A cytochrome c-551 has been characterized from the extreme halophiles, *E. halophila* (Meyer, 1985), *E. halochloris* (Then and Truper, 1983), and *E. abdelmalekii* (Then and Truper, 1984). It has a low redox potential (-59 to 58 mV) which is somewhat lower than that of the green bacterial cytochromes c-555, and is an acidic protein similar to those of other halophiles. Cytochrome c-551 has not been found in the moderate halophiles *E. shaposhnikovii* (Kusche and Truper, 1984) and *E. vacuolata* (Meyer, unpublished). The amino acid sequences of *E. halophila* and *E. halochloris* cytochromes c-551 (Ambler and Meyer, unpublished) have been found to be most closely related to the cytochromes c_5, algal cytochromes c-553, and green bacterial cytochromes c-555. The *Ectothiorhodospira* cytochromes c-551 have the two-residue deletion following the heme ligand methionine that is typical of algal cytochromes c-553, but they also have the cytochrome c_5 tryptophan 35, which is present in the green bacterial cytochrome c-555 as well. Thus, they appear to be intermediate between the green bacterial and algal cytochromes. The *Ectothiorhodospira* cytochromes c-551 uniquely have a four-residue insertion between the N-terminal helix and the heme binding site and a three-residue deletion in the midsection.

9. *Desulfovibrio* cytochromes c-553

Desulfovibrio cytochromes c-553 have a limited distribution in the sulfate-reducing bacteria. They have been found in *D. vulgaris* strain Hildenborough (Bruschi *et al.*, 1972), *D. vulgaris* strain Miyazaki (Yagi, 1979), *Desulfomicrobium baculatus* (formerly *D. desulfuricans*) strain Norway 4 (Fauque *et al.*, 1979), *D. desulfuricans* NCIB 8372 (Eng and Neujahr, 1989), and *D. desulfuricans* Berre-Eau

(Moura *et al.*, 1987). The metabolism of *Wolinella succinogenes* suggests that the 100-mV, 8-kDa cytochrome *c*-552 is also related to the *Desulfovibrio* cytochromes *c*-553 (Moura *et al.*, 1988). The low redox potentials of the *Desulfovibrio* cytochromes *c*-553 (0 to +40 mV) (Bertrand *et al.*, 1982; Bianco *et al.*, 1983; Eng and Neujahr, 1989) probably reflect the anaerobic environment in which the bacteria live.

There are two published amino acid sequences for these proteins and one gene sequence (Bruschi *et al.*, 1972; Nakano *et al.*, 1983; Van Rooijen *et al.*, 1989), which are most similar to a halophilic *Paracoccus* sp. cytochrome *c*-554(548) (35%) and to the first half of *Azotobacter vinelandii* cytochrome c_4 (31%). There is only a single-residue insertion in the midsection relative to *Paracoccus*, but *Desulfovibrio* cytochrome *c*-553 is 1–5 residues shorter than the midsection of the two halves of cytochrome c_4.

The three-dimensional structure of *D. vulgaris* cytochrome *c*-553 shows that portions of the folding pattern are similar to those of other type 1 cytochromes, but there are an equal number of differences (Nakagawa *et al.*, 1990). The glycine and aromatic residues are found in the N- and C-terminal helices as in other type 1 cytochromes *c*. However, the extra helix found in the midsection of *Pseudomonas* cytochrome *c*-551 and cytochrome c_5 is absent in *Desulfovibrio* cytochrome *c*-553. Furthermore, the *Desulfovibrio* sixth-ligand methionine is contained within a helix not present in the other proteins. None of the residues which normally H-bond the heme propionates or the methionine sixth ligand in other type 1 cytochromes *c* are present in the *Desulfovibrio* cytochromes *c*-553.

10. Cytochrome cd_1—nitrite reductase

Cytochrome cd_1 is found in a number of denitrifying bacteria where it functions as a nitrite reductase. Mutants defective in cytochrome cd_1 are also defective in nitrite reduction (Zumft *et al.*, 1988). Cytochrome cd_1 has been reported to occur in *Pseudomonas aeruginosa* (Yamanaka and Okunuki, 1963), *P. stutzeri* (including the former *P. perfectomarinus*) (Cox and Payne, 1973; Zumft *et al.*, 1988), *P. fluorescens* (Coyne *et al.*, 1990), *P. mendocina* and *P. nautica* (Korner *et al.*, 1987), *Alcaligenes faecalis* (Iwasaki and Matsubara, 1971), *Thiobacillus denitrificans* (LeGall *et al.*, 1979), *Roseobacter denitrificans*, formerly *Erythrobacter sp.* (Doi *et al.*, 1989), *Paracoccus denitrificans* (Newton, 1969; Lam and Nicholas, 1969; Timkovich *et al.*, 1982), *Paracoccus halodenitrificans* (Mancinelli *et al.*, 1986), and a halophilic *Paracoccus* species ATCC 12084 (Meyer, unpublished). Cytochrome cd_1 is not the exclusive nitrite reductase in *Pseudomonas* species, as shown by the presence of a copper-containing enzyme in *Pseudomonas aureofaciens* (formerly *P. fluorescens* biotype E) (Zumft *et al.*, 1987). Copper-containing enzymes are found in at least four other species of denitrifying bacteria (Sawada *et al.*, 1978; Masuko *et al.*, 1984; Smith and Tiedje, 1992).

Cytochrome cd_1 has a minimum (subunit) molecular weight of 60,000, and the subunit has two hemes. The nucleotide sequences of cytochromes cd_1 show that the *c*-type heme binding site is near the N-terminus and is homologous to type 1 cytochromes *c* (Silvestrini *et al.*, 1989; Jungst *et al.*, 1991; Smith and Tiedje, 1992). There is a signal peptide, indicating that it is located in the periplasmic space. It can be aligned with cytochromes such as c_4 and c_5, but there is only about 20% similarity at most between these proteins. The sequence of cytochrome cd_1 from *Ps. stutzeri* (Jungst *et al.*, 1991; Smith and Tiedje, 1992) is only 57% identical overall to that of *Ps. aeruginosa* (Silvestrini *et al.*, 1989), and a minimum of seven insertions and deletions are necessary for alignment. It is 26 residues shorter at the N-terminus. In comparison of the *Ps. aeruginosa* and *Ps. stutzeri* heme *c* domains, there is a thirteen-residue insertion between the heme ligands and a two-residue deletion after the heme ligand methionine. There are aromatic residues in the N-terminal and C-terminal helices of the heme *c* domain, and the tryptophan 35 characteristic of cytochrome c_5 is present in the *Ps. stutzeri* protein. There are nine conserved histidines in the heme d_1 domain, which are possible heme binding sites. The N-terminal *c*-type cytochrome sequence in cytochrome cd_1 probably folds in a separate domain of the three-dimensional structure, although this has yet to be demonstrated. *P. aeruginosa* cytochrome cd_1 crystallizes readily, but the structure has not yet been determined (Takano *et al.*, 1979).

The heme d_1 is not covalently bound and is easily lost. The structure of heme d_1 shows that it is not a heme at all, but is a unique iron acryloporphyrindione (Chang and Wu, 1986; Andersson *et al.*, 1990). The heme d_1 is probably the site of nitrite reduction to nitric oxide. A mutant lacking the d_1 heme is inactive in nitrite reduction, although the enzyme is otherwise synthesized and contains the *c*-type heme (Zumft *et al.*, 1988).

Heme *c* is probably where electrons enter from the electron donors, which may be any one or all of *Pseudomonas* cytochromes *c*-551, c_5, c_4, and the copper protein azurin. However, the *Pseudomonas aeruginosa* cytochrome *c*-551 gene is adjacent to the cytochrome cd_1 gene, and they may form an operon (Nordling *et al.*, 1990; Arai *et al.*, 1990), which suggests that cytochrome *c*-551 is the natural electron donor. There are two genes (encoding a tetraheme and a diheme cytochrome) between the cytochromes cd_1 and *c*-551 in *Ps. stutzeri*, and they are separately regulated (Jungst *et al.*, 1991). The tetraheme cytochrome was shown to be necessary for nitrite reduction. There is another small type I cytochrome *c* gene adjacent to the cytochrome *c*-551 gene in both species (Jungst *et al.*, 1991), which has not been characterized as a protein, but it could also be an alternative electron donor to cytochrome cd_1. Electron transfer from the heme *c* to the heme d_1 is relatively slow, which suggests that the distance between hemes is large and that their orientation is unfavorable for electron transfer (Makinen *et al.*, 1983; Silvestrini *et al.*, 1990). It is interesting that the *Paracoccus denitrificans* (Timkovich *et al.*, 1982) and probably the *Roseobacter denitrificans* (formerly *Erythrobacter* sp.) (Doi *et al.*, 1989) cytochromes cd_1 have a different specificity for electron donors than does the *Pseudomonas* enzyme. The former preferably utilize cytochrome c_2 instead of *Pseudomonas* cytochrome *c*-551.

11. Flavocytochromes *c*

Flavocytochromes *c* (FC) are complex proteins containing both heme and flavin subunits. They are found in some species of purple and green sulfur bacteria, where they appear to function as sulfide dehydrogenases. These species include *Chlorobium limicola* (Steinmetz and Fischer, 1981), *C. thiosulfatophilum* (Meyer *et al.*, 1968; Bartsch *et al.*, 1968), *Chromatium vinosum* (Bartsch and Kamen, 1960; Bartsch *et al.*, 1968), *Ch. gracile* (Bartsch, 1978a), and *Thiocapsa roseopersicina* (Truper and Rogers, 1971; Bartsch, 1978a; Zorin and Gogotov, 1981). They are also present in *Pseudomonas putida*, where they function as *p*-cresol (and other *p*-alkyl phenol) dehydrogenases (Hopper and Taylor, 1977; Reeve *et al.*, 1989).

The amino acid sequences of the FC heme subunits have been determined for *Pseudomonas* (McIntire *et al.*, 1986), *Chlorobium* (Van Beeumen *et al.*, 1990), and *Ch. vinosum* (Van Beeumen *et al.*, 1991). The gene sequence of the heme subunit of *Ch. vinosum* has also been determined (Dolata *et al.*, 1993). A three-dimensional structure has been reported for *Pseudomonas* FC (Bellamy *et al.*, 1987; Mathews *et al.*, 1991). All three flavocytochrome *c* heme subunits are homologous to type 1 cytochromes *c*, although not particularly close to any one class. The sequences of the FC heme subunits are not very closely related to one another (25%), and the *Pseudomonas* FC is seven residues shorter than *Chlorobium* FC in the midsection. The *Ch. vinosum* FC heme subunit appears to have been doubled, not unlike *Azotobacter* cytochrome c_4, although the connecting segment is seven residues instead of sixteen. The two halves of the *Ch. vinosum* FC heme subunit are only 11% similar to one another. The first half of the *Ch. vinosum* heme subunit is more similar to the *Chlorobium* subunit (38%) than is the second half.

The three-dimensional structure of the heme subunit of *Pseudomonas* FC (Mathews *et al.*, 1991) is similar to that of other type 1 cytochromes *c* in that it has N- and C-terminal helices, each containing an aromatic residue. Proline 27 appears to H-bond the histidine heme ligand. In the midsection, there is only one helix like that of *Desulfovibrio* cytochrome *c*-553, rather than the two present in *Azotobacter* cytochrome c_5 and *Pseudomonas* cytochrome *c*-551. The heme subunit binds to the flavin subunit at the "front" face, which is where electron transfer is generally thought to occur in type 1 cytochromes *c*. Side chain interactions are not yet well defined at the interface. It is, however, fairly certain that intramolecular electron transfer occurs without dissociation of subunits. This necessitates another site of interaction with the ultimate electron acceptor, which is thought to be an azurin. The nearest approach to the heme in the holoenzyme is at the "bottom" in the region of the heme propionates.

The N-terminal sequences of the *Chlorobium* and *Chromatium* FC flavoprotein subunits show that the covalent flavin binding cysteine is located at position 42 (Van Beeumen *et al.*, 1990, 1991; Dolata *et al.*, 1993), but the covalent flavin binding site of *Pseudomonas* FC is at tyrosine 436 (Mathews *et al.*, 1991). There does not appear to be any similarity at all between the flavin subunits of the phototrophic flavocytochromes *c* and that of *Pseudomonas*. The N-terminal 45 residues of the *Chlorobium* and *Chromatium* FC flavin subunits are actually homologous to a variety of flavoproteins, including succinic dehydrogenase, which has a flavin binding histidine in the same position as cysteine 42 of the phototrophic flavocytochromes *c*.

12. Methanol dehydrogenase–associated cytochrome *c*-550 or c_L

Separate catabolic pathways for utilization of single carbon compounds as well as ethanol, glucose, and glycerol are inducible in the methylotrophic bacteria and acetic acid bacteria. These enzymes generally contain a variation of the PQQ cofactor (McIntire *et al.*, 1991) and interact with high potential electron acceptors (Anthony, 1988, 1992). For example, the copper protein amicyanin is the electron acceptor for the enzyme methylamine dehydrogenase, a triheme cytochrome c is closely coupled to ethanol dehydrogenase, and cytochrome c-550 or c_L is the electron acceptor for methanol dehydrogenase (Anthony, 1986, 1988). Methanol dehydrogenase has 60-kDa and 12-kDa subunits (Harms *et al.*, 1987; Machlin and Hanson, 1988; Nunn *et al.*, 1989; Anderson *et al.*, 1990; Van Spanning *et al.*, 1991). The methanol dehydrogenase–associated cytochrome c-550 is also called cytochrome c_L for its low isoelectric point. Cytochrome c_H (for high isoelectric point) from *Methylobacterium extorquens* is actually a cytochrome c_2, and that from *Methylophilus methylotrophus* is actually related to *Pseudomonas* cytochrome c-551 (Ambler, 1991). Because isoelectric point is not a diagnostic property and leads to misidentification as in the foregoing example, we will use the more convention-al nomenclature based on alpha peak maximum, i.e. c-550 rather than c_L for the methanol dehydrogenase–associated cytochrome c.

The most thoroughly characterized methanol dehydrogenase–associated cyto-chromes c-550 are from *Methylobacterium extorquens* strain AM1 (formerly *Pseudo-monas* AM1) (O'Keeffe and Anthony, 1980) and *Paracoccus denitrificans* (Beardmore-Gray *et al.*, 1982, 1983; Husain and Davidson, 1986), which have molecular weights of 19,000 and have a single heme. The redox potential of this protein is 256 mV, and the isoelectric point is 4.2. The cytochrome gene has been cloned and sequenced from two species, *M. extorquens* (Nunn and Anthony, 1988) and *P. denitrificans* (Van Spanning *et al.*, 1991) where it is known as cytochrome c-551$_i$ (for "inducible"). There are 172 amino acid residues in the mature *M. extorquens* pro-tein and 156 residues in *P. denitrificans*. There are no internal insertions or dele-tions, and the two are 58% identical. It was thought that there is no homology to type 1 cytochromes c (Nunn and Anthony, 1988), but the features of type 1 cytochromes c are definitely present. These include a heme binding site near the N-terminus, a possible sixth ligand methionine near the C-terminus, and a proline which is likely to be H-bonded to the histidine fifth ligand. The methanol dehy-drogenase–associated cytochrome c-550 appears to be most similar to a halophilic *Paracoccus* sp. cytochrome c-554(548) (Ambler *et al.*, 1987a) and to *Azotobacter vinelandii* cytochrome c_4 (Ambler *et al.*, 1984), which have only two to three more residues between heme ligands than does cytochrome c-550. Aromatic residues in the N-terminal and C-terminal helices which are conserved in nearly all type 1 cyto-chromes are present. The relatively low overall similarity to the *Paracoccus* and *Azotobacter* cytochromes (*ca.* 20%) undoubtedly reflects their different functional roles. The methanol dehydrogenase–associated cytochromes c-550 share the ap-proximately 40-residue C-terminal extension with *Thermus thermophilus* cytochrome c-552 (Titani *et al.*, 1985a,b), but they are unique in their approximately 50-residue N-terminal extension. *P. denitrificans* and *M. extorquens* have a cytochrome c-553$_i$ inducible by single carbon compounds in addition to c-551$_i$, which has a similar sequence (Day *et al.*, 1990; Ras *et al.*, 1991). It is slightly larger and has the heme and its ligands in roughly the same location, although there is very low sequence identity (about 18%) and three small insertions are necessary for alignment. The function is unknown, but this protein probably belongs in a separate subclass. This will become clearer as more species are examined.

There are signal peptides in the methanol dehydrogenase–associated cyto-chromes c-550 that direct them to the periplasm, a feature also necessarily present in the two methanol dehydrogenase subunits which have been cloned (Harms *et al.*,

1987; Machlin and Hanson, 1988). *Methylobacter extorquens* methanol dehydrogenase and cytochrome *c*-550 are apparently co-transcribed (Anderson and Lidstrom, 1988), although the methanol dehydrogenase gene in *M. organophilum* is not part of an operon (Machlin and Hanson, 1988). A methanol dehydrogenase–associated cytochrome *c*-550 with properties similar to that of *M. extorquens* has been found in *Methylophilus methylotrophus* (Cross and Anthony, 1980; Beardmore-Gray *et al.*, 1982, 1983), in *Acetobacter methanolicum* (Beardmore-Gray *et al.*, 1982), in *Hyphomicrobium* X (Dijkstra *et al.*, 1988), in *Methylomonas* sp. (DiSpirito *et al.*, 1990), and in *Protaminobacter ruber* (Iba *et al.*, 1985). The methanol dehydrogenase–associated cytochrome *c*-550 is the obligate electron acceptor for methanol dehydrogenase and has been shown to interact with the large subunit of the enzyme through electrostatic association (Chan and Anthony, 1991; Cox *et al.*, 1992).

13. Alcohol dehydrogenase and associated cytochrome subunit

Acetic acid bacteria such as *Acetobacter aceti* and *A. polyoxogenes* contain a membrane-bound alcohol dehydrogenase which uses the PQQ cofactor (Ameyama and Adachi, 1982; Tayama *et al.*, 1989). The enzyme is composed of two subunits of 72 kDa and 44 kDa, both of which are cytochromes *c*. They have relatively high redox potentials of 260 and 340 mV (Ameyama and Adachi, 1982). The larger subunit binds the PQQ cofactor. The soluble PQQ-containing alcohol dehydrogenases from *Pseudomonas aeruginosa* and *Ps. testosteroni* are also 65-kDa cytochromes *c* (Groen *et al.*, 1986; Groen and Duine, 1990). The gene sequences of *Acetobacter aceti* and *A. polyoxogenes* alcohol dehydrogenases show that the 72-kDa subunit has a cytochrome *c* domain fused to the C-terminus and that the adjacent 44-kDa subunit contains three tandem repeats of a small cytochrome *c* (Inoue *et al.*, 1989; Tamaki *et al.*, 1991). The N-terminus of the large subunit of alcohol dehydrogenase is homologous to the methanol dehydrogenases which use the PQQ cofactor but which do not contain heme.

The cytochrome domain in the large subunit and the three domains of the smaller subunit all appear to be related to one another and to the type 1 cytochromes *c*. The heme binding site of each domain is near the N-terminus, and there is a potential methionine sixth ligand in the C-terminal half. There is a proline 10–18 residues after the heme binding site in all four domains which might H-bond the fifth heme ligand histidine, and there are aromatic residues in what would be the N-terminal and C-terminal helices in 2–4 of the domains. The final point of similarity is the presence of a consensus glycine in the presumed N-terminal helix. There is so little similarity otherwise that it is impossible to align them with confidence. Both subunits contain signal peptides directing them to the periplasm.

Three tandem repeats of a type 1 cytochrome is unique. Cytochrome c_4 and *Chromatium* flavocytochrome *c* contain two tandem repeats and are the only other type 1 cytochromes with this characteristic. Cytochrome c_3 from *Desulfuromonas* has three tandem repeats but is probably the result of a deletion of one heme from a twice-doubled ancestral gene. There is no indication that the alcohol dehydrogenase associated 44-kDa cytochrome subunit is any closer to other multiheme cytochromes, such as cytochrome c_4, than to any of the monoheme type 1 cytochromes *c*.

14. *Pseudomonas* nitrite reductase–associated cytochrome *c*

The gene for a small monoheme cytochrome *c* (orf 5) was discovered immediately downstream of the *Pseudomonas* cytochrome c-551 gene in *Ps. stutzeri* and in *Ps. aeruginosa* (Jungst *et al.*, 1991). The gene product of *Ps. stutzeri* orf 5 contains 87 amino acid residues and is 63% identical to that of *Ps. aeruginosa*. The product of this gene has not been characterized as a protein, but it has a signal peptide directing it to the periplasmic space. The soluble cytochromes of *Pseudomonas* have been well characterized; thus, it is possible that this protein was not previously observed because the signal peptide is not cleaved and it remains membrane-bound. It is homologous to the other small type 1 cytochromes from *Pseudomonas* such as cytochrome c-551, c_4, and c_5, but is as divergent as any of these and does not appear to be any closer to one than to another. Its functional role is still unknown. It does not have the glycine and aromatic residues typical of the N-terminal helix, but it does have the proline after the heme which H-bonds the histidine heme ligand and the aromatic residue in the C-terminal helix.

15. *Bacillus* cytochromes *c*

Bacillus is a gram-positive bacterium and has no outer membrane and no periplasmic space, yet it is rich in cytochromes *c*, which are necessarily bound to the outer surface of the cytoplasmic membrane. The *a*-type oxidase in *Bacillus* has a cytochrome *c* gene fused to the C-terminus of the oxidase subunit II, which is the copper-containing subunit homologous to plastocyanin and azurin. However, *Bacillus* has several cytochromes in addition to the oxidase subunit II—cytochrome *c*. Hreggvidsson (1991) obtained partial sequence data for *Bacillus azotoformans* cytochromes c-552 and c-555, which were released by tryptic hydrolysis of membranes. A cytochrome c-550 from *Bacillus subtilis* was isolated by detergent, purified, cloned, and sequenced (Von Wachenfeldt and Hederstedt, 1990a,b). It contains 120 residues, including a large N-terminal membrane anchor. The water-soluble C-terminal fragment of a *B. licheniformis* cytochrome c-552 was sequenced by Van Beeumen (reported in Hreggvidsson, 1991). It is closely related to the *B. subtilis* protein (38% identity) and requires no insertions or deletions for alignment. These cytochromes have a large deletion in the midsection relative to the oxidase cytochrome *c*, which appears to be more closely related to mitochondrial cytochrome *c*. The *B. licheniformis* cytochrome has a glycine and a tyrosine in the N-terminal helix; *B. subtilis* has the tyrosine; all three proteins have the proline which H-bonds the histidine heme ligand; and all have a tryptophan in the C-terminal helix. They are about 10 residues shorter in the midsection than are other small type 1 cytochromes *c* and appear to be most similar to cytochromes c_4.

16. *Bacillus* cytochrome oxidase subunit II–cytochrome *c* fused gene

Cytochrome oxidase purified from the bacteria *Bacillus subtilis* and *Bacillus sp. PS3*, and from *Thermus thermophilus* contain *c*-type heme, and it had been assumed that one of the oxidase subunits was a cytochrome c_1. It was subsequently shown that the oxidase does not contain cytochrome c_1, but that there is a gene fusion between the copper-containing subunit II and a small cytochrome *c* (Sone *et al.*,

1988; Buse *et al.*, 1989). The complete gene sequences of the proteins from the three species confirmed this observation and demonstrated that the cytochrome *c* domain is homologous to the type 1 cytochromes *c* (Ishizuka *et al.*, 1990; Saraste *et al.*, 1991; Mather *et al.*, 1991). The two *Bacillus* heme domains are 56% identical to one another, whereas the *Thermus* heme domain is more divergent, with only 40% identity to the *Bacillus* proteins. The *Thermus* oxidase cytochrome *c* domain also has two insertions or deletions relative to the *Bacillus* proteins. The cytochrome *c* fused to the oxidase is apparently type 1 by virtue of heme attachment near the N-terminus of the domain and a conserved methionine near the C-terminus, which could be the sixth heme ligand. Both a glycine and a phenylalanine occur in the presumed N-terminal helix, and a tyrosine is present in the C-terminal helix, typical of type 1 cytochromes *c*. The oxidase cytochrome appears to be most similar to mitochondrial cytochromes *c* if one assumes an eight-residue deletion just after the heme binding site. The overall similarity is only 20–30% and is not convincing in itself. However, the proline which normally H-bonds the histidine heme ligand is present in the *Bacillus* oxidase cytochromes but not in that of *Thermus*. Basic residues which are conserved in mitochondrial cytochrome *c* and involved in electrostatic binding of cytochrome *c* to oxidase are all substituted in *Thermus* and *Bacillus* oxidase cytochromes except the lysine immediately preceding the methionine sixth heme ligand. The residues which normally H-bond the heme propionates are all substituted. The tyrosine which H-bonds the methionine heme ligand is a tryptophan in the *Bacillus* oxidase cytochromes *c*.

17. Miscellaneous type 1 cytochromes *c*

There are several cytochromes *c* which are obviously type 1, but which have been found in only one species and are not closely related to any of the known subclasses. One or more of these proteins may eventually be found in additional species and therefore may be deserving of recognition as a separate subclass, but for the time being, they will be kept separate. Placement in a miscellaneous category also emphasizes the need for three-dimensional structures before a decision is made concerning their status. *Paracoccus* sp. cytochrome *c*-554(548) (Ambler *et al.*, 1987a), *Thiobacillus neapolitanus* cytochrome *c*-554(547) (Ambler *et al.*, 1985), and *Thermus thermophilus* cytochrome *c*-552 (Titani *et al.*, 1985) were mentioned earlier as showing some similarity to cytochrome c_4. However, these proteins are water-soluble and contain only one heme, whereas cytochrome c_4 is membrane-bound and results from gene doubling. These properties suggest a different functional role.

The cytochrome *c*-555 from *Methylococcus capsulatus* (Ambler *et al.*, 1986) shows some similarity to cytochromes c_5 and to algal cytochrome *c*-553 near the sixth heme ligand methionine and, like algal cytochrome *c*-553, appears to lack the proline which H-bonds the fifth heme ligand histidine. *Bradyrhizobium japonicum* cytochrome *c*-552 (Appleby *et al.*, 1991; Rossbach *et al.*, 1991) also shows similarity to cytochrome c_5. These cytochromes do not have the N-terminal extension and are completely water-soluble. The *Bradyrhizobium fix*O and *fix*P genes are associated with an oxidase and are not closely related to any other cytochromes, although *fix*P is diheme like cytochrome c_4 (Preisig *et al.*, 1993).

B. Type 2 Cytochromes *c*

The most commonly encountered type 2 cytochromes *c* are high-spin and are known as cytochromes *c'*. They were originally characterized in purple phototrophic bacteria where they are widespread, but have since been found in non-phototrophs such as *Paracoccus denitrificans*, *Azotobacter vinelandii*, a *Paracoccus* sp., and an *Alcaligenes* sp. (see Meyer and Kamen, 1982; Goodhew *et al.*, 1990). Some purple bacteria such as *Rps. viridis*, *Rm. vannielii*, *Rps. acidophila*, and *Rp. globiformis* appear to lack cytochromes *c'* entirely. Amino acid sequences are known for about 15 species (Ambler *et al.*, 1981b). The type 2 cytochromes *c* only average about 28% similarity, and it is difficult to position the several small insertions and deletions. The heme is bound near the C-terminus at about position 120, whereas the heme is near the N-terminus of type 1 cytochromes *c*.

The refined high-resolution three-dimensional structure of *R. molischianum* cytochrome *c'* is known (Finzel *et al.*, 1985). Cytochrome *c'* is primarily helical (68%), although there is no obvious relationship to hemoglobin. The structure of *Rhodospirillum rubrum* cytochrome *c'* has also been determined (Yasui *et al.*, 1992). There are four parallel helices arranged in a bundle, which is now recognized as a classic folding pattern for a variety of proteins. Cytochromes *c'* are nearly all dimeric. Only *Rps. palustris* cytochrome *c'* appears to be monomeric. The cytochromes *c'* are more rapidly reduced by free flavin semiquinones, but are more slowly reduced by flavodoxin semiquinone than are type 1 cytochromes *c* (Meyer *et al.*, 1986b). This difference is due to the fact that the heme is more exposed to solvent than in type 1 cytochromes *c*, but it is at the bottom of a deep groove on the surface of the protein, which admits small molecules and excludes other proteins. This structural feature suggests that cytochrome *c'* is designed for reaction with small oxidants and reductants.

There are low-spin examples of type 2 cytochromes *c* which are occasionally found in the same species that produce cytochrome *c'*, and one may think of them as cytochrome *c'* isozymes. They can be recognized by red-shifted UV–visible absorption spectra relative to type 1 cytochromes *c*. Purple bacterial examples are *Rb. sphaeroides* cytochrome c-554 (Meyer and Cusanovich, 1985; Bartsch *et al.*, 1989), *Rps. palustris* cytochromes c-556 and c-554 (Bartsch, 1978a), *Rb. sulfidophila* cytochrome c-556 (Bartsch, 1978a), and *Rps. marina* cytochrome c-558 (Meyer *et al.*, 1990c). The only certain non-phototrophic low-spin type 2 cytochromes *c* are *Agrobacterium tumefaciens* cytochrome c-556 (Van Beeumen *et al.*, 1980) and *Bradyrhizobium japonicum* cytochrome c-555 (Tully *et al.*, 1991). The sixth iron ligand in the low-spin type 2 cytochromes *c* is most likely to be methionine 16 (*R. molischianum* numbering) because it is very near the heme in the three-dimensional structure of *R. molischianum* cytochrome *c'*; nevertheless, methionine is not close enough to bind to the iron in this species. *R. molischianum* cytochrome *c'* is the only type 2 cytochrome *c* which has this methionine and is still high-spin. Whatever small rearrangement of the structure allows this methionine to bind to the iron in some cases but not in others is still unknown.

There is no evidence to suggest that the low-spin type 2 cytochromes *c* are structurally more similar to one another than they are to the high-spin cytochromes

c'. On the contrary, *Rps. palustris* cytochrome c-556 is slightly more similar to *Rps. palustris* cytochrome c' (36%), and *Rb. sphaeroides* cytochrome c-554 is somewhat closer to *Rb. sphaeroides* cytochrome c' (36%) than to other species of type 2 cytochromes c. One insertion/deletion is required for the former alignment and none for the latter. These results suggest that the low-spin proteins have been produced through duplication of the cytochrome c' gene on more than one occasion, and that methionine is merely one of several hydrophobic residues permitted at position 16. Furthermore, the low-spin state of the iron may not have any functional role. High-spin isozymes of type 2 cytochromes should be at least as common or more so than the low-spin isozymes if the spin state does not confer a functional advantage, because there are four hydrophobic residues which have been found at position 16 in addition to methionine. We have not yet found more than one high-spin version of cytochrome c' in any one species, but we have not looked very closely.

The three-dimensional structure of *E. coli* cytochrome b-562 (Mathews *et al.*, 1979) shows that it is primarily helical (64%) and the four helices are arranged in a bundle similar to those in cytochrome c'. The heme fifth ligand histidine 106 is near the C-terminus, and the sixth ligand methionine 7 is near the N-terminus, like the low-spin isozymes of cytochrome c'. These proteins can be aligned with the aid of the three-dimensional structures (Weber *et al.*, 1981), but there is very little sequence similarity (about 15%). However, the location of the heme ligands similar to those of the type 2 cytochromes c can be taken as evidence for homology. There are other proteins which share this structural motif of four helices, such as hemerythrin, ferritin, and tobacco mosaic virus coat protein (Weber and Salemme, 1980), which are not obviously homologous.

The bacterial ferritins contain heme as isolated, which suggests that they actually may be homologous to cytochrome c'. The sequence of *E. coli* bacterioferritin has five histidines and seven methionines which are potential heme ligands (Andrews *et al.*, 1989a,b). The N-terminal sequence of *Nitrobacter winogradskyi* bacterioferritin (Kurokawa *et al.*, 1989) shows that histidine 28 is conserved and along with histidine 130 could be the heme ligands. However, recent studies suggest that both ligands may be methionines in some proteins and histidines in others (Cheesman *et al.*, 1990; Moore *et al.*, 1992). Heme binding is usually non-stoichiometric, although it has been shown that all subunits are capable of binding heme (Kadir and Moore, 1990a). The three-dimensional structure of bacterioferritin is necessary both to tell whether it is structurally related to cytochrome c' and to determine the heme binding residues. It has also been shown that mammalian ferritin is capable of binding heme, although there is no heme in the isolated protein (Kadir and Moore, 1990b). The absence of heme may be a consequence of the relatively harsh isolation procedure. There is no obvious homology between eukaryotic ferritins (Theil, 1987; Ragland *et al.*, 1990) and bacterial ferritins. Methionine 96, histidine 114, and histidine 124 are, however, conserved in the eukaryotic ferritins and are candidates for heme ligands.

C. Type 3 Cytochromes c or Cytochromes c_3

Type 3 cytochromes c are also known as cytochromes c_3 and are present in most, if not all, *Desulfovibrio* species of sulfate-reducing bacteria. They have also been characterized in *Thermodesulfobacterium commune* (Hatchikian *et al.*, 1984) and *Desulfobulbus elongatus* (Samain *et al.*, 1986). There are four hemes in the relatively small 14–15 kDa proteins. Amino acid sequences have been determined for *D. vulgaris* strain Hildenborough (Ambler, 1968), *D. vulgaris* strain Miyazaki (Shinkai *et al.*, 1980), *D. gigas* (Ambler *et al.*, 1969), *D. desulfuricans* strain El Agheila (Ambler *et al.*, 1971), *D. baculatus* (formerly *D. desulfuricans*) strain Norway (Bruschi, 1981), and *D. salexigens* (Ambler, 1973). The gene sequence of *D. vulgaris* strain Hildenborough has also been determined (Voordouw and Brenner, 1986).

A degenerate three-heme, 68-residue, 9-kDa cytochrome c_3 (also called cytochrome c_7), which was isolated from a syntrophic mixture known as *Chloropseudomonas ethylica* (Olson and Shaw, 1969; Shioi *et al.*, 1972) that contains a green sulfur bacterium and a sulfur-reducing bacterium, has been sequenced (Ambler, 1971). The organism from which the cytochrome was derived was identified as *Desulfuromonas acetoxidans* (Probst *et al.*, 1977). It is the second heme in the parent cytochrome which apparently was deleted along with other residues. It is characteristic of the cytochromes c_3 that there is very little conservation of the amino acid sequence except for the heme-binding cysteines and the heme ligands. Consequently, it is difficult to align the sequences and to know even approximately how many insertions and deletions separate them.

The three-dimensional structures of *D. vulgaris* strain Miyazaki (Higuchi *et al.*, 1981, 1984), *D. baculatus* (formerly *D. desulfuricans*) strain Norway 4 (Pierrot *et al.*, 1982), and *D. gigas* (Kissinger, 1989) show that they fold similarly in spite of the large sequence differences. The hemes were numbered arbitrarily and differently in the two published crystallographic studies, but will be numbered sequentially in this review for the sake of clarity. A sequence alignment based on three-dimensional structure was suggested (Higuchi *et al.*, 1981). There is strong evidence from the amino acid sequences that the four-heme cytochromes c_3 evolved by at least one and more likely two successive gene duplications of a smaller protein, although the tetraheme cytochrome we have today folds in a single domain and shows no evidence for doubling in the three-dimensional structure. Sequence evidence for gene doubling includes an unusual four-residue spacer between the cysteines of hemes 2 and 4 of *D. vulgaris* and *D. gigas* cytochromes c_3, but normal two-residue spacing between the cysteines of hemes 1 and 3. *D. baculatus* and *D. salexigens* have the four-residue spacer only in heme 2. The extra two residues at heme 4 apparently were deleted subsequent to duplication.

The redox potentials of the interacting hemes of cytochromes c_3 can be resolved and are very low (about -30 to -400 mV) (DerVartanian *et al.*, 1978; Bruschi *et al.*, 1984; Hatchikian *et al.*, 1984; Samain *et al.*, 1986; Dolla *et al.*, 1987; Benosman *et al.*, 1989; Fan *et al.*, 1990), and the heme ligands are all histidines. The sequence locations of the histidine sixth ligands also support gene doubling, although the ligands are not symmetrically bonded to the four hemes. The sixth ligand of heme 1 is histidine 22 (strain Miyazaki numbering), that of heme 2 is histidine 35, that of heme 3 is histidine 25, and that of heme 4 is histidine 70. There is an apparently vestigial histidine at position 67 in *D. vulgaris* cytochrome c_3, corresponding to histidine 22 in the hypothetical diheme cytochrome, which may have preceded the tetraheme cytochrome in evolution. Histidine 22 would have been the ligand for heme 1, and histidine 25 would have been bonded to heme 2 in the ancestral diheme cytochrome. Hemes 2 and 3 in the tetraheme cytochrome would have changed ligands as a result of gene doubling and rearrangement of the peptide chain.

Desulfovibrio species contain additional proteins with characteristics of cytochromes c_3, first described in *D. gigas* as a 26-kDa, eight-heme cytochrome c_3 (Bruschi *et al.*, 1969). N-terminal sequences of these large-molecular-weight cytochromes c_3 from *D. vulgaris* strain Hildenborough and *D. baculatus* strain Norway 4 show very little similarity to one another and to the previously characterized cytochromes c_3 (Loutfi *et al.*, 1989). There is some confusion about molecular weight, but it has been reported that the 26-kDa cytochromes c_3 from *D. gigas* and *D. baculatus* strain Norway 4 are dimers which can be dissociated only after heme removal (Loutfi *et al.*, 1989). There is also a 41-kDa cytochrome from *D. desulfuricans* ATCC 27774, which is said to have 12 hemes (Liu *et al.*, 1988).

The *D. vulgaris* strain Hildenborough large cytochrome c_3 has a molecular weight of about 75 kDa and has 16 hemes (Higuchi *et al.*, 1987). The gene sequence (Pollock *et al.*, 1991) shows that there are indeed 16 hemes in a 66-kDa protein. The 66-kDa cytochrome c_3 has tandemly repeated domains based on the four-heme motif present in the 14-kDa cytochromes c_3. The first domain is missing the second heme and its associated ligands, reminiscent of the three-heme cytochrome from *Desulfuromonas*. There is an extra heme and approximately 60 residues between domains 3 and 4. The first heme in domain 4 is missing its sixth ligand, and, overall, the 66-kDa protein has one less histidine than required for all 16 hemes, suggesting that one heme is high-spin or has a methionine sixth ligand. There are numerous insertions and deletions in comparison of the four domains with one another and about 20–30% overall identity. There are two-residue spacers between the cysteines of all 16 hemes, which may indicate that this protein is more primitive than the four-heme cytochromes c_3 that were doubled after insertion of two extra residues between cysteines of heme 2.

The deletion of the second heme in domain 1 of the 16-heme cytochrome c_3 suggests that the origin of the three-heme *Desulfuromonas* cytochrome c_3 may have been with a 16-heme cytochrome c_3 rather than with a four-heme protein. When hybridized against the 16-heme cytochrome c_3 DNA, strains of *D. vulgaris*, *D. desulfuricans*, and *D. africanus* tested positive, whereas *D. gigas*, *D. salexigens*, *D. baculatus* Norway 4 and *D. desulfuricans* El Agheila were negative (Pollock *et al.*, 1991). The previous report of a 26-kDa cytochrome c_3 in *D. vulgaris* (Loutfi *et al.*, 1989) was apparently in error, although it may occur in *D. gigas*.

An 84-residue tetraheme cytochrome c_3 has been isolated and sequenced from an untyped purple phototrophic bacterium known as strain H1R (Ambler, 1991). This cytochrome is homologous to but is missing the approximately 20 N-terminal residues of the well-characterized tetraheme cytochromes c_3. The histidine sixth ligands for hemes 1 and 2 are also missing, although there are more than enough histidines in other parts of the sequence to bind to the hemes. The change of ligands in strain H1R cytochrome c_3 shows that it has to fold differently from the others.

D. Type 4 or Tetraheme Reaction Center Cytochromes *c*

The tetraheme photosynthetic reaction center (THRC) cytochrome *c* is distinctly different from any other cytochrome based on its three-dimensional structure

(Deisenhofer *et al.*, 1985). There is clear evidence for two successive gene doublings, such as may have occurred in the four-heme type 3 cytochromes, but there is no other relationship between the structures of these two types of protein. The most obvious difference between types 3 and 4 cytochromes *c* is in their size. Cytochrome c_3 is 14 kDa, whereas the THRC cytochrome is 40 kDa. There is considerable secondary structure in the THRC cytochrome *c* (about 34% helix), with each heme preceded by a helix which contains the heme sixth ligand (for three of the four hemes). Despite clear evidence for gene doubling in the three-dimensional structure, there is no obvious internal sequence homology in the THRC cytochrome *c* (Weyer *et al.*, 1987).

The gene sequence of cytochrome *c*-554 from *Chloroflexus aurantiacus* has been shown to be homologous to the *Rps. viridis* THRC cytochrome *c* (Dracheva *et al.*, 1991). There are a minimum of four insertions and deletions necessary for alignment and only about 20% identity. There is no evidence that the signal peptide is actually cleaved in the *Chloroflexus* THRC cytochrome *c*, and the lipid-binding N-terminal cysteine present in the *Rps. viridis* protein is absent. The gene is not contiguous with the L and M subunits as in *Rps. viridis*, but is separately transcribed. The reaction center genes cloned from *Roseobacter denitrificans* include the first half of the cytochrome, which follows the reaction center M subunit the same as it does in *Rps. viridis* (Liebetanz *et al.*, 1991).

Hemes 1, 2, and 3 of the *Rps. viridis* cytochrome have methionine sixth heme ligands, but heme 4 has a histidine sixth ligand, which is not in the preceding helix, although there is a methionine in that helix which is too far away to bind to the heme. The *Chloroflexus* cytochrome has only four histidines, which provide the fifth heme ligands. Thus, it appears that all four hemes have a methionine sixth ligand. The substitution of the sixth ligand in *Rps. viridis* heme 4 is probably due to folding constraints which followed gene doubling, or may even be an artifact of crystal packing. In this regard, the folding of the two halves of the THRC cytochrome *c* is probably closer to that of the ancestral diheme cytochrome than are the structures of cytochrome c_3 and its presumed diheme predecessor. There is a remote possibility that one of the diheme cytochromes *c* from *Rb. sphaeroides*, *P. stutzeri*, or *P. aeruginosa* could be related to the THRC cytochrome *c*, although the three-dimensional structures of these proteins are necessary to determine relationships because there is virtually no sequence similarity among them.

Each heme in the *Rps. viridis* THRC cytochrome *c* has a different alpha peak maximum and redox potential as follows: heme 1, 553 nm and -60 mV; heme 2, 556 nm and 300 mV; heme 3, 559 nm and 370 mV; and heme 4, 552 nm and 110 mV (Dracheva *et al.*, 1986, 1988; Shopes *et al.*, 1987; Vermeglio *et al.*, 1989; Fritzsch *et al.*, 1989). Generally, cytochromes which have histidine and methionine ligands have higher redox potentials than those with two histidine ligands. However, contrary to expectation, heme 4, which has two histidine ligands, has been assigned a higher potential than heme 1, which has histidine and methionine ligands. Heme 3 is nearest to the reaction center and donates an electron in less than a microsecond. Heme 2 transfers its electron nearly an order of magnitude more slowly, and surprisingly, electron transfer from the low-potential heme 1 is about twice as fast as for heme 3. The redox potentials of the *Chloroflexus* cytochrome *c*-554 hemes can be resolved (300, 220, 120, and 0 mV), but not the alpha peak maxima (Freeman and

Blankenship, 1990). In *Rhodocyclus gelatinosus*, the four hemes have potentials of 320, 300, 130, and 70 mV (Nitschke *et al.*, 1992). Two hemes have alpha peak maxima near 555 nm and two at 551 nm.

Phototrophic bacteria that have the THRC cytochrome c appear to be more common than those that do not. Besides the well-characterized *Rps. viridis*, there is *Ch. vinosum* (Kennel and Kamen, 1971), *Ch. tepidum* (Nozawa *et al.*, 1987), *Th. pfennigii* (Seftor and Thornber, 1984), *Rc. gelatinosus* (Clayton and Clayton, 1978; Fukushima *et al.*, 1988; Nitschke *et al.*, 1992), *Ecto. halochloris* (Engelhardt *et al.*, 1986), *Ectothiorhodospira* sp. (Lefebvre *et al.*, 1984), *Chf. aurantiacus* (Bruce *et al.*, 1982; Freeman and Blankenship, 1990; Meyer *et al.*, 1989), *R. salinarum* (Meyer *et al.*, 1990b), *Rps. marina* (Meyer *et al.*, 1990c), and *R. salexigens* (Meyer *et al.*, 1990a). Photosynthetically incompetent aerobic bacteria, which nevertheless contain bacteriochlorophyll and associated pigments, also have been found to have reaction center cytochrome, namely *Protaminobacter ruber* (Takamiya, 1985), *Roseobacter denitrificans* (formerly *Erythrobacter*) (Takamiya *et al.*, 1987; Liebetanz *et al.*, 1991), and *Pseudomonas radiora* (Nishimura *et al.*, 1989).

The purple bacteria *Rb. sphaeroides*, *Rb. capsulatus*, and *R. rubrum* apparently do not have a THRC cytochrome c; thus, cytochrome c_2 rapidly interacts directly with the reaction center. In *Rps. viridis* and presumably other bacteria which have the THRC cytochrome c, the soluble cytochrome c_2 or other donor cannot interact directly with the reaction centers because access is blocked, and it must transfer electrons to the mediating THRC cytochrome c. In *Rps. viridis*, electron transfer from cytochrome c_2 to reaction center is relatively rapid (Knaff *et al.*, 1991; Meyer *et al.*, 1993). *Rc. gelatinosus* reaction centers can be prepared both with and without bound THRC cytochrome c, and one of the soluble cytochromes (or horse cytochrome c) is as efficient in transferring electrons to one type of reaction center preparation as to the other (Matsuura *et al.*, 1988). Presumably, the type of electron donor to the THRC cytochrome c will influence its evolution. Likewise, the evolution of cytochrome c_2 will be affected by its electron acceptor, whether THRC cytochrome c or the reaction centers themselves.

Although *Rb. sphaeroides* and *Rb. capsulatus* do not have a THRC cytochrome c bound to reaction centers, they apparently induce synthesis of a 45-kDa cytochrome c when grown on DMSO as an electron acceptor, which cross-reacts immunologically with THRC cytochrome c from *Cf. aurantiacus*, which in turn cross-reacts with *Rps. viridis* THRC cytochrome c (McEwan, 1989). There is no known role for this cytochrome in DMSO reduction, and it is not induced by TMAO, which on the other hand induces the same terminal enzyme as DMSO, namely TMAO/DMSO reductase. The gene for the *Rps. viridis* THRC cytochrome c is adjacent to the L and M subunit genes, and it is part of the same operon (Weyer *et al.*, 1987). However, the DMSO cytochrome may be induced fortuitously by DMSO if through some accident of nature the cytochrome gene is moved away from its normal position next to L and M and placed adjacent to the DMSO regulatory element.

What is the function of the THRC cytochrome c, and how or why was it displaced by cytochrome c_2 in *Rb. sphaeroides et al.*? It is possible that the THRC cytochrome c was necessary in early purple bacteria because cytochrome c_2 may have been a late acquisition, and the early electron donor was unable to interact directly with reaction center at a fast enough rate to prevent recombination with reduced quinones. HiPIP and auracyanin may be examples of such inefficient donors. Cytochrome c_2 may have displaced THRC cytochrome c because of mutations in the THRC cytochrome c, which rendered it unable to bind to reaction center, or it may have been the result of a gene rearrangement on the chromosome, which prevented expression of the THRC cytochrome c under photosynthetic growth conditions. These questions will have to await further genetic analyses for resolution.

E. Potentially Distinctive Types of Cytochromes c

There are several cytochromes c that cannot presently be placed in one of the four principal categories described earlier. It is not that these cytochromes are poorly characterized. The proteins treated in this section have been sequenced, yet still defy categorization. They include the large single-heme cytochromes c_1 and f, the monoheme green bacterial reaction center cytochrome, three diheme cytochromes, a tetraheme cytochrome c, and the tetraheme flavocytochrome c from *Shewanella*. Three-dimensional structures of these proteins are absolutely necessary to determine whether there is homology to any of the four principal types of cytochromes described earlier or to one another.

1. Cytochromes c_1 and f

Cytochrome c_1 is a subunit of the mitochondrial cytochrome bc_1 complex, also known as quinol–cytochrome c oxidoreductase (see Trumpower, 1990, for review). The functional role of cytochrome c_1 is analogous to that of cytochrome f, which is a component of the chloroplast cytochrome $b_6 f$ complex, also known as plastoquinol–plastocyanin oxidoreductase (see Hauska *et al.*, 1983, for review). Cytochromes c_1 and f are firmly bound to the membrane, they have similar molecular weights of about 30,000, and they contain a single covalently bound heme. The heme is bound near the N-terminus, and there is a hydrophobic membrane-spanning segment near the C-terminus. Nevertheless, sequence homology between cytochromes c_1 and f is not obvious, and three-dimensional structures are necessary to determine whether they are related.

Cytochrome c_1 has a redox potential (ca. 230 mV) similar to its reaction partner, soluble mitochondrial cytochrome c (ca. 260 mV). Cytochromes c_1 have been characterized in bacteria as well as in eukaryotes—for example, *Paracoccus denitrificans* (Ludwig *et al.*, 1983) and *Rhodobacter sphaeroides* (Yu *et al.*, 1986). Amino acid or nucleotide sequences are known for bovine (Wakabayashi *et al.*, 1980), human (Nishikimi *et al.*, 1987), yeast (Sadler *et al.*, 1984), *Neurospora crassa* (Romisch *et al.*, 1987), *Euglena gracilis* (Mukai *et al.*, 1989), *Paracoccus denitrificans* (Kurowski and Ludwig, 1987), *Rps. viridis* (Verbist *et al.*, 1989), *Bradyrhizobium japonicum* (Thony-Meyer *et al.*, 1989), two strains of *Rhodobacter capsulatus* (Davidson and Daldal, 1987a,b; Gabellini and Sebald, 1986), and *Rhodospirillum rubrum* (Majewski and Trebst, 1990). The *Bradyrhizobium* cytochrome c_1 and cytochrome b are synthesized as a polyprotein, which is post-translationally cleaved into the individual subunits (Thony-Meyer *et al.*, 1989, 1991). The known gene sequences contain 20- to 70-residue leader sequences which are eventually cleaved following transport of the protein across the membrane. A 15- to 21-residue C-terminal hydrophobic segment which is flanked by basic residues apparently anchors the mature cytochrome to the membrane. This segment is located at bovine positions 207–222.

The heme is bound to a pair of cysteines near the N-terminus at bovine positions 37 and 40 as in other c-type cytochromes. *Euglena* and *Crithidia* are exceptions in that the first heme-binding cysteine is missing (Mukai *et al.*, 1989; Priest and Hajduk, 1992). The soluble mitochondrial cytochromes c from *Euglena* and *Crithidia* are also missing this same cysteine (Pettigrew, 1972, 1973). There is a conserved methionine in the C-terminal half of the protein at bovine position 160, which may be the sixth heme ligand. The locations of the heme and its ligands suggest a relationship to the type 1 cytochromes c, but the much larger size and lack of obvious homology necessitate three-dimensional structural analysis to determine whether a relationship actually exists. A number of insertions and deletions are necessary to align the mitochondrial and bacterial cytochromes c_1, which makes it even more difficult to decide whether there is any similarity to mitochondrial cytochrome c or bacterial cytochrome c_2. Horimoto *et al.* (1991) used principal component analysis to detect overall similarities between cytochromes c and c_1. In support of this contention, there is a glycine and an aromatic residue positioned the same as those of the N-terminal helix in mitochondrial cytochrome c. There is also a conserved proline at position 111 and an arginine at position 120, which might correspond to pro 30 and arg 38 in mitochondrial cytochrome c. If they are in fact homologous, then there would be nearly 60 residues inserted in this region. Likewise, there is a conserved tryptophan at position 192, which might correspond to the aromatic residue in the C-terminal helix of mitochondrial cytochrome c. If so, there would be a 15-residue insertion between the methionine heme ligand and the trptophan.

Another interesting aspect of cytochrome c_1 is the association with a so-called "hinge" protein which is thought to facilitate electron transfer from cytochrome c_1 to cytochrome c (Kim *et al.*, 1987). The amino acid sequences of these 9–17 kDa proteins from bovine (Wakabayashi *et al.*, 1982) and yeast (Van Loon *et al.*, 1984) sources are not very similar to one another, but they share a characteristically large number of acidic residues. *Paracoccus* cytochrome c_1 is larger than all the others (450 versus 240–280 residues) and is unusual in having a very acidic N-terminal 150-residue segment (Kurowski and Ludwig, 1987). There is no obvious homology to the "hinge" proteins, but it appears that the acidic N-terminus could fill the same role as the acidic "hinge" proteins in other cytochrome bc_1 complexes. The *R. capsulatus* cytochrome bc_1 complex contains a normal cytochrome c_1, but it also has an additional 14-kDa peptide, which may be related to the peptide which is present at the N-terminus of the *Paracoccus* cytochrome c_1. The obvious implication is that the site of interaction on cytochrome c_1 for the basic cytochrome c may also be basic and requires intervention of the acidic peptide to overcome electrostatic repulsion. Although many of the bacterial cytochromes c_2 which interact with cytochrome c_1 are acidic, they all have positive charge at the site of electron transfer similar to that of mitochondrial cytochrome c (Meyer et al., 1984).

Cytochromes f have been most thoroughly characterized in the cruciferous plants where they can be readily solubilized without gross alteration of properties (Gray, 1978). They have also been isolated from the green alga *Scenedesmus acutus* (Bohme et al., 1980a) and the cyanobacterium *Spirulina platensis* (Bohme et al., 1980b). The molecular weights of cytochromes f are similar to those of the cytochromes c_1 (*ca.* 30,000), but the redox potentials are considerably higher (*ca.* 380 mV). The gene sequences of cytochromes f from several plants (Willey et al., 1984), from the green alga *Chlamydomonas reinhardtii* (Bertsch and Malkin, 1991; Buschlen et al., 1991; Matsumoto et al., 1991), and the cyanobacteria *Nostoc* sp. (Kallas et al., 1988), *Synechococcus* sp. (Widger, 1991), and *Synechocystis* sp. (Mayes and Barber, 1991) have been determined. The heme is bound near the N-terminus to cysteines at positions 56 and 59 of the spinach protein. Only two insertions and deletions are required to align *Nostoc* with the plant proteins, and, overall, the structure of cytochrome f appears to be more conserved than that of cytochrome c_1. However, the methionine which is thought to be the sixth heme ligand in cytochrome c_1 is absent in cytochrome f. Although there is a conserved methionine at spinach position 318 (which is only two residues removed from the C-terminus), there is considerable controversy about whether the sixth heme ligand is actually a methionine. Several recent papers suggest that the sixth ligand is a lysine (Davis et al., 1988; Rigby et al., 1988; Simpkin et al., 1989), of which nine are conserved.

Rather than a single hydrophobic segment near the C-terminus of cytochrome f, which might anchor the mature protein to the membrane, there are three segments of 14, 15, and 20 residues. The last 20-residue segment, located at spinach positions 286–306, is more hydrophobic than the others and is flanked by basic residues. It is probably analogous to the hydrophobic segment in cytochrome c_1.

The question of the existence of a "hinge" protein may be a moot point, because cytochrome f normally interacts with the highly acidic copper protein plastocyanin and appears to have a large positive charge at the site of electron transfer (Qin and Kostic, 1992). Even in algae, where a small cytochrome c-553 is interchangeable with plastocyanin, that cytochrome c-553 generally is strongly negatively charged. A few cyanobacteria, such as *Anabaena*, have a basic plastocyanin and cytochrome c-553. In these cases, the sites of electron transfer are also basic, as opposed to the plant and the majority of algal proteins which have negatively charged sites of electron transfer (Meyer et al., 1987). Perhaps there is an acidic "hinge" protein in *Anabaena*, or else acidic residues may have been coordinately substituted for basic ones in cytochrome f.

2. *Rb. sphaeroides* diheme cytochrome c-551

Rhodobacter sphaeroides cytochrome c-551 is a 16-kDa protein, which contains two hemes (Meyer and Cusanovich, 1985). It has not been found in any other species. The sequence shows that it is not related to any known cytochrome (Van Beeumen, 1991). Although it is about the same size as cytochrome c_4 (138 residues), the two heme binding sites appear not to have been produced by gene doubling or by gene fusion. One heme is near the N-terminus (at cysteine positions 24 and 27), and the other is near the C-terminus (at cysteine positions 125 and 128). The cytochrome has a low redox potential (-254 mV) and presumably has two histidine ligands to each heme. There are potential histidine ligands at positions 6, 51, and 106.

The protein appears to be split in two by endogenous proteolysis without alteration of properties, which means either that proteolysis does not cause the protein to unfold, or that the two halves fold in separate domains. It is possible that this cytochrome is representative of an unrelated type of cytochrome c, but there is also a possibility that it may be distantly related to the tetraheme reaction center cytochrome c or one of the two diheme cytochromes described later. The three-dimensional structure is necessary to determine whether it is related to a known protein. The low helix content of 23% (measured by circular dichroism) suggests that it is not related to the other multiheme proteins such as reaction center cytochrome c.

3. *P. stutzeri* diheme cytochrome *c*-552

Pseudomonas stutzeri (formerly *P. perfectomarinus*) cytochrome c-552 has a molecular weight of 26,000 and contains two hemes (Liu *et al.*, 1981). A preliminary sequence has been reported (Denariaz *et al.*, 1989), and the gene was discovered as part of the cytochrome cd_1 gene cluster between a tetraheme cytochrome c gene and the *Pseudomonas* cytochrome c-551 gene (Jungst *et al.*, 1991). A homolog of this cytochrome was recently found in the purple phototrophic bacterium *Rhodocyclus tenuis* by amino acid sequence determination (Van Beeumen, 1991). The *P. stutzeri* and *Rc. tenuis* cytochromes are about 39% similar. The hemes are not situated the same as in *Rb. sphaeroides* cytochrome c-551 (at *P. stutzeri* cysteine positions 41 and 44, and at 128 and 131), and the two proteins are probably not related. *P. stutzeri* and *Rc. tenuis* cytochromes have an unusual three-residue spacer between the two cysteines that bind the second heme, which has not been observed elsewhere. The cytochrome is irreversibly denatured at low pH, which may partially explain why it has been observed in only two species.

The *P. stutzeri* protein has one high-potential heme ($+174$ mV) and one low-potential heme (-180 mV) (Liu *et al.*, 1981), whereas both hemes appear to be low-potential in the *Rc. tenuis* protein. There are conserved histidines (at *P. stutzeri* positions 33 and 239) and methionines (at positions 120 and 146), which could be heme ligands. Endogenous proteolysis splits the *P. stutzeri* cytochrome near the N-terminus and causes dramatic changes in properties. One heme becomes high-spin, presumably through loss of one of its heme ligands, and the protein takes on cytochrome c peroxidase activity. The segment which is removed contains one of the conserved histidine residues (at position 33), and it can be assumed that this histidine is a heme ligand. There is no obvious similarity to the diheme *Ps. aeruginosa* cytochrome c peroxidase described next.

4. Bacterial cytochrome *c* peroxidase

Bacterial cytochrome c peroxidase (BCCP) was discovered by Lenhoff and Kaplan (1956) in *Pseudomonas aeruginosa*. The early literature was reviewed by

Meyer and Kamen (1982). Since then, the amino acid sequence of the *Pseudomonas* BCCP has been determined (Ronnberg *et al.*, 1989). There are 302 amino acid residues and two heme-binding sites, at cysteines 52 and 54 and at 177 and 180. BCCP is labile to purification and storage, resulting in cleavage of the peptide chain at position 202 (Ronnberg, 1987). Microheterogeneity at several positions suggests that there is more than one gene for BCCP. There is no obvious relationship to any other cytochrome for which the sequence is known, except for an approximately 44-kDa soluble cytochrome *c*-556 purified from *Rhodobacter capsulatus*, which also has BCCP activity (Hanlon *et al.*, 1992). The sequence of this latter protein is approximately 45% similar to *Pseudomonas* BCCP (Van Beeumen, 1991).

It is remotely possible that BCCP may be the result of a tandem gene duplication of a type 1 cytochrome *c*. The second heme domain is preceded by a glycine and a phenylalanine which could be in an N-terminal helix. A proline follows both hemes and could H-bond the histidine heme ligand. There is a conserved aromatic residue which could be in a C-terminal helix that follows a conserved methionine (potential heme ligand). The three-dimensional structure is necessary to establish whether this speculation has any validity.

The enzyme is generally distributed in *Pseudomonas* species and recently has been reported in *Paracoccus denitrificans* (Goodhew *et al.*, 1990). The specificity of the *Paracoccus* BCCP is quite different from that of *P. aeruginosa* BCCP in that it is most active with cytochrome c_2 instead of with *Pseudomonas* cytochrome *c*-551. One heme of BCCP is low-spin and has a high redox potential, whereas the other has a low redox potential and is high-spin at room temperature and low-spin at low temperature (Ellfolk *et al.*, 1983; Villalain *et al.*, 1984; Foote *et al.*, 1985). Presumably one heme has histidine and methionine coordination, and the other has just one histidine at room temperature or two histidines at low temperature. Methionines 95 and 254 appear to be conserved, and one of them could be the sixth ligand to the high-redox-potential heme. It is not yet known whether there are any conserved histidines, which might be the distal histidines for the low-potential heme, although Ronnberg *et al.* (1989) speculate that it is histidine 240.

5. *Pseudomonas stutzeri* tetraheme cytochrome *c* gene

A tetraheme cytochrome *c* gene was discovered immediately downstream of the cytochrome cd_1 gene in *Pseudomonas stutzeri* (Jungst *et al.*, 1991). It encodes a 201-residue protein, which is larger than the tetraheme cytochromes c_3 but smaller than the tetraheme reaction center cytochrome *c*. It does not appear to be related to either of these tetraheme cytochromes, and there is no obvious similarity to any monoheme cytochrome. It has a 25-residue N-terminal hydrophobic segment, which may anchor it to the membrane. Deletion of this gene results in loss of ability to utilize nitrite, although cytochrome cd_1 is still synthesized and active. This demonstrates that the tetraheme cytochrome is an obligate component of the electron transport pathway to nitrite. However, the gene product has not yet been characterized. We have found a homolog of the *Ps. stutzeri* tetraheme cytochrome *c* gene upstream of the flavocytochrome *c* operon in *Chromatium vinosum* (Dolata *et al.*, 1993). The protein corresponding to this gene has not yet been observed. There is 49% identity between the *Chromatium* and *Pseudomonas* proteins, and only a single-residue deletion is required for alignment. *Chromatium* is not a denitrifier, and cytochrome cd_1 has not been observed in this species. Thus, the tetraheme cytochrome c cannot fill the same functional role as in *Ps. stutzeri*.

There is evidence for gene duplication in the tetraheme cytochrome c, although there are very few identical residues in the two halves. If one- and three-residue deletions are postulated, then there are two conserved methionines and two conserved histidines between the first and second hemes and the third and fourth hemes, which could be sixth ligands for the four hemes. This suggests that two hemes have a high redox potential and that two of them have a relatively low potential.

6. Green bacterial reaction center cytochrome c

The green sulfur phototrophic bacteria such as *Chlorobium* contain a reaction center similar to plant photosystem I (Buttner *et al.*, 1992) and unlike the reaction center of purple bacteria. Nevertheless, it contains a bound cytochrome c, the gene sequence of which shows that it contains a single heme and is 18 kDa in size (Okkels *et al.*, 1992). The N-terminal half is hydrophobic and provides the membrane anchor, whereas the heme is bound to the hydrophilic segment. There is no obvious similarity to other cytochromes, although there is a methionine 26 amino acid residues downstream of the heme binding site which could be the sixth ligand. Consensus residues of the N-terminal helix in type 1 cytochromes are absent in the *Chlorobium* cytochrome, but it is possible that some of the residues beyond the methionine fit the consensus for the C-terminal helical region of type 1 cytochromes. For example, a phenylalanine three residues after the methionine matches an aromatic residue in the algal cytochromes c_6, and the sequence IXXWL is approximately the same distance from the methionine as in type 1 cytochromes such as algal cytochrome c_6. If the *Chlorobium* cytochrome is related to type 1 cytochromes, then it has experienced an approximately 15-residue deletion between the heme binding site and the methionine relative to other small bacterial cytochromes such as cytochromes c_5 and c_6.

7. *Shewanella* flavocytochrome c or fumarate reductase

Fumarate reductase is normally a membrane-bound protein, but is soluble in some bacteria and in yeast. The mitochondrial fumarate reductase contains a flavoprotein subunit, a ferredoxin subunit, and a membrane-spanning b-type cytochrome subunit which anchors the other two subunits to the membrane. The gene sequence of the soluble fumarate reductase from *Shewanella putrefaciens* shows that the 571-residue protein is composed of two domains, an N-terminal tetraheme cytochrome c domain of 117 residues and a C-terminal flavoprotein domain of 454 residues homologous to the flavoprotein subunit of the membrane-bound fumarate reductases (Pealing *et al.*, 1992). The gene contains a cleavable signal sequence which shows that the protein is periplasmic. The four hemes have low redox potentials, two of which are near -204 mV and two of which are near -320 mV, suggesting that the heme ligands are all histidines, such as found in the cytochromes c_3. There is no obvious similarity to any other c-type cytochromes, including the three known kinds of tetraheme cytochromes such as the *Desulfovibrio* cytochromes c_3 (Voordouw and Brenner, 1986), the purple bacterial reaction center cytochromes c (Weyer *et al.*, 1987), and the *Ps. stutzeri* NirT gene (Jungst *et al.*, 1991). Furthermore, there is no strong evidence for gene doubling such as is apparent in the other tetraheme cytochromes. The observation that all four hemes are low potential does suggest that there could be a distant relationship to the cytochromes c_3. When the *Shewanella* FR heme domain is aligned with cytochrome c_3, it can be seen that histidines corresponding to the sixth ligands for hemes 3 and 4 may be present, but not those for hemes 1 and 2. However, two additional histidines are in the same locations as found for the low-potential cytochrome from *Rhodopseudomonas* sp. strain H1R (Ambler, 1991). If these two proteins are homologous to the type 3 cytochromes, then they must fold differently to accommodate the substitution of the ligands for two of the hemes.

III. Role of Gene Transfer and of Isozymes

Isozymes were briefly discussed in the introduction, but they will be covered in more depth here. Isozymes differ from pseudogenes in that the latter are not normally expressed and presumably have no function. They can and probably have replaced functional cytochrome c either in whole or in part after sustaining a number of mutations not possible in the functional protein. Isozymes may have functional roles which are similar to those of the principal functional cytochrome c, but they may accept mutations at a more rapid rate if they are not essential. To the extent that they do fill some functional role, they would accept mutations more slowly than pseudogenes. Isozymes appear to be less common than pseudogenes in animals, but much less is known of the relative numbers and importance of these two types of genes in bacteria. Cytochrome c_2 isozymes were found in *R. molischianum* (Flatmark *et al.*, 1970) and are 34% different from one another (Ambler *et al.*, 1979a). *R. molischianum* iso-1 cytochrome c_2 is missing the otherwise conserved tryptophan 59 of mitochondrial cytochrome c and has a lower redox potential than iso-2 (290 vs. 380 mV). It is not known whether it is functional. *R. molischianum* is a strict anaerobe and presumably does not have an oxidase which would interact with cytochrome c_2. There may therefore be only a single role for cytochrome c_2 (as mediator between the bc_1 complex and reaction center).

Another isozyme of cytochrome c_2 was found in *Rb. sphaeroides* when the functional protein was experimentally deleted from the genome (Donohue *et al.*, 1988). At first, the bacteria would not grow phototrophically without the deleted wild-type cytochrome c_2, but they spontaneously regained that ability. They were then found to produce an isozyme of cytochrome c_2 which was called iso-2 cytochrome c_2 (Fitch *et al.*, 1989). The properties of iso-2 cytochrome c_2 are not particularly different from those of iso-1, and it is likely that it is as effective as iso-1 in coupling photosynthetic electron transfer. The sequence of iso-2 cytochrome c_2, however, shows that it is much more divergent than we would expect for an isozyme (Rott *et al.*, 1993; Van Beeumen *et al.*, unpublished). Iso-2 cytochrome c_2 is expressed in wild-type cells at a very low level under either aerobic or photosynthetic growth conditions (Rott *et al.*, 1992). The normal function in wild-type cells is unknown.

In contrast to *Rb. sphaeroides*, the functional cytochrome c_2 gene of *Rb. capsulatus* was deleted, but the bacteria were still able to grow photosynthetically (Daldal *et al.*, 1986). There is no soluble cytochrome c_2 in the mutant, and it appeared that cytochrome c_2 was unnecessary for photosynthetic growth, unlike *Rb. sphaeroides*. Based on spectral evidence, it appeared that cytochrome c_1 may donate electrons directly to reaction center (Prince *et al.*, 1986). An alternative explanation is that there is a membrane-bound version of cytochrome c_2 in *Rb. capsulatus* such as is found in *Bradyrhizobium* (Bott *et al.*, 1991). In fact, Jenney and Daldal (1993) report a membrane-bound cytochrome *c* gene in *Rb. capsulatus* which complements the photosynthesis minus mutants of *Rb. sphaeroides*.

There are five related cytochromes which occur in *Pseudomonas* and related species. Cytochrome cd_1 has a well-defined activity as nitrite reductase, but we do not know the functional roles of cytochromes *c*-551, c_5, or c_4 in *Pseudomonas* and *Azotobacter*, although they appear to be interchangeable *in vitro* (Cheddar *et al.*, 1989). They have similar redox potentials, and they have a similar lack of charged side chains near the edge of the heme exposed to solvent where electron transfer is expected to occur most rapidly. The cytochrome *c*-551 and orf 5 genes are located next to that for cytochrome cd_1, which may indicate that they are coordinately synthesized and are natural reaction partners *in vivo*. *Azotobacter* does not denitrify and does not have cytochrome cd_1; therefore, the role of cytochrome *c*-551 in this species may be different than in *Pseudomonas*. Cytochromes *c*-551, c_4, and c_5, as well as orf 5, may be thought of as isozymes, although they are not as close in sequence as one might expect isozymes to be. Cytochromes c_4 and c_5 are normally membrane-bound, and that may prevent them from functioning interchangeably with the soluble cytochrome *c*-551 *in vivo*. Cytochrome *c*-551 is evolving relatively slowly in *Azotobacter* (Ambler, 1973), *Pseudomonas* (Ambler, 1974; Ambler and Wynn, 1973), *Hydrogenobacter* (Sanbongi *et al.*, 1989), and *Nitrosomonas* (Miller and Nicholas, 1986). Only a few cytochrome c_4 and c_5 sequences are known, so we do not yet know whether they are evolving at the same or different rates as cytochrome *c*-551. Because cytochromes c_4 and c_5 are membrane-bound, we do not know whether all three are generally present in all species which produce cytochrome *c*-551, or whether they are expressed at the same level. More sequence data and genetic studies are needed to clarify their relationships.

What is the evidence for gene transfer, and how would one recognize its occurrence? A good discussion of this topic was published by Krawiec and Riley (1990). A protein which is common in one family and which is also found in an apparently unrelated species could be taken as preliminary but not sufficient evidence for gene transfer. For example, cytochrome cd_1 is generally present in the denitrifying pseudomonads, but is also present in *Paracoccus*, a species which produces cytochrome c_2 but none of the other *Pseudomonas* cytochromes or azurin. We can postulate gene transfer of cytochrome cd_1 to *Paracoccus*, but cannot prove it. On the other hand, perhaps it was the cytochrome c_2 which may have been transferred from purple bacteria (in which it is commonly found) to *Paracoccus* because the cytochrome c_2 is most similar to that of *Rb. capsulatus* (Ambler *et al.*, 1981b). However, cytochrome *c'* and BCCP are present in both *Rb. capsulatus* (Van Beeumen, 1991) and *Paracoccus* (Goodhew *et al.*, 1990), thus lending support to the possibility for transfer of cytochrome cd_1 to *Paracoccus*. The best evidence for gene transfer would be a protein of nearly identical sequence in two otherwise unrelated species. This has not yet been observed. If the gene sequences were known, the base composition of the suspected foreign gene might be found to be different from that of the organism as a whole (Krawiec and Riley, 1990).

Cytochromes *c*, found in some purple bacteria, are more closely related to *Pseudomonas* cytochrome *c*-551 than to other cytochromes (Ambler *et al.*, 1979b). Cytochrome c_2 has not been found in these species, which include *Rc tenuis*, *Rc. purpureus*, and *Rc. gelatinosus*. We do not know whether the cytochrome *c*-551 was transferred to the phototrophic bacteria from *Pseudomonas*, whether the photosynthetic operon was transferred to *Pseudomonas* in the opposite direction, or whether *Pseudomonas* arose by deletion of photosynthetic genes. In any case, it may be assumed that the phototrophic cytochrome *c*-551 has a different role from that in *Pseudomonas* for no other reason than it appears to be evolving at a different rate. Because the cytochrome *c*-551 is more variable in the purple bacteria than in the pseudomonads, it is more likely that the postulated gene transfer was from *Pseudomonas* to the purple bacteria, in which it is being molded to fill a new role.

Chloroflexus is generally thought to be a green phototrophic bacterium because it contains the *Chlorobium* chlorophylls arranged in *Chlorobium* vesicles (Pierson and Castenholz, 1974). On the other hand, it is capable of fully aerobic growth, utilizing organic compounds which none of the green sulfur bacteria are capable of. It also lacks the gram-negative cell wall typical of green sulfur bacteria. Even more remarkably, it lacks the ferredoxin-reducing photosystem of *Chlorobium* and instead has a quinone-reducing photosystem very much like that of the purple bacteria (Pierson and Thornber, 1983; Shiozawa *et al.*, 1987). The gene sequence of the *Chloroflexus* reaction center (Ovchinnikov *et al.*, 1988a,b) is homologous to that of purple bacteria, and there is even a tetraheme reaction center cytochrome (Freeman and Blankenship, 1990; Dracheva *et al.*, 1991) such as found in *Rps. viridis*. Therefore, this appears to be a case of gene transfer of light-harvesting pigments from *Chlorobium* to an aerobic bacterium, which already had the purple bacterial reaction center. How large a DNA fragment might have been transferred is unknown, but it is probably a lot larger than a cytochrome *c* gene. It is possible that the purple bacterial reaction center genes were also transferred to *Chloroflexus*, but this is a more difficult question to resolve at present. The reaction center is composed of at least four polypeptides with a combined molecular weight of about 100,000, which may make the gene cluster less likely to be transferred than the genes encoding the light-harvesting proteins, whose combined molecular weight may be a lot smaller. Perhaps *Chloroflexus* is a little of both green and purple bacteria, but no closer to one than to the other.

HiPIP and cytochrome *c'* are commonly found in purple bacteria, and they are also present in a halophilic denitrifying *Paracoccus* sp. (Hori, 1961; Tedro *et al.*, 1977; Ambler *et al.*, 1987a). There are no other similarities to purple bacteria such as might be expected if the *Paracoccus* were a purple bacterium which had lost its photosynthetic capacity. Thus, we may assume that the genes for HiPIP and cytochrome c' were transferred to *Paracoccus*. It appears to be coincidence that the genes for both proteins may have been transferred together, but neither gene has yet been cloned, and it is not known whether these genes may be linked.

Cytochrome c' has been found in *Azotobacter* (Yamanaka and Imai, 1972), but not in the various denitrifying pseudomonads, which are closely related to *Azotobacter*. Because cytochrome c' is commonly found in purple bacteria, we may assume that the cytochrome c' gene was transferred from purple bacteria to *Azotobacter*.

There are a number of aerobic bacteria which contain bacteriochlorophyll but which are incapable of photosynthetic growth. For example, *Roseobacter denitrificans* (formerly *Erythrobacter*) contains bacteriochlorophyll and has a cytochrome c_2 closely related to that of *Rb. capsulatus* (Okamura *et al.*, 1987). The photosynthetic reaction center and a tetraheme reaction center cytochrome c are also like those of purple bacteria (Takamiya *et al.*, 1987; Liebetanz *et al.*, 1991). This similarity appears to be due to loss of an essential component in a purple bacterium which prevents photosynthetic growth, but it could just as well be a case of transfer of an incomplete photosynthetic pathway to an aerobic bacterium. The presence of photosynthetic pigments in a symbiotic *Rhizobium* sp. also appears to be a case of gene transfer (Evans *et al.*, 1990).

It should be fairly obvious that gene transfer can and does occur in bacteria. It is also postulated to occur between prokaryotes and eukaryotes (Doolittle *et al.*, 1990; Bork and Doolittle, 1992). It does not appear to be very common, but is not easy to recognize and is even more difficult to prove. In any case, gene transfer probably affects the evolution of cytochrome c in bacteria as much as do isozymes and pseudogenes, although it may be necessary to determine complete genome sequences of a number of bacteria before actual instances of gene transfer can be accepted.

IV. Influence of Function on Evolution

In evolutionary studies, there is a tendency to think of proteins and nucleic acids as records of a slow mutational process which occurs over millions of years of time without constraints on either the type or extent of change allowed. There are even those who do not permit consideration of function to cloud their perception of relationships based on apparent mutational change, perhaps to avoid preconceived notions. It was not deemed necessary to consider function as an important variable, because the function appears to be conserved in animals and because of the relative success in using molecular methods of sequence analysis to show relationships among cytochromes of higher animals, which more or less correspond to the fossil record. On the other hand, bacterial cytochromes are more variable than eukaryotic cytochromes, and bacterial evolutionary trees have the appearance of a starburst. This result occurs because there is a limit to variation which is determined by function, and most bacterial cytochromes that have been examined have reached that limit of change (Meyer *et al.*, 1986a).

We know that proteins require certain amino acids for function. Mitochondrial cytochrome c requires the two cysteines, 14 and 16, to bind the heme, histidine 18 and methionine 80 for iron ligation, and approximately 6–9 basic residues for interaction with its reaction partner. There are also residues required for maintenance of structure such as proline 30, which H-bonds the histidine heme ligand, and arginine 38 and tryptophan 59, which H-bond the rear heme propionate. Tyrosine 52 H-bonds the front heme propionate, and tyrosine 67 H-bonds the methionine heme ligand, etc. "Slowly" and "rapidly" evolving proteins, therefore, reflect the greater or lesser fraction of residues which are required for maintenance of three-dimensional structure and function. The function of a protein dictates not only the extent of mutational change allowed in that protein, but also the type of change. When we say that hemoglobin and cytochrome *c* are evolving at different rates, we mean that more variability is allowed in one than in the other without impairment of function. This apparent difference in rate of change can affect our ability to align sequences. Only the heme ligands and one of the cysteines are absolutely invariant in mitochondrial cytochromes *c* and bacterial cytochromes c_2, but there are a number of residues which are conserved in 8 out of 10 species that aid in alignment. Mutagenesis has shown that these latter residues may in some cases confer only a slight advantage to the species, because substitutions do not significantly alter growth rate of the organism, biosynthesis of the protein, its stability, its redox potential, or its ability to function *in vivo* or *in vitro* (Hampsey *et al.*, 1986).

One often sees literature references to rapidly evolving proteins or genes being used for comparison of closely related organisms, and slowly evolving genes used for distantly related organisms. However, one should get the same result using either type of protein or gene, providing that such proteins can be aligned accurately and that a sufficiently large sample is considered. We are referring not to changes in the rate of evolution of a particular protein with time, but to the overall amount of change observed in different proteins. Hypothetical examples may be used to illustrate the point. A slowly evolving protein such as histone may have a limit of variability of about 20% difference. If two species of this protein were found to differ by 15%, then the effective difference would be 15/20 = 75%. A moderately slowly evolving protein might have a limit to change of about 40% difference. If two species of this protein were found to differ by 30%, then the effective difference would still be 75%. For a moderately rapidly evolving protein such as cytochrome *c* which has a limit of change of 60%, a 45% difference between any two species would represent an effective 75% difference. A fairly rapidly evolving protein such as hemoglobin has a limit of change of about 80%, therefore 60% actual difference would represent 75% effective difference. The extent of variation of fibrinopeptides is apparently limited only by amino acid composition; thus, 71% actual difference with a limit of 95% difference represents the same 75% effective difference as shown in the above examples. What would prevent the use of histones for evolutionary studies would be the relatively small statistical sampling size rather than the apparent rate of change. That is, both histones and fibrinopeptides should give the same limited result. Although 16s and 23s rRNA would be in the same category as histones for evolutionary studies, they are much larger and provide a greater sample size than does histone. The best proteins for evolutionary studies are those that show the greatest limit to change, but that are large and can still be unambiguously aligned. Preferably, one would use several such proteins to increase the sample size. The whole mitochondrial genome should be sufficiently large for such purposes, although its rapid rate of change relative to cytoplasmic proteins in animals and maternal inheritance may limit its usefulness. It is obvious that if one were to compare two homologous cytochromes c which had different functions and different limits to change as described above, then the outcome would be erroneous.

In bacteria, larger than normal sequence differences suggest that there may be changes in function, such as for the typical *Pseudomonas* cytochromes which occur in a few species of purple phototrophic bacteria (Ambler *et al.*, 1979b; Van Beeumen, 1991). The function in either case is still unknown. Simple divergent mutations, such as are observed with eukaryotes, have not been observed with these bacteria except when working with strains of a single species. Rather, each species has diverged to the maximum extent. Stated another way, before there are sufficient phenotypic differences between phototrophic bacteria or pseudomonads to allow us to recognize a distinct species, the proteins have already changed to the maximum extent. This will probably be found to be true for most known bacteria. The exceptions are the species of medical importance. They are characterized more by the disease or illness they cause than by their own properties. Thus, medically important species of a particular genus or family appear to be more numerous and are often more closely related to one another than are the soil bacteria. Until consistent statistically proven molecular methods of classification are adopted, this will continue to be the case.

A. Interaction of Cytochromes c with A-Type Oxidases

Reaction of mitochondrial cytochrome c with the bovine a-type oxidase complex, the cytochrome bc_1 complex, and yeast cytochrome c peroxidase has been thoroughly studied, with the conclusion that basic residues on cytochrome c have been identified as contributing to the reaction mechanisms (Staudenmayer *et al.*, 1977; Ferguson-Miller *et al.*, 1978; Pettigrew, 1978; Speck *et al.*, 1979; Konig *et al.*, 1980). Chemical modification has identified about nine basic residues at positions 8, 13, 25, 27, 72, 73, 79, 86, and 87 as being in the vicinity of the site of oxidation and reduction of cytochrome c near the exposed heme edge. These residues are generally invariant in mitochondrial cytochromes c, with the exception of *Tetrahymena pyriformis*, which has substituted all but two of these basic residues by neutral or acidic residues. As expected, *Tetrahymena* cytochrome c is not very effective in its interaction with bovine oxidase. Presumably, compensating changes have occurred in the *Tetrahymena* oxidase. There are generally 4–8 basic residues in the vicinity of the site of oxidation of the cytochromes c_2, and they all show positive charge in their interactions *in vitro* with bovine oxidase (and with the negatively charged flavin, FMN), despite the fact that some cytochromes c_2 have an overall negative charge (Errede and Kamen, 1978; Meyer *et al.*, 1984).

Cytochrome c_2 in purple bacteria is known primarily as the electron donor to the photosynthetic reaction center. Positive charges on the surface of this protein near the site of electron transfer are also considered to be important in this system (Rickle and Cusanovich, 1979; Overfield *et al.*, 1979; Rosen *et al.*, 1983). However, cytochromes c_2 are generally poor electron donors to bovine oxidase, in spite of the presence of the requisite positive electrostatic field (Errede and Kamen, 1978), but that is mainly due to the high redox potentials of these proteins (average 340 mV for cytochrome c_2 vs. 260 mV for mitochondrial cytochrome c). Presumably the bacterial oxidases would have a much higher redox potential. It is not known what factors are most important in determining the redox potential of cytochromes, but H-bonding interactions with the heme ligands, buried charges, packing of hydrophobic residues against the heme, and exposure of heme to solvent must all contribute. Not all purple bacteria are capable of aerobic growth. In fact, there are only a few purple bacteria which are facultatively capable of fully aerobic growth and presumably have such an oxidase which has a correspondingly high redox potential. Nonetheless, it has been reported that at least some bacteria contain a membrane-bound cytochrome c, which is thought to be the immediate electron donor to oxidase instead of cytochrome c_2 (Hudig and Drews, 1983; Berry and Trumpower, 1985).

An a-type oxidase has been characterized from *Rb. sphaeroides* (Shapleigh and Gennis, 1992; Hosler *et al.*, 1992; Cao *et al.*, 1991, 1992). Although it is active with horse cytochrome c, the natural substrate appears to be a membrane-bound cytochrome c (Hosler *et al.*, 1992). A *Paracoccus* oxidase has likewise been characterized (Raitio *et al.*, 1987). There is evidence of a second oxidase in *Rb. sphaeroides*, *Paracoccus*, and *Bradyrhizobium* (Raitio *et al.*, 1990; Bott *et al.*, 1992; Shapleigh *et al.*, 1992). The gene for the membrane-bound cytochrome c which is the electron donor to oxidase in *Bradyrhizobium* has been sequenced and shown to be homologous to mitochondrial cytochrome c and bacterial cytochrome c_2 (Bott *et al.*, 1991).

There are o-type cytochrome oxidases in addition to the well-known a-type (known as cytochromes aa_3), and these undoubtedly have different requirements for interaction with cytochromes c. Heme o does not have the formyl group of heme a, but retains the farnesyl side chain (Puustinen and Wikstrom, 1991). O-type oxidases are presumably the more common type in facultatively aerobic purple bacteria. They have not yet been purified, and there are no sequence data available. Quinol oxidases, such as those thoroughly characterized in *E. coli* (Green *et al.*, 1988; Chepuri *et al.*, 1990) may also be present in purple bacteria. Therefore, it is not obvious that a-type oxidases influence the evolution of cytochrome c_2 to a very great extent.

It may be concluded from this analysis that basic residues in cytochrome c must be conserved for reaction with most a-type oxidases, and that residues, which would affect the redox potential of cytochrome c, must also be conserved. Because a-type oxidases are relatively rare in purple phototrophic bacteria, such oxidases would have less influence on the evolution of cytochrome c_2 than they would have on mitochondrial cytochrome c in eukaryotes.

B. Interaction with Photosynthetic Reaction Centers

The purple bacterial cytochromes c_2 do not all interact directly with the reaction center. Instead, the presence of an intervening membrane-bound tetraheme reaction center cytochrome appears to be the rule rather than the exception. *Rps. viridis* possesses the most notable example of an intervening reaction center cytochrome subunit (Deisenhofer *et al.*, 1985). The highest-potential heme of the *Rps. viridis* reaction center cytochrome is at least as efficient an electron donor to the photo-oxidized special pair chlorophyll as is cytochrome c_2 in *Rb. sphaeroides* (Dracheva *et al.*, 1986). The *Rps. viridis* cytochrome c_2 appears to be required to couple the cytochrome bc_1 complex to reaction center cytochrome, and the rate constant for reaction between cytochrome c_2 and reaction center cytochrome is about 10^6 M^{-1} s^{-1} at infinite ionic strength (Shill and Wood, 1984; Meyer *et al.*, 1993). Cytochrome c_2 forms a complex with reaction center at low ionic strength, and the limiting rate constant is 270–1300 s^{-1} (Knaff *et al.*, 1991; Meyer *et al.*, 1993). The positive charge on cytochrome c_2 interacts with negative charge on the reaction center cytochrome. Horse cytochrome c is at least as effective, if not more so, than cytochrome c_2 as an electron donor to *R. rubrum* or *Rb. sphaeroides* reaction centers (Rickle and Cusanovich, 1979), although it is less effective in *Rps. viridis*. Perhaps that is because of the intervening bound cytochrome in the latter but not the former.

There is no soluble cytochrome c_2 in *Chloroflexus*, *Ch. vinosum*, or *Rc. gelatinosus*, all of which have the membrane-bound tetraheme reaction center cytochrome. *Chloroflexus* instead contains a small copper protein called auracyanin (Trost *et al.*, 1988; McManus *et al.*, 1992), and *Ch. vinosum* and *Rc. gelatinosus* produce a high-redox-potential ferredoxin known as HiPIP (Bartsch, 1978b), which could be the electron donors to the reaction-center cytochrome. The exact role of the reaction-center cytochrome is unknown, but it may be required to mediate reaction with donors other than cytochrome c_2. It has, however, been reported that *Ch. vinosum* has a minor and labile soluble cytochrome c, which has characteristics of a cytochrome c_2 that include efficient electron transfer to the reaction center (Van Grondelle *et al.*, 1977; Tomiyama *et al.*, 1983; Gray *et al.*, 1983). The identity of this protein has not been confirmed by sequence determination, however. One could speculate that cytochrome c_2 was a late acquisition by purple bacteria which proved to be more efficient than auracyanin or HiPIP and eventually replaced them as donor to reaction center. The tetraheme reaction-center cytochrome may then have become redundant as well, if the cytochrome c_2 could interact directly and efficiently with the special pair chlorophyll.

Evolutionary pressures on cytochrome c_2 are obviously different if it reacts with the reaction center directly, rather than with the intervening RC cytochrome. These pressures would include both surface charges and residues affecting redox potential. The same factors apply to the RC cytochrome if cytochrome c_2 is the electron donor, or alternatively if auracyanin or HiPIP serve as electron donors. We may eventually detect structural differences corresponding to the two roles just described, but at present there are insufficient data. In the case of certain cytochromes c_2, the evolutionary pressure could include the effects of oxidase, reductase, nitrite reductase, and reaction center all at once. If the interaction with reaction center is a recently acquired function for cytochrome c_2, then it may not yet have contributed much to the evolution of cytochrome c_2.

C. Interaction with Nitrite Reductase

Cytochrome c_2 is presumably the electron donor to the cytochrome cd_1 type of nitrite reductase in *Paracoccus denitrificans*, with which horse cytochrome c has only 14% and *Pseudomonas* cytochrome c-551 has only 0.3% as much activity as cytochrome c_2 (Timkovich *et al.*, 1982). *Roseobacter denitrificans* (formerly *Erythrobacter*) strain Och 114 cytochrome cd_1 is probably also specific for cytochrome c_2 (Doi *et al.*, 1989). *Pseudomonas aeruginosa* cytochrome cd_1 is specific for cytochrome c-551, but it has 2% activity with azurin and 8% activity with *Paracoccus* cytochrome c_2 (Timkovich *et al.*, 1982). The cytochrome c-551 gene is adjacent to that for cytochrome cd_1 in *P. aeruginosa*, and near it in *Ps. stutzeri*, which suggests along with the activity just noted that it is the natural electron donor (Silvestrini *et al.*, 1989; Jungst *et al.*, 1991). Strains of *Rb. sphaeroides* have a copper-containing nitrite reductase, which also appears to utilize *Rb. sphaeroides* cytochrome c_2 and horse cytochrome c as electron donors (Sawada *et al.*, 1978). The *Pseudomonas aureofaciens* (formerly *P. fluorescens* biotype E) copper-containing nitrite reductase uses azurin as electron donor (Zumft *et al.*, 1987). The copper-containing enzymes from *Achromobacter cycloclastes* and *Alcaligenes faecalis* S6 utilize pseudoazurin as electron donor, whereas *Pseudomonas* cytochrome c-551 is inactive (Kakutani *et al.*, 1981; Liu *et al.*, 1986).

It is expected that the evolution of both nitrite reductase and its electron donors will be affected by their interaction. As the specificity of the enzyme changes, so, too, will its evolutionary development. In *Rb. sphaeroides*, the evolution of cytochrome c_2 will be constrained by its interaction with oxidase, reductase, reaction centers, and nitrite reductase. Thus, one should not expect simple divergent evolution except where these constraints are held constant. It should be admitted that denitrification is relatively rare in phototrophic bacteria, and that nitrite reductase should not significantly affect the evolution of cytochromes c_2 as a whole. Because the nitrite reductases can utilize a variety of electron donors in different species, it appears that such adaptation is relatively simple and does not require a great many mutations. One could therefore look upon it as providing relatively weak functional constraints on the structure of cytochrome c_2.

D. Interaction with the bc_1 Complex

Those factors which have been identified as contributing to establishment of specificity for electron transfer will be influenced to varying degrees by each of the reactions in which a particular cytochrome may participate. As mentioned before, charge distribution, redox potentials, and surface topography will all require certain conserved residues in particular species, which will influence the evolution of that cytochrome in one direction or the other. As the reaction partners of cytochrome c remain constant, so will the evolutionary pressures, but as reaction partners change, so will the structure. Although cytochrome c_2 has many more identified roles than does mitochondrial cytochrome c, the extent of variation does not seem to be that much greater. Perhaps it is because one particular role which appears to be universally present is much more demanding of the structural integrity of cytochrome c or c_2 than are the others, namely interaction with the cytochrome bc_1 complex. If the bc_1 complex is most important in directing structural variation of cytochrome c_2, then we would expect a more or less constant rate of change in all those species in which the two are both present. This expectation has to be established by future experiments. Although the bc_1 complex has been identified in only a few bacterial species, it is expected to be present in most. For example, it is believed that all phototrophic bacteria contain a bc_1 complex. The genes for the bc_1 complex have only been found in four species, but there is spectral evidence for others (see Knaff, 1993, for a review).

V. Role of Gene Fusion and Post-translational Processing

In gram-negative bacteria, the soluble cytochrome c_2 is periplasmic and, by analogy to mitochondria, is thought to interact with oxidase on the outside of the cytoplasmic membrane. However, a super-complex containing the subunits of both the bc_1 complex and the a-type oxidase has been isolated from *Paracoccus*, which also contains an additional 22-kDa cytochrome c subunit (Berry and Trumpower, 1985). The complex oxidizes quinol in the absence of soluble cytochrome c_2, and the reaction is not stimulated by added cytochrome c_2. The existence of this complex raises some doubt as to whether the soluble cytochrome c_2 is the natural electron donor to oxidase in *Paracoccus*. A second oxidase gene for subunit 1 which is adjacent to the cytochrome c_2 gene in *Paracoccus* further complicates the picture (Raitio et al., 1990). Perhaps only one of the two oxidases utilizes the soluble cytochrome c_2. An a-type oxidase-associated, membrane-bound, 20-kDa cytochrome c_2 was cloned and sequenced from *Bradyrhizobium japonicum* (Bott et al., 1991). It differs from soluble cytochromes c_2 in that the signal peptide is not cleaved following transport across the membrane. The *Paracoccus* cytochrome described earlier may be similar. It may be speculated that the membrane location of this cytochrome restricts its functional role to this specific reaction.

These experiments raise the additional question of what is the electron donor to oxidase in gram-positive bacteria, which have no outer membrane and, therefore, no periplasmic space or soluble periplasmic cytochromes. One could assume that a cytochrome c is either anchored to the outside of the membrane, as in the *Paracoccus* and *Bradyrhizobium* examples, or that it is not necessary, as reported for *E. coli* (Anraku and Gennis, 1987) or for *Sulfolobus acidocaldarius* (Anemuller and Schafer, 1990). It has been found that a cytochrome c accompanies the solubilized a-type oxidase in *Bacillus* PS3, *Bacillus firmus*, and *Bacillus* YN 2000 (Sone and Yanagita, 1982; Kitada and Krulwich, 1984; Qureshi et al., 1990), and it therefore appears that a cytochrome c may be present as electron donor to oxidase in at least some gram-positive as well as gram-negative bacteria.

A super-complex similar to that isolated from *Paracoccus* was found in *Bacillus* PS3, which contained a 21-kDa cytochrome c subunit in addition to cytochrome c_1 (29 kDa) and the oxidase cytochrome c–containing subunit 2 (38 kDa) (Sone et al., 1987). The gene encoding the latter subunit contains a cytochrome c gene fused to the copper protein subunit 2 gene (Ishizuka et al., 1990). A similar gene fusion was discovered in *B. subtilis* (Saraste et al., 1991). *Thermus thermophilus*, which is gram-negative, also has a cytochrome c–containing oxidase like that of *Bacillus* (Yoshida et al., 1984), in which the cytochrome c is fused to the copper-containing oxidase subunit 2 (Buse et al., 1989; Mather et al., 1991).

A membrane-bound cytochrome *c* can be expected to evolve differently than a soluble cytochrome even if their reaction partners (cytochrome oxidase) are the same. For example, the lysine residues in cytochrome *c*, which are involved in forming an electrostatic complex with oxidase, may no longer be necessary if the cytochrome *c* is tethered to the oxidase. Furthermore, the mobility of the cytochrome *c* is reduced by membrane binding, thus restricting its ability to interact with other potential reaction partners.

VI. Quantification of Cytochrome Evolution

Evolutionary trees may be constructed from matrices of sequence differences, which are obtained from pairwise comparison of protein or gene sequences. Insertions and deletions are used to maximize identity when aligning two proteins, and penalties are sometimes added to prevent excessive use of such gaps. That is, a gap is usually not allowed unless it results in several new identities. Multiple sequence alignment is a recent innovation, which further restricts the use of gaps (e.g., Murata *et al.*, 1985; Feng and Doolittle, 1987; Lipman *et al.*, 1989). Ideally, one would use three-dimensional structures to determine experimentally the number and location of gaps, but there is a limit to the number of crystal structures available, and rearrangements may still prevent firm conclusions.

Once an alignment is chosen and sequence differences tabulated, there are a number of ways in which to construct evolutionary trees. This is where a great deal of activity is centered in the study of evolution. However, sequence differences must be converted to numbers of mutations, because only the latter are proportional to elapsed time. There are more mutations than there are sequence differences because parallel and back mutations are silent. This is usually considered to be a small and unimportant correction which does not alter the topology of the resultant tree. This assumption is more or less true where there is only a small amount of change (comparison of higher eukaryotes), but where there is a large amount of change (comparison of microorganisms), parallel and back mutations, collectively known as convergent mutations, become highly significant. In fact, the numbers of mutations increase roughly exponentially with increasing percentage sequence divergence. At some point, divergence and convergence are balanced, and there is no more apparent change in the sequences. Where divergence and convergence are equally balanced, the number of mutations which have occurred approaches infinity and cannot reliably be estimated (Meyer *et al.*, 1986a). At this point, the underlying assumptions break down, and trees are no longer valid. The problem is in estimating this limit for any protein so that the appropriate correction for convergence may be applied.

For those who attempt to apply a correction for convergence, it is often assumed that the only limitation on the amount of sequence difference is that imposed by amino acid composition, which is about 90–95% difference for cytochromes *c*. Of course, we know that two proteins rarely achieve a level of divergence limited only by composition, because a certain number of residues are required for maintenance of structure and function as discussed earlier. Either these residues cannot be substituted, or there are only a few possibilities allowed at these positions. The limit to change in a protein sequence is, therefore, indirectly established by the function rather than by the amino acid composition. The function of the protein is thus a very important variable in evolutionary studies, which has generally been ignored. This practice is not surprising for bacterial cytochromes especially, because the functions of these proteins are usually unknown or poorly defined. The limit to change thus has to be determined empirically for each protein of defined function. The problem of assigning a limit to change was recognized early in the study of cytochrome c evolution (Margoliash and Smith, 1965; Margoliash and Fitch, 1968), where a limit to change of 75–80% was assumed for mitochondrial cytochromes c, and an exponential correction appropriate for this percentage applied. This important lesson seems to have been forgotten or ignored in many recent studies. However, we have since learned that the average limit to change is smaller than originally assumed and is closer to 60%. Valid evolutionary trees may be constructed using data which represent only a small amount of change relative to this limit (and therefore require only a small correction for convergence), such as for higher eukaryotes. On the other hand, an evolutionary tree resulting from use of highly divergent sequences, such as the bacterial cytochromes c_2, usually has the appearance of a shellburst, and all species appear to be equidistant from one another. Trees which have this appearance may be assumed to be invalid.

The limit to change in protein sequences may be estimated from a plot of the frequency of the data in the matrix of percentage difference (Meyer *et al.*, 1986a). For bacterial cytochromes c_2 and c', this curve approximates a normal distribution, the mean of which is the limit to change, about 60% for mitochondrial cytochrome c and cytochrome c_2 and about 75% for cytochrome c'. The early estimates of 75–80% difference for mitochondrial cytochrome c were thus much too high. For proteins of unknown function, such a normal distribution would suggest that the function is the same for all proteins compared. A bimodal distribution would suggest that at least some of the proteins have a different function or that some species are truly closer to one another than are others. For bacterial proteins, the conditions for construction of valid evolutionary trees based on difference matrices are rarely met, and some other means of quantifying relationships must be found.

All-inclusive trees (Dayhoff and Schwartz, 1981; Dickerson, 1980b) which were constructed without consideration of the magnitude of convergence or the varied functions of type 1 bacterial cytochromes lack validity. Attempts to combine one flawed tree with another to obtain an improved tree (Dayhoff and Schwartz, 1981) also lack credibility, as pointed out by Demoulin (1979). As emphasized throughout this review, trees which include only a single subclass of cytochrome that has a well-defined function, such as cytochrome c_2, are still invalid for the most part. Therefore, one cannot hope to combine several functionally unrelated proteins in a single tree and draw any meaningful conclusions. The amount of bacterial data which can be used to construct valid trees from matrices of sequence differences is so small a fraction of the total as to be uninteresting. We believe that analysis of insertions and deletions eventually will provide an improved measure of relatedness among bacterial species.

Insertions and deletions established by comparison of three-dimensional structures may be used to quantify relationships among bacterial proteins which have

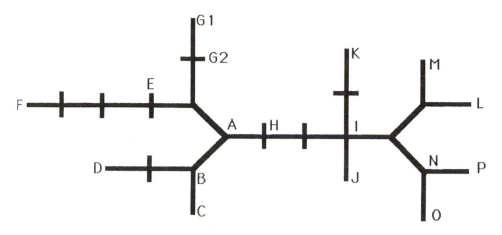

Figure 2.1

Evolutionary relationships among mitochondrial cytochromes *c* and cytochromes c_2
based on internal insertions and deletions. Each node represents a gap: (A) *Rps. viridis*,
Rm. vannielii, *Rps. acidophila*, and *Nitrobacter agilis* cytochromes c_2; (B) most eukaryotic
mitochondrial cytochrome *c*; (C) *Euglena gracilis* and *E. viridis* mitochondrial cytochrome *c*;
(D) *Tetrahymena pyriformis* mitochondrial cytochrome *c*; (E) *Rp. globiformis* cytochrome c_2;
(F) *R. salexigens* cytochrome c_2; (G1) *R. molishianum* and *R. fulvum* iso-1 cytochromes c_2;
(G2) *R. molischianum* and *R. fulvum* iso-2 cytochromes c_2; (H) *Agrobacterium tumefaciens*,
Rb. sphaeroides iso-2, and *Rps. marina* cytochromes c_2; (I) *R. rubrum* and *Aq. itersonii*
cytochromes c_2; (J) *R. photometricum* cytochrome c_2; (K) *Rps. palustris* cytochrome c_2; (L)
Rb. sphaeroides iso-1 cytochrome c_2; (M) *Rb. sulfidophila* cytochrome c_2; (N) *Rb. capsulatus*
and *Rb.* sp. TJ12 cytochromes c_2; (O) *Paracoccus denitrificans* cytochrome c_2; (P)
Roseobacter denitrificans (formerly *Erythrobacter*) OcH 114 cytochrome c_2. "Large"
cytochromes c_2 are on the right side of the figure (I to P) and "medium" on the left (A to G).

reached a limit of amino acid substitution. For proteins which have diverged only a
small amount, insertions and deletions are rare and easy to place, but as two sequences approach a limit of change, gaps become common and difficult to place
with certainty. The bacterial cytochromes c_2 have reached a limit of change in most
species examined (about 60% average difference). A number of insertions and deletions have occurred in this family, and many have been placed with the aid of crystal structures. Because gaps occur less commonly than amino acid substitutions,
convergence is less of a problem. Furthermore, back mutations may not result in
precise elimination of a gap, and a record of its occurrence might still be visible
if, for example, only two residues of a three-residue insertion were subsequently
deleted.

Insertions and deletions in mitochondrial cytochrome *c* and cytochrome c_2 can
be quantified and tabulated in a matrix in the same way as sequence differences.
This analysis gives a better picture of relationships for divergent species than does
the standard evolutionary tree, as shown in Fig. 2.1, although it is not very useful for
closely related species. Additionally, one can see that there are two groups of cytochromes c_2 which differ in the number of shared gaps. Large cytochromes c_2 have

shared three- and eight-residue insertions and a single-residue deletion relative to the small cytochromes c_2. These three gaps are unlikely to have occurred simultaneously and represent a major division of the cytochromes c_2. This conclusion is not obvious from inspection of the matrix of sequence differences. For example, *Rps. palustris* cytochrome c_2 is often placed with *Rps. viridis*, *Rm. vannielii*, and *Rps. acidophila* cytochromes c_2 based on the matrix of percentage sequence differences, but insertions and deletions place *Rps. palustris* c_2 nearer to *R. rubrum* and *R. photometricum* cytochromes c_2. As a caveat, *Rp. globiformis* and *Euglena* share the same insertions and deletions relative to *Rps. viridis* c_2, although we assume they were independent, parallel events. Furthermore, there are some cytochromes c_2 which have only the three-residue insertion. These may represent either intermediate or parallel changes. We have not yet been able to apply this analysis to any other class of cytochromes, but we expect to be able to do so when additional three-dimensional structures become available.

May we assume that the evolution of cytochrome c is the same as that of the species in which it is found? Only to the extent that the cytochrome has the same function in all species compared and that there are no isozymes or pseudogenes. *Rb. sphaeroides* has both large and medium cytochrome c_2 isozymes (Fitch *et al.*, 1989; Rott and Donohue, 1990; Rott *et al.*, 1992, 1993). Which one is representative of the species? Only the large cytochrome c_2 is expressed under photosynthetic conditions. It may be assumed to be representative, but it could also be a recent acquisition which replaced the medium cytochrome c_2. In this regard, we may compare *Rb. sphaeroides* with *Rb. capsulatus*, two species which are closer than the average pair of purple bacteria and which both have a functional large cytochrome c_2. *Rb. capsulatus* does not appear to have the medium isozyme, and we may thus assume that it is the medium cytochrome c_2 which was recently acquired by *Rb. sphaeroides*. This analysis is complicated by the presence of a membrane-bound cytochrome c in *Rb. capsulatus* (Jenney and Daldal, 1993). Thus, comparison of the evolution of several unrelated proteins should give us a better idea of relationships among species.

Do relationships inferred from comparison of cytochrome c_2 match those obtained from some other protein such as cytochrome c'? To the limited extent that the comparison can be made, the answer is yes. Cytochromes c' from *Rb. sphaeroides* and *Rb. capsulatus* are significantly closer to each other than average, and so are the large cytochromes c_2 (Ambler *et al.*, 1979a, 1981b). We observe the same with *R. rubrum* and *R. photometricum*. Some species do not have a cytochrome c', and they all have a functional medium cytochrome c_2. Other species lack the membrane-bound reaction center cytochrome, and they all have a functional large cytochrome c_2. Beyond such qualitative statements, we cannot be more precise. We do not know precisely where to place insertions and deletions in cytochrome c' because there are only two known crystal structures. We now have three structures for the high-potential ferredoxin known as HiPIP, but need still more. Therefore, we cannot at present quantitatively analyze insertions and deletions the same way as in cytochrome c_2. Furthermore, those species which have cytochrome c_2 usually do not have HiPIP. Eventually we will have the sequences of additional proteins for comparison, such as reaction center, cytochrome bc_1 complex, rubisco, and nitrogenase, and then we will be able to compare several gene products. At present, 16s rRNA partial or complete sequences are known for many bacterial species which utilize cytochrome c_2.

To what extent does the evolution of cytochrome c_2 match that of rRNA? Dickerson (1980a) and Woese et al. (1980) thought the correspondence was good. However, we are much more cautious for the following reasons. Many of the foregoing limitations to analysis of the evolution of cytochrome c_2 also apply to rRNA, although there are pronounced differences. There should be a limit to change in rRNA where divergence and convergence are approximately balanced. Trees for rRNA would have no validity in this region. We can expect a smaller amount of change in rRNA and more rapid convergence because there are only four bases as opposed to 20 amino acids, and because approximately half the bases will be H-bonded in AU and CG pairs. In fact, this limit appears to be near 40% difference for 16s rRNA (Martens et al., 1987). That is much smaller than for cytochrome c_2 (60%) and suggests that under the best of circumstances, conclusions based on rRNA sequences should be more limited than those based on cytochromes c. In most comparisons, partial rather than complete rRNA sequences have been determined. This results in an error not encountered with proteins, which is based on the poor correlation between partial and complete sequences. In fact, the correlation (Woese, 1987; Martens et al., 1987) is so weak as to be nearly useless. Using partial rRNA sequences is comparable to using amino acid compositions of proteins for evolutionary analysis. Because there are so few complete sequences, we cannot yet compare rRNA evolution to that for cytochrome c_2 and draw meaningful conclusions.

One cannot analyze insertions and deletions in rRNA with confidence because the three-dimensional structure of the ribosome is unknown. In fact, using the sequences alone, insertions and deletions in rRNA are far more difficult to place than are those of proteins because there are only four bases. An attempt has been made to extend the validity of rRNA trees by use of what is called signature analysis. This simply involves looking for conserved bases within a group of presumably related species. This strategy is highly subjective and does not have the same authority as analysis of conserved features in cytochromes c, which have established roles based on three-dimensional structures. Nevertheless, those who analyze rRNA sequences have a large following in spite of its shortcomings. To their credit, the similarity apparent between Rb. sphaeroides and Rb. capsulatus or between Rb. capsulatus and Paracoccus denitrificans cytochromes c_2 has also been noted from comparison of rRNA (Fox et al., 1980). However, a relationship between Chromatium vinosum and E. coli has been reported (Fox et al., 1980) and is almost certainly erroneous. Provided one stays within the narrow window of reliability of matrix data, one can draw valid conclusions, but when dealing with bacteria, the ideal is rarely achieved.

VII. Acknowledgment

This work was supported in part by a grant from the National Institutes of Health, GM 21277.

VIII. References

Almassy, R. J., & Dickerson, R. E. (1978) *Proc. Natl. Acad. Sci. USA 75*, 2674–2678.

Amati, B. B., Goldschmidt-Clermont, M., Wallace, C. J. A., & Rochaix, J. D. (1988) *J. Mol. Evol. 28*, 151–160.

Ambler, R. P. (1968) *Biochem. J. 109*, 47P.

Ambler, R. P. (1971) *FEBS Letts. 18*, 351–353.

Ambler, R. P. (1973) *Systematic Zoology 22*, 554–565.

Ambler, R. P. (1974) *Biochem. J. 137*, 3–14.

Ambler, R. P. (1982) in *From Cyclotrons to Cytochromes* (N. O. Kaplan & A. B. Robinson, eds.), Academic Press, New York, pp. 263–280.

Ambler, R. P. (1991) *Biochim. Biophys. Acta 1058*, 42–47.

Ambler, R. P., & Daniel, M. (1991) *Biochem. J. 274*, 825–831.

Ambler, R. P., & Murray, S. (1973) *Biochem. Soc. Trans. 1*, 162–164.

Ambler, R. P., & Taylor, E. (1973) *Biochem. Soc. Trans. 1*, 166–168.

Ambler, R. P., & Wynn, M. (1973) *Biochem. J. 131*, 485–498.

Ambler, R. P., Bruschi, M., & LeGall, J. (1969) *FEBS Letts. 5*, 115–117.

Ambler, R. P., Bruschi, M., & LeGall, J. (1971) *FEBS Letts 18*, 347–350.

Ambler, R. P., Daniel, M., Hermoso, J., Meyer, T. E., Bartsch, R. G., & Kamen, M. D. (1979a) *Nature 278*, 659–660.

Ambler, R. P., Meyer, T. E., & Kamen, M. D. (1979b) *Nature 278*, 661–662.

Ambler, R. P., Meyer, T. E., Kamen, M. D., Schichman, S. A., & Sawyer, L. (1981a) *J. Mol. Biol. 147*, 351–356.

Ambler, R. P., Bartsch, R. G., Daniel, M., Kamen, M. D., McLellan, L., Meyer, T. E., & Van Beeumen, J. (1981b) *Proc. Natl. Acad. Sci. USA 78*, 6854–6857.

Ambler, R. P., Daniel, M., Melis, K., & Stout, C. D. (1984) *Biochem. J. 222*, 217–227.

Ambler, R. P., Meyer, T. E., Trudinger, P. A., & Kamen, M. D. (1985) *Biochem. J. 227*, 1009–1013.

Ambler, R. P., Dalton, H., Meyer, T. E., Bartsch, R. G., & Kamen, M. D. (1986) *Biochem. J. 233*, 333–337.

Ambler, R. P., Daniel, M., McLellan, L., Meyer, T. E., Cusanovich, M. A., & Kamen, M. D. (1987a) *Biochem. J. 248*, 365–371.

Ambler, R. P., Meyer, T. E., Cusanovich, M. A., & Kamen, M. D. (1987b) *Biochem. J. 246*, 115–120.

Ameyama, M., & Adachi, O. (1982) *Methods in Enzymology 89*, 450–457.

Anderson, D. J., & Lidstrom, M. E. (1988) *J. Bacteriol. 170*, 2254–2262.

Anderson, D. J., Morris, C. J., Nunn, D. N., Anthony, C., & Lidstrom, M. E. (1990) *Gene 90*, 173–176.

Andersson, L. A., Loehr, T. M., Wu, W., Chang, C. K., & Timkovich, R. (1990) *FEBS Letts. 267*, 285–288.

Andrews, S. C., Smith, J. M. A., Guest, J. R., & Harrison, P. M. (1989a) *Biochem. Biophys. Res. Comm. 158*, 489–496.

Andrews, S. C., Harrison, P. M., & Guest, J. R. (1989b) *J. Bacteriol. 171*, 3940–3947.

Anemuller, S., & Schafer, G. (1990) *Eur. J. Biochem. 191*, 297–305.

Anthony, C. (1986) *Adv. Microbial Physiol. 27*, 113–210.

Anthony, C. (1988) in *Bacterial Energy Transduction* (C. Anthony, ed.), Academic Press, New York, pp. 293–316.

Anthony, C. (1992) *Biochim. Biophys. Acta 1099*, 1–15.

Anraku, Y., & Gennis, R. B. (1987) *TIBS 12*, 262–266.

Appleby, C. A., James, P., & Hennecke, H. (1991) *FEMS Microbiol. Letts. 83*, 137–144.

Arai, H., Sanbongi, Y., Igarashi, Y., & Kodama, T. (1990) *FEBS Letts. 261*, 196–198.

Arnason, U., & Johnsson, E. (1992) *J. Mol. Evol. 34*, 493–505.

Baba, M. L., Darga, L. L., Goodman, M., & Czelusniak, J. (1981) *J. Mol. Evol. 17*, 197–213.

Bartsch, R. G. (1978a) in *The Photosynthetic Bacteria* (R. K. Clayton and W. R. Sistrom, eds.), Plenum Press, New York, pp. 249–279.

Bartsch, R. G. (1978b) *Methods in Enzymol. 53*, 329–340.

Bartsch, R. G., & Kamen, M. D. (1960) *J. Biol. Chem. 235*, 825–831.

Bartsch, R. G., Meyer, T. E., & Robinson, A. B. (1968) in *Structure and Function of Cytochromes* (K. Okunuki, M. D. Kamen, and I. Sekuzu, eds.), University Park Press, Baltimore, pp. 443–451.

Bartsch, R. G., Ambler, R. P., Meyer, T. E., & Cusanovich, M. A. (1989) *Arch. Biochem. Biophys.* **271**, 433–440.

Beardmore-Gray, M., O'Keeffe, D. T., & Anthony, C. (1982) *Biochem. J.* **207**, 161–165.

Beardmore-Gray, M., O'Keeffe, D. T., & Anthony, C. (1983) *J. Gen. Microbiol.* **129**, 923–933.

Bellamy, H. D., Mathews, F. S., McIntire, W. S., & Singer, T. P. (1987) in *Flavins and Flavo-proteins* (D. E. Edmondson and D. B. McCormick, eds.), Walter de Gruyter, New York, pp. 673–676.

Benning, M. M., Wesenberg, G., Caffrey, M. S., Bartsch, R. G., Meyer, T. E., Cusanovich, M. A., Rayment, I., & Holden, H. M. (1991) *J. Mol. Biol.* **220**, 673–685.

Benosman, H., Asso, M., Bertrand, P., Yagi, T., & Gayda, J. (1989) *Eur. J. Biochem.* **182**, 51–55.

Berry, E. A., & Trumpower, B. L. (1985) *J. Biol. Chem.* **260**, 2458–2467.

Berry, M. J., George, S. J., Thomson, A. J., Santos, H., & Turner, D. L. (1990) *Biochem. J.* **270**, 413–417.

Bertrand, P., Bruschi, M., Denis, M., Gayda, J. P., & Manca, F. (1982) *Biochem. Biophys. Res. Comm.* **106**, 756–760.

Bertsch, J., & Malkin, R. (1991) *Plant Mol. Biol.* **17**, 131–133.

Bhatia, G. E. (1981) Ph.D. Thesis, University of California at San Diego.

Bianco, P., Haladjian, J., Loutfi, M., & Bruschi, M. (1983) *Biochem. Biophys. Res. Comm.* **113**, 526–530.

Bilous, P. T., Cole, S. T., Anderson, W. F., & Weiner, J. H. (1988) *Mol. Microbiol.* **2**, 785–795.

Bohm, R., Sauter, M., & Bock, A. (1990) *Mol. Microbiol.* **4**, 231–243.

Bohme, H., Brutsch, S., Weithmann, G., & Boger, P. (1980a) *Biochim. Biophys. Acta* **590**, 248–260.

Bohme, H., Pelzer, B., & Boger, P. (1980b) *Biochim. Biophys. Acta* **592**, 528–535.

Bork, P., & Doolittle, R. F. (1992) *Proc. Natl. Acad. Sci. USA* **89**, 8990–8994.

Bott, M., Ritz, D., & Hennecke, H. (1991) *J. Bacteriol.* **173**, 6766–6772.

Bott, M., Preisig, O., & Hennecke, H. (1992) *Arch. Microbiol.* **158**, 335–343.

Bruce, B. D., Fuller, R. C., & Blankenship, R. E. (1982) *Proc. Natl. Acad. Sci. USA* **79**, 6532–6536.

Brune, D. C. (1989) *Biochim. Biophys. Acta* **975**, 189–221.

Bruschi, M. (1981) *Biochim. Biophys. Acta* **671**, 219–226.

Bruschi, M., LeGall, J., Hatchikian, E. C., & Dubourdiu, M. (1969) *Bull. Soc. Fr. Physiol. Veg.* **15**, 381–390.

Bruschi, M., LeGall, J., and Bovier-Lapierre, G. (1972) *Biochem. Biophys. Acta* **271**, 48–60.

Bruschi, M., Loutfi, M., Bianco, P., & Haladjian, J. (1984) *Biochem. Biophys. Res. Comm.* **120**, 384–389.

Buschlen, S., Choquet, Y., Kuras, R., & Wollman, F. A. (1991) *FEBS Letts.* **284**, 257–262.

Buse, G., Hensel, S., & Fee, J. A. (1989) *Eur. J. Biochem.* **181**, 261–268.

Bushnell, G. W., Louie, G. V., & Brayer, G. D. (1990) *J. Mol. Biol.* **214**, 585–595.

Buttner, M., Xie, D. L., Nelson, N., Pinther, W., Hauska, G., & Nelson, N. (1992) *Proc. Natl. Acad. Sci. USA* **89**, 8135–8139.

Cao, J., Shapleigh, Gennis, R., Revzin, A., & Ferguson-Miller, S. (1991) *Gene* **101**, 133–137.

Cao, J., Hosler, J., Shapleigh, J., Revzin, A., & Ferguson-Miller, S. (1992) *J. Biol. Chem.* **267**, 24273–24278.

Carter, D. C., Melis, K. A., O'Donnell, S. E., Burgess, B. K., Furey, W. F. Jr., Wang, B. C., & Stout, C. D. (1985) *J. Mol. Biol.* **184**, 279–295.

Chan, H. T. C., & Anthony, C. (1991) *Biochem. J.* **280**, 139–146.

Chang, C. K., & Wu, W. (1986) *J. Biol. Chem.* **261**, 8593–8596.

Cheddar, G., Meyer, T. E., Cusanovich, M. A., Stout, C. D., & Tollin, G. (1989) *Biochemistry* **28**, 6318–6322.

Cheesman, M. R., Thomson, A. J., Greenwood, C., Moore, G. R., & Kadir, F. (1990) *Nature* **346**, 771–773.

Chepuri, V., Lemieux, L., Au, D. C. T., & Gennis, R. B. (1990) *J. Biol. Chem.* **265**, 11185–11192.

Clayton, B. J., & Clayton, R. K. (1978) *Biochim. Biophys. Acta* **501**, 470–477.

Cohn, C. L., Hermodson, M. A., & Krogmann, D. W. (1989a) *Arch. Biochem. Biophys.* **279**, 219–226.

Cohn, C. L., Sprinkle, J. R., Alam, J., Hermodson, M., Meyer, T. E., & Krogmann, D. W. (1989b) *Arch. Biochem. Biophys.* **270**, 227–235.

Cole, S. T. (1987) *Eur. J. Biochem.* **167**, 481–488.

Cox, C. D., Jr., & Payne, W. J. (1973) *Can. J. Microbiol.* **19**, 861–872.

Cox, J. M., Day, D. J., & Anthony, C. (1992) *Biochim. Biophys. Acta* **1119**, 97–106.

Coyne, M. S., Arunakumari, A., Pankratz, H. S., & Tiedje, J. M. (1990) *J. Bacteriol. 172*, 2558–2562.

Cross, A. R., & Anthony, C. (1980) *Biochem. J. 192*, 421–427.

Cusanovich, M. A., & Bartsch, R. G. (1969) *Biochim. Biophys. Acta 189*, 245–255.

Daldal, F., Cheng, S., Applebaum, J., Davidson, E., & Prince, R. C. (1986) *Proc. Natl. Acad. Sci. USA 83*, 2012–2016.

Davidson, E., & Daldal, F. (1987a) *J. Mol. Biol. 195*, 13–24.

Davidson, E., & Daldal, F. (1987b) *J. Mol. Biol. 195*, 25–29.

Davis, D. J., Frame, M. K., & Johnson, D. A. (1988) *Biochim. Biophys. Acta 936*, 61–66.

Day, D. J., Nunn, D. N., & Anthony, C. (1990) *J. Gen. Microbiol. 136*, 181–188.

Dayhoff, M. O., & Schwartz, R. M. (1981) *Ann. N.Y. Acad. Sci. 361*, 92–104.

Deisenhofer, J., Epp, O., Miki, K., Huber, R., & Michel, H. (1985) *Nature 318*, 618–624.

Demoulin, V. (1979) *Science 205*, 1036–1038.

Denariaz, C. M., Liu, M. Y., Payne, W. J., LeGall, J., Marquez, L., Dunford, H. B., & Van Beeumen, J. (1989) *Arch. Biochem. Biophys. 270*, 114–125.

DerVartanian, D. V., & LeGall, J. (1974) *Biochim. Biophys. Acta 346*, 79–99.

DerVartanian, D. V., Xavier, A. V., & LeGall, J. (1978) *Biochimie 60*, 321–325.

Dickerson, R. E. (1980a) *Nature 283*, 210–212.

Dickerson, R. E. (1980b) *Scientific American 242*, 136–153.

Dijkstra, M., Frank, J., Van Wielink, J. E., & Duine, J. A. (1988) *Biochem. J. 251*, 467–474.

DiSpirito, A. A., Lipscomb, J. D., & Lidstrom, M. E. (1990) *J. Bacteriol. 172*, 5360–5367.

Doi, M., Shioi, Y., Morita, M., & Takamiya, K. (1989) *Eur. J. Biochem. 184*, 521–527.

Dolata, M. M., Van Beeumen, J. J., Ambler, R. P., Meyer, T. E., & Cusanovich, M. A. (1993) *J. Biol. Chem. 268*, 14426–14431

Dolla, A., Cambillau, C., Bianco, P., Haladjian, J., & Bruschi, M. (1987) *Biochem. Biophys. Res. Comm. 47*, 818–823.

Donohue, T. J., McEwan, A. G., & Kaplan, S. (1986) *J. Bacteriol. 168*, 962–972.

Donohue, T. J., McEwan, A. G., Van Doren, S., Crofts, A. R., & Kaplan, S. (1988) *Biochemistry 27*, 1918–1925.

Doolittle, R. F., Feng, D. F., Anderson, K. L., & Alberro, M. R. (1990) *J. Mol. Evol. 31*, 383–388.

Dracheva, S. M., Drachev, L. A., Zaberezhnaya, S. M., Konstantinov, A. A., Semenov, A. Y., & Skulachev, V. P. (1986) *FEBS Letts. 205*, 41–46.

Dracheva, S. M., Drachev, L. A., Konstantinov, A. A., Semenov, A. V., Skulachev, V. P., Arutjunjan, A. M., Shuvalev, V. A., & Zaberezhnaya, S. M. (1988) *Eur. J. Biochem. 171*, 253–264.

Dracheva, S., Williams, J. C., Van Driessche, G., Van Beeumen, J. J., & Blankenship, R. E. (1991) *Biochemistry 30*, 11451–11458.

Ellfolk, N., Ronnberg, M., Aasa, R., Andreasson, L. E., & Vanngard, T. (1983) *Biochim. Biophys. Acta 743*, 23–30.

Eng, L. H., & Neujahr, H. Y. (1989) *Arch. Microbiol. 153*, 60–63.

Engelhardt, H., Engel, A., and Baumeister, W. (1986) *Proc. Natl. Acad. Sci. USA 83*, 8972–8976.

Errede, B., & Kamen, M. D. (1978) *Biochemistry 17*, 1015–1027.

Evans, M. J., & Scarpulla, R. C. (1988) *Proc. Natl. Acad. Sci. USA 85*, 9625–9629.

Evans, W. R., Fleischman, D. E., Calvert, H. E., Pyati, P. V., Alter, G. M., & Rao, N. S. S. (1990) *Appl. Environ. Microbiol. 56*, 3445–3449.

Fan, K., Akutsu, H., Kyogoku, Y., & Niki, K. (1990) *Biochemistry 29*, 2257–2263.

Fauque, G., Bruschi, M., & LeGall, J. (1979) *Biochem. Biophys. Res. Comm. 86*, 1020–1029.

Feng, D. F., & Doolittle, R. F. (1987) *J. Mol. Evol. 25*, 351–360.

Ferguson-Miller, S., Brautigan, D. L., & Margoliash, E. (1978) *J. Biol. Chem. 253*, 149–159.

Finzel, B. C., Weber, P. C., Hardman, K. D., & Salemme, F. R. (1985) *J. Mol. Biol. 186*, 627–643.

Fischer, U. (1989) *Z. Naturforsch. 44c*, 71–76.

Fitch, J., Cannac, V., Meyer, T. E., Cusanovich, M. A., Tollin, G., Van Beeumen, J., Rott, M. A., & Donohue, T. J. (1989) *Arch. Biochem. Biophys. 271*, 502–507.

Flatmark, T., Dus, K., DeKlerk, H., & Kamen, M. D. (1970) *Biochemistry 9*, 1991–1996.

Foote, N., Peterson, J., Gadsby, P. M. A., Greenwood, C., & Thomson, A. J. (1985) *Biochem. J. 230*, 227–237.

Fox, G. E., Stackebrandt, E., Hespel, R. B., Gibson, J., Maniloff, J., Dyer, T. A., Wolfe, R. S., Balch, W. E., Tanner, R. S., Magrum L. J., Zablen, L. B., Blakemore, R., Gupta, R., Bonen, L., Lewis, B. J., Stahl, D. A., Luehrson, K. R., Chen, K. N., & Woese, C. R. (1980) *Science 209*, 457–463.

Freeman, J., & Blankenship, R. E. (1990) *Photosynthesis Res.* **23**, 29–38.

Fritzsch, G., Buchanan, S., & Michel, H. (1989) *Biochim. Biophys. Acta* **977**, 157–162.

Fukushima, A, Matsuura, K. Shimada, K., & Satoh, T. (1988) *Biochim. Biophys. Acta* **933**, 399–405.

Gabellini, N., & Sebald, W. (1986) *Eur. J. Biochem.* **154**, 569–579.

Gadsby, P. M. A., Hartshorn, R. T., Moura, J. J. G., Sinclair-Day, J. D., Sykes, A. G., & Thomson, A. J. (1989) *Biochim. Biophys. Acta* **994**, 37–46.

Gaul, D. F., Gray, G. O., & Knaff, D. B. (1983) *Biochim. Biophys. Acta* **723**, 333–339.

Goodhew, C. F., Wilson, I. B. H., Hunter, D. J. B., & Pettigrew, G. W. (1990) *Biochem. J.* **271**, 707–712.

Goosen, N., Horsman, H. P. A., Huinen, R. G. M., & Van de Putte, P. (1989) *J. Bacteriol.* **171**, 447–455.

Gray, G. O., Gaul, D. F., & Knaff, D. B. (1983) *Arch. Biochem. Biophys.* **222**, 78–86.

Gray, J. C. (1978) *Eur. J. Biochem.* **82**, 133–141.

Gray, K. A., Davidson, E., & Daldal, F. (1992) *Biochemistry* **31**, 11864–11873.

Green, G. N., Fang, H., Lin, R. J., Newton, G., Mather, M., Georgiou, C. D., & Gennis, R. B. (1988) *J. Biol. Chem.* **263**, 13138–13143.

Grisshammer, R., Wiessner, C., and Michel, H. (1990) *J. Bacteriol.* **172**, 5071–5078.

Groen, B. W., & Duine, J. A. (1990) *Methods in Enzymology* **188**, 33–39.

Groen, B. W., Van Kleef, J. A. G., & Duine, J. A. (1986) *Biochem. J.* **234**, 611–615.

Hampsey, D. M., Das, G., & Sherman, F. (1986) *J. Biol. Chem.* **261**, 3259–3271.

Hanlon, S. P., Holt, R. A., & McEwan, A. G. (1992) *FEMS Microbiol. Letts.* **97**, 283–288.

Harms, N., DeVries, G. E., Maurer, K., Hoogendijk, J., & Stouthamer, A. H. (1987) *J. Bacteriol.* **169**, 3969–3975.

Hatchikian, E. C., Papavassiliou, P., Bianco, P., & Haladjian, J. (1984) *J. Bacteriol.* **159**, 1040–1046.

Hauska, G., Hurt, E., Gabellini, N., & Lockau, W. (1983) *Biochim. Biophys. Acta* **726**, 97–133.

Hauska, G., Nitschke, W., & Herrmann, R. G. (1988) *J. Bioenerg. Biomemb.* **20**, 211–228.

Hennig, B. (1975) *Eur. J. Biochem.* **55**, 167–183.

Higuchi, Y., Kusunoki, M., Yasuoka, N., Kakudo, M., & Yagi, T. (1981) *J. Biochem.* **90**, 1715–1753.

Higuchi, Y., Kusunoki, M., Matsuura, Y., Yasuoka, N., & Kakudo, M. (1984) *J. Mol. Biol.* **172**, 109–139.

Higuchi, Y., Inaka, K., Yasuoka, N., & Yagi, T. (1987) *Biochim. Biophys. Acta* **911**, 341–348.

Holton, R. W., & Myers, J. (1967) *Biochim. Biophys. Acta* **131**, 362–374.

Hopper, D. J., & Taylor, D. G. (1977) *Biochem. J.* **167**, 155–162.

Hori, K. (1961) *J. Biochem.* **50**, 440–449.

Horimoto, K., Suzuki, H., & Otsuka, J. (1991) *Prot. Seq. Data Anal.* **4**, 33–42.

Hosler, J. P., Fetter, J., Tecklenburg, M. M. J., Espe, M., Lerma, C., & Ferguson-Miller, S. (1992) *J. Biol. Chem.* **267**, 24264–24272.

Hreggvidsson, G. O. (1991) *Biochim. Biophys. Acta* **1058**, 52–55.

Hudig, H., & Drews, G. (1983) *FEBS Letts.* **152**, 251–255.

Hunter, D. J. B., Brown, K. R., & Pettigrew, G. W. (1989) *Biochem. J.* **262**, 233–240.

Husain, N., & Davidson, V. L. (1986) *J. Biol. Chem.* **261**, 8577–8580.

Iba, K., Takamiya, K., & Sato, K. (1985) *Plant Cell Physiol.* **26**, 117–122.

Inoue, S., Hiroyoshi, T., Matsubara, H., & Yamanaka, T. (1984) *Biochim. Biophys. Acta* **790**, 188–195.

Inoue, S., Matsubara, H., & Yamanaka, T. (1985) *J. Biochem.* **97**, 947–954.

Inoue, S., Inoue, H., Hiroyoshi, T., Matsubara, H., & Yamanaka, T. (1986) *J. Biochem.* **100**, 955–965.

Inoue, T., Sunagawa, M., Mori, KA., Imai, C., Fukuda, M., Takagi, M., & Yano, K. (1989) *J. Bacteriol.* **171**, 3115–3122.

Ishizuka, M., Machida, K., Shimada, S., Mogi, A., Tsuchiya, T., Ohmori, T., Souma, Y., Gonda, M., & Sone, N. (1990) *J. Biochem.* **108**, 866–873.

Iwasaki, H., & Matsubara, T. (1971) *J. Biochem.* **69**, 847–857.

Jenney, F. E., Jr., & Daldal, F. (1993) *EMBO J.* **12**, 1283–1292.

Johnson, M. S., Sutcliffe, M. J., & Blundell, T. L. (1990) *J. Mol. Evol.* **30**, 43–59.

Jungst, A., Wakabayashi, S., Matsubara, H., & Zumft, W. G. (1991) *FEBS Letts.* **279**, 205–209.

Kadir, F. H. A., & Moore, G. R. (1990a) *FEBS Letts.* **271**, 141–143.

Kadir, F. H. A., & Moore, G. R. (1990b) *FEBS Letts.* **276**, 81–84.

Kagimoto, K., McCarthy, J. L., Waterman, M. R., & Kagimoto, M. (1988) *Biochem. Biophys. Res. Comm.* **155**, 379–383.

Kakutani, T., Watanabe, H., Arima, K., & Beppu, T. (1981) *J. Biochem.* **89**, 463–472.

Kallas, T., Spiller, S., & Malkin, R. (1988) *Proc. Natl. Acad. Sci. USA* **85**, 5794–5798.

Kemmerer, E. C., Lei, M., & Wu, R. (1991) *J. Mol. Evol.* **32**, 227–237.

Kennel, S. J., & Kamen, M. D. (1971) *Biochim. Biophys. Acta* **253**, 153–156.

Kim, C. H., Balny, C., & King, T. E. (1987) *J. Biol. Chem.* **262**, 8103–8108.

Kim, I. C., & Nolla, H. (1986) *Biochem. Cell Biol.* **64**, 1211–1217.

Kissinger, C. R. (1989) Ph.D. Thesis, University of Washington.

Kitada, M., & Krulwich, A. T. (1984) *J. Bacteriol.* **158**, 963–966.

Knaff, D. B. (1993) *Photosynthesis Res.* **35**, 117–133.

Knaff, D. B., Willie, A., Long, J. E., Kriauciunas, A., Durham, B., & Millett, F. (1991) *Biochemistry* **30**, 1303–1310.

Koga, H., Yamaguchi, E, Matsunaga, K., Aramaki, H., & Horiuchi, T. (1989) *J. Biochem.* **106**, 831–836.

Konig, B. W., Osheroff, N., Wilms, J., Muijsers, A. O., Dekker, H. L., & Margoliash, E. (1980) *FEBS Letts.* **111**, 395–398.

Korner, H., Frunzke, K., Dohler, K., & Zumft, W. G. (1987) *Arch. Microbiol.* **148**, 20–24.

Korszun, Z. R., & Salemme, F. R. (1977) *Proc. Natl. Acad. Sci. USA* **74**, 5244–5247.

Kortluke, C., Horstmann, K., Schwartz, E., Rohde, M., Binsack, R., & Friedrich, B. (1992) *J. Bacteriol.* **174**, 6277–6289.

Krawiec, S., & Riley, M. (1990) *Micro. Rev.* **54**, 502–539.

Krogmann, D. W. (1991) *Biochim. Biophys. Acta* **1058**, 35–37.

Kurokawa, T., Fukumori, Y., & Yamanaka, T. (1989) *Biochim. Biophys. Acta* **976**, 135–139.

Kurowski, B., & Ludwig, B. (1987) *J. Biol. Chem.* **262**, 13805–13811.

Kusche, W. H., & Truper, H. G. (1984) *Z. Naturforsch.* **39c**, 894–901.

Lam, Y., & Nicholas, D. J. D. (1969) *Biochim. Biophys. Acta* **180**, 459–472.

Lederer, F., Simon, A. M., & Verdiere, J. (1972) *Biochem. Biophys. Res. Comm.* **47**, 55–58.

Lefebvre, S., Picorel, R., Cloutier, Y., & Gingras, G. (1984) *Biochemistry* **23**, 5279–5288.

LeGall, J., Payne, W. J., Morgan, T. V., & DerVartanian, D. (1979) *Biochem. Biophys. Res. Commun.* **87**, 355–362.

Leitch, F. A., Brown, K. R., & Pettigrew, G. W. (1985) *Biochim. Biophys. Acta* **808**, 213–218.

Lenhoff, H. M., & Kaplan, N. O. (1956) *J. Biol. Chem.* **220**, 967–982.

Liebetanz, R., Hornberger, U., & Drews, G. (1991) *Mol. Microbiol.* **5**, 1459–1468.

Limbach, K. J., & Wu, R. (1985a) *Nucleic Acids Res.* **13**, 617–630.

Limbach, K. J., & Wu, R. (1985b) *Nucleic Acids Res.* **13**, 631–644.

Lipman, D. J., Altschul, S. P., & Kececioglu, J. D. (1989) *Proc. Natl. Acad. Sci. USA* **86**, 4412–4415.

Liu, M. C., Peck, H. D., Jr., Payne, W. J., Anderson, J. L., Dervartanian, D. V., & LeGall, J. (1981) *FEBS Letts.* **129**, 155–160.

Liu, M. C., Costa, C. Coutinho, I. B., Moura, J. J. G., Moura, I., Xavier, A. V., & LeGall, J. (1988) *J. Bacteriol.* **170**, 5545–5551.

Liu, M. Y., Liu, M. C., Payne, W., & LeGall, J. (1986) *J. Bacteriol.* **166**, 604–608.

Louie, G. V., & Brayer, G. D. (1990) *J. Mol. Biol.* **214**, 527–555.

Louie, G. V., Hutcheon, W. L. B., & Brayer, G. D. (1988) *J. Mol. Biol.* **199**, 295–314.

Loutfi, M., Guerlesquin, F., Bianco, P., Haladjian, J., & Bruschi, M. (1989) *Biochem. Biophys. Res. Comm.* **159**, 670–676.

Ludwig, B., Suda, K., & Cerletti, N. (1983) *Eur. J. Biochem.* **137**, 597–602.

Machlin, S. M., & Hanson, R. S. (1988) *J. Bacteriol.* **170**, 4739–4747.

Majewski, C., & Trebst, A. (1990) *Mol. Gen. Genet.* **224**, 373–382.

Makinen, M. W., Schichman, S. A., Hill, S. C., & Gray, H. B. (1983) *Science* **222**, 929–931.

Mancinelli, R. L., Cronin, S., & Hochstein, L. I. (1986) *Arch. Microbiol.* **145**, 202–208.

Margoliash, E., & Fitch, W. M. (1968) *Ann. N.Y. Acad. Sci.* **151**, 359–381.

Margoliash, E., & Smith, E. L. (1965) in *Evolving Genes and Proteins* (V. Bryson & H. J. Vogel, eds.), Academic Press, New York, pp. 221–242.

Martens, B., Spiegl, H., & Stackebrandt, E. (1987) *System. Appl. Microbiol.* **9**, 224–230.

Masuko, M., Iwasaki, H., Sahurai, T., Suzuki, S., & Nakahara, M. (1984) *J. Biochem.* **96**, 447–454.

Mather, M. W., Springer, P., & Fee, J. A. (1991) *J. Biol. Chem.* **266**, 5025–5035.

Mathews, F. S., Bethge, P. H., & Czerwinski, E. W. (1979) *J. Biol. Chem.* **254**, 1699–1706.

Mathews, F. S., Chen, Z., Bellamy, H. D., & McIntire, W. S. (1991) *Biochemistry 30*, 238–247.

Matsumoto, T., Matsuo, M., & Matsuda, Y. (1991) *Plant Cell Phys. 32*, 863–872.

Matsuura, K., Fukushima, A., Shimada, K., & Satoh, T. (1988) *FEBS Letts. 237*, 21–25.

Matsuura, Y., Takano, T., & Dickerson, R. E. (1982) *J. Mol. Biol. 156*, 389–404.

Mayes, S. R., & Barber, J. (1991) *Plant Mol. Biol. 17*, 289–293.

McEwan, A. G. (1989) Symposium on Molecular Biology of Membrane-Bound Complexes in Phototrophic Bacteria (Cogdell, R., Drews, G., van Grondelle, R., Hunter, N., Kaplan, S., Melandri, B. A., Kondratieva, E. N., Niederman, R. A., Vignais, P., and Zuber, H., eds.)

McIntire, W. S., Singer, T. P., Smith, A. J., & Mathews, F. S. (1986) *Biochem. 25*, 5975–5981.

McIntire, W. S., Wemmer, D. E., Chistoserdov, A., & Lidstrom, M. E. (1991) *Science 252*, 817–824.

McManus, J. D., Brune, D. C., Han, J., Sanders-Loehr, J., Meyer, T. E., Cusanovich, M. A., Tollin, G., & Blankenship, R. E. (1992) *J. Biol. Chem. 267*, 6531–6540.

Merchant, S., & Bogorad, L. (1987) *J. Biol. Chem. 262*, 9062–9067.

Meulenberg, J. J. M., Sellink, E., Riegman, N. H., & Postma, P. W. (1992) *Mol. Gen. Genet. 232*, 284–294.

Meyer, T. E. (1985) *Biochim. Biophys. Acta 806*, 175–183.

Meyer, T. E., & Cusanovich, M. A. (1985) *Biochim. Biophys. Acta 807*, 308–319.

Meyer, T. E., & Kamen, M. D. (1982) *Adv. Prot. Chem. 35*, 105–212.

Meyer, T. E., Bartsch, R. G., Cusanovich, M. A., & Mathewson, J. H. (1968) *Biochim. Biophys. Acta 153*, 854–861.

Meyer, T. E., Kennel, S. J., Tedro, S. M., & Kamen, M. D. (1973) *Biochim. Biophys. Acta 292*, 634–643.

Meyer, T. E., Watkins, J. A., Przysiecki, C. T., Tollin, G., & Cusanovich, M. A. (1984) *Biochem. 23*, 4761–4767.

Meyer, T. E., Cusanovich, M. A., & Kamen, M. D. (1986a) *Proc. Natl. Acad. Sci. USA 83*, 217–220.

Meyer, T. E., Cheddar, G., Bartsch, R. G., Getzoff, E. D., Cusanovich, M. A., & Tollin, G. (1986b) *Biochemistry 25*, 1383–1390.

Meyer, T. E., Cusanovich, M. A., Krogmann, D. W., Bartsch, R. G., & Tollin, G. (1987) *Arch. Biochem. Biophys. 258*, 307–314.

Meyer, T. E., Tollin, G., Cusanovich, M. A., Freeman, J. C., & Blankenship, R. E. (1989) *Arch. Biochem. Biophys. 272*, 254–261.

Meyer, T. E., Fitch, J. C., Bartsch, R. G., Tollin, G., & Cusanovich, M. A. (1990a) *Biochim. Biophys. Acta 1016*, 364–370.

Meyer, T. E., Fitch, J., Bartsch, R. G., Tollin, D., & Cusanovich, M. A. (1990b) *Biochim. Biophys. Acta 1017*, 118–124.

Meyer, T. E., Cannac, V., Fitch, J., Bartsch, R. G., Tollin, D., Tollin, G., & Cusanovich, M. A. (1990c) *Biochim. Biophys. Acta 1017*, 125–138.

Meyer, T. E., Bartsch, R. G., Cusanovich, M. A., & Tollin, G. (1993) *Biochemistry 32*, 4719–4726.

Miller, D. J., & Nicholas, J. D. (1986) *Biochem. Int. 12*, 167–172.

Mills, G. (1991) *J. Theor. Biol. 152*, 177–190.

Mittal, S., Zhu, Y. Z., & Vickery, L. E. (1988) *Arch. Biochem. Biophys. 264*, 383–391.

Montgomery, D. L., Leung, D. W., Smith, M., Shalit, P., Faye, G., & Hall, B. D. (1980) *Proc. Natl. Acad. Sci. USA 77*, 541–545.

Moore, G. R., Cheesman, M. R., Kadir, F. H. A., Thomson, A. J., Yewdall, S. J., & Harrison, P. M. (1992) *Biochem. J. 287*, 457–460.

Moura, I., Fauque, G., LeGall, J., Xavier, A. V., & Moura, J. J. G. (1987) *Eur. J. Biochem. 162*, 547–554.

Moura, I., Liu, M. Y., Costa, C., Liu, M. C., Pai, G., Xavier, A. V., LeGall, J., Payne, W. J., & Moura, J. J. G. (1988) *Eur. J. Biochem. 177*, 673–682.

Mukai, K., Wakabayashi, S., & Matsubara, H. (1989) *J. Biochem. 106*, 479–482.

Murata, M., Richardson, J. S., & Sussman, J. L. (1985) *Proc. Natl. Acad. Sci. USA 82*, 3073–3077.

Murphy, M. E. P., Nall, B. T., & Brayer, G. D. (1992) *J. Mol. Biol. 227*, 160–176.

Nakagawa, A., Higuchi, Y., Yasuoka, N., Katsube, Y., & Yagi, T. (1990) *J. Biochem. 108*, 701–703.

Nakano, K., Kikumoto, Y., & Yagi, T. (1983) *J. Biol. Chem. 258*, 12409–12412.

Newton, N. (1969) *Biochim. Biophys. Acta 185*, 316–331.

Nishikimi, M., Suzuki, H., Ohta, S., Sakurai, T., Shimomura, Y., Tanaka, M., Kagawa, Y., & Ozawa, T. (1987) *Biochem. Biophys. Res. Comm. 145*, 34–39.

Nishimura, Y., Musaka, S., Iizuka, H., & Shimada, K. (1989) *Arch. Microbiol.* *152*, 1–5.

Nitschke, W., Agalidis, I., & Rutherford, A. W. (1992) *Biochim. Biophys. Acta* *1100*, 49–57.

Nordling, M., Young, S., Karlsson, B. G., & Lundberg, L. G. (1990) *FEBS Letts.* *259*, 230–232.

Nozawa, T., Trost, J. T., Fukada, T., Hatano, M., McManus, J. D., & Blankenship, R. E. (1987) *Biochim. Biophys. Acta* *894*, 468–476.

Nugent, J. H. A., Gratzer, W., & Segal, A. W. (1989) *Biochem. J.* *264*, 921–924.

Nunn, D. N., & Anthony, C. (1988) *Biochem. J.* *256*, 673–676.

Nunn, D. N., Day, D., & Anthony, C. (1989) *Biochem. J.* *260*, 857–862.

Ochi, H., Hata, Y., Tanaka, N., Kakudo, M., Sakurai, T., Aihara, S., & Morita, Y. (1983) *J. Mol. Biol.* *166*, 407–418.

Okamoto, Y., Minami, Y., Matsubara, H., & Sugimura, Y. (1987) *J. Biochem.* *102*, 1251–1260.

Okamura, K., Miyata, T., Iwanaga, S., Takamiya, K., & Nishimura, M. (1987) *J. Biochem.* *101*, 957–966.

O'Keeffe, D. T., & Anthony, C. (1980) *Biochem. J.* *192*, 411–419.

Okkels, J. S., Kjaer, B., Hansson, O., Svendsen, I., Moller, B. L., & Scheller, H. V. (1992) *J. Biol. Chem.* *267*, 21139–21145.

Olson, J. M., & Shaw, E. K. (1969) *Photosynthetica* *3*, 288–290.

Ovchinnikov, Y. A., Abdulaev, N. G., Shmuckler, B. E., Zargarov, A. A., Kutuzov, M. A., Telezhinskaya, I. N., Levina, N. B., & Zolotarev, A. S. (1988a) *FEBS Letts.* *232*, 364–368.

Ovchinnikov, Y. A., Abdulaev, N. G., Zolotarev, A. S., Shmuckler, B. E., Zargarov, A. A., Kutuzov, M. A., Telezhinskaya, I. N., & Levina, N. B. (1988b) *FEBS Letts.* *231*, 237–242.

Overfield, R. E., Wraight, C. A., & DeVault, D. (1979) *FEBS Letts.* *105*, 137–142.

Palmer, G., & Reedijk, J. (1991) *Eur. J. Biochem.* *200*, 599–611.

Pealing, S. L., Black, A. C., Manson, F. D. C., Ward, F. B., Chapman, S. K., & Reid, G. A. (1992) *Biochemistry* *31*, 12132–12140.

Pettigrew, G. W. (1972) *FEBS Letts.* *22*, 64–66.

Pettigrew, G. W. (1973) *Nature* *241*, 531–533.

Pettigrew, G. W. (1974) *Biochem. J.* *139*, 449–459.

Pettigrew, G. W. (1978) *FEBS Letts.* *86*, 14–16.

Pettigrew, G. W., & Brown, K. R. (1988) *Biochem. J.* *252*, 427–435.

Pettigrew, G. W., & Moore, G. R. (1987) Cytochromes *c* Biological Aspects, Springer-Verlag, New York.

Pettigrew, G. W., Leaver, J. L., Meyer, T. E., & Ryle, A. P. (1975) *Biochem. J.* *147*, 291–302.

Pierrot, M., Haser, R., Frey, M., Payan, F., & Astier, J. P. (1982) *J. Biol. Chem.* *257*, 14341–14348.

Pierson, B. K., & Castenholz, R. W. (1974) *Arch. Microbiol.* *100*, 5–24.

Pierson, B. K., & Thornber, J. P. (1983) *Proc. Natl. Acad. Sci. USA* *80*, 80–84.

Pollock, W. B. R., Loutfi, M., Bruschi, M., Rapp-Giles, B. J., Wall, J. D., & Voordouw, G. (1991) *J. Bacteriol.* *173*, 220–228.

Preisig, O., Anthamatten, D., & Hennecke, H. (1993) *Proc. Natl. Acad. Sci. USA* *90*, 3309–3313.

Priest, J. W., & Hajduk, S. L. (1992) *J. Biol. Chem.* *267*, 20188–20195.

Prince, R. C., Davidson, E., Haith, C. E., & Daldal, F. (1986) *Biochemistry* *25*, 5208–5214.

Probst, I., Bruschi, M., Pfennig, N., & LeGall, J. (1977) *Biochem. Biophys. Acta* *460*, 58–64.

Puustinen, A., & Wikstrom, M. (1991) *Proc. Natl. Acad. Sci. USA* *88*, 6122–6126.

Qin, L., & Kostic, N. M. (1992) *Biochemistry* *31*, 5145–5150.

Qureshi, M. H., Yumoto, I., Fujiwara, T., Fukumori, Y., & Yamanaka, T. (1990) *J. Biochem.* *107*, 480–485.

Ragland, M., Briat, J. F., Gagnon, J., Laulhere, J. P., Massenet, O., & Teil, E. C. (1990) *J. Biol. Chem.* *265*, 18339–18344.

Raitio, M., Jalli, T., & Saraste, M. (1987) *EMBO J.* *6*, 2825–2833.

Raitio, M., Pispa, J.M., Metso, T., & Saraste, M. (1990) *FEBS Letts.* *261*, 431–435.

Ramseier, T. M., Winteler, H. V., & Hennecke, H. (1991) *J. Biol. Chem.* *266*, 7793–7803.

Ras, J., Reijnders, W. N. M., Van Spanning, R. J. M., Harms, N., Oltmann, L. F., & Stouthamer, A. H. (1991) *J. Bacteriol.* *173*, 6971–6979.

Reeve, C. D., Carver, M. A., & Hopper, D. J. (1989) *Biochem. J.* *263*, 431–437.

Rickle, G. K., & Cusanovich, M. A. (1979) *Arch. . Biochem. Biophys.* *197*, 589–598.

Rigby, S. E. J., Moore, G. R., Gray, J. C., Gadsby, P. M. A., George, S. J., & Thomson, A. J. (1988) *Biochem. J.* *256*, 571–577.

Romisch, J., Tropschug, M., Sebald, W., & Weiss, H. (1987) *Eur. J. Biochem.* *164*, 111–115.

Ronnberg, M. (1987) *Biochim. Biophys. Acta* *916*, 112–118.

Ronnberg, M., Kalkkinen, N., & Ellfolk, N. (1989) *FEBS Letts.* *250*, 175–178.

Rosen, D., Okamura, M. Y., Abresch, E. C., Valkirs, G. E., & Feher, G. (1983) *Biochemistry* *22*, 335–341.

Rossbach, S., Loferer, H., Acuna, G., Appleby, C. A., & Hennecke, H. (1991) *FEMS Microbiol. Letts.* *83*, 145–152.

Rott and Donohue (1990) *J. Bacteriol.* *172*, 1954–1961.

Rott, M. A., Fitch, J., Meyer, T. E., & Donohue, T. J. (1992) *Arch. Biochem. Biophys.* *292*, 576–582.

Rott, M. A., Witthuhn, V. C., Schilke, B. A., Soranno, M., Ali, A., & Donohue, T. J. (1993) *J. Bacteriol.* *175*, 358–366.

Russell, P. R., & Hall, B. D. (1982) *Mol. Cell. Biol.* *2*, 106–116.

Sadler, I., Suda, K., Schatz, G., Kaudewitz, F., & Haid, A. (1984) *EMBO J.* *3*, 2137–2143.

Samain, E., Albagnac, G., & LeGall, J. (1986) *FEBS Letts.* *204*, 247–250.

Sanbongi, Y., Ishii, M., Igarashi, Y., & Kodama, T. (1989) *J. Bacteriol.* *171*, 65–69.

Sanbongi, Y., Yang, J. H., Igarashi, Y., & Kodama, T. (1991) *Eur. J. Biochem.* *198*, 7–12.

Sandmann, G. (1986) *Arch. Microbiol.* *145*, 76–79.

Santos, H., & Turner, D. L. (1988) *Biochim. Biophys. Acta* *954*, 277–286.

Saraste, M., Metso, T., Nakari, T., Jalli, T., Lauraeus, M., & Van der Oost, J. (1991) *Eur. J. Biochem.* *195*, 517–525.

Sawada, E., Satoh, T., & Kitamura, H. (1978) *Plant and Cell Physiol.* *19*, 1339–1351.

Sawyer, L., Jones, C. L., Damas, A. M., Harding, M. M., Gould, R. O., & Ambler, R. P. (1981) *J. Mol. Biol.* *153*, 831–835.

Scarpulla, R. C. (1984) *Mol. Cell. Biol.* *4*, 2279–2288.

Schirmer, T., Bode, W., Huber, R., Sidler, W., & Zuber, H. (1985) *J. Mol. Biol.* *184*, 257–277.

Seftor, R. E. B., & Thornber, J. P. (1984) *Biochim. Biophys. Acta* *764*, 148–159.

Self, S. J., Hunter, C. N., & Leatherbarrow, B. J. (1990) *Biochem. J.* *265*, 599–604.

Serrano, A., Gimenez, P., Scherer, S., & Boger, P. (1990) *Arch. Microbiol.* *154*, 614–618.

Shapleigh, J. P., & Gennis, R. B. (1992) *Mol. Microbiol.* *6*, 635–642.

Shapleigh, J. P., Hill, J. J., Alben, J. O., & Gennis, R. B. (1992) *J. Bacteriol.* *174*, 2338–2343.

Shill, D. A., & Wood, P. M. (1984) *Biochim. Biophys. Acta* *764*, 1–7.

Shinkai, W., Hase, T., Yagi, T., & Matsubara, H. (1980) *J. Biochem.* *87*, 1747–1756.

Shioi, Y., Takamiya, K., & Nishimura, M. (1972) *J. Biochem.* *71*, 285–294.

Shiozawa, J. A., Lottspeich, F., & Feick, R. (1987) *Eur. J. Biochem.* *167*, 595–600.

Shopes, R. J., Levine, L. M. A., Holten, D., & Wraight, C. A. (1987) *Photosyn. Res.* *12*, 165–180.

Silvestrini, M. C., Galeotti, C. L., Gervais, M., Schinina, E., Barra, D., Bossa, F., & Brunori, M. (1989) *FEBS Letts.* *254*, 33–38.

Silvestrini, M. C., Tordi, M. G., Musci, G., & Brunori, M. (1990) *J. Biol. Chem.* *265*, 11783–11787.

Simpkin, D., Palmer, G., Devlin, F. J., McKenna, M. C., Jensen, G. M., & Stephens, P. J. (1989) *Biochemistry* *28*, 8033–8039.

Smith, G. B., & Tiedje, J. M. (1992) *Appl. Environ. Microbiol.* *58*, 376–384.

Smith, M., Leung, D. W., Gillam, S., & Astell, C. R. (1979) *Cell* *16*, 753–761.

Sneath, P. H. A. (1989) *System. Appl. Microbiol.* *12*, 15–31.

Sodergren, E. J., & DeMoss, J. A. (1988) *J. Bacteriol.* *170*, 1721–1729.

Sone, N., & Yanagita, Y. (1982) *Biochim. Biophys. Acta* *682*, 216–226.

Sone, N., Sekimachi, M., & Kutoh, E. (1987) *J. Biol. Chem.* *262*, 15386–15391.

Sone, N., Yokoi, F., Fu, T., Ohta, S., Metso, T., Raitio, M., & Saraste, M. (1988) *J. Biochem.* *103*, 606–610.

Speck, S. H., Ferguson-Miller, S., Osheroff, N., & Margoliash, E. (1979) *Proc. Natl. Acad. Sci. USA* *76*, 155–159.

Sprinkle, J. R., Hermodson, M., & Krogmann, D. W. (1986) *Photosyn. Res.* *10*, 63–73.

Staudenmayer, N., Ng, S., Smith, M. B., & Millett, F. (1977) *Biochemistry* *16*, 600–604.

Steinmetz, M. A., & Fischer, U. (1981) *Arch. Micribiol.* *130*, 31–37.

Steinmetz, M. A., & Fischer, U. (1982a) *Arch. Microbiol.* *131*, 19–26.

Steinmetz, M. A., & Fischer, U. (1982b) *Arch. Microbiol.* *132*, 204–210.

Steinmetz, M. A., Truper, H. G., & Fischer, U. (1983) *Arch. Microbiol.* *135*, 186–190.

Swank, R. T., & Burris, R. H. (1969) *Biochim. Biophys. Acta* *180*, 473–489.

Takamiya, K. (1985) *Arch. Microbiol.* *143*, 15–19.

Takamiya, K., Iba, K., & Okamura, K. (1987) *Biochim. Biophys. Acta* *890*, 127–133.

Takano, T., & Dickerson, R. E. (1981) *J. Mol. Biol.* *153*, 79–115.

Takano, T., Dickerson, R. E., Schichman, S. A., & Meyer, T. E. (1979) *J. Mol. Biol.* **133**, 185–188.

Tamaki, T., Fukaya, M., Takemura, H., Tayama, K., Okamura, H., Kawamura, Y., Nishiyama, M., Horinouchi, S., & Beppu, T. (1991) *Biochim. Biophys. Acta* **1088**, 292–300.

Tanaka, M., Haniu, M., Yasunobu, K. T., & Kimura, T. (1973) *J. Biol. Chem.* **248**, 1141–1157.

Tanaka, M., Haniu, M., Yasunobu, K. T., Dus, K., & Gunsalus, I. C. (1974) *J. Biol. Chem.* **249**, 3689–3701.

Tanaka, Y., Fukumori, Y., & Yamanaka, T. (1982) *Biochim. Biophys. Acta* **707**, 14–20.

Tarr, G. E., & Fitch, W. M. (1976) *Biochem. J.* **159**, 193–199.

Tayama, K., Fukuya, M., Okumura, H., Kawamura, Y., & Beppu, T. (1989) *App. Microbiol. Biotech.* **32**, 181–185.

Tedro, S. M., Meyer, T. E., & Kamen, M. D. (1977) *J. Biol. Chem.* **252**, 7826–7833.

Theil, E. C. (1987) *Ann. Rev. Biochem.* **56**, 289–315.

Then, J. & Truper, H. G. (1983) *Arch. Microbiol.* **135**, 254–258.

Then, J. & Truper, H. G. (1984) *Arch. Microbiol.* **139**, 295–298.

Thony-Meyer, L., Stax, D., & Hennecke, H. (1989) *Cell* **57**, 683–697.

Thony-Meyer, L., James, P., & Hennecke, H. (1991) *Proc. Natl. Acad. Sci. USA* **88**, 5001–5005.

Timkovich, R., & Dickerson, R. E. (1976) *J. Biol. Chem.* **251**, 4033–4046.

Timkovich, R., Dhesi, R., Martinkus, K. J., Robinson, M. K., & Rea, T. M. (1982) *Arch. Biochem. Biophys.* **215**, 47–58.

Tissieres, A. (1956) *Biochem. J.* **64**, 582–589.

Titani, K., Ericsson, L. H., Hon-Nami, K., & Miyazawa, T. (1985a) *Biochem. Biophys. Res. Comm.* **128**, 781–787.

Titani, K., Ericsson, L. H., Hon-Nami, K., & Miyazawa, T. (1985b) *Biochem. Biophys. Res. Comm.* **128**, 808–813.

Tomiyama, Y., Doi, M., Takamiya, K., & Nishimura, M. (1983) *Plant Cell Physiol.* **24**, 11–16.

Trost, J. T., McManus, J. D., Freeman, J. C., Ramakrishna, B. L., & Blankenship, R. E. (1988) *Biochemistry* **27**, 7858–7863.

Trumpower, B. L. (1990) *Microbiol. Rev.* **54**, 101–129.

Truper, H. G., & Rogers, L. H. (1971) *J. Bacteriol.* **108**, 1112–1121.

Tully, R. E., Sadowsky, M. J., & Keister, D. L. (1991) *J. Bacteriol.* **173**, 7887–7895.

Van Beeumen, J. (1991) *Biochim. Biophys. Acta* **1058**, 56–60.

Van Beeumen, J., Ambler, R. P., Meyer, T. E., Kamen, M. D., Olson, J. M., & Shaw, E. K. (1976) *Biochem. J.* **159**, 757–774.

Van Beeumen, J., Tempst, P., Stevens, P., Bral, D., Van Damme, J., & De Ley J. (1980) *Protides of Biol. Fluids* **28**, 69–74.

Van Beeumen, J., Van Bun, S., Meyer, T. E., Bartsch, R. G., & Cusanovich, M. A. (1990) *J. Biol. Chem.* **265**, 9793–9799.

Van Beeumen, J. J., DeMol, H., Samyn, B., Bartsch, R. G., Meyer, T. E., Dolata, M. M., & Cusanovich, M. A. (1991) *J. Biol. Chem.* **266**, 12921–12931.

Vanfleteren, J. R. Evers, E. A. I. M., Van de Werken, G., & Van Beeumen, J. J. (1990) *Biochem. J.* **271**, 613–620.

Van Grondelle, R., Duysens, L. N. M., Van der Wel, J. A., Van der Wal, H. N. (1977) *Biochim. Biophys. Acta* **461**, 188–201.

Van Loon, A. P. G. M., De Groot, R. J., De Haan, M., Dekker, A., & Grivell, L. A. (1984) *EMBO J.* **3**, 1039–1043.

Van Rooijen, G. J. H., Bruschi, M., & Voordouw, G. (1989) *J. Bacteriol.* **171**, 3575–3578.

Van Spanning, R. J. M., Wansell, C., Harms, N., Oltmann, L. F., & Stouthamer, A. H. (1990) *J. Bacteriol.* **172**, 986–996.

Van Spanning, R. J. M., Wansell, C. W., DeBoer, T., Hazelaar, M. J., Anazawa, H., Harms, N., Oltmann, L. F., & Stouthamer, A. H. (1991) *J. Bacteriol.* **173**, 6948–6961.

Verbist, J., Lang, F., Gabellini, N., & Oesterhelt, D. (1989) *Mol. Gen. Genet.* **219**, 445–452.

Vermeglio, A., Richaud, P., & Breton, J. (1989) *FEBS Letts.* **243**, 259–263.

Viebrock, A., & Zumft, W. G. (1988) *J. Bacteriol.* **170**, 4658–4668.

Villalain, J., Moura, I., Liu, M. C., Payne, W. J., LeGall, J., Xavier, A. V., & Moura, J. J. G. (1984) *Eur. J. Biochem.* **141**, 305–312.

Virbasius, J. V., & Scarpulla, R. C. (1988) *J. Biol. Chem.* **263**, 6791–6796.

Von Wachenfeldt, C., & Hederstedt, L. (1990a) *J. Biol. Chem.* **265**, 13939–13948.

Von Wachenfeldt, C., & Hederstedt, L. (1990b) *FEBS Letts.* **270**, 147–151.

Voordouw, G., & Brenner, S. (1986) *Eur. J. Biochem.* **159**, 347–351.

Wakabayashi, S., Matsubara, H., Kim, C. H., Kawai, K., & King, T. E. (1980) *Biochem. Biophys. Res. Comm.* **97**, 1548–1554.

Wakabayashi, S., Takeda, H., Matsubara, H., Kim, C. H., & King, T. E. (1982) *J. Biochem.* **91**, 2077–2085.

Weber, P. C., & Salemme F. R. (1980) *Nature* **287**, 82–84.

Weber, P. C., Salemme, F. R., Mathews, F. S., & Bethge, P. H. (1981) *J. Biol. Chem.* **256**, 7702–7704.

Weyer, K. A., Lottspeich, F., Gruenberg, H., Lang, F., Oesterhelt, D., & Michel, H. (1987) *EMBO J.* **6**, 2197–2202.

Widger, W. R. (1991) *Photosyn. Res.* **30**, 71–84.

Willey, D. L., Howe, C. J., Auffret, A. D., Bowman, C. M., Dyer, T. A., & Gray J. C. (1984) *Mol. Gen. Genet.* **194**, 416–422.

Woese, C. R. (1987) *Microbiol. Revs.* **51**, 221–271.

Woese, C. R., Gibson, J., & Fox, G. E. (1980) *Nature* **283**, 212–214.

Woolley, K. J. (1987) *Eur. J. Biochem.* **166**, 131–137.

Woolley, K. J., & Athalye, M. (1986) *Biochem. Biophys. Res. Comm.* **140**, 808–813.

Yagi, T. (1979) *Biochim. Biophys. Acta* **548**, 96–105.

Yamanaka, T., & Imai, S. (1972) *Biochem. Biophys. Res. Comm.* **46**, 150–154.

Yamanaka, T., & Okunuki, K. (1963) *Biochim. Biophys. Acta* **67**, 379–393, 407–416.

Yamanaka, T., DeKlerk, H., & Kamen, M. D. (1967) *Biochim. Biophys. Acta* **143**, 416–424.

Yamanaka, T., Nagano, T., Shoji, K., & Fukumori, Y. (1991) *Biochim. Biophys. Acta* **1058**, 48–51.

Yasui, M., Harada, S., Kai, Y., Kasai, N., Kusunoki, M., & Matsuura, Y. (1992) *J. Biochem.* **111**, 317–324.

Yoshida, T., Lorence, R. M., Choc, M. G., Tarr, G. E., Findling, K. L., & Fee, J. A. (1984) *J. Biol. Chem.* **259**, 112–123.

Yu, L., Dong, J. H., & Yu, C. A. (1986) *Biochim. Biophys. Acta* **852**, 203–211.

Zorin, N. A., & Gogotov, I. N. (1981) *Biokhimiya* **45**, 1497–1502 (English translation, pp. 1134–1138).

Zumft, W. G., Gotzmann, D. J., & Kroneck, P. M. H. (1987) *Eur. J. Blochem.* **168**, 301–307.

Zumft, W. G., Dohler, K., Korner, H., Lochelt, S., Viebrock, A., & Frunzke, K. (1988) *Arch. Microbiol.* **149**, 492–498.

PART

II

Structural Studies

3

Structural Studies of Eukaryotic Cytochromes c

GARY D. BRAYER
MICHAEL E. P. MURPHY

Department of Biochemistry and Molecular Biology
University of British Columbia

I. Overview

Eukaryotic cytochromes c have long been the subjects of structural analyses using x-ray diffraction techniques. At the current time, a total of five such proteins have been elucidated to sufficiently high resolution to allow for detailed comparative analyses. Much more importantly, these structures have proven central to the process of understanding the mechanism of cytochrome c–mediated electron transfer events. The tertiary structures known are those for yeast iso-1, yeast iso-2, tuna, horse, and rice cytochromes c. A sixth structure involving the protein from bonito has also been solved, but at a lower resolution, and therefore will not be discussed in detail herein. Coordinate sets for all of these proteins either are or will shortly be available through the Protein Data Bank (Bernstein *et al.*, 1977).

One of the first cytochromes c to be subjected to x-ray structural analyses was the oxidized form of horse heart cytochrome c (Dickerson *et al.*, 1967). These studies utilized the multiple isomorphous replacement approach and culminated in a 2.8-Å resolution structure determination (Dickerson *et al.*, 1971). This work was the first to define the general folding characteristics of the mitochondrial cytochromes c. As it turns out, further analysis of the horse protein was abandoned in favor of tuna cytochrome c, for which better-quality crystals could be grown (Takano *et al.*, 1973).

Subsequent work resulted in the determination of high-resolution structures for tuna cytochrome c in both oxidation states (Takano and Dickerson, 1981a,b). Coincident with this work was the elucidation of the structure of cytochrome c from bonito (Tanaka *et al.*, 1975; Matsuura *et al.*, 1979). Both the tuna and bonito structures clearly demonstrated the subtle nature of structural changes occurring between oxidation states and set the stage for more comprehensive mechanistic studies by other investigators. The next cytochrome c structure to be completed was that of the oxidized rice protein (Ochi *et al.*, 1983). This was followed by reports detailing the structure of reduced yeast iso-1-cytochrome c at low (Louie *et al.*, 1988a) and then high (Louie and Brayer, 1990) resolution, as well as the structure of reduced yeast iso-2-cytochrome c (Murphy *et al.*, 1992).

Of all the cytochrome c structures elucidated to date, the reduced form of the yeast iso-1 protein has been determined to the highest resolution (1.23 Å). Crystals of this protein are of superb quality and should allow extension of this structure to even higher resolution. A further advantage of the yeast system is that the iso-1 protein can be resolved in both oxidation states using isomorphous crystalline material. This has significantly enhanced our ability to detect the subtle changes occurring on going from one oxidation state to the other. The structure of oxidized yeast iso-1-cytochrome c has also been completed at high resolution and has led to a series of new proposals concerning the mechanism of electron transfer mediated by cytochromes c (Berghuis and Brayer, 1992).

The yeast system has also opened up cytochrome c to scrutiny by site-directed mutagenesis techniques. There are now more than 35 mutant yeast iso-1-cytochromes c that have undergone structural analyses in our laboratory (Louie *et al*, 1988b; Louie and Brayer, 1989; Murphy *et al.*, 1993, 1995; Berghuis *et al.*, 1994a,b; and Rafferty *et al.*, 1995). These studies are directed at a wide variety of objectives including understanding heme environmental factors, determining the role of internally bound waters, elucidating redox partner complexation interactions, determining the factors involved in polypeptide chain folding, and detailing the structural factors governing the mechanism by which electron transfer takes place.

A crucial element in the successful determination of the wild-type and mutant structures of yeast iso-1-cytochrome c has been the development of a novel crystallization methodology (Leung *et al.*, 1989). Without "hair seeding," progress in elucidating yeast protein structures would be exceedingly slow in contrast to the rather rapid rate at which structures can presently be determined. This method has also proven essential in growing crystals of the iso-2 protein from yeast, which was resistant to crystallization by all other conventional methods attempted. The structure of yeast iso-2-cytochrome c has now been completed, along with that of a composite protein having polypeptide chain loops from both the yeast iso-1 and iso-2 proteins (Murphy *et al.*, 1992). Finally, the original structure of horse heart cytochrome c has been recently redetermined to high resolution (1.9 Å) in our laboratory to resolve some unsettled issues surrounding this protein (Bushnell *et al.*, 1990).

The scope of the present work is limited to a discussion of those five eukaryotic cytochromes c for which high-resolution three-dimensional structures are presently available. This discussion includes a brief summary of primary sequence variance, followed by a more detailed discussion of the available crystallization methods.

Following this, each protein will be comparatively analyzed in terms of its three-dimensional structure, using as a basis of discussion the structure of yeast iso-1-cytochrome *c*, which has been determined to the highest resolution. Finally, the results of structural studies of the tuna and yeast iso-1 proteins in both oxidation states will be detailed, with special emphasis on how these relate to mechanistic proposals to explain the process of electron transfer.

II. Sequence Variance

Cytochrome *c* is a protein ubiquitous to all eukaryotic organisms. Since it is relatively abundant and readily isolated, the sequences of many such proteins have been determined. In the past, this extensive inventory of sequence data has been used to trace the phylogeny of the eukaryotes (Dickerson, 1980a,b). As Figure 3.1 illustrates, comparison of the available sequences shows that eukaryotic cytochromes *c* are highly conserved. Indeed, of the 103 to 115 residues present, there are 21 amino acids which are invariant and an additional 11 which are always one of two amino acids (Hampsey *et al.*, 1988; Louie, 1990). Only 14 amino acid positions are highly variable, having a variability index greater than 15 (Wu and Kabat, 1970).

An alignment of the sequences of those cytochromes *c* for which high-resolution three-dimensional atomic structures are available, is presented in Table 3.1. These include the two cytochromes from yeast (iso-1 and iso-2), as well as those from tuna, horse and rice. The amino acid numbering scheme used herein is taken from Table 3.1, which is based on the sequence of tuna cytochrome *c*. Note that this means the N-terminal extensions of the yeast and rice cytochromes *c* have been assigned negative numbers. The pairwise sequence identity between all five cytochromes *c* varies from 51 to 84%.

III. Crystallization Techniques

The process of characterizing the three-dimensional structures of eukaryotic cytochromes *c* has utilized a variety of crystallization methods. Summarized here are detailed methods used to obtain diffraction quality crystals for the five eukaryotic cytochromes *c* for which high-resolution three-dimensional structures have been successfully elucidated. In addition, a description of the approach taken to crystallize and study specifically mutated variants is also provided.

A. Yeast Iso-1-cytochrome *c*

1. Reduced state

Wild-type yeast (*Saccharomyces cerevisiae*) iso-1-cytochrome *c* was purchased from Sigma Chemical Co. (St. Louis) and used without further purification. The initial screening that determined conditions suitable for the formation of crystals has been described by Sherwood and Brayer (1985). Optimal conditions for use in the

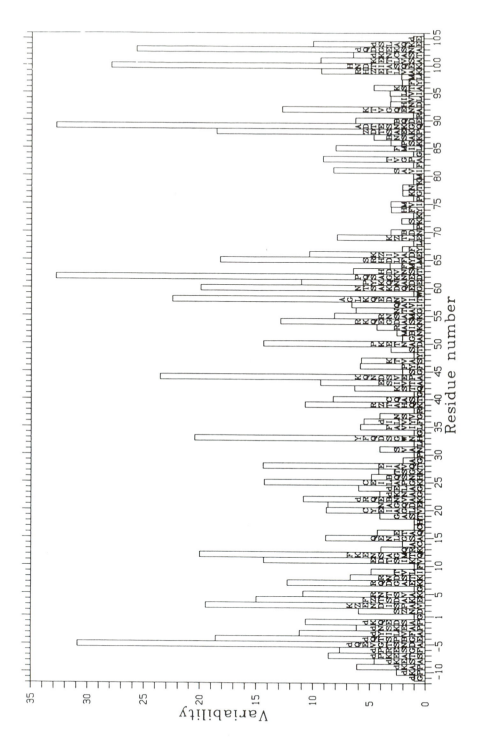

Figure 3.1

Degree of amino acid sequence variability along the polypeptide chain of eukaryotic cytochromes *c*. At each position, the variability (Wu and Kabat, 1970) is calculated as the number of different amino acids observed to occur at this position, divided by the frequency of occurrence of the most common amino acid (whose identity is given at the very bottom of the histogram). Other amino acids which occur at each position have also been listed along each histogram bar. Note that a lower-case "d" indicates that the designated residue is deleted in some species. The sequences of 92 eukaryotic cytochromes *c* (Hampsey *et al.*, 1988, Louie, 1990) have been included in this analysis.

Table 3.1

Sequence alignment for yeast iso-1, yeast iso-2, tuna, horse, and rice cytochromes *c*.

```
        -9              1                10                    30
Iso-1   - - - T E F K A G S A K K G A T L F K T R C L Q C H T V E K G G P H K V G
Iso-2   A K E - T G F K P G S A K K G A T L F K T R C C Q C H T I E E G G P N K K V G
Tuna    - - - - - - G F - - G D V A K G K K T F V Q K C A Q C H T V E N G G K H K V G
Horse   - - - - - - G D - - G D V E K G K K I F V Q K C A Q C H T V E K G G K H K T G
Rice    - A S F S E A P P G N P K A G E K I F K T K C A Q C H T V D K G A G H K Q G

         30            40              50                    60
Iso-1   P N L H G I F G R H S G Q A E G Y S Y T D A N I K K N V L W D E N N M S E
Iso-2   P N L H G L F G R H S G Q V K G Y S Y T D A N I K K N V K W D E D S M S E
Tuna    P N L W G L F G R K T G Q A E G Y S Y T D A N K S K G I V W N E D T L M E
Horse   P N L H G L F G R K T G Q A P G F T Y T D A N K N K G I T W K E E T L M E
Rice    P N L N G L F G R Q S G T T P G Y S Y S T A D K N M A V I W E E N T L Y D

         70            80              90                    100
Iso-1   Y L T N P K K Y I P G T K M A F G G L K K E K D R N D L I T Y L K K A C E -
Iso-2   Y L T N P K K Y I P G T K M A F G G L K K E K D R N D L I T Y M T K S A T K -
Tuna    Y L E N P K K Y I P G T K M I F A G I K K K G E R Q D L V A Y L K S A T S -
Horse   Y L E N P K K Y I P G T K M I F A G I K K K T E R E D L I A Y L K K A T N E
Rice    Y L L N P K K Y I P G T K M V F P G L K J P Q E R A D L I S Y L K E A T S -
```

The sequences of yeast iso-1 (Smith *et al.*, 1979), yeast iso-2 (Montgomery *et al.*, 1980), tuna (Kreil, 1965), horse (Margoliash *et al.*, 1961) and rice (Mori and Morita, 1980) cytochromes *c* have been aligned to maximize the amino acid homology present. The single-letter code is used to identify amino acids, and residue numbering is based on the sequence of tuna cytochrome *c*. Note that the single-letter code "J" is used to denote ε-N-trimethyl lysine residues. Boxes enclose regions of sequence identity in all five cytochromes *c*. For this purpose lysine and ε-N-trimethyl lysine have been considered equivalent. Statistical data concerning homology relationships can be found in Table 3.4.

hanging drop method (McPherson, 1990) were determined to be the equilibration of 5 μL of protein solution (25 mg/mL) containing 78% saturated (3.17 M) ammonium sulfate, 40 mM dithiothreitol, and 0.1 M sodium phosphate (adjusted to pH 6.2 using ammonium hydroxide or acetic acid) against 1.0 mL of 92% saturated (3.74 M) ammonium sulfate in the same buffer (Louie et al., 1988a).

Although large crystals of iso-1-cytochrome c could be obtained in this manner, many attempts did not yield crystals, apparently because of the absence of nucleation sites. For this reason, free interface diffusion (Salemme, 1972), in which the formation of crystal nuclei is favored by the protein initially experiencing a supersaturating concentration of precipitant, was attempted and found to be the most reliable method for producing the large crystals required for collection of high-resolution data (Louie and Brayer, 1990). Crystallization conditions were selected based on the optimized conditions determined using the hanging drop vapor diffusion method; however, the final ammonium sulfate concentration (90%) was chosen to be slightly lower than the 92% used in the vapor diffusion experiments. This lower final concentration would be expected to favor slower crystal growth and less loss of protein through precipitate formation. A typical trial, carried out in a melting point capillary (1.5-mm inner diameter), involved the layering of 5 μL of a concentrated protein solution (~200 mg/mL) in a buffer containing 0.1 M sodium phosphate (pH 6.2) and 40 mM dithiothreitol, on top of 45 μL of 100% saturated (4.06 M) ammonium sulfate in the same buffer.

A fair amount of turbidity formed initially at the interface between the protein and saturated ammonium sulfate solutions in the free interface experiment. The resultant precipitate would then sink into the column of ammonium sulfate; however, as the two solutions mixed, much of the precipitate was observed to redissolve. Crystal growth initiated at nuclei that had formed within the initial precipitate and generally occurred at the original interface or along the trail made by this precipitate as it fell into the column of ammonium sulfate. Crystals grew to a maximal size of 0.9 × 0.6 × 0.5 mm in about two weeks. At the end of this period, the solution had become completely clear (the initial protein solution was bright red) with all the protein incorporated into the crystals present. In general, very little precipitate remained. Crystal symmetry and unit cell parameters for reduced yeast iso-1-cytochrome c are summarized in Table 3.2a.

2. Oxidized state

To enhance comparisons of yeast iso-1-cytochrome c between the reduced and oxidized states, this protein was crystallized isomorphously in both oxidation states and refined to high resolution using the same techniques (Berghuis and Brayer, 1992). This feature allowed for careful comparative analyses in the absence of potential interfering factors such as different crystallizing conditions, packing arrangements or refinement strategies which could obscure the relatively small conformational changes occurring between oxidation states.

Yeast iso-1-cytochrome c for these crystallizations was isolated from recombinant yeast as described by Pielak et al. (1985, 1986). It should be noted that this protein had a threonine residue at position 102 (see Table 3.1) in place of a cysteine

Table 3.2a
Crystal growth and unit cell parameters for eukaryotic cytochromes c.

Cytochrome c	Crystal growth conditions[a]	Space group	Unit cell	Resolution[b]	References
Yeast iso-1					
(a) Reduced	92% A.S., 0.1 M phosphate, PH 6.2, 40 mM DTT	$P4_32_12$	$a = b = 36.46$ Å $c = 136.86$ Å	1.2 Å	Sherwood and Brayer, 1985 Louie et al., 1988a Louie and Brayer, 1990
(b) Oxidized	92% A.S., 0.1 M phosphate, pH 6.2, 30 mM NaNO_3	$P4_32_12$	$a = b = 36.47$ Å $c = 137.24$Å	1.9 Å	Leung et al., 1989 Berghuis and Brayer, 1992
Yeast iso-2 (reduced)	92% A.S., 0.1 M phosphate, pH 6.0, 0.3 M NaCl, 40 mM DTT	$P4_32_12$	$a = b = 36.44$ Å $c = 137.86$ Å	1.9 Å	Leung et al., 1989 Murphy et al., 1992
Tuna					
(a) Reduced	85% A.S., excess ascorbate, pH7.5	$P2_12_12$	$a = 34.44$ Å $b = 87.10$ Å $c = 37.33$ Å	1.5 Å	Takano and Dickerson, 1981a
(b) Oxidized	50% A.S., 15% NaNO_3, pH 7.0, 1.0 M ammonium phosphate	$P4_3$	$a = b = 74.42$ Å $c = 36.30$ Å	1.8 Å	Swanson et al., 1977 Takano and Dickerson, 1981b
Horse (oxidized)	94% A.S., 0.1 M phosphate, 0.75 M NaCl, pH 6.4	$P4_3$	$a = b = 58.34$ Å $c = 41.83$ Å	1.9 Å	Bushnell et al., 1990
Rice (oxidized)	3.6 M A.S., pH 6.0	$P6_1$	$a = b = 43.78$ Å $c = 110.05$ Å	1.5 Å	Ochi et al., 1983

[a] A.S. refers to ammonium sulfate concentration, DTT refers to dithiothreitol.
[b] Resolution refers to the maximum resolution of the diffraction data used to determine the structure of each protein.

to eliminate protein dimerization during purification and subsequent manipulations (Cutler *et al.*, 1987). This feature was essential to keeping the protein in its monomer form under oxidizing conditions. Small crystals (0.2 × 0.1 × 0.05 mm) of the oxidized protein were grown using the hanging drop method and employing a hair seeding technique (Leung *et al.*, 1989). The conditions under which crystal growth was observed were similar to those previously described for the reduced form of yeast iso-1-cytochrome (Sherwood and Brayer, 1985; Louie *et al.*, 1988a), with the exception that the buffer contained 30 mM sodium nitrate instead of a reducing agent. To obtain larger crystals, macro seeding was performed, which after 7 to 10 days resulted in crystals suitable for high-resolution x-ray diffraction analyses (∼0.5 × 0.4 × 0.2 mm). To ensure complete oxidation of the protein, crystals were transferred into a solution containing 20 mM potassium ferricyanide prior to mounting for diffraction studies. Spectroscopic analyses of crystalline material confirmed that the iso-1-cytochrome *c* present was in the oxidized form. Observed crystallographic parameters are shown in Table 3.2a.

B. Yeast Iso-2-cytochrome *c* (Reduced)

The protein used was prepared as described by Nall and Landers (1981). Yeast iso-2-cytochrome *c* turned out to be resistant to crystallization using the techniques described for its iso-1 protein counterpart and could not be crystallized using the full range of other conventional techniques available (McPherson, 1990). Indeed, crystallization required the use of a specially designed hair seeding technique developed for this protein in our laboratory (Leung *et al.*, 1989).

The hair seeding method can be summarized as follows. A seeding solution was prepared by crushing, with a drawn glass fiber, crystals of yeast iso-1-cytochrome *c* in about 3 μL of the seed crystal's precipitating solution. Next, a "hair stick" was passed through the seeding solution and then gently stroked through a protein droplet containing the iso-2 protein. During this procedure, the protein droplet was positioned on a siliconized glass coverslip. The seeded iso-2 protein droplet was then assembled into a conventional hanging drop vapor diffusion plate and allowed to equilibrate against a reservoir containing 1 mL of precipitating solution. Crystals of iso-2-cytochrome *c* were found to grow in the hanging droplet only along the path of the original hair stroke.

The 5 μL hanging droplet of iso-2-cytochrome *c* contained 20 mg of protein per mL in 65% saturated ammonium sulfate, 0.1 M sodium phosphate, 0.3 M sodium chloride, and 0.04 M dithiothreitol and was adjusted to pH 6.0. Before seeding, the protein solution was centrifuged for three minutes to remove debris and precipitate. The 1 mL reservoir solution was in the same buffer with 88 to 92% saturated ammonium sulfate. Using the hair seeding technique, crystals of iso-2-cytochrome *c* could be reproducibly grown overnight and continued to increase in size for about two weeks. The largest crystals grown to date measure 0.45 × 0.45 × 0.25 mm. Once iso-2-cytochrome *c* crystals were available, these were used to form the seeding solution described above, in place of iso-1 protein crystals. The use of iso-2 protein crystals in this way considerably improved the subsequent crystal growth obtained.

The hair stick used was constructed by attaching a 1- to 2-cm length of human hair to the end of an application stick. Individual hairs obtained for this purpose were used without any further modification. However, the effectiveness of the hair seeding process did vary considerably with hair samples from different sources. We believe the hair stick acts by picking up small crystal fragments from the seeding solution, which are then deposited in the protein droplet and function as nucleation sites for crystal growth. It is notable that the transferred seeds are too small to be visible when the hair stick is examined at 80× magnification under a microscope.

As discussed previously, these seeds were added to droplets well below their supersaturation points. We cannot account for the apparent resistance of these microscopic seeds to dissolution. Nonetheless, this phenomenon has been observed before. For example, past experiments by both Koeppe *et al.* (1975) and Thaller *et al.* (1981) have made use of this unique property of seed crystals to preferentially dissolve precipitated protein in order to grow larger crystals using the pulsed diffusion and repeated seeding techniques, respectively. These results and our own work suggest that seed crystals can maintain their integrity for substantial periods of time under unfavorable conditions and thereby serve as growth nucleation points when conditions of supersaturation are subsequently restored.

Often, it was found that the seeding process was too effective, resulting in large numbers of relatively small crystals. However, strict control of the number of initial nucleation sites resulted in the growth of much larger crystals. Several successful nucleation control strategies were found. These include diluting the seeding solution; passing the hair stick successively through several protein droplets, thereby decreasing the number of seeds transferred to later drops, and decreasing the length of the stroke through the protein droplet, or alternatively only dipping the point of the hair stick into this solution at a few selected points.

Crystallization parameters and unit cell data for crystals of yeast iso-2 cytochrome *c* are summarized in Table 3.2a.

C. Yeast Iso-1/Iso-2 Cytochrome *c* Mutants

Yeast cytochrome *c* proteins having various types of amino acid substitutions have proven much less cooperative in forming diffraction-quality crystals. Both the hanging drop and free interface diffusion methods, employing similar conditions to those used to grow wild-type crystals, are only rarely effective. This problem can be alleviated to a great degree using the hair seeding technique discussed earlier for yeast iso-2 cytochrome *c* (Leung *et al.*, 1989). Thus, hair seeding, often followed by further macroscopic seeding techniques, is our laboratory's primary method for the growth of large crystals of mutant yeast cytochrome *c* proteins. One major advantage of this approach has been that all mutant protein crystals have so far turned out to be isomorphous to those of the wild-type yeast cytochromes *c*, thereby greatly easing the task of subsequent structure determinations. Conditions used in the various mutant protein crystallizations range around those for the wild-type proteins. Although the structures of mutants of yeast cytochromes *c* are not discussed in this work, a list of those described in the literature is given in Table 3.2b with associated references.

Table 3.2b
Completed structural studies of mutant yeast cytochromes c.

Protein	References
Asn52Ala Ile75Met	Rafferty *et al.*, 1995
Asn52Ile Asn52Ile/Tyr67Phe	Langen *et al.*, 1992; Berghuis *et al.*, 1994b
Tyr67Phe	Berghuis *et al.*, 1994a
Pro71Ala Pro71Ile Pro71Ser Pro71Val	Murphy *et al.*, 1995
Phe82Gly	Louie and Brayer, 1989
Phe82Ser	Louie *et al.*, 1988b
B-2036	Murphy *et al.*, 1992
RepA2 RepA2(Val20)	Murphy *et al.*, 1993

D. Tuna Cylochrome c

1. Reduced state

This protein is obtained from tuna heart muscle and purified as outlined in Swanson *et al.*, 1977. Crystals were obtained from a 10% protein solution having a five fold excess of ascorbate and in 85% saturated ammonium sulfate. This solution was adjusted to a pH of 7.5 with NH_4OH and then loosely sealed in a glass vial to allow for slow evaporation. Maximal crystal growth was observed after about one month.

2. Oxidized state

As for the reduced state, crystals were obtained from $\sim 10\%$ protein solutions. However, in this case, no reducing agent was present, and the protein was initially dissolved in a 1.0 M ammonium phosphate buffer which was also 15% saturated in ammonium sulfate and had its pH adjusted to 7.0 (Swanson *et al.*, 1977). Then, powdered $NaNO_3$ was slowly added to $\sim 15\%$ saturation, followed by addition of powdered ammonium sulfate to a final saturation of $\sim 50\%$. This procedure led to the growth of microcrystals. The resultant solution was stored in loosely sealed vials. Slow evaporation led to the growth of diffraction-quality crystals after about one month. Crystallographic parameters for both the reduced and oxidized tuna cytochromes c are outlined in Table 3.2a.

E. Horse Cytochrome *c* (Oxidized)

This protein was crystallized at room temperature (22 °C) using conventional hanging drop techniques (Bushnell *et al.*, 1990). The protein sample was purchased from Sigma Chemical Co. (St. Louis). The starting well solution (1 mL) was composed of 94% saturated ammonium sulfate, 0.75 M sodium chloride, and 0.1 M sodium phosphate and was adjusted to pH 6.4. The initial droplet solution (5 μL) was similar except for only 70% ammonium sulfate saturation and the presence of 20 mg protein/mL. Relevant crystallographic parameters are summarized in Table 3.2a.

F. Rice Cytochrome *c* (Oxidized)

This protein was obtained from the embryos of *Oryza sativa L* (Ochi *et al.*, 1983) and crystallized from a solution containing 2% (w/v) protein, 3.6 M ammonium sulfate, and adjusted to pH 6.0. The resultant hexagonal bipyramid crystals grew to a size suitable for x-ray analyses after three months. Crystallographic parameters are summarized in Table 3.2a.

G. Bacterial Cytochromes *c*

Although bacterial cytochromes *c* are not discussed in this work, for completeness Table 3.2c lists the crystallization conditions used in successful structure determinations of these proteins.

IV. High-Resolution Cytochrome *c* Tertiary Structures[1]

Currently the most highly refined structure of a cytochrome *c* is that of the reduced yeast iso-1 protein at 1.2-Å resolution (Table 3.2a; Louie and Brayer, 1990). This structure includes not only an accurate placement of all non-hydrogen protein, heme, and water atoms (internal and external), but also tentative assignments of hydrogen atom positions. Given its high degree of refinement, the yeast iso-1 cytochrome *c* structure will be used in this section as the focus for a discussion of the comparative similarities and differences with other known cytochrome *c* structures. These include the iso-2 protein from yeast and those from tuna, horse, and rice. All of these latter structures have also been determined to relatively high resolutions (1.5–2.0 Å). Rather than repeat the many similarities between these cytochromes *c*, the follow-

---

[1] The amino acid numbering scheme used herein is based upon the alignment in Table 3.1. The nomenclature used for amino acid and heme atoms follows the conventions of the Protein Data Bank (Bernstein *et al.*, 1977). A schematic representation of the heme group of cytochrome *c* with atom labels is illustrated in Figure 3.2. Cytochrome *c* atomic coordinates used herein were obtained from the following sources: reduced yeast iso-1 (Louie and Brayer, 1990; Protein Data Bank file 1YCC); oxidized yeast iso-1 (Berghuis and Brayer, 1992; Protein Data Bank file 2YCC); reduced yeast iso-2 (Murphy *et al.*, 1992; Protein Data Bank file 1YEA); reduced tuna (Takano and Dickerson, 1981a; Protein Data Bank file 5CYT); oxidized tuna (Takano and Dickerson, 1981b; Protein Data Bank file 3CYT); oxidized horse (Bushnell *et al.*, 1990); and oxidized rice (Ochi *et al.*, 1983; Protein Data Bank file 1CCR).

Table 3.2c

Crystal growth and unit cell parameters for bacterial type I cytochromes c.

Cytochrome c	Crystal growth conditions[a]	Space group	Unit cell	Resolution[b]	References
Azotobacter vinelandii c_5	1.2 M A.S., 0.4 M phosphate, pH 6.9	C2	$a = 45.73$ Å, $b = 37.56$ Å $c = 42.55$ Å, $\beta = 111.30°$	2.5 Å	Stout, 1978 Carter *et al.*, 1985
Desulfovibro vulgaris c_{553}	75% (v/v) ethanol, 50 mM Tris HCl, pH 7.3	P2$_1$2$_1$2$_1$	$a = 52.90$ Å, $b = 68.10$ Å $c = 34.90$ Å	1.8 Å	Bando *et al.*, 1979 Higuchi *et al.*, 1984
Rhodobacter capsulatus c_2	3.5 M A.S., 0.25 M NaCl, 50 mM phosphate, pH 7.5	R32	$a = b = 100.03$ Å $c = 162.10$ Å	2.5 Å	Holden *et al.*, 1987 Benning *et al.*, 1991
Rhodospirillum rubrum c_2 Reduced (Oxidized)	80% A.S., pH 5.8	P2$_1$2$_1$2$_1$	$a = 32.22$ Å, $b = 37.36$ Å $c = 84.62$ Å	1.68 Å (2.0 Å)	Salemme *et al.*, 1973 Bhatia, 1981
Paracoccus denitrificans c_2	95% A.S. 1.0 M NaCl, pH 7.5	P2$_1$2$_1$2$_1$	$a = 42.70$ Å, $b = 82.17$ Å $c = 31.56$ Å	2.5 Å	Timkovich and Dickerson, 1972 Timkovich and Dickerson, 1976
Pseudomonas aeruginosa c_{551} Oxidized (Reduced)	60% A.S., 1 M NaCl, 50 mM phosphate, pH 5.6 to 5.7 (1% ascorbate)	P2$_1$2$_1$2$_1$	$a = 29.47$ Å, $b = 49.30$ Å $c = 49.74$ Å	1.6 Å (1.6 Å)	Matsuura *et al.*, 1982

[a] A.S. refers to ammonium sulfate concentration.
[b] Resolution refers to the maximum resolution of the diffraction data used to determine the structure of each protein.

114

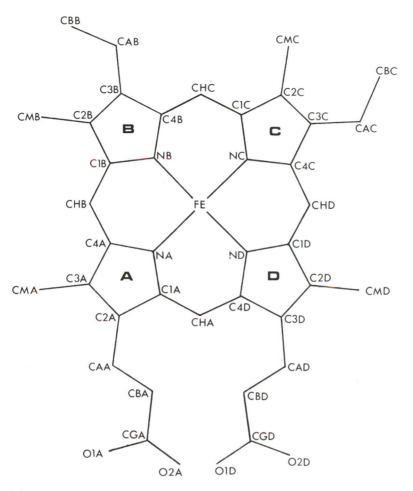

Figure 3.2
A schematic representation of the atomic skeleton of the heme group of cytochrome c and the atom labeling convention used herein, which follows that of the Protein Data Bank (Bernstein *et al.*, 1977). Each of the individual pyrrole rings has also been labelled with its letter code. Thioether bonds are made to the side chains of Cys14 and Cys17 through heme atoms CAB and CAC, respectively. Pyrrole rings C and D are near the solvent-exposed edge of the heme, whereas pyrrole rings A and B are buried within the interior of cytochrome c.

ing discussion of yeast iso-1-cytochrome c will consider these, whereas subsequent headings under the other cytochromes c will highlight their unique structural characteristics.

A. Yeast Iso-1-cytochrome c (Reduced State)

1. Introduction

In yeast (*Saccharomyces cerevisiae*), as in other eukaryotes, cytochrome c is part of the respiratory electron transport chain and acts as an electron shuttle along the inner mitochondrial membrane. Two isozymes of cytochrome c occur in yeast. These two proteins (the iso-1 and iso-2 forms) are believed to have equivalent physiological roles (Mattoon and Sherman, 1966). Of the cytochrome c present in yeast, the relative isozymic amounts are dependent on several metabolic factors (Laz *et al.*, 1984; Dumont *et al.*, 1990). In general, yeast iso-1-cytochrome c constitutes from 75 to 95% of the total cytochrome c present (Sels *et al.*, 1965).

Although they are components of the mitochondrion, yeast iso-1 and iso-2 cytochromes c are encoded by nuclear genes. The biosynthesis of these proteins involves the following steps: translation of the apo-proteins on cytosolic ribosomes; removal of the amino-terminal methionine residue of the polypeptide chain through the action of an amino peptidase (Tsunasawa *et al.*, 1985); trimethylation of the ε-nitrogen of Lys-72 (DiMaria *et al.*, 1979); and covalent attachment of protoheme IX to Cys 14 and Cys 17 through the action of a specific heme lyase (Dumont *et al.*, 1988). This final step has been shown to be concomitant with translocation of cytochrome c across the outer mitochondrial membrane. As part of the folding process, His18 and Met80 become ligands to the heme iron atom. Further discussion of cytochrome c biosynthesis and folding is provided in the chapters by Jordi and De Kruijff and by Nall in this volume (chapters 12 and 4, respectively). The fully processed yeast iso-1 and iso-2 cytochromes c contain 108 and 112 amino acids, respectively, in a single polypeptide chain. One unique feature is that both proteins have extensions at their N-terminal ends, in comparison to vertebrate cytochromes c, which are typically 103 or 104 amino acids in length (Table 3.1).

The versatility of the yeast system offers several advantages over others in the study of electron transfer reactions. As a result, extensive genetic and biochemical analyses related to the function and expression of the yeast cytochromes c have been carried out. One important feature is the ease with which mutants can be generated *in vivo* (Hampsey *et al.*, 1986) in contrast to other eukaryotic systems. Furthermore, yeast is amenable to genetic manipulation methods, and it is possible to design and construct mutant cytochromes c to probe the various functional properties of this protein (Smith, 1986). Indeed, this latter feature allows the yeast system to be used to investigate more general questions, such as the salient factors governing the relationship between amino acid sequences, polypeptide chain folding, and the resultant expression of biological activity.

2. Polypeptide chain conformation

The yeast iso-1-cytochrome c molecule has the typical cytochrome c fold with the polypeptide chain organized into a series of α-helices and fairly extended loop structures (Dickerson, 1980a). The protein fold envelops the heme prosthetic group within a hydrophobic pocket formed by a shell of polypeptide chain one to two residues thick. The polypeptide chain backbone, heme group, and heme ligands of yeast iso-1-cytochrome c are illustrated in Figure 3.3a. Additional illustrations in comparable orientations for yeast iso-2, tuna, horse, and rice cytochromes c are also shown in Figure 3.3. Rotation matrices and translation vectors to bring all of these cytochrome c structures to a common orientation are shown in Table 3.3. A structural equivalence and sequence identity matrix for all these cytochromes is presented in Table 3.4.

Table 3.3
Rotation matrices and translation vectors applied to achieve structural superposition of yeast iso-2, tuna, horse and rice cytochromes c onto yeast iso-1-cytochrome c[a].

Rotation matrix	Translation vector

a. Yeast Iso-2

$$\begin{bmatrix} 0.99977 & 0.01278 & 0.01711 \\ -0.01251 & 0.99980 & -0.01563 \\ -0.01731 & 0.01541 & 0.99973 \end{bmatrix} \qquad \begin{bmatrix} -0.24324 \\ 0.26410 \\ -0.35243 \end{bmatrix}$$

b. Tuna

$$\begin{bmatrix} 0.27960 & 0.38527 & 0.87943 \\ -0.70408 & -0.54047 & 0.46062 \\ 0.65277 & -0.74797 & 0.12014 \end{bmatrix} \qquad \begin{bmatrix} -7.79026 \\ 12.86496 \\ -7.61622 \end{bmatrix}$$

c. Horse

$$\begin{bmatrix} 0.54089 & -0.20725 & 0.81516 \\ 0.76967 & 0.51281 & -0.38032 \\ -0.33920 & 0.83311 & 0.43688 \end{bmatrix} \qquad \begin{bmatrix} 11.08282 \\ 19.46846 \\ -17.30244 \end{bmatrix}$$

d. Rice

$$\begin{bmatrix} -0.67674 & -0.00102 & -0.73622 \\ 0.58985 & 0.59767 & -0.54302 \\ -0.44058 & -0.80174 & -0.40386 \end{bmatrix} \qquad \begin{bmatrix} 22.80838 \\ 22.47102 \\ 35.40743 \end{bmatrix}$$

[a] The 412 main chain atoms common to all five cytochromes c (residues 1 to 103) were overlaid in a pairwise fashion by a least-squares procedure to obtain the indicated operations. The application of the rotation matrix and translation vector to superimpose the coordinates (x_0, y_0, z_0) onto those of yeast iso-1-cytochrome c is as follows:

$$\begin{bmatrix} \text{Rotation} \\ \text{matrix} \end{bmatrix} \begin{bmatrix} x_0 \\ y_0 \\ z_0 \end{bmatrix} + \begin{bmatrix} V \\ e \\ c \end{bmatrix} = \begin{bmatrix} x_0' \\ y_0' \\ z_0' \end{bmatrix}$$

118

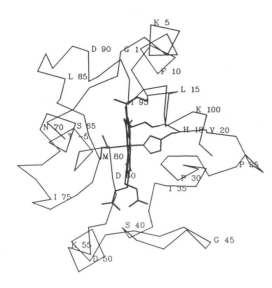
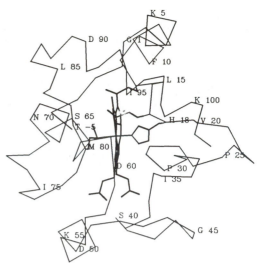

A

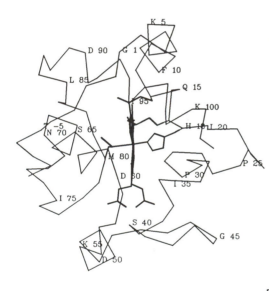
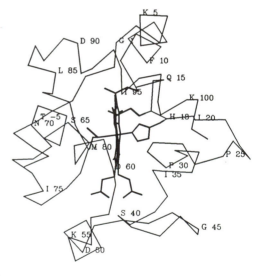

B

Figure 3.3

Stereo drawings showing the conformations of the main chains of (a) yeast iso-1, (b) yeast iso-2, (c) tuna, (d) horse, and (e) rice cytochromes *c*. In each case, the heme group and the heme ligands formed by the side chains of His18 and Met80 are also illustrated, as are the two covalent thioether heme bonds to Cys14 and Cys17 (shown in thick lines). The alpha-carbon atoms of every fifth residue are labeled with sequence numbers according to the alignment presented in Table 3.1. Also indicated is the one-letter amino acid code for each labeled amino acid. The view presented here is edge-on to the heme group, with that portion of the molecule containing the exposed heme edge directly facing the viewer. The matrices of Table 3.3 have been used to obtain the common viewpoint presented.

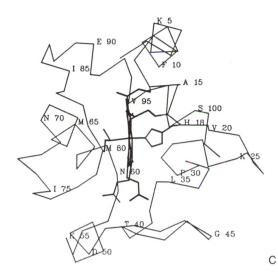
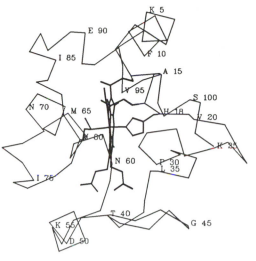

C

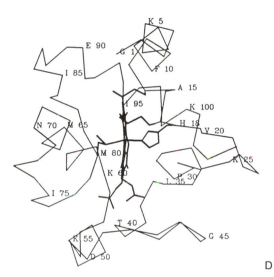
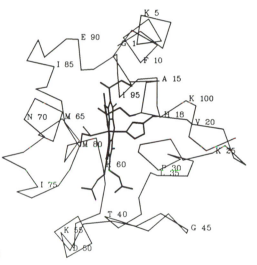

D

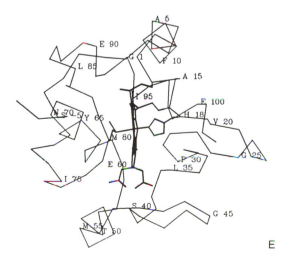
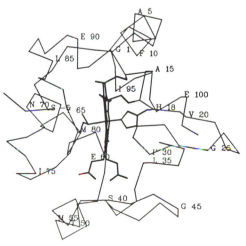

E

Table 3.4

Structural equivalence and amino acid sequence identity matrix for yeast iso-1, yeast iso-2, tuna, horse, and rice cytochromes c^a.

	Yeast iso-1 (reduced) (108)	Yeast iso-2 (reduced) (112)	Tuna (reduced) (103)	Horse (oxidized) (104)	Rice (oxidized) (111)
Yeast iso-1 (reduced)	—	91 (84%)	63 (61%)	64 (62%)	62 (57%)
Yeast iso-2 (reduced)	0.30 (0.41)	—	59 (57%)	60 (58%)	57 (51%)
Tuna (reduced)	0.40 (0.54)	0.51 (0.66)	—	85 (83%)	61 (59%)
Horse (oxidized)	0.48 (0.54)	0.57 (0.67)	0.43 (0.51)	—	64 (62%)
Rice (oxidized)	0.40 (0.51)	0.47 (0.63)	0.45 (0.61)	0.49 (0.60)	—

[a] The total number of amino acids in each protein is given below its name in the heading. The upper triangular portion of the matrix contains the number of chemically identical residues found in common between each pair of proteins aligned in Table 3.1. The percentage of the number of identical residues in the smaller protein is given in parentheses.

The lower triangular portion of the matrix contains the results of least-squares structural overlap comparisons based on the atomic positions of main-chain atoms common to all five cytochromes c (412 atoms, residues 1 to 103). For each pair of proteins, the resultant average deviations in these common atoms is listed. The root mean square deviation values are indicated in parentheses.

A plot of all main chain torsion angles (ϕ, ψ) for iso-1-cytochrome c is shown in Figure 3.4. In this plot, there is a clustering of residues with the α-helical conformation ($-60°$, $-40°$) and of glycine residues that belong to type II β-turns ($85°$, $5°$). All non-glycine residues fall within or near the allowed regions, except for Lys27 ($-121°$, $-127°$). Lys27 is the first residue in a tight turn that has the conformation of a reversed classical γ-turn (Milner-White et al., 1988; Nemethy and Printz, 1972), in which the amide group of residue i (Lys27) hydrogen bonds the carbonyl group of residue $i + 2$ (Gly29). The additional hydrogen bonds made by the $i + 2$ residue (Gly29 N to Cys17 O) and the side chain of the i residue (Lys27 NZ to Leu15 O) may explain the stability of this unusual structural feature. Of note is the lowered stability of mutant iso-1-proteins that possess side chains (Leu, Gln, Trp, and Tyr) at position 27 (Hickey et al., 1988) that are incapable of forming the latter interaction. Figure 3.5 illustrates the conformation of the reversed classical γ-turn involving residues 27 to 29 in yeast iso-1-cytochrome c.

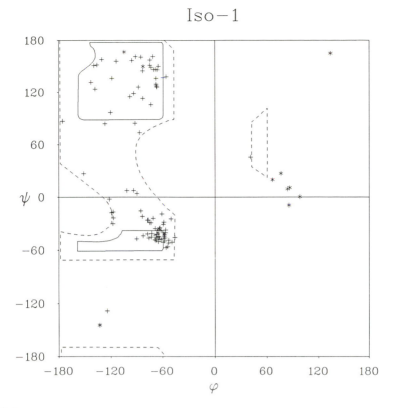

Figure 3.4
Ramachandran plot showing the main chain ϕ and ψ dihedral torsion angles of yeast iso-1-cytochrome c. Glycine residues are denoted by " $*$ "; all other residues are denoted by " $+$ ". Main chain conformations for an alanyl residue (having an N–CA–C angle, τ, of 112°) in poly(L-alanine), which are fully allowed, and allowed assuming shortened permissible minimum interatomic contact distances (Ramakrishnan and Ramachandran, 1965), are enclosed by continuous and broken lines, respectively.

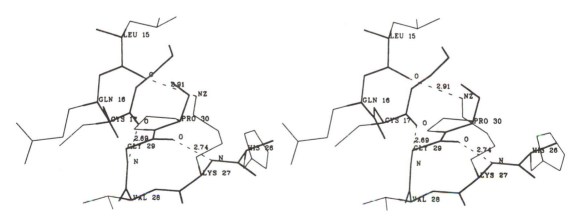

Figure 3.5
A stereo drawing illustrating the conformation of a tight reversed classical γ-turn in yeast iso-1-cytochrome c involving residues 27 to 29 and the hydrogen bond interactions formed by these residues. The polypeptide main chain is shown in thick lines, and side chains in thin lines. Hydrogen bonds present between Lys27 N and Gly29 O, Cys17 O and Gly29 N, and Lys27 NZ and Leu15 O are shown as broken lines, and their corresponding hydrogen-bond lengths (in angstroms) are indicated.

A detailed analysis of the secondary structural elements present in yeast iso-1-cytochrome c is presented in Table 3.5. The average main chain torsion angles for all residues having an α-helical conformation are in good agreement with those observed in other highly refined protein structures (Sheriff *et al.*, 1987; McGregor *et al.*, 1987). However, some distortions from ideal α-helical geometry do occur. For example, the N-terminal α-helix contains a kink at Ala7 and Thr8, with both the Ala7 O–Lys11 N and Thr8 N–Lys4 O hydrogen bonds being long. This distortion accommodates hydrogen bonding by the side chains of two nearby threonine residues to the carbonyl groups of residues in the preceding turn of the helix (Thr8 OG1–Lys5 O and Thr12 OG1–Thr8 O). The C-terminal end of this helix also possesses an unusual distortion in which the carbonyl groups of residues 12, 13, and 14 are directed markedly away from the helix axis, and the carbonyl group of residue 10 accepts a hydrogen bond from the amide groups of both residues 14 and 15. This distortion appears to be necessary to accommodate the formation of a thioether bond from Cys14 to the heme group.

The helix encompassing residues 60 to 70 is distorted towards a 3_{10} helix at both its N- and C-terminal ends. The carbonyl group of the first residue of this helix is involved in hydrogen bonding with the $i + 3$ and $i + 4$ main chain amide groups (Asp60 O to Asn63 N and Met64 N). The last turn of this helix also assumes a 3_{10} helix hydrogen-bonding pattern (Thr69 N hydrogen bonds Glu66 O, and Asn70 N hydrogen bonds Tyr67 O). This latter distortion may be caused by Thr69, whose OG1 atom forms a hydrogen bond with Ser65 O. Also, in the middle of this helix, Glu66 N and Asn62 O do not form the expected hydrogen bond, whereas Asn63 O forms bonds with both Glu66 N and Tyr67 N.

A threonine residue also appears to cause a distortion in the C-terminal α-helix (residues 87 to 102), which is kinked at Thr96 to allow the OG1 atom of this residue to form a hydrogen bond with the carbonyl group of residue 92. This results in the amide nitrogen atoms of both residues 96 and 97 being involved in bifurcated hydrogen bonds with the carbonyl groups of residues 92 and 93, and with those of residues 93 and 94, respectively. Beginning at residue 99, the C-terminal helix takes on main chain torsion angles characteristic of a 3_{10} helix. This αC1 distortion (Baker and Hubbard, 1984) results in Lys100 N hydrogen bonding with the carbonyl groups of residues 96 and 97, and Ala101 N forming a bifurcated hydrogen bond to the carbonyl groups of residues 97 and 98. This allows the side chain of Lys99 to form a hydrogen bond with the carbonyl group of residue 96, and it also facilitates the hydrogen-bonded interaction of the two most C-terminal residues with the rest of the molecule, so that Cys102 O forms a hydrogen bond with Gly34 N and the C-terminal carboxyl group of residue 103 forms a salt-bridge to His33 NE2.

The other two helices present in yeast iso-1-cytochrome c (residues 49 to 55 and 70 to 75; see Table 3.5) are shorter and less well formed, having only approximate α-helical geometry. Overall, the α-helices in iso-1-cytochrome c are typical of those found in other high-resolution protein structures in that their conformations do not strictly adhere to that of the idealized α-helix (Arnott and Dover, 1967). In the present study, a particularly notable feature is that the presence of threonine residues within a helix causes substantial distortion of regular α-helical geometry, and in two of the helices, threonine marks the end of these secondary structures. The propensity

Table 3.5
Secondary structural elements present in yeast iso-1-cytochrome *c*.

Secondary structural element	Residues involved	Main-chain torsion angles (°)[a]
α-Helix	2–14	$(-64.7, -40.8)$
β-Turn (type II)	21–24	$(-65, 128), (84, 8)$
γ-Turn	27–29	$(-121, -127), (-57, -50), (-104, 164)$
β-Turn (type II)	32–35	$(-55, 138), (66, 19)$
β-Turn (type II)	35–38	$(-64, 126), (86, -11)$
β-Turn (type II)[b]	43–46	$(-82, 147), (87, 9)$
α-Helix	49–55	$(-65.5, -37.8)$
α-Helix	60–70	$(-63.6, -38.6)$
α-Helix	70–75	$(-66.3, -40.0)$
β-Turn (type II)	75–78	$(-66, 128), (99, 0)$
α-Helix	87–102	$(-63.5, -42.2)$
Overall:		
α-Helix	43 residues	$(-64.3, -40.3)$
β-Turn (type II)	5 turns	$(-66.4, 133.4), (84.4, 5.0)$

[a] The main chain torsion angles listed refer to the average (ϕ, ψ) angles for residues in each α-helical segment (excluding terminal residues), and to $(\phi, \psi)_2$ and $(\phi, \psi)_3$, respectively, for residues in each type II β-turn. For the γ-turn, all three of $(\phi, \psi)_1$, $(\phi, \psi)_2$ and $(\phi, \psi)_3$ are given. Overall averages for those classes with more than one occurrence are listed.
[b] Mediated through a water molecule.

for threonine to cause distortions in α-helices has been observed in other protein structures (McGregor *et al.*, 1987; Poulos *et al.*, 1987).

Aside from the helical segments described, there is little other regular secondary structure present in the iso-1-cytochrome *c* molecule. Residues 37 to 40 and 57 to 59 form a very short, highly distorted two-stranded anti-parallel β-sheet (Figure 3.6). This secondary structure element contains only three interstrand main chain to main chain hydrogen bonds. There are five turns of the β-type II class possessing average main chain torsion angles typical of this class of secondary structure (Table 3.5, and see Richardson, 1981). A modified β-turn occurs at residues 43 to 46, in which Ala43 O forms a hydrogen bond with a water molecule instead of with Tyr46 N, as would be expected in a classical type II β-turn. This altered conformation appears to be designed to facilitate the hydrogen bonding of Glu44 O to His26 NE2, as well as the packing of the side chain of Tyr46 against the pyrrolidine ring of Pro30. Other regions of the polypeptide chain can be described as being in a coil conformation.

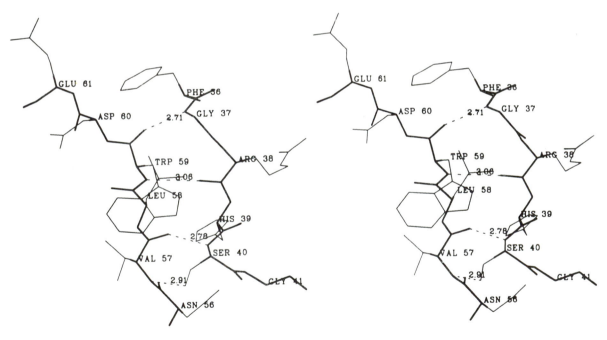

Figure 3.6

A stereo-drawing showing the conformation of a short segment of distorted two-stranded antiparallel β-sheet in yeast iso-1-cytochrome *c*, which involves residues 37 to 40 and 57 to 59. The polypeptide main chain is shown in thick lines, and side chains in thin lines. The four interstrand hydrogen bonds that occur (Gly37 N–Trp59 O, Arg38 O–Trp59 N, Ser40 N–Val57 O, and Ser40 OG–Val57 N) are shown as thin, broken lines, and the bond distances (Å) involved have been indicated.

3. Conformations of side chains

A stereo drawing showing the placement of all the side chains of yeast iso-1-cytochrome c is presented in Figure 3.7a. Similar drawings for yeast iso-2, tuna, horse, and rice cytochromes c are also illustrated in Figure 3.7. All of these structures are drawn having the same direction of view, which is identical to that presented in Figure 3.3. Several analyses of the preferred values for side-chain dihedral angles in proteins have been published (Ponder and Richards, 1987; Summers *et al.*, 1987; McGregor *et al.*, 1987). The side-chain conformations observed in iso-1-cytochrome c show trends similar to those noted in these studies. The histogram in Figure 3.8 shows the distribution of side-chain dihedral torsion angles observed for those cases where a staggered conformation is expected. The overall rms deviation of these torsion angles (for sp^3 tetrahedral carbon atoms) from the optimal staggered conformation (i.e. $\pm 60°$, $180°$) is $19.2°$. Of the 16 lysine residues in yeast iso-1-cytochrome c, a number of these side chains, or parts thereof, were poorly defined in our structural analysis. If a total of 29 uncertain torsional angle values from 13 such side chains are excluded (usually beyond the CD or CE carbon atoms), then the overall rms deviation quoted above falls to $16.5°$.

As Figure 3.8 shows, there is a strong avoidance of the g^- conformation ($+60°$) for χ_1 angles, with only 12 of 85 present having this conformation. For residues within α-helices, this avoidance is absolute except for Thr8 and Thr12, which are part of the N-terminal α-helix and whose conformations are influenced by side chain hydrogen bonding interactions as described earlier. In general, for threonine and serine residues, the g^- conformation is not disfavored with respect to g^+ and t (McGregor *et al.*, 1987). Six of the 12 side chains in iso-1-cytochrome c that have the g^- conformation at χ_1 are threonine or serine residues. The remaining six residues are all glutamate and lysine. For these residues, χ_2 strongly favors the t conformation if χ_1 is g^- (four out of six cases). For aromatic residues, a very narrow distribution of χ_1 torsion angles about preferred values is observed: $180°$ (rms. $\Delta = 10°$) for phenylalanine and tyrosine; and $-60°$ (rms $\Delta = 4.6°$) for histidine and the single tryptophan present. In five out of seven occurrences, isoleucine and valine residues have the most preferred conformation with the hydrogen atom of the CB carbon being *gauche* to both the main chain N and C atoms. Of the sulfur-containing side chains, all three cysteine residues are g^+. The two methionine residues — Met64 in an α-helix and Met80, which forms a heme ligand — are tg^-g^- and ttt, respectively. Overall, the conformations of the majority of the side chains in iso-1-cytochrome c are consistent with established torsional angle preferences and are in agreement with side chain rotamer templates compiled in other studies (Ponder and Richards, 1987; McGregor *et al.*, 1987).

4. Heme conformation and interactions

The protoporphyrin IX heme prosthetic group of yeast iso-1-cytochrome c is buried within a hydrophobic pocket formed by the polypeptide chain. Only five atoms of the heme (CMC, CBC, CAC, CHD, and CMD), which are positioned on the front edge of this group in the orientation of Figure 3.3, are exposed to external

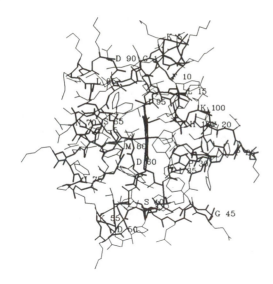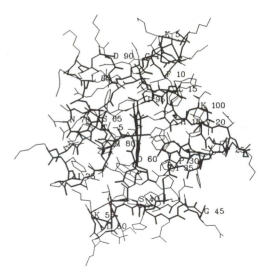

A

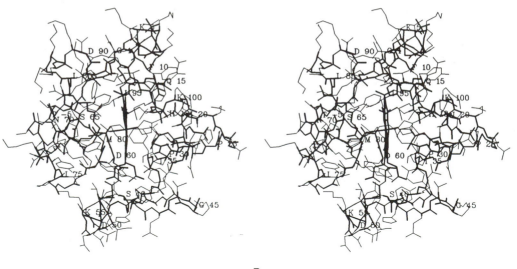

B

Figure 3.7

Stereo-drawings illustrating side chain placements in (a) yeast iso-1, (b) yeast iso-2, (c) tuna, (d) horse, and (e) rice cytochromes *c*. Side chain groups are drawn with thin bonds, while the polypeptide chain backbone is drawn with thick lines. The heme group and ligand bonds are also drawn with thick lines. For clarity, every fifth alpha-carbon atom is labeled with its one-letter amino acid code designation and sequence number (Table 3.1). The orientation of view is the same as that in Figure 3.3.

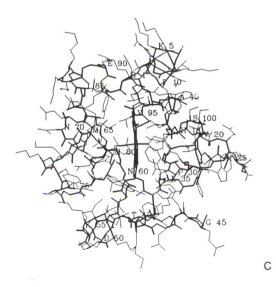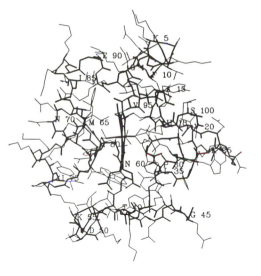

C

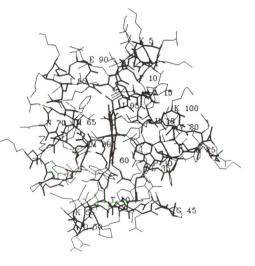

D

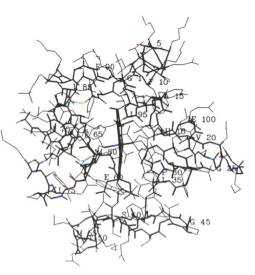

E

128

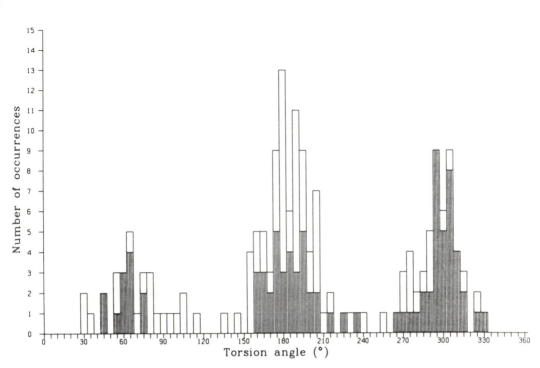

Figure 3.8
Distribution of side chain dihedral torsion angles in yeast iso-1-cytochrome c. Only those angles for which a staggered conformation is expected to be most favored have been included (168 angles in total). The distribution of all χ_1 torsion angles, excluding those belonging to proline residues (85 angles in total), is shown by the shaded bars.

solvent. This is particularly evident in the space-filling diagram presented in Figure 3.9. A compilation of heme atom accessibility in yeast iso-1, yeast iso-2, tuna, horse, and rice cytochromes c is provided in Table 3.6. In total, only about 9.5% of the total surface area of the heme of yeast iso-1-cytochrome c is exposed to solvent.

The heme group is covalently attached to the polypeptide chain *via* two thioether bonds to the side chains of Cys14 and Cys17. The bond geometry of these thioether linkages is typical for carbon–sulfur single bonds, with the average CB–SG–C_{heme} angle being 101.2°.

The heme group itself is not absolutely planar but is distorted into a saddle shape with the front and rear edges of the heme deviating toward the right, and the top and bottom edges toward the left of the molecule in the direction of view presented in Figures 3.3 and 3.7. Deviations of the planes of each pyrrole ring with respect to the plane of the porphyrin ring as a whole and with respect to the plane of the Fe-coordination pyrrole nitrogen plane, are shown in Table 3.7. Pyrrole rings B and C are both involved in thioether linkages to the polypeptide chain and show the greatest deviations with respect to the porphyrin ring plane as a whole. These results suggest that covalent bond attachment of the heme to the polypeptide chain is an important factor in establishing the observed heme conformation.

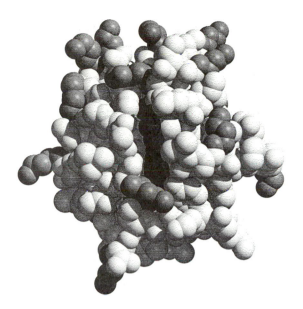

Figure 3.9

A space-filling representation of all the non-hydrogen atoms of yeast iso-1-cytochrome *c*. Heme group atoms are drawn in black, and in the direction of view presented, this group's solvent-exposed edge directly faces the viewer. Only ~9.5% of the total heme surface is exposed to solvent. Dark-grey atoms are associated with positively charged lysine and arginine residues that encircle the exposed heme edge and are believed to be responsible for complexation with redox partners. White spheres represent all other surface atoms. The molecular orientation presented here is equivalent to that in Figures 3.3 and 3.7.

Table 3.6
Heme solvent accessibilities[a].

Cytochrome *c* structure	Yeast iso-1 (reduced)	Yeast iso-2 (reduced)	Tuna (reduced)	Horse (oxidized)	Rice (oxidized)
1. Solvent-accessible heme atoms and surface area exposed (Å^2)[b]					
CHD	2.9	0.0	4.6	0.0	0.0
CMC	9.4	8.5	8.2	8.3	10.4
CAC	3.7	0.0	1.9	5.2	0.0
CBC	18.0	20.9	23.5	17.3	12.8
CMD	10.4	11.7	10.1	3.9	8.1
2. Total heme exposure (Å^2)	44.4	41.4	48.3	34.7	31.3
3. % heme surface area exposed	9.5	8.9	10.5	7.4	6.8
4. % protein surface contributed by heme atoms	0.9	0.8	1.0	0.7	0.6

[a] Computations were done using the method of Connolly (1983) with a probe radius of 1.4 Å.
[b] See Figure 3.2 for heme atom labeling conventions.

The heme iron coordinate bond distances are also listed in Table 3.7 and are typical of those found in small-molecule six-coordinate iron–porphyrin complexes (Scheidt, 1978). The out-of-plane coordinate bonds involving the His18 and Met80 ligands show only slight deviations from being perpendicular to the pyrrole nitrogen plane. The Met80 ligand bond is in the R configuration in all eukaryotic cytochromes c of known structure (Mathews, 1985). The imidazole ring of His18 lies almost perpendicular (99°) to the pyrrole nitrogen plane of the heme. Its orientation, in which a 46.5° angle is made between the imidazole ring of this side chain and a vector drawn through the NA and NC pyrrole nitrogen atoms, is such that the CD2 and CE1 atoms make minimal contacts with the heme pyrrole nitrogen atoms.

Table 3.7
Heme conformation and ligand geometry in yeast iso-1-cytochrome c[a].

I. Angular deviations (°) between both the plane normals of individual pyrrole rings and the heme coordinate bonds, and both the pyrrole nitrogen plane and the porphyrin ring plane[b]

(a) Pyrrole ring[c]	Pyrrole N plane	Porphyrin ring plane
A	9.4 (12.6)	6.7 (8.3)
B	11.1 (14.1)	11.9 (12.5)
C	8.8 (9.6)	9.8 (13.8)
D	8.2 (12.8)	6.0 (10.0)
(b) Heme coordinate bonds		
Fe–His18 NE2	2.1 (7.2)	3.3 (6.5)
Fe–Met80 SD	4.9 (3.3)	7.5 (5.4)

II. Heme iron coordination bond distances (Å)

His18 NE2	1.99 (2.01)
Met80 SD	2.35 (2.43)
Hem NA	1.97 (1.97)
Hem NB	2.00 (1.98)
Hem NC	1.99 (2.02)
Hem ND	2.00 (2.05)

III. Heme propionate hydrogen bond interactions[d]

Heme atom	Hydrogen bond partners and distances (Å)
O1A	Tyr48 OH 2.83(2.83), Wat121 2.81(2.85), Wat168 2.85(2.87)
O2A	Gly41 N 3.21(2.60), Asn52 ND2 3.34(3.54), Trp59 NE1 3.09(3.43), Wat121 4.01(3.34)
O1D	Thr49 OG1 2.64(2.79), Thr78 OG1 2.90(3.07), Lys79 N 3.17(2.67)
O2D	Thr49 N 2.94(2.75)

[a] Values in parentheses are for the oxidized form of the protein.

[b] Each pyrrole ring plane is defined by nine atoms, which include the five ring atoms plus the first carbon atom bonded to each ring carbon. The porphyrin ring plane is defined by the five atoms in each of the four pyrrole rings, the four bridging methine carbon atoms, the first carbon atom of each of the eight side-chains, and the heme iron (33 atoms in total). The pyrrole N plane is defined by only the four pyrrole nitrogen atoms. These two latter planes differ in orientation by an angle of 2.6° in the reduced protein and 4.3° in the oxidized form.

[c] See Figure 3.2 for a schematic representation of the heme labeling conventions used herein.

[d] Distances are provided for the heme O2A to Wat121 (reduced state) and Asn52 ND2 (oxidized state) interactions for the sake of comparison, even though these are too long to represent hydrogen bonds.

The two heme propionate groups are buried within the protein matrix and form a number of hydrogen bonds with nearby polar groups (Table 3.7). Thus, while internally positioned, these propionate groups occupy a local environment that is clearly hydrophilic in character. Figure 3.10 illustrates the placement of these groups and the hydrogen bonding observed. In addition to the hydrogen bonds listed in

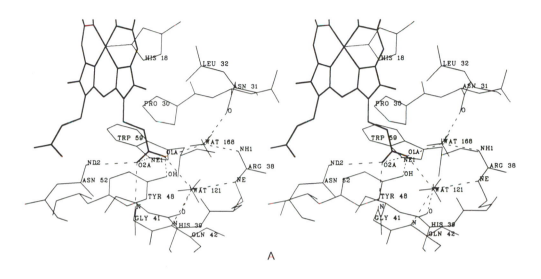

A

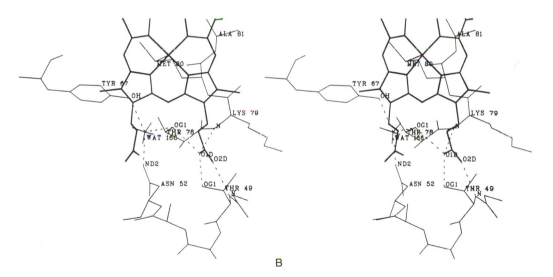

B

Figure 3.10

Local environment and hydrogen-bonding interactions observed for the heme propionate groups attached to (a) pyrrole ring A and (b) pyrrole ring D in yeast iso-1-cytochrome *c* (see Figure 3.2 for heme atom nomenclature used). Heme and propionate groups are drawn with thick lines, and the surrounding polypeptide chain with thinner lines. Hydrogen bonds are illustrated with thin broken lines. Groups involved in forming hydrogen bonds to heme propionate carboxyl oxygen atoms, along with the hydrogen bond lengths, are detailed in Table 3.7.

Table 3.7, a charged interaction between the heme propionate carboxyl atom O1A and the guanidinium group of Arg38 is formed. As shown in Figure 3.10, this interaction is mediated by two water molecules.

5. Hydrogen bonding

A summary of all intramolecular hydrogen bonding interactions present in yeast iso-1, yeast iso-2, tuna, horse, and rice cytochromes c is presented in Table 3.8. In the first section of Table 3.8, hydrogen bonds found in common in all these cytochromes are listed and subdivided into three classes: those involving only main chain atoms, those between main chain and side chain atoms, and those in which only side chain atoms are involved. Following sections of Table 3.8 list other hydrogen bonds occurring in each of these cytochromes c. Of the 213 main chain groups that could potentially form hydrogen bonds in yeast iso-1-cytochrome c, 96 make hydrogen bonds to other main chain atoms in the molecule. An additional 33 main chain groups make intramolecular hydrogen bonds only to side chain atoms. In total, 61% of main chain atoms potentially capable of doing so form intramolecular hydrogen bonds, and in only 23 of the 108 residues does a main chain atom not directly form such a hydrogen bond. In addition, 31 carbonyl oxygen atoms and nine amide nitrogen atoms are each involved in two hydrogen bond interactions. Only 24 (11% of the total) main chain groups (10 amide nitrogen and 14 carbonyl oxygen atoms) do not form hydrogen bonds to other protein atoms or to solvent molecules.

The extent of hydrogen bonding involving side chain atoms is even greater. Yeast iso-1-cytochrome c possesses 62 polar side chains which bear 91 atoms capable of participating in hydrogen bonding. Hydrogen-bonded interactions are not observed for only two side chains, and for only six polar atoms. The two amino acid residues found not to be involved in hydrogen bonding are both surface lysine residues (residues 4 and 100), the side chains of which are directed into the solvent. Failure to observe hydrogen bonding in these cases is likely due to the substantial mobility of these lysine residues and of their associated solvent molecules. Considering that side chain oxygen atoms are capable of participating in two (carboxyl and amide) or three (hydroxyl) hydrogen bonds each and that many side chain nitrogen atoms can act as hydrogen bond donors to two (Gln NE2; Asn ND2; and Arg NH1, 2) or three (Lys NZ) acceptors, yeast iso-1-cytochrome c contains a total of 197 side chain hydrogen-bonding functionalities. Of these, 131 (67%) are observed to participate in hydrogen bonds. Of the 119 hydrogen bonds formed by side chain atoms, 45 are intramolecular, with the remainder involving solvent molecules or other protein molecules in the crystal lattice.

Overall, 90.1% of all main chain and side chain polar groups form well-defined hydrogen-bond interactions. The saturation of hydrogen bonding potential, considering all possible hydrogen bonding functionalities, is 69.4% (360 of 519). This high degree of hydrogen bonding present in iso-1-cytochrome c is typical of that found in other protein structures (Baker and Hubbard, 1984) and undoubtedly contributes significantly to the stability of the folded state of this protein. The lack of fulfillment of hydrogen bonding potential that is present in the iso-1-cytochrome c structure is due, in most cases (particularly for main chain atoms), to the rendering of some potential sites inaccessible to favorable hydrogen-bonding interactions by the fold of the protein.

Table 3.8
Hydrogen bonds in yeast iso-1, yeast iso-2, tuna, horse and rice cytochromes c^a.

I. Hydrogen bonds common to all five cytochromes c

A. Main/main[b]

6 N–2 O	34 N–102 O	67 N–63 O	93 N–89 O
10 N–6 O	35 N–32 O	68 N–64 O	94 N–90 O
11 N–7 O	38 N–35 O	70 N–67 O	95 N–91 O
12 N–8 O	40 N–57 O	74 N–70 O	96 N–92 O
14 N–10 O	53 N–49 O	75 N–71 O	97 N–93 O
15 N–10 O	54 N–50 O	78 N–75 O	98 N–94 O
17 N–14 O	59 N–38 O	85 N–68 O	99 N–95 O
24 N–21 O	64 N–60 O	91 N–87 O	101 N–97 O
29 N–17 O	65 N–61 O	92 N–88 O	102 N–98 O
32 N–19 O			

B. Main/side

2 N–Asp93 OD1	Thr19 OG1–25 O	49 N–Hem O2D	79 N–Hem O1D
His18 ND1–30 O	41 N–Hem O2A	52 N–Thr/Ser49 OG	80 N–Thr78 OG1

C. Side/Side

Tyr48 OH–Hem O1A	Trp59 NE1–Hem O2A	Tyr67 OH–Met80 SD[c]	Thr78 OG1–Hem O1D
Thr/Ser49 OG–Hem O1D			

II. Other hydrogen bonds found in yeast iso-1-cytochrome c (reduced)

A. Main/main

7 N–3 O	18 N–14 O	55 N–52 O	73 N–70 O
8 N–4 O	27 N–29 O	69 N–65 O	82 N–80 O
9 N–5 O	37 N–59 O	69 N–66 O	100 N–96 O
13 N–9 O	48 N–46 O		

B. Main/side

(−3) N–Thr(−5) OG1	Lys27 NZ–15 O	43 N–Tyr48 OH	Lys79 NZ–47 O
1 N–Thr96 OG1	31 N–His26 ND1	Tyr46 OH–28 O	Lys86 NZ–69 O
5 N–Ser2 OG	Asn31 ND2–21 O	Lys55 NZ–74 O	Arg91 NH2–85 O
Thr8 OG1–5 O	His33 ND1–20 O	57 N–Ser40 OG	Arg91 NH2–86 O
Thr12 OG1–8 O	33 N–Asn31 OD1	62 N–Asp60 OD1	Thr96 OG1–92 O
Thr12 OG1–9 O	Arg38 NH1–33 O	63 N–Asp60 OD1	Lys99 NZ–96 O
20 N–Glu21 OE1	Arg38 NH2–33 O	Thr69 OG1–65 O	Cys102 SG–98 O
His26 NE2–44 O	Ser40 OG–52 O	73 N–Asn70 OD1	

C. Side/side

Thr(−5)OG1–Glu61 OE1	Asn31 ND2–Thr19 OG1	Asn52 ND2–Hem O2A	Tyr74 OH–Asn63 OD1
Thr(−5)OG1–Asn62 OD1	Thr49 OG1–Hem O1D	Asn63 ND2–Asp60 OD2	Arg91 NE–Ser65 OG

III. Other hydrogen bonds found in yeast iso-2-cytochrome c (reduced)

A. Main/main

(−5) N–(−9) O	10 N–7 O	55 N–51 O	92 N–89 O
(−4) N–(−8) O	13 N–10 O	56 N–53 O	96 N–93 O
7 N–3 O	18 N–14 O	66 N–62 O	100 N–96 O
8 N–4 O	27 N–29 O	69 N–66 O	100 N–97 O
9 N–5 O	37 N–34 O	70 N–66 O	103 N–100 O

(continued)

Table 3.8 *(continued)*
Hydrogen bonds in yeast iso-1, yeast iso-2, tuna, horse and rice cytochromes c^a.

B. Main/side

(−9) N–Glu66 OE2	Lys27 NZ–16 O	Tyr46 OH–28 O	73 N–Asn70 OD1
Thr(−5) OG1–(−9) O	Asn31 ND2–21 O	Ser47 OG–47 O	Lys79 NZ–47 O
Thr(−5) OG1–(−8) O	31 N–Asn26 OD1	Lys55 NZ–74 O	Lys86 NZ–69 O
1 N–Thr96 OG1	His33 ND1–20 O	57 N–Ser40 OG	Lys86 NZ–83 O
5 N–Ser2 OG	33 N–Asn31 OD1	62 N–Asp60 OD1	Arg91 NH2–85 O
Thr8 OG1–5 O	Arg38 NH1–33 O	Ser63 OG–58 O	Arg91 NH2–86 O
Thr12 OG1–9 O	Arg38 NH2–33 O	63 N–Asp60 OD1	Thr96 OG1–92 O
Thr12 OG1–8 O	Ser40 OG–52 O	Thr69 OG1–65 O	Lys103 NZ–103 O
Lys27 NZ–15 O	43 N–Tyr48 OH	Thr69 OG1–66 O	

C. Side/side

Thr19 OG1–Asn31 ND2	Asn52 ND2–Hem O2A	Lys86 NZ–Asn70 ND2	Arg91 NE–Ser65 OG
Thr49 OG1–Hem O1D	Tyr74 OH–Ser63 OG		

IV. Other hydrogen bonds found in tuna cytochrome c (reduced)

A. Main/main

7 N–3 O	18 N–14 O	55 N–52 O	73 N–70 O
8 N–4 O	28 N–17 O	66 N–62 O	100 N–96 O
9 N–5 O	37 N–59 O	69 N–66 O	101 N–98 O
13 N–9 O	46 N–43 O		

B. Main/side

Thr9 OG1–5 O	33 N–Asn31 OD1	Ser54 OG–51 O	Lys79 NZ–47 O
16 N–Gln16 OE1	Arg38 NH1–31 O	Lys55 NZ–74 O	Lys86 NZ–69 O
20 N–Glu21 OE1	Thr40 OG1–52 O	60 N–Thr63 OG1	Lys86 NZ–83 O
21 N–Glu21 OE1	Gln42 NE2–40 O	Thr63 OG1–58 O	Arg91 NE–85 O
His26 NE2–44 O	43 N–Tyr48 OH	63 N–Asn60 OD1	Arg91 NH2–85 O
31 N–His26 ND1	Tyr46 OH–28 O	Thr63 OG1–60 O	Thr102 OG1–98 O
Asn31 ND2–21 O	Ser54 OG–50 O		

C. Side/side

Lys13 NZ–Glu90 OE1	Asn31 AD2–Thr19 OG1	Thr49 OG1–Hem O1D	Thr63 OG1–Asn60 AD1
Thr19 OG1–Asn31 AD2	Arg38 NH1–Hem O1A	Asn52 AD1–Hem O2A	Lys99 NZ–Asn61 AD2

V. Other hydrogen bonds found in horse cytochrome c (oxidized)

A. Main/main

13 N–9 O	37 N–59 O	66 N–62 O	91 N–88 O
18 N–15 O	55 N–52 O	69 N–65 O	100 N–96 O

B. Main/side

2 N–Asp2 OD1	Thr40 OG1–57 O	57 N–Thr40 OG1	Lys86 NZ–69 O
Thr19 OG1–19 O	Thr40 OG1–40 O	60 N–Thr63 OG1	Lys86 NZ–83 O
His26 NE2–44 O	Lys53 NZ–49 O	Thr63 OG1–60 O	Arg91 NE–85 O
31 N–His26 ND1	Lys55 NZ–74 O	73 N–Asn70 OD1	Thr102 OG1–98 O
His33 ND1–20 O	56 N–Thr40 OG1		

C. Side/side

Lys5 NZ–Asp2 OD2	Arg38 NH1–Hem O1A	Thr49 OG1–Hem O2D	Lys53 NZ–Asp50 OD2
Thr19 OG1–Asn31 ND2	Thr49 OG1–Hem O1D	Asn52 ND2–Hem O2A	Lys99 NZ–Glu61 OE2

VI. Other hydrogen bonds found in rice cytochrome c (oxidized)

A. Main/main

(−4) N–(−7) O	13 N–9 O	53 N–50 O	69 N–66 O
(−3) N–(−6) O	18 N–14 O	55 N–51 O	73 N–70 O
7 N–3 O	27 N–29 O	56 N–53 O	74 N–71 O
8 N–4 O	37 N–59 O	66 N–62 O	103 N–100 O
9 N–5 O			

Table 3.8 (*continued*)

B. Main/side			
(−8) N–Glu−4 OE1	Asn31 ND2–21 O	57 N–Ser40 OG	Gln89 NE2–(−1) O
4 N–Asn2 OD1	Asn33 OD1–20 O	60 N–Thr63 OG1	Gln89 NE2–89 O
5 N–Asn2 OD1	33 N–Asn31 OD1	61 N–Glu61 OE1	Arg91 NH2–65 O
Thr12 OG1–9 O	Ser40 OG–52 O	Thr63 OG1–58 O	Arg91 NH1–86 O
Thr12 OG1–8 O	Thr42 OG1–39 O	Thr63 OG1–60 O	Ser96 OG–92 O
His26 NE2–44 O	43 N–Tyr48 OH	73 N–Asn70 OD1	Thr102 OG1–98 O
Lys27 NZ–15 O	Tyr46 OH–28 O	Lys79 NZ–47 O	Ser103 OG–99 O
31 N–His26 ND1	Lys53 NZ–48 O	87 N–Glu90 OE2	

C. Side/side			
Lys22 NZ–Asn33 ND2	Thr43 OG1–Tyr48 OH	Lys73 NZ–Asn70 ND2	Lys99 NZ–Glu61 OE2
Arg38 NH2–Thr43 OG1	Ser49 OG–Hem O1D		

[a] Hydrogen bonds were defined by the following criteria: a H ... A distance <2.60 Å, 2.70 Å, or 3.05 Å where A is an oxygen, nitrogen, or sulfur atom, respectively: a D-H ... A angle >120°; and a C-A ... H angle >90°. In the tuna coordinate set, atoms of the side-chain amide groups of asparagines and glutamines were not distinguished and were simply labeled AD1, AD2, and AE1, AE2, respectively. That labeling is retained in this table. Hydrogen bonds were analyzed for both conformations of these amide groups for the tuna structure and Asn33 of the rice structure. The imidazole ring of His33 of the horse structure was rotated 180° to optimize hydrogen bond interactions.

[b] Where main chain atoms are listed, only the residue number is given.

[c] As discussed elsewhere, the nature of this interaction is likely oxidation state–dependent.

6. Solvent structure

During the structural refinement of yeast iso-1-cytochrome *c*, a total of 116 water molecules were identified. Of these, six water molecules occupy sites that are inaccessible to a 1.4 Å probe sphere used to delimit the envelope enclosing the volume of the molecule (four of these six are buried in the protein interior, whereas two occupy depressions in the protein surface). The thermal factors of all observed water molecules range from 10.1 to 62.2 Å² with a mean of 37.1 Å². The majority of the water molecules populate the first solvation layer of the protein molecule, with 93% of those identified being within 4 Å of the protein surface. The overall average distance from each water molecule to the nearest protein atom is 3.1 Å. A more detailed analysis of the hydrogen bonding of surface water molecules and their organization into networks is presented in Louie and Brayer (1990). The solvent structure about yeast iso-1-cytochrome *c* also contains a well-ordered sulfate ion located at the N-terminal end of the helical segment containing residues 2 to 14 (Table 3.5), to which it forms a number of hydrogen bonds.

In total, there are four water molecules completely buried within the interior of yeast iso-1-cytochrome *c*. One of these, Wat110 is positioned adjacent to the His18 ligand (see Figure 12a in Louie and Brayer, 1990). The other three are in close proximity to the heme group. As shown in Figure 3.10b, Wat166 is hydrogen-bonded to atoms belonging to three buried side chains (Asn52 ND2, Tyr67 OH, and Thr78 OG1). Together, Wat121 and Wat168 act as a bridge between the heme propionate atom O1A and the guanidinium group of Arg38. In addition (Figure 3.10a), these water molecules also form hydrogen bonds to other internal main chain atoms (His39 O and Gln42 N with Wat121, and Asn31 O with Wat168). Table 3.9 lists all those water molecules conserved in the structures of yeast iso-1, yeast iso-2, tuna, horse, and rice cytochromes c. Of the four completely internal water molecules present in yeast iso-1-cytochrome c, only Wat121 and Wat166 have counterparts in the vertebrate cytochromes c. As discussed later, both of these water molecules appear to play important functional roles in the electron transfer mechanism of cytochrome c.

7. Polypeptide chain mobility

Refined thermal factors can be used to assess the relative flexibility of various regions of the yeast iso-1-cytochrome c molecule. The overall average for all non-hydrogen atoms is 16.5 $Å^2$. Several stretches of polypeptide chain backbone are associated with low thermal factors. The majority of these are involved in forming the heme pocket. They include amino acid residues packing directly against the heme (residues 14 to 18, 28 to 32, and 78 to 82) and three α-helical regions (residues 2 to 14, 60 to 70 and 87 to 102) that form part of the heme pocket. The heme itself is held tightly within the molecule, having an average thermal factor of 5.3 $Å^2$. Those regions of the polypeptide chain that are further removed from the heme and that occupy surface positions have much higher thermal factors. These include the N-terminal region (residues -5 to -1); residues 36 to 39, which form a β-turn (Table 3.5); residues 51 to 61, which include a small helix and adjoining coil region; and residues 73 to 75, which are toward the end of a short helical segment. A composite plot of the average thermal factors of yeast iso-1, yeast iso-2, tuna, horse, and rice cytochromes c along the course of the polypeptide chain is shown in Figure 3.11.

8. Other distinctive features

a. *N-terminus*   Typical of fungal cytochromes c the polypeptide chain of yeast iso-1-cytochrome c has an extension at its N-terminus (5 amino acids for this protein; see Table 3.1). Residues -5 to -1 have ϕ, ψ values characteristic of an extended conformation (Schulz and Schirmer, 1979). Notably, Gly1 adopts an unusual conformation ($\phi = -132°$, $\psi = -146°$; accessible only to glycine residues), which causes a bend in the polypeptide chain at this position. As a result, the N-terminal residues of yeast iso-1-cytochrome c are projected parallel to the rear surface of the molecule and make few interactions with the body of the molecule (Figure 3.3). The most prominent contacts are made by the side chain of Phe(-3), which occupies a small surface pocket formed by the atoms Glu61 CG and CD, Asn92 CA, and Ile95 CB and CG2. The lack of intramolecular interactions involving this N-terminal segment is reflected in high thermal factors for residues -5 to -1.

b. *ε-N-Trimethyl lysine 72*   The cytochromes c of fungi and plants are trimethylated at the epsilon nitrogen of lysine 72 (Paik *et al.*, 1989). This residue is specifically modified by the enzyme S-adenosylmethionine: protein-lysine N-methyltransferase

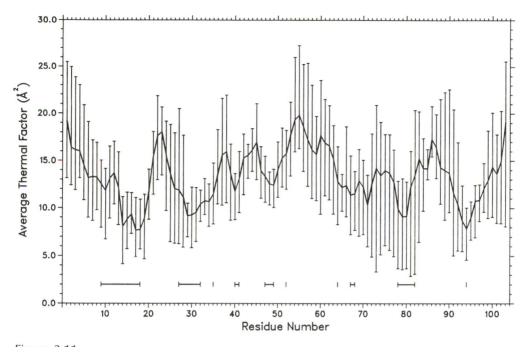

Figure 3.11
A plot of the overall average thermal factors of main chain atoms for those amino acids
common to yeast iso-1, yeast iso-2, tuna, horse, and rice cytochromes c. In constructing this
plot, the thermal factors for all structures were initially normalized to those of yeast
iso-1-cytochrome c. The vertical bars indicate the range of thermal factors observed at a
given amino acid position. This analysis can be used as a measure of the relative flexibility of
different regions of the polypeptide chain. The lower horizontal bars delineate those amino
acids within 8 Å of the heme porphyrin ring (see Table 3.7 for definition). It is clear that the
heme moiety has a marked influence on the mobility of adjacent polypeptide chain, with
nearby residues showing considerably less flexibility than those more remote.

(DiMaria et al., 1979). Various roles, generally relating to involvement in forming
intermolecular interactions, have been proposed for the trimethylated lysine residue.
Polastro et al. (1978) have shown that the methylated iso-1-cytochrome c species
binds more tightly to mitochondria than does the unmethylated species, possibly
through an increased affinity for cytochrome c oxidase (Holzschu et al., 1987).
Farooqui et al. (1981) have found that methylated iso-1-cytochrome c is degraded
intracellularly at a decreased rate, likely as a result of the indirect protection from
proteolytic attack afforded by increased binding to mitochondria. Methylation of
lysine 72 has also been suggested to be required for transport of newly synthesized
cytochrome c from the cytoplasm to the inner mitochondrial membrane (DiMaria et
al., 1979; Paik et al., 1980). Finally, the presence of trimethylated lysine 72 (Tml72)
has been shown to result in an increased binding affinity of iso-1-cytochrome c for
cytochrome b_2 (Holzschu et al., 1987). Despite these findings, Holzschu et al. (1987)

have shown that replacement of Tml72 by an arginine residue has only minor effects on the properties of yeast iso-1-cytochrome c. Thus, the functional significance of lysine 72 trimethylation is not yet firmly established.

As illustrated in Figure 3.7, the side chain of Tml72 has essentially a fully extended conformation and projects directly towards the front of the molecule. The bulky and aliphatic trimethylamine has one of its methyl groups packed against Ala81. Relative to other lysine side chains in iso-1-cytochrome c, that of Tml72 is fairly rigidly positioned (average thermal factor 20.7 Å^2)

Structural studies, coupled with other functional studies, indicate two possible roles for trimethylation of lysine 72. First, the aliphatic portion of the side chain of Tml72, in occupying a position near the Met80 ligand, may contribute to maintaining the integrity of the heme pocket. This would be consistent with the lowered stability of the Arg-72 variant iso-1-cytochrome c (Holzschu et al., 1987). The more hydrophilic side chain of arginine would be unlikely to pack against residue 81 or to adopt the same rigid conformation observed for Tml72. Secondly, the trimethylamine group may form complementary interactions with other proteins. Holzschu et al. (1987) have shown that the Arg72 variant of iso-1-cytochrome c has ~4-fold higher K_m for binding to cytochrome b_2. Additional evidence is provided by NMR studies of the interaction between cytochrome b_5 and cytochrome c from the yeast Candida krusei, which show that the linewidth of the trimethyl group of Tml72 is considerably broadened upon bimolecular complexation (Eley and Moore, 1983). The indicated decrease in mobility of Tml72 suggests that this residue forms part of the protein–protein interface.

c. Cysteine 102   One unusual feature of yeast iso-1-cytochrome c is the presence of a cysteine at position 102 in the sequence. This residue is positioned on the inward-facing side of the C-terminal α-helix (Figure 3.7; Table 3.5). The sulfhydryl group is inaccessible to solvent and occupies a predominantly hydrophobic pocket formed by the side chains of Val20, Leu32, Ile35, Phe36, and Leu98. The χ_1 torsion angle of $-55°$ places the SG atom in a position where it can form a hydrogen bond to the carbonyl group of Leu98.

Because the side chain of Cys102 is normally directed into the interior of the molecule, covalent dimerization of iso-1-cytochrome c via intermolecular disulfide bond formation (Bryant et al., 1985) would require at least a partial unfolding of the C-terminal helices of the two molecules involved. Since the C-terminal helix is an important structural component of the heme pocket, dimerization would be expected to destabilize the overall fold of the protein. These considerations are consistent with experiments which show that dimerization increases the susceptibility of iso-1-cytochrome c to denaturation by heat, guanidine hydrochloride, urea, and acid (Bryant et al., 1985); alkaline isomerization, and imidazole and azide binding to the heme iron (Saigo, 1986); and digestion by proteases (Motonaga et al., 1965). From measurements of the same susceptibilities, Bryant et al. (1985) and Motonaga et al. (1965) have found that reaction of Cys102 with iodoacetate or thiosulfite also destabilizes iso-1-cytochrome c. Inspection of the structure of yeast iso-1-cytochrome c indicates that the pocket occupied by the SG atom of Cys102 is compact, and thus derivatization of the sulfhydryl group with bulky substituents would perturb the native conformation of this region of the molecule. It is also notable that Moench and Satterlee (1989) have shown, using NMR techniques, that both dimerization and modification with 5,5′-dithiobis(2-nitrobenzoate) cause significant structural changes in the iso-1-cytochrome *c* molecule, which in turn perturb the heme environment.

B. Structural Conservation between Yeast Iso-1 and Other Cytochromes *c*

As is evident from Figure 3.1, the sequences of eukaryotic cytochromes *c* are highly conserved. Indeed, some 20% of the amino acids are invariant. As Tables 3.1 and 3.4 indicate, the sequence identities between those five cytochromes *c* for which high-resolution structures are known varies from a low of 51% between yeast iso-2 and rice cytochromes *c* to a high of 85% between the two yeast isozymes. Two trends emerge from a closer look at these data. First, those side chains positioned at the protein surface account for the vast majority of sites at which amino acid variability occurs. Second, sequence homology is highest towards the C-terminal end of the polypeptide chain and for those amino acids which either form a part of the heme environment or are near the solvent-exposed heme edge.

Not unexpectedly, the high degree of sequence homology expresses itself in a high degree of structural conservation among yeast iso-1, yeast iso-2, tuna, horse, and rice cytochromes *c*. This is clearly illustrated in Figures 3.3 and 3.7, which show the structures of all five cytochromes *c* in comparable orientations based on the rotation matrices of Table 3.3. Reference to Table 3.4 shows the average deviations between main chain atoms in these structures varies from a low of 0.30 Å between the yeast isozymes to a high of 0.57 Å between yeast iso-2 and horse cytochromes *c*. The overall average deviation between all pairwise comparisons in Table 3.4 is 0.45 Å. This is a remarkably small number for proteins from such a wide range of sources.

The high degree of structural similarity shared by these cytochromes *c* is further emphasized in the composite plot shown in Figure 3.12. Here, the overall average deviation of main chain atoms in yeast iso-2, tuna, horse, and rice cytochromes *c* from yeast iso-1-cytochrome *c* is presented as a function of residue number. A further indication of the range of pairwise deviations between these cytochromes *c* is given by the vertical bars. Although there are small regions of differences, which will be discussed in a following section (particularly in the amino terminal half of the molecule), one is struck by the remarkable overall similarity in amino acid positioning in all five cytochromes *c*.

As Figure 3.11 shows, structural homology also extends to polypeptide chain mobility; that is, there is the tendency for the same residues in all five cytochromes *c* to have the same relative thermal motion. Furthermore, as the horizontal bars in Figure 3.11 clearly indicate, the heme group has a dominating influence on the dynamics of the cytochrome *c* molecule, and there is a strong correlation between close positioning to this group and low observed thermal factors. Comparison of Figures 3.11 and 3.12 also shows that regions having higher polypeptide chain mobilities are those that tend to have the largest deviations in three-dimensional structure between cytochromes *c*. These stretches of polypeptide chain also show a strong tendency to be in exposed locations on the protein surface.

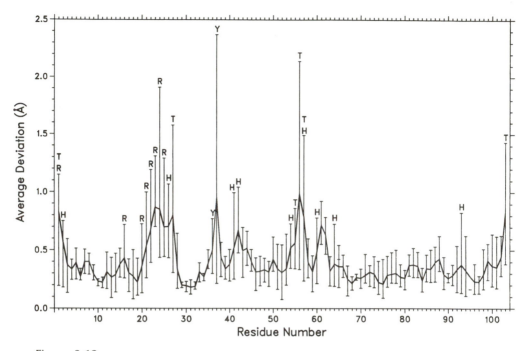

Figure 3.12

A plot of the overall average deviations of main chain atoms of yeast iso-2, tuna, horse, and rice cytochromes c from those of yeast iso-1-cytochrome c along the course of the polypeptide chain. Only residues 1 to 103, which are common to all five cytochromes c, are represented. The vertical bars represent the range of individual pairwise average deviations of main chain atoms between yeast iso-1 and the other cytochromes c. Structures with pairwise deviations both larger than 0.7 Å and 30% greater than the overall average deviations of all structures are labeled at the particular residues involved (Y = yeast iso-2, T = tuna, H = horse and R = rice cytochromes c).

The strong sequence and structural homology observed is also expressed in a high degree of conservation in hydrogen bonding patterns between all five cytochromes c. As shown in Table 3.8, a total of 37 main chain to main chain, eight side chain to main chain, and five side chain to side chain hydrogen bonds are conserved in all these cytochromes c. This represents 48% of all the hydrogen bonds in yeast iso-1-cytochrome c. Hydrogen bond conservation is highest in those secondary structural elements that form the heme crevice or for residues directly in the heme pocket and interacting with heme groups.

Coincident with the strong structural conservation in the heme pocket is the similarity in heme positioning in cytochromes c. This is further reflected in a similar presentation of the exposed edge of the heme on the protein surface. As Table 3.6 indicates, overall heme exposure averages about 9%, and there is a clear pattern of heme atoms involved. Two internally bound water molecules, 121 and 166 of the yeast iso-1 protein, are also highly conserved in cytochromes c (Table 3.9), and as discussed herein are likely to be important in the function of this protein.

Table 3.9
Conserved water molecules found in yeast iso-1, yeast iso-2, tuna, horse, and rice cytochromes *c*[a].

Residue numbers of conserved water molecules					Average thermal factor	Conserved hydrogen bonds
Iso-1	Iso-2	Tuna	Horse	Rice		
121	118	26	125	115	17.8	39 O, 42 N, Arg38 guanidino (not in iso-2), Hem O1A
122	115	1	128	117	14.0	79 O, 81 N
166	106	3	112	131	18.3	Asn(Asp)52 ND2(OD2), Tyr67 OH, Thr78 OG1 (not in horse)
196	142	4	142	116	21.4	Phe36 N (not in iso-1)

[a] A water molecule was considered conserved if the corresponding water in the other overlapped structures was less than 2.0 Å away using yeast iso-1-cytochrome *c* as the reference structure. A further constraint was the presence of conserved water-protein hydrogen bonds. Hydrogen bond exceptions are noted in parentheses.

C. Unique Structural Features of Yeast Iso-2, Tuna, Horse, and Rice Cytochromes *c*

1. Polypeptide chain conformations

Despite the high degree of structural similarity among all five high resolution cytochrome *c* structures, there are points of substantial structural differences. These differences are depicted in Figure 3.12. A clear example of structural dissimilarity involves the conformations at both the N and C-terminal ends in all five cytochromes *c*. Many of these differences undoubtedly reflect the surface positioning of these termini and a greater flexibility to assume alternative conformations. In the cases of the yeast iso-1, yeast iso-2, and rice proteins, there are large extensions to the polypeptide chain at the N terminus (Table 3.1) whose conformations strongly influence the positioning of residues 1 to 3 (Louie *et al.*, 1988a). Horse cytochrome *c* is also unique in having a one-residue extension at the C-terminal end of the polypeptide chain. This additional residue is found in a highly solvent-exposed position (Figure 3.3) and has only a relatively small effect on the conformations of the preceding amino acid residues.

Also showing a higher degree of variance is the surface polypeptide chain loop consisting of residues 22 to 27 (Figure 3.3), which includes a type II β-turn involving residues 21 to 24 (Table 3.5). This loop is positioned differently in rice cytochrome *c* (Figure 3.12). Apparently, this results from having an alanine residue at position 24 (see Table 3.1) that would spatially conflict with nearby residues in the other cytochromes *c*. In yeast iso-1, yeast iso-2, tuna, and horse cytochromes *c*, residue 24 is a glycine. Thus, this region in rice cytochrome *c* has a more open conformation, which substantially lengthens the β-turn hydrogen bond between residues 21 and 24. Residues 24 to 26, in particular, take on a completely different conformation in the rice protein. These conformational changes are mediated by a $\sim 180°$ flip of the peptide bond between Ala24 and Gly25, as well as the unique presence of a glycine at residue 25 in rice cytochrome c (Table 3.1), which adopts a conformation inaccessible to amino acids with side chains.

A smaller deviation in polypeptide chain conformation in this area (residues 26 to 27 in particular) is observed in horse and tuna cytochromes c. This deviation appears to arise from the unique presence of proline at sequence position 25 in the yeast iso-1 and iso-2 cytochromes c (lysine in both the horse and tuna proteins). This rotationally constrained amino acid has a somewhat different ψ angle, which alters the course of the main chain about residues 26 and 27. Beyond this local perturbation, the polypeptide chains of all five cytochromes c return to a homologous course (Figure 3.12).

As Figure 3.12 illustrates, a unique structural feature occurs in yeast iso-2 cytochrome c about residue 37. In all five cytochromes c, this residue is a glycine (Table 3.1); however, in all but the yeast iso-2 protein, this residue packs against the adjacent amino acid at position 58, which is generally hydrophobic in nature. One potential explanation for the aberrant conformation observed in yeast iso-2-cytochrome c is the presence of a charged lysine residue at position 58 in this protein (Murphy et $al.$, 1992).

Another example of a localized conformational difference is observed about residues 56 and 57. These surface residues are displaced further into solvent in the horse and tuna proteins, apparently as a result of the replacements of Ser40 and Val57 found in yeast iso-1, yeast-iso-2, and rice cytochromes c, with threonine and isoleucine, respectively. These larger side chains lead to spatial conflicts between themselves and Trp59, which results in the bulging of the region about residues 56 and 57 out to a more solvent-exposed position (Figure 3.3). Further disruption of the conformation of this region of the polypeptide chain is avoided by the presence of a glycine at residue 56 in the horse and tuna cytochromes c (asparagine in yeast iso-1 and iso-2 cytochromes c, and alanine in the rice protein). Gly56 restores the path of the polypeptide chain to that found in yeast iso-1, yeast iso-2, and rice cytochromes c by adopting a conformation not allowed for amino acids having side chains.

The conformational differences observed about residues 22 to 27 and 56 to 57 emphasize the importance of glycine residues in providing the polypeptide chain with the necessary flexibility to accommodate bulky side chain substitutions. In the approximately 90 known sequences of eukaryotic cytochromes c, the occurrence of a residue with a side chain (generally alanine) at position 24 is almost always accompanied by a glycine at position 25. In addition, where a threonine (instead of serine) occurs at position 40 and a residue larger than valine (most often isoleucine) occurs at position 57, then a glycine is found at position 56.

2. Arginine-38

Arginine-38 is an invariant residue in the sequences of eukaryotic cytochromes c (Figure 3.1, Table 3.1). What is unusual about this residue is the variety of conformations its side chain assumes in various species of cytochrome c to form a salt-bridge interaction with the heme propionate group extending from pyrrole ring A. Shown in Figure 3.13 are the conformations of this residue in all five cytochrome *c* structures. In each case, a distinctive conformation and unique mode of interaction between the guanidinium group and the heme propionic acid is observed.

In yeast iso-1 and iso-2 cytochromes *c*, two internal water molecules are located adjacent to Arg38 (Figures 3.13a and b). In the yeast iso-1 protein, the NH1 and NE atoms of Arg38 form hydrogen bonds to these two water molecules, which are in turn hydrogen-bonded to the heme carboxyl group O1A atom. In yeast iso-2-cytochrome *c*, only one of these internal water molecules directly bridges between Arg38 and the adjacent heme propionate. In rice, the guanidinium group of Arg38 interacts end-on with three bridging water molecules, with both NH1 and NH2 donating hydrogen bonds (Figure 3.13e). In tuna and horse cytochromes *c*, Arg38 NE hydrogen bonds a lone bridging water molecule, while NH1 forms a direct hydrogen bond to the O1A heme atom (Figures 3.13c and d).

Many of the differences observed in Arg38 conformation appear to be due to local amino acid substitutions. In yeast iso-1 and iso-2, the guanidinium group of Arg38 cannot occupy the buried position found in the tuna and horse proteins because the replacement of Leu35 by an isoleucine residue would place the CG1 atom of this alternative side chain too close to the terminal NH1 atom. The further replacement of an alanine for a valine at position 43 in yeast iso-2 cytochrome *c* could account for the differences observed in this region between the yeast isozymes. In rice cytochrome *c*, the external positioning of the guanidinium group of Arg38 seems to be due to the unique hydrogen bond formed between Arg38 NH2 and the OG1 atom of Thr43. In all the other cytochromes *c*, residue 43 is either alanine or valine.

These results demonstrate that conserved or invariant residues do not necessarily carry out their role in an identical manner in different members of a protein family, and that there is flexibility in the formation of their intramolecular interactions. In addition, the intricate interplay between the conserved residue and compensatory changes nearby in the protein environment make difficult the assignment of definitive roles to conserved residues based on the inspection of just one member of the protein family. Nonetheless, the different modes by which a conserved residue can carry out its role provide clues to the specific function of that residue.

V. Oxidation State–Coupled Conformational Changes in Cytochrome *c*

Insight into the nature of oxidation state–dependent conformational changes occurring during cytochrome c–mediated electron transfer events has proven surprisingly elusive. Since early work in the first half of this century (Keilin, 1930), this critical biological process has been studied by a wide variety of techniques (see reviews by Margoliash and Schejter, 1966; Dickerson and Timkovich, 1975; Salemme, 1977; and Mathews, 1985). An abundance of experimental evidence indicates that structural differences do exist between the oxidized and reduced forms of cytochrome *c*. For example, studies suggest that the oxidized state has a significantly increased adiabatic compressibility (Eden *et al.*, 1982). In addition, the radius of gyration as observed by small-angle x-ray scattering is larger for the oxidized protein (Trewhella *et al.*, 1988).

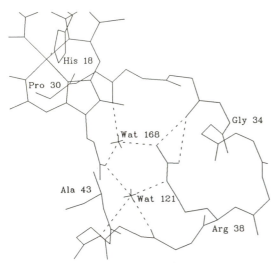

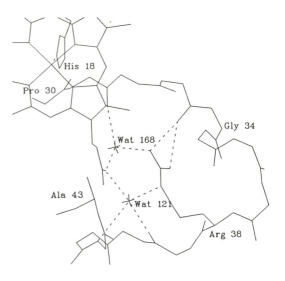

A

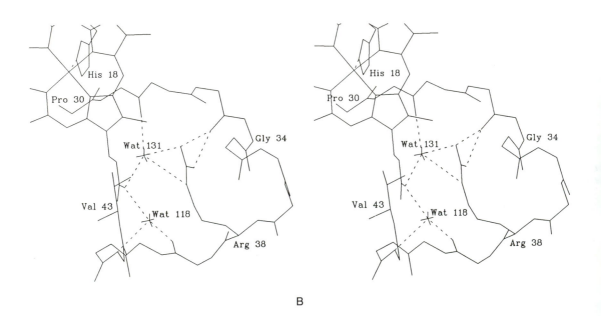

B

Figure 3.13

Comparison of the conformations of the invariantly conserved residue Arg38 in (a) yeast iso-1, (b) yeast iso-2, (c) tuna, (d) horse and (e) rice cytochromes *c*. In all diagrams, the full structures of Arg38 and residue 43 are shown, along with all water molecules hydrogen-bonded to the adjacent heme propionate. Only main chain atoms are drawn for other amino acids in the polypeptide chain segment composed of residues 30 to 42. All hydrogen bonds are illustrated as dashed lines.

144

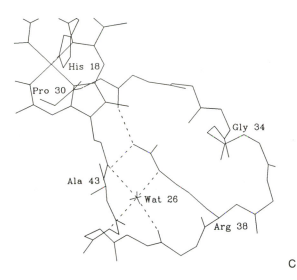
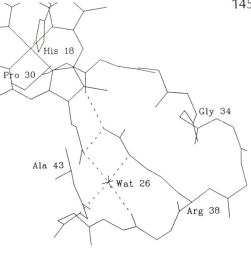

C

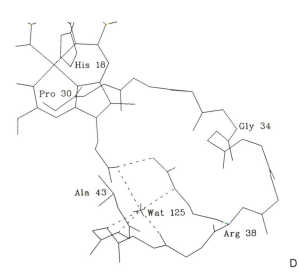
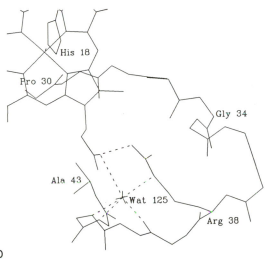

D

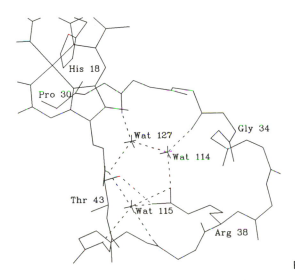
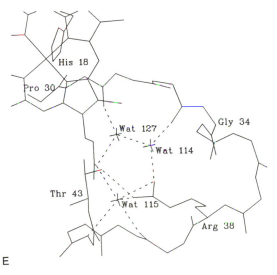

E

The oxidized state of cytochrome c is also more susceptible to proteolytic digestion (Nozaki *et al.*, 1958) and thermal and chemical denaturation (Butt and Keilin, 1962; Dickerson and Timkovich, 1975). Furthermore, an increase in hydrogen exchange rates for the oxidized state of cytochrome c has been measured (Ulmer and Kagi, 1968; Liu *et al.*, 1989). Also, second-derivative amide I infrared spectroscopy suggests that there are differences in the secondary structure between the oxidation states of cytochrome c (Dong *et al.*, 1992). And finally, a differential chemical modification technique (Bosshard and Zurrer, 1980) and one-dimensional NMR studies (Moore, 1983; Williams *et al.*, 1985) provide further indications that the reduced and oxidized forms of cytochrome c have distinctly different conformations. Nonetheless, the origin and nature of these unique properties remains poorly characterized from a structural perspective.

A. Tuna Cytochrome c

The first three-dimensional structural study to address the issue of oxidation state–dependent conformational changes was completed by Takano and Dickerson (1980, 1981a,b). Although isomorphous crystals of tuna cytochrome c in both oxidation states could not be obtained, the reduced and oxidized proteins could be crystallized individually in different space groups. In the case of the oxidized protein, there were two molecules in the crystallographic asymmetric unit.

These studies showed that the most striking differences between oxidation states occur in the vicinity of an internally bound water molecule which is positioned adjacent to the heme group. This water molecule is conserved in all cytochrome c structures determined to date (Table 3.9), and the equivalent water molecule in yeast iso-1-cytochrome c can be seen in Figure 3.10b (Wat166). Upon cytochrome c oxidation, this water molecule, which forms hydrogen bonds to the side chains of Asn52, Tyr67, and Thr78, was found to shift position in a direction towards the heme iron atom. This movement was accompanied by shifts in the side chains of Asn52 and Tyr67. The trigger for this pattern of structural changes was postulated to be the attraction of the internally buried water molecule to the increased positive charge on the oxidized heme iron.

The oxidation state–dependent conformational changes observed for tuna cytochrome c do not seem sufficient to account for the majority of differences observed by other methods. It seems apparent that the structural changes induced are for the most part extremely subtle, requiring very accurate structural analyses to be detected. Unfortunately, the inability to obtain isomorphous crystals of tuna cytochrome c in the same space group would tend to obscure such subtle details. Furthermore, it is possible that differences in the chemical and physical properties of the two oxidation states arise from differences in dynamic behavior. This is an issue that would be difficult to address under the non-isomorphous conditions of the tuna cytochrome c structural analyses. As will be subsequently discussed, more recent studies with the yeast cytochrome c system indicate that such factors appear to play a critical role in oxidation state stabilization and redox partner interactions.

B. Cross-Species Comparisons

There are now available high-resolution structures for three cytochromes *c* in the reduced state (yeast iso-1, yeast iso-2, tuna) and four in the oxidized state (yeast iso-1, tuna, horse, and rice). As illustrated in Figure 3.14, the position of the buried conserved internal water molecule found adjacent to the heme in these proteins (Wat166 in yeast iso-1, Table 3.9) is highly dependent upon oxidation state. Those water molecules found in oxidized structures are on average shifted by 1.2 Å in a direction towards the heme iron atom. The water to heme iron atom distance decreases on average from 6.6 Å to 5.9 Å. Also shown in Figure 3.14 are the conformations of the side chains of Asn52, Tyr67, and Thr78. With only a few exceptions, these amino acid residues are conserved in the sequences of all eukaryotic cytochromes *c*. Between oxidation states, corresponding shifts are observed in the side chains of Asn52 and Tyr67. These shifts, along with that of the centrally buried water molecule, represent the largest consistent dislocations observed on comparison between oxidation states across the yeast iso-1, yeast iso-2, tuna, horse, and rice cytochromes *c*. However, it

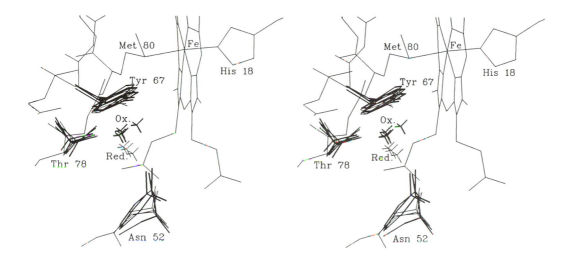

Figure 3.14

A composite stereo diagram showing the positions of a conserved internally bound water molecule (Wat166 in yeast iso-1-cytochrome *c*; see Table 3.9) found in all eukaryotic cytochromes *c* for which three-dimensional structures have been completed. Those water molecule positions drawn with dark lines are from proteins in the oxidized state, while those drawn with thin lines are from reduced proteins. The three amino acid side chains to which this water molecule is hydrogen-bonded (Asn52, Tyr67, and Thr78) are also drawn in the same manner. For clarity, the hydrogen bonds present have not been drawn. This internal water shifts on average 1.2 Å in a direction towards the heme iron atom in the oxidized state of cytochrome *c*. Corresponding shifts occur in the side chains of Asn52 and Tyr67. Also shown for perspective are additional polypeptide chain, heme atoms, and heme ligands from the yeast iso-1-cytochrome *c* structure. As discussed herein, this water molecule may play a role in the electron transfer mechanism of cytochrome *c*.

must be pointed out that other systematic conformational changes may be present but obscured in these analyses by structural dissimilarity arising from amino acid sequence differences.

The results obtained are essentially identical with those observed for the reduced/oxidized pair of structures determined for tuna cytochrome c (Takano and Dickerson, 1981a,b). The consistency of these observations across all the available cytochrome c structures from various sources strongly implicates the observed internally bound water molecule as having an important role in the mechanism of cytochrome c. This conclusion is supported by the close proximity of the heme iron atom, as well as the cross-species conservation of the surrounding amino acid residues to which this internal water molecule is hydrogen-bonded.

C. The Two Oxidation States of Yeast Iso-1-cytochrome c

High-resolution three-dimensional structural analyses of yeast iso-1-cytochrome c have been completed in both oxidation states using isomorphous crystalline material and similar structural determination methodologies (Berghuis and Brayer, 1992). This feature of these structural analyses allows for a comprehensive comparison to be made in the absence of other potential interfering factors such as different crystallizing conditions, packing arrangements or refinement strategies which could obscure the relatively small conformational changes that occur between oxidation states. These studies show oxidation state–dependent changes are expressed for the most part in terms of adjustments to heme structure, movement of internally bound water molecules, and segmental thermal parameter changes along the polypeptide chain, rather than as explicit polypeptide chain positional shifts, which are found to be minimal.

1. Polypeptide chain conformations

A plot of the observed average positional deviations between corresponding main chain and side chain atoms of oxidized and reduced yeast iso-1-cytochrome c is shown in Figure 3.15. The conformational changes observed are small, and it is notable that none of the hydrogen bond interactions between main chain atoms are lost upon change in oxidation state. The overall average value for main chain atom differences in all residues is 0.31 Å. Note that Thr(-5) and Glu(-4) are substantially disordered in both structures of yeast iso-1-cytochrome c, and the differences observed are therefore likely due to differential fits to the same poor electron density, rather than being a function of oxidation state. This disorder is apparent from the large thermal B values (~ 50–55 Å^2) assigned to these residues in both proteins.

Two conformational changes do appear to result from differing oxidation state. The larger involves Gly84 (average deviation of 0.75 Å) and results in the formation of a new hydrogen bond in the oxidized protein (length 2.63 Å) between the side chain of Arg13 (NH1 moves ~ 1.0 Å) and the carbonyl oxygen atom of Gly84 (moves ~ 1.1 Å; see Berghuis and Brayer, 1992). In the reduced protein, both of these groups are associated with surface solvent molecules (Louie and Brayer, 1990). A second

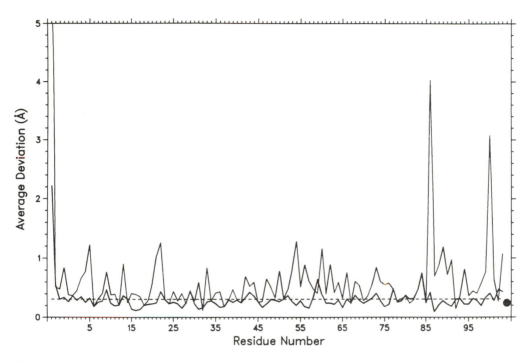

Figure 3.15

A comparison of the average positional deviations between main chain (thick line) and side chain (thin line) atoms of the oxidized and reduced forms of yeast iso-1-cytochrome *c*. The horizontal dashed line represents the overall value of 0.31 Å observed for the average deviation between all main chain atoms. The filled dark circle at position 104 represents the overall average deviation for all heme atoms of 0.24 Å.

conformational change in the main chain of Trp59 (average deviation 0.64 Å) is likely the result of a lengthening in the oxidized protein, of the hydrogen bond between the side chain of this residue and the nearby heme propionate group (from 3.1 to 3.4 Å). This is reflected in an increase in the average thermal factor value for the side chain of Trp59 from 15 to 21 Å^2. Associated with this movement is the reorientation of the side chain of Asp60 (overall average shift 1.2 Å) and a shift in Wat-124 to which both this residue and Trp59 are hydrogen bonded. Other large side chain displacements in Figure 3.15 are associated with poorly defined surface residues and would appear to result from positional uncertainty rather than from change in oxidation state.

For both oxidation states, comparable values are observed for the overall average thermal factor for all atoms in the polypeptide chain of yeast iso-1-cytochrome *c* (16.4 Å^2 and 16.5 Å^2; reduced and oxidized, respectively). However, as evident from Figure 3.16, comparison of thermal factors along the course of the polypeptide chain reveals substantive differences. In the oxidized protein, the most significant decrease in main chain atom thermal factors involves Arg38, a residue having a water medi-

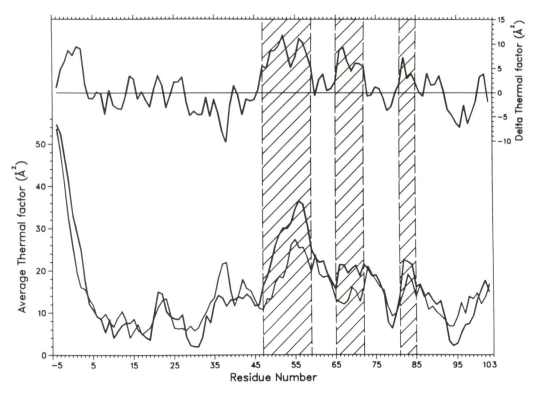

Figure 3.16
Plots of the average thermal factors of main chain atoms along the course of the polypeptide chain in oxidized (thick line) and reduced (thin line) yeast iso-1-cytochrome c. The upper plotted line (axis designation at top right) shows the observed differences between the mean main chain thermal factors in the two oxidation states. Three regions of polypeptide chain with significantly higher thermal factors in the oxidized protein (residues 47–59, 65–72, and 81–85) have been highlighted with cross-hatched boxes.

ated interaction with the pyrrole ring A propionate group (Figure 3.13). Also affected are the two immediately adjacent residues Gly37 and His39. The side chain atoms of Arg38 exhibit an even larger $\sim 14 \text{ Å}^2$ lowering of average thermal factor values in the oxidized state.

In total, four regions of polypeptide chain have significantly higher thermal factors in the oxidized form of yeast iso-1-cytochrome c. The significance of the first, which involves the N-terminus, is questionable because this region is disordered and poorly resolved in electron density maps. The three other segments of polypeptide chain affected (residues 47–59, 65–72 and 81–85) are particularly interesting because they appear to be related to oxidation state–dependent conformational changes in internal water structure and the heme group. As Figure 3.16 illustrates, those portions of polypeptide chain involved are sharply delineated, with maximal increases in thermal factors in the oxidized state being observed for Asn52, Tyr67, and Phe82.

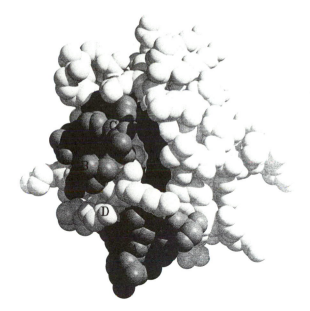

Figure 3.17

A space-filling representation of yeast iso-1-cytochrome *c* showing the location of the three segments of polypeptide chain (drawn with gray spheres) that have main chain thermal factors which are significantly larger in the oxidized protein. A general indication of the location of the individual affected polypeptide chain segments is provided by the letters A–C, corresponding to residues 47–59, 65–72, and 81–85, respectively. Heme atoms are shown with black spheres, while white spheres indicate regions of polypeptide chain with comparable or lower thermal factors in oxidized yeast iso-1-cytochrome *c*. Although contiguously linked through interactions in the protein interior, the three segments of polypeptide chain having higher thermal factors are subdivided at the protein surface by the strand of polypeptide chain composed of residues 73 to 80 (labeled D).

Figure 3.17 shows that these three segments of polypeptide chain are located to the Met80 ligand side of the heme group.

2. Heme structure

High-resolution studies of reduced yeast iso-1-cytochrome *c* have shown that the heme group is substantially distorted from planarity into a saddle shape (see Section IV.A.4). In the oxidized protein the type of distortion observed is similar, but considerably more pronounced suggesting the degree of heme planarity is dependent on oxidation state (Table 3.7). None of the heme iron coordinate distances is significantly different between oxidation states, although the largest deviation observed is a lengthening of the Met80 ligand bond. Overall, the average displacement of side chain atoms of Met80 is 0.36 Å on conversion from the reduced to the oxidized states, with the largest shifts observed for the CB ($\Delta 0.55$ Å) and CE ($\Delta 0.47$ Å) carbon atoms.

152

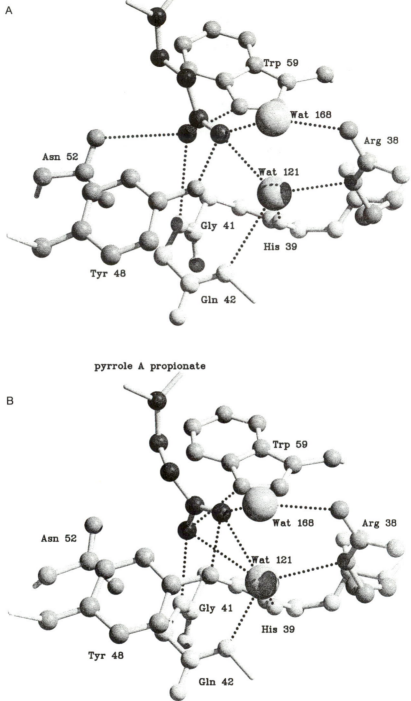

Figure 3.18

Drawings of the region about the pyrrole ring A propionate group in (a) reduced and (b) oxidized yeast iso-1-cytochrome *c*, illustrating the positional shifts and altered hydrogen-bonding patterns observed. The pyrrole ring A propionate group is highlighted with dark-shaded balls. Hydrogen bonds are indicated by thin dashed lines. The two internally bound water molecules, Wat121 and 168, which mediate the interaction of Arg38 with this heme propionate, are shown with larger spheres.

These displacements result in a small 7° change in the torsion angle composed of the CE and SD atoms of this side chain and the heme Fe and NA atoms. It is notable that a substantive increase in thermal factors for the side chain of Met80 is also observed. In reduced yeast iso-1-cytochrome c, the average side chain thermal factor is 5 Å^2, whereas it increases to 12 Å^2 in the oxidized protein. In both structures, the average main chain thermal factor remains constant at 10 Å^2. In contrast, no significant oxidation state–dependent change in ligand bond distance or thermal parameters is observed for His18. However, the orientation of the His18 side chain may be a function of oxidation state in that a 46.7° angle is found between the imidazole ring plane and a vector drawn through the NA and NC pyrrole nitrogen atoms in reduced yeast iso-1-cytochrome c, whereas this value is 55.8° in the oxidized protein.

Another difference between the heme structures of oxidized and reduced cytochrome c is the positioning of the pyrrole ring A propionate group. As evident from Table 3.7 and Figure 3.18, not all groups that hydrogen-bond to this propionate can accommodate the observed positional shifts in the oxidized state. In particular, much weaker hydrogen bonds are made to Asn52 and Trp59, and both of these side chains have increased thermal factors. Also affected is the position of the internally bound water molecule, Wat121, which is conserved in all eukaryotic cytochrome c structures determined to date (Table 3.9). Notably, the side chain of Arg38, whose interaction with the propionate group is mediated by Wat121 and Wat168, shows no significant positional displacement in the oxidized protein.

3. The internal water, Wat166

The conserved and internally bound water molecule, Wat166 (Table 3.9), undergoes a large shift in position in response to oxidation state. As already discussed, this is apparently a characteristic feature of all eukaryotic cytochromes c (see Figure 3.14). In oxidized yeast iso-1-cytochrome c, Wat166 moves 1.7 Å almost directly towards the heme iron atom, closing the distance between these groups from 6.6 Å to 5.0 Å (Figure 3.19). Most affected by this movement is Asn52, to which Wat166 no longer forms a hydrogen bond. The hydrogen bond between Asn52 and the pyrrole ring A propionate group is also lost (Table 3.7).

As illustrated in Figure 3.20a, Wat166 in reduced yeast iso-1-cytochrome c is tightly constrained into a small spherical cavity having a volume of ~10 Å^3. To accommodate Wat166 movement, the densely packed protein matrix must undergo small concerted conformational adjustments, the largest of these involving the side chain of Asn52 (average side chain atom displacement of 0.47 Å). As shown in Figure 3.20b, the resultant cavity surrounding Wat166 in the oxidized protein is significantly expanded (~25 Å^3) and is highly asymmetrical, with Wat166 occupying only that portion of the available volume closest to the heme iron atom. This leaves considerable free volume available to amino acids adjacent to Wat166, and they are therefore conformationally less constrained.

Although the hydrogen atoms of Wat166 cannot be resolved experimentally, the nature of the surrounding hydrogen-bonded network makes it possible to speculate on the orientation of the dipole moment of Wat166 (Berghuis and Brayer, 1992).

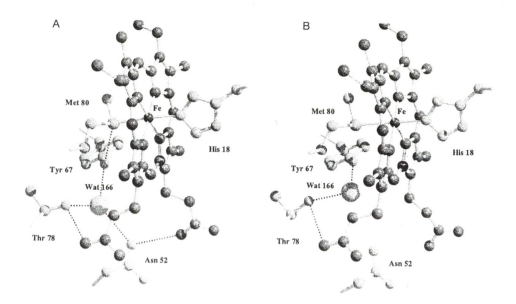

Figure 3.19

Heme and polypeptide chain structure about the internally bound water molecule Wat166 in (a) reduced and (b) oxidized yeast iso-1-cytochrome *c*. Heme ligand interactions are indicated by thin white bonds, whereas hydrogen bonds are shown by thin black dashed lines. Wat166 is found in all eukaryotic cytochrome *c* structures completed to date. In the reduced protein it interacts with three highly conserved residues (Asn52, Tyr67, and Thr78). In oxidized yeast iso-1-cytochrome *c*, the hydrogen bond to Asn52 is broken, and Wat166 shifts 1.7 Å to be within 5.0 Å of the now charged heme iron atom. Wat166 is also a central feature in interactions with three segments of polypeptide chain exhibiting higher thermal parameters in the oxidized protein (see Figure 3.16).

As illustrated in Figure 3.21a, in the reduced protein the positive end of the modelled dipole of Wat166 is pointing in the general direction of the heme, suggesting that it serves to help stabilize this uncharged and electron-rich state. The placement proposed for the dipole moment of Wat166 in the oxidized protein (Figure 3.21b) has its negative end pointed almost directly towards the now positively charged heme iron atom. Given the shorter Wat166 to heme iron atom distance (5.0 Å) in oxidized yeast iso-1-cytochrome *c*, coupled with the reorientation of the Wat166 dipole, one would expect Wat166 to provide a significant contribution toward stabilizing the positive charge on the heme iron atom.

The loss of the Tyr67 OH to Met80 SD hydrogen bond in the oxidized state (see Figure 3.21) could be expected to be an additional stabilizing feature in two ways. First, it would tend to make the Met80 heme ligand less electron-withdrawing, a situation favoring stabilization of the positively charged heme iron atom. Second, the new hydrogen bond interaction from Wat166 to the side chain of Tyr67 assists proper

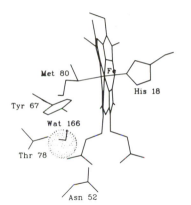
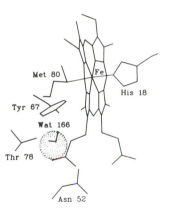

A

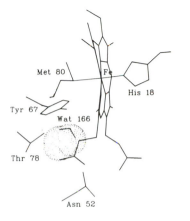
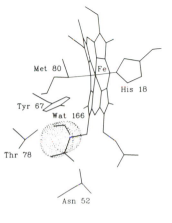

B

Figure 3.20
Stereo drawings of the internal cavity occupied by Wat166 in the (a) reduced and (b) oxidized forms of yeast iso-1-cytochrome c. Also drawn are the nearby residues to which Wat166 is hydrogen-bonded, and the adjacent heme group and heme ligands. In the reduced protein, the volume of the cavity containing Wat166 is ~ 10 Å^3, which expands to ~ 25 Å^3 in oxidized yeast iso-1-cytochrome c.

orientation of the dipole moment of Wat166 adjacent to the heme group to stabilize maximally the positive charge resident there. That the hydrogen bond to the side chain of Met80 is lost is supported by the observed overall higher thermal factors of both the side chains of Met80 and Tyr67 in the oxidized state. Collectively, these dipole modeling studies suggest a major role for Wat166 in stabilizing the oxidized and reduced forms of cytochrome c through the differential orientation of dipole moment, shift in distance to the heme iron atom, and alterations in the surrounding hydrogen bonded network.

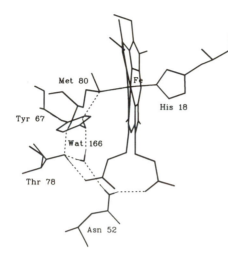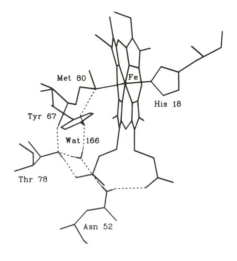

A

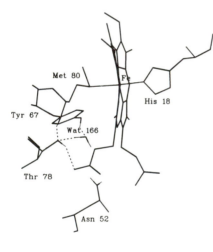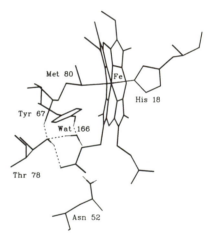

B

Figure 3.21

Stereo-drawings showing the immediate vicinity about Wat166 in (a) reduced and (b) oxidized yeast iso-1-cytochrome *c*. Also shown is the hypothetical placement of hydrogen atoms for the purpose of discerning the dipole moment of Wat166 in the presence and absence of a positive charge at the heme iron atom. The positions of hydrogen atoms (thin lines; only those involved in hydrogen bonds are drawn) were calculated based on the geometry of attached non-hydrogen groups (thick lines) and the hydrogen bonding network present (dashed lines). These modeling studies support the premise that Wat166 is a factor in stabilizing the two oxidation states of cytochrome *c*.

4. Focal points for oxidation state–dependent structural alterations

A compilation of significant structural differences observed on going from the reduced to oxidized states of yeast iso-1-cytochrome c is presented in Table 3.10. There appear to be three points about which these are focused. These involve the pyrrole ring A propionate group, the internal water molecule Wat166, and the Met80 ligand. All three of these features are structurally linked through intermediary interactions between either segments of polypeptide chain or amino acid side chains.

Table 3.10
Structural changes observed upon oxidation of yeast iso-1-cytochrome c[a].

A. *Positional displacements of polypeptide chain (see Figure 3.15).*
   1. Movement of Arg13 and Gly84 to form a hydrogen bond.
   2. Lengthening of the interaction between Trp59 and the heme pyrrole ring A propionate group, with an associated change in Asp60.

B. *Thermal factor parameters of main chain atoms (see Figure 3.16).*
   1. Lower values observed for residues 37–39, focussed at Arg38
      —side chain of Arg38 also has reduced values.
   2. Higher values found for three polypeptide chain segments
      (a) residues 47–59, focused at Asn52
      (b) residues 65–72, focused at Tyr67
      (c) residues 81–85, focused at Phe82
      —All three side chains of Asn52, Tyr67 and Phe82 show higher mobility.

C. *Heme structure and ligands.*
   1. Increased distortion of heme planarity (see Table 3.7).
   2. Readjustment of the pyrrole ring A propionate group with a realignment of hydrogen bonding interactions (see Figure 3.18).
   3. Movement (CB and CE atoms) and higher thermal parameters for the Met80 side chain.
   4. Rotation of the imidazole ring plane of His18.

D. *Internal water structure.*
   1. Large displacement of Wat166 towards the heme iron atom, coupled with a change in hydrogen bonded interactions (see Figure 3.19).
   2. Wat166 movement is facilitated by shifts in the protein matrix to enlarge the available internal cavity space (see Figure 3.20).
   3. Reorientation of the dipole of Wat166 to favor stabilization of the charged heme iron atom (see Figure 3.21).

E. *Hydrogen bond interactions (see Figures 3.18, 3.19, 3.21; Table 3.7).*
   1. Stronger: Gly41 N–Heme O2A
   2. Weaker: Trp59 NE1–Heme O2A
   3. Lost: Asn52 ND2–Heme O2A
                 Asn52 ND2–Wat166
                 Met80 SD–Tyr67 OH
   4. New: Wat121–Heme O2A
                 Gly84 O–Arg13 NH1
                 Asp60 OD1–Wat124

---

[a] A detailed assessment of these features can be found in Berghuis and Brayer, 1992.

They are, furthermore, localized for the most part on the Met80 ligand side of the heme group, adjacent to the solvent exposed edge of this moiety (Figure 3.17). As a whole, the amino acids involved also form a large portion of the protein surface implicated in being central to the complexation interactions formed with electron transfer partners (Salemme, 1976; Poulos and Kraut, 1980; Lum *et al.*, 1987; Pelletier and Kraut, 1992).

It is instructive to examine each focal point for structural change in oxidized yeast iso-1-cytochrome *c* in light of the conformational shifts observed and their probable role in stabilizing this state of the protein.

a. *Heme pyrrole ring A propionate group*   Upon oxidation there is an apparent change in the relationship between the heme iron atom and the pyrrole ring A propionate, leading to a conformational adjustment in this latter group (see Figure 3.18) and realignment of the hydrogen bonds that it forms (Tables 3.7 and 3.10). This change includes stronger interactions made to Gly41 N and Wat121 and weaker links to Asn52 ND2 and Trp59 NE1 (Figures 3.18 and 3.19). The movement of this heme propionate also appears to elicit a shift in the main and side chain atoms of Trp59 and a reorientation in the side chain of Asp60. In addition, the lengthening of the Trp59 to heme propionate hydrogen bond and the release of the hydrogen bond to Asn52 lead to larger thermal parameters for the side chains of both residues and appear to be associated with the overall greater mobility of the polypeptide chain segment composed of residues 47–59 (Figure 3.16). This segment of polypeptide chain is located immediately adjacent to both internally buried heme propionate groups (Figures 3.3 and 3.7). Maximal increases in thermal parameters occur about Asn52, which is centrally located in a small helix composed of residues 49–55 (Table 3.5).

Another residue in this vicinity affected by change in oxidation state is Arg38, whose side chain forms a salt bridge to the heme pyrrole ring A propionate group. This interaction is mediated through the two internally bound water molecules, Wat121 and Wat168. It is notable that the main chain atoms of Trp59 and Arg38 are hydrogen-bonded to one another, again illustrating how the various affected structural features are coupled through a web of interactions. However, in contrast to Trp59, the main and side chain atoms of Arg38 are positionally stable and experience a large decrease in thermal parameters in the oxidized protein.

Precisely what triggers the conformational changes about the pyrrole ring A propionate group and how these serve to stabilize the oxidized state of the protein is uncertain. Nonetheless, there are two conformational states in this region, corresponding to the two oxidation states of the heme iron atom. It is likely that the intricate network of hydrogen bonds present in this region facilitates these changes, perhaps by oxidation state–dependent, differential delocalization of the negative charge of the ionized heme propionate group (Barlow and Thornton, 1983; Singh *et al.*, 1987). Alternatively, direct through-space interaction of the positively charged heme iron atom and the negatively charged pyrrole ring A propionate group may be the primary factor involved, and the observed conformational changes serve to enhance this interaction (Moore, 1983).

Differential through-space interactions between the positively charged side chain of Arg38 and the heme iron atom are also possible. Replacement of arginine at position 38 by other amino acids is known to substantially lower the reduction potential of cytochrome *c* (Proudfoot and Wallace, 1987; Cutler *et al.*, 1989), indicating the ability of this amino acid to directly affect the functional properties of the heme. Finally, it is conceivable that the increased distortion of heme planarity in the oxidized state (Table 3.7) leads to a need for conformational adjustments that are focused in the area of the pyrrole ring A propionate group. In this view, the adjustments to the surrounding protein matrix may be designed simply to accommodate the alternative positions of this propionate group. It is notable that structurally similar changes have been observed in yeast iso-1-cytochrome *c* variants having lowered reduction potentials and apparently hybrid oxidation state conformations (Louie *et al.*, 1988b; Louie and Brayer, 1989).

b. *Wat166*   A second focal point for oxidation state-dependent structural changes is Wat166. This internally buried water molecule is directly linked to shifts about the pyrrole ring A propionate group through interactions with the side chain of Asn52 (Figure 3.19a). In the oxidized state, the Wat166 to Asn52 hydrogen bond is lost (Table 3.7), but comparable interactions are retained by Wat166 to Tyr67 and Thr78, despite its positional shift towards the heme iron atom. Two segments of polypeptide chain (residues 47–59 and 65–72) which have higher thermal factors in the oxidized protein (Figure 3.16) are located adjacent to the Wat166 internal cavity. One of these (residues 47–59) has already been discussed with respect to oxidation state–induced conformational changes about the pyrrole ring A propionate group. The additional factors of the loss of the Wat166 to Asn52 hydrogen bond and the presence of an enlarged internal cavity nearby also undoubtedly contribute to the increased mobility of this polypeptide chain segment. Residues 65–72 are located for the most part at the C-terminal end of an alpha-helical structure composed of residues 60 to 70 (Figures 3.3 and 3.7). For this segment of polypeptide chain, maximal increases in thermal parameters are associated with the main and side chain atoms of Tyr67, a residue which directly hydrogen-bonds to Wat166 in both oxidation states (Figure 3.21). It is notable that elimination of the Wat166 cavity by the replacement of Asn52 with an isoleucine residue leads to significantly increased stabilization of the oxidized form of yeast iso-1-cytochrome *c* (Das *et al.*, 1989; Hickey *et al.*, 1991).

The trigger for Wat166 movement in the oxidized state of cytochrome *c* and for the subsequent cascade of related conformational changes, appears to be an electric field–induced response to the oxidation state of the heme. The shift of Wat166 could serve several functions. It results in direct stabilization of the positively charged heme iron atom by Wat166, which appears to be orientationally fixed by surrounding hydrogen-bond contacts to enhance this effect (Figure 3.21). Through the side chain of Asn52, Wat166 is also physically linked to conformational changes about the pyrrole ring A propionate group and may help mediate these adjustments. Wat166 movement also appears to be a primary factor in the increased mobility of nearby polypeptide chain segments which form a part of the putative surface binding site of electron transfer partners, and in this way could potentially influence binding in an oxidation state–dependent manner (Figure 3.17). Finally, as discussed elsewhere, its positioning may result in modification of the strength of the Met80 SD–heme iron atom ligand interaction and in this way play an integral role in the electron transfer process.

    c. *The Met80 ligand region*   The region about the Met80 heme ligand and Phe82 forms a third focus for oxidation state–dependent conformational changes. Coincident with observations showing that oxidized cytochrome c has a more open structure, and the present study, which suggests this is likely localized on the Met80 side of the heme group (Figure 3.17), are data indicating a substantive weakening of the Met80 SD sulfur to heme iron atom ligand bond in the oxidized state (see reviews by Dickerson and Timkovich, 1975; Williams *et al.*, 1985). This is supported by our observation that the side chain of Met80 has an increased overall thermal factor in oxidized yeast iso-1-cytochrome c. This change in thermal factor values is very localized in that neither the heme group average thermal factor (5.3 Å^2 reduced; 5.6 Å^2 oxidized) nor that of the His18 ligand is affected. Also apparent is some displacement of side chain atoms and a small change in the torsion bond angle between the Met80 side chain and the heme iron atom. However, bond weakening does not appear to cause a lengthening of the Met80 SD–heme iron bond (Korzun *et al.*, 1982). This observation is supported by the current study, where although this bond distance shows the largest deviation among heme iron coordinate bonds (Table 3.7), the values obtained in both oxidation states are within experimental error of each other.

    An essential parameter in the regulation of the relatively high reduction potential of cytochrome c (+ 290 mV for yeast iso-1-cytochrome c; Rafferty *et al.*, 1990) is the strong withdrawing power that the Met80 ligand exerts on the heme iron atom (see reviews of Marchon *et al.*, 1982; Mathews, 1985). Our results and those of others suggest that relative to the reduced state, the Met80 ligand in the oxidized protein is weakened and exerts less of an electron-withdrawing effect, a situation favoring stabilization of the positively charged heme iron atom. An important factor in modulating Met80 ligand bond strength is likely to be the interaction this group has with the side chain of Tyr67 (Figure 3.21a). This hydrogen bond could be expected to increase the electron-withdrawing strength of the Met80 ligand and thereby increase the heme reduction potential further. This is clearly illustrated by the ca. 50 mV drop in reduction potential observed when this hydrogen bond is eliminated by mutation of Tyr67 to a phenylalanine (Berghuis *et al.*, 1994a) or by semisynthetic substitution of Tyr67 by fluorophenylalanine (Frauenhoff and Scott, 1992).

    As discussed earlier, in the oxidized state it appears that the Met80 SD to Tyr67 OH hydrogen bond is broken (Figure 3.21b). This change would facilitate stabilization of oxidized cytochrome c through two mechanisms. First, breakage of this bond would diminish the electron-withdrawing power of the Met80 ligand. Second, the side chain of Tyr67 would become available to bind Wat166 immediately adjacent to the heme, oriented so that its dipole moment is directed to stabilize maximally the positive charge resident on the heme iron atom. Thus Wat166 movement and the placement of the Tyr67 hydrogen bond provide a convenient mechanism for modulating the electron-withdrawing power of the Met80 ligand according to the oxidation state of the heme group. It is also notable that through interaction with Wat166, the Met80 ligand is physically linked to other oxidation state–dependent changes not only about Wat166, but in the more remote region around the pyrrole ring A propionate group.

Perturbations about the Met80 ligand also appear to be associated with the greater mobility observed in the adjacent polypeptide chain segment composed of residues 81–85. Maximal increases in thermal factors are localized at Phe82, a residue implicated in the electron transfer mechanism of cytochrome *c* (Nocek *et al.*, 1991; Louie *et al.*, 1988b; Louie and Brayer, 1989). The side chain of this residue is positioned parallel to the plane of the heme near its solvent-exposed edge and is in van der Waals contact with the side chain of Met80 (Figure 3.7). It is also positioned in that portion of the cytochrome *c* surface believed to form the contact face in complexes with electron transfer partners (Poulos and Kraut, 1980; Lum *et al.*, 1987). A small change in the orientation of the phenyl group of Phe82 with respect to the heme plane is observed in the oxidized state. It has been postulated that the side chain of Phe82 may undergo very large motions in complexes with electron transfer partners (Wendoloski *et al.*, 1987), though this suggestion has not proved amenable to experimental verification as yet (Burch *et al.*, 1990).

Another conformational change in this region is the formation of a new hydrogen bond between Gly84 O and Arg13 NH1. Arg13 has also been implicated as important in electron-transfer complexation interactions (Margoliash and Bosshard, 1983; Satterlee *et al.*, 1987; Lum *et al.*, 1987). This new hydrogen bond has also been observed in mutant yeast iso-1-cytochrome *c* structures which mimic the oxidized state (Louie *et al.*, 1988b; Louie and Brayer, 1989). Given its apparent importance to the electron transfer process, it is not unexpected that the region about Phe82 is sensitive to oxidation state changes about the Met80 ligand and, through this latter group, to other affected areas of the protein.

5. Mechanistic implications

Taken together, our results suggest there is a concerted and cooperative deployment of interconvertible structural changes which facilitates stabilization of the alternative oxidation states of the heme group of cytochrome *c*. Based on these observations and the surface presentation of oxidation state–dependent conformational changes, it is possible to propose a trigger mechanism for electron transfer events mediated by cytochrome *c*. As illustrated in Figure 3.17, those residues about the three focal points of conformational readjustments are localized to the Met80 ligand side of the heme, adjacent to the solvent-exposed edge of this group. In terms of the outer surface of cytochrome *c*, these changes are centrally positioned in that region of the surface believed to form complementary complexation interactions with redox partners. One rather unusual aspect of the surface presentation of those residues conformationally adjusted according to oxidation state is the unperturbed nature of the polypeptide chain composed of amino acids 73–80, which bisects this region.

It is notable that the 73–80 polypeptide chain segment in cytochromes *c* has the most highly conserved sequence, with almost complete absence of natural variance (Figure 3.1). The major secondary structural feature formed by this portion of polypeptide chain is a Type II β-turn composed of residues 75–78 (Table 3.5), which includes the side chain of Thr78, which forms a hydrogen bond to the oxidation state–sensitive Wat166. As already described, various interactions interconnect Wat166 with the other two focal points of oxidation state–dependent change.

Given its central location within the putative contact zone with redox partners, its high degree of sequence conservation, its unique response to oxidation state and its linkage through Thr78 to the three focal points of conformational change, it seems possible that the 73–80 segment of polypeptide chain acts to trigger the alternative structural changes required to stabilize the two oxidation states of cytochrome c. This peptide segment is appropriately positioned to carry out this triggering function by virtue of its surface location next to the solvent-exposed edge of the heme, a likely route for an electron to travel to and from the heme group (Pelletier and Kraut, 1992; Beratan et al., 1992). One might think of residues 73–80, or perhaps more specifically the 75–78 β-turn, as a binary push button contact switch operated by protein–protein contacts with redox partners, leading to the generation of the necessary conditions required to facilitate electron transfer.

Precisely how this proposed push-button switch might function cannot be determined in detail on the basis of our current studies. This arises from only being able to resolve the two endpoints of the oxidation reaction, the reduced and oxidized cytochrome c structures, not the intermediary transition state conformation (or possibly conformations). Nonetheless, the current results suggest that the proposed push-button switch would retain the same spatial positioning in the two end oxidation states, but once pushed by protein–protein contacts in a complex with a redox partner, would act to promote changes in protein conformation that would facilitate electron transfer in either direction. One role of differential mobility in the region about residues 73–80 could be that of mediating these changes by providing the necessary conformational flexibility. This proposal is analogous to the type of switching carried out in a conventional electrical circuit having a push-button contact switch that reversibly alters the direction of current flow.

It must be emphasized, as diagrammatically illustrated in Figure 3.22, that the structural details of the intermediary conformational states through which cytochrome c must pass on going between alternative oxidation states are missing from our understanding of this process. Also missing is any information on the potential role played by redox partner complexation interactions beyond the proposed global recognition and switching events. While it seems likely that the intermediary state is some hybrid of the fully oxidized and reduced cytochrome c structures, it is entirely possible that redox partner complexation results in transient conformations (of either residues 73–80 or elsewhere) that are not evident in the two end states. Such surface transitions have been suggested from the results of other studies (Koloczek et al., 1987; Weber et al., 1987; Hildebrandt and Stockburger, 1989). Indeed, it could very well be that our minimalist three-state scheme may be too simple and that multiple transition states span the gap between oxidized and reduced cytochrome c.

Our results do suggest a number of approaches to investigating possible intermediary states occurring during electron transfer. For example, an obvious target for future structure–function–mutagenesis studies is the protein segment composed of residues 73 to 80. Of particular interest here would be the effect of three classes of replacements: those involving residues that outwardly face onto the surrounding

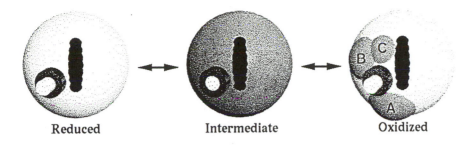

Reduced   Intermediate   Oxidized

Figure 3.22

A schematic representation of a proposed push-button switching mechanism for electron transfer events in yeast iso-1-cytochrome c. Upon complexation with redox partners, residues 73–80, or perhaps more specifically 75–78, alter their conformation so as to facilitate electron transfer and the necessary concomitant structural readjustments. After electron transfer has taken place and the oxidation state–dependent conformational changes have occurred, cytochrome c disengages from the complexed redox partner, and residues 75–78 return to their original conformation. The exposed heme edge is shown in black in this figure, whereas areas labeled A, B, and C correspond to those segments of polypeptide chain which become more mobile in the oxidized state of cytochrome c (see Figure 3.17). Residues 73–80 are represented by the push-button switch.

solvent and presumably contact the surfaces of complexed redox proteins (Lys73, Tyr74, Pro76, Gly77, Lys79); those residues that contact Wat166 or line its internal cavity (Ile75, Thr78, Met80); and those residues likely to be instrumental in setting the folded conformation of this region of the polypeptide chain (Pro76, Gly77). In this last group, Gly77 is a particularly interesting residue, because it is not only at the third position of the type II β-turn, but it also adopts ϕ, ψ angles not accessible to other amino acids with side chains.

At this point only limited attention has been directed at the role of the 73–80 segment. The available studies show that mutation of residues in this region leads to destabilization of cytochrome c and low biological activity (Boon *et al.*, 1979; Wood *et al.*, 1988; ten Kortenaar *et al.*, 1985; Wallace, *et al.*, 1989). These results are in accord with the mechanistic proposals discussed herein.

VI. Acknowledgments

The authors are indebted to past and present laboratory co-workers including Albert Berghuis, Terence Lo, Gordon Louie and Gordon Bushnell, whose results for yeast iso-1 and horse cytochromes c are described herein. We also thank Jacque Fetrow, Grant Mauk, George McLendon, Barry Nall, Fred Sherman, and Michael Smith, as well as members of their laboratories, for assistance in various aspects of these studies. This work was supported by the Medical Research Council of Canada.

VII. References

Arnott, S., & Dover, S. D. (1967) *J. Mol. Biol.* **30**, 209–212.

Baker, E. N., & Hubbard, R. E. (1984) *Prog. Biophys. Mol. Biol.* **44**, 97–179.

Bando, S., Matsuura, Y., Tanaka, N., Yasouka, N., Kakudo, M., Yamanaka, T., & Inokuchi, H. (1979) *J. Biochem. (Tokyo)* **86**, 269–272.

Barlow, D. J., & Thornton, J. M. (1983) *J. Mol. Biol.* **168**, 867–885.

Benning, M. M., Wesenberger, G., Caffrey, M. S., Bartsch, R. G., Meyer, T. E., Cusanovich, M. A., Rayment, I., & Holden, H. M. (1991) *J. Mol. Biol.* **220**, 673–685.

Beratan, D. N., Onuchic, J. N., Winkler, J. R., and Gray, H. B. (1992). *Science* **258**, 1740–1741.

Berghuis, A. M., & Brayer, G. D. (1992). *J. Mol. Biol.* **223**, 259–976.

Berghuis, A. M., Guillemette, J. G., Smith, M., and Brayer, G. D. (1994a) *J. Mol. Biol.* **235**, 1326–1341.

Berghuis, A. M., Guillemette, J. G., McLendon, G., Sherman, F., Smith, M., and Brayer, G. D. (1994b) *J. Mol. Biol.* **236**, 786–799.

Bernstein, F. C., Koetzle, T. F., Williams, G. J. B., Meyer, E. F., Bruce, M. D., Rodgers, J. R., Kennard, O., Shimanouchi, T., & Tasumi, M. (1977) *J. Mol. Biol.* **112**, 535–542.

Bhatia, G. E. (1981). Ph.D. Thesis, University of California at San Diego.

Boon, P. J., Van Raay, A. J. M., Tesser, G. I., and Nivard, R. J. F. (1979) *FEBS Lett.* **108**, 131–135.

Bosshard, H. R., & Zurrer, M. (1980) *J. Biol. Chem.* **255**, 6694–6699.

Bryant, C., Stottmann, J. M., & Stellwagen, E. (1985) *Biochemistry* **24**, 3459–3464.

Burch, A. M., Rigby, S. E. J., Funk, W. D., MacGillivary, R. T. A., Mauk, M. R., Mauk, A. G., & Moore, G. R. (1990) *Science* **247**, 831–833.

Bushnell, G. W., Louie, G. V., & Brayer, G. D. (1990) *J. Mol. Biol.* **214**, 585–595.

Butt, W. D., & Keilin, D. (1962) *Proc. Royal Soc. London* **B512**, 429–458.

Carter, D. C., Melis, K. A., O'Donnel, S. E., Burgess, B. K., Furey, W. F., Wang, B.-C., & Stout, C. D. (1985) *J. Mol. Biol.* **184**, 279–295.

Connolly, M. L. (1983) *Science* **221**, 709–713.

Cutler, R. L., Pielak, G. J., Mauk, A. G., & Smith, M. (1987) *Protein Eng.* **1**, 95–99.

Cutler, R. L., Davies, A. M., Creighton, S., Warshel, A., Moore, G. R., Smith, M., & Mauk, A. G. (1989) *Biochemistry* **28**, 3188–3197.

Das, G., Hickey, D. R., McLendon, D., McLendon, G., & Sherman, F. (1989) *Proc. Natl. Acad. Sci. (USA)* **86**, 496–499.

Dickerson, R. E. (1980a) *Sci. Am.* **242**, 99–110.

Dickerson, R. E. (1980b) in *The Evolution of Protein Structure and Function*, pp. 173–202, Academic Press, New York.

Dickerson, R. E., & Timkovich, R. (1975) in *The Enzymes* (Boyer, P. D., ed.), 3rd Ed., Vol. 11, Part A, pp. 397–547, Academic Press, New York.

Dickerson, R. E., Kopka, M. L., Borders, C. L., Vamum, J., & Weinzierl, J. E. (1967) *J. Mol. Biol.* **29**, 77–95.

Dickerson, R. E., Takano, T., Eisenberg, D., Kallai, O. B., Samson, L., Cooper, A., & Margoliash, E. (1971) *J. Biol. Chem.* **246**, 1511–1532.

DiMaria, P., Polastro, E., DeLange, R. J., Kim, S., & Paik, W. K. (1979) *J. Biol. Chem.* **254**, 4645–4652.

Dong, A., Huang, P., & Caughey, W. S. (1992) *Biochemistry* **31**, 182–189.

Dumont, M., Ernst, J. F., & Sherman, F. (1988) *J. Biol. Chem.* **263**, 15928–15937.

Dumont, M., Matthews, A., Nall, B., Baim, S., Eustice, D., & Sherman, F. (1990) *J. Biol. Chem.* **265**, 2733–2739.

Eden, D., Matthew, J. B., Rosa, J. J., & Richards, F. M. (1982) *Proc. Natl. Acad. Sci. (USA)* **79**, 815–819.

Eley, C. G. S. & Moore, G. R. (1983) *Biochem. J.* **215**, 11–21.

Farooqui, J., DiMaria, P., Kim, S., & Paik, W. K. (1981) *J. Biol. Chem.* **256**, 5041–5045.

Frauenhoff, M. M., & Scott, R. A. (1992) *Proteins: Struct. Func. Genet.* **14**, 202–212.

Hampsey, D. M., Das, G., & Sherman, F. (1986) *J. Biol. Chem.* **261**, 3259–3271.

Hampsey, D. M., Das, G., & Sherman, F. (1988). *FEBS Lett.* **231**, 275–283.

Hickey, D. R., McLendon, G., & Sherman, F. (1988) *J. Biol. Chem.* **263**, 18298–18305.

Hickey, D. R., Berghuis, A. M., Lafond, G., Jaeger, J. A., Cardill, T. S., McLendon, D., Das, G., Sherman, F., Brayer, G. D., & McLendon, G. (1991) *J. Biol. Chem.* **266**, 11686–11694.

Higuchi, Y., Kusunoki, M., Matsuura, Y., Yasuoka, N., & Kakudo, M. (1984) *J. Mol. Biol.* **172**, 109–139.

Hildebrandt, P., & Stockburger, M. (1989) *Biochemistry* **28**, 6722–6728.

Holden, H. M., Meyer, T. E., Cusanovich, M. A., Doldale, F., & Rayment, I. (1987) *J. Mol. Biol.* **195**, 229–231.

Holzschu, D., Principio, L., Conklin, K. T., Hickey, D. R., Short, J., Rao, R., McLendon, G., & Sherman, F. (1987) *J. Biol. Chem.* **262**, 7125–7131.

Keilin, D. (1930) *Proc. Roy. Soc.* **B106**, 418–444.

Koeppe, R. E., Stroud, R. M., Pena, V. A., & Santi, D. V. (1975) *J. Mol. Biol.* **98**, 155–160.

Koloczek, H., Horie, T., Yonetanni, T., Anni, H., Maniaia, G., & Vanderkooi, J. M. (1987) *Biochemistry* **26**, 3142–3148.

Korzun, Z. R., Moffat, K., & Cusanovich, M. A. (1982) *Biochemistry* **21**, 2253–2258.

Kreil, G. (1965) *Hoppe-Seyler's Z. Physiol. Chem.* **340**, 86–87.

Langen, R., Brayer, G. D., Berghuis, A. M., McLendon, G., Sherrnan, F., and Warshel, A. (1992) *J. Mol. Biol.* **224**, 589–600.

Laz, T., Pietras, D., & Sherman, F. (1984) *Proc. Natl. Acad. Sci. (USA)* **81**, 4475–4479.

Leung, C. J., Nall, B. T., & Brayer, G. D. (1989) *J. Mol. Biol.* **206**, 783–785.

Liu, G., Crygon, C. A., & Spiro, T. G. (1989) *Biochemistry* **28**, 5046–5050.

Louie, G. V. (1990) Ph.D. Thesis, University of British Columbia.

Louie, G. V., & Brayer, G. D. (1989) *J. Mol. Biol.* **210**, 313–322.

Louie, G. V., & Brayer, G. D. (1990) *J. Mol. Biol.* **214**, 527–555.

Louie, G. V., Hutcheon, W. L., & Brayer, G. D. (1988a) *J. Mol. Biol.* **199**, 295–314.

Louie, G. V., Pielak, G. J., Smith, M., & Brayer, G. D. (1988b) *Biochemistry* **27**, 7870–7876.

Lum, V. R., Brayer, G. D., Louie, G. V., Smith, M., & Mauk, A. G. (1987) in *Protein Structure, Folding and Design* (Oxender, D., ed.), pp. 143–150, Alan Liss, New York.

Marchon, J. C., Mashiko, T., & Reed, C. A. (1982) in *Electron Transport and Oxygen Utilization* (Ho, C., ed.), pp. 67–72, Elsevier, Amsterdam.

Margoliash, E., & Bosshard, H. R. (1983) *Trends Biochem. Sci.* **35**, 316–320.

Margoliash, E., & Schejter, A. (1966) *Adv. Protein Chem.* **21**, 113–286.

Margoliash, E., Smith, E., Kreil, G., & Tuppy, H. (1961) *Nature (London)* **192**, 1125–1127.

Mathews, F. S. (1985) *Prog. Biophys. Mol. Biol.* **46**, 1–56.

Matsuura, Y., Hata, Y., Yamaguchi, T., Tanaka, N., & Kakudo, M. (1979) *J. Biochem. (Tokyo)* **85**, 729–737.

Matsuura, Y., Takano, T., & Dickerson, R. E. (1982) *J. Mol. Biol.* **156**, 389–409.

Mattoon, J. R., & Sherman, F. (1966) *J. Biol. Chem.* **241**, 4330–4338.

McGregor, M. J., Islam, S. A., & Sternberg, M. J. E. (1987) *J. Mol. Biol.* **198**, 295–310.

McPherson, A. (1990) *Eur. J. Biochem.* **189**, 1–23.

Milner-White, E. J., Ross, B. M., Ismall, R., Belhadj-Mostefa, K., & Poet, R. (1988) *J. Mol. Biol.* **204**, 777–782.

Moench, S. J., & Satterlee, J. D. (1989) *J. Biol. Chem.* **264**, 9923–9931.

Montgomery, D., Leung, D., Smith, M., Shalit, P., Faye, G., & Hall, B. (1980) *Proc. Natl. Acad. Sci. (USA)* **77**, 541–545.

Moore, G. R. (1983) *FEBS Lett.* **161**, 171–175.

Mori, E., & Morita, Y. (1980) *J. Biochem. (Tokyo)* **87**, 249–266.

Motonaga, K., Katano, H., & Nakanishi, K. (1965) *J. Biochem. (Tokyo)* **57**, 29–33.

Murphy, M. E. P., Nall, B. T., & Brayer, G. D. (1992) *J. Mol. Biol.* **227**, 160–176.

Murphy, M. E. P., Fetrow, J. S., Burton, R. E., & Brayer, G. D. (1993) *Protein Sci.* **2**, 1429–1440.

Murphy, M. E. P., Komar-Panicucci, S., Sherman, F., McLendon, G., & Brayer, G. D. (1995) in preparation.

Nall, B. T., & Landers, T. A. (1981) *Biochemistry* **20**, 5403–5411.

Nemethy, G., & Printz, M. P. (1972) *Macromolecules* **5**, 755–758.

Nocek, J. M., Stemp, E. D. A., Finnegan, M. G., Koshy, T. I., Johnson, M. K., Margoliash, E., Mauk, A. G., Smithl, M., & Hoffman, B. M. (1991) *J. Am. Chem. Soc.* **113**, 6822–6831.

Nozaki, M., Mutzushima, H., Horio, T., & Okunuki, K. (1958) *J. Biochem.* **256**, 673–676.

Ochi, H., Hata, Y., Tanaka, N., Kakudo, M., Sakuri, T., Achara, S., & Morita, Y. (1983) *J. Mol. Biol.* **166**, 407–418.

Paik, W. K., Polastro, E., & Kim, S. (1980) in *Current Topics in Cellular Regulation* (Horecker, B. L., & Stadtman, E. R., eds.), Vol. 16, pp. 87–111, Academic Press, New York.

Paik, W. K., Cho, Y.-B., Frost, B., & Kim, S. (1989) *Biochem. Cell. Biol.* **67**, 602–611.

Pelletier, H., and Kraut, J. (1992) *Science* **258**, 1748–1755.

Pielak, G. J., Mauk, A. G., & Smith, M. (1985) *Nature (London)* **313**, 152–154.

Pielak, G. J., Oikawa, K., Mauk, A. G., Smith, M., & Kay, C. M. (1986) *J. Am. Chem. Soc.* **108**, 2724–2727.

Polastro, E., Deconinck, M. M., Devogel, M. R., Mailier, E. L., Looze, Y., Schneck, A. G., & Leonis, J. (1978) *FEBS Lett.* **86**, 17–20.

Ponder, J. W., & Richards, F. M. (1987) *J. Mol. Biol.* **193**, 775–791.

Poulos, T. L., & Kraut, J. (1980) *J. Biol. Chem.* **255**, 10322–10330.

Poulos, T. L., Sheriff, S., & Howard, A. J. (1987) *J. Biol. Chem.* **262**, 13881–13884.

Proudfoot, A. E. I. & Wallace, C. J. A. (1987) *Biochcm. J.* **248**, 965–967.

Rafferty, S. P., Pearce, L. L., Barker, P. D., Guillemette, G., Kay, C. M., Smith, M., & Mauk, A. G. (1990) *Biochemistry* **29**, 9365–9369.

Rafferty, S. P., Guillemette, J. G., Berghuis, A. M., Brayer, G. D., Smith, M., & Mauk, A. G. (1995) submitted for publication.

Ramakrishnan, C., & Ramachandran, G. N. (1965) *Biophys. J.* **5**, 909–933.

Richardson, J. S. (1981) *Adv. Protein Chem.* **34**, 167–339.

Saigo, S. (1986) *J. Biochem. (Tokyo)* **100**, 157–165.

Salemme, F. R. (1972) *Arch. Biochem. Biophys.* **151**, 533–539.

Salemme, F. R. (1976) *J. Mol. Biol.* **102**, 563–568.

Salemme, F. R. (1977) *Ann. Rev. Biochem.* **46**, 299–329.

Salemme, F. R., Freer, S. T., Xuong, N. H., Alden, R. A., & Kraut, J. (1973) *J. Biol. Chem.* **248**, 3910-3921.

Satterlee, J. D., Moench, S. J., & Erman, J. E. (1987) *Biochim. Biophys. Acta* **912**, 87–97.

Scheidt, W. R. (1978) in *The Porphyrins* (Dolphin, D., ed.), Vol. III, Part A, pp. 463–511, Academic Press, New York.

Schulz, G. E., & Schirmer, R. H. (1979) in *Principles of Protein Structure*, Springer-Verlag, New York.

Sels, A., Fukuhara, G., Pere, G. E, & Onimski, P. (1965) *Bioch. Biophys. Acta* **95**, 486–502.

Sheriff, S., Hendrickson, W. A., & Smith, J. L. (1987) *J. Mol. Biol.* **197**, 273–296.

Sherwood, C., & Brayer, G. D. (1985) *J. Mol. Biol.* **185**, 209–210.

Singh, J., Thornton, J. M., Snarey, M., & Campbell, S. F. (1987) *FEBS Lett.* **224**, 161–171.

Smith, M. (1986) *Phil. Trans. R. Soc. Lond.* **A317**, 295–304.

Smith, M., Leung, D. W., Gillam, S., Astell, C. R., Montgomery, D. L., & Hall, B. D. (1979) *Cell* **16**, 753–761.

Stout, C. D. (1978) *J. Mol. Biol.* **126**, 105–108.

Summers, N. L., Carlson, W. D., & Karplus, M. (1987) *J. Mol. Biol* **196**, 175–198.

Swanson, R., Trus, B. L., Mandel, N., Mandel, G., Kallai, O. B., & Dickerson, R. E. (1977) *J. Biol. Chem.* **252**, 759–775.

Takano, T., & Dickerson, R. E. (1980) *Proc. Natl. Acad. Sci. (USA)* **77**, 6371–6375.

Takano, T., & Dickerson, R. E. (1981a) *J. Mol. Biol.* **153**, 79–94.

Takano, T., & Dickerson, R. E. (1981b) *J. Mol. Biol.* **153**, 95–115.

Takano, T., Kallai, O. B., Swanson, R., & Dickerson, R. E. (1973) *J. Biol. Chem.* **248**, 5234–5255.

Tanaka, N., Yamane, T., Tsukihara, T., Ashida T., & Kakudo, M. (1975) *J. Biochem. (Tokyo)* **77**, 147–162.

ten Kortenaar, P. B. W., Adams, P. S. H. M., & Tesser, G. I. (1985) *Proc. Natl. Acad. Sci., U.S.A.* **82**, 8279–8283.

Thaller, C., Weaver, L. H., Eichele, G., Wilson, E., Karlsson, R., & Jansonuis, J. N. (1981) *J. Mol. Biol.* **147**, 465–469.

Timkovich, R., & Dickerson, R. E. (1972) *J. Mol. Biol.* **72**, 199–202.

Timkovich, R., & Dickerson, R. E. (1976) *J. Biol. Chem.* **251**, 4033–4046.

Trewhella, J., Carlson, V. A. P., Curtis, E. H., & Heidorn, D. B. (1988) *Biochemistry* **27**, 1121–1125.

Tsunasawa, S., Stewart, J. W., & Sherman, F. (1985) *J. Biol. Chem.* **260**, 5382–5391.

Ulmer, D. D., & Kagi, J. H. R. (1968) *Biochemistry* **7**, 2710–2717.

Wallace, C. J. A., Mascagni, P., Chait, B. T., Collawn, J. F., Paterson, Y., Proudfoot, A. E. I., & Kent, S. B. H. (1989) *J. Biol. Chem.* **264**, 15199–15209.

Weber, C., Michel, B., & Bosshard, H. R. (1987) *Proc. Natl. Acad. Sci. (USA)* **84**, 6687–6691.

Wendoloski, J. J., Matthew, J. B., Weber, P. C., & Salemme, F. R. (1987) *Science* **238**, 794–797.

Williams, G., Moore, G. R., & Williams, R. J. P. (1985) *Comments Inorg. Chem.* **4**, 55–98.

Wood, L. C., Muthukrishnan, K., White, T. B., Ramdas, L., & Nall, B. T. (1988) *Biochemistry* **27**, 8554–8561.

Wu, T. T., & Kabat, E. A. (1970) *J. Exp. Med.* **132**, 211–250.

4

Cytochrome *c* Folding and Stability

BARRY T. NALL

Department of Biochemistry
University of Texas Health Science Center

I. Introduction: Cytochrome *c* as a Model System for Protein Folding

One of the most fundamental of all properties of a protein is the ability to fold spontaneously to a unique three-dimensional structure (Creighton, 1990; Kim and Baldwin, 1982, 1990). The objective of studies of protein folding is to understand the physical mechanisms by which amino acid sequence determines protein structure and stability. Cytochrome *c* has played an important role in efforts to decipher the folding code. Debate remains about the proper definition of the protein folding problem. What are the questions? What might constitute a solution? Regardless, most agree that the problem will have been solved if, with knowledge of a protein's amino acid sequence, we could make the following predictions:

1. Will the polypeptide fold, and if so, what will be the thermodynamic stability of the folded structure relative to the unfolded state?

2. Are disulfide bonds or prosthetic groups required for overall stability? Do they facilitate folding rates? When sulfhydryls are present, are they paired in

the folded protein, and what pairings are needed by intermediates during the process of folding?

3. What are the properties of noncovalent but highly structured intermediates in folding? What constitutes the rate-limiting step for folding, and how do the structures of partially folded proteins differ before and after the transition state? What does the transition state look like?

4. What is the three-dimensional arrangement of the atoms of the polypeptide chain to within, say, three angstroms resolution?

The first three points are the preliminaries and constitute the immediate objectives of current research. The last problem, prediction of detailed structures from sequence, is a long-term objective constituting the "Holy Grail" of the folding problem.

Although recognized as important, protein folding has not always been a fashionable area of research. Early studies of structural transitions in polypeptides and proteins generated hope that the problem would be solved quickly, but these hopes were soon dashed as the complexity of the problem became apparent. It is only with the advent of powerful new experimental tools that studies of folding have been viewed as likely to result in rapid progress. Fortunately, cytochrome c is particularly well suited for study with some of the most important of these new tools, site-directed mutagenesis and high-field multidimensional NMR spectroscopy. The former allows changes in sequence of a protein to be made at will, while the latter provides a new means of determining the structural and dynamic consequences of sequence changes. Improvements in the efficiency and reliability of traditional x-ray crystallographic methods of structure determination have made it possible to determine high-resolution structures for numerous mutant proteins. There have also been important advances in the theory of protein behavior, although the most successful applications of computational methods have focused more on properties of fully folded proteins than on the process of protein folding.

The objectives of this review are to discuss a range of questions relating to protein folding and to describe some of the approaches to obtaining answers. The focus will be primarily, but not entirely, on folding of cytochrome c. Some important subjects relating to cytochrome c have been largely neglected in order to maintain an emphasis on the protein folding process. For example, no attempt has been made to review the wide range of folded forms of cytochrome c together with their respective redox and heme ligation states, except as these forms relate to the protein folding process.

A. Advantageous Physical Properties

The wide range of biophysical techniques used to investigate folding places stringent requirements on the behavior of suitable model proteins. Since many physical methods are insensitive, the protein must be easily purified in large quantity and be highly water soluble. Particularly important (and correspondingly rare) is that the monomeric form of the unfolded protein remains soluble at high protein concentrations. In addition, the protein should be small and monomeric, since large multi-domain proteins have complex unfolding transitions in which different domains unfold at distinct stages with varying degrees of interdependence. Other important requirements are the presence of intrinsic, sensitive, optical probes of conformation, and convenient assays of biological function. Cytochrome *c* meets most needs of the protein-folding researcher. It is structurally and thermodynamically a single-domain protein with highly soluble folded and unfolded forms.

The yeast cytochrome *c* system, highly refined for studies by classical genetics (Sherman and Stewart, 1978), molecular cloning, and site-directed mutagenesis (Hampsey *et al.*, 1986; Pielak *et al.*, 1985), is among the most versatile for *in vivo* and *in vitro* studies of protein structure and function. A unique feature is the ease with which genetic or *in vivo* studies of cytochrome *c* can be compared to bio-physical work. The covalently attached heme is folded deep in the active site and allows detection and detailed spectroscopic comparisons of normal and mutant cytochromes *c* in both living cells and the test tube. Many mutant protein systems have been genetically engineered (when necessary) to allow detection of functional variants *in vivo*, but few other systems allow detection and quantitation of both active and inactive proteins in living cells (Sherman *et al.*, 1968). Many of the first mutant forms of yeast cytochrome *c* were isolated by spectroscopic screening rather than by functional assays or genetic selection. Over the years visible spectroscopy has continued to play an important role in the further development of the yeast cytochrome *c* system.

Cytochrome *c* has retained its place as a protein of choice for development of methods for atomic resolution biological structure determination (see Chapter 5, by Pielak *et al.*, this volume). NMR spectral assignments are more extensive than for any other protein of comparable size. Full 2-D NMR assignments have been reported for both redox states of two distantly related mitochondrial forms (Feng *et al.*, 1990, 1989; Gao *et al.*, 1990; Wand *et al.*, 1989), and there are partial assignments for homologs from numerous other organisms (Cookson *et al.*, 1978; Williams *et al.*, 1985). Few protein families have been studied as thoroughly by x-ray crystallography. High-resolution structures have been obtained for several species variants, including both isozymes of yeast cytochrome *c*. The structure of reduced yeast iso-1 at 1.23 Å is one of the most highly refined structures available for any protein, and several structures have been reported for mutant forms of yeast cytochromes *c* (see Chapter 3, by Brayer and Murphy, this volume).

B. Sequence Comparisons and the Folding Code

An approach to deciphering the folding code is to compare three-dimensional structures for as many different amino acid sequences as possible. One objective is to identify related sequences which fold to the same three-dimensional structure (Dickerson, 1972, 1980). Current means of sequence comparisons, however, sometimes fail to find sequence similarity for what turn out to be structurally similar proteins.

Another objective is to determine which sequences fold to similar kinds of secondary structure in structurally different proteins. This is especially difficult. Nevertheless, by emphasizing specific aspects of sequence information, predictive schemes have been developed for locating secondary structure (Chou and Fasman, 1978; Levitt, 1978; Lim, 1974). All such schemes are inaccurate (Kabsch and Sander, 1983), but they are better than pure chance. The crux of the problem is that short- and long-range tertiary interactions often tip the scales in favor of otherwise less likely secondary structures. There are some exciting new suggestions for dealing with the interplay between secondary and tertiary structure, particularly with regard to identifying sequences encoding helical "caps" that define helix endpoints (Presta and Rose, 1988; Richardson and Richardson, 1988). Developing tests of the validity of these new approaches remains an active area of research for both experimentalists and theoreticians.

Cytochrome c is the classic example of a protein family in which polypeptides with a range in sequence identity fold to the same structure. Sequence comparisons within and between protein families have been valuable in defining evolutionary relationships between organisms and between proteins (see Chapter 2, by Meyer, this volume), as well as for determining which amino acids are required for function. Although it is implicit in such comparisons, there has been less emphasis on the folding information provided by sequence comparisons. Sequence comparisons give important clues about how folding information is encoded. The existence of protein sequence families shows that the folding code is highly degenerate and implies that "context" is important in determining the tertiary structure of a specific sequence segment.

C. Mutant Forms: Evaluation of Roles of Specific Sites in Folding and Stability

Some of the most exciting genetic experiments in folding have been those in which "saturation mutagenesis" is used to determine the density with which folding information is distributed among different sites in a protein sequence (Bowie and Sauer, 1989; Reidhaar and Sauer, 1988). In these experiments the coding region of a protein gene is heavily mutagenized *in vitro* and reintroduced to the cell. Functional variants are isolated and sequenced to determine for each site in the polypeptide chain which amino acid replacements are compatible with functional chain folding. Sites in which only one or a few amino acids occur are assumed to carry the bulk of the folding information. Genetic experiments of this type have been valuable for pinpointing important locations in proteins of unknown structure for which there are few natural sequence variants. For the cytochrome c protein family, the large number of variants provides similar information. Thus, the distribution of folding information along the cytochrome c sequence is known, at least to a first approximation. The major challenge is to determine the biophysical form in which sequence-encoded folding information is expressed during folding. This will require detailed characterization of proteins in which the folding signals have been gently perturbed so that specific physical steps can be identified without changes in the overall folding process.

II. Investigations of Stability

A. Global Stability

An understanding of the sequence determinants of global stability is not necessarily a solution to the folding problem. For example, do mutations that enhance folding information always lead to more stable proteins? Can a sequence that lacks the thermodynamic stability to fold still encode three-dimensional structural information? Are folding and stability determinants distinct? Is the folding information distributed along a sequence in exactly the same manner as the determinants of stability? An example of a mutant protein in which folding information is perturbed more than stability is a protein that folds to a qualitatively different conformation. Such a protein, though stable, encodes a different tertiary fold. Another example is a mutant protein with equal or enhanced stability that folds and unfolds more slowly than the normal protein. Stability is retained, but the information needed to achieve the folded state is less than optimal. The distinction is that mutations that alter folding information perturb intermediate structures and rates of folding. In the extreme case, the pathway to a particular structure is completely blocked (infinitely slow) and an alternative structure is attained. The question of the relationship between folding determinants and stability determinants is a mechanistic one. Two extremes are the sequential model of folding (Kim and Baldwin, 1982) and the puzzle model (Harrison and Durbin, 1985). The sequential model assumes that specific intermediates are populated which direct folding to the native state in a reasonable amount of time. Puzzle assembly envisions a process in which local segments of native-like structure form rapidly and are assembled to a folded protein in a haphazard manner, much as one assembles pieces of a jigsaw puzzle. Puzzle assembly allows only native-like structure in folding intermediates and does not have a specific transition state. If one path to the folded state is blocked, the protein simply chooses another equally rapid path. In this case mutations can block folding only by introducing thermodynamic instability. Sequential folding postulates unique structures for folding intermediates and transition states. Mutations disrupting the transition state block folding without necessarily affecting global stability. Mechanistic investigations of folding of mutant proteins seek to characterize folding transition states by mutational perturbation. These studies assume that rate-determining aspects of folding follow a sequential pathway involving unique intermediate species (Dill, 1985).

Understanding stability is important. For puzzle assembly, stability is everything, and understanding the sequence determinants of stability provides a solution to the folding problem. For a sequential mechanism, a solution to the folding problem involves understanding how sequence determines the relative stabilities of folding intermediates and folding transition states. Probably both models are correct in that they give descriptions of different aspects of folding. Some aspects of puzzle assembly must be operative in the early stages of folding, where the vast array of unfolded species is narrowed down to a more manageable set of species.

To be a proper measure of protein stability, it is necessary that an unfolding transition be measured under conditions of high reversibility. In some instances the reversibility requirement determines the choice of experimental methods for measuring stability. The least ambiguous means of determining stability is by thermal unfolding with a differential scanning microcalorimeter (DSC) (Privalov, 1979), because DSC measures the thermodynamic properties of the transition directly. Unfortunately, DSC unfolding transitions for many proteins are not reversible at the high protein concentrations and elevated temperatures required. Fully reversible transitions can be measured for a much wider range of proteins using denaturant-induced unfolding with guanidine hydrochloride or urea, but to calculate structural stability from denaturant-induced unfolding transitions, it is necessary to make mechanistic assumptions about the character of the transition. The usual assumption is that the unfolding transition is two-state (Brandts, 1969): The only species populated at appreciable levels within the transition zone are fully folded and fully unfolded forms.

1. Thermal unfolding

a. Differential scanning microcalorimetry   Fortunately, ferricytochrome c exhibits a highly reversible unfolding transition under conditions needed for differential scanning calorimetry. The most extensive work to date is with horse cytochrome c. Thermal unfolding of horse cytochrome c is increasingly irreversible as pH is raised towards neutrality. Scanning calorimetry studies of the pH and NaCl dependence of the unfolding of horse cytochrome c have been carried out below pH 3.2, where the transition is at least 80% reversible (Potekhin and Pfeil, 1989). An important result of these studies is the detection of two folded forms of cytochrome c with distinct thermodynamic properties. The two folded species differ in the number of H^+ and Cl^- bound and in the transitional heat capacity changes (ΔC_p^d) on unfolding. The overall reaction scheme is as follows:

$$N + 5.6\,H^+ + 1.7\,Cl^- \overset{K_1}{\leftrightarrow} N^*H^+_{4.4}Cl^-_{1.7} + 1.2\,H^+ \overset{K_2}{\leftrightarrow} DH^+_{5.6} + 1.7\,Cl^-$$

where N is the low-salt form of cytochrome c, N* a high-salt form that binds 4.4 protons and 1.7 Cl^- ions, and D the denatured (unfolded) form of the protein. The equilibrium constants K_1 and K_2 describe equilibria between $N \leftrightarrow N^*$ and $N^* \leftrightarrow D$, respectively. The high-salt form, N*, is believed to be the "molten-globule" form detected by NMR spectroscopy (Ohgushi and Wada, 1983). The thermodynamic behavior of the high-salt form of cytochrome c is hardly that expected for a molten globule. A defining feature of a molten-globule state is that it lacks a cooperative unfolding transition (Shakhnovich and Finkelstein, 1982), but the N* state is shown to unfold to D with high cooperativity.

A series of studies have been carried out on yeast iso-1-MS,[1] yeast iso-2, and three hybrid (or composite) proteins in which segments from both iso-1 and iso-2

---

[1] Iso-1-MS is yeast iso-1-cytochrome c in which the free sulfhydryl of Cys 102 has been modified by addition of an $-S-CH3$ group.

are combined (Liggins, Sherman, Mathews, and Nall, 1994). Yeast cytochromes *c* are somewhat better suited to scanning calorimetry studies and have unfolding transitions that are greater than 96% reversible at pH 6 (Liggins, J., unpublished). Compared to iso-1, iso-2 has four extra amino terminal residues in addition to 17 other changes in sequence (Ernst *et al.*, 1982). The three composites retain the amino and carboxy terminal segments of iso-1 but differ at a total of 4, 8, or 10 internal sequence positions, respectively. Comparisons of the five proteins by scanning calorimetry provides a stability test of the interchangeability of segments of two closely related proteins. If global stability is dominated by short-range forces (between residues close in the linear sequence), then the composites should have intermediate stability compared to the normal isozymes. Results show that the two parental isozymes have transition midpoints differing by 3.8°, with iso-2 more thermally stable than iso-1-MS. The composites, with transition midpoints 1–3.2 °C lower than that of iso-1-MS, are less stable than either of the parental isozymes. This finding suggests that the stability determinants of some proteins may be dominated by interaction of protein segments distant in the linear sequence (i.e., dominated by long-range forces).

b. Optical probes and two-state models   A variety of groups in cytochrome *c* provide optical probes that can be used to measure unfolding transitions. For yeast cytochrome *c*, these include Tyr-46, Tyr-48, Trp-59, Tyr-67, Tyr-74, Tyr-97, and a heme which is covalently attached to Cys-14 and Cys-17 through thioether bonds. The tyrosine side chains distributed throughout the structure provide ultraviolet absorbance probes of hydrophobic core formation on folding. Absorbance of the heme in the visible region provides a specific probe of folding and active-site formation (heme ligation), as well as a measure of biological function (redox state). The 695-nm absorbance band monitors a specific event in construction of the active site, the formation of the methionine 80 sulfur to heme iron coordinate bond (Shechter and Saludjian, 1967). Absorbance changes at 695 nm provide a simple but direct means of monitoring a range of tightly coupled reactions which include methionine 80-heme ligation, opening of the heme crevice, and global unfolding (Kaminsky *et al.*, 1973). Since fluorescence is intrinsically highly temperature-dependent, it is not as useful as absorbance for monitoring temperature-induced unfolding transitions. The tryptophan–heme donor-quencher pair in cytochrome *c*, however, is an exquisitely sensitive measure of overall heme–tryptophan distance (Tsong, 1974), especially for the unfolded protein. Thus, fluorescence is an excellent means of monitoring denaturant-induced unfolding (Tsong, 1976).

Thermal unfolding of several normal and mutant variants of cytochromes *c* have been measured and stability differences estimated with a two-state thermodynamic model of unfolding. Examples include a thermostable cytochrome *c*-552 (Nojima *et al.*, 1978) and cytochromes *c* from horse, cow, and *Candida krusei* (Kawaguchi and Noda, 1977). Representative transitions for yeast cytochromes *c* (Mathews, Zuniga, Sherman, and Nall, unpublished) are shown (Figure 4.1a,b) as ΔA vs. T plots and as derivative plots, $d\Delta A/dT$ vs. T. In Table 4.1, estimates of the thermodynamic parameters describing the transitions are given, along with observed reversibilities of the transitions. When interpreting stabilities of mutant proteins, it

174

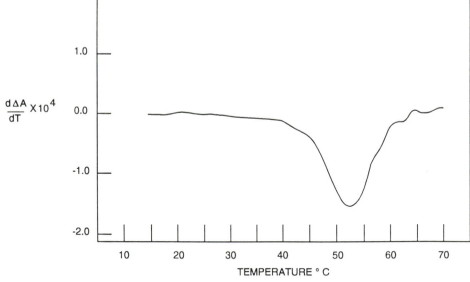

Figure 4.1

Thermal unfolding transition for yeast iso1-MS cytochrome c monitored by (A) absorbance changes and (B) $d\Delta A/dt$ changes at 287 nm (Zuniga and Nall, unpublished). Cooling to below the transition zone followed by reheating shows that unfolding is 96% reversible. Conditions are 0.1 M sodium phosphate, pH 6.0. The thermodynamic parameters describing the transition, listed in Table 4.1, are estimated as described in Table 4.1.

Table 4.1

Two-state thermodynamic parameters for the thermal unfolding transitions of normal and mutant yeast cytochromes c.[a]

Protein[b]	Reversibility %	T_m °C	Transition width, $\Delta T_{1/2}$ °C	$\Delta\Delta G(T_m°)$ kcal/mol	$\Delta H'$ kcal/mol	$\Delta S'$ cal/mol-K
Iso-1-MS	96	52.3 ± 0.3	9.0 ± 0.3	0	82 ± 2	250 ± 7
Iso-2	101	54.2 ± 0.3	9.8 ± 0.3	-0.44 ± 0.04	75.7 ± 0.5	232 ± 1
A7S Iso-1-MS	93	49.3 ± 0.4	9.1 ± 0.3	0.7 ± 0.1	79 ± 2	244 ± 7
L9T Iso-1-MS	101	42.8 ± 0.3	10.5 ± 0.5	2.0 ± 0.1	66 ± 3	207 ± 10
L9H Iso-1-MS	87	37.5 ± 0.3	12.6 ± 1	2.5 ± 0.2	53 ± 4	170 ± 14
L9S Iso-1-MS	103	36.3 ± 0.3	12.7 ± 0.3	2.70 ± 0.06	52 ± 1	168 ± 4
L9F Iso-1-MS	71	40.7 ± 0.3	11.6 ± 1	2.2 ± 0.1	59 ± 5	188 ± 15
H33L Iso-1-MS	99	41.6 ± 0.4	12.2 ± 0.3	1.90 ± 0.09	56 ± 1	178 ± 3
H33A Iso-1-MS	89	48.3 ± 0.3	9.8 ± 0.4	0.89 ± 0.03	73 ± 3	227 ± 8
H33S Iso-1-MS	97	48.5 ± 0.3	9.5 ± 0.3	0.87 ± 0.03	74.9 ± 0.5	233 ± 2
E66W Iso-1-MS	95	43.5 ± 0.3	12.0 ± 0.8	1.6 ± 0.1	58 ± 4	183 ± 12
P71V Iso-1-MS	99	43.2 ± 0.3	13.3 ± 0.4	1.49 ± 0.07	52 ± 1	165 ± 5
P71T Iso-1-MS	91.2	32.1 ± 0.5	18.9 ± 1	2.3 ± 0.2	34 ± 2	112 ± 6
P71I Iso-1-MS	93	43.0 ± 0.6	13.3 ± 0.7	1.50 ± 0.03	52 ± 3	165 ± 9
P71S Iso-1-MS	96	30.1 ± 0.4	18.5 ± 1	2.5 ± 0.2	34 ± 2	113 ± 8
Y97L Iso-1-MS	94	28.8 ± 0.3	14.9 ± 0.4	3.3 ± 0.1	42 ± 1	140 ± 4
Iso-1-AM	96	49.0 ± 0.3	9.9 ± 0.3	0.73 ± 0.08	72 ± 2	224 ± 6
P76G Iso-2[c]	101	50.5 ± 1	16 ± 4	0.24 ± 0.15	48 ± 12	147 ± 35
Iso-2/Iso-1-MS Composite: CYC1-136-B	99	49.3 ± 0.5	9.5 ± 0.5	0.7 ± 0.1	76 ± 4	235 ± 12

(*continued*)

Table 4.1 (*continued*)

Protein[b]	Reversi-bility %	T_m °C	Transition width, $\Delta T_{1/2}$ °C	$\Delta\Delta G(T_m°)$ kcal/mol	$\Delta H'$ kcal/mol	$\Delta S'$ cal/mol-K
Iso-2/Iso-1-MS Composite: CYC1-158-B	97	49.0 ± 0.3	12.5 ± 0.3	0.58 ± 0.03	58 ± 1	178 ± 3
Iso-2/Iso-1-MS Composite: CYC1-136-C	99	48.9 ± 0.3	11.2 ± 0.3	0.66 ± 0.04	64 ± 2	198 ± 6

[a] Thermal transitions are monitored by absorbance changes at 287 nm in 0.1 M Na phosphate, pH 6.0. The primary data, $A(287)$ vs. T, are converted to derivative form, dA/dT vs. T. T_m is defined as the temperature at which $dA/dT = 0$, i.e., the maximum excursion of the $|dA/dT|$ vs. T peak. $\Delta T_{1/2}$ is the width in °C at half height of the $|dA/dT|$ vs. T peak. $\Delta H'$, $\Delta S'$, and $\Delta\Delta G$ are calculated from T_m and $\Delta T_{1/2}$ as described elsewhere (Dumont et al., 1990). $\Delta H'$ and $\Delta S'$ are given at T_m', the transition midpoint for each particular protein, while $\Delta\Delta G$ for all proteins is given at $T = T_m°$, the transition midpoint for iso-1-MS.

[b] Iso-1 and iso-2 are the two normal yeast cytochrome c isozymes. Iso-2 has four additional amino terminal residues not found in iso-1 and differs at 17 other positions in the amino acid sequence. The vertebrate cytochrome c numbering system is used to denote amino acid positions to facilitate comparison between members of the cytochrome c family. Thus, negative integers are used to denote the additional residues on the amino terminus of iso-1 and iso-2. Point mutations are denoted by XNY, where X is the normal amino acid, Y the replacement amino acid, and N the location in the sequence. Three composite proteins are listed which are composed of segments from both iso-1 and iso-2. The composites are denoted by genetic allele (Ernst et al., 1982). Taking iso-1 as the standard, the three composites have four, eight, and 10 positions, respectively, at which the iso-1 amino acid is replaced by the amino acid found at the same position in iso-2. Composite alleles may be described as multiple point mutant forms of iso-1-MS as follows: CYC1-136-B = K54N, L58K, N62D, N63S Iso-1-MS; CYC1-158-B = H26N, A43V, E44K, K54N, L58K, N62D, N63S, G83A Iso-1-MS; CYC1-136-C = L15Q, V20I, K20E, H26N, A43V, E44K, K54N, L58K, N62D, N63S, Iso-1-MS. Iso-1 and the composite proteins have a Cys at position 102 which must be blocked for reversible folding in the absence of disulfide dimerization. Iso-1-MS is iso-1 treated with methyl methane thiosulfonate (Ramdas et al., 1986) to replace the Cys-102 sulfhydryl proton with an –S–CH$_3$ group. Iso-1-AM is iso-1 treated with iodoacetamide (Zuniga and Nall, 1983) to replace the sulfhydryl proton by a –CONH$_2$ group.

[c] The thermal unfolding of P76G iso-2 is broad and asymmetric, indicating that the transition is not a two-state process. Methods used to obtain the thermodynamic parameters assume a two-state transition, so these parameters are rough estimates (at best) for P76G iso-2.

is useful to have a measure of the extent to which mutant proteins fold to similar or different structures. Shown in Table 4.2 are estimates, taken from ultraviolet second-derivative spectroscopy, of the degree of exposure of tyrosine side chains to solvent. These give a rough measure of the extent of mutational perturbation of the native structure. A detailed structural interpretation of changes in tyrosine exposure is not possible, but differences indicate a change in the tyrosine environment which may be related to solvent access to the hydrophobic core.

Table 4.2
Estimation of average exposure of tyrosines to solvent from second derivative
spectrophotometry.[a]

Protein	r_n	α	Average number exposed Tyr	Temperature, °C	Reference
Iso-1-MS	0.72 ± 0.01	0.378 ± 0.004	1.89 ± 0.02	5	
	0.74 ± 0.04	0.39 ± 0.015	1.95 ± 0.08	20	Ramdas *et al.*, 1986
Iso-2	0.754 ± 0.003	0.390 ± 0.001	1.95 ± 0.01	5	
A7S Iso-1-MS	0.74 ± 0.01	0.383 ± 0.004	1.92 ± 0.02	5	
L9T Iso-1-MS	0.74 ± 0.02	0.384 ± 0.006	1.92 ± 0.03	5	
L9H Iso-1-MS	0.84 ± 0.02	0.423 ± 0.007	2.11 ± 0.04	5	
L9S Iso-1-MS	0.818 ± 0.009	0.413 ± 0.003	2.07 ± 0.02	5	
L9F Iso-1-MS	0.80 ± 0.02	0.406 ± 0.007	2.03 ± 0.04	5	
H33L Iso-1-MS	0.77 ± 0.02	0.397 ± 0.005	1.99 ± 0.03	5	
H33A Iso-1-MS	0.76 ± 0.01	0.392 ± 0.004	1.96 ± 0.02	5	
H33S Iso-1-MS	0.71 ± 0.01	0.376 ± 0.004	1.88 ± 0.02	5	
E66W Iso-1-MS[b]	0.613 ± 0.004	0.340 ± 0.003	1.70 ± 0.02	5	
P71V Iso-1-MS	0.74 ± 0.01	0.390 ± 0.003	1.95 ± 0.02	20	Ramdas *et al.*, 1986
P71T Iso-1-MS	0.83 ± 0.01	0.420 ± 0.003	2.10 ± 0.02	20	Ramdas *et al.*, 1986
P71I Iso-1-MS	0.79 ± 0.01	0.400 ± 0.003	2.00 ± 0.02	20	Ramdas *et al.*, 1986
P71S Iso-1-MS	—	—	—	—	—
Y97L Iso-1-MS	0.681 ± 0.004	0.376 ± 0.002	1.50 ± 0.01	5	
Iso-1-AM	0.75 ± 0.01	0.388 ± 0.004	1.94 ± 0.02	5	
P76G Iso-2	0.920 ± 0.02	0.450 ± 0.008	2.25 ± 0.04	5	

(*continued*)

Table 4.2 (*continued*)

Protein	r_n	α	Average number exposed Tyr	Temper- ature, °C	Reference
Iso-2/Iso-1-MS Composite: CYC1-136-B	0.727 ± 0.008	0.381 ± 0.003	1.90 ± 0.02	10	
Iso-2/Iso-1-MS Composite: CYC1-158-B	0.69 ± 0.01	0.368 ± 0.004	1.84 ± 0.02	10	
Iso-2/Iso-1-MS Composite: CYC1-136-C	0.653 ± 0.002	0.354 ± 0.001	1.77 ± 0.01	10	

[a] The average exposure of tyrosine side chains to solvent is estimated from second derivative spectra according to Ragone *et al.* (1984). Values for r_n are ratios of peak to valley heights in the second derivative spectra of the folded proteins in 0.1 M Na phosphate, pH 6.0. Ragone *et al.* have shown that r_n is a measure of the polarity of the environment of the tyrosine side chains and the tyrosine/tryptophan ratio. The average degree of exposure of a tyrosine to solvent in the folded state is $\alpha = (r_n - r_a)/(r_u - r_a)$, where $r_i = (A_i x + B_i)/(C_i x + 1)$. x is the molar ratio of tyrosine to tryptophan in the protein, and A_i, B_i, and C_i are parameters that depend on the polarity of the tyrosine environment. Yeast cytochromes c have five tyrosines, so $x = 5$ and the ratio for fully exposed tyrosines (unfolded protein) is calculated as $r_u = 2.44$ by using parameters appropriate to water: $A = 0.21$, $B = 0.66$, and $C = -0.06$. The ratio for fully buried tyrosines, $r_a = -0.325$, is similarly estimated using parameters appropriate to ethylene glycol: $A = -0.18$, $B = 0.64$, and $C = -0.04$. Since there are five tyrosines in yeast iso-1 and iso-2, the average number of exposed tyrosines is 5 α.

[b] The E66W iso-1-MS mutant protein has one additional tryptophan residue, so the molar ratio of tyrosine to tryptophan is 2.5. Both α and the average number of exposed tyrosines are calculated as described in footnote (a), using $x = 2.5$.

2. Denaturant-induced unfolding

Denaturant-induced unfolding of cytochrome c is a reversible process (Stellwagen, 1968) and allows quantitative comparisons of stability. Some of the first measurements of stability differences for proteins with similar sequences are those obtained by denaturant-induced unfolding of natural variants of cytochrome c (Knapp and Pace, 1974; McLendon and Smith, 1978). For closely related cytochromes c, it is difficult to interpret the observed small changes in stability: There is little correlation between differences in side chain hydrophobicity and differences in stability. This may be because most of the differences between closely related proteins are conservative changes at highly variable sites on the protein surface. The larger stability differences observed between distantly related cytochromes c are more consistent with expectations from hydrophobicity differences (Nall and Landers, 1981). In turn, sequence differences between distantly related proteins tend to be more drastic with some changes at partially buried sites. Denaturant-induced unfolding is less accurate than thermal unfolding for measuring small stability differences. Errors in stability measurements determined by denaturant-induced unfolding are generally in the range of 0.2–1.0 kcal/mol, while those for (well-behaved) thermal transitions are much smaller.

Estimation of thermodynamic parameters from denaturant-induced unfolding transitions requires a choice of models for the unfolding process (Pace, 1990). Solution occurs at two levels. First, a chemical mechanism must be chosen that relates the significantly populated states of the protein, e.g., two-state or multistate. For small globular proteins, two-state mechanisms are used because there is little hope of extracting meaningful thermodynamic parameters for complex multistate mechanisms. Fortunately, scanning calorimetry and other tests show that a two-state mechanism is an excellent approximation for equilibrium unfolding transitions for small globular proteins (Privalov, 1979). Second, it is necessary to choose a model for the chemical interactions between the folded and unfolded forms of the protein and solvent or denaturant. Several models of this type have been described (Tanford, 1969). The simplest, which has a basis in statistical thermodynamics, assumes a linear relationship between the free energy of unfolding and the denaturant concentration (Schellman, 1978).

The major advantage of denaturant-induced unfolding is that high reversibility can be maintained over extended periods of time. For time-intensive measures of protein properties, this is all-important. For example, Figure 4.2 shows the denaturant-induced unfolding of iso-2 cytochrome *c* measured by 1-D NMR spectroscopy. Refolding of the unfolded samples after storage for a month at 5 °C gave a folded protein with a 1-D NMR spectrum essentially the same as that of freshly prepared folded samples. Thermal unfolding of the same protein has also been monitored by NMR spectroscopy. While many of the basic features of the transition are similar to those of the denaturant-induced transition, there is significant irreversibility after only a few hours at high temperature (Ramdas, L., unpublished). Regardless, the fact that an unfolding transition can be measured at the high protein concentrations needed for NMR spectroscopy is testimony to the remarkable solubility of both folded and unfolded forms of cytochrome *c*.

Polyacrylamide gel electrophoresis systems can be used to measure protein stability (Creighton, 1979). The partially denaturing acrylamide gel analysis system developed by Creighton has been shown to be effective not only for measuring stability, but also as a means of estimating rates for slow folding reactions (Creighton, 1980). Gel systems allow rapid screening of equilibrium and kinetic properties for mutant proteins to find those worthy of more detailed kinetic and structural characterization (Figure 4.3a,b). We have used formamide gradient gel electrophoresis to test reversibility and the two-state mechanism for unfolding of various forms of yeast cytochromes *c* (Figure 4.4). Unlike other kinds of denaturant-induced unfolding techniques, differential measurements of stability are possible with gels. Partially denaturing gels in which native and mutant proteins are electrophoresed together can detect stability differences of less than 0.1 kcal/mol and can give stability estimates for irreversibly unfolded species (Zuniga and Nall, unpublished) such as the reduced form of cytochrome *c* (Fig 4.5).

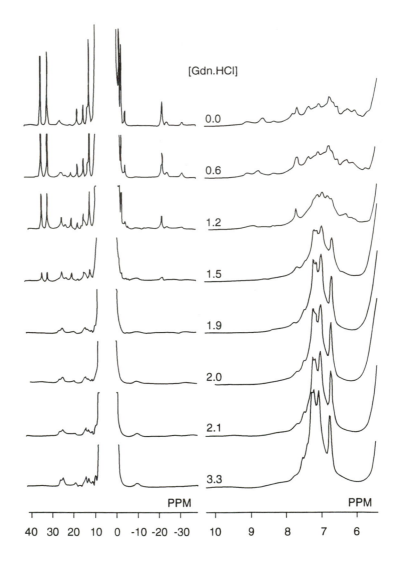

[Gdn.HCl]

0.0

0.6

1.2

1.5

1.9

2.0

2.1

3.3

PPM   PPM

40 30 20 10 0 -10 -20 -30   10   9   8   7   6

Figure 4.2

Guanidine hydrochloride–induced unfolding of yeast iso-2-cytochrome c monitored by proton nuclear magnetic resonance (NMR) spectroscopy (Nall, unpublished). Spectra on the left show the paramagnetically shifted proton resonances of the heme–heme ligand complex that make up the active site of cytochrome c. The spectra on the right show the aromatic ring proton resonances of His, Trp, Tyr, and Phe. Resonance positions are indicated as PPM relative to TSP, sodium 3-(trimethylsilyl)tetradeuteriopropionate. Other conditions are 0.1 M sodium phosphate, pD = 7.2, and the indicated concentrations of deuterated guanidine hydrochloride, Gdn.HCl.

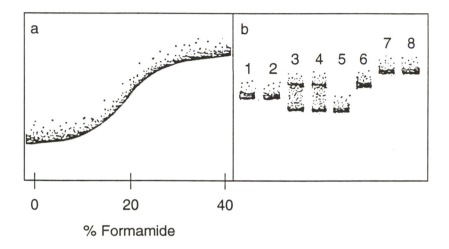

Figure 4.3
Diagrammatic illustration of a polyacrylamide slab gel method of screening the folding
properties of mutant proteins. (A) Electrophoretic pattern expected when the native or
unfolded wild-type protein is electrophoresed as a thin line into a slab gel containing (left
to right) a 0 to 40% gradient of formamide as a denaturant. The pattern shown is for the
case where the folding–unfolding equilibrium is fast compared to the electrophoresis time.
Note that the transition midpoint for the wild-type protein is about 20% formamide.
(B) A partially denaturing gel (20% formamide) used to compare folding properties of mutant
and wild-type proteins at the transition midpoint of the wild-type protein. In different lanes
the protein has been loaded as: (1) wild-type folded protein; (2) wild-type unfolded protein;
(3), (4) folded and unfolded mutant protein (respectively), where the overall stability of the
mutant protein is close to that of the wild-type protein, but where the rate of the folding-
unfolding reaction is comparable to the electrophoresis time; (5), (6) folded and unfolded
mutant protein (respectively), where the stability is similar to that of the wild-type protein
but where the rate of folding–unfolding is slow compared to the electrophoresis time; (7), (8)
folded and unfolded mutant protein, where the overall stability of the mutant protein is less
than that of the wild-type protein and where the folding–unfolding equilibrium is fast
compared to the electrophoresis time.

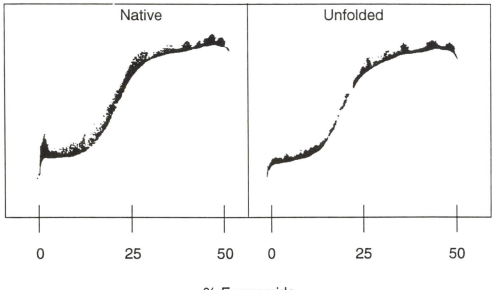

% Formamide

Figure 4.4

Formamide gradient denaturing gel of the unfolding transition for yeast iso-1-AM cytochrome c. A long thin line of iso-1-AM cytochrome c is loaded as a native (low formamide) protein (left) and as an unfolded (high formamide) protein (right). Electrophoresis is from the top towards the bottom of the gel for 2 hours at 400 volts. The gel contains 11% acrylamide and a 0 to 50% formamide gradient (left to right) perpendicular to the direction of electrophoresis. Other conditions are 0.05 M bis–tris acetate, pH 7.2, 2 °C. Identical gel patterns are observed for electrophoresis of folded and unfolded protein, showing that the folding–unfolding rates are fast on the (2-hour) electrophoretic time scale. The single-step sigmoidal shapes show that the electrophoretically detected unfolding transitions are in close accord with a two-state unfolding mechanism. The two-state model can be used to obtain unfolding free energies from the gel patterns (Goldenberg and Creighton, 1984). The free energy of unfolding for iso-1-AM is estimated to be $\Delta G_d^\circ = -3.4$ kcal/mol in the absence of formamide at 2 °C (Zuniga & Nall, unpublished). This is compared to $\Delta G_d^\circ = -2.1$ kcal/mol at 20 °C from the guanidine hydrochloride–induced unfolding transition (Zuniga and Nall, 1983).

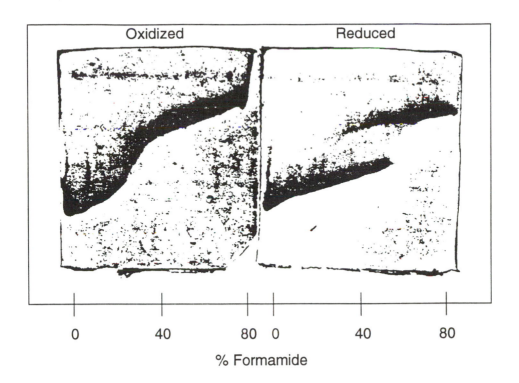

0   40   80   0   40   80

% Formamide

Figure 4.5

Comparison of the stabilities of the oxidized and the reduced forms of iso-1-AM cytochrome *c* by formamide gradient gel electrophoresis (Zuniga and Nall, unpublished). Folded oxidized iso-1-AM is loaded on the gel on the left, and folded reduced iso-1-AM on the right. Electrophoresis is for 3 hours at 200 volts through an 11% acrylamide gel containing (from left to right) a 0 to 80% formamide gradient. Other conditions are 0.05 M bis–tris acetate, pH 7.2, 10 °C. Unfolding of the reduced protein is not fully reversible because the heme oxidizes on unfolding. Nevertheless, a "shadow" of the the unfolding transition is observed, although it is not known whether this transition is the same as for the (hypothetical) reversible transition for the reduced protein. A two-state analysis of the gels (Goldenberg and Creighton, 1984) gives $\Delta G_d° = -2.7$ kcal/mol for oxidized iso-1-AM and $\Delta G_d° = -6.6$ kcal/mol for reduced iso-1-AM at 10 °C in the absence of formamide. The stability difference between the reduced and oxidized forms is estimated to be $\Delta\Delta G_d°$ (red–ox) $= -3.9$ kcal/mol at 10 °C, pH 7.2. Through a thermodynamic cycle, the difference in unfolding free energies is known to be equal to the difference in the redox free energies for the folded and unfolded forms of the cytochrome *c*-heme ligand complex (Pfeil, 1981). This provides a direct link between structure and function: Global stability differences are equal to differences in electron binding affinities.

3. pH-induced changes in conformation and folding

Conformations of both folded and unfolded forms of cytochrome c depend on pH (Babul and Stellwagen, 1971, 1972; Saigo, 1986; Tsong, 1975, 1977; Saigo, 1981). There are at least three folded forms that differ in heme ligation state. X-ray crystallographic structures have been solved for both redox states of the form present at neutral pH but have not been obtained for other folded conformations. The form with Met-80 and His-18 as heme ligands occurs near neutral pH and is the dominant species in the range of pH 5–8. At high pH (above pH 8–10), the dominant species is the alkaline conformation (see Chapter 9, by Wilson and Greenwood, this volume). The structure of this form is not known, but loss of the 695-nm absorbance band and a paramagnetically shifted NMR signal, both assigned to the Met-80 heme ligand complex, show that Met-80 is no longer a heme ligand. Since the electronic state of the heme remains low-spin, an intrinsic protein ligand, probably an ε-amino group of Lys (Gadsby *et al.*, 1987; Stellwagen *et al.*, 1975; Wallace, 1984; Wilgus and Stellwagen, 1974), is believed to displace Met-80 as a heme ligand. The low-pH form has reduced stability, but anion binding stabilizes the folded structure (Stellwagen and Babul, 1975; Tsong, 1976). The neutral-pH native conformation is a single species, but there is evidence of conformational heterogeneity for both the acid and alkaline conformations. Globular high-spin (Dyson and Beattie, 1982; Robinson *et al.*, 1983) or "molten-globule" acid forms (Goto *et al.*, 1990a,b) have been reported. There is evidence from NMR spectroscopy of additional alkaline species (Hong and Dixon, 1989), but some of these forms could be high pH–induced unfolded forms.

At high denaturant concentrations (guanidine hydrochloride or urea), the protein conformation approximates a "random coil" polypeptide. Both the alkaline and neutral-pH unfolded forms have low-spin hemes, indicating that protein heme ligands are retained in unfolded alkaline and neutral species. For the fully unfolded protein at acid pH, a high-spin heme state suggests that the ligands are water molecules (Stellwagen and Babul, 1975). The nature of the unfolded protein heme ligands at neutral and high pH is not known with certainty, but the best guess is that one or both are likely to be His imidazole side chains (Babul and Stellwagen, 1971). There may be heme ligation differences between unfolded forms at neutral pH and high pH for yeast iso-2 cytochrome c, since there are changes in the paramagnetically shifted proton resonances in the NMR spectrum on raising pH (Osterhout *et al.*, 1985).

The pH dependence of the guanidine hydrochloride-induced unfolding transition for horse cytochrome c has been studied in detail by Tsong (1976). A less detailed study shows similar pH effects for yeast iso-2 cytochrome c (Osterhout *et al.*, 1985). There is very little pH dependence for unfolding near neutral pH, but large decreases in stability at acid pH (below pH 5.0). The decreased stability is correlated with the presence of high-spin forms of cytochrome c in which water displaces the usual heme ligands of Met-80 and His-18. This suggests that the decreased stability results from rupture of the heme coordinate bonds. The loss of the heme ligands is caused by protonation of protein side chains, probably the His-18 imidazole and a heme propionate (Hartshorn and Moore, 1989; Stellwagen and Babul, 1975). A quantitative, structure-based understanding of pH-induced stability changes is complicated by the lack of more detailed information for folded and unfolded forms of cytochrome *c* with nonnative heme ligation.

Heterogeneity of the heme ligand state of guanidine hydrochloride-unfolded cytochrome *c* has been investigated by NMR spectroscopy (Muthukrishnan and Nall, 1991). The experimental approach involves titrating the unfolded protein with deuterated imidazole. Heme ligand resonances are identified by observation of the paramagnetic region and the histidine ring C_2H region of the proton NMR spectrum of unfolded cytochrome *c*. Deuterated imidazole has no proton NMR spectrum of its own, and as an extrinsic heme ligand displaces the intrinsic ligands when the imidazole concentration exceeds the effective (local) concentration of the intrinsic ligands. On displacement, resonances of the intrinsic ligands disappear from the paramagnetic region of the spectrum and reappear in the diamagnetic spectral region. Titrations have been carried out on members of the mitochondrial cytochrome *c* family that contain different numbers of His residues. These include cytochromes *c* from horse (3 His), tuna (2 His), yeast iso-1-MS (4 His), and yeast iso-2 (3 His). A correlation is found between the number of His residues in a particular variety of cytochrome *c* and the number of proton resonances that appear in the His ring C_2H region of the NMR spectrum at saturating deuterated imidazole (Figure 4.6). For unfolded iso-2 in the absence of imidazole, observation of a single broad resonance in the diamagnetic His C_2H region of the spectrum suggests that protons on all three His side chains are subject to paramagnetic shifts and/or resonance broadening. This suggests that all His side chains participate as heme ligands, at least some of the time. Therefore, the unfolded state of cytochrome *c* is heterogeneous and the degree of heterogeneity depends on the number of His residues present in the particular type of cytochrome *c*. Whenever there are more than two His side chains in cytochrome *c*, the unfolded form of the protein is an equilibrium mixture of forms with different His side chains as the heme ligands, i.e., the available His residues take turns as heme ligands. Ligation of the heme by His in the unfolded protein alters the configurational entropy of the unfolded protein and changes the stability of the unfolded states relative to the folded state. Configurational entropy differences and heme ligand heterogeneity must be taken into account when comparing stability or folding mechanisms for natural and mutational variants of cytochrome *c* with different numbers of His residues.

4. Decoding the sequence requirements for stability and folding

a. *Highly conserved sites*   For the mitochondrial cytochrome *c* family, there is little doubt that conserved locations in the sequence carry the bulk of the folding information. Surveys of the cytochrome *c* protein family show that there is a wide range in the tolerance of sequence sites for amino acid substitutions. Using the approximation that changes at one site do not influence which residues are allowed at other sites, it is possible to calculate the total number of all possible protein sequences that will fold (Reidhaar and Sauer, 1990; Yockey, 1977). The first step in this process is to estimate the number of amino acids allowed at a given site from the number of different amino acids observed at that location in functional mutant

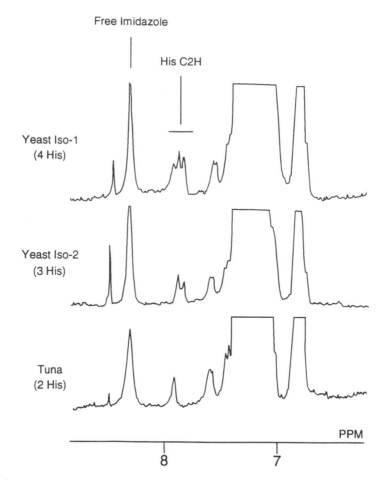

Free Imidazole

His C2H

Yeast Iso-1
(4 His)

Yeast Iso-2
(3 His)

Tuna
(2 His)

PPM

8   7

Figure 4.6

Nuclear magnetic resonance (NMR) spectra of unfolded cytochrome c in the presence of imidazole (Muthukrishnan and Nall, 1991). Spectra of iso-1-MS, iso-2, and tuna cytochromes c are shown in the presence of 3.5 M deuterated guanidine hydrochloride and 150 mM imidazole. These three members of the mitochondrial cytochrome c family have 4, 3, and 2 His residues, respectively. The high concentration of guanidine hydrochloride is sufficient to fully unfold the proteins to an unstructured state containing His side chains as intrinsic heme ligands at the fifth and sixth coordination positions. In the absence of imidazole (not shown), there is a greatly reduced resonance intensity in the His C_2H ring proton region: The resonances are shifted and broadened by the heme paramagnetism. At a concentration of 150 mM, the extrinsic deuterated imidazole binds to the heme iron atom, displacing intrinsic heme ligands. The C_2H ring proton resonances from the displaced His side chains appear in the expected spectral region. At saturating deuterated imidazole, the number of resonances is one less than the total number of His side chains in a given cytochrome c variant. Three conclusions are drawn: (1) for unfolded cytochromes c with more than two His residues, the His side chains take turns being the heme ligands; (2) at high imidazole concentration, one His (probably His 18) remains as a heme ligand, but all others are displaced; (3) there is a heterogeneous heme ligand state for unfolded cytochromes c with more than two His residues. Resonance positions are indicated for an imidazole proton and for the His C_2H ring protons. The imidazole is fully deuterated at the 99% level and is not seen in the proton NMR spectra at low imidazole concentrations. However, the 1% contaminating protonated imidazole shows up as a strong resonance in the proton NMR spectrum at high imidazole concentrations.

or natural variant sequences. The product for all sequence sites of the number of allowed residues at each site gives the total number of sequences capable of folding to the cytochrome *c* structure. It is estimated that a cytochrome *c*–like fold will result for 2.8×10^{61} out of the total of 2.5×10^{131} possible sequences (Yockey, 1977). Calculations of this kind using sequences from protein families give a measure of the degeneracy of the folding code and are valuable for designing mutagenesis experiments to further decipher folding information (Bowie *et al.*, 1990; Bowie and Sauer, 1989; Lim and Sauer, 1989; Reidhaar and Sauer, 1988, 1990).

b. *Saturation mutagenesis*   For cytochrome *c*, there is a long history of use of mutagenesis to investigate the sequence requirements of folding and stability *in vivo*. Results from this work have tended to be in line with expectations of surveys of the cytochrome *c* sequence family (Hampsey *et al.*, 1986, 1988), in that mutation of evolutionarily conserved residues leads to nonfunctional proteins more often than do substitutions at evolutionarily variable sites. Surprisingly, most conserved sites can be mutated with partial retention of *in vivo* function. This observation may indicate that the cytochrome *c* family has a broad distribution of sequences capable of folding and imparting partial function, but a smaller subset of sequences encoding fully functional proteins. A complication is that there are gradations of function so that distinctions between functional and nonfunctional proteins are rather arbitrary.

Sauer and co-workers have applied methods of saturation mutagenesis to identify critical sites for folding and function of lambda repressor and ARC repressor (Bowie and Sauer, 1989; Reidhaar and Sauer, 1988). In other work, the context dependence of substitutions that allow successful packing of the lambda repressor hydrophobic core has been investigated (Lim and Sauer, 1989). Typically two to four sequence locations are heavily mutated at once, and functional mutants are recovered. Results from such studies show that in the hydrophobic core, the allowed residues at one location depend on which amino acids are present at other core sites. Hydrophobic core variants that give stable functional proteins retain the chemical properties of the core residues (hydrophobic only) and maintain tight limits on the cumulative volume of the core residues. Steric limitations on side-chain packing are also important in determining whether a particular mutant sequence will fold to a functional protein. Analogous experiments have not been reported for cytochrome *c*.

B. Local or Segmental Stability: Independent Folding Modules

One of the fundamental questions in protein folding is the extent to which the conformation of a protein segment is encoded locally or determined by the "context" provided by (sequentially) distant segments of polypeptide. If conformation is largely determined by local sequence, then it should be possible to decipher much of the folding information by studies of the preferred conformations of short peptides and protein segments. Presumably, the preferred conformations would be the same as those of the protein segment in the intact protein. An encouraging case in which this has been observed is ribonuclease A, where the same residues have been shown to be helical in a protein fragment, the free S peptide, and the fully folded protein

(Kim and Baldwin, 1984). This finding shows that a helix stop signal is encoded by short-range interactions within the local sequence. On the other hand, there are also instances where the solution structure of a protein segment differs from that found in the native protein (Dyson *et al.*, 1985).

A sequence might encode a structure but still lack the thermodynamic stability to fold. For example, a sequence may encode a structure that has a global free energy minimum, but if the depth of the minimum is small compared to thermal energy (RT), the peptide remains unfolded. Experimentally, sequences encoding thermodynamically unstable structures will be difficult to distinguish from sequences for which there is no global minimum. The problem of determining which peptides retain folding information but lack stability, and which have neither folding information or stability, is difficult. For entire proteins, the thermodynamic balance between folded and unfolded forms is tenuous. For peptides with folding information, thermodynamic stability may be the exception rather than the rule. It is important to find ways of enhancing the stability of encoded structures. One way is with nonphysiological solvents (Nelson and Kallenbach, 1986), but it is necessary to show that these solvents induce native-like rather than aberrant interactions.

a. *Loops and turns*   Careful inspection of known protein structures suggests that stable structures are composed of smaller folding units (Rose, 1979, 1985; Rose and Roy, 1980; Yuschok and Rose, 1983). One algorithm detects local regions of high packing density which have features expected for structural elements. In most cases, the high-density regions correspond to well-known local folding motifs, such as loops, turns, sheets, and helices (Zehfus and Rose, 1986). Since cytochrome *c* is constructed almost entirely of loops and helices, it is a good system for testing the idea of modular assembly of proteins. Loops have been identified by computational methods, and modeling studies suggest ways in which hybrid forms of cytochrome *c* might be constructed by enlarging or deleting loops (Leszczynski [Fetrow] and Rose, 1986). An exciting recent series of genetic experiments shows that stable chimeric cytochrome *c* analogs can be constructed that resemble predicted structures (Fetrow *et al.*, 1989).

b. *Helices and helical ends*   Locating helices by quantifying the sequence determinants of helical stability is an important first step in structure prediction. The classic method describes overall stability using helix–coil transition theory (Poland and Scheraga, 1970; Zimm and Bragg, 1959) and extracts residue-specific stability parameters from host–guest experiments (Scheraga, 1978). The host–guest experiments, in effect, determine the average contributions to helix stability for each type of amino acid by measuring the helix–coil transitions for model polypeptides. Unfortunately, results of this approach have not yet allowed accurate prediction of helix locations. The problem may be twofold. First, residue-specific contributions to stability give an estimate of the dependence of stability on amino acid composition, but they give no information on context (or sequence) dependence of stability for sequences of the same composition (Padmanabhan *et al.*, 1990). Second, there may be qualitatively different contributions to helical stability in proteins compared to model polypeptides, especially for the residues at helix ends (Strehlow and Baldwin,

1989). Stabilities measured in high-molecular-weight model polypeptides are made on systems in which most helical residues are within the interior of the helix, where backbone H-bonds are made to both preceding and following residues. For short helices found in globular proteins, the situation is different. Half or more of the residues are near the helix ends, where backbone H-bonding needs must be met by nonhelical residues (Presta and Rose, 1988).

Predictions of helix stability should be more accurate as values for parameters improve and sequence effects become better understood. Initial results showed small differences in the contributions to helix stability for most residues, so predicted differences in helix stabilities were always small. It is now known that Ala and Leu have "*s*" values, equilibrium constants for adding one helical residue to the end of an existing helix (Poland and Scheraga, 1970), that are much larger than expected (Marqusee *et al.*, 1989; Padmanabhan *et al.*, 1990). The helix–coil parameters for other amino acids may also require revision. A wider range of stability parameters for individual amino acids leads to larger differences in calculated helix stabilities. Eventually, improvements in the parameters describing helix–coil transitions should help in successfully predicting the locations of helical regions in proteins.

An exciting new means of locating helices within protein sequences (Presta and Rose, 1988) focuses on flanking residues instead of the helical residues themselves. The scheme is based on the observations that the average length of a helix in a globular protein is very short, so that most residues in the helix are close to one end or the other. Since each of the backbone H-bonds reaches ahead (on the N-terminus) or behind (on the C-terminus) by four residues, the four residues at each end of the helix will have backbone H-bonding requirements not met by other helical residues. These requirements must be satisfied by protein side chains and nonhelical backbone H-bonds from flanking residues or by solvent. The approach is to determine all the N-terminal (or C-terminal) flanking sequences capable of assuming nonhelical side-chain/backbone conformations that provide H-bond acceptors (donors) for the four N-terminal (C-terminal) helical residues. Such sequences make up N-terminal or C-terminal helical boundaries which are probable start and end points for helices. Tests of the scheme show that boundary sequences occur at helix ends with high probability in a wide range of proteins, including cytochrome *c*.

C. Fragment Complementation as a Bridge between Global and Local Stability

Model peptide systems give insight into short-range determinants of local structure while investigations of stability and mechanism for intact proteins give a broad outline of the folding process. Folding by fragment complementation is of increasing importance as a means of bridging the gap between peptide and whole-protein systems. Several studies of the interaction of protein fragments to form folded structures have been performed. Recently a two-peptide fragment complex stabilized by a disulfide bond has been used to model a folding intermediate for bovine pancreatic trypsin inhibitor (Oas and Kim, 1988). Other workers have carried out detailed fragment complementation studies for staphylococcal nuclease A and cytochrome *c*

(Taniuchi *et al.*, 1986). Two regions have been identified at which the polypeptide can be cleaved without loss of the cytochrome *c* fold: between residues 23 and 25 and between residues 39 and 55 (Hantgan and Taniuchi, 1977; Juillerat *et al.*, 1980; Parr *et al.*, 1978). The allowed cleavage regions correspond to segments missing in evolutionary variant members of the cytochrome *c* family (Taniuchi *et al.*, 1986). With breakpoints within the allowed regions, three-fragment complexes of cytochrome *c* have been shown to form stable folded structures (Juillerat *et al.*, 1980). Fragment complementation has also been used to investigate stability differences for mutant complexes in which key residues are varied (Juillerat and Taniuchi, 1986). Fragment systems are valuable in assessing the importance of long-range tertiary interactions to the stability of folding intermediates and transition states. The ability to vary fragment concentration gives the experimentalist a way to test the affects of specific tertiary interactions on folding. An exciting example of a successful application of this approach is the inhibition of folding of dihydrofolate reductase by a peptide fragment (Hall and Frieden, 1989).

III. Mechanistic Studies of Folding

A reaction as complex as folding of a protein must involve a variety of specific chemical processes. Hydrogen bonds and salt bridges are formed, and perhaps broken and interchanged during the folding process. Solvent-induced hydrophobic associations may occur, or even a general hydrophobic collapse to a somewhat less than fully ordered "globule" may ensue. Polypeptide chains might become entangled and have to extricate themselves prior to further folding. Intricate shapes and structures may be constructed only to partially or fully unravel on transformation into other more stable forms. The objective of much experimental work on the process of folding has been to obtain direct evidence for some of these expectations.

A. Kinetic Features of Folding

1. Folding to the native conformation

The refolding kinetics of the yeast cytochromes *c* are very similar to those of the distantly related horse cytochrome *c* (Ikai *et al.*, 1973; Tsong, 1973), suggesting that the qualitative features of protein folding reactions are conserved within a protein family (Nall and Landers, 1981). Fast and slow phases are observed, and both (absorbance-detected) phases give native enzyme as a product (Nall, 1983). The fast phase is relatively large and constitutes 70–80% of the amplitude for absorbance-detected folding and 80–90% for fluorescence-detected folding. Assuming that the proline model (Brandts *et al.*, 1975) explains the existence of the slow-folding phases, the large fast-phase amplitude makes cytochrome *c* an excellent system for studies of folding reactions free from *cis–trans* isomerization. Slightly different kinetics are found when different physical probes are used to monitor refolding rates. In particular, fluorescence-detected slow folding is about 10-fold faster ($\tau = 10$–20 s) than absorbance-detected slow folding ($\tau = 100$–200 s) (Nall, 1983). Differences in

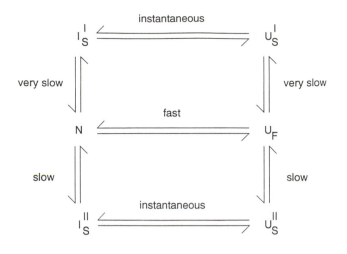

Figure 4.7

Minimal kinetic mechanism for unfolding of yeast iso-2-cytochrome c at neutral pH
(Osterhout and Nall, 1985). U_f are the fast-folding species, I^I_S are the absorbance-detected
slow-folding species, and I^{II}_S are the fluorescence-detected slow-folding species. The U_f species
have the absorbance and fluorescence properties of the unfolded protein. I^I_S are species with
the fluorescence of the native protein but the absorbance of the unfolded protein. I^{II}_S are
species with the fluorescence of the unfolded protein but the absorbance of the native protein.
For refolding, both "I" species are formed during the mixing time of the stopped-flow
experiments and give rise to (undetected) changes in fluorescence (I^I_S) or absorbance (I^{II}_S).
The species present immediately following mixing can be calculated from refolding
amplitudes. I^I_S is estimated to be 19%, I^{II}_S is 8%, and U_f is 73% of the nonnative species.

The U species are the unfolded forms of the protein. U^I_S and U^{II}_S are the unfolded
species responsible for the absorbance-detected or fluorescence-detected slow folding phases,
respectively. The relative populations of U species after unfolding has gone to completion are
estimated to be 19% U^I_S, 8% U^{II}_S, and 73% U_f.

rates of fast folding are also observed, but only at high and low pH (Nall *et al.*,
1988).

For yeast cytochromes c, the absorbance-detected slow phase is about 10-fold
slower than fluorescence-detected slow folding. Refolding monitored by 695-nm absorbance, believed to be an indicator of native protein formation, shows only the
slower ("absorbance-detected") of the two slow phases. Presumably, the fluorescence-
detected slow phase monitors conversion of partly folded species to fully native
species. The fluorescence-detected slow folding intermediates are reducible and have
the absorbance properties of the native protein, but have intermediate fluorescence
properties. Similarly, the absorbance-detected phase monitors conversion of inactive
species with native-like fluorescence but intermediate absorbance to fully native
species (Figure 4.7). The properties of both slow phases are consistent with involve-

ment of proline isomerization. The activation enthalpies are in the expected range: $\Delta H^{\ddagger} = 21$ kcal/mol for fluorescence and 27 kcal/mol for absorbance. In addition, double jump-kinetic experiments, which measure the rates of formation of slow-refolding forms of unfolded proteins, show that both species are generated slowly in the unfolded protein with slightly different rates: τ_{DJ}(fluor) $= 24$ s, τ_{DJ}(abs) $= 114$ s (Osterhout and Nall, 1985).

2. Native-like intermediates and folding to alkaline or mutant conformations

Folding to mutant (or alkaline) conformations occurs in a surprising manner: Native or native-like species appear as transient intermediates in folding to the alkaline form (Nall, 1986). In the first (fast-phase) step of folding, an intermediate is formed that has the absorbance spectrum and ascorbic acid reducibility of native cytochrome c. Following formation of the native-like intermediate, a slow conformational change occurs leading to the mutant (alkaline) conformation. The second, slow conformational change can be induced in the absence of overall folding by pH jumps. Folding to a mutant conformation via a native-like state has been observed for both iso-1 and iso-2 (at high pH), for three different replacements of Pro71, and for a Pro76 to Gly76 mutant (Nall, 1986; Ramdas and Nall, 1986; Wood et al., 1988). These results show that folding, at least some of the time, prefers to pass through specific states with structural features (heme ligation state) differing from those of the final product of folding. This suggests the existence of unique species which direct folding, perhaps along a sequential pathway (Kim and Baldwin, 1982).

3. Heme ligands and folding

The heme–ligand complex makes up the active site of cytochrome c and is the focus of physical studies of cytochrome c function. Heme ligation also influences the kinetic and equilibrium properties of folding. At or above neutral pH, unfolded cytochrome c retains heme ligands, probably His imidazole side chains (Babul and Stellwagen, 1971). The heme coordinate bonds in the unfolded protein close loops of polypeptide chain to produce a cross-linked "random coil" unfolded state. During refolding, the His side chains are replaced by His-18 and Met-80, the ligands of the native structure.

Hence, ligand exchange is necessary for folding, but current evidence suggests that it is not rate-limiting. Fast folding reactions for cytochrome c are in the same time range as other small globular proteins, so fast folding rates are probably governed by the same (unknown) physical processes as for non-heme proteins. Folding studies of site-directed mutants and catalysis by the enzyme peptidyl prolyl cis–trans isomerase show that the slow folding phases are generated by proline isomerization (Lin et al., 1988; Wood et al., 1988; Veeraraghavan and Nall, 1994). However, heme ligation may be a factor in determining the number of kinetic phases in folding. Refolding of horse cytochrome c from guanidine hydrochloride occurs with three kinetic phases, but refolding in the presence of extrinsic ligands (such as imidazole) occurs with only one kinetic phase (Brems and Stellwagen,

1983). One explanation of this observation is that relatively fast heme ligation reactions are coupled to slow chain configurational changes that may involve proline imide bond isomerization. However, all three kinetic phases remain for refolding of imidazole–cytochrome *c* from urea (Myer, 1984). The different results for urea-unfolded vs. guanidine hydrochloride–unfolded protein are surprising. For similar (random coil) unfolded states, the number of kinetic phases is expected to be independent of the mode of denaturation. At present, the involvement of heme ligation in folding of horse cytochrome *c* is not fully understood, and the experimental results concerning this issue are puzzling.

For yeast iso-2-cytochrome *c*, three kinetic phases (two slow phases and one fast phase) are observed in the stopped-flow time range for refolding at neutral pH. Amplitudes and rates of refolding are independent of the initial pH from pH 3 to above pH 9 (Osterhout and Nall, 1985). This observation shows that the kinetic properties of refolding are independent of the heme ligation state in the unfolded protein (no ligands at acid pH; His ligands at neutral and higher pH). The number of fast kinetic phases in folding to the native conformation depends on the final pH over a similar range. Near neutral pH, the rate of fast folding appears to be the same when monitored by absorbance or fluorescence spectroscopy. At higher and lower pH, there are two fast folding reactions, with absorbance-detected fast folding occurring in a slightly faster time range than fluorescence-detected fast folding. The rates of both fast folding reactions pass through broad minima near neutral pH, suggesting involvement of ionizable groups in rate-limiting steps (Nall *et al.*, 1988). At high pH there is the additional slow kinetic phase that involves conversion of the native (or native-like) conformation to a nonnative alkaline form (Nall, 1986). Other properties of iso-2 folding differ from those reported for horse cytochrome *c*. For iso-2, distinct slow phases are detected by absorbance and fluorescence. Slow folding rates differ by as much as 20-fold at neutral pH and show little change with increasing pH. The difference between the rates of absorbance-detected and fluorescence-detected slow folding decreases as pH is lowered and approaches zero at pH 4 (Nall *et al.*, 1988). The slight differences between fast folding rates detected by fluorescence and absorbance are significant only well above or below neutral pH.

B. Proline Isomerization and Folding Rates

In unstructured polypeptide chains, the amino acids are linked to one another by amide bonds in which the amide proton and carbonyl oxygen are always *trans*. Prolines, however, have an imide linkage and the carbonyl oxygen, and the C_α proton can be either *cis* or *trans* to one another. In most cases the folded protein requires a unique proline conformer for each proline-accommodating site in the tertiary structure. Both *cis* and *trans* prolines have been found in native proteins by x-ray crystallography, although the trans form is much more common (Brandts *et al.*, 1975; Frommel and Preissner, 1990). The *trans* form of the proline imide bond is probably two to four times more common than the *cis* form for both folded proteins and unstructured polypeptides. The conversion between *cis* and *trans* forms of imide linkages is slow (time constants of 10–1000 s) and has a large activation enthalpy (20–25 kcal/mol). As first pointed out by Brandts (Brandts *et al.*, 1975), these and other properties of the slow kinetic phases in protein folding reactions are similar to those of proline imide bond isomerization. A model emerged to explain a major part of the kinetic complexity of folding reactions, the presence of the slow kinetic phases in protein refolding:

1. Native proteins require specific isomeric states for each proline, *cis* or *trans*.

2. In the unfolded protein, the conformational constraints on a proline are relieved, and, from molecule to molecule, a particular proline occurs in both *cis* and *trans* forms.

3. For a protein molecule to refold quickly, all prolines must be in the native format dictated by the tertiary structure.

4. Polypeptide chains with one or more prolines in a nonnative format refold slowly as proline imide bonds isomerize prior to or along with refolding.

For refolding conditions where native protein structure is very stable, the slow folding process is more complicated than expected from the simple form of the proline isomerization model. This complication arises because partially folded structures form (Kim and Baldwin, 1980; Schmid and Baldwin, 1979) which can strongly influence rates of chemically slow events such as isomerization. In some cases, "native-like" structures occur which are fully active enzymes (Cook *et al.*, 1979). These native-like structures with incorrect proline isomers can be thought of as transient forms of "mutant" proteins.

1. Identification of crucial prolines

Since many of the properties of the slow folding phases are those expected for reactions involving proline isomerization, considerable effort has been devoted to determining which prolines generate which slow phases. We have investigated slow folding of point mutants in which a proline is replaced by other amino acids. Yeast iso-1 and iso-2 cytochromes *c* have prolines at identical locations, with the exception of position (-1),[2] where a proline occurs in iso-2 but not in iso-1. Since the slow folding behavior of iso-1 and iso-2 is essentially the same, Pro(-1) does not appear to play a significant role in slow folding. Of the remaining prolines, those at positions 30, 71, and 76 are conserved among all members of the mitochondrial cytochrome *c* family, while Pro25 occurs at a location where non-proline residues are also allowed.

Proteins with replacements of Pro71 retain both the absorbance-detected and fluorescence-detected slow folding phases (Ramdas and Nall, 1986; White *et al.*,

---

[2] The vertebrate cytochrome *c* numbering is used to denote amino acid positions to facilitate comparison between members of the cytochrome *c* family. Iso-1 has five additional) amino-terminal residues, and iso-2 has nine additional amino-terminal residues compared to vertebrate cytochromes *c*. Both iso-1 and iso-2 have one residue less on the carboxy terminus. Thus, the vertebrate numbering of iso-1 and iso-2 starts at positions -5 and -9, respectively, and extends to position 103 (Hampsey, 1986, #6; Dickerson, 1972, #3). For example, Pro-76 in the vertebrate numbering system corresponds to Pro-85 in the iso-2 numbering system.

1987). Thus, Pro71, which occurs between two short helical regions in the native protein, does not appear to block folding as detected by either fluorescence or absorbance probes. Another possibility is that local sequence constraints keep Pro71 in the native *trans* format, even in the unfolded protein. A mutant protein in which Pro25 is replaced by Gly also retains both slow phases, so Pro25, found in a turn region, does not block refolding (MacKinnon, C., unpublished). Replacement of Pro76 leads to loss of the absorbance-detected phase, while fluorescence-detected slow folding remains largely unchanged (Wood *et al.*, 1988). So the absorbance-detected phase is assigned to (generated by) Pro76. N52I, P30A iso-2, a double mutant protein that replaces proline 30 with alanine, retains both the absorbance-detected and fluorescence-detected phases (McGee, unpublished). Thus, all prolines can be replaced, one by one, without eliminating the fluorescence-detected phase. So what generates the fluorescence-detected slow phase? Of course, the fluorescence-detected phase may be generated by some other process, perhaps heme ligand exchange. However, recent experiments that use the enzyme peptidyl prolyl *cis–trans* isomerase to catalyze slow folding slow that fluorescence-detected slow folding involves proline isomerization (Veeraraghavan and Nall, 1994). At present, we believe that the fluorescence-detected slow phase involves isomerization of two or more proline residues.

2. Simplified folding mechanism

The main result of replacing proline residues with other amino acids is that the kinetic mechanism of folding is simplified: Slow folding phases are eliminated. This occurrence allows focus on the fast folding phases. Physical interactions that simply modulate the slow folding kinetics may play a dominant role in fast folding. It is important to learn more about the fundamental processes limiting rates of fast folding. One important approach is through studies of site-directed mutants. On the other hand, folding of medium- to high-molecular-weight proteins almost always involves slow phases and thus, presumably, proline isomerization. For large proteins with many prolines or for small proteins containing *cis* isomers, essentially all folding involves isomerization. Thus, it is also important to investigate further the interplay between the structure of intermediate species kinetically trapped by non-native proline isomers and the resulting slow kinetic phases.

3. Proline replacement and alkaline conformational change

Proline replacement mutants have led to surprises with regard to folding to mutant or alternative folded conformations. Replacements of conserved prolines, at least those at positions 71 and 76, depress the pK of a pH-induced conformational change, which occurs near pH 9 in normal cytochromes c but near pH 7 in many mutant proteins (Nall *et al.*, 1989; Pearce *et al.*, 1989; Ramdas and Nall, 1987). Thus, slightly above neutral pH the mutant proteins fold to mutant (or alkaline) conformations rather than native or native-like structures. Little is known about the alkaline conformational state of cytochrome c, except that Met 80 is no longer a heme ligand, since the 695-nm absorbance band assigned to Met 80 (Shechter and Saludjian, 1967) is lost. Surprisingly, the alkaline conformation of iso-2 is more stable towards denaturant-induced unfolding than the native protein (Osterhout *et al.*, 1985). For horse cytochrome *c*, hydrogen–deuterium exchange studies suggest that the alkaline form has much but not all of the same secondary structure as native cytochrome *c* (Dohne *et al.*, 1989).

4. Implications for folding of large proteins (with more prolines) *in vivo*

It is interesting to ask whether proline imide isomerization is important for folding in the cell. There is little direct experimental evidence one way or the other, but there is some circumstantial evidence. Creighton has suggested that isomerization-limited folding of large proteins with many prolines may be very slow (Creighton, 1978): Half-times of the order of 10 minutes are estimated for a protein containing 20 prolines. Thus, in some cases, folding rates are expected to be slower than rates of protein synthesis and comparable to bacterial doubling times. Moreover, unfolded proteins sometimes have cellular half-lives shorter than the slowest folding rates. So how do proteins fold in a cell? Is proline isomerization circumvented? Is isomerization catalyzed? There are four codons for proline (CCX, where X = U, C, A, or G), so one possibility is that there are specific codons for *cis* and *trans* prolines. If so, the isomeric state of a proline needed by folded proteins might be specified during protein synthesis. As long as protein synthesis was faster than isomerization, folding would proceed at a rapid (fast-phase) rate. This possibility has been tested by inspecting proline codon preferences for proteins of known tertiary structure where the isomeric requirements of the native protein are known (Stickle, Rose, Nall, and Hardies, unpublished). The results show that there is no significant correlation between the use of particular codons and the isomeric state of prolines in folded proteins. Another possibility is that there are intracellular catalysts of isomerization and folding. This suggestion is strongly supported by the discovery of a peptidyl prolyl isomerase (PPI) (Fischer and Bang, 1985; Fischer *et al.*, 1984, 1989; Lang *et al.*, 1987). Moreover, the enzyme has been shown to speed slow folding *in vitro*. Another family of proteins, the chaperonins, help in folding of some proteins, but do so by a distinctly different mechanism. Apparently, chaperonins facilitate intramolecular folding reactions by binding partly folded or misfolded proteins and blocking nonspecific intermolecular aggregation (Freedman, 1992).

5. Catalysis by prolyl isomerase

Veeraraghavan and Nall (1994) have investigated catalysis of folding of yeast iso-2-cytochrome *c* by human peptidyl prolyl isomerase (PPI). PPI catalysis of the absorbance-detected phase (Pro-76) is observed but has a much lower k_{cat}/K_m than does PPI catalysis of isomerization in an unstructured peptide substrate. Slow folding of a mutant protein is catalyzed better than is slow folding of the native protein, suggesting that mutational destabilization of structure in a folding intermediate enhances the rate of catalyzed folding. The fluorescence-detected slow phase is catalyzed somewhat more efficiently than absorbance-detected slow folding. At present, the working model is that structure in folding intermediates modulates catalysis by PPI.

C. Flow Quench/HX Analysis of Structure of Folding Intermediates

Characterization of intermediates in protein folding reactions is difficult. Intermediates are poorly populated at equilibrium because of the cooperativity of unfolding transitions. In kinetic experiments, intermediates are highly populated, but only transiently. This limits methods for characterizing intermediates to fast techniques such as spectrophotometry or fluorometry. Unfortunately, most "fast" methods give limited structural information. An exception is amide proton–deuterium exchange (HX) (Englander and Kallenbach, 1984). HX combined with rapid mixing techniques and 2-D NMR spectroscopy gives high-resolution information on a fast time scale. This approach is based on the fact that H-bonding protects an amide proton from exchange with deuterated solvent, so exchange measurements give the location and strength of H-bonds in proteins. HX and stopped-flow mixing are used together to exchange-label exposed amides in transient structural intermediates in folding. Conversely, H-bonded amide protons in folding intermediates are protected from exchange. Subsequent analysis of the extent and patterns of labeling by NMR spectroscopy gives the rate of formation of H-bonded structure at known locations as the protein folds. The folding of cytochrome *c* is one of the first applications of this method (Roder *et al.*, 1988). The results show that the N-terminal and C-terminal helices form early in folding with similar rates. Tertiary H-bonds and shorter interior helices form more slowly. The combination of H-exchange labeling and NMR spectroscopy promises to become a standard method of determining the order in which H-bonds are formed during folding.

IV. Summary

Cytochrome *c* remains a favorite of the biophysical chemist for study of problems in protein structure and function. In particular, renewed interest in protein folding has stimulated new studies of cytochrome *c* in areas as diverse as genetics, 2-D NMR spectroscopy, and x-ray crystallography. The challenge is to use newly acquired experimental tools to make significant progress in understanding the biophysical mechanisms by which tertiary structure is encoded in sequence. Most of the prerequisites are in place: High-resolution three-dimensional structures, mutant proteins, NMR spectroscopy assignments, and kinetic mechanisms. Detailed, quantitative information on structure and folding of normal and mutant proteins is rapidly becoming available. The biophysical character of the protein folding process is yielding to experiment.

V. Acknowledgments

This review would not have been possible without research support from the National Institutes of Health (GM 32980, RR 05043) and the Robert Welch Foundation (AQ 838). One section of this review (III.B) has been revised from an earlier report (MacKinnon *et al.*, 1990) with the permission of the publishers. The studies of folding of yeast cytochromes c carried out in my laboratory are the result of the dedicated efforts of numerous research colleagues. Special thanks is due Efrain H. Zuniga, who prepared many of the figures and who carried out much of the unpublished work presented in this review.

VI. References

Babul, J., & Stellwagen, E. (1971) *Biopolymers 10*, 2359–2361.

Babul, J., & Stellwagen, E. (1972) *Biochemistry 11*, 1195–1200.

Bowie, J. U., & Sauer, R. T. (1989) *Proc. Natl. Acad. Sci. USA 86*, 2152–2156.

Bowie, J. U., Reidhaar, O. J., Lim, W. A., & Sauer, R. T. (1990) *Science 247*, 1306–1310.

Brandts, J. F. (1969) in *Structure and Stability of Biological Macromolecules* (Timasheff, S., & Fassman, G. D., eds.), pp. 213–290, Marcel Dekker, Inc., New York.

Brandts, J. F., Halvorson, H. R., & Brennan, M. (1975) *Biochemistry 14*, 4953–4963.

Brems, D. N., & Stellwagen, E. (1983) *J. Biol. Chem. 258*, 3655–3660.

Chou, P. Y., & Fasman, G. D. (1978) *Adv. Enzymol. 47*, 45–148.

Cook, K. H., Schmid, F. X., & Baldwin, R. L. (1979) *Proc. Natl. Acad. Sci. USA 76*, 6157–6161.

Cookson, D. J., Moore, G. R., Pitt, R. C., Williams, R. J., Campbell, I. D., Ambler, R. P., Bruschi, M., & LeGall, J. (1978) *Eur. J. Biochem. 83*, 261–275.

Creighton, T. E. (1978) *J. Mol. Biol. 125*, 401–406.

Creighton, T. E. (1979) *J. Mol. Biol. 129*, 235–264.

Creighton, T. E. (1980) *J. Mol. Biol. 137*, 61–80.

Creighton, T. E. (1990) *Biochem. J. 270*, 1–16.

Dickerson, R. E. (1972) *Sci. Am. 226*, 58–72.

Dickerson, R. E. (1980) *Sci. Am. 242*, 136–153.

Dill, K. (1985) *Biochemistry 24*, 1501–1509.

Dohne, S. M., Elove, G. A., Roder, H., & Nall, B. T. (1989) *Biophys. J. 55*, 557a.

Dumont, M. D., Mathews, A. J., Nall, B. T., Baim, S. B., Eustice, D. C., & Sherman, F. (1990) *J. Biol. Chem. 265*, 2733–2739.

Dyson, H. J., & Beattie, J. K. (1982) *J. Biol. Chem. 257*, 2267–2273.

Dyson, H. J., Cross, K. J., Houghten, R. A., Wilson, I. A., Wright, P. E., & Lerner, R. A. (1985) *Nature 318*, 480–483.

Englander, S. W., & Kallenbach, N. R. (1984) *Quart. Rev. Biophys. 16*, 531–625.

Ernst, J. F., Stewart, J. W., & Sherman, F. (1982) *J. Mol. Biol. 161*, 373–394.

Feng, Y., Roder, H., Englander, S. W., Wand, A. J., & Di Stefano, D. L. (1989) *Biochemistry 28*, 195–203.

Feng, Y., Roder, H., & Englander, S. W. (1990) *Biophys. J. 57*, 15–22.

Fetrow, J. S., Cardillo, T. S., & Sherman, F. (1989) *Proteins 6*, 372–381.

Fischer, G., & Bang, H. (1985) *Biochim. Biophys. Acta 828*, 39–42.

Fischer, G., Bang, H., & Mech, C. (1984) *Biomed. Biochim. Acta 43*, 1101–1111.

Fischer, G., Wittmann, L. B., Lang, K., Kiefhaber, T., & Schmid, F. X. (1989) *Nature 337*, 476–478.

Freedman, R. B. (1992) in *Protein Folding* (Creighton, T. E., ed.), pp. 455–539, W. H. Freeman and Company, New York.

Frommel, C., & Preissner, R. (1990) *FEBS 277*, 159–163.

Gadsby, P. M., Peterson, J., Foote, N., Greenwood, C., & Thomson, A. J. (1987) *Biochem. J. 246*, 43–54.

Gao, Y., Boyd, J., Williams, R. J. P., & Pielak, G. J. (1990) *Biochemistry 29*, 6994–7003.

Goldenberg, D. P., & Creighton, T. E. (1984) *Anal. Biochem. 138*, 1–18.

Goto, Y., Calciano, L. J., & Fink, A. L. (1990a) *Proc. Natl. Acad. Sci. USA 87*, 573–577.

Goto, Y., Takahashi, N., & Fink, A. L. (1990b) *Biochemistry 29*, 3480–3488.

Hall, J. G., & Frieden, C. (1989) *Proc. Natl. Acad. Sci. USA 86*, 3060–3064.

Hampsey, D. M., Das, G., & Sherman, F. (1986) *J. Biol. Chem. 261*, 3259–3271.

Hampsey, D. M., Das, G., & Sherman, F. (1988) *FEBS Lett. 231*, 275–283.

Hantgan, R. R., & Taniuchi, H. (1977) *J. Biol. Chem. 252*, 1367–1374.

Harrison, S. C., & Durbin, R. (1985) *Proc. Natl. Acad. Sci. USA 82*, 4028–4030.

Hartshorn, R. T., & Moore, G. R. (1989) *Biochem. J. 258*, 595–598.

Hong, X., & Dixon, D. W. (1989) *FEBS Lett. 246*, 105–108.

Ikai, A., Fish, W. W., & Tanford, C. (1973) *J. Mol. Biol. 73*, 165–184.

Juillerat, M. A., & Taniuchi, H. (1986) *J. Biol. Chem. 261*, 2697–2711.

Juillerat, M., Parr, G. R., & Taniuchi, H. (1980) *J. Biol. Chem. 255*, 845–853.

Kabsch, W., & Sander, C. (1983) *FEBS Lett. 155*, 179–182.

Kaminsky, L. S., Miller, V. J., & Davison, A. J. (1973) *Biochemistry 12*, 2215–2221.

Kawaguchi, H., & Noda, H. (1977) *J. Biochem. 81*, 1307–1317.

Kim, P. S., & Baldwin, R. L. (1980) *Biochemistry 19*, 6124–6129.

Kim, P. S., & Baldwin, R. L. (1982) *Ann. Rev. Biochem. 51*, 459–489.

Kim, P. S., & Baldwin, R. L. (1984) *Nature 307*, 329–334.

Kim, P. S., & Baldwin, R. L. (1990) *Ann. Rev. Biochem. 59*, 631–660.

Knapp, J. A., & Pace, C. N. (1974) *Biochemistry 13*, 1289–1294.

Lang, K., Schmid, F. X., & Fischer, G. (1987) *Nature 329*, 268–270.

Leszczynski (Fetrow), J. F., & Rose, G. D. (1986) *Science 234*, 849–855.

Levitt, M. (1978) *Biochemistry 17*, 4277–4285.

Liggins, J. R., Sherman, F., Mathews, A. J. & Nall, B. T. (1994) *Biochemistry 33*, 9209–9219.

Lim, V. I. (1974) *J. Mol. Biol. 88*, 873–894.

Lim, W. A., & Sauer, R. T. (1989) *Nature 339*, 31–36.

Lin, L. N., Hasumi, H., & Brandts, J. F. (1988) *Biochim. Biophys. Acta 956*, 256–266.

MacKinnon, C., Veeraraghavan, S., Kreider, I., Allen, M. J., Liggins, J. R., & Nall, B. T. (1990) in *Chemical Aspects of Enzyme Biotechnology* (Baldwin, T. O., Raushel, F. M., & Scott, A. I., eds.), pp. 53–64, Plenum Press, New York and London.

Marqusee, S., Robbins, V. H., & Baldwin, R. L. (1989) *Proc. Natl. Acad. Sci. USA 86*, 5286–5290.

McLendon, G., & Smith, M. (1978) *J. Biol. Chem. 253*, 4004.

Muthukrishnan, K., & Nall, B. T. (1991) *Biochemistry 30*, 4706–4710.

Myer, Y. P. (1984) *J. Biol. Chem. 259*, 6127–6133.

Nall, B. T. (1983) *Biochemistry 22*, 1423–1429.

Nall, B. T. (1986) *Biochemistry 25*, 2974–2978.

Nall, B. T., & Landers, T. A. (1981) *Biochemistry 20*, 5403–5411.

Nall, B. T., Osterhout, J. J., & Ramdas, L. (1988) *Biochemistry 27*, 7310–7314.

Nall, B. T., Zuniga, E. H., White, T. B., Wood, L. C., & Ramdas, L. (1989) *Biochemistry 28*, 9834–9839.

Nelson, J., & Kallenbach, N. (1986) *Proteins 1*, 211–217.

Nojima, H., Hon-nami, K., Oshima, T., & Noda, H. (1978) *J. Mol. Biol. 122*, 33–42.

Oas, T. G., & Kim, P. S. (1988) *Nature 336*, 42–48.

Ohgushi, M., & Wada, A. (1983) *FEBS Lett. 164*, 21.

Osterhout, J. J., & Nall, B. T. (1985) *Biochemistry 24*, 7999–8005.

Osterhout, J. J., Jr., Muthukrishnan, K., & Nall, B. T. (1985) *Biochemistry 24*, 6680–6684.

Pace, C. N. (1990) *TIBTECH 8*, 93–98.

Padmanabhan, S., Marqusee, S., Ridgeway, T., Laue, T. M., & Baldwin, R. L. (1990) *Nature 344*, 268–270.

Parr, G. R., Hantgan, R. R., & Taniuchi, H. (1978) *J. Biol. Chem. 253*, 5381–5388.

Pearce, L. L., Gartner, A. L., Smith, M., & Mauk, A. G. (1989) *Biochemistry 28*, 3152–3156.

Pfeil, W. (1981) *Mol. Cell. Biochem. 40*, 3–28.

Pielak, G., Mauk, A. G., & Smith, M. (1985) *Nature 313*, 152–154.

Poland, D., & Scheraga, H. (1970) *Theory of Helix–Coil Transitions in Biopolymers* (first ed.), Academic Press, New York.

Potekhin, S., & Pfeil, W. (1989) *Biophys. Chem. 34*, 55–62.

Presta, L. G., & Rose, G. D. (1988) *Science 240*, 1632–1641.

Privalov, P. L. (1979) *Adv. Protein Chem. 33*, 167–241.

Ragone, R., Colonna, G., Balestrieri, C., Servillo, L., & Irace, G. (1984) *Biochemistry 23*, 1871–1875.

Ramdas, L., & Nall, B. T. (1986) *Biochemistry 25*, 6959–6964.

Ramdas, L., & Nall, B. T. (1987) *J. Cell. Biochem. Supplement 11C*, 218.

Ramdas, L., Sherman, F., & Nall, B. T. (1986) *Biochemistry 25*, 6952–6958.

Reidhaar, O. J., & Sauer, R. T. (1988) *Science 241*, 53–57.

Reidhaar, O. J. F., & Sauer, R. T. (1990) *Proteins 7*, 306–316.

Richardson, J. S., & Richardson, D. C. (1988) *Science 240*, 1648–1652.

Robinson, J. J., Strottmann, J. M., & Stellwagen, E. (1983) *J. Biol. Chem. 258*, 6772–6776.

Roder, H., Elove, G. A., & Englander, S. W. (1988) *Nature 335*, 700–704.

Rose, G. D. (1979) *J. Mol. Biol. 134*, 447–470.

Rose, G. D. (1985) *Methods Enzymol. 115*, 430–440.

Rose, G. D., & Roy, S. (1980) *Pro. Natl. Acad. Sci. USA 77*, 4643–4647.

Saigo, S. (1981) *J. Biochem. 89*, 1977–1980.

Saigo, S. (1986) *J. Biochem. (Tokyo) 100*, 157–165.

Schellman, J. A. (1978) *Biopolymers 17*, 1305–1322.

Scheraga, H. A. (1978) *Pure Appl. Chem. 50*, 315–324.

Schmid, F. X., & Baldwin, R. L. (1979) *J. Mol. Biol. 135*, 199–215.

Shakhnovich, E. I., & Finkelstein, A. V. (1982) *Dokl. Akad. Nauk S.S.S.R. 267*, 1247.

Shechter, E., & Saludjian, P. (1967) *Biopolymers 5*, 788–790.

Sherman, F., & Stewart, J. W. (1978) in *Biochemistry and Genetics of Yeast* (Bacila, M., Horecker, B. L., & Stoppani, A. D. M., eds.), p. 273, Academic Press, New York.

Sherman, F., Stewart, J. W., Parker, J. H., Inhaber, E., Shipman, N. A., Putterman, G. J., Gardisky, R. L., & Margoliash, E. (1968) *J. Biol. Chem. 243*, 5446–5456.

Stellwagen, E. (1968) *Biochemistry 7*, 2893–2898.

Stellwagen, E., & Babul, J. (1975) *Biochemistry 14*, 5135–5140.

Stellwagen, E., Babul, J., & Wilgus, H. (1975) *Biochim. Biophys. Acta 405*, 115–121.

Strehlow, K. G., & Baldwin, R. L. (1989) *Biochemistry 28*, 2130–2133.

Tanford, C. (1969) *Adv. Prot. Chem. 24*, 1–95.

Taniuchi, H., Parr, G. R., & Juillerat, M. A. (1986) *Methods Enzymol. 131*, 185–217.

Tsong, T. Y. (1973) *Biochemistry 12*, 2209–2214.

Tsong, T. Y. (1974) *J. Biol. Chem. 249*, 1988–1990.

Tsong, T. Y. (1975) *Biochemistry 14*, 1542–1547.

Tsong, T. Y. (1976) *Biochemistry 15*, 5467–5473.

Tsong, T. Y. (1977) *J. Biol. Chem. 252*, 8778–8780.

Veeraraghavan, S. & Nall, B. T. (1994) *Biochemistry 33*, 687–692.

Wallace, C. J. (1984) *Biochem. J. 217*, 601–604.

Wand, A. J., Di Stefano, D. L., Feng, Y., Roder, H., & Englander, S. W. (1989) *Biochemistry 28*, 186–194.

White, T. B., Berget, P. B., & Nall, B. T. (1987) *Biochemistry 26*, 4358–4366.

Wilgus, H., & Stellwagen, E. (1974) *Proc. Natl. Acad. Sci. USA 71*, 2892–2894.

Williams, G., Moore, G. R., Porteous, R., Robinson, M. N., Soffe, N., & Williams, R. J. P. (1985) *J. Mol. Biol. 183*, 409–428.

Wood, L. C., White, T. B., Ramdas, L., & Nall, B. T. (1988) *Biochemistry 27*, 8562–8568.

Yockey, H. P. (1977) *J. Theor. Biol. 67*, 377–398.

Yuschok, T. J., & Rose, G. D. (1983) *Int. J. Pept. Protein. Res. 21*, 479–484.

Zehfus, M. H., & Rose, G. D. (1986) *Biochemistry 25*, 5759–5765.

Zimm, B. H., & Bragg, J. K. (1959) *J. Chem. Phys. 31*, 526–535.

Zuniga, E. H., & Nall, B. T. (1983) *Biochemistry 22*, 1430–1437.

Spectroscopic Properties

5

Nuclear Magnetic Resonance Studies of Class I Cytochromes c

GARY J. PIELAK
DOUGLAS S. AULD
STEPHEN F. BETZ
SHARON E. HILGEN-WILLIS
L. LILIANA GARCIA

Department of Chemistry
University of North Carolina

I. Introduction

Many studies of the nuclear magnetic resonance (NMR) of cytochromes c have appeared since the 56.4-MHz proton spectrum of horse protein was published by Kowalsky in 1962. The purpose of this article is to provide the researcher with a feeling for the various NMR spectra of cytochrome c and facile access to the literature on the subject. Papers published up to February 1995 are included here. This review is limited to Class I cytochromes c. [Some assignments in Class II cytochromes c have been presented by Moore et al. (1982).] Members of Class I are low-spin, contain a single heme, and possess His and Met as iron ligands, with the Met ligand occurring near the C-terminus (Pettigrew and Moore, 1987). Earlier reviews at least partially devoted to NMR studies of cytochromes c include those by Wüthrich (1970, 1976), McDonald and Phillips (1973), Levine et al. (1979), Moore et al. (1983), Williams et al. (1985c), Senn and Wüthrich (1985), Williams (1988, 1989), MacKenzie et al. (1992a), and, especially, Moore and Pettigrew (1990).

Three sections follow this introduction. A discussion of the properties of cytochromes c that make them good candidates for NMR spectroscopy is presented in Section II. Section III deals with the assignment of resonances to specific nuclei. The final section deals with the use of the assignments to provide information about structure, dynamics, and function. When more or complementary information is to be found in other places in the review, the relevant section numbers are given.

In this review IUPAC-IUB (1970, 1985) nomenclature for amino acids and the heme (Commission on the Nomenclature of Biological Chemistry, 1960) is used (Figure 5.1). The use of numeral superscripts to denote substituent groups of the heme was suggested by Bonnett (1978), and the use of letter superscripts to specify protons on the propionic acid side chains is based on Fig. 2 of Chau *et al.* (1990). Many studies used other nomenclatures for the heme, so a concordance is given in Table 5.1. The nomenclature wherein Greek letters are used in a systematic way for every amino acid is accepted by the IUPAC-IUB, but is not the preferred one for the aromatic amino acids. Except for the aromatic amino acids, this nomenclature is equivalent to that used in the Brookhaven Data Bank of x-ray crystallographic data (Bernstein *et al.*, 1977). For His, the IUPAC recommends using τ and π to name the ring nitrogens (Figure 5.1). The preferred nomenclature for the benzoid side chains of Tyr and Phe and for the indole ring of Trp (Figure 5.1) involves arabic numerals. For Phe and Tyr, the *ortho* positions are referred to as 2 and 6, and the *meta* positions are referred to as 3 and 5. The term "heteronucleus," as used in this review, refers to any nucleus besides the proton. Protons are named for the heavy atom to which they are attached.

The residue numbering system used here is based on the sequences of higher eukaryotic cytochromes c (Moore and Pettigrew, 1990). The thioether linkages to the heme emanate from cysteines 14 and 17, and the ligands to the iron are residues 18 and 80. Therefore, plant and fungal cytochromes c, which possess extensions at the N-terminus, begin with negative residue numbers. Prokaryotic cytochromes c are referred to using a numbering system unique to each protein. Unnatural protein variants are denoted using the one-letter code (IUPAC-IUB Joint Commission on Biochemical Nomenclature, 1985), with the wild-type residue listed first, followed by the primary sequence position, and the variant residue. Multiple changes are separated by a semicolon (e.g., F82Y; C102T). The *de facto* wild-type *Saccharomyces cerevisiae* iso-1-cytochrome c, the C102T variant (Cutler *et al.*, 1987), is structurally identical to the wild-type protein (Gao *et al.*, 1990b, 1991a; Berghuis and Brayer, 1992), but is more amenable to biophysical studies (Betz and Pielak, 1992) because removal of the sole free cysteine stops intermolecular dimer formation.

An extensive review of the NMR experiment is beyond the scope of this review. The reader is referred to the book by Ernst *et al.* (1987) for a rigorous account of modern NMR techniques, the book by Derome (1987) for a largely nonmathematical treatment of modern NMR, and the books by Freeman (1987) and Homans (1989) for an explanation of NMR jargon. Review articles by Bax (1989) and Ernst (1992), the books by Wüthrich (1976, 1986), and Volumes 176 and 177 of *Methods in Enzymology* are other excellent sources for information on protein NMR.

Wild-type cytochromes c from more than 20 sources have been studied by NMR. However, there are only two higher eukaryotic, two lower eukaryotic, and

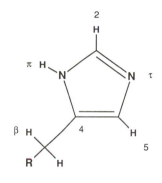

A

B

Figure 5.1
(A) The structure of His and Trp annotated with the IUPAC-IUB nomenclature used in this article (IUPAC-IUB Joint Commission on Biochemical Nomenclature, 1985). (B) The structure of the heme annotated with the IUPAC nomenclature used in this article (Commission on the Nomenclature of Biological Chemistry, 1960).

Table 5.1
A concordance of His, Trp, and heme nomenclatures.

IUPAC-IUB[a]	IUPAC-IUB[b]	Old nomenclature
Nomenclature for His		
τ	$\varepsilon 2$	3
2	$\varepsilon 1$	2
π	$\delta 1$	1
4	γ	5
5	$\delta 2$	4
Nomenclature for Trp		
1	$\varepsilon 1$	
2	$\delta 1$	
3	γ	
3a	$\delta 2$	
4	$\varepsilon 3$	
5	$\zeta 3$	
6	$\eta 2$	
7	$\zeta 2$	
7a	$\varepsilon 2$	

Heme nomenclature

IUPAC-IUB[c]	Brookhaven[d]	Old nomenclature[e]	
2^1	CMB	methyl 1	
3^1	CAB ring B	thioether 2 methine	ring 1
3^2	CBB	thioether 2 methyl	
5	CHC	meso α	
7^1	CMC	methyl 3	
8^1	CAC ring C	thioether 4 methine	ring 2
8^2	CBC	thioether 4 methyl	
10	CHD	meso β	
12^1	CMD	methyl 5	
13^{1a}	CAD ring D	propionate 6 (outer) β or α	ring 3
13^{1b}	CAD	propionate 6 (outer) β' or α'	
13^{2a}	CBD	propionate 6 (outer) α or β	
13^{2b}	CBD	propionate 6 (outer) α' or β'	
15	CHA	meso γ	

Table 5.1 (*continued*)

IUPAC-IUB[a]	IUPAC-IUB[b]	Old nomenclature	
17^{1a}	CAA	propionate 7 (inner) β or α	
17^{1b}	CAA	propionate 7 (inner) β' or α'	
17^{2a}	CBA ring A	propionate 7 (inner) α or β	ring 4
17^{2b}	CBA	propionate 7 (inner) α' or β'	
18^{1}	CMA	methyl 8	
20	CHB	meso δ	

[a] IUPAC-IUB Joint Commission on Biochemical Nomenclature, 1985; this is the preferred system.
[b] IUPAC-IUB Commission on Biochemical Nomenclature, 1970.
[c] Commission on the Nomenclature of Biological Chemistry (1960); Bonnett, 1978: Chau *et al.*, 1990.
[d] Louie *et al.*, 1988a.
[e] *Cf.* Keller and Wüthrich, 1978b; Moore and Williams, 1984.

four prokaryotic cytochromes *c* whose proton resonances have been extensively assigned (Table 5.2). The eukaryotic proteins are from horse, tuna, and the yeast *Saccharomyces cerevisiae* (the C102T variant of iso-1-, and wild-type iso-2-cytochromes; see notes added in proof). The prokaryotic proteins are from *Pseudomonas aeruginosa* and *Pseudomonas stutzeri* (ferrocytochromes *c*-551), the photoheterotroph *Rhodobacter capsulatus* (cytochrome c_2), and *Desulfovibrio vulgaris* Hildenborough (cytochrome *c*-553). Many ^{13}C assignment are available for both oxidation states of the horse protein (Gao *et al.*, 1990a; Santos and Turner, 1992) and for reduced *Desulfovibrio vulgaris* Hildenborough cytochrome *c*-553. Furthermore, essentially all the backbone ^{15}N resonances have been assigned for the *Rhodobacter capsulatus* (Gooley *et al.*, 1990) and *Pseudomonas aeruginosa* (Timkovich, 1990) proteins. Complete listings of assignments can be obtained from the BioMagResBank (Department of Biochemistry, University of Wisconsin, Madison 53706-1569; Ulrich *et al.*, 1989). References to earlier assignments for these proteins are given in the text and listed in Table 5.2. References to resonance assignments for other cytochromes *c* are given in Table 5.3. Procedures for the preparation of ferro- and ferricytochromes *c* for proton NMR spectroscopy are given by Pielak *et al.* (1988a).

Table 5.2
Cytochromes c for which extensive proton assignments are available.[a]

	Amides	α	Side chains	Ref.[b]
	Higher eukaryotes			
	Horse			
Ferro	99/100	103/104	nearly complete	a
Ferri	99	103	nearly complete	b
	Tuna			
Ferro	0/99	~14/103	partial	c
Ferri	71	71	partial	d, e
	Lower eukaryotes			
	C102T variant of *Saccharomyces cerevisiae* iso-1			
Ferro	100/104	105/108	nearly complete	f
Ferri	100	104	nearly complete	f
	***Saccharomyces cerevisiae* iso-2**			
Ferro	0/107	0/112	very few	g
Ferri	75	73	partial	h
	Prokaryotes			
	***Pseudomonas aeruginosa* c-551**			
Ferro	76/76	79/82	nearly complete	i
Ferri	0	1	very few, see Table III	j
	***Pseudomonas stutzeri* c-551**			
Ferro	76/76	82/82	nearly complete	k
Ferri	none	none	very few	l
	***Rhodobacter capsulatus* c_2**			
Ferro	106/111	115/116	nearly complete	m
Ferri	nearly complete	?	?	n
	***Desulfovibrio vulgaris* Hildenborough c-553**			
Ferro	76/77	77/78	nearly complete	o
Ferri	0	0	none	

[a] Fractional numbers indicate the number of protons assigned over the number in the protein. The numbers given for α protons do not take into account the fact that Gly residues possess two such protons.

[b] a, Wand *et al.*, 1989; b, Feng *et al.*, 1989; c, Moore *et al.*, 1985; d, Gao *et al.*, 1989; e, Williams *et al.*, 1985a; f, Gao *et al.*, 1990b; g, Senn *et al.*, 1983a; h, Garcia, 1991; i, Chau *et al.*, 1990, Detlefsen *et al.*, 1990, Timkovich, 1991; j, Leitch *et al.*, 1984, but see III.L, Moratal *et al.*, 1993; k, Cai *et al.*, 1992; l, Leitch *et al.*, 1984; m, Gooley *et al.*, 1990; n, referred to as Zhao *et al.*, cited as unpublished in Gooley *et al.*, 1991; o, Medvedeva *et al.*, 1993, Marion and Guerlesquin, 1992; Senn *et al.*, 1983b, Cookson *et al.*, 1978.

Table 5.3
References to nonexchangeable-proton assignments for various cytochromes c.

Candida krusei

Cookson *et al.* (1978): His-18[r], Phe-82[r*], Met-80[r]

Boswell *et al.* (1982): Thr-19[o], Ile-20[r], trimethyllysine-72[r,o], Phe-82[r,o]

Robinson *et al.* (1983): His-39[r], Trp-59[r,o], Tyr-74[r,o]

Williams *et al.* (1985b): Ala-15[o] Thr-19[o], Ala-43[o], Ala-51[o], Ala-81[o], Ala-101[o]

Moore *et al.* (1985): Ala-15[r], Thr-19[r], Val-28[r], Leu-32[r], Ala-43[r], Thr-49[r], Ala-51[r], Thr-63[r], Leu-68[r], Ile-75[r], Thr-78[r], Leu-98[r], Ala-101[r*]

Luntz *et al.* (1989): Leu-68[o]

Chicken

Moore and Williams (1980f): Ile-57[r,o]

Cow

Moore and Williams (1980f): Ile-57[r,o]

Wand and Englander (1985): Ser-47[r]

Crithidia oncopelti **cytochrome** c**-557**

Keller *et al.* (1979): spectra presented, but only heme assignments reported

Cryptococcus

Sibel'dina *et al.* (1976): Met-80[o]

Desulfovibrio desulfuricans **cytochrome** c**-553**

Senn *et al.* (1983b): His ligand[r], Met ligand[r,o]

Dog

Moore and Williams (1980f): Ile-57[r,o]

Donkey

Moore and Williams (1980f): Ile-57[r,o], Thr-89[r]

Drosophila melanogaster

Luntz *et al.* (1989): Leu-68[o]

Euglena gracilis c**-552**

Keller *et al.* (1977): Met-56[r]

Keller and Wüthrich (1981): Phe-6[r], His-14[r], Leu-27[r], Trp-59[r], Tyr-74[r], Trp-83[r]

Guanaco

Robinson *et al.* (1983): Ile-57[r,o]

Halotolerant *micrococcus* **cytochrome** c**-554**

Cookson *et al.* (1978): Met ligand[r]

(continued)

210

Table 5.3 (*continued*)
References to nonexchangeable-proton assignments for various cytochromes *c*.

Kangaroo

Robinson *et al.* (1983): Ile-57[r,o]

Lamprey

Williams *et al.* (1985b): Val-3°, Val-9°, Val-11°, Thr-19° Val-20°, Ala-23°, Thr-28°, Leu-35°, Thr-40°, Ala-43°, Thr-49°, Ala-51° Ile-57°, Val-58°, Thr-63°, Leu-64°, Val-66°, Leu-68°, Thr-78°, Ile-81°, Ala-83°, Ile-85°, Leu-94°, Ile-95°, Ala-96°, Leu-98°, Thr-102°

Neurospora crassa

Boswell *et al.* (1982): Thr-19°, Trimethyllysine-72[r,o], Phe-82[r,o]

Eley *et al.* (1982b): Phe-10[r,o], ala-15°, Leu-32[r], Tyr-46°, Tyr-48°, Ile-57[r,o], Trp-59[o,r], Tyr-74[r,o], Met-80[r,o]

Pig

Sibel'dina *et al.* (1976): Met-80°

Pigeon

Moore and Williams (1980f): Ile-57[r,o]

Pseudomonas aeruginosa **cytochrome *c*-551**

Moratal *et al.* (1993): Cys-12°, Gly-24°, Ile-48°, Trp-56°, Met-61°
(for the reduced protein, see Table 5.2)

Pseudomonas mendocina **cytochrome *c*-551**

Leitch *et al.* (1984): Phe-7[r], His-47[r], trp-56[r], Trp-77[r]

Rabbit

Moore and Williams (1980f), Robinson *et al.* (1983): Ile-57[r,o]

Rat

Robinson *et al.* (1983): Ile-57[r,o]

Luntz *et al.* (1989): Ala-15°, Thr-19°, Leu-68°, Phe-82°

Rhodomicrobium vannielii **cytochrome *c*₂**

Moore *et al.* (1984): Trp-59[r], methionine ligand[o]

Rhodospirillum fulvum **iso-1-cytochrome *c*₂**

Cookson *et al.* (1978): Tyr-43[r*], Met-75[r], Phe-79[r*]

Rhodospirillum rubrum **cytochrome *c*₂**

Cookson *et al.* (1978) Tyr-48[r*], Met-91[r], Phe-93[r*]

Smith (1979): Leu-32[r], His-42[r], Trp-62[r], Tyr-70[r], Met-91[r,o]

Senn and Wüthrich (1983): His-18[r,o]

Yu and Smith (1990a): Phe-20[r], Pro-30[r], Trp-62[r,o], Phe-93[r]

Table 5.3 (*continued*)

***Rhodospirillum viridis* cytochrome c_2**

Cookson *et al.* (1978): Tyr-47^{r*}, Met-68^r, trp-58^{r*}, Phe-81^{r*}

Synechococcus lividus

Katz and Crespi (1972): spectra presented, but no assignments given

Turtle

Stellwagen and Shulman (1973): spectra presented, but no assignments reported

N.B.—References to cytochromes *c* that have been more extensively assigned are given in Table 5.2. Superscripts r and o refer to the reduced and oxidized states, respectively. Many of the assignments are not complete. The assignments listed here are as reported in the references cited. A superscript * indicates that this assignment should be treated with caution (III.L). Note that the assignments of residue 46 and Tyr-48 in both oxidation states have been revised (Keller and Wüthrich, 1981; Arean *et al.*, 1988), as have the assignments of Thr-9 and Ala-101 (Arean *et al.*, 1988) in the oxidized protein. Note that assignments for Tyr-48 equivalents in ferrocytochromes c_2, Tyr-48 in *Rhodospirillum rubrum*, Tyr-47 in *Rhodopseudomonas viridis*, and Tyr-43 in *Rhodospirillum fulvum* iso-1-, in Cookson *et al.* (1978) need to be revised. Trp assignments in cytochromes *c*-551 and other prokaryotic cytochromes *c* have been revised (Chau *et al.*, 1990), as have the assignments of Phe-82 (Boswell *et al.*, 1980) and its ferrocytochrome c_2 equivalents (i.e., Phe-93 in *Rhodospirillum rubrum*, Phe-81 in *Rhodopseudomonas viridis*, and Phe-79 in *Rhodospirillum fulvum* iso-1-cytochrome c_2; Cookson *et al.*, 1978).

II. Properties of Cytochromes *c* Relevant to the NMR Experiment

Why has cytochrome *c* been so extensively studied by NMR techniques? The answer involves the suitability of the protein to the NMR experiment. NMR is not the most sensitive of analytical techniques; even at the current proton resonance frequencies of 500–750 MHz, sample concentrations in the millimolar range are required. With a molecular weight for eukaryotic cytochromes *c* of approximately 13,000 g mol^{-1} and a sample volume of 0.5 mL, this translates to between 10 and 50 mg of protein. This is much more protein than is necessary for other biophysical techniques such as fluorescence, circular dichroism (cd), and absorption spectroscopies. The fact that Sigma Chemical Company (St. Louis) sells cytochrome *c* from several species has made this molecule available to spectroscopists who prefer not to isolate proteins. Cytochromes *c* can also be isolated from almost any tissue or organism (Brautigan *et al.*, 1978). The high positive charge of many cytochromes *c* makes their isolation and purification by ion-exchange chromatography straightforward, and the red color of the proteins makes them easy to identify on chromatographic columns. This high positive charge also aids in preventing aggregation, which is important for reasons discussed later.

High-resolution proton NMR spectroscopy is sensitive to the size of the molecule under study in two related ways. As the number of amino acids in the protein increases, so does the number of resonances and the possibility that overlapping resonances will obviate assignment. Complicating this is the increase in the linewidth of individual resonances with increasing molecular size as measured by the decrease in T_2, the spin–spin relaxation time, as a function of τ_c, the correlation time. It should be noted that aggregation of protein molecules and increased viscosity, both of which can result from the high sample concentration required for NMR, have the same effect as increasing the molecular size. Cytochrome c exhibits a correlation time of $\sim 5 \times 10^{-9}$ s (Andersson et al., 1979; Senn et al., 1984), and typical methyl resonances have a width at half-height of approximately 15 Hz in high-field instruments. Horse, tuna, and Saccharomyces cerevisiae iso-1-cytochrome c do not aggregate even at concentrations approaching 20 mM.

Studies involving protein stability, the activation energy of amide proton exchange, the flip rate of aromatic residues, etc., require acquisition of NMR spectra as a function of temperature. For the titration of ionizable functional groups and for controlling the exchange rate of labile protons, it is important to know the extremes of pH that can be explored. In this review, pH values are uncorrected for the deuterium isotope effect and represent pH meter readings. The correction factor is $p^2H = pH + 0.4$ (Glasoe and Long, 1960). There is, however, an isotope effect on pK that often makes correction unnecessary (Schowen and Schowen, 1982). Cytochromes c maintain their native form and yield high-quality NMR spectra over a wide range of temperature and pH (see Sections III.N.2 and III.N.3, respectively).

Variants of cytochrome c generated by in vitro mutagenesis or semisynthesis exhibit a continuum of stabilities ranging from not stable enough to isolate in large quantities (Sorrell et al., 1989; Garcia et al., 1992), to moderately stable (Pearce et al., 1989; Auld et al., 1993; Pielak et al., 1995) to more stable than the wild-type protein (Das et al., 1989; Luntz et al., 1989, but see Schejter et al., 1992; Section IV.J.3). Before beginning a detailed study of any new cytochrome c, it is important to determine that the pH and temperature to be used are consistent with conditions that favor the native state (Pearce et al., 1989; Sections III.N.2 and III.N.3).

When spectra of cytochromes c are compared to those of other, non–heme-containing proteins, it is noted that resonances of cytochrome c exhibit greater chemical shift dispersion. This increased dispersion is brought about by the ring-current shift of the heme (Section III.B.1) and, for the oxidized protein, paramagnetic shifts caused by the interaction of proton spins with the spin of the unpaired electron on the iron (Section III.B.2). This increased dispersion can be considered an asset because it relieves crowding of resonances. It can be considered a liability because it invalidates the use of random-coil chemical shifts (Bundi and Wüthrich, 1979) for making initial guesses concerning the type of proton (i.e., amide, methyl, etc.) that gives rise to resonances (see Sections III.B.1 and III.B.2).

The ability to express economically large quantities of isotopically enriched eukaryotic cytochromes c is desirable or essential for some NMR studies involving nuclei other than protons. The assignment of ^{13}C and ^{15}N (Section III.K.2) and the determination of the relaxation parameters for ^{13}C and ^{15}N (Lipari and Szabo, 1982; Section IV.K) are examples where isotopic enrichment is useful because of the low receptivity of these nuclei.

It might be imagined that the cloned genes of eukaryotic cytochromes *c* could be used to produce gram quantities of the proteins via heterologous expression in the work-horse bacterium *Escherichia coli*. Bacterial cytochromes *c* have been expressed in *E. coli*, but yields are low (McEwan *et al.*, 1989; Self *et al.*, 1990; von Wachenfeldt and Hederstedt, 1990). Expression of eukaryotic cytochromes *c* in bacteria leads to production of the apoprotein, perhaps because the heterologous gene does not have a bacterial signal sequence and the bacterial heme lyase (the enzyme that attaches the heme) does not recognize the eukaryotic apoprotein. Isotopic enrichment of higher eukaryotic cytochromes *c* could be achieved by expressing the protein from a cloned gene in a lower eukaryote capable of performing the various post-translational modifications. Using the yeast *Saccharomyces cerevisiae* cytochromes *c* as an example, the synthesis of the protoporphyrin, the insertion of the iron, the removal of the N-terminal Met, the transport of the apoprotein into the mitochondrion, the covalent insertion of heme into apocytochrome *c*, and the trimethylation of Lys-72 require a plethora of gene products. Unless one of these steps is entirely limiting, many gene products must be over-expressed for holocytochrome *c* to be over-expressed. The expression of rat (Scarpulla and Nye, 1986; Luntz *et al.*, 1989; Clements *et al.*, 1989), tuna, pigeon, and horse (Hickey *et al.*, 1991b) cytochromes *c* in yeast have been reported, but to our knowledge there is no report of isotopic enrichment via heterologous expression of a cloned eukaryotic cytochrome *c*. Expression of isotopically enriched eukaryotic cytochromes *c* may present a more serious economic problem than a technical one unless overproducing strains can be obtained. Although acceptable yields of cytochromes *c* can be obtained by fermentation of microorganisms on rich media, use of defined media (required for incorporation of heteronuclei) often greatly reduces the yield. Even if mass production of labeled protein turns out to be an expensive proposition, all is not bleak because much progress has been made in assigning ^{13}C [ferro- and ferri- horse cytochrome *c* (Gao *et al.*, 1990a; Santos and Turner, 1992; Turner and Williams, 1993), *Desulfovibrio vulgaris* ferrocytochrome *c*-553 (Medvedeva *et al.*, 1993)] and ^{15}N resonances [*Pseudomonas aeruginosa* ferrocytochrome *c*-551 (Timkovich, 1990)] at natural abundance (Section III.K.2).

There are several reports of isotopic enrichment of lower eukaryotic cytochromes *c* expressed in homologous systems. Some of the first assignments of ^{13}C resonances were made for *Neurospora crassa* cytochrome *c* grown on labeled Met (Eakin *et al.*, 1975a). Nall and Zuniga (1990) labeled *Saccharomyces cerevisiae* iso-2-cytochrome *c* (Section IV.A). Simplification of the proton spectrum of *Synechococcus lividus* cytochrome *c* resulting from growing this blue-green alga on D_2O with or without supplementation by proton-containing amino acids was reported by Katz and Crespi (1972), although no assignments were reported. There is also a report (Kayushin *et al.*, 1975) of deuterium enrichment of the cytochrome *c* from the yeast *Cryptococcus*, but details are sketchy. Reports concerning the expression of labeled (^{15}N and/or ^{13}C) prokaryotic cytochromes *c* include those by Stockman *et al.* (1989)

for *Anabaena 7120* cytochrome *c*-553, Gooley *et al.* (1991a) and MacKenzie *et al.* (1992b) for *Rhodobacter capsulatus* cytochrome c_2, and Yu and Smith (1990a) for *Rhodospirillum rubrum* cytochromes c_2 (Section III.K.2).

III. Assignment of Resonances to Nuclei

To exploit the information available from NMR spectra of proteins, resonances must be assigned to their cognate nuclei. This section deals primarily with proton assignments. Assignment of ^{13}C, ^{15}N, ^{19}F, and ^{57}Fe resonances is discussed in Section III.K. The ability to make proton assignments hinges on three NMR observables: secondary shift, spin–spin coupling, and the nuclear Overhauser effect (NOE). The ability to assign *completely* the proton spectrum of a protein the size of cytochrome *c* hinges on the observation of exchangeable protons.

A. Exchangeable Protons

In aqueous solution, protons of hydroxyl, amide, amine, carboxylate, sulfhydryl, and guanidino groups exchange with protons from water and are called exchangeable or labile protons. These protons exchange rapidly with bulk water in the denatured form of proteins (Section IV.B). The exchange rate is pH-dependent, with a minimum at acidic pH. The rate is dramatically reduced if these protons are involved in multiple, consecutive hydrogen bonds or are buried in the protein interior (Section IV.B; Englander and Kallenbach, 1984; Englander and Mayne, 1992). Nevertheless, in folded proteins the exchange rate for many labile protons is still fast enough that NMR spectra are usually acquired in 90% H_2O/10% D_2O (the D_2O being required to lock the spectrometer) and, in many cases, at a slightly acidic pH. The range of rate constants for the exchange of labile protons in the N-terminal helix of horse cytochrome *c* is >10 h^{-1} to 10^{-5} h^{-1} (Wand *et al.*, 1986), which is typical of the range found for other cytochromes *c* (Section IV.B; Marmorino *et al.*, 1993). Detecting proton resonances whose concentration is on the order of millimolar in a sample containing 100 M protons (from the solvent water) is quite a challenge, but with current instrumentation and ingenious methods for solvent suppression (Hore, 1989), this type of experiment is now routine. Early reports of cytochrome *c* spectra containing resonances from exchangeable protons (Section IV.B) include Katz and Crespi (1972), Stellwagen and Shulman (1973) (although assignments from this study should be treated with caution; see Section III.L), McDonald and Phillips (1973), and Patel and Canuel (1976).

B. Secondary Shift

For aqueous samples, the standard reference for chemical shift, δ, is 2,2-dimethyl-2-silapentane-5-sulfonate, which is abbreviated "DSS." Its methyl proton resonance is defined as 0 parts per million (ppm). Many studies use 1,4-dioxane, whose resonance appears at 3.74 ppm relative to DSS, as a secondary standard because several proton resonances of cytochromes *c* are found near 0 ppm. Protons that resonate at lower field than DSS are assigned positive values of δ. Changes in chemical shift are referred to as $\Delta\delta$. Throughout this article, "upfield" refers to shifts to higher field and corresponds to a decrease in the value of δ and a negative value of $\Delta\delta$. "Downfield" refers to shifts to lower field and corresponds to an increase in the value of δ and a positive value for $\Delta\delta$.

The frequency at which a nucleus resonates contains information about both its chemical nature and its environment. For instance, amide protons resonate at 7 to 9 ppm, and methyl and other aliphatic protons resonate near 1 to 2 ppm, because of the electronegativity of the nucleus to which the proton is directly bonded. Bundi and Wüthrich (1979) give a list of chemical shifts for amino acid protons in random-coil peptides. These are appropriately referred to as random-coil chemical shifts. Deviations from the random-coil chemical shifts are referred to as secondary or structure-dependent shifts. Secondary shifts cause a greater dispersion of chemical shifts in a protein than would be expected from the summation of the random-coil chemical shifts of the constituent amino acids (McDonald and Phillips, 1967). Although secondary shifts are useful because they shift proton resonances outside of the "envelope" of random-coil chemical shifts, they can also be a liability, as they make a mockery of the process of using random-coil chemical shifts to facilitate assignment.

Secondary shifts are brought about by local diamagnetic, paramagnetic, and electrostatic fields. Chemical exchange can also lead to secondary shifts, but, as the name implies, this source of chemical shift dispersion results from the presence of two forms of the protein in chemical equilibrium. The first three topics are discussed next, while a discussion of the chemical exchange is delayed until the sections on electron self-exchange (III.F) and the flipping of aromatic residues (IV.A).

1. Ring-current shift

Chemical shifts of protons in cytochrome c are modulated by local diamagnetic fields. Diamagnetic shifts contain angular and distance (i.e., structural) information because of their anisotropic nature. Most commonly these fields arise because of the presence of nearby aromatic groups (reviewed by Perkins, 1982).

The ring-current shift, $\Delta\delta_R$, arises from interaction of a proton with the current loop produced by an aromatic moiety. There are several models for calculating the ring-current shift (reviewed by Perkins, 1982). At large distances from the ring center, all these models reduce to the dipolar equation of Pople (1956):

$$\Delta\delta_R = -10^6 \frac{ne^2a^2}{6\pi mc^2 r^3} \quad (1 - 3\cos^2\theta), \qquad (5.1)$$

where n is the number of circulating electrons, e is the electronic charge, m is the electronic mass, c is the velocity of light, a is the radius of the ring, θ is the angle between the proton in question and the center of the ring (with 90° defined as the ring plane), and r is the distance between the center of the ring and the proton. A resonance is shifted upfield relative to its position in the absence of the diamagnetic field if the proton is above or below the plane of the aromatic moiety and within a cone defined by $\theta = 54.7°$ [i.e., $(1 - 3\cos^2\theta) = 0$] and downfield if it is outside the cone and nearer the plane. Note that unlike paramagnetic shift (Section III.B.2), the diamagnetic shift is independent of temperature (Section IV.F).

The semiclassical Johnson-Bovey (1958) equation, however, yields a better fit to experimentally observed ring-current shifts for cytochrome c than does Equation (5.1) (Perkins, 1982). Calculations of $\Delta\delta_R$ for cytochrome c using this and other models are described by Perkins (1980, 1982), Senn *et al.* (1984), and Cross and Wright (1985).

The heme is the largest aromatic moiety in the protein and plays a major role in increasing chemical shift dispersion. Heme-induced ring-current shifts can be very large. For instance, the ϵ-methyl group of Met-80 in the ferrocytochromes c, because of its location directly over the heme, experiences a ring-current shift of over 4 ppm, relative to its chemical shift in a random polypeptide. This is approximately three times the ring-current shift brought about by the aromatic side chain of Phe.

2. Paramagnetic shift

Cytochromes c can exist in two oxidation states. In the reduced (ferro) state, the heme iron is diamagnetic (low-spin d^6). In the oxidized (ferri) state (see Section III.N.1 for a discussion of non-native states), the iron is paramagnetic with one unpaired electron (low-spin d^5). This electron affects the chemical shifts of nearby protons via contact (a.k.a. Fermi contact or scalar) and pseudocontact (a.k.a. dipolar) interactions. Oxidized cytochrome c is completely in the low-spin state at neutral pH and room temperature (Ångström *et al.*, 1982).

The paramagnetic shift is defined as the chemical shift of a particular resonance in the oxidized state of cytochrome c minus the chemical shift of the resonance from this same proton in the reduced state. Paramagnetic shift is related to the hyperfine shift. The hyperfine shift is defined as the chemical shift of a nucleus in a paramagnetic molecule minus the chemical shift of the same nucleus in the isostructural diamagnetic molecule. The key word is "isostructural." For many protons of cytochrome c, the paramagnetic shift and the hyperfine shift are identical. However, because there is a small conformational change between the two oxidation states, the paramagnetic shift contains the hyperfine shift plus a term corresponding to the effect of the conformational change on the chemical shift (Section IV.F).

A complete and authoritative set of articles concerning paramagnetic shift is found in the book *NMR of Paramagnetic Molecules: Principles and Application*, edited by La Mar, Horrocks, and Holm (Academic Press, New York, 1973). Other discussions of paramagnetic shifts are given by Wüthrich (1970, 1976), Satterlee (1986, 1990), Bertini *et al.* (1989), Turner and Williams (1993), and Turner (1993). Clayden *et al.* (1987) provide a review of paramagnetic shift specifically devoted to cytochrome c.

a. *Contact shift* For heme proteins, contact shifts result from a portion of the unpaired electron density of the Fe^{3+} residing on the nucleus of interest. Because electron density is transferred through bonds, contact shift is restricted to protons that are bound to the heme (i.e., heme and ligand protons). A short and simple review is given by Satterlee (1990), and a more detailed, but simple, review is given by Jesson (1973). The magnitude of the contact shift for simple heme models is described by

$$\Delta\delta_{c} = -\frac{(A/h)g\beta S(S+1)}{3\hbar\gamma_{N}kT},\qquad (5.2)$$

where A/h represents the contact interaction constant and has the units of frequency, g is the rotationally averaged electron g-factor, β is the Bohr magneton, h is Planck's constant, S is the electronic spin ($\frac{1}{2}$ for native ferricytochrome c), $\hbar$ is Planck's constant over 2π, γ_{N} is the magnetogyric ratio, k is Boltzmann's constant, and T is absolute temperature. Note that like nuclear coupling constants (J), A can be either positive or negative. For instance, for protons 5, 10, 15, and 20 (the meso protons) of the heme (Figure 5.1), A is positive, but for the heme methyl groups, A is negative (see Wüthrich, 1976). This explains why the heme meso protons are shifted upfield (Keller et al., 1976; Moore et al., 1977; Kimura et al., 1981) and the methyl protons are shifted downfield in ferricytochrome c (Table 5.4). Turner (1993), focusing mostly on ^{13}C, calculates contact shifts for the heme of horse and tuna cytochromes c and compares them to the experimentally determined values. The agreement is rather good considering the simplifications that must be made. The author also explains the curious temperature dependence of heme chemical shifts.

Even though electrons and protons are both spin $\frac{1}{2}$ species, the usual splitting of proton resonances is not observed in the NMR spectrum because the rate of electron-spin relaxation is much greater than the difference in frequency between the individual components of the notional doublet. By analogy to the flip rates of benzoid moieties (Section IV.A) and the chemical shift of fast-exchanging amide protons (Section V.B), this situation results in the observation of a single average resonance whose frequency is determined by the population difference between the electronic spin states (Jesson, 1973). The effect of electron spin on the quality of EPR and proton NMR data is discussed in Section III.N.1.

The contact interaction term (A) is modulated by the extent to which the unpaired electron is found out on the heme. Both theory and experiment place approximately 2% of the unpaired electron in the p_z orbital of each of the heme methyl carbons 7^1 and 18^1, and approximately 0.5% on each of the heme methyl carbons 2^1 and 12^1 of eukaryotic cytochromes c (Wüthrich, 1976; Keller and Wüthrich, 1978b; Turner, 1993; see Section III.G for a discussion of some prokaryotic cytochromes c). The magnitude of the paramagnetic shift for heme methyls 7^1 and 18^1 is approximately 30 ppm (Table 5.4). Therefore, contact shifts can be much larger than even heme-induced ring-current shifts. In addition to the contact shift, heme and ligand protons are also subject to significant pseudo-contact shifts (vide infra). In fact, the contribution of the pseudo-contact shift to the proton NMR spectra of hemes and heme proteins can be equal in magnitude, but opposite in sign, to the contact shift (Shulman et al., 1971).

How or whether this asymmetric distribution of electrons is related to electron transfer between physiological electron transfer partners is not known. As stated by Turner (1993), the energy difference that maintains this asymmetry is much less than

Table 5.4
Heme assignments.[a]

	Reduced			Oxidized		
	Pseudomonas aeruginosa			*Pseudomonas aeruginosa*		
Proton(s)	Horse[b]	C102T[f]	c-551[h]	Horse[b]	C102T[f]	c-551
2^1	3.52	3.49	3.69	7.4	7.80	24.8[h]
3^1	5.23	5.23	5.97	−1.28[d]	−0.8	4.02[l]
3^2	1.49	1.45	1.87	−2.1	−2.23	2.7[h]
5	9.32	9.33	9.87			8.92[k]
7^1	3.85	3.87	3.76	33.2	31.58	13.4[h]
8^1	6.36	6.37	6.18	2.18[d]	2.09	−0.09[m]
8^2	2.61	2.55	2.44	3.1	2.92	−0.1[h]
10	9.60	9.70	9.36	−0.78[d]	−0.24[g]	−0.99[l]
12^1	3.61	3.55	3.32	10.5	10.82	32.1[h]
13^{1a}	4.19[c,d]	4.22[c,g]	4.25[c,i]	−1.47[c,d]	−1.45[c,g]	13.73[c,l]
13^{1b}	3.84[c,e]	4.00[c,g]	4.57[c,i]	1.76[c,d]		10.24[c,l]
13^{2a}	2.67[c,e]	2.68[c,g]	2.67[c,i]	2.77[c,d]	2.72[c,g]	1.70[l]
13^{2b}		2.99[c,g]	3.42[c,i]	1.17[c,d]		0.15[l]
15	9.63	9.61	9.42	8.02[d]		6.32[l]
17^{1a}	4.17[c,d]	4.11[c]	4.62[c,i]	11.4[c,j]	12.93[c,l]	5.45[l]
17^{1b}	3.64[c,d]	3.60[c]	3.92[c,i]	18.6[c,j]	14.96[c,g]	1.22[l]
17^{2a}	3.16[c,e]	3.37[c]	3.35[c,i]	1.7[c,j]	1.43[c,g]	0.28[l]
17^{2b}	2.64[c,e]	2.70[c]	2.72[c,i]	−0.2[c,j]	−0.14[c,g]	−0.02[l]
18^1	2.18	2.30	3.42	33.9	34.58	17.8[h]
20	9.04	9.13	9.24			−2.75[l]

[a] These data were collected under a variety of temperatures and pH's that are given in the original references.

[b] Unless otherwise noted, these assignments are from Keller and Wüthrich (1978b).

[c] Stereospecific assignments are not assumed.

[d] Feng *et al.* (1990b).

[e] Chau *et al.* (1990). These assignments are tentative.

[f] The C102T variant of yeast iso-1-cytochrome *c*. Unless otherwise stated, these assignments are from Gao *et al.* (1990) and are in excellent agreement with those for the wild-type protein (Senn *et al.*, 1983a).

[g] Gao *et al.* (1991b).

[h] These resonances were identified by Keller *et al.* (1976) and assigned by Keller and Wüthrich (1978a).

[i] Chau *et al.* (1990); Detlefsen *et al.* (1990).

[j] Santos and Turner (1987).

[k] Chemical shifts for the meso protons in the oxidized protein were proposed by Keller *et al.* (1976), and one was observed by Moore *et al.* (1977).

[l] Moratal *et al.* (1993).

[m] Timkovich (1991).

the energy available upon the binding of cytochrome c to redox partners. In fact, the binding of redox partners to cytochrome c is known to affect the chemical shifts of certain heme resonances (Section IV.I). See notes added in proof.

b. *Pseudo-contact shift*   Pseudo-contact shift arises because of charge–dipole interactions between the unpaired electron and various protons. It possesses both angular and distance dependencies (i.e., it is anisotropic) and is in many ways analogous to the diamagnetic shift (Horrocks, 1973; Section III.B.1). In the absence of anisotropic effects (i.e., in isotropic systems) there is no pseudo-contact shift.

The pseudo-contact shift for an axially symmetric system is given by (Kurland and McGarvey, 1970)

$$\Delta\delta_{pc} = \frac{\beta^2 S(S+1)}{9kTr^3}(g_{//}{}^2 - g_\perp{}^2)3\cos^2\theta - 1. \tag{5.3}$$

$g_{//}$ and $g_\perp$ are the principal components of the g-tensor in the plane of the heme and along the z-axis, respectively, and β is the Bohr magneton [other constants are defined in Equation (5.2)]. The distance from the iron to the proton in question is r. Equation (5.3) has the same geometric form as Equation (5.1), which describes the ring-current shift, but it is important to note the sign of the combined constants, which appears in front of the geometric term. For porphyrin systems, $g_{//}$ is greater than $g_\perp$, and the leading constants are positive, but these leading terms are negative for the ring-current shift. Therefore, for the same arrangement of protons, ring-current and pseudo-contact shifts will have opposite signs. That is, protons experiencing pseudo-contact interactions that lie within the cone defined by the angle $54.7°$ will exhibit downfield shifts and protons closer to the plane of the heme will exhibit upfield shifts.

The situation for cytochromes c is a bit more complex. For the case of axial symmetry discussed earlier, there are only two principal orientations of the g-tensor (in the plane, $g_\perp$, and along the z-axis, $g_{//}$). However, there are three orientations in the protein, g_x, g_y, and g_z. Cytochromes c exhibit rhombic symmetry (Mailer and Taylor, 1972). Equation (5.4) describes the pseudocontact shift for the case of rhombic symmetry:

$$\Delta\delta_{pc} = \frac{\beta^2 S(S+1)}{9kTr^3}[g_{ax}(3\cos^2\theta - 1) + 1.5\,g_{eq}\sin^2\theta\cos^2 2\phi], \tag{5.4}$$

where g_i is the anisotropy of the effective g-values and, specifically, $g_{ax} = g_z{}^2 - \frac{1}{2}(g_x{}^2 + g_y{}^2)$ and $g_{eq} = g_x{}^2 - g_y{}^2$. g_x, g_y, and g_z are the principal components of the g-tensor. The remaining terms are described in Equation (5.3). The x- and y-axes are in the heme plane, and the z-axis is perpendicular to it. The angle θ is measured from the positive z-axis. The angle ϕ is measured clockwise in the x–y plane from the positive x-axis as viewed from the positive z-axis (see Williams *et al.*, 1985b). Note that Equation (5.4) becomes equivalent to Equation (5.3) when g_x equals g_y. Comparison of Equations (5.3) and (5.4) reveals that although the complexity of Equation (5.4) is greater, its sensitivity to structure has also increased because of the introduction of ϕ.

Although the x- and y-axes are defined earlier as being in the heme plane, this is not enough information. For Equation (5.4) to be useful, the orientation of the axes with respect to the heme plane must be known. Mailer and Taylor (1972) defined the orientation of the g-tensor by examining the electron paramagnetic resonance (EPR) spectrum of crystalline horse ferricytochrome c where the orientation of the molecule in the crystal was known from x-ray diffraction studies. A line through heme nitrogens 22 and 24 defines the x-axis, and a line through nitrogens 21 and 23 defines the y-axis (Figure 5.1). It is also possible to determine the orientation by fitting experimentally determined paramagnetic shifts to pseudocontact shifts calculated using the coordinates from a crystal structure and Equation (5.4). These calculations are done at several different orientations of the principal axes of the g-tensor with respect to the molecular axis. The best fit between the calculated and observed values should occur at the correct orientation (Section IV.F).

Note from Equations (5.2)–(5.4) that there should be an inverse dependence of $\Delta \delta_c$ and $\Delta \delta_{pc}$ on temperature such that at high temperatures the observed paramagnetic shifts should diminish. That is, at high temperature the chemical shift of protons belonging to the oxidized protein should approach those of the reduced protein, assuming there is no difference in structure or dynamics between the two states (Section IV.F). However, this simple behavior is not always observed, especially for protons that experience significant contact effects (Moore and Williams, 1980d; Ångström et al., 1982; Turner and Williams, 1993; Turner, 1993). The cause of this complex temperature-dependent behavior can be ascribed to very small changes between the reduced and oxidized protein, to the failure of Equations (5.3) and (5.4) to account for admixtures of the excited and ground states (i.e., the g-tensor and the hyperfine coupling constant are temperature-dependent), and to the presence of small amounts of high-spin forms (Section III.N.2). Note that secondary shifts brought about by aromatic moieties (III.B.1), hydrogen bonds, and electrostatic effects (III.B.3) are independent of temperature. Turner and Williams (1993; Section IV.F), using both proton and ^{13}C data, put this last fact to good use to separate paramagnetic and diamagnetic terms and to show, independent of crystallographic data, that there are subtle differences in structure between the reduced and oxidized states of horse cytochrome c.

A fine illustration of contact and pseudo-contact shifts is the difference in electronic structure of the heme of Pseudomonas aeruginosa cytochrome c-551 compared to that found in eukaryotic cytochromes c (Section III.G). The use of pseudo-contact shifts to examine structural and dynamic differences between reduced and oxidized cytochrome c and between the crystalline and solution states is discussed in Section IV.F. The possibility of using pseudo-contact shifts to examine unnatural variants of cytochromes c is discussed in Section IV.J.2.

3. Shift due to electrostatic interactions, the carbonyl bond, and hydrogen bonding

These sources of secondary shift are common to all proteins and are, in general, smaller than those brought about by the sources discussed earlier. When a non-ionizable proton is within approximately 3 Å of a titratable proton and in the correct orientation, the chemical shift of the former will exhibit a titration curve with the pK of the latter. A prime example of this is the pK of 4.5 observed for the position 2 proton of Trp-77, presumably because of its proximity to the carboxylate side chain of Glu-4 of *Pseudomonas aeruginosa* cytochrome c-551 (Chau *et al.*, 1990). This type of secondary shift is dependent on both distance and orientation and is due to the effect of the electric field on the diamagnetic shielding of the nucleus under observation.

Carbonyl groups contribute to both electrostatic and dipolar fields. Protons in the plane of the bond are shifted upfield. Because many hydrogen bonds in proteins involve carbonyl groups and amide protons of the protein backbone, it comes as no surprise that hydrogen bonds give rise to secondary shifts. The secondary shift due to hydrogen bonding can be greater than 1 ppm for an amide proton. Wagner *et al.* (1983) have developed an empirical r^{-3} relationship between the length of a hydrogen bond and the chemical shift of amide and α proton resonances that make up the bond. The relationship between the chemical shift of backbone protons and secondary structure is also discussed by Dalgarno *et al.* (1983), Szilàgyi and Jardetzky (1989), Gao *et al.* (1990b, 1991c), Williamson (1990), and Augspurger *et al.* (1992). The reader is referred to Perkins (1982) for a more detailed account of the effects discussed in this section.

C. Spin–Spin Coupling

To identify residue types, the spin–spin coupling (a.k.a. *J*-coupling) must be examined. Nonidentical protons separated by three bonds or less possess coupling constants on the order of 1 to 20 Hz. These coupling constants are manifested in the multiplet structure observed in proton spectra. These couplings can be observed in one-dimensional (1-D) spectra when resonances are well resolved from the protein envelope and may be useful for determining what type of proton is being observed. For instance, a doublet may be observed for the methyl group(s) of an Ala, Val, Leu, and the γ_2 methyl of Ile, and a triplet may be observed for the δ methyl group of Ile. Unfortunately, inspection of the 1-D spectrum does not identify the resonance that causes the multiplet. However, if the multiplet is saturated by application of a selective pulse, the resonance to which it is coupled can be identified by the collapse of its multiplicity. This method works well when the resonances to be decoupled are well resolved. Results become confusing when resonances can no longer be selectively decoupled because other nearby resonances are affected by the decoupling pulse. Much of the early work on determining proton assignments in cytochromes c relied on selective decoupling. The classic series of papers by Moore and Williams (1980a–f) is a fine example of the use of 1-D methods. However, the complex nature of the proton spectrum of a protein the size of cytochrome c makes it impossible to assign all resonances to specific nuclei by 1-D techniques.

D. Two-Dimensional (2-D) NMR; the COSY Experiment

The idea of 2-D NMR is to split the frequency scale into two dimensions, defined as F_1 and F_2, and then induce some sort of correlation between resonances. The correlation is spin–spin coupling (Section III.C) in COSY-type experiments, and the nuclear Overhauser effect (NOE, Section III.H) and/or chemical exchange (Sections

III.F and IV.A) in NOESY-type experiments. Most common homonuclear 2-D experiments exhibit symmetry about a diagonal where F_1 equals F_2. This diagonal is no more than the 1-D spectrum. It does not look like a typical 1-D spectrum because to represent three dimensions of data (two frequency domains and signal intensity) in two-dimensions the intensities are represented as contours so that the spectrum can be viewed from above without closely spaced cross peaks obscuring each other. In the COSY experiment, any two nondegenerate protons that are separated by three or fewer chemical bonds are correlated with each other. This correlation leads to a cross peak with frequency coordinates of the chemical shift of the two J-coupled nuclei. Because magnetization transfers between progressively and regressively coupled transitions are phase shifted relative to one another by 180°, COSY cross peaks in phase-sensitive experiments exhibit antiphase behavior, where the separation between components is related to the coupling constants. See Derome (1987) for a simplified discussion of the COSY experiment.

E. Other 2-D Experiments Involving Spin–Spin Coupling

There are several useful derivatives of the COSY experiment. By inserting a second transfer step tuned to the coupling constant of interest, it is possible to obtain relayed COSY spectra. These spectra contain cross peaks associated with protons separated by four or fewer bonds, providing that the intervening heteronuclei bear a proton (e.g., coupling is not observed between the β protons of Phe, Tyr, His, or Trp and the position 2 or 6 proton). Similarly, a double-relayed COSY spectrum exhibits cross peaks from protons separated by five or fewer bonds. It is important to note that relayed and double-relayed experiments contain not only the relayed cross peaks, but also COSY cross peaks.

Another experiment involving scalar coupling is TOCSY (Braunschweiler and Ernst, 1983; closely related to HOHAHA, Bax and Davis, 1985). This experiment relies on a spin-locking pulse to transfer magnetization throughout an entire spin system. The longer the spin lock, the more magnetization is transferred. Therefore, the equivalent of COSY, relayed-COSY, etc., experiments can be obtained simply by varying the length of time that the spin lock is applied. These experiments offer two potential advantages over experiments of the COSY type. First, cross peaks from TOCSY spectra are all phased in the same way, unlike the phase-sensitive COSY experiment which yields antiphase cross peaks. This gives a theoretical enhancement in sensitivity. Second, in a TOCSY-type experiment, longer-order magnetization transfer can be investigated by increasing only one parameter, the length of the spin lock, rather than requiring a total change in pulse sequence as for COSY-type experiments.

F. Electron Exchange and Transfer

Electron self-exchange occurs when ferro- and ferricytochrome c are mixed and these individual species exchange electrons. The measurement of this exchange using NMR was pioneered by Redfield and Gupta (1971). The concepts of paramagnetic shift and electron self-exchange can be brought together to describe another weapon in the arsenal of techniques for making proton assignments. Assuming the rate of electron exchange is much slower than the difference in chemical shift between the resonances of identical protons in the two oxidation states (the paramagnetic shift), then for a resonance that is saturated in the reduced protein, magnetization may be transferred to the oxidized state, and the resonance in the oxidized protein will be partially saturated and *vice versa*. The main impediment to electron self-exchange is that cytochrome *c* molecules are prevented from interacting by their net positive charges (Gupta *et al.*, 1972). Therefore, the rate of self-exchange can be increased by masking these charges by increasing the salt concentration. The rate of electron transfer can also be adjusted by varying the ratio of the two oxidation states. Saturation transfer competes with the cross relaxation that causes the NOE (Section III.H), but conditions can be found where the rate of electron transfer is approximately a factor of 10 faster than cross relaxation.

Saturation transfer experiments have been used to cross-assign hyperfine-shifted resonances in oxidized cytochrome *c* (see Moore *et al.*, 1985). A 2-D version of saturation transfer using the NOESY pulse sequence has been applied to cross-assign resonances in tuna cytochrome *c* (Boyd *et al.*, 1984). Spectra obtained from such an experiment contain both NOE and exchange cross peaks. These two kinds of cross peaks can be separated by collecting separate NOESY spectra for the reduced and oxidized samples. The optimization of 1-D and 2-D exchange experiments is discussed by Feng *et al.* (1990b) and Turner (1985), respectively.

Saturation transfer or inversion recovery experiments (Forsèn and Hoffman, 1963) can also be used to determine directly the rate of electron self-exchange (Redfield and Gupta, 1971; Gupta *et al.*, 1972; Dixon *et al.*, 1989; Concar *et al.*, 1991b; Whitford *et al.*, 1991b; for a review see Dixon and Hong, 1990) and the rate of electron exchange between physiological electron transfer proteins (Concar *et al.*, 1991a; Whitford *et al.*, 1991a). If the self-exchange rate is fast enough, line broadening alone can be used to determine the rate constant (Timkovich *et al.*, 1988; Kimura *et al.*, 1981). A detailed discussion of these studies is beyond the scope of this review, but it is important to note that there are quantitative differences between results from studies where electron transfer between cytochrome *c* and cytochrome b_5 (Section IV.I.3) is assessed using saturation transfer (Concar *et al.*, 1991a) and where electron transfer is assessed using photoinduced methods with ruthenium-modified cytochrome b_5 (Willie *et al.*, 1992, 1993) or stopped-flow mixing and native cytochrome b_5 (Eltis *et al.*, 1991).

The TOCSY experiment (Section III.E) can also be exploited to cross-assign protons in the reduced and oxidized states (Feng and Roder, 1988; Feng *et al.* 1991). In this experiment, exchange cross peaks, *J*-coupled cross peaks, and the diagonal have the same sign, whereas NOE cross peaks have the opposite sign. *J*-coupled and exchange cross peaks can be separated by collecting separate TOCSY spectra for the reduced and oxidized samples.

A complete study of decoupling and saturation transfer experiments leads to a list of amino acid types and their assignments. The situation is much like amino acid analysis. That is, amino acid analysis gives the number of a particular type of amino acid residue present in the protein, but not where each is in the sequence. However, before we discuss sequence-specific assignments, the assignment of heme resonances will be reviewed.

G. Heme Assignments and the Chirality of the Met Ligand Sulfur

Resonances belonging to the heme and its ligands were the first to be assigned to specific protons because of the heme's diamagnetic and paramagnetic properties (Wüthrich, 1969; McDonald *et al.*, 1969). A list of heme assignments for a higher eukaryotic cytochrome *c* (horse), a lower eukaryotic cytochrome *c* (the C102T variant of *Saccharomyces cerevisiae* iso-1-cytochrome *c*), and a prokaryotic cytochrome *c* (*Pseudomonas aeruginosa* cytochrome *c*-551) is given in Table 5.4. Although assignments for the protons directly attached to the heme ring (5, 10, 15, 20) in the oxidized state of *Pseudomonas aeruginosa* cytochrome *c*-551 have been reported (Keller and Wüthrich, 1978a; Chao *et al.*, 1979), these should be considered to be preliminary. It should be noted that Chao *et al.* (1979) confused the assignment of the heme methyl groups of cytochrome *c*-551. The assignment of heme proton resonances of cytochrome *c*-557 from the protozoan *Crithidia oncopelti* has been reported by Keller *et al.* (1979). Although this protein contains only one thioether linkage via carbon 8^1, the chemical shifts are very similar to those reported for higher eukaryotic cytochromes *c*. Other cytochromes *c* for which heme assignments are available include ferrocytochromes *c*-553 from the cyanobacterium *Anabaena 7120* (Zehfus *et al.*, 1990) and from the anaerobe *Desulfovibrio vulgaris* (Marion and Guerlesquin, 1992; Medvedeva *et al.*, 1993). Moench *et al.* (1992) present a table of proton chemical shifts for horse, tuna, and yeast iso-1- and iso-2-cytochromes *c* in the presence and absence of cytochrome *c* peroxidase. Finally, Timkovich (1991) has reported proton and ^{13}C assignments for horse ferricytochrome *c*, *Pseudomonas aeruginosa* ferricytochrome *c*-551, bis(cyano)iron(III) protoporphyrin IX, and the free-base (i.e., iron-free) porphyrin.

The pattern of heme methyl group proton chemical shifts for the eukaryotic ferricytochromes *c* [$(7^1, 18^1) > (2^1, 12^1)$] is reversed for cytochromes *c*-551, as are the signs of chemical shifts of thioether methyl protons ($3^2, 8^2$). The chemical shifts of the directly attached heme methyl protons are controlled primarily by contact shifts (Section III.B.2.a), so this reversal must be related to a redistribution of the unpaired electron density found on the heme carbons. On the other hand, the chemical shift of the thioether methyl protons is controlled primarily by the pseudo-contact shift (Shulman *et al.*, 1971; Section III.B.2.b). The reversal of sign reflects a rotation of the principal axes of the *g*-tensor between cytochromes *c*-551 and eukaryotic cytochromes *c* (Keller and Wüthrich, 1978a; Senn *et al.*, 1980) and correlates with reversal of the chirality of the Met ligand sulfur as measured by a change in sign of the Cotton effect at 695 nm (Senn *et al.*, 1980; *vide infra*).

The chirality of the Met ligand sulfur can be determined for the reduced proteins by comparison of NOEs between protons of the Met ligand and the heme (Senn and Wüthrich, 1983; Senn *et al.*, 1983a,b). The absolute configuration about the Met ligand is obtained from the crystal structure. By calibrating NOE and CD observations for a cytochrome *c* for which a crystal structure is available, the absolute configuration of the met ligand can be determined for a cytochrome *c* whose crystal structure is unknown. In this way it has been determined that all eukaryotic cytochromes *c* studied to date, as well as *Rhodospirillum rubrum* cytochrome c_2 and cytochromes *c*-552 from *Euglena gracilis* and *Spirulina maxima*, possess *R*-chirality, whereas *Pseudomonas aeruginosa*, *Pseudomonas stutzeri*, *Pseudomonas mendocina*, and other prokaryotic cytochromes *c*-551 possess *S*-chirality (Senn and Wüthrich,

1983). The stereochemistry of this sulfur has also been determined by a synthesis of NOE measurements with van der Waals constraints and ring-current shift calculations (Senn *et al.*, 1984). For both these classes, the chirality of the sulfur is independent of oxidation state. Cytochromes *c*-553 represent a third class, because they exhibit different chiralities in the ferro- and ferri states (Senn *et al.*, 1983b; Marion and Guerlesquin, 1992). Energetically, however, it seems unlikely that this third class is real. In summary, most cytochromes *c* exhibit *R*-chirality, cytochromes *c*-551 exhibit *S*-chirality, and cytochromes *c*-553 exhibit *S*-chirality in the oxidized state and *R*-chirality in the reduced state. Senn and Wüthrich (1985) present a more detailed review of these results.

The heme of all cytochromes *c* possesses two propionic acid side chains. These are referred to as 13 and 17 in IUPAC nomenclature and as 6 and 7 in older nomenclature (Table 5.1). In Brookhaven nomenclature, these propionates are described as emanating from carbons C3D and C2A (but see Table 5.1 for the designation of each proton-bearing heavy atom). These side chains are usually called the inner (17) and outer propionates (13), because in eukaryotic cytochromes *c* the outer propionate is relatively solvent-exposed and the inner propionate is not. Partial proton assignments for these side chains in both redox forms of the heme undecapeptide were determined by Kimura *et al.* (1981). The proton assignments for the inner propionic acid side chain of the heme of *Pseudomonas aeruginosa* ferrocytochrome *c*-551 were determined by Leitch *et al.* (1984). These assignments were confirmed and extended by Chau *et al.* (1990), Detlefsen *et al.* (1990), and Cai and Timkovich (1991). Assignments for the outer propionate of *Pseudomonas aeruginosa* ferricytochrome *c*-551 by Leitch *et al.* (1984) have been reassigned to the inner propionate (Timkovich and Cai, 1993; Moratal *et al.*, 1993). Assignments for the propionates of *Pseudomonas stutzeri* ferrocytochrome *c*-551 were reported by Cai *et al.* (1992). Chau *et al.* also proposed assignments for this side chain in horse ferrocytochrome *c*. Feng *et al.* (1990b) have assigned the propionic acid side chains in both oxidation states of horse cytochrome *c*. Assignment of the inner propionic acid side chains of oxidized tuna cytochrome *c* were reported by Moore and Williams (1984), but note that the α and β designations for these protons are reversed with respect to the scheme of Feng *et al.* (1990b). The inner propionic acid side chains have been assigned in wild-type *Saccharomyces cerevisiae* iso-1-cytochrome *c*, a Cys-102 derivative, the Cys-102 covalent dimer (Moench and Satterlee, 1989), the C102T variant, the F82S;C102T variant (Gao *et al.*, 1991b), and the Cys-102-methylated derivative (Busse *et al.*, 1990). Partial assignments for both heme propionates have been reported for *Rhodospirillum rubrum* ferro- and ferricytochrome c_2 by Yu and Smith (1990a). Complete assignment of both propionates in *Desulfovibrio vulgaris* ferrocytochrome *c*-553 was reported by Marion and Guerlesquin (1992). All the assignments discussed here are internally consistent, although stereospecific assignments have not been reported in all cases.

H. The Nuclear Overhauser Effect (NOE)

Fitting resonance assignments into the primary structure involves exploitation of the nuclear Overhauser effect. Once an equilibrium population of proton nuclear spins has been perturbed, there are only a limited number of ways it can relax.

The present discussion focuses on relaxation via dipole–dipole interactions with surrounding protons. The reader is referred to the book by Noggle and Schirmer (1972) for detailed information.

Simple proton dipolar relaxation, however, is not the only mechanism for relaxation. For residues very near the heme in oxidized cytochrome c, relaxation attributable to interactions between the unpaired electron and nuclear spins is also important (Section III.B). Electron exchange between ferro- and ferri- proteins (III.F), as well as the flipping of benzoid side chains (IV.A), can also affect relaxation.

Ignoring for the moment the relaxation mechanisms involving the unpaired electron (Section III.B.2) and chemical exchange (III.F, IV.A, IV.B), when a proton is excited it may transfer some of its spin to other nearby protons by dipolar coupling. This is the basis of the NOE. For protons, the distances over which NOEs may be observed range from the sum of the van der Waals radii of two protons (approximately 2 Å), out to approximately 5 Å. The strength of the NOE is modulated by the size and mobility of the molecule. Because this article is limited to cytochromes c, there is no need to deal with the effect of molecular size, except to say that for proteins the size of cytochrome c the proton NOE is negative with respect to the NOE for a small molecule, and that this negative NOE is near its maximum value.

Regions of differential mobility have a considerable effect on the NOE. If all protons on cytochrome c were rigidly bound so that they all moved with the correlation time of protein, the strength of the NOE could be described as proportional to the inverse sixth power of the separation between protons. The effect of an increase in mobility is to diminish the strength of the NOE. NOEs from mobile residues, such as Lys side chains which protrude into solution, are either very weak or not observed. In the oxidized protein, NOEs involving some of the heme protons and protons attached to the ligands are also diminished by relaxation of these protons by the unpaired electron. It is important to note that the absence of an NOE is not a positive indication that two protons are greater than 5 Å apart. Neither is the presence of an NOE necessarily a positive indication that protons are within 5 Å of each other, because spin diffusion can lead to the observation of NOEs from protons that are separated by more than this distance. One-dimensional NOE experiments played an important role in assigning the heme resonances, determining the stereochemistry of the Met-80 sulfur ligand (Section III.G), and making amino acid assignments based on data from crystal structures.

I. Sequence-Specific Assignments (Wüthrich, 1986)

Amide protons on successive amino acid residues are usually less than 5 Å apart. Therefore, NOEs may be observed from the amide proton of residue i to the amide proton of residues $i + 1$ and $i - 1$. Pro residues, which lack an amide proton, and other residues whose amide protons are in fast exchange with solvent water (Section IV.B), break this pattern. The trail of NOEs leading from one residue to the next is referred to as connectivity, and the general technique of using NOE data to "walk" along the backbone of the protein is referred to as sequence-specific assignment.

With a trail of amide–amide crosspeaks in hand, the identities of their α and side-chains resonances are obtained from inspection of COSY, or other J-coupled spectra.

J. The Determination of Higher-Order Structure

1. Secondary structure (Wüthrich, 1986)

At the next level of protein structure, different elements of secondary structure give rise to distinct patterns of NOE connectivities. For instance, a NOE between the α proton of residue i and the amide proton of residue $i + 3$ is indicative of an α-helix. The tighter 3_{10} helix yields more i, $i + 2$ α-amide NOEs. Using this approach, the secondary structures of tuna, horse, the C102T variant of yeast iso-1-cytochrome c, *Pseudomonas aeruginosa* cytochrome c-551, *Pseudomonas stutzeri* cytochrome c-551, *Rhodobacter capsulatus* cytochrome c_2, and *Desulfovibrio vulgaris* Hildenborough cytochrome c-553 have been defined (Table 5.2). It should be noted that for the tuna protein only i, $i + 1$ connectivities were reported. The N- and C-terminal helices of all three eukaryotic proteins are found to be relatively un-distorted, whereas this helix exhibits significant 3_{10}-character in the *Rhodobacter capsulatus* protein (Gooley *et al.*, 1990). In the horse protein, the fifties and sixties helices are also found to be undistorted. A short helix between residues 72 and 75 was also observed for the horse protein. For the C102T variant, distortions were observed for the fifties, sixties, and seventies helices, with the sixties helix exhibiting 3_{10} character. It is interesting to note that the region around the sixties helix is also the least conserved region of eukaryotic cytochromes c (Moore and Pettigrew, 1990) and that the chemical shifts of amide protons in this region also exhibit the most variability (Gao *et al.* 1989, 1990b). These findings are in reasonable agreement with those from x-ray crystallographic studies (Dickerson *et al.*, 1971; Takano and Dickerson, 1981a,b; Louie *et al.*, 1988a; Louie and Brayer, 1990; Bushnell *et al.*, 1990; Berghuis and Brayer, 1992). The sixties helix and the C-terminal helix show a 3_{10}-to α-helix transition in the full-fledged solution structure of horse ferrocytochrome c (Qi *et al.*, 1994b; Section IV.F).

2. Tertiary structure

By examining NOEs between units of secondary structure, it is possible to de-termine the way these units interact to yield the native protein. Before the advent of 2-D NMR techniques, the comparison of distances obtained from the crystal struc-tures of cytochromes c to the observation of NOEs between two protons was used as a tool for assignment. Two-dimensional techniques yield a plethora of NOEs (Wüthrich, 1986). Using techniques of distance geometry and/or molecular dynamics, it is possible to calculate the structure of a protein based on the presence and strength of NOEs. Recently, the first tertiary cytochrome c structures derived from only NMR techniques, those of ferrocytochrome c-551 from *P. aeruginosa* (Detlefsen *et al.*, 1991), *Pseudomonas stutzeri* (Cai *et al.*, 1992), ferrocytochrome *c* from *Desulfovibrio vulgaris* (Blackledge *et al.*, 1995), and horse ferrocytochrome *c* (Qi *et al.*, 1994b; Section IV.F) have been reported.

A partial list of the extent of the assignment of amino acid proton resonances of various cytochromes *c*, and the references to these assignments, are given in Tables 5.2 and 5.3. Assignments for heme proton resonances are given in Table 5.4.

K. Assignment of Heteronuclei

This section is divided into two parts based on the ease of detection of nuclei.

1. ^{19}F

^{19}F has a sensitivity of approximately 80% that of the proton and a high natural abundance. Therefore, there is no problem with detection. However, fluorine is not present in native cytochromes *c*, so it must be introduced by chemical modification. Smith and Millett (1980) examined the ^{19}F spectra of horse cytochrome *c* trifluoracetylated at individual Lys residues, and the effect of addition of cytochrome *c* peroxidase on these resonances and their longitudinal relaxation (Section IV.I.4). The interaction of ^{19}F-labeled horse cytochrome *c* with phospholipids was reported by Staudenmayer *et al.* (1976) and is briefly discussed in Section IV.I.7.

2. ^{13}C, ^{15}N, and ^{57}Fe

^{13}C, ^{15}N, and ^{57}Fe have low sensitivities with respect to the proton (1.6%, 0.1%, and 0.003%, respectively) and are present at a low natural abundance (1.1%, 0.4%, and 2.2%, respectively). However, these nuclei are natural components of proteins. Therefore, enrichment of the protein in this type of nucleus can be used to examine the properties of the unmodified forms. As discussed in Section II, a possible problem concerning assignment of insensitive heteronuclei resonances in eukaryotic cytochromes *c* is that it is not easy to produce the large quantities of labeled material required. A short but excellent discussion of ^{13}C NMR spectroscopy of proteins is given by Allerhand *et al.* (1973).

Several studies have been performed using chemically modified cytochromes *c* where the reagent used to modify the protein contained ^{13}C. Eakin *et al.* (1975b) examined the extent of carboxymethylation of horse cytochrome *c* upon reaction with ^{13}C-labeled bromoacetic acid and showed that the Met-80 sulfur was available for reaction in the presence of CN^-, which disrupts the Met-80–Fe bond. Procedures for the derivatization of Lys and Tyr residues with ^{13}C-labeled acetyl imidazole, along with spectra for the modified species, were reported by Nieman *et al.* (1982), although neither mechanistic nor physiological effects of this probe were discussed. Stellwagen *et al.* (1977) reported studies in which horse cytochrome *c* was reacted with ^{13}C-enriched *O*-methylisourea, causing the conversion of Lys to homoarginine. Because one Lys residue, presumed by the authors to be Lys-79, reacts much more slowly than the other 18, differential modification was performed. The binding of cytochrome *c* peroxidase to the differentially modified cytochrome *c* was also examined, and these results are discussed in Section IV.I.4. Studies on *Saccharomyces cerevisiae*, *Paracoccus denitrificans*, *Pseudomonas aeruginosa*, *Rhodospirillum rubrum*, *Rhodopseudomonas capsulatus*, and *Rhodopseudomonas sphaeroides* cytochromes *c* modified with ^{13}C-enriched *O*-methylisourea were reported by Kennelly *et al.* (1981). These authors were able to assign tentatively the ^{13}C resonances from Lys-27 and Lys-29 (numbering based on higher eukaryotic cytochromes *c*). The reaction of these derivatives with various cytochrome *c* oxidases is discussed in Section IV.H.1.

Until recently, the assignment of ^{13}C resonances in unmodified cytochrome *c* has been restricted to a few nonprotonated aromatic carbons (Oldfield and Allerhand, 1973; Oldfield *et al.*, 1975; Wilbur and Allerhand, 1977). Although Oldfield *et al.* (1975) suggested that one Tyr residue of horse ferrocytochrome *c* begins to ionize at pH 9.5, Boswell *et al.* (1983) used ^{13}C NMR to probe the ionization of Tyr residues and concluded that all have p*K*s greater than 11. A limited number of ^{13}C assignments were determined by direct ^{13}C–^{1}H 2-D correlation (Santos and Turner, 1985, 1986). Burch *et al.* (1990) put these assignments to good use in the examination of the complex between cytochrome *c* and cytochrome b_5 (Section IV.I.3). Except for the experiments by Schejter *et al.* (1978), Wooten *et al.* (1981; Section III.N.3), and Spooner and Watts (1992) (Section IV.I.7), which involved the post-translational insertion of a ^{13}C label at the methyl group of Met-80, all ^{13}C studies of cytochrome *c* save one involved the nucleus at natural abundance. Nall and Zuniga (1990) investigated Tyr ring flipping in iso-2-cytochrome *c* from *Saccharomyces cerevisiae* grown on minimal media enriched with [3,5-^{13}C]-labeled Tyr (Section IV.A).

Recently, the strategy of indirect detection has been applied to the ^{13}C assignment problem in horse cytochrome *c* (Gao *et al.*, 1990a) using 2-D ^{13}C–^{1}H experiments. The technique of indirect detection offers a considerable increase in sensitivity over direct methods. In the study just cited, a 5 mM protein sample was used, and the time required to acquire a 2-D data set was approximately 24 h. Data from the nearly complete proton assignment of the horse protein (Wand *et al.*, 1989; Feng *et al.*, 1989; Feng and Roder, 1988; Feng *et al.*, 1991; Gao *et al.*, 1990b) were used to assign ^{13}C resonances. Gao *et al.* (1990a) were able to assign 205 of the 437 proton-bearing carbons of the reduced protein. These assignments include all the Ala, Val, Thr, and Gly carbons, all 48 methyl carbons, all proton-bearing aromatic carbons, and 92 α-carbons. Using different techniques, Santos and Turner (1992) have assigned many of the same ^{13}C resonance in the horse ferrocytochrome *c* (Section III.L). Additionally, these authors report ^{13}C assignments in the oxidized protein. With these assignments in hand, it is now perhaps possible to study the dynamics of cytochrome *c* via determination of relaxation parameters (Section IV.K). Timkovich (1991) has reported ^{13}C assignments for the heme of horse ferricytochrome *c* and *Pseudomonas aeruginosa* ferricytochrome *c*-551, as well for bis(cyano)iron(III) protoporphyrin and free-base porphyrin. There is also a report (Stockman *et al.*, 1989) of preliminary ^{13}C assignments for an enriched sample of *Anabaena 7120* ferrocytochrome *c*-553.

There are fewer reports of ^{15}N assignments in cytochrome *c*. Gooley *et al.* (1990) have reported the essentially complete assignment of backbone ^{15}N resonances of *Rhodobacter capsulatus* ferrocytochrome c_2. They used these assignments to determine the exchange rate of backbone amide protons (Section IV.B). Gooley and

MacKenzie (1990) and Gooley *et al.* (1991a, 1992) have also used $^1H-^{15}N$ 2-D methods to examine the structure and dynamics of the P35A variant of this protein (Section IV.J.1). Nitrogen-15 resonances from the side chain of the His ligand (His-18), Trp-62, His-42, and one Lys residue, as well as the N-terminal Glu and two Pro residues (Pro-30 and Pro-74 or Pro-85), have been assigned in *Rhodospirillum rubrum* cytochrome c_2 (Yu and Smith, 1988; 1990a,b). There is also a report (Stockman *et al.*, 1989) of preliminary ^{15}N assignments for an enriched sample of *Anabaena 7120* ferrocytochrome c-553. Timkovich used the amide-proton assignments of *Pseudomonas aeruginosa* cytochrome c-551 (Chau *et al.*, 1990) to assign the ^{15}N resonances using indirect detection. This work shows unequivocally that ^{15}N resonances can be assigned at natural abundance. Although not a part of the native protein, the binding of ^{15}N-enriched CN^- to ferricytochrome c was investigated by Behere *et al.* (1986). These authors showed that the ^{15}N chemical shift cytochrome c–CN complexes varied over a range of 10 ppm for 11 eukaryotic cytochromes c, indicating incredible phylogenetic conservation of the heme environment.

The ^{57}Fe NMR resonance of horse ferrocytochrome c has been reported by Baltzer (1987). To perform this experiment, the biologically incorporated Fe was removed and ^{57}Fe reincorporated. The chemical shift of the ^{57}Fe is radically different from that reported for the carbon monoxide derivative of myoglobin, indicating ^{57}Fe NMR is an extremely sensitive probe of the heme environment. However, even at 90% enrichment, this experiment required 24 h of acquisition, over which time 2.6×10^6 transients were acquired. These results have been confirmed and extended by Chung *et al.* (1990). Deuterium NMR studies are not within the scope of this review, but some studies whose results are briefly discussed within the text include those by Wooten *et al.* (1981; Section III.N.3) and Spooner and Watts (1991a,b; IV.I.7).

L. Revision of Earlier Assignments

The assignment problem is a difficult one, and mistakes are bound to occur. It is gratifying to note that in almost all cases the corrections were published by the group that made the original assignment. Except where noted, the following discussion refers to proton assignments. Some assignments in the N-terminal helix of the C102T variant of *Saccharomyces cerevisiae* iso-1-cytochrome c prior to the work of Gao *et al.* (1990b) have been revised. Assignments for the benzoid protons of residues 46 and 48 were revised by Keller and Wüthrich (1981) and Arean *et al.* (1988). Assignment of the benzoid protons of Phe-82 and Phe-10 were reversed prior to the reports of Boswell *et al.* (1980a) and Keller and Wüthrich (1981). Assignment of Ala-101 and Thr-9 in ferricytochromes c were reversed prior to the report by Arean *et al.* (1988). Assignment of Trp-56 and Trp-77 of *Pseudomonas aeruginosa* ferrocytochrome c-551 were reversed (Moore *et al.*, 1977) prior to the work of Chau *et al.* (1990). Detlefsen *et al.* (1990) have shown that Moore *et al.* (1977) reversed a γ and β proton of Met-61 for this protein. There is also a discrepancy in the assignment of the α protons for Pro-3, Pro-62, and Pro-63 between the studies of Chau *et al.* (1990) and Detlefsen *et al.* (1990), which was resolved by Cai and Timkovich (1991). Timkovich also showed that some of the significant differences in chemical shift between the two studies were due to differences in pH. Several heme assignments for *Pseudomonas aeruginosa* ferricytochrome *c*-551 reported by Timkovich (1991) have been revised by Cai and Timkovich (1993). Finally, an unresolved discrepancy involves the fact that although Detlefsen *et al.* (1990) report the assignment of the η proton of Arg-47 of *Pseudomonas aeruginosa* ferrocytochrome *c*-551, Cai and Timkovich (1991) have been unable to confirm this assignment.

The early and valiant attempt by Stellwagen and Shulman (1973) to assign exchangeable protons should be treated with caution in the light of more recent efforts (Table 5.2). Iso-2-cytochrome *c* has yet to be fully assigned (Garcia, 1991), but Nall and Zuniga (1990) compared their assignments for the five Tyr residues with the corresponding residues in horse cytochrome *c*. With the exception of Tyr-97, their assignments agree with the horse assignments to within 0.17 ppm. If the 3,5 and 2,6 proton assignments are switched for horse cytochrome *c* (Eley *et al.*, 1982b; Wand *et al.*, 1989; Arean *et al.*, 1988), the chemical shifts for Tyr-97 are within 0.02 ppm for the two proteins. It should be kept in mind that different species are being compared, and this could reflect a difference in environment rather than an error in assignment. Santos and Turner (1992) report differences in ^{13}C assignments for one methyl group on Leu-68, one methyl group on Leu-94, and the $\zeta 3$ carbon of Trp-59 in horse ferrocytochrome *c* compared to Gao *et al.* (1990a). As stated by Santos and Turner, their assignments are preferred because they used a cross-assignment strategy, comparing shifts in the reduced and oxidized forms, and because their shifts are more consistent with diamagnetic shifts for these types of carbons. Turner and Williams (1993) have revised the stereospecific assignments for Leu-64 and Leu-94. Also note that reference 30 in Gao *et al.* (1990a) should be Santos and Turner (1992).

In general, any reference to an assignment published before the date of the report that revised it should be treated with caution. A case of disputed chemical shifts in which it is not clear which assignments are correct is that of Tyr-48 in yeast iso-1-ferrocytochrome *c* (Gao *et al.*, 1990b; Thurgood *et al.*, 1991a). Finally, assignments for the oxidized form of the N52I;C102A variant in the middle portion of Table 1 on page 346 of Gao *et al.* (1992) are clearly incorrect. We are currently reassigning this protein.

M. Ionization of His Side Chains

All eukaryotic cytochromes *c* contain a buried His residue at position 18. The τ nitrogen of His-18 is one of the heme ligands, and the π proton is hydrogen-bonded to the carbonyl of Pro-30 (Dickerson *et al.*, 1971; Louie and Brayer, 1990). Many cytochromes *c*, including those from horse, tuna, *Candida krusei*, and one isoform (iso-1-) from *Saccharomyces cerevisiae* also contain a His residue at position 26. The τ nitrogen of His-26 is involved in a hydrogen bond with the carbonyl oxygen of Glu-44 (Dickerson *et al.*, 1971), and the π nitrogen is hydrogen-bonded to the main chain amide of Asn-31 (Louie and Brayer, 1990). Neither His-18 nor His-26 titrates between pH 5 and pH 8. From experiments on tuna cytochrome *c*, Cohen and Hayes (1974) estimate that the pK of His-26 is approximately 3.5. A partial list of pKs of titratable His residues is presented in Table 5.5. Note that despite the fact that the

Table 5.5
The pK of His residues in eukaryotic cytochromes c at approximately 25°C.

Source	Residue number	Oxidation state	pK	Reference
Horse	33	reduced	6.5 ± 0.05	Dobson *et al.*, 1975
	33	reduced	6.54 ± 0.02	Cohen and Hayes, 1974
	33	oxidized	6.4 ± 0.05	Dobson *et al.*, 1975
	33	oxidized	6.41 ± 0.02	Cohen *et al.*, 1974
	26[a]	oxidized denatured	6.71	Muthukrishnan and Nall, 1991
	33[a]	oxidized denatured	6.40	Muthukrishnan and Nall, 1991
	39[a]	oxidized denatured	6.56	Muthukrishnan and Nall, 1991
Candida krusei				
	33	oxidized	6.74 ± 0.01	Cohen and Hayes, 1974
	33[a]	oxidized	6.65 ± 0.05	Robinson *et al.*, 1983
	39	reduced	7.3 ± 0.05	Robinson *et al.*, 1983
	39	oxidized	6.56 ± 0.09	Cohen and Hayes, 1974
Saccharomyces cerevisiae				
	33	reduced	6.7 ± 0.05	Robinson *et al.*, 1983
	33	oxidized	6.8 ± 0.05	Robinson *et al.*, 1983
	39	reduced	7.2 ± 0.05	Robinson *et al.*, 1983
	39	oxidized	6.7 ± 0.2	Robinson *et al.*, 1983

[a] Tentative.

imidazole of His-33 is involved in hydrogen bonds with Val-20 and Glu-103 in yeast iso-1-cytochrome c (Louie and Brayer, 1990), this His exhibits a normal pK. Moench *et al.* (1991) have suggested that the ionization state of His-33 influences the chemical shift of heme methyl 7^1, the effect of ionization being transmitted via the hydrogen bond between His-33 and the C-terminal helix. The pKs of the His residues, with the exception of His-18, in denatured *Saccharomyces cerevisiae* iso-2-cytochrome c have been tentatively determined by Muthukrishnan and Nall (1991; Section III.N.4; Table 5.5), as have the pKs in horse apocytochrome c (Cohen *et al.*, 1974; Snel *et al.*, 1991; Section IV.I.7). Both Muthukrishnan and Nall (1991) and Snel *et al.*

(1991) observed that the amino acid sequence in the region of each His affects pK. For His-33 and His-39, there appears to be little change in pK between the folded and denatured forms, but there is a large change for His-26. This observation should be helpful in unraveling the thermodynamics of denaturation as a function of pH (Pfeil and Privalov, 1976; Hawkes *et al.*, 1984; Cohen and Pielak, 1994). The pK of histidine residues is also central to the assignment of pKs to the heme propionates (Section IV.G).

N. Non-native Forms

NMR spectra of cytochromes c have been examined at extremes of temperature and pH that lead to the observation of non-native forms. Derivatives of the protein where the Met ligand of the heme has been replaced with other ligands such as CN^- and N_3^- have also been studied using NMR, but these forms are not reviewed here. The main purpose of this section is to describe some of the properties of non-native forms of cytochrome c in enough detail so that these forms can be recognized and, if desired, avoided. NMR studies of chemically prepared apocytochrome c are discussed in Sections III.M and IV.I.7.

1. Properties of high-spin and low-spin Fe(III) porphyrins

Before describing the conditions that result in non-native forms, a discussion of the properties of high- and low-spin Fe(III) porphyrin species is presented. Except for some heme and heme–ligand resonances, proton resonances from native ferricytochromes c are not significantly broadened by the unpaired electron, because the electronic spin–spin relaxation time is short relative to the NMR time scale. However, some of the non-native forms of cytochrome c are high-spin d^5, $S = \frac{5}{2}$. Note that this increases the magnitude of $\Delta\delta_c$ (Section III.B.2.a) even in the absence of any other structural or dynamic change. High-spin character lengthens the electronic relaxation time, broadening the proton resonances from these forms. Note, however, that spin $\frac{5}{2}$ species often exhibit smaller pseudo-contact shifts (Section III.B.2.b) because the five unpaired electrons tend to distribute themselves more evenly among the five d-orbitals. This makes the electron distribution more isotropic, g_x and g_y are similar, and the bracketed term in Equation (5.4) decreases.

$S = \frac{1}{2}$ species yield narrow proton resonances at room temperature, but very broad EPR spectra. This explains why EPR data for cytochromes c are acquired at low temperature. On the other hand, $S = \frac{5}{2}$ species give rise to broadened and highly shifted proton resonances, but yield relatively sharp EPR spectra at high temperature. See La Mar (1973) and Wüthrich (1976) for excellent discussions of these points.

2. The effect of temperature

In degassed solutions, horse ferrocytochrome c is reported to exhibit one well-ordered structure yielding high-quality spectra to at least 97°C at pH 7 (Moore and Williams, 1980c,d; Arean *et al.*, 1988). Remarkably, the recovery of protein after spectroscopy at that temperature is high (G. Moore, personal communication). This observation bodes well for thermodynamic studies of ferrocytochromes c (Betz and Pielak, 1992; Bixler *et al.*, 1992; Hilgen-Willis *et al.*, 1993) using NMR. Horse ferricytochrome c is reported to exhibit one well-ordered structure up to 77°C at pH 5.25 (Moore and Williams, 1980d). This is in reasonable agreement with thermodynamic data for bovine ferricytochrome c (Privalov and Khechinashvili, 1974), but the temperature dependence of contact and pseudo-contact shifts [Equations (5.2) and (5.4)] complicates interpretation in terms of the two-state hypothesis. Also note that over these temperature ranges, there are changes in the dynamics of benzoid side chains (Section IV.A). Most other wild-type cytochromes c are not as robust as horse cytochrome c, but they are certainly stable for more than 24 h up to temperatures of approximately 30°C. The temperature dependence of the proton spectrum of horse N^ϵ-maleyl-ferricytochrome c was investigated by Boswell *et al.* (1983), who showed that this derivative goes through a low-spin form missing the Met-80 ligand between approximately 40°C and 50°C and becomes a random-coil polypeptide at 57°C.

Ångström *et al.* (1982) used NMR to examine the magnetic susceptibility of horse cytochrome c as a function of temperature and pH. At neutral pH and temperatures above 42°C, they observed a new species that closely resembles the alkaline-isomerized form (Section III.N.3). However, below pH 5.5, a transition to a high-spin form of the protein was noted as the temperature was raised above 57°C. The authors calculate that at this pH, approximately 9% of the protein is in the high-spin form at 69°C. This form is in fast exchange with the low-spin form. In many studies of labile protons, the pH of the protein sample is reduced to below 6. Therefore, depending on the source of the protein, significant amounts of the high-spin form may be present under these conditions, leading to degradation of spectral quality. The high-spin/low-spin equilibrium just described at least partly explains the complex temperature dependence of paramagnetically shifted resonances (Moore and William, 1980d). Other causes for the complex temperature dependence are discussed in Section III.B.2.

It is interesting to note that at pH 5.4 and high temperature, the resonance attributable to the methyl protons of Met-80 is still present, yet the 695-nm band is essentially absent. Because the 695-nm band is indicative of the intact Met-80–Fe bond and the increase in magnetic susceptibility is indicative of the high-spin form of the protein, Ångström *et al.* (1982) suggest there may be a change in the orientation of the Met-80–Fe bond prior to its dissociation.

3. The effect of pH

Using optical spectroscopy, Theorell and Akesson (1941) described five states of horse ferricytochrome c at low ionic strength, depending on the pH. State I exists below pH 0.4, and state II between pH 0.4 and pH 2.5. States I and II are reported to be high-spin (Gupta and Koenig, 1971; Babul and Stellwagen, 1972; Myer and Saturno, 1990), but Morishima *et al.* (1977) have reported NMR evidence that suggests State II is low-spin. State III exists between pH 2.5 and pH 9.4 and is the state to which the majority of this review is dedicated. State IV exists between pH

9.4 and 12.8. Like state III, this form is low-spin, and the transition between states III and IV is referred to as the alkaline isomerization or the alkaline transition (see Chapter 19 by Wilson and Greenwood, this volume). State V exists above pH 12.8 and is high-spin.

Horse ferrocytochrome *c* is reported to exhibit one well-ordered structure yielding high-quality spectra from pH 3.5 to 12.5 at 27°C (Moore and Williams, 1980c). The most complete NMR study of the pH dependence of horse ferricytochrome *c* is that of Morishima *et al.* (1977), who cite references to earlier work in this area. Morishima *et al.* (1977) observed changes in the proton spectra at pH 2.5, 4, 9, 11, and 12.5 at 22°C, thus subdividing state V of Theorell and Akesson (1941). As determined by proton NMR, horse ferrocytochrome *c* denatures below pH 4.4 at 57°C and below pH 3.8 at 27°C. At alkaline pH, the protein precipitates at pH 10.4, 77°C, and denatures at pH 13, 8°C (Moore and Williams, 1980c). Clearly, for most cytochromes *c* a wide range of temperatures and pHs can be explored in the absence of irreversible denaturation. Both the alkaline and acid transitions were studied using horse cytochrome *c* into which either ^2H or ^{13}C had been ingeniously incorporated using chemical methods (Wooten *et al.*, 1981). Ångström *et al.* (1982) observed a dependence of the magnetic susceptibility of horse ferricytochrome *c* on pH between pH 5 and 7 with an apparent pK of 6.2. It is interesting to speculate (as Ångström *et al.* have) that this pK may be related to that observed by Burns and LaMar (1979; Section IV.D).

Acid denaturation at high ionic strength or in the presence of concentrated HCl results in the formation of a novel form of the protein. Unlike denaturation at low ionic strength, which results in a fully extended protein (*vide ultra*), this high-spin form is compact, exists between pH 2 and 3, and has the properties expected for a "molten globule" (Goto *et al.*, 1990). Because of its nature, the molten globule state is not precisely defined, but, generally speaking, it possesses a compact form containing secondary structural units found in the native form, but lacks most native tertiary interactions (Goto *et al.*, 1990; Dill and Shortle, 1991; Ptitsyn, 1992). In other words, the molten globule exists between the native structure and the fully denatured form and has been proposed as a kinetic folding intermediate (but see Sosnick *et al.*, 1994; Section IV.C). Direct evidence that a molten globule–like state is populated during the folding of apomyoglobin has been reported by Jennings and Wright (1993). Jeng *et al.* (1990) have used NMR to probe the molten globule state of cytochrome *c*, and Sosnick *et al.* (1994) have examined the folding kinetics of the molten globule of ferricytochrome *c*. Their techniques and results are discussed in Sections IV.B and IV.C, respectively. See notes added in proof.

In the alkaline isomerized state, certain heme resonances and the resonances of the Met-80 ligand disappear (and other resonances appear) with a pK of between 8 and 9, depending on the source of the cytochrome *c*. Ferrer *et al.* (1993) have shown that either of two Lys residues can replace the Met-80 sulfur as the ligand to the heme iron, and that one of these is Lys-79. Like the native protein, the product of alkaline isomerization is also low-spin, and therefore its spectra exhibit narrow resonances. From other studies, it has been shown that the pK of the alkaline isomerization is actually a summation of a microscopic pK of 11 and a conformational change with a pK of approximately -2 (Davis *et al.*, 1974). However, the situation appears to be even more complex. In an NMR study by Hong and Dixon (1989), two forms of the alkaline isomerized forms were observed. The chemical shift of resonances for heme methyls 7^1, 12^1, and 18^1 were reported for both isomers. The isomers are both low-spin and are in slow exchange on the NMR time scale. Hong *et al.* (1989) report values of ΔH and ΔS for the isomerization, as well as the activation parameters and pH dependencies for the forward and reverse reactions. As pointed out by Hong *et al.* (1989), optical and NMR studies of alkaline isomerization do not detect exactly the same species. It also appears that the alkaline isomerized form is inactive in electron self-exchange, although the rate of self-exchange in a sample containing no salt is inhibited by 50% of its maximum at pH 10.5 (Gupta *et al.*, 1972).

An important point is that certain amino acid substitutions lower the apparent pK of the alkaline transition of *Saccharomyces cerevisiae* iso-1-cytochrome *c* from 9 to as low as 7 (Pearce *et al.*, 1989). These pKs have been determined by optical spectroscopy for several unnatural variants of *Saccharomyces cerevisiae* iso-1-cytochrome *c* altered at position 82. Pearce *et al.* (1989) find that certain mutations alter the microscopic pK while others, at the same position, change the conformational pK. They also present an extensive review of the literature on the alkaline isomerization of cytochromes *c*. The appearance of a species resembling the alkaline isomerized form at pH 6 has also been reported for the P76G variant of yeast iso-2-cytochrome *c* (Wood *et al.*, 1988). The ability of variations in amino acid sequence to alter the amount of protein in the physiologically important state at neutral pH should be borne in mind when choosing conditions for studies of unnatural variants.

In summary, there are a plethora of non-native forms that can be studied by NMR. These forms differ in structure, dynamics, and spin state. In studying unnatural variants of cytochrome *c*, it is important to remember that the substitutions may change the conditions under which these forms will be observed.

4. Heme peptides

The heme octapeptide comprises residues 14–21, including the pendant heme, and is produced by proteolysis of cytochrome *c*. Smith and McLendon (1981) analyzed the proton NMR spectrum of the heme octapeptide of horse cytochrome *c* in the presence and absence of a sixth ligand to the iron (methionine, pyridine, and CN^-). Spectra were analyzed in terms of contact and dipolar shifts. These authors concluded that differences between spectra of the heme octapeptide–methionine complex and native oxidized cytochrome *c* probably result from the orientation of the Met ligand forced upon it by the protein. Proton resonances of the heme undecapeptide (residues 11–21) were assigned by Kimura *et al.* (1981). One problem with working with heme peptides is that, like free heme, they tend to aggregate in aqueous solution (Othman *et al.*, 1993).

5. The effect of guanidinium chloride and urea

High concentrations of urea or guanidinium chloride denature cytochrome *c* (Brems *et al.*, 1982; Hickey *et al.*, 1991a; Betz and Pielak, 1992; Bixler *et al.*, 1992; Hilgen-Willis *et al.*, 1993). Hartshorn and Moore (1989; Section IV.G) examined the

1-D proton NMR spectrum of horse ferricytochrome *c* in urea. At pH 4 and 6 M urea, the spectrum is that expected for a random coil, but at pH 7 and 8 M urea the authors claim that the protein is not fully denatured. Muthukrishnan and Nall (1991) used proton NMR to study denatured *Saccharomyces cerevisiae* iso-2-ferricytochrome *c* in the presence of 3.5 M guanidinium chloride. The denatured form is low-spin, yet the sulfur atom of Met-80 is no longer coordinated to the iron, indicating that some other strong-field ligand or ligands have replaced His-18 and/or Met-80. Upon addition of a large excess of deuterated imidazole, new resonances indicative of his C2 protons are observed. Iso-2-cytochrome *c* contains three His residues; however, even at large excesses of added imidazole, only two new resonances are observed in the denatured form. The authors suggest that in the absence of added imidazole, the natural ligand, His-18, remains coordinated in the denatured form and that the other two His residues "share" the sixth coordination site. The resonances of the coordinated His residues are in the paramagnetic region of the spectrum and were not observed. Addition of imidazole is postulated to displace the other two His residues, whose resonances are now found at the chemical shift expected for His in a (diamagnetic) random coil. Interestingly, the displaced His residues have chemical shifts that differ by approximately 0.1 ppm. By making several assumptions, the investigators estimated their individual p*K*s and tentatively assigned the C2 resonances to specific residues. By determining the amount of imidazole that must be added to displace half the His residues, the effective concentration of His relative to the heme in the denatured protein was calculated. The authors find the effective concentration is approximately 10 mM. These experiments were also performed on tuna and methylated *Saccharomyces cerevisiae* iso-1-cytochrome *c*. The results of this study are consistent with the idea that this denatured cytochrome *c* is nearly a random coil.

In agreement with the results of Muthukrishnan and Nall (1991), Elöve *et al.* (1994) have shown that in 4.5 M guanidinium chloride, the amide proton of His-18 in horse ferricytochrome *c* is substantially protected from exchange with solvent water (Section IV.B). The effect of non-native histidine ligation on the folding kinetics of horse ferricytochrome *c* folding is discussed in Section IV.C.

IV. The Use of Assignments

A. Flip Rate of Benzoid Rings

If the side chain of a Tyr or Phe residue is firmly fixed within the anisotropic environment of a protein, then all the benzoid protons should possess unique shifts. If the ring is flipped by 180° about the β–γ bond, the position 2 and 6 protons exchange places, as do the position 3 and 5 protons. If this rate of flipping is very slow compared to the difference in chemical shift between them, then they will exhibit unique shifts. However, if the rate of flipping is increased so that it is very fast compared to the difference in chemical shift, then the protons experience an average environment and the individual resonances of protons 2 and 6 coalesce into a single resonance at the average value of their individual chemical shifts. The same is true for the resonances of protons 3 and 5. In between these two extremes, the individual resonances become broad (exchange broadening). The flip rate of benzoid protons in cytochrome c was first discussed by Dobson et $al.$ (1975) and Moore and Williams (1975), and first measured by Campbell et $al.$ (1976).

The rate of rotation of methyl groups is much greater than the difference between the chemical shift of the three protons. Therefore, all methyl groups give rise to only one proton resonance with an integrated intensity of three protons. The symmetric side-chain methyl protons of Val and Leu might also be expected to flip like those of Phe and Tyr, resulting in averaging of their chemical shifts. However, flipping Val and Leu side chains is more complex. Although the side chain is symmetric, a 180° rotation about the α–β bond for Val and the β–γ bond of Leu changes the environment of the prochiral methine proton. Therefore, distinct resonances are observed for each of the prochiral methyl protons whether or not they flip. Although the side chains of most Val and Leu residues in eukaryotic cytochromes c are immobile, the amino acid at position 57 (Ile or Val) is an exception (Robinson et $al.$, 1983; Section IV.I.6).

Using logic strictly analogous to cross-assignment of resonances between the ferro- and ferri- state (i.e., that the rate of cross relaxation is much less than the rate of flipping), a saturation transfer experiment can be performed to determine the flip rate of slow-flipping benzoid side chains. Sometimes benzoid side chains can be coaxed from the slow-flipping regime to the fast-flipping regime by increasing the temperature without denaturing the protein. In these cases, measurement of the flip rate as a function of temperature will yield the activation parameters for ring flipping (Campbell et $al.$, 1975; Wüthrich and Wagner, 1975; Nall and Zuniga, 1990). If the transition between slow and fast flipping does not occur within the allowed range of temperatures, then limits may still be placed on the rate of flipping by simulation of spectra. In general, a slow flip rate is approximately 1–10 s^{-1} and a fast flip rate is 10^3–10^4 s^{-1}. For slow-flipping aromatics in horse cytochrome c, $\Delta H^\dagger$ and $\Delta S^\dagger$ at 25 °C are found to be approximately 25 kcal mol^{-1} and 25 cal mol^{-1} K^{-1}, respectively (Campbell et $al.$, 1976).

Nall and Zuniga (1990) have made a detailed investigation of Tyr ring flips in $Saccharomyces$ $cerevisiae$ iso-2-cytochrome c. The protein was isolated from yeast grown on minimal media supplemented with unenriched amino acids except for Tyr. [3,5-^{13}C]-labeled Tyr was supplemented to allow the use of isotope-edited NMR spectroscopy to study the ring-flip motion of the five Tyr residues in this protein. Spectra were obtained for the reduced and oxidized protein over a temperature range of 5–55 °C for the reduced protein and 5–45 °C for the oxidized protein. Only resonances for the reduced protein were assigned because of the quality and complexity of the spectra for the oxidized protein.

For the reduced protein, Nall and Zuniga (1990) found that Tyr-48, Tyr-46, and Tyr-67 are in slow exchange at 20 °C (two resonances observed for each ring) and in fast exchange at 50 °C (total of three "averaged" resonances observed). For the remaining two Tyr residues, Tyr-74 and Tyr-97, one resonance each was observed over the temperature range examined. The observation of just one resonance for each ring does not necessarily mean the rings are rapidly flipping. The one resonance could be due to the combination of chemical shift degeneracy and slow-flipping rings. The scheme outlined for the five Tyr residues was verified by saturation-transfer NOE difference spectroscopy.

Eyring plots were constructed using rate constants estimated from spectral simulations of the reduced protein and the activation parameters determined. Tyr-48 and Tyr-46 have roughly the same activation parameters: $\Delta H^\dagger = 28$ kcal mol^{-1} and $\Delta S^\dagger = 42$ cal mol^{-1} K^{-1}. Tyr-67 exhibits larger values: $\Delta H^\dagger = 36$ kcal mol^{-1}, $\Delta S^\dagger = 72$ cal mol^{-1} K^{-1}. With the exception of Tyr-97, there is good agreement between horse and iso-2-cytochromes *c* for the Tyr flip rates. Tyrosine-97 seems to be in fast exchange at low temperature in iso-2-cytochrome *c*, but in slow exchange at low temperature in horse cytochrome *c*. However, as discussed earlier, Tyr-97 of iso-2-cytochrome *c* could be in the slow-flipping regime if the chemical shift of the 3 and 5 protons are degenerate. Auld *et al.* (1993) have shown that substitution of Phe-10 in iso-1-cytochrome *c* appears to cause the ring of Tyr-97 to change from the slow-flipping to the fast-flipping regime (Section IV.J.2).

In summary, NMR studies of eukaryotic cytochromes *c* show that the paired aromatic residues, Phe-10 and Tyr-97, exhibit relatively slow flip rates, (Moore and Williams, 1980c; Eley *et al.*, 1982b; Arean *et al.*, 1988). The fact that this pair of residues flips at all might seem remarkable, because inspection of the crystal structure reveals that each member of the pair inhibits the movement of the other. This is also true for the paired, slow-flipping residues Tyr-48 and Phe-/Tyr-46 (Moore and Williams, 1980c). Although some assignments have been revised (Boswell *et al.*, 1980b; Keller and Wüthrich, 1981; Arean *et al.*, 1988; Section III.L), the conclusions concerning ring mobility of all the residues except Phe-10 and Phe-82 are unaffected. The assignment of Phe-10 and Phe-82 have been corrected (Boswell *et al.* 1980b) since they were originally proposed by Moore and Williams (1980a). This is an important correction because Phe-10 is a slow-flipping residue while Phe-82 is fast-flipping. In the C102T variant of *Saccharomyces cerevisiae* iso-1-cytochrome *c*, Phe-(−3) of the N-terminal extension also flips rapidly (Pielak *et al.*, 1988a). It is also important to note that the dynamics of benzoid rings is conserved, with the possible exception of Tyr-97, in all eukaryotic cytochromes *c*, and is the same in both oxidation states in all cases examined. The benzoid residues at positions 36, 74 and 82 exhibit fast-flipping behavior, and Tyr-67 exhibits slow-flipping behavior.

It is interesting that in the crystal structure there does not appear to be enough room to allow flipping of Phe-82. This observation, taken together with the fact that the paired aromatics 10/97, and 46/8 flip, albeit at a much slower rate, is direct evidence for protein breathing. The nature of weakly polar aromatic–aromatic interactions has been discussed by Burley and Petsko (1988). For two aromatic rings, a favorable interaction is thought to arise from the electrostatic attraction of the π-electrons and partially negatively charged carbon atoms of one ring with the partially positively charged protons of the other. An analogous weakly polar interaction has been proposed between aromatic protons and the electron lone pairs of sulfur in methionine, cysteine, and cystine and between aromatic residues and amino groups. Note, however, that slow-flipping rings are not a requirement for weakly polar interactions (Serrano *et al.*, 1991).

Turning to prokaryotic cytochromes *c*, Gooley and MacKenzie (1990) and Gooley *et al.* (1991a) have examined the flip rates of Phe-51 and Tyr-53 in the wild-type form and P35A variant of ferrocytochrome c_2 from the prokaryote *Rhodobacter capsulatus* (Section IV.J.1.). The dynamics of benzoid rings in other ferrocytochromes c_2 has been examined by Cookson *et al.* (1978). Note, however, the assignments for these aromatic residues have been revised (III.L). Thus, residues listed as homologs to the eukaryotic residues Phe-82 and Tyr-48 in Cookson *et al.*, are actually homologs to Phe-10 and Tyr-46, respectively.

A comparison of the dynamics of homologous aromatic amino acids between eukaryotic and other prokaryotic cytochromes *c* is presented in Section IV.E. There are several reports concerning qualitative aspects of ring dynamics in *Pseudomonas aeruginosa* and *P. stutzeri* ferrocytochromes c-551 and, with the exception of Tyr-27, the reports are in complete agreement (Chau *et al.*, 1990; Detlefsen *et al.*, 1990; Cai and Timkovich, 1991).

B. Exchangeable Protons

One theory for explaining the exchange of labile protons states that for buried labile protons to exchange, the protein must partially unfold to expose these protons to the bulk solvent (Hvidt and Nielsen, 1966; for reviews, see Englander and Kallenbach, 1984; Englander and Mayne, 1992). Therefore, a great deal of information about protein structure, stability, and dynamics is available from studies of labile proton exchange (Sections III.N.3; IV.F; IV.I.3; IV.J.4).

McDonald and Phillips (1973) demonstrated that approximately 80% of the labile protons of ferrocytochrome *c* exchange with deuterons from D_2O after one hour at room temperature and pH 5. Patel and Canuel (1976) demonstrated that the exchange of labile protons is base-catalyzed (as it is for proteins in general) and that there can be a several-hundred-fold increase in exchange rate when cytochrome *c* is oxidized (Section III.A). Moore *et al.* (1980a) identified one of these as the indole proton of Trp-59.

The reader may wonder why all exchangeable protons are not detectable in 90% H_2O/10% D_2O. In the discussion about ring flipping in IV.A, it was stated that if benzoid rings are in the fast-flipping regime (i.e., exchange is fast with respect to the difference in chemical shift), then the chemical shifts of the individual 2, 6 (and 3, 5) protons appear at the average of the individual chemical shifts. When there is a difference in the populations of the fast-exchanging species, as there is between exchangeable protons from the protein (on the order of millimolar) and protons from water (~ 100 M), then the resonance of the exchanging species appears at the *weighted* average of the chemical shift in the two states. The difference between the chemical shift of an amide proton and water protons is approximately 10^3 Hz when the experiment is carried out using a 500-MHz spectrometer. Such a small difference in chemical shift and such a large difference in populations result in the resonance for the fast-exchanging labile protons appearing at essentially the chemical shift of the water protons.

The theory of amide proton exchange rates was developed by Hvidt and Nielsen (1966). Cytochrome *c* and most other proteins appear to follow the so-called EX2 mechanism of proton exchange:

$$N(\text{H}) \underset{k_\text{f}}{\overset{k_\text{u}}{\rightleftarrows}} N'(\text{H}) \xrightarrow[\text{D}_2\text{O}]{k_\text{c}} N'(\text{D}) \underset{k_\text{f}}{\overset{k_\text{u}}{\leftrightarrows}} N(\text{D}). \qquad (5.5)$$

The observed rate constant for exchange, k_{ex}, is related to the frequency of fluctuations of the protein backbone in the vicinity of the proton. For most proteins, k_{ex} is related to the equilibrium constant for unfolding of the protein in the region around the labile proton (local unfolding) through the relationship

$$k_{ex} = (k_u/k_f) \times k_c = K_{op} \times k_c. \tag{5.6}$$

Given k_{ex} and k_c, K_{op} can be calculated. To a first approximation, k_c depends only on the local amino acid sequence, and values of k_c are available (Molday $et\ al.$, 1972; Roder $et\ al.$, 1985a,b; Bai $et\ al.$, 1993; Connelly $et\ al.$, 1993). k_c/k_{ex} is the protection factor, P. The free energy for local unfolding can be calculated from the relationship $\Delta G_{op} = -RT \ln K_{op} = RT \ln P$. A problem with this type of experiment is that the error in the protection factors can be as high as 50%. Therefore, only very large differences in protection factors can be interpreted with confidence. Results from experiments where amide proton exchange rates for identical protons are compared between protein variants or between oxidation states must be interpreted with caution for at least two reasons. First, amino acid substitutions may affect k_c (Gooley $et\ al.$, 1992; Section IV.J.IV). Second, if the reduced sample is contaminated with oxidized protein, electron self-exchange (III.F) will cause k_{obs} values to be contaminated by contributions from the oxidized protein.

Wand $et\ al.$ (1986) have measured the exchange rate of protons in the N-terminal helix of horse cytochrome c. These authors find that exchange rates decrease sharply between the regions comprising residues 5–7 and 8–15. Amide protons from residues 5–7 exchange with rate constants of greater than 10 h^{-1}, decreasing to 0.63 h^{-1} for residue 8, and between 10^{-4} and 10^{-2} h^{-1} for residues 9–15. This gradient of exchange rates is in accord with the statistical mechanics of α-helices, which predicts "fraying" at the end of helices (Zimm and Bragg, 1959). The activation energies for proton exchange are approximately 10 ± 2 kcal mol^{-1}. Only residues 14 and 15 of the N-terminal helix exhibit a redox state–induced change in exchange rates, with rates increasing approximately 10-fold for the oxidized protein. For $Rhodospirillum\ rubrum$ cytochrome c_2, the exchange rates of the π proton of the His ligand and the indole proton of Trp-62 have been followed as a function of oxidation state (Yu and Smith, 1988, 1990b). The half-time for exchange in these protons in the reduced state is on the order of a month, whereas the half-time in the oxidized state is on the order of hours.

Marmorino $et\ al.$ (1993) have investigated amide proton exchange in both oxidation states of the C102T variant of $Saccharomyces\ cerevisiae$ iso-1-cytochrome c (I). These authors find that slowly exchanging backbone protons tend to lack solvent-accessible surface area, possess backbone hydrogen bonds, and be present in regions of regular secondary structure as well as in Ω-loops. Furthermore, there is no correlation between k_{obs} and the distance from a backbone amide nitrogen to the nearest solvent-accessible atom. These observations are consistent with the local unfolding model. Comparisons of the free energy change for denaturation, ΔG_d, at 298 K (Cohen and Pielak, 1994) to the free energy change for local unfolding, ΔG_{op}, at 298 K for the oxidized protein suggest that certain conformations possessing higher free energy than the denatured state are detected at equilibrium under conditions that favor the native state. The observation of species higher in free energy than the denatured state under equilibrium conditions was also reported for ribonuclease A by Mayo and Baldwin (1993). Reduction of the C102T variant results in a general increase in ΔG_{op}. Comparisons of ΔG_d to ΔG_{op} for the reduced protein show that the most open states of the reduced protein possess more structure than its chemically denatured form. This persistent structure in high-energy conformations of the reduced form appears to involve the axially coordinated heme. The results of Marmorino *et al.* (1993) are in general agreement with those of Wand *et al.* (1986), except that the exchange rate of certain amide protons in the N-terminal helix of the yeast protein is much faster than that determined for the horse protein (Wand *et al.*, 1986). This may be due to an N-terminal extension which is absent from higher eukaryotic cytochromes *c*. It must be noted, however, that small amounts of contaminating oxidized protein in reduced samples complicate interpretation (*vide ultra*).

The exchange rates for many of the labile protons of *Rhodobacter capsulatus* ferrocytochrome c_2 have been reported by Gooley *et al.* (1990). Like the C102T variant of iso-1-cytochrome *c*, the exchange rate of certain amide protons in the N-terminal helix of the yeast protein is faster than that determined for the horse protein (Wand *et al.*, 1986). There is also a pattern observed for the exchange rates in this region. Residues 7, 10, and 13 exhibit slower exchange rates than the rest of the helix. Gooley *et al.* (1990) cite this $i, i + 3$ pattern in exchange rates, along with the observation of $i, i + 2$ proton amide–amide and proton amide–α NOEs, as evidence for 3_{10}-character in this helix. Gooley *et al.* (1991b) went on to examine the exchange rate of essentially all amide protons as a function of oxidation state. These authors found that there is essentially no redox-state dependence for the N- and C-terminal helices, but there is a difference in rates for the helix between residues 82 and 87 (there is no homologous helix in eukaryotic cytochromes c). These authors have used these redox-dependent exchange rates to rationalize changes in formal (a.k.a. reduction) potential brought about by amino acid substitutions (Gooley *et al.*, 1992; Section IV.J.4).

Timkovich *et al.* (1992) have investigated amide proton exchange in *Pseudomonas aeruginosa* ferrocytochrome *c*-551 using both 2-D methods and saturation transfer. To identify amide protons that are lost before the first time point could be obtained, the authors exploited the fact that exchange cross peaks between the amides in question and bulk water can be observed in HOHAHA spectra acquired in H_2O. The regions of slowest exchange are found at the end of the forties helix and within the C-terminal helix. The observation of slow exchange rates at the extreme end of a helix is surprising, given the fraying argument presented earlier. One observation concerning amide proton exchange rates, which is common to all cytochromes *c* examined to date, is that some of the most slowly exchanging protons are present in the C-terminal helix. Only in horse cytochrome *c*, however, does it appear that exchange in the N-terminal helix is also very slow.

Amide proton exchange in horse ferricytochrome *c* has been studied in tetrahydrofuran (THF) by Wu and Gorenstein (1993). The work was aimed at understanding the role that water plays in maintaining the active form of proteins. Experiments were performed in THF/1% D_2O buffer, where the protein sample exists as a two-phase particle suspension. Like the study of the molten globule state

(Section III.N.3; Jeng *et al.*, 1990; *vide ultra*), samples sufficient for a 2-D homonuclear ^1H NMR experiment were removed at log-spaced time points. For each sample, the THF was removed by freeze-drying and the dried protein was stored in $N_2(1)$, The proton occupancy of each sample was then assessed by redissolving the protein in D_2O buffer, pH 5.1, and acquiring a TOCSY spectrum. Exchange rates for both the protein and a model peptide are slower in THF/1% H_2O than in D_2O. The exchange rate constant for the model peptide was then used along with the k_{obs} values to evaluate protection factors for local unfolding in THF/1% H_2O via Equation (5.6). The results are consistent with the idea that the protein remains folded in the nonaqueous solvent, although structural changes may have occurred. Specifically, it was found that residues in the C-terminal helix that are packed against the N-terminal helix in the native protein in aqueous solution are not protected to the same degree in THF. This suggests that in the nonaqueous medium the interface between the two helices is destabilized, perhaps by repacking.

As noted in Section III.N.3, when horse cytochrome *c* is exposed to low pH at high ionic strength a compact high-spin species is formed (Goto *et al.*, 1990). This form has been suggested to be a molten globule and is therefore of intense interest. Given its high-spin nature, how can amide proton exchange rates be determined? In a clever set of experiments, Jeng *et al.* (1990) first transferred the native protein in H_2O to a D_2O solution, pH 2.2, containing 1.5 M NaCl. At times ranging from 2 min to 500 h, a sample was removed and conditions changed so as to return the protein to its native state. A COSY spectrum was acquired for each time point, and the decrease in cross peak volume was used to determine the exchange rate of the amide protons. This allows the exchange rate of the acid-denatured form to be studied without putting the high-spin form into the spectrometer. The protection factors were then compared to those obtained for the native protein (*vide ultra*). As is evident from the observation of slowly exchanging amide protons, the N-, C-, and sixties helices are still present in this form of the protein. On the other hand, many of the amide protons involved in hydrogen bonds that determine the tertiary structure of the native protein are absent. The amide proton of Leu-85, which is involved with a tertiary hydrogen bond with the carbonyl of Leu-68, is also slowly exchanging, suggesting that this form of the protein, although lacking tertiary structure, is still compact. Taken together, these data provide good evidence that this form of cytochrome *c* represents a molten globule. Finally, Elöve *et al.* (1994) have examined amide proton exchange for horse ferricytochrome *c* in the presence of guanidinium chloride as a function of pH (Section III.N.5).

C. The Folding of Horse Cytochrome *c*

Activity in determining the path (or as is more likely, paths) for protein folding is intense, and because cytochrome *c* is so well-behaved, its folding has been extensively studied. Roder *et al.* (1988) have used NMR to follow ferricytochrome *c* from a random coil to its native state. First, cytochrome *c* is dissolved in D_2O containing guanidinium chloride where, depending on pH (Section III.N.5; Muthukrishna and Nall, 1991; and see later discussion), the protein approximates a random coil. The denaturant is then quickly diluted with an H_2O-containing buffer to a concentration well below that which causes denaturation, and the protein begins to fold. This is called the initial folding time, t_f. Assuming the rate of exchange between labile protons of the denatured protein and water is much slower than folding, hydrogen bonds, which are taken as an indication of folding, that form during this period will have trapped deuterons. After various times, the pH of the refolding reaction is quickly brought to approximately 9, causing the exchange rate of amide deuterons not involved in structural hydrogen bonds to increase dramatically. This is the quench time, t_q. Finally, the pH of the reaction is adjusted to 5, and the protein reduced by addition of a D_2O buffer containing ascorbate. The ferro- form is preferred because it is known that many amide protons exchange more slowly in ferrocytochrome c (Patel and Canuel, 1976; Wand *et al.*, 1986).

A magnitude-mode COSY spectrum of samples from each time point is obtained, and the well-resolved NH-α cross peaks are integrated. [A magnitude-mode spectrum is used because a pure phase COSY cross peak has an integrated intensity of zero. However, Radford *et al.* (1992) have used phase-sensitive spectra and summed the absolute value of each component of a crosspeak.] If the initial time allowed for folding is short, few deuterons are captured, and the resulting spectrum contains mostly protons. As the initial time allowed for folding is increased, more and more deuterons are captured, and the intensity of the NH-α cross peaks decreases. Therefore, each sample represents a snapshot of the degree of folding just before the quench. A plot of the integral of each cross peak as a function of the folding time allows the degree of folding of individual parts of the protein to be followed.

When the folding reaction is carried out at neutral pH, two patterns are observed. First, residues in the N- and C-terminal helices are protected early on (40% in 30 ms), suggesting that these helices are formed first in the folding process. Second, amide protons from the sixties and seventies helices and a tertiary hydrogen bond from the indole proton of Trp-59 are not protected to the same degree until much later (*ca.* 40% in 3 s). Therefore, it appears that the folding of horse cytochrome c begins with the formation and interaction of the N- and C-terminal helices, after which other processes, such as Pro isomerization and exchange of heme ligands, become rate-determining. See notes added in proof.

A caveat pertinent to all quenched-flow studies is that only interactions that are both along the folding pathway and ultimately present in the native state are observed. Thus, protons that exchange too quickly in the native state cannot be used as probes. Additionally, information about non-native interactions that occur along the pathway is unavailable. Thus, when interpreting results from quenched-flow studies, it is important to note that interactions observed as a function of folding time are relevant to the folding pathway, but non-native interaction(s) that impede renaturation (causing the intermediate to build up), and non-native interactions that facilitate renaturation, are not observed.

The data of Roder *et al.* (1988) also provide indirect evidence that more than one pathway for folding is available. Twenty percent of the molecules fold completely within the dead time of the device, while another 60% accumulate in the early intermediate. A reasonable explanation for these observations follows. The intermediate possesses unfavorable interactions that must rearrange for folding to continue, while 20% of the molecules do not possess this unfavorable interaction and fold quickly.

An estimate of the number of species present at any one time, as well as an estimate of their stability, can be obtained by a modification of the quenched-flow experiment where protection at a fixed t_f and t_q is followed as a function of the quench pH (Elöve and Roder, 1991). Protected protons in each species present at t_f are assumed to exchange via a two-state model:

$$S(^1H) \xrightleftharpoons{K_{d,s}} D(^1H) \xrightarrow[{}^2H_2O]{k_c} D(^2H) \xrightleftharpoons{(K_{d,s})^{-1}} S(^2H), \qquad (5.7)$$

where D is the exchange-competent state, and S is used instead of "native" to differentiate S, which may be an intermediate, from the native state. As discussed earlier, for most proteins under most conditions k_c is rate-limiting. This same concept can be used to determine $K_{d,s}$ in quenched-flow experiments. Proton occupancy P (pH) as a function of pH at a constant t_f is given by

$$P(\text{pH}) = 1 - \exp[-K_{d,s}(k_{OH} 10^{(\text{pH}-pKw)})t_q], \qquad (5.8)$$

where the term in parentheses describes the pH dependence of k_c (Molday *et al.*, 1972; Roder *et al.*, 1985a,b; Bai *et al.*, 1993).

Data from such studies are interpreted in a manner analogous to studies of folding equilibria and kinetics using different probes. If only one species is present at t_f, the sigmoidal curve of occupancy vs. time for every amide proton is identical. If, on the other hand, multiple species are present, different amide protons will behave differently. Furthermore, if plots of pH versus occupancy exhibit only one sigmoid, or multiple components are well resolved, inspection of Equation (5.8) reveals that $\Delta G_{d,s}$ for this species relative to the denatured state can be calculated from $K_{d,s}$. Also, as can be seen from Equation (5.8), the same information is available from studies of occupancy vs. t_q (Radford *et al.*, 1992). To obtain more information about the number of species present, and to estimate the stability of the helix–helix interaction in the early intermediate, Elöve and Roder (1991) examined the effect of quench pH on occupancy after 30 ms of renaturation. Twenty percent of the molecules possess protected N-, C-, and sixties helices; 40% are present as the early intermediate (protected N- and C-helices only); and 20% are unfolded. Furthermore, they showed that the data are consistent with a stability for the paired helices of approximately 3 kcal mol^{-1}. The fast-refolding form probably possesses certain critical Pro residues in the *trans* (native) conformation and possesses Met-80 as the sixth heme ligand. [All species possess His-18 as a heme ligand (Muthukrishnan and Nall, 1991; Section III.N.5).]

Quenched-flow/NMR data have been combined with stopped-flow data to describe the intermediates in more detail (Elöve *et al.*, 1992). The intermediates possess a His as the sixth ligand, which must be replaced by the sulfur of Met-80 before folding can be completed (Roder and Elöve, 1994; Elöve *et al.*, 1994). The slowest-refolding forms probably possess critical Pro residues in *cis* (non-native) conformations. Information about the number of different species present at any time can also be obtained using quenched-flow/mass spectrometry (Miranker *et al.*, 1993). A short and readable account of quenched-flow/NMR and quenched-flow/mass spectrometry is given by Englander (1993).

To determine the effect of histidine ligation to the Met-80 site, folding from guanidine chloride was examined as a function of initial and final pH and in the presence of imidazole (Roder and Elöve, 1994; Elöve *et al.*, 1994; III.N.5). The results suggest that above a pH of ∼6, non-native histidines are coordinated to the Met-80 site. Thus, the intermediate possessing protection within the N- and C-terminal helices is observed near neutral pH because it is trapped by non-native histidine ligation. Nevertheless, quenched-flow NMR data obtained below pH 6 suggest that an intermediate with protected N- and C-terminal helices is still formed, but, at this pH, it is a minor species. It is also interesting to note that Hartshorn and Moore (1989; Section III.N.5) suggest that the structure of the urea-denatured protein also changes with pH.

Sosnick *et al.* have examined the kinetics of folding of horse ferricytochrome *c* starting with the denatured protein at pH 2 [a value less than that used by Elöve *et al.* (1994)] and jumping the pH to either 4.9 or 6.2. Seventy percent of each amide proton is protected with a time constant of ∼15 ms, independent of the final condition (the other 30% probably undergo rate-determining proline isomerization). Under these conditions, no kinetic intermediates are observed, and therefore the protein folds in a two-state process. These authors also examined the folding kinetics of the molten globule (III.N.3) and found that folding occurs in less than 3 ms, independent of whether the folding reaction was carried out at pH 4.9 or 6.2. The finding that the molten globule to native transition is fast suggests that the molten globule of cytochrome *c* occurs after the transition state. Such behavior has not been observed for other molten globules (Creighton, 1994). As Sosnick *et al.* (1994) point out, however, the molten globule of cytochrome *c* is more highly structured than the molten globule of most proteins.

A complete understanding of the role of the interaction between the N- and C-terminal helices in the kinetics of folding must await two more types of experiments. The first type of experiment involves examination of variants with substitutions at the interface between the two helices (Auld and Pielak, 1991; Fredericks and Pielak, 1993) using stopped- and quenched-flow techniques. The second type of experiment involves a detailed kinetic analysis of orthologous cytochromes *c*. If, for instance, horse and yeast iso-1-cytochrome *c* fold by different paths, then the importance of specific intermediates would seem to be diminished. Such findings would suggest that there are many low-energy paths to the native state, and small changes in sequence can dictate which path is taken.

D. Heterogeneity in the Environment of Heme Methyl Groups

For horse ferricytochrome *c*, Burns and La Mar (1981) have observed by proton NMR that heme methyl 7^1 exists in two different conformations. In aqueous solution at neutral pH and room temperature, these two conformations are in fast exchange with respect to their difference in chemical shift. Therefore, a resonance appears at the average of their chemical shifts. By decreasing the temperature to below 0 °C in 20% methanol, the two conformations can be brought into slow exchange. Under these conditions a group or groups with a pK of 7 were found to control the equilibrium between the two conformations. Burns and La Mar (1981) suggested that Phe-82, which is near the heme 7^1 methyl, can exist in two conformations, and that these conformations are reflected in the changes of the methyl proton resonance. They also suggested that His-26, the amino acid at position 33 and/or position 46, may control this conformational change. As Ångström et al. (1982) and Moench et al. (1991) point out, the suggestion that this pK is linked to the ionization of His-26 is unlikely because this residue is believed to possess a pK below 4 (Cohen and Hayes, 1974; Section III.M). Further evidence for conformational heterogeneity in the environs of Phe-82, based on the effect of pH and ionic strength (Section IV.H.1), has been reported by Moench et al. (1991). These authors also suggest that the Phe-82 is less constrained in *Saccharomyces cerevisiae* iso-1- than it is in other, higher eukaryotic cytochromes c.

In a study of horse cytochrome c modified by addition of a 4-carboxy-3, 5-dinitrophenyl group to Lys-13, Falk et al. (1981) suggest that a salt bridge between the side chains of the heme methyl 7^1 proximal amino acid residues 13 and 90 controls this transition. It has also been suggested by Falk and Ångström (1983) and Satterlee et al. (1987) that this salt bridge may be important for the interaction of cytochrome c with cytochrome c oxidase and cytochrome c peroxidase, respectively (Sections V.I.1; IV.I.4). Also related to these studies is the observation by Ångström et al. (1982) of a pK of 6.2 for the pH dependence of the magnetic susceptibility of horse ferricytochrome c. It may be possible to resolve these possibilities by examination of unnatural variants at positions 82 (Pielak et al., 1985; Inglis et al., 1991; Hilgen and Pielak, 1991), 13 (Hazzard et al., 1988), 46, and 33.

Splitting of the heme methyl 18^1 resonance has been observed for *Rhodospirillum rubrum* cytochrome c_2 (Yu and Smith, 1990b). The apparent contradiction caused by the observation of splitting of different heme methyl resonances in prokaryotic vs. eukaryotic cytochromes c was for a time taken as evidence that there might be some error in assignment (Burns and La Mar, 1979). However, this now seems unlikely, and other mechanisms must be investigated. The underlying mechanism leading to the splitting of the heme methyl 18^1 resonance is more complex than the splitting of the heme methyl 7^1 resonances described earlier, but does involve two forms in slow exchange, with the ratio of the forms being pH-dependent.

E. Comparison of Eukaryotic and Prokaryotic Cytochromes c

The most extensive sets of proton assignments for prokaryotic cytochromes c are those reported for *Pseudomonas aeruginosa* cytochrome c-551 (Chau et al., 1990; Detlefsen et al., 1990, 1991; Timkovich and Cai, 1993); for *P. stutzeri* ferro-cytochrome c (Cai et al., 1992); and for *Rhodobacter capsulatus* cytochrome c_2 (Gooley et al., 1990, 1991b). Because cytochrome c-551 is only 82 residues long, a complete comparison with higher eukaryotic cytochromes c is not possible. Nevertheless, Chau *et al.* (1990) have analyzed several patterns involving chemical shifts and the dynamics of benzoid moieties.

Chemical shifts for the structurally homologous protons are, as expected, similar between cytochromes *c*-551 and its eukaryotic cousins. These residues in cytochrome *c*-551 and eukaryotic proteins, respectively, include the His ligand to the heme iron (residues 16 and 18), Gly-24 and -29, Pro-25 and -30, Ile-48 and Leu-68, Trp-56 and 59, the Met ligand to the heme iron (residues 61 and 80), and Val-78 and Leu-98. The mobility of benzoid side chains (Section IV.A) in *Pseudomonas aeruginosa* cytochrome *c*-551 resembles in some respects the mobility found in its higher eukaryotic cousin. For instance, Phe-7 in cytochrome *c*-551 is homologous to Phe-10 in eukaryotic cytochromes *c* in that they both exhibit separate resonances (i.e., slow flipping) and similar chemical shifts. Although there is no sequence identity in this region, Phe-34 and Tyr-27 of *c*-551 are homologs of the eukaryotic residues Tyr-48 and Phe-36, respectively, in the sense that they appear to fill similar structural roles. Both Tyr-27 and Phe-36 exhibit fast-flip behavior. However, Phe-34 appears to be in the fast-flip regime, whereas its eukaryotic homolog, Tyr-48, flips slowly. Chau *et al.* (1990) suggest that because of its smaller size, cytochrome *c*-551 is unable to supply the protein bulk required to inhibit the flipping of Phe-34.

The homologs of the eukaryotic residues Phe-/Tyr-46, Tyr-48, Tyr-67, Phe-82, and Tyr-97 in *Rhodobacter capsulatus* ferrocytochrome c_2 (Phe-51, Tyr-53, Phe-98, and Tyr-110) show very similar ring-flip dynamics in both proteins (Gooley *et al.*, 1990, 1991a). However, Phe-10 in eukaryotic cytochromes *c* is in the slow-flipping regime, whereas this residue is rapidly flipping in the protein from *Rhodobacter capsulatus*. Comparison of flip rates for other ferrocytochromes c_2 to their eukaryotic homologs has been reported by Cookson *et al.* (1978), but note that significant revisions in assignment have been made (III.L; IV.A; Table 5.3).

Comparison of prokaryotic and eukaryotic cytochromes *c* with respect to the p*K* of the inner propionic acid side chain is discussed in Section IV.G, and the effect of oxidation state on the exchange rate of the π proton of the ligand His of *Rhodospirillum rubrum* cytochrome c_2 is discussed in Section IV.B.

F. Redox-State–Dependent Structural Changes and Differences between the Crystalline and Solution States

Given a crystal structure, the pseudo-contact shift may be calculated [Williams *et al.* 1985a; Feng *et al.* 1990a; Gao *et al.*, 1991a; Equation (5.4)] from the magnitude and orientation of the three principal values of the *g*-tensor with respect to the atomic coordinates. These parameters are available from EPR data on crystalline horse cytochrome *c* (Mailer and Taylor, 1972). Alternatively, the crystal coordinates plus the paramagnetic shift data can be used to search for the *g*-tensor parameters. These will yield as output both best-fit calculated shift values and *g*-tensor parameters. These data can be compared with existing EPR and susceptibility data. The observed and calculated values will agree very precisely only if the crystal and solution structures are identical in both oxidation states. Differences will then reveal changes between oxidized and reduced forms in the crystal and in solution, or they will reveal differences between crystal and solution states. It should be borne in mind that crystalline proteins contain a large amount of water. In fact, crystals of *Saccharomyces cerevisiae* iso-1-cytochrome *c* are 30% water (Sherwood and Brayer, 1985). Therefore, to a first approximation, gross conformational changes are not expected between the structure of proteins in crystals and in solution.

Changes in conformation between the ferro- and ferri- state will give rise to changes in a proton's chemical shift because of changes in relationship of the proton with respect to the unpaired electron and/or changes in the relationship with respect to diamagnetic and/or electrostatic fields (Section III.B). A change in the spatial relationship between a proton and the unpaired electron will lead to discrepancies between calculated and observed pseudo-contact shifts.

Williams *et al.* (1985a) reported a comparison of calculated and observed paramagnetic shifts for tuna cytochrome *c*. This study was performed before extensive proton assignments were available, and only 21 resonances were used to optimize the *g*-tensor. Nevertheless, the agreement between calculated and observed values was impressive. An important assumption made by these authors was that there was no change in solution structure between the ferro- and ferri- structures, and that any discrepancy between calculated and observed pseudocontact shifts must be due to changes between the solution and crystal structures. These authors concluded that there are changes between the solution and crystal structures in and around the C-terminal helix, near the position 3 thioether linkage to the heme, and around the heme propionic acid side chains. Evidence for changes near the thioether was supported by NOE data.

With the availability of many more proton assignments from interpretation of 2-D NMR spectra of horse (Wand *et al.*, 1989; Feng *et al.*, 1989, 1991; Feng and Roder, 1988), the C102T variant of yeast iso-1-cytochrome *c* (Gao *et al.*, 1990b) and, to a lesser extent, tuna cytochrome *c* (Gao *et al.*, 1989), and the availability of high-resolution crystal structures for these proteins (Bushnell *et al.*, 1990: Louie and Brayer, 1990; Takano and Dickerson, 1981a,b), more detailed analyses of paramagnetic shift data for these eukaryotic cytochromes *c* were undertaken. As discussed later, ^{13}C chemical shifts (Santos and Turner, 1992) have also been used to examine the horse protein. It must be stressed that the calculated and observed pseudo-contact shifts agree remarkably well for all three proteins. This shows that there is very little change between both the solution and crystal structures and between the solution structure of the oxidized and reduced proteins.

Feng *et al.* (1990a) used a sophisticated statistical screen to find discrepancies between calculated and observed shifts which are due to changes in structure either between solution and crystal structures, or between oxidized and reduced solution structures. This screen involves calculating the minimum change in proton coordinates that would be required to bring the calculated and observed shifts into agreement. If the required movement is much greater than the estimated error in the proton coordinates, then the discrepancy between calculated and observed shifts is due to changes in structure either between the oxidized and reduced proteins in solution, or between the solution and crystal structures. Residues captured by this screen were clustered with respect to primary structure, occurring mainly between 38 and 43 and between 50 and 70. The regions between residues 38–43 and 50–60 are known to interact with the heme propionates and are involved in the conformational change described by Takano and Dickerson (1981b). Amide protons are, in general, overrepresented in the forties and fifties regions. This fact, taken together with the observation by Wagner *et al.* (1983) that there is a close correlation between the chemical shift of amide protons and the dimensions of hydrogen bonds, strongly suggests that changes in hydrogen bonding and other diamagnetic shifts, rather than gross conformational reorganization, accompanies the change in oxidation. Discrepancies observed from residue 60 to 70 may represent changes in diamagnetic shift due to rearrangement of aromatic protons, although changes between solution and crystal structures cannot be ruled out. A similar study on the C102T variant of *Saccharomyces cerevisiae* iso-1-cytochrome *c* by Gao *et al.* (1991a) yielded essentially identical results, suggesting that the redox-dependent conformational changes are similar for all eukaryotic cytochromes *c*.

Turner and Williams (1993) combined ^{13}C and proton assignments in their analysis of pseudo-contact shifts for horse cytochrome *c*. They also used a different screen, involving the maximum change in pseudo-contact shift given a 0.05 nm change in atomic coordinates, to find nuclei whose paramagnetic shift does not fit the calculated pseudo-contact shift. Their results are in accord with changes observed by crystallography and by other NMR studies. Specifically, the authors find a region near Trp-59 where calculated pseudo-contact shifts and observed paramagnetic shifts fit poorly. These authors also exploit the fact that, in the absence of temperature-dependent changes in structure, diamagnetic shifts are independent of temperature [Equation (5.1), Sections III.B.1, III.B.3], whereas pseudo-contact shifts should exhibit an inverse dependence [Equation (5.4), Section III.B.2.b]. The observation that nuclei whose calculated and observed shifts do not match possess typical temperature dependencies strongly suggests that the mismatch is caused by changes in diamagnetic fields which accompany the change in oxidation state. Although the change in the paramagnetic shift upon raising the temperature from 30°C to 50°C is linearly related to the size of paramagnetic shift at 30°C, the slope is greater than that predicted by Equation (5.2). Part of the anomalous temperature dependence is attributed to mixture of the excited and ground states of the iron. The remainder is attributed to thermal expansion of the protein.

Timkovich and Cai (1993) performed pseudo-contact shift calculations on *Pseudomonas aeruginosa* cytochrome *c*-551, the only prokaryotic cytochrome *c* that has been extensively assigned in both oxidation states. These authors report only small deviations between calculated and observed shifts. Unlike the eukaryotic proteins, almost all the protons which exhibit large deviations between calculated and observed pseudo-contact shifts are in regions of large pseudo-contact shift gradients. Thus, the structures of the reduced and oxidized form of this protein appear to be even more similar than the reduced and oxidized forms of eukaryotic cytochromes *c*.

Results reported on variants generated at positions 52 (Das *et al.*, 1989) and 67 (Luntz *et al.*, 1989) using *in vitro* mutagenesis are of interest with respect to the just-described differences in hydrogen bonding near the propionate and the buried water molecules. Preliminary NMR data were reported (Sections IV.J.1, IV.J.3), and detailed studies are eagerly awaited. The Y67F (Luntz *et al.*, 1989) and N52I (Hickey *et al.*, 1991a) variant proteins appear to be more stable than the wild-type protein, but for the Y67F variant, interpretation of such results are complex (IV.J.2; IV.J.3; Schejter *et al.*, 1992; Frauenhoff and Scott, 1992). A preliminary spectral analysis of NOESY spectra has appeared for the Y67F variant protein (Luntz *et al.*, 1989), showing that the interaction between Leu-68 and heme methyl 12^1 has changed in the variant protein (IV.J.3). It will be interesting to look at these proteins again now that it is possible to examine directly the protons of buried water molecules (Qi *et al.*, 1994a; IV.F).

The difference in paramagnetism between ferro- and ferri-cytochromes c complicates comparison between the two oxidation states. This is because discrepancies between calculated and observed pseudo-contact shifts could be due to changes in position with respect to the unpaired electron and/or changes in diamagnetic shift. If the oxidized and reduced proteins were both diamagnetic, comparison might be simplified. Co^{3+}-substituted cytochrome c has the same overall charge as ferricytochrome c, but unlike ferricytochrome c, Co^{3+}-substituted cytochrome c is diamagnetic (d^6). Therefore, a comparison of ferrocytochrome c and cobalticytochrome c could, in theory, yield information about changes in diamagnetic shift without the complication of paramagnetic shift. There are several assumptions inherent in the last statement, and these are discussed by Moore *et al.* (1980c), wherein the authors compare the 1-D spectra of several metal-substituted cytochromes c, including cobalticytochrome c. Superficially, at least, ferrocytochrome c and cobalticytochrome c appear very similar. However, with all the new information available from the total assignment of several cytochromes c, cobalticytochrome c should be reinvestigated.

Examination of the redox-state dependence of labile proton exchange (Section IV.B) is not complicated by arguments involving paramagnetic effects and differences between crystal and solution. Patel and Canuel (1976) have demonstrated that labile protons of ferrocytochrome c exchange much more slowly than those of the oxidized protein (IV.B). Wand *et al.* (1986), Gooley *et al.* (1990), and Marmorino *et al.* (1993) have identified several of these differentially changing protons. The difference in exchange rates as a function of oxidation state may reflect the fact that cytochromes c bind reduced heme more tightly than oxidized heme, as reflected in their positive formal potentials or differences in the locally unfolded structures. The x-ray scattering experiments of Trewhella *et al.* (1988) show that the radius of gyration and the maximum linear dimension of horse cytochrome c increase by more than 1 Å upon oxidation. However, interpretation of these and some other results in terms of a conformational or dynamic change should be treated with caution, in light of the results of Feng and Englander (1990; IV.H.1). Further work in this area, especially studies of variant proteins, will lead to interesting new ideas about the function of cytochromes c.

It is now possible to compare directly the solution structure of horse ferrocytochrome c (Qi *et al.*, 1994b) to the crystal structure (Bushnell *et al.*, 1990). As expected, the two structures are quite similar. The small differences are discussed next.

In the crystal, the side chain of Ile-81 obstructs solvent access to heme methyl 7^1. In solution, this methyl possesses significant solvent-accessible surface because the side chain of Ile-81 crosses the heme edge. Comparison of the crystal to the solution structure also reveals that the benzoid ring of Phe-82 is ~ 2 Å closer to the heme iron than it is in the crystal. The authors also state that in solution, the side chain of Ile-81 moves upon oxidation, and they suggest that this may be important for the mechanism of electron transfer. It is important to note, however, that many lower eukaryotic cytochromes c possess an Ala at position 81, and that methyl 7^1 is significantly exposed in the crystal structure of both oxidized and reduced yeast iso-1-cytochrome c (Berghuis and Brayer, 1992). Another difference between the crystal and solution forms involves additional interactions between the N- and C-terminal helices in solution. These interactions include a backbone–backbone hydrogen bond between the amide hydrogen of Lys-2 and the carbonyl oxygen of Asp93, and a salt bridge between the side chain of Lys-5 and Asp-93. In terms of absences, no hydrogen bonds involving the heme propionates are observed in solution, and all but one of the Type-II turns in the crystal do not fall easily into this category in solution, despite the fact that the turns are well defined by NOEs.

Although a description of the solution structure of horse ferricytochrome c has yet to appear, Qi *et al.* (1994a) have determined the position of the buried water molecules in both oxidation states. These experiments involve three-dimensional homonuclear ^{1}H spectroscopy that combines TOCSY and NOESY spectra (Otting *et al.*, 1991). The waters observed in these spectra are in fast exchange with solvent water and have residence times of a least a few hundred picoseconds.

The position and movement of these buried waters is important because comparison of reduced and oxidized crystal structures of yeast iso-1- (Berghuis and Brayer, 1992) and tuna cytochromes c (Takano and Dickerson, 1981a,b) shows that a particular water found in both structures exhibits a nearly identical redox-state–induced change. This conserved movement, and the fact that this water is common to other crystal structures of ferricytochromes c, suggests that the redox-state shift of the water is functionally important. With three exceptions, the positions of these waters change neither with oxidation state nor between crystal and solution.

Two of the three exceptions involve buried waters that are observed in the solution structure that are not observed in the crystal. One of these, Wat-6, is near a turn and is therefore close to solvent. The other, Wat-5, is buried in the heme crevice. Wat-5 may affect the reorganization energy for electron transfer.

The third exception involves a water molecule, called Wat-1 by Qi *et al.* (1994a) and Wat-166 by Berghuis and Brayer (1992), located on the Met-80 side of the heme. In the reduced protein, Wat-1/Wat-166 is hydrogen bonded to the side chains of Asn-52, Tyr-67, and Thr-78. This water moves in response to a change in redox state, but the movement is different in solution from that in the crystal. Examination of crystal structures show that upon oxidation this water moves 1.6 Å closer to the heme iron. In solution it moves in the opposite direction: toward the surface of the protein, 3.7 Å farther away from the iron. As expected because of the different direction and magnitude of movement, there are large differences in hydrogen bonding of this water between the solution and crystal structure of the oxidized proteins. Thus, electron transfer mechanisms and proposals regarding the reduction potential of cytochrome c that implicate movement of Wat-166 must be reconsidered.

G. The Ionization of Heme Propionic Acids

Although all cytochromes c exhibit some pH dependence of their formal potentials, eukaryotic cytochromes c show a much less marked dependence than cytochromes c such as *Pseudomonas aeruginosa* cytochrome c-551. It has been proposed that one of the main determinants of the pH dependence is the ionization of the position 17-propionic acid side chain of the heme. In cytochromes c such as *Pseudomonas aeruginosa* c-551, the pK of approximately 7 for the formal potential follows the ionization of the side chain (Moore *et al.*, 1980b; Cai and Timkovich, 1992), whereas in higher eukaryotic cytochromes c the formal potential and the ionization state of this carboxylic acid remain unchanged between pH 5.0 and 8.8 (Moore *et al.*, 1984). For certain eukaryotic cytochromes c, such as *Saccharomyces cerevisiae*, the pH dependence of the formal potential appears to be influenced by a His at position 39 (Moore *et al.*, 1984). The pKs of the propionic acids in free heme have been estimated to be 4.8 and 5.6 (Phillips, 1963), but, except for a qualitative study by Chau *et al.* (1990), we are unaware of any experimentally determined value, probably because the free heme aggregates in aqueous solution. It has been proposed that in eukaryotic cytochromes c, the nearby Arg-38, which is invariant in eukaryotic cytochromes c, decreases the pK of the inner propionic acid (17; see Table 5.1 and Figure 5.1) to below 5 (Moore *et al.*, 1984). However, when Arg-38 of yeast iso-1-cytochrome c was changed to Ala, His, Lys, Leu, Asn, or Gln using site-directed mutagenesis, the chemical shift of the inner propionic acid 17^1 protons changed little relative to the wild-type protein (Cutler *et al.*, 1989). Davies *et al.* (1993) examined the pH dependence of the reduction potential and the resonances from the inner propionic acid upon changing Tyr-48 and Trp-59, as well as Arg-38. Although the reduction potential and the pK of the alkaline transition is decreased in the Y48F, Y48F; W59F, and R38A; Y48F; W59F variants, the pK of the propionic acid seems to remain unchanged. The fact that a wide range of amino acid substitutions at the conserved positions which interact with the inner propionic acid affect the pH dependence of the formal potential (at least at pH values below the alkaline transition) indicates that current ideas on the pK of the inner propionic acid side chain in mitochondrial cytochromes c require revision. In summary, the pK of the inner propionic acid of higher eukaryotic cytochromes c remains unknown. As pointed out by Davies *et al.* (1993), either the pK is less than 5, but substitutions are insufficient to perturb it, or the pK is greater than 7.5, and substitutions only served to increase it further. In a clever titration study (which did not directly involve NMR), Hartshorn and Moore (1989; Section III.N.5) suggest that the inner propionic acid possesses a pK of >9 and that the outer propionic acid possesses a pK of <4.5.

Examinations of variants of *Saccharomyces cerevisiae* iso-1-cytochrome c containing multiple substitutions at positions 38 and 39 might clear up some of the confusion regarding the outer propionic acid. NMR studies of the pH dependence of the N52I variant protein (Das *et al.*, 1989; Hickey *et al.*, 1991a; Gao *et al.*, 1992) and the Y67F variant (Luntz *et al.*, 1989) might also prove useful because these residues are near the inner propionate and are involved in the structural reorganiza-

tion between ferro- and ferricytochrome c (Takano and Dickerson, 1981b; Feng
et al., 1990a; Gao *et al.*, 1991a). The reader may feel that the outer propionate
has been neglected. This side chain appears to be important for the interaction of
cytochromes c with the mitochondrial membrane (Section IV.I.7).

Turning to those cytochromes c that exhibit a variation of formal potential with
pH, the chemical shift of the heme propionates of *Pseudomonas aeruginosa* ferrocytochrome c-551 was studied by Chau *et al.* (1990), who found that the pH dependence of the heme propionate resonances was less in the protein than in free heme
(25% v/v dimethyl sulfoxide/D_2O) and that the direction and magnitude of the
changes in chemical shifts as a function of pH were ambiguous. The Trp-56 indole
proton and the Arg-47 ϵ proton are each hydrogen-bonded to a different oxygen of
the inner propionate in the crystal structure (Matsuura *et al.*, 1982, and references
therein). Neither the exchangeable side-chain protons of Arg-47 nor the τ or π
protons of His-47 in *Pseudomonas stutzeri* ferrocytochrome c-551 (Cai and Timkovich, 1992) are observable, even in the absence of presaturation of the residual
water resonance, suggesting that these protons are in fast exchange. [This result is
somewhat unsettling, as Detlefsen *et al.* (1990) have reported the assignment of this
Arg η proton.] Nevertheless, the indole proton exhibits a pK of approximately 7,
suggesting that the pK represents the ionization of the inner heme propionate. These
results were summarized and extended by Cai and Timkovich (1992), who examined
both *Pseudomonas aeruginosa* and *P. stutzeri* ferrocytochrome c. The latter protein
possesses a His at position 47, and observation of the chemical shift of its C-2 and
C-5 protons as a function of pH yields a pK of approximately 8.2. On the other
hand, observation of the indole proton of Trp-56 yields a pK of approximately 3.
Observation of the amide proton of Val-55, which is hydrogen-bonded to the outer
propionate in both proteins, yields a pK of approximately 7. The authors present a
model for the ionization of the inner propionic acid residues of these two proteins
based on the fact that His is a weaker base than Arg. The inner propionate
in *Pseudomonas stutzeri* ferrocytochrome c-551 is assigned a pK of 3, and that in
Pseudomonas aeruginosa ferrocytochrome c-551 is assigned a pK of 7.2 (Cai and
Timkovich, 1992). The model for *Pseudomonas stutzeri* cytochrome c-551 is in
agreement with scheme C in Figure 11 of Leitch *et al.* (1984), and, despite appear-
ances, the data for the native proteins are in general agreement with the ideas of Yu
and Smith (1990a).

In their study of the *Rhodospirillum rubrum* cytochrome c_2, which is a member
of the same class as *Pseudomonas stutzeri* cytochrome c-551, Yu and Smith (1990b)
examined the chemical shift of the inner propionic acid side chain as a function of
pH in the oxidized state of the protein. pKs of 6.3 and 8.7 were observed. These
were assigned to the pK of His-42 (which is hydrogen-bonded to the inner propionate) and a conformational change that results in the expulsion of the His
ligand, respectively. It should be noted that the pK of His-42 could not be independently measured in the oxidized protein, and its value was inferred from experiments on the reduced protein (Yu and Smith, 1990a). Moore *et al.* (1984) assigned a
pK of 6.3 to the inner propionic acid of the closely related *Rhodomicrobium vannielii*
ferrocytochrome c_2. Yu and Smith (1990b) also reported the pK for the N-terminus.
There is a controversy concerning the number of pKs required to fit the pH depen-

dence of the reduction potential for cytochromes c_2 and the identities of the ionizing species. The reader is directed to the original references (Yu and Smith, 1990a,b; Moore et al., 1984).

H. The Interaction of Cytochromes c with Small Ions Other Than Protons and Hydroxide

1. Diamagnetic ions

Although there have been several studies that used NMR to probe ion binding (Section IV.H.2; Stellwagen and Shulman, 1973; Andersson et al., 1979), probably the most precise measurements of binding constants were made using other techniques (Gopal et al., 1988). It is important to note that many small anions bind to cytochromes c. This binding may affect the studies wherein the binding of other molecules is monitored in the presence of a background electrolyte. Cacodylate binds to horse, tuna, and cow cytochromes c with an association constant much less than 100 M^{-1} at 25 °C and pH 7 (Gopal et al., 1988) and is the electrolyte of choice (Barlow and Margoliash, 1966).

Trewhella et al. (1988) used x-ray scattering to measure changes in the size of cytochrome c as a function of both oxidation state and NaCl concentration. Results from this study indicate that upon oxidation at low ionic strength, both the radius of gyration and the maximum linear dimension of horse cytochrome c increased by greater than 1 Å. Furthermore, increasing the concentration of NaCl ameliorated the increase in size of the oxidized form. It is tempting to interpret these observations in terms of redox-dependent conformational and dynamic changes (Section IV.F). Note that these experiments were performed in a background buffer containing 5 mM phosphate.

In light of this result, Feng and Englander (1990) investigated the effect of increasing NaCl concentrations on the proton NMR spectrum of horse ferro- and ferricytochrome c in a background buffer of 5 mM phosphate (the same concentration as that used by Trewhella et al., 1988). They found that increasing the NaCl concentration from 0 to 200 mM had only a very small effect on chemical shift. For the reduced protein, only the chemical shift of the Glu-61 amide proton was changed by greater than 0.1 ppm. (This proton exhibits a large secondary shift and therefore is very sensitive to even minute structural changes.) For the oxidized protein, significant changes were noted for protons from Glu-61, Ala-83, Lys-87, Lys-88, and Thr-89. With the exception of Glu-61, these residues form one of the anion binding sites of cytochrome c (Arean et al., 1988; Section IV.H.2) Taken together with the results from studies of the pseudo-contact shift (III.B.2.b; IV.4), it appears that there is no substantial structural change between ferro- and ferricytochromes c. Feng and Englander (1990) explain the results of the x-ray scattering experiments (Trewhella et al., 1988), as described in the next paragraph.

The size increase noted upon oxidation and its diminution with increasing NaCl concentration from the study of x-ray scattering may be explained by the presence of phosphate in this experiment. At low ionic strength, phosphate is more tightly bound to the oxidized form, and a net increase in size was observed. As the ionic strength was increased, NaCl effectively competed with phosphate for the binding site on cytochrome c, and the increase in size was not observed. Gopal *et al.* (1988) showed that phosphate binds to oxidized horse cytochrome c with a binding constant of approximately 1600 M^{-1}, and to the reduced form with a binding constant of approximately 500 M^{-1}. Feng and Englander seem to have been unaware of the results of Gopal *et al.* (1988), but their argument is only strengthened by the 1988 study.

There is a further set of experiments that could be performed which would shed light on this matter. Gopal *et al.* (1988) demonstrated that the pattern of ion binding to horse ferro- versus ferricytochrome c is reversed for the tuna protein. Therefore, if the increase in molecular size of horse cytochrome c upon oxidation is caused by increased avidity of phosphate for this form, then the results should be reversed if the same study were conducted using the tuna protein. It is also important to note that the conclusions from other studies which demonstrate significant differences between the two redox states of cytochromes c are still valid (Sections IV.B, IV.F); and references cited by Feng and Englander, 1990). A caveat discussed by Feng and Englander (1990) involves the possibility that in studies which examined differences between the two redox forms performed in the absence of phosphate, the binding of other salts may mimic the effect of phosphate.

Ragg and Moore (1984) examined the binding of the diamagnetic transition-metal complex [Co(CN$_6$)]$^{3-}$ to horse ferro- and ferricytochrome c using ^{59}Co NMR. Two binding sites were observed with association constants of $2.0 \pm 0.6 \times 10^3$ M^{-1} and $1.5 \pm 0.5 \times 10^2$ M^{-1} at pH 7.3, 25 °C, and an ionic strength of 0.07 M (KCl). The dependence of binding on ionic strength was also investigated. In competition experiments against the cobalt complex, the binding of [Ru(CN)$_6$]$^{4-}$ was determined to be less than 300 M^{-1} at an ionic strength of 0.12 M (KCl), 25°C, pH 7.3. Moench *et al.* (1991) have investigated the effect of ionic strength on the environment of Phe-82 in horse, tuna, and yeast iso-1-cytochromes c using both NaCl and KNO$_3$. These results suggest that changes in ionic strength (and pH; Section IV.D) cause a small conformational change in this region of the protein, and that the yeast protein is more susceptible to the effects of salt than are the other two proteins. The results also exemplify the complexity of such studies in that the effects observed depend upon the identity of the salt, the identity of the protein, and the resonances that are followed. The binding of other diamagnetic species, especially phosphates, has been studied indirectly via their ability to displace paramagnetic probes (*vide infra*).

2. Paramagnetic ions

The most extensive series of experiments on the binding of transition metal complexes to cytochromes c have emanated from R. J. P. Williams' laboratory (Williams *et al.*, 1982; Eley *et al.*, 1982a; Williams and Moore, 1984; Arean *et al.*, 1988). These investigators used paramagnetic difference spectroscopy (PDS) to identify binding sites for both cationic and anionic complexes. The idea behind PDS experiments is that paramagnetic transition metal complexes such as [Cr(CN)$_6$]$^{3-}$, [Cr(en)$_3$]$^{3+}$, and various Fe(III) and Gd(III) complexes selectively broaden proton resonances in the vicinity of the binding site with an r^{-6} distance dependence. [These highly symmetric ions have long electronic relaxation times (Section III.N.1).] In the PDS experiment, a 1-D spectrum is first acquired. Then a small volume of a concentrated solution of the paramagnetic complex is added, and a second spectrum is acquired. When the two spectra are subtracted, resonances that are far from the metal ion binding site cancel. Resonances near the complex ion binding site, however, are broadened in the presence of the complex and are observed as narrow resonances in the difference spectrum. After resonances from the affected residues are assigned, a search of the crystallographic coordinates is made for positively and negatively charged residues in the vicinity that might form the binding site. PDS has been used to identify four anion and four cation binding sites in horse and tuna cytochrome c. The cation binding sites encompass the C-terminal carboxylate, (Asp-2, Asp-93), Glu-44, and (Glu-66, Glu-69). The anion binding sites encompass (Lys-5, Lys-87, Lys-88), (Lys-7, Lys-25, Lys-27), (Lys-13, Lys-72, Lys-79, Lys-86), and Lys-99. In the course of studies to characterize electron self-exchange within complexes of horse cytochrome c (Whitford *et al.*, 1991b), Concar *et al.* (1991 b,c) used $[Cr(CN)_6]^{3-}$ to show that the binding site for polyphosphates, specifically hexametaphosphate and tripolyphosphate, includes Lys-13, Lys-86, and Lys-87. Eley *et al.* (1982a) showed that $[Fe(CN)_6]^{3-}$ binds to trimethyllysine-72 of *Candida krusei* ferricytochrome c with an association constant of 140 ± 15 M^{-1} at 27°C and pH 7.

I. The Interaction of Cytochromes c with Proteins and Phospholipids

From a biological standpoint, the most important interactions of cytochromes c are those with protein redox partners and the mitochondrial membrane (Tzagoloff, 1982; Pettigrew and Moore, 1987). Several hypothetical models for complexes of cytochrome c with its redox partners have been proposed (e.g., Salemme, 1976; Poulos and Kraut, 1980; Poulos and Finzel, 1984; Matthew *et al.*, 1983). These are based on model-building studies using the x-ray crystallographic coordinates of the individual proteins. A molecular dynamics study of the model complex between cytochrome c and cytochrome b_5 has been reported (Wendoloski *et al.*, 1987). Until recently, no x-ray crystal structures for these complexes have been reported, owing to difficulties in obtaining suitable crystals (Poulos *et al.*, 1987). Pelletier and Kraut (1992) have now reported crystal structures for cytochrome c peroxidase complexed to both horse and *Saccharomyces cerevisiae* iso-1-cytochromes c.

NMR has been used to investigate the interaction of cytochrome c with cytochrome b_5, L-lactate cytochrome c reductase, cytochrome c peroxidase, cytochrome c oxidase, flavodoxin, and plastocyanin. All of these studies, with one exception, are based on the assumption that the protein complex is in fast exchange (compared to the chemical shift time scale; Section IV.A) with the uncomplexed proteins. This appears to be a valid assumption because individual protons (usually hyperfine-shifted heme resonances of cytochrome c) exhibit unique chemical shifts in the complex, and the chemical shifts of these protons change as a function of the ratio of the two proteins (Gupta and Redfield, 1973). The exception to the fast-exchange situation is the study of the covalent complex formed between cytochrome c and cytochrome c peroxidase (Moench *et al.*, 1987; Section IV.I.4). Another indicator of complex formation is an increase in linewidth when the two proteins are mixed. This broadening is due to the increased correlation time (Section II) of the complex compared to its individual constituents. In the discussion that follows, it should be borne in mind that although physiologically important protein–protein complexes have been studied, these complexes are by necessity catalytically incompetent. NMR can, however, be used to determine rates of electron transfer (III.F).

1. Cytochrome c oxidase

Investigation of the complex between cytochrome c and cytochrome c oxidase by NMR is hampered by the low solubility of the oxidase. Even when the solubility is increased by addition of detergents, the large size of the molecules hampers observation of proton resonances because of the increase in linewidth (II). However, the effect of complex formation on the value of T_1 of the heme methyl resonances has been investigated by Falk and Ångström (1983). These authors determined T_1 as a function of added cytochrome c oxidase up to a mole ratio of cytochrome c oxidase to cytochrome c of 0.07. The longitudinal relaxation of resonances belonging to methyl groups 18^1 and 12^1 shows a nonlinear dependence on the amount of cytochrome c oxidase added, whereas the heme methyl 7^1 resonance exhibits linear behavior. This nonlinear dependence of T_1 is also observed for cytochrome c by itself when Lys-13 was modified (Falk et al., 1981; Falk and Ångström, 1983). The nonlinear behavior of Lys-13–modified cytochrome c has been suggested by these same authors to be caused by the dissociation of the salt bridge between Lys-13 and Glu-90. Falk and Ångström (1983) conclude that the interaction of cytochrome c with cytochrome c oxidase causes a disruption of this salt bridge. Kennelly et al. (1981) examined the reaction of cytochrome c oxidases from cow, *Paracoccus denitrificans*, and *Pseudomonas aeruginosa* with several prokaryotic and eukaryotic cytochromes c in which Lys residues had been converted to homoarginine by reaction with ^{13}C-enriched O-methylisourea. Although the reactions with the oxidases was not followed by NMR, this study reports the ^{13}C NMR spectra of the derivatives (Section II.K.2). The modifications had only a small effect on activity. In fact, in one case a modified protein was more active than the cognate native protein.

2. L-Lactate cytochrome c reductase

This reductase, found in yeast, is composed of two domains, one containing flavodoxin and the other, a cytochrome b_2 unit (reviewed by Chapman et al., 1991). The b_2 unit contains a cytochrome b_5–like moiety. An NMR study was undertaken by Thomas et al. (1987) to characterize the interaction of the holoenzyme (and the cytochrome b_2 unit) from the yeast *Hansenula anomala* with the homologous cytochrome c. This cytochrome c contains three methylated Lys residues at positions 55, 72, and 73. The Lys residues at positions 72 and 73 are trimethylated, and Lys-55 is a mixture of mono- and dimethyllysine. Methylated Lys residues are common to all plant, algal, and fungal cytochromes c whose amino acid sequences have been determined. These methyl groups are freely rotating in solution, having a much shorter correlation time than the bulk of the protein, as evidenced by the narrowness of the resonances. The binding of cytochrome *c* to either the core protein or the cytochrome b_5–containing domain was monitored via the change in chemical shift of a hyperfine-shifted heme resonance of the oxidized cytochrome *c* (probably heme methyl 18^1) as a function of the amount of the other protein component. Binding was also monitored via chemical shift changes of the methylated Lys residues. Changes in chemical shift of the hyperfine-shifted resonance as a function of the cytochrome *c*:holoenzyme or cytochrome *c*:core protein ratio indicated that a complex was formed, but the stoichiometry appeared to be less than one. Interpretation of results from the titration of the holoenzyme is dangerous because it could not be followed beyond a mole ratio of holoenzyme to cytochrome *c* of 0.4—a result of the low solubility of the former. When complex formation was monitored via chemical shift changes of trimethyllysines-72 and 73, the stoichiometry of the complex appeared to be less than 0.1 core protein/cytochrome *c*. Furthermore, the linewidth of the trimethyllysine residues was not affected significantly by complex formation, suggesting that the correlation time of this residue is not affected by complex formation. These results concerning the dynamics of the trimethylated Lys residues upon complex formation are in conflict with results from the study of the interaction of cytochrome *c* with the tryptic fragment of cytochrome b_5 (Eley and Moore, 1983; *vide ultra*) and should be interpreted with caution.

3. Cytochrome b_5

Eley and Moore (1983) have studied the interaction of a tryptic fragment of bovine cytochrome b_5 with horse and *Candida krusei* ferricytochromes *c*. Changes in chemical shift of heme methyl 18^1 of oxidized cytochrome *c* as a function of the ratio of cytochrome *c* to cytochrome b_5 indicated that binding does take place. Eley and Moore (1983) assumed, rather than proved, that a one-to-one complex was formed. However, the results of this study are in agreement with the study of Mauk *et al.* (1982), who used optical spectroscopy to demonstrate a one-to-one complex.

On the other hand, Miura *et al.* (1980) concluded in their NMR study of complex formation that a two-to-one cytochrome *c*:cytochrome b_5 complex was formed. Furthermore, Whitford *et al.* (1990) observed, in addition to the binary complex, a ternary complex consisting of two molecules of cytochrome *c* and one molecule of cytochrome b_5. They also report binding constants of approximately 10^5 M^{-1} and 10^4 M^{-1}, respectively, but few details on how these values were obtained are presented. In a further experiment where the ionic strength was varied between 20 and 150 mM phosphate at a fixed concentration of the two proteins, the authors noted three distinct species as defined by the linewidth of the cytochrome b_5 heme 12^1 methyl group—two different forms of the dimer and free cytochrome b_5. There seems no way to reconcile the results of Whitford *et al.*, Mauk *et al.*, and Miura *et al.* As discussed by the first group, it is possible to reconcile the results of Whitford *et al.* with those of Mauk *et al.*, if it is assumed that the binding of the second cytochrome *c* and the rearrangement of the binary complex do not induce any further changes in optical properties at the wavelength investigated. It is also possible that the higher protein concentration used by Whitford *et al.* enhances formation of the ternary complex.

Eley and Moore (1983) also noted an increase in linewidth upon complex formation. In this study, the linewidth of the methyl resonance from trimethyllysine-72 of the *Candida krusei* protein was observed to increase upon complex formation, suggesting that the mobility of the methyl groups is restricted in the complex. The authors went on to explore the residues involved in the interaction domain between the two proteins. The binding domain was explored by addition of $[Cr(en)_3]^{3+}$ to a one-to-one mixture of cytochrome c and cytochrome b_5. The experiment is a variation of the PDS experiment discussed in Section IV.H.2. If binding of $[Cr(en)_3]^{3+}$ to either protein is disturbed by formation of the cytochrome c/cytochrome b_5 complex, then resonances normally broadened by their proximity to a chromium ion are now unaffected and hence disappear from the difference spectrum. There are many assumptions inherent in such an experiment, including that there are only minor chemical shift differences upon interaction of proteins, and that the interaction of $[Cr(en)_3]^{3+}$ with the individual protons does not disturb the protein complex (discussed by Hartshorn *et al.*, 1987). Therefore, results cannot be interpreted in a quantitative manner. Nevertheless, results from this experiment show that Ile-81, Phe-82, Ala-83, and Ile-85 are involved in the interaction of cytochrome c with cytochrome b_5. This is in accord with the model-building study of Salemme (1976).

In a molecular dynamics simulation of the complex between horse cytochrome c and cytochrome b_5, Wendoloski *et al.* (1987) observed that the side chain of cytochrome c residue Phe-82 changes position upon complex formation. This proposed complex may be thought of as the activated complex. The conformational change moves the benzoid side chain of Phe-82 to a more central position between the two hemes in the hypothetical complex, implying a crucial role for Phe-82 in electron transfer. Burch *et al.* (1990) tested this proposal using NMR. These authors showed that the ^{13}C and proton NMR spectra of the cytochrome c-cytochrome b_5 complex can be simulated by the simple summation of the individual spectra. The proton resonances of the benzoid side chain of Phe-82 were specifically identified in the complex, and their chemical shifts were shown to be unaffected by complex formation. Assuming that a change in environment would result in a change in chemical shift, Burch *et al.* (1990) concluded that no appreciable conformational change occurs upon complex formation. In a related study, Whitford *et al.* (1990) examined changes in the proton chemical shifts of the Phe-82 side chain and examined NOESY spectra of the one-to-one cytochrome c:ferricytochrome b_5 complex. These authors also conclude that there is no change in the environment of Phe-82 upon formation of the complex (but see Section IV.I.3).

Does this observation invalidate the proposal of Wendoloski *et al.* (1987)? As Burch *et al.* (1990) indicated, NMR measures some average structure representing the collision complex. Because the simulated structure would represent the activated complex, it may be that this complex is formed rarely and transiently. Such a complex would not be detected by NMR.

The most important point is that even if a conformational change does occur, it is not directly coupled to the rate of electron transfer. This is because changing Phe-82 of *Saccharomyces cerevisiae* iso-1-cytochrome c does not drastically alter the rate of electron (or perhaps energy) transfer in reactions with itself, with cytochrome b_5, and with Zn-substituted cytochrome c peroxidase (Concar *et al.*

1991a,b; Willie *et al.*, 1993; Everest *et al.*, 1991). Furthermore, neither the steady-state rate of the cytochrome *c*/cytochrome *c* peroxidase reaction (Pielak *et al.*, 1985) nor the ability of yeast harboring the position-82 alleles as their only source of cytochrome *c* to grow on nonfermentable carbon sources (Pielak *et al.*, 1985; Inglis *et al.*, 1991; Hilgen and Pielak, 1991) are dramatically affected by the residue at position 82.

Another portion of the study by Burch *et al.* involved examination of the methyl proton resonances of modified Lys residues of horse cytochrome *c*. The protein was modified because the resonances of the 19 unmodified Lys side chains are overlapping multiplets and could not be assigned by 1-D NMR. Making the assumption that the chemical modification induces negligible changes in the structure of cytochrome *c* (which is not strictly valid—*vide* Falk *et al.*, 1981) or its complex with cytochrome b_5, the authors compared the proton chemical shifts of methyl groups attached to the Lys residues in the presence and absence of cytochrome *c*. They noted that six of the methyl groups experience changes in chemical shift upon complex formation, compared to only four proposed in Salemme's model (1976). It may seem strange that the lack of change in the chemical shift of aromatic protons of Phe-82 was used to argue against a conformational change, yet a change in chemical shift for the modified Lys residues was used to argue for their involvement in complex formation. As discussed by Burch *et al.* (1990), electrostatic interaction (Section III.B.3) between the modified Lys residues and the surface carboxylates of cytochrome b_5 in the absence of a conformational change would produce the alteration in chemical shift upon complex formation. The involvement of more Lys residues than proposed by Salemme (1976) is not surprising when it is remembered that the former is a static model and the latter is dynamic. That is, there may be several modes of binding with equally low energies, and it is the ensemble average of these binding modes that is observed by NMR. The observation of two different first-order rate constants for the reaction of ruthenium-modified cytochrome b_5 with cytochrome *c* (Willie *et al.*, 1992, 1993) supports this notion, but note that in saturation transfer studies using the native proteins (Concar *et al.*, 1991a), only one rate constant is observed (Section III.F). Finally, as pointed out by Burch *et al.* (1990), the amino acid sequence of their recombinant lipase-solubilized bovine liver cytochrome b_5, as well as the sequence used in the previous modeling studies, contains three errors.

4. Cytochrome *c* peroxidase

The interaction of cytochrome *c* with *Saccharomyces cerevisiae* cytochrome *c* peroxidase (CCP) has been studied by NMR in both noncovalent (Gupta and Yonetani, 1973; Stellwagen *et al.*, 1977; Smith and Millett, 1980; Satterlee *et al.*, 1987; Moench *et al.*, 1992; Yi *et al.*, 1992) and covalent (Moench *et al.*, 1987, 1992) complexes. It is clear from the proton data that the proteins form a one-to-one complex, and that cytochrome *c* resonances from the 18^1 and 7^1 methyl protons of the heme, the Met-80 ligand, and one of the 17^2 propionic acid side-chain protons of the heme are affected by complex formation. These changes in chemical shift need not represent large structural reorganizations. The pyrrole ring to which the

7^1 methyl group is attached is exposed to solvent in free cytochrome c. Because these proton resonances are sensitive to contact and pseudo-contact shift (Section III.B.2), a very small conformational change could affect not only the chemical shift of the 7^1 methyl protons, but also, via the d_{xz} orbital of the iron (axes defined in III.B.2.b), the chemical shift of the 17^2 and 18^1 protons.

Nearly all NMR data on complex formation published to date have been interpreted with respect to the hypothetical model of Poulos and Kraut (1980; Poulos and Finzel, 1984). The largest difference between the model and the crystallographically determined structures (Pelletier and Kraut, 1992) is that in the former the two hemes are essentially coplanar, and in the latter there is an angle of $60°$ between them. A figure in which the hypothetical complex is superimposed on the crystal structure of the complex is given by Poulos and Fenna (1994). Other differences involve the residues that are in contact with each other in the complexes. NMR studies have focused on the region around pyrrole ring B (Table 5.1) and Phe-82 of cytochrome c. These regions are important in both the Poulos and Kraut model complex and the crystallographically determined complex. Therefore, NMR observations that support the former also support the latter.

Satterlee *et al.* (1987) performed a titration of cytochrome c with cytochrome c peroxidase, keeping the total amount of protein constant. They noted an increase in the linewidth of cytochrome c heme methyl 7^1 and 18^1 of approximately 2.5 times in the complex relative to the free protein. Using the observed changes in chemical shift of heme methyl 18^1, Satterlee *et al.* (1987) determined the off rate for cytochrome c peroxidase in the complex to be 1133 ± 120 s^{-1}. As discussed by Moench *et al.* (1992), care must be taken not to overinterpret such results, because the exchange kinetics are dependent on the concentration of the complex. See notes added in proof.

In a simple and elegant experiment, Yi *et al.* (1992, 1993) examined the NMR spectra of cytochrome c:CCP-CN complexes where the sixth ligand to the iron in CCP is cyanide. This has the effect of making CCP low-spin, thus sharpening the proton resonances within the complex (Section III.B.2.a). Under these conditions, changes in chemical shift of greater than 0.1 ppm are observed for the resonances of heme protons 7^1 and 18^1 of CCP-CN upon formation of the protein complex. This result demonstrates that the binding of cytochrome c to CCP-CN alters the heme environment within CCP-CN even though the heme of CCP is buried within the protein. Other results from Yi *et al.* confirm those from the study of the native complex (Moench *et al.*, 1992; *vide infra*), in that changes in chemical shift are greater for homologous than for heterologous complexes. That different complexes are formed between homologous and heterologous partners is clearly shown in the crystallographic study (Pelletier and Kraut, 1992), but note that similar first-order rate constants are observed for electron transfer from *Saccharomyces cerevisiae* iso-1-ferrocytochrome c to Zn-substituted CCP (Everest *et al.*, 1991) and from horse ferrocytochrome c to Zn-substituted CCP (Wallin *et al.*, 1991).

Just when it seemed safe to suggest that formation of physiological (but inactive) protein complexes involving cytochrome c does not affect the environment of Phe-82 (Section IV.I.3), Moench *et al.* (1992) showed that the change in chemical shifts for its ring protons upon forming a one-to-one complex with CCP is different for different cytochromes c. That is, changes in the chemical shift of the Phe-82 4 proton and heme 7^1 methyl protons upon formation of the homologous noncovalent cytochrome c:CCP complexes (i.e., *Saccharomyces cerevisiae* cytochromes c) are greater than the changes observed upon formation of heterologous complexes (i.e., the tuna or horse protein). Furthermore, the changes in chemical shift for the Phe-82 and 7^1 protons are linearly correlated, and the direction of the change in chemical shift for Phe-82 is consistent with the idea that the side chain moves away from the heme. It is important to note that reorganization of the Phe-82 side chain upon complex formation is not observed in the crystallographic study (Pelletier and Kraut, 1992). Phe-82 is, however, in van der Waals contact with CCP in the homologous complex, but not in the heterologous complex. These observations serve to point out a strength and a weakness of protein proton NMR. Clearly NMR can detect small changes in environment, but interpretation of these changes in terms of specific structures remains difficult.

In the study by Stellwagen *et al.* (1977), a differentially ^{13}C-modified cytochrome c was used to investigate binding. The authors found that addition of CCP to the modified cytochrome c resulted in an increase in the ^{13}C linewidth of the modified Lys residues, consistent with complex formation, and that certain (unassigned) resonances were shifted by a fraction of a part per million. As noted by the authors, however, the presence of H_2O in the D_2O-containing sample also affects linewidth. In the study by Smith and Millett (1980), the ^{19}F chemical shifts of cytochromes c individually trifluoroacetylated at Lys-13, Lys-79, and Lys-87 were examined in the presence and absence of CCP. Aside from the expected line broadening, no change in chemical shift was observed upon complex formation. The T_1 of these ^{19}F resonances were determined in both the oxidized complex and in the dithionite-reduced (diamagnetic) complex. No change in T_1 was noted for any of the modified proteins upon reduction. [This result is surprising because even the methyl groups of trimethyllysine-72 in *Saccharomyces cerevisiae* iso-1-cytochrome c exhibit a measurable paramagnetic shift (Gao *et al.*, 1989).] The invariance of T_1 was used to determine a minimum distance of at least 15 to 20 Å between the two heme irons. This is consistent with both the hypothetical model of Poulos and Kraut (25 Å; 1980; Poulos and Finzel, 1984) and the crystal structure of the complex (26.5 Å; Pelletier and Kraut, 1992).

In the covalent complex (Moench *et al.*, 1987), the linewidth of resonances is approximately three times as great as that in the noncovalent complex because in the covalent complex the two proteins can no longer be in fast exchange. The changes in chemical shift of the heme methyl and propionic acid protons listed earlier are almost identical for the covalent and non-covalent complexes. Moench *et al.* (1992) compared the proton spectra of the covalent CCP:horse cytochrome c and CCP:*Saccharomyces cerevisiae* iso-1-cytochrome c complexes. The covalent complex involving iso-1-cytochrome c exhibits many different and partially resolved hyperfine-shifted resonances, and thus there appear to be many different environments for these protons. The NMR spectrum of the covalent complex involving horse cytochrome c does not exhibit this heterogeneity. In both cases, many different types of complexes are formed, and these complexes possess different numbers of crosslinks. Therefore, interpretation of this result is not straightforward, except to state that the homologous complex exhibits a more complex spectrum than does the heterologous complex. There is no evidence for multiple conformations in either the homologous or heterologous complexes studied using crystallography (Pelletier and Kraut, 1992). Interpretation of this result in terms of the cross-linking studies is not clear. However, two different first-order rate constants are observed for electron transfer between ferrocytochromes c and Zn-substituted CCP (Wallin *et al.*, 1991; Everest *et al.*, 1991) and between ruthenium-modified ferrocytochromes c and compound I of CCP (Hahm *et al.*, 1992).

The position 7^1 methyl group of cytochrome c is in close proximity to the side chain of Lys-13 (of horse or tuna cytochrome c). Data from chemical modification studies and proposals from computer graphics simulations suggest that this Lys is involved in the electrostatic interaction between cytochrome c and CCP. No such interaction, however, is found in the crystal structure of the complex (Pelletier and Kraut, 1992). The side chain of Lys-13 in tuna cytochrome c may form a salt bridge with the side-chain carboxylate of Glu-91 in solution (Takano and Dickerson, 1981a). In horse cytochrome c, the interaction would be between Lys-13 and Glu-90, and in yeast iso-1-cytochrome c, the interaction would be between Arg-13 and Asp-90. Satterlee *et al.* (1987) suggest that interruption of this interaction, with resulting rearrangement of the Lys side chain so that it can interact with an oppositely charged residue on CCP, would probably change the environment of heme methyl 7^1, thus explaining the change in chemical shift observed for the complex. Falk *et al.* (1981) and Falk and Angstrom (1983) expressed a similar view (Section IV.D). An *intramolecular* interaction between Lys-13 and Glu-90 is observed in the crystal structure of the horse cytochrome c:CCP *complex*, but not in the yeast cytochrome c:CCP complex (Pelletier and Kraut, 1992). On the other hand, the crystallographically determined structure does possess intermolecular van der Waals interactions between Arg-13 and Ile-9 of yeast iso-1-cytochrome c and the side chain of Tyr-39 of CCP. In the heterologous complex, the interaction between residues 9 and 39 is present, but the interaction between residues 13 and 39 is absent. This may explain the phylogenetic effects of complex formation observed for the heme methyl and position 82 protons.

Comparing results from NMR studies of different complexes, the heme methyl 18^1 resonance is not reported to shift upon formation of the L-lactate cytochrome c reductase–cytochrome c complex (Section IV.I.2; Thomas *et al.*, 1987). It is also interesting to note that for the interaction of cytochrome c with bovine cytochrome b_5, a salt bridge between Lys-13 and a carboxylate on cytochrome b_5 is also proposed (Salemme, 1976). However, in the complex, the cytochrome c heme methyl 7^1 resonance is virtually unaffected.

A preliminary report has appeared describing the effect of complex formation on amide proton exchange rates (IV.B) in cytochrome c (McLendon, 1991). The author claims that the data support the Poulos and Kraut model (1980; Poulos and Finzel, 1984), but few details are given. See notes added in proof.

Finally, it is important to note that the observation of changes in chemical shift of hyperfine-shifted methyl resonances upon binding another molecule does not necessarily mean that a specific complex is formed. Boswell *et al.* (1980b) observed chemical shift changes very similar to those obtained for the interaction of cytochrome c with CCP upon addition of polyglutamate to oxidized cytochrome c.

5. Nonphysiological interactions

Although the interaction of eukaryotic cytochromes c with either the photo-synthetic blue copper protein plastocyanin (King *et al.*, 1985) or flavodoxin from *Azotobacter vinelandii* (Hazzard and Tollin, 1985) are not physiologically significant, they do serve as important models for physiological interactions. A model for the electrostatic interaction of cytochrome c with flavodoxin has been proposed based on crystallographic studies (Matthew *et al.*, 1983). For both proteins, complex formation can be monitored by observing the change in chemical shift of various hyperfine-shifted heme resonances as a function of the ratio of the protein to cytochrome c. In both cases the heme 18^1 methyl resonance shifts upfield as the complex is formed. The pattern for the plastocyanin complex is in complete accord with effects observed in the complexes of cytochrome c with CCP and with cytochrome b_5, as discussed earlier. The pattern of other changes in hyperfine chemical shift changes for the flavodoxin complex is somewhat smaller and different from the cases mentioned earlier. For the plastocyanin complex, the change in chemical shift was used to show that a one-to-one complex was formed, while for both the plastocyanin and the flavodoxin complexes, the increase in linewidth as a function of complex formation was used to determine the one-to-one stoichiometry. The off-rates of plastocyanin and flavodoxin from their complexes with cytochrome c were calculated to be greater than 10^3 s^{-1} and approximately 10^2 s^{-1}, respectively. Gd^{3+} binds to the negatively charged patch of plastocyanin. For the plastocyanin:cytochrome c complex, King *et al.* (1985) measured the paramagnetic relaxation enhancement of the water resonance as a function of complex formation to show that the binding constant is approximately 1.5×10^4 M^{-1} at pH 6.

6. Antibodies

Moore and Williams first presented experimental evidence for the correlation of antigenicity with highly mobile regions of protein structure in 1980 (f). In particular, Williams and Moore (1985) and their collaborators focused on the region of eukaryotic cytochromes c around residue 57. The chemical shift of the amino acid side chain at this position in eukaryotic cytochromes c (either Ile or Val) exhibits both complex pH- and temperature-dependent behavior. Because the temperature-dependent changes in chemical shift occur in the reduced protein, they cannot be due to a paramagnetic effect (Sections III.B.1, II.B.2). The resonance of the δ protons also broadens with increasing pH below that which results in the alkaline transition (III.N.3). Other resonances in the vicinity of position 57, particularly Trp-59, do not exhibit this broadening, further suggesting a localized effect. The temperature dependence of the chemical shift is suggested by Robinson *et al.* (1983) to be caused by changes in the mobility of the Ile-57 side chain as a function of temperature under the influence of the ring-current shift brought about by Tyr-74 [i.e., changes in r and θ of Equation (5.1)]. The data also suggest that the pH dependence of the chemical shift is caused by the ionization of Lys-55 (III.B.3). The pH dependence of the linewidth is suggested to be caused by several different conformations for the amino acids surrounding and including Ile-57. These fast-exchanging conformations cause a slight conformation-dependent shift for the resonances of Ile-57, resulting in line broadening. Moore and Williams (1980f) had previously noted a correlation between the heterogeneity of the region around position 57 and its importance as an epitope for antibody binding (Urbanski and Margoliash, 1977). The effect of changing Trp-59 and Tyr-48 to Phe on the chemical shift of Val-57 in yeast iso-1-ferricytochrome c has been discussed by Thurgood *et al.* (1991a; IV.J.1).

Paterson *et al.* (1990) have reported a detailed analysis of the residues involved in the binding of a monoclonal antibody to horse ferricytochrome c (see Chapter 11 by Paterson, this volume). The experimental protocol involved determination by proton NMR of the differential exchange rate of backbone amide protons (Section IV.B). The rationale behind the protocol is related to that used to probe cytochrome c folding (IV.C). The antibody was covalently bound to a polymer matrix and mixed with cytochrome c in a proton-containing buffer. The reaction mixture was transferred to a column, excess unbound cytochrome c was removed, and the column was washed with a D_2O-containing buffer. All manipulations after this point were made in D_2O-containing buffers at 8 °C. At various times, ranging from 1 h to 20 days after the addition of the D_2O containing buffer, the cytochrome c was removed from the column by elution with a buffer of pH 2.5. The pH of the eluent was then quickly adjusted to pH 7.5, and the cytochrome c reduced with ascorbate and concentrated. If the bound antibody retarded the exchange rate of amide protons, then these amides would have retained the protons that were present before the addition of D_2O. If exchange rates were not retarded, then the amide protons would have been replaced by deuterons. A magnitude-mode COSY spectrum was then acquired for each time point. Integration of the proton α-amide cross peaks as a function of the time the cytochrome c and antibody were in contact provided the rate constants for amide proton exchange. These were divided by the rate constants determined in the absence of antibody, and a "protection ratio" for each amide proton was calculated.

Protection ratios of greater than 3 were observed for 12 residues, and ratios greater than 100 were observed for Gly-37, Arg-38, Trp-59, Lys-100, and Ala-101. As Paterson *et al.* (1990) point out, this experiment will miss residues important for antibody binding if their amide-proton exchange rates are very slow in the absence of antibody. The surface defined by the protected residues is a noncontiguous patch with a solvent-exposed surface area of 750 Å^2, comprising portions of the 60s helix, the C-terminal helix, and an Ω-loop (residues 37–59). Note that Ile-57, discussed earlier, lies within the patch defined by this study.

Using identical experimental procedures, Mayne *et al.* (1992) have examined the binding of two other monoclonal antibodies to horse cytochrome c. The epitope for these antibodies is centered on the Ω-loop mentioned earlier. A difference between the results reported by Paterson *et al.* and those reported by Mayne *et al.* is that in the former study, only amide protons in the region of binding are affected, whereas in the latter study, there are "knock-on" effects. That is, the protection ratios for amide protons outside the epitope are affected to a greater degree compared to the earlier study. In the absence of antibody, local unfolding is less favorable for the 60s and C-terminal helices than for the Ω-loop. Given this observation, the authors suggest that distortion of the Ω-loop upon antibody binding is more likely to affect the exchange rates in other parts of the protein (via cooperative interactions) than is antibody binding to the 60s and the C-terminal helices.

7. Phospholipids, membranes, and micelles

Although cytochrome *c* is an extrinsic membrane protein, its interaction with the phospholipid bilayer may be important for its function. Cardiolipin is a major component of the mitochondrial membrane and is required for the function of cytochrome *c* oxidase (Robinson and Capaldi, 1977). Whether this requirement is due to interaction of the oxidase with the membrane or due to a requirement for the binding of cytochrome *c* to the oxidase is not known. Cardiolipin–phosphatidylcholine vesicles decrease the formal potential of cytochrome *c* by over 60 mV (Kimelberg and Lee, 1970), and cytochrome *c* exhibits specific binding to cardiolipin in these vesicles (Brown and Wüthrich, 1977). These data show that the interaction between cardiolipin and cytochrome *c* is an important but underexplored aspect of the chemistry of cytochrome *c*.

Brown and Wüthrich (1977) investigated the interaction of cytochrome *c* with 1 : 4 cardiolipin–phosphatidylcholine vesicles. This study dealt exclusively with the effect of cytochrome *c* on the NMR and EPR spectra of the lipid, and is therefore beyond the scope of this article. However, the results are summarized because they bear on a study of the effect of vesicles on the proton NMR spectrum of the protein discussed in the next paragraph. Brown and Wüthrich demonstrated that cytochrome *c* interacts with the vesicle mainly through electrostatic interactions which do not perturb the interior structure of the vesicles. Using Met-65–spin-labeled horse cytochrome *c*, the authors demonstrated that the region around Met-65 interacts with the phospholipid bilayer. This region is far removed from the heme crevice (but is in the vicinity of the heme propionates), suggesting that the crevice points away from the membrane. They also presented evidence that suggests cytochrome *c* causes phase segregation of cardiolipin. That is, cardiolipin is preferentially located on the surface of the vesicles in an immobilized boundary. However, the study of Devaux *et al.* (1986) found no evidence for segregation when cytochrome *c* interacts with phosphatidylcholine–phospatidylserine vesicles. This discrepancy may be indicative of a unique interaction between cardiolipin and cytochrome *c*.

The interaction between horse cytochrome *c* and cardiolipin/phosphatidylcholine vesicles was examined using proton NMR (Soussi *et al.*, 1990). In this fascinating study, cardiolipin vesicles were added to a mixture of cytochrome *c* and phosphatidylcholine vesicles. In control experiments, the authors showed that addition of phosphatidylcholine liposomes has only a minimal effect on the linewidth of hyperfine-shifted resonances of oxidized cytochrome *c*, indicating that there is no specific interaction with phosphatidylcholine. However, addition of cardiolipin vesicles to a mixture of cytochrome *c* and phosphatidylcholine vesicles causes a dramatic increase in linewidth. The authors state that this broadening is due to the binding of cytochrome *c* to large aggregates which occurs via fusion of cardiolipin and phosphatidylcholine vesicles. Upon titration of the cytochrome *c*–phosphatidylcholine mixture with cardiolipin, the resonance of the heme methyl 12^1 protons disappears, and a new resonance appears further downfield. The resonance at 11.4 ppm, belonging to proton 17^1 of the inner propionic acid side chain, disappears at the same rate as that of 12^1, and a new resonance appears downfield. The authors interpret these results in terms of a cardiolipin-induced conformational or dynamic change near the heme position 13, the outer propionic acid side chain. This is in agreement with the study of Brown and Wüthrich (1977) discussed earlier.

The interaction of horse cytochrome c specifically and individually trifluoro-acetylated at Lys-22, Lys-25, and (to a lesser specificity) Lys-86 with mitochondrial phospholipids was investigated using ^{19}F by Staudenmayer et al. (1976). The complex of the protein and the phospholipid was insoluble, obviating observation of individual protein proton resonances due to their restricted motion in the complex. However, the rotational motion of the trifluoromethyl groups appears to be much less restricted (not surprising because these groups are on the periphery of the molecule), and ^{19}F resonances were observed. Of course these resonances were significantly broadened (short T_2). It is difficult to derive concrete conclusions from these experiments, except to say that they indicate binding to phospholipids.

Spooner and Watts (1991a,b) have examined the interaction of horse cytochrome c with cardiolipin bilayers using ^2H and ^{31}P NMR both in solution and in the solid state. Deuterium was incorporated into the ε nitrogens of Lys residues, the C2 position of His residues, and into the backbone amides. Interaction of these proteins with the bilayer increases the mobility of the methyl groups and the His residues and increases the exchange rate of backbone amide deuterons. These results suggest that the interaction of the protein with the bilayer results in the reversible denaturation of the protein. Examination of the ^{31}P resonances of the lipid head-groups indicate that cytochrome c profoundly increases headgroup relaxation, independent of the oxidation state of the iron. Further studies by these authors (Spooner and Watts, 1992) using horse cytochrome c, wherein the Met-80 and Met-65 methyl groups had been labeled with ^{13}C, are consistent with the idea that cardiolipin partially unfolds the protein. These results suggest that interaction of cytochrome c with cardiolipin increases the amount of high-spin protein. In summary, cardiolipin appears to be acting like a detergent or denaturant (Spooner and Watts, 1991b), leading to conformational changes in cytochrome c which ultimately result in something like reversible denaturation. However, the bound form is not completely denatured, and Spooner and Watts (1992) compare it with the molten-globule state (III.N.3). If this situation is physiologically relevant, its implications for biological electron transfer are fascinating.

The interaction of apocytochrome c with sodium dodecyl sulfate micelles has been studied using the technique of photochemically induced dynamic nuclear polarization proton NMR (photo-CIDNP) by Snel et al. (1991). In this technique (Kaptein, 1982), a dye is added to the protein, and the solution is exposed to a laser flash. The photoexcited dye interacts reversibly with the aromatic residues of the protein. The back reaction causes nuclear-spin polarization in the protons of the aromatic residues. Collection of proton NMR data in the presence and absence of the flash yields the photo-CIDNP spectrum, which contains information about the aromatic residues of the protein. Snel et al. (1991) observed that photo-CIDNP enhancements are reduced in the micelles, indicating that the aromatic residues of apocytochrome c interact with micelles at the interface. Using conventional proton NMR, these authors also showed that interaction of the apoprotein with micelles increases the pK of histidine residues (Section III.M) from approximately 6 to 8.1. This observation is in accord with the idea that association of the apoprotein with the negatively charged surface of the vesicles should increase the pK of histidine.

J. Modified Cytochromes *c*

Although there is probably no substitute for the determination of the complete tertiary structure of a variant cytochrome *c*, this would seem unnecessary for variants whose conformation varies little from that of the wild-type protein. The sticking point is the determination of whether the change is local or global. This section describes methods that can help locate and describe changes in tertiary structure, whether they be small or large.

1. Chemical shifts

Before the advent of directed mutagenesis (Hutchinson *et al.*, 1978), naturally occurring variants of cytochrome *c* were used to assign resonances to protons from specific residues (*cf.* Moore and Williams, 1980f). The concept of using natural variants stems from the reasonable assumption that all cytochromes *c* have very similar folds, so that assignments can be made by comparing 1-D spectra from two cytochromes *c* that differ by one (or several) amino acids. An example of a single amino acid substitution is provided by horse and donkey cytochromes *c*, wherein a Thr of horse cytochrome *c* is changed to a Ser in the donkey protein. One-dimensional NMR difference spectroscopy suggested only very small structural perturbations between these two proteins (Moore and Williams, 1980f). A further example involves horse and cow cytochrome *c*. By comparing their COSY spectra, Wand and Englander (1985) were able to assign the spin system of Ser-47 in cow cytochrome *c*. Although this type of approach is useful, data from 1-D difference NMR and comparison of COSY spectra are difficult to interpret because very small changes in structure can cause significant changes in chemical shift. In other words, environmental effects, which cause the dispersion of chemical shift, although easy to observe, are usually difficult to interpret in a quantitative manner. Examination of chemical shifts yields information about the type of proton and something about its environment. Changes in chemical shift between the wild-type and variant proteins indicate some change in structure. To obtain detailed information, the resonances must be assigned to their cognate amino acid.

Wagner *et al.* (1983) have described a correlation between the dimensions of a hydrogen bond and the amide and α protons of the residues involved. The difference in chemical shift of main-chain amide protons between horse and yeast iso-1-cytochromes *c* has recently been examined by Gao *et al.* (1990b), who conclude that there is a subtle difference between the hydrogen bonding in the region 55 to 70 of these two proteins. Gao *et al.* (1989) also analyzed the difference in chemical shifts between horse and tuna cytochromes *c*, and again found a large deviation in the chemical shift of amide protons in this region. This region also exhibits a large variation in amino acid sequence among cytochromes *c* of various origins (Cutler *et al.*, 1987, and references therein; Moore and Pettigrew, 1990). It is interesting to note that the region between 60 and 70 also exhibits large discrepancies between calculated and observed pseudo-contact shifts (Section IV.F).

Single-site protein variants, whether generated by nature or *in vitro*, offer great promise for the study of cytochrome *c* by NMR because any changes in spectra must be due to the change caused by the substituted residue. An early NMR study of the unnatural cytochromes *c* semisynthetic M65Hse;Y67L, and M65Hse;Y74L (Hse, homoserine), was reported by Eley *et al.* (1982b). These authors concluded that substitution of residues 74 and 67 of horse cytochrome *c* with Leu does not greatly affect the structure. A small change in structure for cytochrome *c* chemically modified at specific Lys residues (Falk *et al.*, 1981; Falkand Ångström, 1983) is discussed in Section IV.D. Moench and Satterlee (1989) have studied the change in chemical shift of the heme and several amino acid resonances of *Saccharomyces cerevisiae* iso-1-cytochrome *c* either upon modification of Cys-102 with 5-5′-dithiobis(2-nitrobenzoate) or methyl methanethiosulfonate, or through intermolecular disulfide formation (Busse *et al.*, 1990). These authors conclude that these modifications cause a small conformational change and suggest that the oxidation state of the Cys-102 side chain plays a physiological role in yeast. However, yeast bearing the gene for iso-1-cytochrome *c* in which the codon for Cys-102 has been replaced by the codon for Thr exhibit no change in phenotype (Cutler *et al.*, 1987).

The advantage of variants constructed by *in vitro* mutagenesis is that changes can be placed at any position and any or all of the 19 variants at any position can be made. (Whether or not the variant proteins can be expressed in quantities sufficient for study by NMR is another matter.) Studies on variants at Phe-82, show that upon substitution of Phe with nonaromatic residues, the chemical shifts of the heme protons near position 82 move downfield, indicative of the loss of diamagnetic shift (Pielak *et al.*, 1988b). In a preliminary study of a cytochrome *c* variant in which Tyr-67 was replaced by Phe, there was no change in chemical shift or in the quantity or intensity of NOEs (but see Section IV.J.3) observed for the side-chain protons of Ala-15, Thr-19, and Phe-82 (Luntz *et al.*, 1989). This observation suggests that no change in structure occurred in that region of the protein, which is quite distant from the site of mutation. Wood *et al.* (1988) have reported the 1-D spectra of both the reduced and oxidized states of the variant of yeast iso-2-cytochrome *c* in which the evolutionarily invariant Pro-76 has been replaced by Gly. Although the 1-D spectra of the wild-type protein and the variant are essentially identical (but see IV.J.3), it is interesting that vestiges of the alkaline isomerized form were noted in the downfield region of the oxidized variant at pH 6, suggesting that the pK of the alkaline isomerization is lowered significantly (III.N.3).

Gooley and co-workers (1990, 1991a, 1992) have compared the P35A variant of *Rhodobacter capsulatus* cytochrome c_2 to the wild-type protein using heteronuclear ^{1}H–^{15}N 2-D NMR. The carbonyl of the strictly conserved Pro at position 35 is hydrogen-bonded to the π proton of heme ligand His-17. Comparison of the NOE spectra indicates that the conformations of the wild-type and variant proteins are indistinguishable. Their study did show, however, that the amide proton of Gly-34, and the π proton of His-17, which are involved in separate hydrogen bonds in the wild-type protein, exchange approximately 100 times more rapidly in the variant protein. This would seem to indicate destabilization of these hydrogen bonds. $^3J_{\alpha NH}$, ^{1}H, and ^{15}N chemical shift data argue, however, that these hydrogen bonds are intact in the variant. These results can be reconciled if the replacement of pro by ala causes a solvent-accessible cavity or if the mutation causes a local increase in flexibility, which, by reducing steric hindrance, increases solvent accessibility. These hypotheses are supported, although not distinguished, by the observation that the ring flip rates (Section IV.A) of the nearby residues Phe-51 and Tyr-53 are increased in the variant protein. These authors also examined the amide proton exchange rates for the P35A and Y75F variant proteins (Gooley *et al.*, 1992; IV.J.4).

Other studies of variant cytochromes c in which changes in chemical shifts are discussed include Thurgood *et al.* (1991a,b), Greene *et al.* (1993), and Auld *et al.* (1993). Thurgood *et al.* (1991a) examined the changes in diamagnetic shift experienced by Val-57 (Sections IV.A, IV.I.6) upon substitution of Phe-48 and Trp-59 with Phe and show that for the β-proton and one of the methyl groups, the observed changes are consistent with the expected change in ring-current field (Perkins, 1980, 1982; III.B.I). The work of Greene *et al.* is discussed in Section IV.J.3. The report of Auld *et al.* (1993) deals with comparisons between changes in chemical shift and changes in paramagnetic shift and is discussed in the next section.

In summary, changes in chemical shift between variants and the wild-type protein are indicative of structural or dynamic changes, but interpretation of these changes is difficult. If changes in chemical shift are restricted to amide (or other exchangeable protons), this is a good indication that there has been (at least) a change in hydrogen bonding. To pinpoint structural alterations, changes in NOEs, paramagnetic shifts, and amide proton exchange rates should also be examined. To determine dynamic alterations, the flip rate of aromatic protons, changes in the exchange rate of amide protons, and the relaxation of ^{13}C and ^{15}N (Section IV.K) should be examined.

2. Paramagnetic shifts

It has been shown using the comparison of calculated pseudo-contact and observed paramagnetic shift that there is a very good fit between the solution and crystal structure of cytochrome c (Williams *et al.*, 1985b; Feng *et al.*, 1990a; Section IV.F). Thus, any change in paramagnetic shift between the wild-type protein and a variant will represent a change in structure. Complications arise because paramagnetic shifts alone do not indicate which oxidation state is affected. It is possible to localize the change to one oxidation state by looking at the difference in chemical shift between the wild-type and the variant protein in the same oxidation state. Analysis of differences in chemical shifts and NOEs in both the reduced and oxidized proteins, as well as differences in paramagnetic shift between the C102T and F82S;C102T variants of *Saccharomyces cerevisiae* iso-1-cytochrome c, have been reported by Gao *et al.* (1991b). There is evidence for small conformational differences in both oxidation states of the double variant near position 82. Differences in structure are more evident in the oxidized forms of the variants. These differences extend to distant parts of the protein. It appears that the oxidized double variant has undergone a small rearrangement of several regions of the protein which are linked by a hydrogen-bond network. It was shown that the rearrangement involves hydrogen bonds associated with the heme propionates and associated water molecules. The conclusions from NMR data are similar to those from examination of the crystal structures of the reduced forms of wild-type protein and the F82S variant (Louie *et al.*, 1988b).

Auld *et al.* (1993) examined the effect of changing Phe-10 in *Saccharomyces cerevisiae* iso-1-cytochrome c to methionine, and the methionine analog, *S*-methyl cysteine (Cys$_{SMe}$). Phe-10 is involved in a weakly polar interaction (Section IV.A; Burley and Petsko, 1988) with Tyr-97. The authors examined the structure and stability of the two double variants to determine if the aromatic–aromatic interaction in the wild-type protein is replaced by a weakly polar sulfur–aromatic interaction in the double variants. Because reduction of the Cys$_{SMe}$ side chain is concomitant with reduction of the heme iron, this derivative could only be examined in the oxidized state.

Comparing chemical shifts of identical protons, it was noted that approximately one-third of the differences were outside the error of the determination of two chemical shifts. Most of the protons that gave rise to the anomalous chemical shift differences are located near the site of the mutation. This was true for comparisons of both variants to the wild-type protein. When the F10M;C102T and F10C$_{SMe}$; C102T variants were compared directly, it was observed that the two variants are more similar to each other than they are to the wild-type protein. Thus, it appears that the substitutions change the wild-type structure in the same way.

To locate these changes, the paramagnetic shifts of the F10M;C102T and the C102T variants were compared. Nearly 80% of the protons whose chemical shift differences are anomalous do not experience anomalous changes in paramagnetic shift. This indicates that many of the differences in chemical shift are caused by changes in diamagnetic fields. Most of the protons that exhibit anomalous differences in paramagnetic shift are located near the site of the mutation. However, changes were also observed in the region of the protein that interacts with the heme propionates. Because similar observations were made for the F82S;C102T and N52I;C102T variants, it appears that the region of the protein in the vicinity of the propionates is especially malleable. In fact, this malleability was predicted by Chothia and Lesk (1985). Analysis of NOESY data for both proteins indicates that both sulfur-containing side chains are in position to form a weakly polar interaction with Tyr-97.

Several observations led the authors to conclude that the side chain of Tyr-97 is in the fast-flipping regime in the F10M;C102T and F10C$_{SMe}$;C102T variants. First, only a single strong COSY cross peak is observed between the δ and ε proton resonances of Tyr-97 in both oxidation states of the F10M;C102T variant and in the oxidized form of the F10C$_{SMe}$;C102T variant. Second, strong NOEs were observed between both the 2,6 and 3,5 protons of Tyr-97 and its β protons in both oxidation states of the F10M;C102T variant and in the oxidized form of F10C$_{SMe}$;C102T variant.

The F10M;C102T and F10C$_{SMe}$;C102T variants are 2–3 kcal mol^{-1} less stable than iso-1-cytochrome c at 300 K. Comparison of the stabilities of the F10M;C102T and F10C$_{SMe}$;C102T variants allowed evaluation of the potential weakly polar interaction between the additional sulfur atom of the F10C$_{SMe}$;C102T variant and the aromatic moiety of Tyr-97. The F10C$_{SMe}$;C102T variant is 0.7 ± 0.3 kcal mol^{-1} more stable than the F10M;C102T protein. The increased stability is explained by the difference in hydrophobicity of the sulfur-containing side chains. The authors concluded that any weakly polar interaction between the additional sulfur and the aromatic ring is too weak to detect or is masked by destabilizing contributions to the free energy of denaturation. Examination of paramagnetic shifts for other variant proteins should prove a powerful tool for locating and describing structural changes.

3. NOEs

A feature of structures determined by distance geometry or molecular dynamics from NOE data is their low resolution compared to those derived from x-ray diffraction data. Deviations between pairs of distance geometry structures may be 2 Å for backbone atoms. Furthermore, local features may be poorly defined, and large root-mean-square deviations between dihedral angles in pairs of NMR structures are often observed. One reason for the large deviations is the lack of stereospecific assignments for prochiral protons. The situation may be better for solution structures of cytochromes *c*, because stereospecific assignments are available from paramagnetic shift data (Feng *et al.*, 1990a). However, given an x-ray structure of the wild-type protein, NMR will be expected to provide an accurate definition of structural change due to the substitution of amino acid residues.

Although knowledge of the differences in chemical shift between the wild-type and variant protein is essential, changes in chemical shift, except in exceptional circumstances (such as amide protons; Wagner *et al.*, 1983; Gao *et al.*, 1990b), are difficult to interpret in terms of precise structural changes. However, comparison of data from COSY spectra between wild-type and variant proteins yield a chemical shift map by correlating changes in chemical shift with the spin system of the amino acid whose resonances experience the shift. This provides knowledge of chemical shift changes for resonances from corresponding residues between the two proteins, which may then be used together with interresidue NOEs from protons on the corresponding residues in the wild-type and variant proteins to identify structural changes. A combination of overlayed and difference 2-D NOESY spectra can therefore be quickly analyzed to indicate structural perturbations.

The preceding ideas have been used to examine structural changes occurring between the C102T variant of yeast iso-1-cytochrome *c* and the F82Y;C102T double variant (Pielak *et al.*, 1988b; Greene *et al.*, 1993). It appears that structural changes are restricted to portions of the protein near position 82, and that Leu-85 has been rearranged because of the insertion of a phenolic hydroxyl group on going from Phe-82 to Tyr-82. Greene *et al.* (1993) observed resonances from the two η protons of Arg-13 in the F82Y;C102T variant. These protons are geminally coupled and exhibit NOEs to the hydroxyl proton of Tyr-82. Geminal coupling of Arg η protons is not observed in the C102T protein and does not appear to have been reported for any protein. These observations are interpreted in terms of a novel hydrogen bond between the side chain of Arg-13 and the hydroxyl group of Tyr-82, which was predicted from inspection of the crystal structure of the reduced wild-type protein (Louie *et al.*, 1988a).

One-dimensional NOE studies were performed on the P76G variant of yeast iso-2-ferrocytochrome *c* (Wood *et al.*, 1988). NOEs resulting from irradiation of proton resonances between -1 and -4 ppm (Met-80 protons) and from irradiation of the meso protons of the heme (5, 10, 15, 20; Figure 5.1) in both the wild-type protein and the variant were compared. The only difference noted involved loss of a NOE between the Met-80 β' proton and the heme 15 meso proton (Section IV.J.1). As Wood *et al.* (1988) suggest, this loss could be the result of an increase in the distance between these two protons resulting from a slight reorganization of the

Met-80 side chain, or could result from changes in spin diffusion between the wild-type protein and the variant.

In their preliminary study of the NOESY spectrum of the Y67F variant of rat cytochrome c, Luntz et al. (1989; Section IV.F) observed changes in the NOE between one Leu-68 methyl group and the 2^1 and 12^1 heme methyl groups. They suggest that in the variant protein, the side chain of Leu-68 moves in such a way that it occupies the position vacated by an internal water molecule (IV.F). However, it now appears that this buried water molecule is present in the Y67F variant (Frauenhoff and Scott, 1992). Furthermore, although this substitution increased the stability of the iron–Met-80 sulfur bond, it decreased the free energy of denaturation relative to the wild-type protein (Schejter et al., 1992). Similar discrepancies between melting temperatures and free energies of denaturation for variant cytochromes c have been reported by Betz and Pielak (1992). In their examination of the F82S;C102T variant of iso-1-cytochrome c, Gao et al. (1991b) concluded that NOEs may not be as useful in identifying conformational changes as changes in pseudo-contact shift (III.B.2.b; IV.J.2). Finally, Auld et al. (1993; IV.J.2), in their examination of the F10M;C102T and F10C$_{SMe}$;C102T variants, showed that the sulfur-containing side chains at position 10 are close to the side chain of Tyr-97.

4. Amide proton exchange rates

Gooley et al. (1992) and MacKenzie et al. (1992b) have examined amide proton exchange for the Y75F and P35A protein variants of *Rhodobacter capsulatus* ferrocytochrome c_2. The Y75F mutation decreases the formal potential by 59 mV. For helix III, which spans residues 79 to 88, in the Y75F variant, increases in exchange rates for the reduced form of the variant mimic the changes observed when the reduced and oxidized forms of the wild-type protein are compared (Gooley et al., 1991b). Thus, the unfolding of this region appears to influence the formal potential. Furthermore, the increases in this region upon substituting Tyr by Phe are uniform, strongly suggesting that this helix is involved in a cooperative unfolding unit. Changes in exchange rate for the P35A variant are more localized, primarily involving residues near the site of the mutation. Rationalization of changes in exchange rates in terms of changes in formal potential is problematic in the case of the P35A variant, because there is only an 8 mV decrease in formal potential. As the authors point out, this observation is better rationalized in terms of changes in the hydrophobic core or creation of a cavity which alters solvent access in the native state (Section IV.J.1). The authors then interpret the differences in the free energy for local unfolding determined from exchange experiments with differences in the free energy of denaturation from chemical denaturation. It is unlikely, however, that the free energies of denaturation from chemical denaturation experiments on the reduced proteins (Caffrey et al., 1991) are valid, because the reduced form of the denatured protein is expected to possess a large and negative formal potential, making denaturation irreversible (Bixler et al., 1992; Hilgen-Willis et al., 1993) except under strictly anaerobic conditions (Cohen and Pielak, 1995). Nevertheless, the conclusion that substitutions which adversely affect the reduced protein relative to the oxidized protein will decrease the formal potential of the protein variant is supported by the amide proton exchange data, provided that substitutions do not affect the formal potential of the denatured protein. This line of reasoning may also help explain the decreased formal potentials observed for variants of yeast iso-1-cytochrome c upon substitution of Trp-59 and Tyr-48 (Section IV.G; Davies *et al.*, 1993). Furthermore, if the denatured state is unchanged, the reverse of this argument may explain the increase in formal potential for the variants such as N52I;C102T of yeast iso-1-cytochrome c (Hickey *et al.*, 1991a). Such explanation should be borne in mind when interpreting formal potentials in terms of the polarity arguments of Kassner (1972), because changes in polarity upon substituting specific side chains may not alter the potential as expected if the substitutions affect local or global unfolding.

K. Heteronuclear Relaxation

The relaxation of nuclear spins contains important information about protein dynamics (Lipari and Szabo, 1982). However, such measurements for protons are not very useful because there are too many nearby protons that influence the relaxation of the proton in question. The spin one-half nuclei ^{13}C and ^{15}N offer the advantage that dipolar interactions of *directly attached protons* dominate the relaxation of these nuclei. However, ^{13}C and ^{15}N have a low receptivity relative to the proton (Section III.K.2). Therefore, enrichment is desirable, if not essential. Clore *et al.* (1990) have presented an excellent example of the use of the ideas of Lipari and Szabo (1982) to examine the backbone dynamics of interleukin-1β via ^{15}N-proton NMR. As discussed in Section III.K.2, extensive ^{13}C assignments at natural abundance for horse ferrocytochrome c (Gao *et al.*, 1990) and ^{15}N assignments for *Pseudomonas aeruginosa* ferrocytochrome c-551 at natural abundance (Timkovich, 1990) are available, but whether the required high-quality T_1 and T_2 and heteronuclear NOE data can be obtained without isotopic enrichment is still questionable. *Rhodobacter capsulatus* cytochrome c_2, whose backbone ^{15}N resonances have been completely assigned using an enriched sample (Gooley *et al.*, 1990, 1991b, 1992), can be subjected to these types of experiments. Studies comparing the backbone dynamics of ferro- and ferricytochromes c are eagerly awaited.

V. Acknowledgments

D.S.A. was supported by a Department of Education Fellowship. S.F.B. was supported by a Department of Education Fellowship and a North Carolina Board of Governors' Fellowship in Science and Technology. G.J.P. thanks Dr. Geoffrey R. Moore and Professors Michael Smith, F. R. S., A. Grant Mauk, and Robert J. P. Williams, F. R. S. for moral support and encouragement, and Bryan Fine for ferreting out some of the references. Financial support for this work was provided by the National Institutes of Health (GM42501) and by donors of the Petroleum Research Fund, administered by the American Chemical Society (21597-G).

VI. References

Allerhand, A., Childers, R. F., & Oldfield, E. (1973) *Annu. N.Y. Acad. Sci.* **222**, 764–777.

Andersson, T., Thulin, E., & Forsén, S. (1979) *Biochemistry* **18**, 2487–2493.

Ångström, J., Moore, G. R., & Williams, R. J. P. (1982) *Biochim. Biophys. Acta* **703**, 87–94.

Arean, C. O., Moore, G. R., Williams, G., & Williams, R. J. P. (1988) *Eur. J. Biochem.* **173**, 607–615.

Augspurger, J., Pearson, J. G., Oldfield, E., Dykstra, C. E., Park, K. D., & Schwartz, D. (1992) *J. Magn. Res.* **100**, 342–357.

Auld, D. S., & Pielak, G. J. (1991) *Biochemistry* **30**, 8684–8690.

Auld, D. S., Young, G. B., Saunders, A. J., Doyle, D. F., Betz, S. F., & Pielak, G. J. (1993) *Prot. Sci.* **2**, 2187–2197.

Babul, J., & Stellwagen, E. (1972) *Biochemistry* **11**, 1195–1200.

Bai, Y., Milne, J. S., Mayne, L., & Englander, S. W. (1993) *Proteins* **17**, 75–86.

Baltzer, L. (1987) *J. Am. Chem. Soc.* **109**, 3479–3481.

Barlow, G. H., & Margoliash, E. (1966) *J. Biol. Chem.* **241**, 1473–1477.

Bax, A. (1989) *Annu. Rev. Biochem.* **58**, 223–256.

Bax, A., & Davis, D. G. (1985) *J. Magn. Res.* **63**, 207–213.

Behere, D. V., Ales, D. C., & Goff, H. M. (1986) *Biochim. Biophys. Acta* **871**, 285–292.

Berghuis, A. M., & Brayer, G. D. (1992) *J. Mol. Biol.* **223**, 959–976.

Bernstein, F. C., Koetzle, T. F., Williams, G. J. B., Meyer, E. F., Brice, M. D., Rodgers, J. R., Kennard, O., Shimanouchi, T., & Tasumi, M. (1977) *J. Mol. Biol.* **112**, 535–542.

Bertini, I., Banci, L., & Luchinat, C. (1989) *Methods Enzymol.* **177**, 246–263.

Betz, S. F., & Pielak, G. J. (1992) *Biochemistry* **31**, 12337–12344.

Bixler, J., Bakker, G., & McLendon, G. (1992) *J. Am. Chem. Soc.* **114**, 6938–6939.

Blackledge, M. J., Medvedeva, S., Poncin, M., Guerlesquin, F., Bruschi, M., & Marion, D. (1995) *J. Mol. Biol.* **245**, 661–681.

Bonnett, R. (1978) *The Porphyrins* (Dolphin, D., ed.) **1**, 1–27.

Boswell, A. P., McClune, G. J., Moore, G. R., Williams, R. J. P., Pettigrew, G. W., Inubishi, T., Yonetani, T., & Harris, D. E. (1980a) *Biochem. Soc. Trans.* **8**, 637–638.

Boswell, A. P., Moore, G. R., Williams, R. J. P., Chien, J. C. W., & Dickinson, L. C. (1980b) *J. Inorg. Biochem.* **13**, 347–352.

Boswell, A. P., Eley, C. G. S., Moore, G. R., Robinson, M. N., Williams, G., Williams, R. J. P., Neupert, W. J., & Hennig, B. (1982) *Eur. J. Biochem.* **124**, 289–294.

Boswell, A. P., Moore, G. R., Williams, R. J. P., Harris, D. E., Wallace, C. J. A., Bocieck, S., & Welti, D. (1983) *Biochem. J.* **213**, 679–686.

Boyd, J., Moore, G. R., & Williams, G. (1984) *J. Magn. Res.* **58**, 511–516.

Braunschweiler, L., & R. R. Ernst (1983) *J. Magn. Res.* **53**, 521–528.

Brautigan, D. L., Ferguson-Miller, S., & Margoliash, E. (1978) *Methods Enzymol.* **53**, 128–164.

Brems, D. N., Cass, R., & Stellwagen, E. (1982) *Biochemistry* **21**, 1488–1493.

Brown, L. R., & Wüthrich, K. (1977) *Biochim. Biophys. Acta* **468**, 389–410.

Bundi, A., & Wüthrich, K. (1979) *Biopolymers* **18**, 285–297.

Burch, A. M., Rigby, S. E. J., Funk, W. D., MacGillvray, R. T. A., Mauk, M. R., Mauk, A. G., & Moore, G. R. (1990) *Science* **247**, 831–833.

Burley, S. K., & Petsko, G. A. (1988) *Adv. Protein Chem.* **39**, 125–189.

Burns, P. D., & La Mar, G. N. (1979) *J. Am. Chem. Soc.* **101**, 5844–5846.

Burns, P. D., & La Mar, G. N. (1981) *J. Biol. Chem.* **256**, 4934–4939.

Bushnell, G. W., Louie, G. V., & Brayer, G. D. (1990) *J. Mol. Biol.* **214**, 585–595.

Busse, S. C., Moench, S. J., & Satterlee, J. D. (1990) *Biophys. J.* **58**, 45–51.

Caffrey, M. S., Daldal, F., Holden, H. M., & Cusanovich, M. A. (1991) *Biochemistry* **30**, 4119–4125.

Cai, M., & Timkovich, R. (1991) *Biochem. Biophys. Res. Commun.* **178**, 309–314.

Cai, M., & Timkovich, R. (1992) *FEBS Lett.* **311**, 213–216.

Cai, M., & Timkovich, R. (1993) *Biochemistry* **32**, 11516–11523.

Cai, M., Bradford, E. G., & Timkovich, R. (1992) *Biochemistry* **31**, 8603–8612.

Campbell, I. D., Dobson, C. M., & Williams, R. J. P. (1975) *Proc. R. Soc. Lond. Ser. B* **189**, 503–509.

Campbell, I. D., Dobson, C. M., Moore, G. R., Perkins, S. J., & Williams, R. J. P. (1976) *FEBS Lett.* **70**, 96–100.

Chao, Y.-Y. H., Bersohn, R., & Aisen, P. (1979) *Biochemistry 18*, 774–779.

Chapman, S. K., White, S. A., & Reid, G. A. (1991) *Adv. Inorg. Chem. 36*, 257–301.

Chau, M.-H., Cai, M. L., & Timkovich, R. (1990) *Biochemistry 29*, 5076–5087.

Chothia, C., & Lesk, A. M. (1985) *J. Mol. Biol. 182*, 151–158.

Chung, J., Lee, H. C., & Oldfield, E. (1990) *J. Magn. Res. 90*, 148–157.

Clayden, N. J., Moore, G. R., & Williams, G. (1987) *Metal Ions in Biological Systems 21*, 187–227.

Clements, J. M., O'Connell, L. I., Tsunasawa, S., & Sherman, F. (1989) *Gene 83*, 1–14.

Clore, G. M., Driscoll, P. C., Wingfield, P. T., & Gronenborn, A. M. (1990) *Biochemistry 29*, 7387–7401.

Cohen, D. S., & Pielak, G. J. (1994) *Prot. Sci. 3*, 1253–1260.

Cohen, D. S., & Pielak, G. J. (1995) *J. Am. Chem. Soc. 117*, 1675–1677.

Cohen, J. S., & Hayes, M. H. (1974) *J. Biol. Chem. 249*, 5472–5477.

Cohen, J. S., Fisher, W. R., & Schechter, A. N. (1974) *J. Biol. Chem. 249*, 1113–1118.

Commission on the Nomenclature of Biological Chemistry (1960) *J. Amer. Chem. Soc. 82*, 5545–5584, see page 5582.

Concar, D. W., Whitford, D., Pielak, G. J., & Williams, R. J. P. (1991a) *J. Amer. Chem. Soc. 113*, 2401–2406.

Concar, D. W., Whitford, D., & Williams, R. J. P. (1991b) *Eur. J. Biochem. 199*, 553–560.

Concar, D. W., Whitford, D., & Williams, R. J. P. (1991c) *Eur. J. Biochem. 199*, 569–574.

Connelly, G. P., Bai, Y., Jeng, M.-F., & Englander, S. W. (1993) *Proteins 17*, 87–92.

Cookson, D. J., Moore, G. R., Pitt, R. C., Williams, R. J. P., Campbell, I. D., Ambler, R. P., Bruschi, M., & Le Gall, J. (1978) *Eur. J. Biochem. 83*, 261–275.

Creighton, T. E. (1994) *Structural Biology 1*, 135–138.

Cross, K. J., & Wright, P. E. (1985) *J. Magn. Res. 64*, 220–231.

Cutler, R. L., Pielak, G. J., Mauk, A. G., & Smith, M. (1987) *Protein Eng. 1*, 95–99.

Cutler, R. L., Davies, A. M., Creighton, S., Warshel, A., Moore, G. R., Smith, M., & Mauk, A. G. (1989) *Biochemistry 28*, 3188–3197.

Dalgarno, D. C., Levine, B. A., & Williams, R. J. P. (1983) *Bioscience Reports 3*, 443–452.

Das, G., Hickey, D. R., McLendon, D., McLendon, G., & Sherman, F. (1989) *Proc. Natl. Acad. Sci. USA 86*, 496–499.

Davies, A. M., Guillemette, J. G., Smith, M., Greenwood, C., Thurgood, A. G. P., Mauk, A. G., & Moore, G. R. (1993) *Biochemistry 32*, 5431–5435.

Davis, L. A., Schejter, A., & Hess, G. P. (1974) *J. Biol. Chem. 249*, 2624–2632.

Derome, A. E. (1987) *Modern NMR Techniques for Chemistry Research*, Pergamon, Oxford.

Detlefsen, D. J., Thanabal, V., Pecoraro, V. L., & Wagner, G. (1990) *Biochemistry 29*, 9377–9386.

Detlefsen, D. J., Thanabal, V., Pecoraro, V. L., & Wagner, G. (1991) *Biochemistry 30*, 9040–9046.

Devaux, P. F., Hoatson, G. L., Favre, E., Fellmann, P., Farren, B., MacKay, A. L., & Bloom, M. (1986) *Biochemistry 25*, 3804–3812.

Dickerson, R. E., Takano, T., Eisenberg, D., Kallai, O. B., Samson, L., Cooper, A., & Margoliash, E. (1971) *J. Biol. Chem. 246*, 1511–1535.

Dill, K. A., & Shortle, D. (1991) *Annu. Rev. Biochem. 60*, 795–825.

Dixon, D. W., & Hong, X. (1990) *Adv. in Chem. Ser. 226*, 161–179.

Dixon, D. W., Hong, X., & Woehler, S. E. (1989) *Biophys. J. 56*, 339–351.

Dobson, C. M., Moore, G. R., & Williams, R. J. P. (1975) *FEBS Lett. 51*, 60–65.

Eakin, R. T., Morgan, L. O., & Matwiyoff, N. A. (1975a) *Biochem. J 152*, 529–535.

Eakin, R. T., Morgan, L. O., & Matwiyoff, N. A. (1975b) *Biochemistry 14*, 4538–4543.

Eley, C. G. S., & Moore, G. R. (1983) *Biochem. J. 215*, 11–21.

Eley, C. G. S., Moore, G. R., Williams, G., & Williams, R. J. P. (1982a) *Eur. J. Biochem. 124*, 295–303.

Eley, C. G. S., Moore, G. R., Williams, R. J. P., Neupert, W., Boon, P. J., Brinkhof, H. H. K., Nivard, R. J. F., & Tesser, G. I. (1982b) *Biochem. J. 205*, 153–165.

Elöve, G. A., & Roder, H. (1991) in *Protein Refolding* (Georgiou, G., & De Bernardez-Clark, E., eds.), ACS Symposium Series *470*, 50–63.

Elöve *et al.* (1991) p. 83.

Elöve, G. A., Chaffotte, A. F., Roder, H., & Goldberg, M. E. (1992) *Biochemistry 31*, 6876–6883.

Elöve, G. A., Bhuyan, A. K., & Roder, H. (1994) *Biochemistry 33*, 6925–6935.

Eltis, L. D., Herbert, R. G., Barker, P. D., Mauk, A. G., & Northrup, S. H. (1991) *Biochemistry 30*, 3663–3674.

Englander, S. W. (1993) *Science 262*, 848–849.

Englander, S. W., & Kallenbach, N. R. (1984) *Q. Rev. Biophys. 16*, 521–655.

Englander, S. W., & Mayne L. (1992) *Annu. Rev. Biophys. Biomol. Struct.* **21**, 243–265.

Ernst, R. R. (1992) *Angew. Chem. Intl. Ed. Engl.* **31**, 805–823.

Ernst, R. R., Bodenhausen, G., & Wokaun, A. (1987) *Principles of Nuclear Magnetic Resonance in One and Two Dimensions*, Oxford University Press, Oxford.

Everest, A. M., Wallin, S. A., Stemp, E. D. A., Nocek, J. M., Mauk, A. G., & Hoffman, B. M. (1991) *J. Am. Chem. Soc.* **113**, 4337–4338.

Falk, K.-E., & Ångström, J. (1983) *Biochim. Biophys. Acta* **722**, 291–296.

Falk, K.-E., Jovall, P. A., & Ångström, J. (1981) *Biochem. J.* **193**, 1021–1024.

Feng, Y., & Englander, S. W. (1990) *Biochemistry* **29**, 3505–3509.

Feng, Y., & Roder, H. (1988) *J. Magn. Res.* **78**, 597–602.

Feng, Y., Roder, H., Englander, S. W., Wand, A. J., & Di Stefano, D. L. (1989) *Biochemistry* **28**, 195–203.

Feng, Y., Roder, H., & Englander, S. W. (1990a) *Biochemistry* **29**, 3505–3509.

Feng, Y., Roder, H., & Englander, S. W. (1990b) *Biophys. J.* **57**, 15–22.

Feng, Y., Wand, A. J., Roder, H., & Englander, S. W. (1991) *Biophys. J.* **59**, 323–328.

Ferrer, J. C., Guillemette, J. G., Bogumil, R., Inglis, S. C., Smith, M., & Mauk, A. G. (1993) *J. Am. Chem. Soc.* **115**, 7507–7508.

Forsèn, S., & Hoffman, R. A. (1963) *J. Chem. Phys.* **39**, 2892–2901.

Frauenhoff, M. M., & Scott, R. A. (1992) *Prot. Struct. Func. Genet.* **14**, 202–212.

Fredericks, Z. L., & Pielak, G. J. (1993) *Biochemistry* **32**, 929–936.

Freeman, R. (1987) *A Handbook of Nuclear Magnetic Resonance*, Wiley, New York.

Gao, Y., Lee, A. D. J., Williams, R. J. P., & Williams, G. (1989) *Eur. J. Biochem.* **182**, 57–65.

Gao, Y., Boyd, J., & Williams, R. J. P. (1990a) *Eur. J. Biochem.* **194**, 355–365.

Gao, Y., Boyd, J., Williams, R. J. P., & Pielak, G. J. (1990b) *Biochemistry* **29**, 6994–7003.

Gao, Y., Boyd, J., Pielak, G. J., & Williams, R. J. P. (1991a) *Biochemistry* **30**, 1928–1934.

Gao, Y., Boyd, J., Pielak, G. J., & Williams, R. J. P. (1991b) *Biochemistry* **30**, 7033–7040.

Gao, Y., Veitch, N. C., & Williams, R. J. P. (1991c) *J. Biomol. NMR* **1**, 457–471.

Gao, Y., McLendon, G., Pielak, G. J., & Williams, R. J. P. (1992) *Eur J. Biochem.* **204**, 337–352.

Garcia, L. L. (1991) Ph.D. Dissertation, University of North Carolina at Chapel Hill.

Garcia, L. L., Fredericks, Z., Sorrell, T. N., & Pielak, G. J. (1992) *New J. Chem.* **16**, 629–632.

Glasoe, P. K., & Long, F. A. (1960) *J. Phys. Chem.* **64**, 188–190.

Gooley, P. R., & MacKenzie, N. E. (1990) *FEBS Lett.* **260**, 225–228.

Gooley, P. R., Caffrey, M. S., Cusanovich, M. A., & MacKenzie, N. E. (1990) *Biochemistry* **29**, 2278–2290.

Gooley, P. R., Caffrey, M. S., Cusanovich, M. A., & MacKenzie, N. E. (1991a) *Eur. J. Biochem.* **196**, 653–661.

Gooley, P. R., Zhao, D., & MacKenzie, N. E. (1991b) *J. Biomol. NMR* **1**, 145–154.

Gooley, P. R., Caffrey, M. S., Cusanovich, M. A., & MacKenzie, N. E. (1992) *Biochemistry* **31**, 443–450.

Gopal, D., Wilson, G. S., Earl, R. A., & Cusanovich, M. A. (1988) *J. Biol. Chem.* **263**, 11652–11656.

Goto, Y., Calciano, L. J., & Fink, A. L. (1990) *Proc. Natl Acad. Sci. USA* **87**, 573–577.

Greene, R. M., Betz, S. F., Hilgen-Willis, S. E., Auld, D. S., Fencl, J. B., & Pielak, G. J. (1993) *J. Inorg. Biochem.* **51**, 663–676.

Gupta, R. K., & Koenig, S. H. (1971) *Biochem. Biophys. Res. Commun.* **45**, 1134–1143.

Gupta, R. K., & Yonetani, T. (1973) *Biochim. Biophys. Acta* **292**, 502–508.

Gupta, R. K., Koenig, S. H., & Redfield, A. G. (1972) *J. Magn. Res.* **7**, 66–73.

Hahm, S., Durham, B., & Millet, F. (1992) *Biochemistry* **31**, 3472–3477.

Hartshorn, R. T., & Moore, G. R. (1989) *Biochem. J.* **258**, 595–598.

Hartshorn, R. T., Mauk, A. G., Mauk, M. R., & Moore, G. R. (1987) *FEBS Lett.* **213**, 391–395.

Hawkes, R., Grutter, M. G., & Schellman, J. (1984) *J. Mol. Biol.* **175**, 195–212.

Hazzard, J. T., & Tollin, G. (1985) *Biochem. Biophys. Res. Commun.* **130**, 1281–1286.

Hazzard, J. T., McLendon, G., Cusanovich, M. A., Das, G., Sherman, F., & Tollin, G. (1988) *Biochemistry* **27**, 4445–4451.

Hickey, D. R., Berghuis, A. M., Lafond, G., Jaeger, J. A., Cardillo, T. S., McLendon, D., Das, G., Sherman, F., Brayer, G. D., & McLendon, G. (1991a) *J. Biol. Chem.* **266**, 11686–11694.

Hickey, D. R., Jayaraman, K., Goodhue, C. T., Shah, J., Fingar, S. A., Clements, J. M., Hosokawa, Y., Tsunasawa, S., & Sherman, F. (1991b) *Gene* **105**, 73–81.

Hilgen, S. E., & Pielak, G. J. (1991) *Protein Eng.* **4**, 575–578.

Hilgen-Willis, S. Bowden, E. F., & Pielak, G. J. (1993) *J. Inorg. Biochem.* **51**, 649–653.

Homans, S. W. (1989) *A Dictionary of Concepts in NMR* (revised paperback edition), Oxford University Press, Oxford.

Hong, X., & Dixon, D. W. (1989) *FEBS Lett.* **246**, 105–108.

Hore, P. J. (1989) *Methods Enzymol.* **176**, 64–77.

Horrocks, W. De W., Jr. (1973) in *NMR of Paramagnetic Molecules: Principles and Applications* (La Mar, G. N., Horrocks, W. De W., Jr., & Holm, R. H., eds.), pp. 127–177, Academic Press, New York.

Hutchinson, C. A. III, Phillips, S., Edgell, M. H., Gillam, S., Jahnke, P., & Smith, M. (1978) *J. Biol. Chem.* **258**, 6551–6560.

Hvidt A., & Nielsen, S. O. (1966) *Adv. Protein. Chem.* **21**, 287–386.

Inglis, S. C., Guillemette, J. G., Johnson, J. A., & Smith, M. (1991) *Protein Eng.* **4**, 569–574.

IUPAC-IUB Commission on Biochemical Nomenclature (1970) *J. Biol. Chem.* **245**, 6489–6497.

IUPAC-IUB Joint Commission on Biochemical Nomenclature (1985) *J. Biol. Chem.* **260**, 14–41.

Jeng, M.-F., Englander, S. W., Elöve, G. A., Wand, A. J., & Roder, H. (1990) *Biochemistry* **29**, 10433–10437.

Jeng, M.-F., Englander, S. W., Pardue, K., Rogalsky, J. S., & McLendon, G. (1994) *Structural Biology* **1**, 234–238.

Jennings, P. A., & Wright, P. E. (1993) *Science* **262**, 892–896.

Jesson, J. P. (1973) in *NMR of Paramagnetic Molecules: Principles and Applications* (La Mar, G. N., Horrocks, W. De W., Jr., & Holm, R. H., eds.), pp. 1–52, Academic Press, New York.

Johnson, C. E., & Bovey, F. A. (1958) *J. Chem. Phys.* **29**, 1012–1014.

Kaptein, R. (1982) in *Biological Magnetic Resonance* (Berliner, L. J., & Reuben, J., eds.), Vol. 4, pp. 145–188, Plenum, New York.

Kassner, R. J. (1972) *Proc. Natl. Acad. Sci. U.S.A.* **69**, 2263–2267.

Katz, J. J., & Crespi, H. L. (1972) *Pure and Applied Chem.* **32**, 221–250.

Kayushin, L. P., Sibel'dina, L. A., Lazareva, A. V., Okon, M. S., & Kutyshenko, V. P. (1975) *Studia Biophysica* **51**, 221–227.

Keller, R. M., & Wüthrich, K. (1978a) *Biochem. Biophys. Res. Commun.* **83**, 1132–1139.

Keller, R. M., & Wüthrich, K. (1978b) *Biochim. Biophys. Acta* **533**, 195–208.

Keller, R. M., & Wüthrich, K. (1981) *Biochim. Biophys. Acta* **668**, 307–320.

Keller, R. M., Wüthrich, K., & Pecht, I. (1976) *FEBS Lett.* **70**, 180–184.

Keller, R. M., Wüthrich, K., & Schejter, A. (1977) *Biochim. Biophys. Acta* **491**, 409–415.

Keller, R. M., Picot, D., & Wüthrich, K. (1979) *Biochim Biophys. Acta* **580**, 259–265.

Kennelly, P. J., Timkovich, R., & Cusanovich, M. A. (1981) *J. Mol. Biol.* **145**, 583–602.

Kimelberg, H. K., & Lee, C. P. (1970) *J. Membr. Biol.* **2**, 252–262.

Kimura, K., Peterson, J., Wilson, M., Cookson, D. J., & Williams, R. J. P. (1981) *J. Inorg. Biochem.* **15**, 11–25.

King, G. C., Binstead, R. A., & Wright, P. E. (1985) *Biochim. Biophys. Acta* **806**, 262–271.

Kowalsky, A. (1962) *J. Biol. Chem.* **237**, 1807–1819.

Kurland, R. J., & McGarvey, B. R. (1970) *J. Magn. Res.* **2**, 286–301.

La Mar, G. N. (1973) in *NMR of Paramagnetic Molecules: Principles and Applications* (La Mar, G. N., Horrocks, W. De W., Jr., & Holm, R. H., eds.), pp. 85–126, Academic Press, New York.

Leitch, F. A., Moore, G. R., & Pettigrew, G. W. (1984) *Biochemistry* **23**, 1831–1838.

Levine, B. A., Moore, G. R., Ratcliffe, R. G., & Williams, R. J. P. (1979) *International Review of Biochemistry: Chemistry of Macromolecules IIA—Simple Macromolecules* (Offord, R. E., ed.), Vol. 24, pp. 77–141, University Park, Baltimore.

Lipari, G., & Szabo, A. (1982) *J. Amer. Chem. Soc.* **104**, 4546–4559.

Louie, G. V., & Brayer, G. D. (1990) *J. Mol. Biol.* **214**, 527–555.

Louie, G. V., Hutcheon, W. L. B., & Brayer, G. D. (1988a) *J. Mol. Biol.* **199**, 295–314.

Louie, G. V., Pielak, G. J., Smith, M., & Brayer, G. D. (1988b) *Biochemistry* **27**, 7870–7876.

Luntz, T. L., Schejter, A., Garber, E. A. E., & Margoliash, E. (1989) *Proc. Natl. Acad. Sci. USA* **86**, 3524–3528.

MacKenzie, N. E., Gooley, P. R., & Hardaway, L. A. (1992a) *Annu. Rev. Physiol.* **54**, 749–773.

MacKenzie, N. E., Gooley, P. R., & Zhao, D. (1992b) in *Synthesis and Applications of Isotopically Labelled Compounds 1991* (Buncel, E. & Kabalka, G. W., eds.), pp. 138–143, Elsevier, Amsterdam.

Mailer, C., & Taylor, C. P. S. (1972) *Can. J. Biochem.* **50**, 1048–1055.

Marion, D., & Guerlesquin, F. (1992) *Biochemistry* **31**, 8171–8179.

Marmorino, J. L., Auld, D. S., Betz, S. F., Doyle, D. F., Young, G. B., & Pielak, G. J. (1993) *Prot. Sci.* **2**, 1966–1974.

Marmorino, J. L., & Pielak, G. J. (1995) *Biochemistry* **34**, 3140–3143.

Matsuura, Y., Takano, T., & Dickerson, R. E. (1982) *J. Mol. Biol.* **156**, 389–409.

Matthew, J. B., Weber, P. C., Salemme, F. R., & Richards, F. M. (1983) *Nature (London)* **301**, 169–171.

Mauk, M. R., Reid, L. S., & Mauk, A. G. (1982) *Biochemistry* **21**, 1843–1846.

Mauk, M. R., Ferrer, J. C., & Mauk, A. G. (1994) *Biochemistry* **33**, 126009–126014.

Mayne, L., Paterson, Y., Cerasoli, D., & Englander, S. W. (1992) *Biochemistry* **31**, 10678–10685.

Mayo, S. L., & Baldwin, R. L. (1993) *Science* **262**, 873–876.

McDonald, C. C., & Phillips, W. D. (1967) *J. Amer. Chem. Soc.* **89**, 6332–6341.

McDonald, C. C., & Phillips, W. D. (1973) *Biochemistry* **12**, 3170–3186.

McDonald, C. C., Phillips, W. D., & Vinogradov, S. N. (1969) *Biochem. Biophys. Res. Commun.* **36**, 442–449.

McEwan, A. G., Kaplan, S., & Donohue, T. J. (1989) *FEMS Lett.* **59**, 253–258.

McLendon, G. (1991) *Structure and Bonding* **75**, 159–174.

Medvedeva, S., Simorre, J.-P., Brutscher, B., Guerlesquin, F., & Marion, D. (1993) *FEBS Lett.* **333**, 251–256.

Miranker, A., Robinson, C. V., Radford, S. E., Aplin, R. T., & Dobson, C. M. (1993) *Science* **262**, 896–900.

Miura, R., Sugiyama, T., Akasaka, K., & Yamano, T. (1980) *Biochem. Int.* **1**, 532–538.

Moench, S. J., & Satterlee, J. D. (1989) *J. Biol. Chem.* **264**, 9923–9931.

Moench, S. J., Satterlee, J. D., & Erman, J. E. (1987) *Biochemistry* **26**, 3821–3826.

Moench, S. J., Shi, T.-M., & Satterlee (1991) *Eur. J. Biochem.* **197**, 631–641.

Moench, S. J., Chroni, S., Lou, B.-S., Erman, J. E., & Satterlee, J. D. (1992) *Biochemistry* **31**, 3661–3670.

Molday, R. S., Englander, S. W., & Kallen, R. G. (1972) *Biochemistry* **11**, 150–158.

Moore, G. R., & Pettigrew, G. W. (1990) *Cytochromes c: Evolutionary, Structural and Physico-chemical Aspects*, Springer-Verlag, Berlin.

Moore, G. R., & Williams, R. P. P. (1975) *FEBS Lett.* **53**, 334–338.

Moore, G. R., & Williams, R. J. P. (1980a) *Eur. J. Biochem.* **103**, 493–502.

Moore, G. R., & Williams, R. J. P. (1980b) *Eur. J. Biochem.* **103**, 503–512.

Moore, G. R., & Williams, R. J. P. (1980c) *Eur. J. Biochem.* **103**, 513–521.

Moore, G. R., & Williams, R. J. P. (1980d) *Eur. J. Biochem.* **103**, 523–532.

Moore, G. R., & Williams, R. J. P. (1980e) *Eur. J. Biochem.* **103**, 533–541.

Moore, G. R., & Williams, R. J. P. (1980f) *Eur. J. Biochem.* **103**, 543–550.

Moore, G. R., & Williams, G. (1984) *Biochim. Biophys. Acta* **788**, 147–150.

Moore, G. R., Pitt, R. C., & Williams, R. J. P. (1977) *Eur. J. Biochem.* **77**, 53–60.

Moore, G. R., De Aguiar, A. B. V. P., Pluck, N. D., & Williams, R. J. P. (1980a) in *Biochemical and Medical Aspects of Tryptophan Metabolism* (Hayaishi, O., Ishimura, Y., & Kido, R., eds.), pp. 83–94, Elsevier/North-Holland, Amsterdam.

Moore, G. R., Pettigrew, G. W., Pitt, R. C., & Williams, R. J. P. (1980b) *Biochim. Biophys. Acta* **590**, 261–271.

Moore, G. R., Williams, R. J. P., Chien, J. C. W., & Dickson, L. C. (1980c) *J. Inorg. Biochem.* **12**, 1–15.

Moore, G. R., McClune, G. J., Clayden, N. J., Williams, R. J. P., Alsaadi, B. M., Ångström, J., Ambler, R. P., Van Beeumen, J., Tempst, P., Bartsch, R. G., Meyer, T. E., & Kamen, M. D. (1982) *Eur. J. Biochem.* **123**, 73–80.

Moore, G. R., Ratcliffe, R. G., & Williams, R. J. P. (1983) *Essays Biochem.* **19**, 142–195.

Moore, G. R., Harris, D. E., Leitch, F. A., & Pettigrew, G. W. (1984) *Biochim. Biophys. Acta* **764**, 331–342.

Moore, G. R., Robinson, M. N., Williams, G., & Williams, R. J. P. (1985) *J. Mol. Biol.* **183**, 429–446.

Moratal, J. M., Donaire A., Salgado, J., Jiménez, H. R., Castells, J., & Piccioli, M. (1993) *FEBS Lett.* **324**, 305–308.

Morishima, I., Ogawa, S., Yonezawa, T., & Iizuka, T. (1977) *Biochim. Biophys. Acta* **495**, 287–298.

Muthukrishnan, K., & Nall, B. T. (1991) *Biochemistry* **30**, 4706–4710.

Myer, Y. P., & Saturno, A. F. (1990) *J. Prot. Chem.* **9**, 379–387.

Nall, B. T., & Zuniga, E. H. (1990) *Biochemistry* **29**, 7576–7584.

Nieman, R. A., Gust, D., & Cronin, J. R. (1982) *Anal. Biochem.* **120**, 347–350.

Noggle, J. H., Schirmer, R. E. (1972) *The Nuclear Overhauser Effect: Chemical Applications*, Academic Press, New York.

Oldfield, E., & Allerhand, A. (1973) *Proc. Natl. Acad. Sci. USA* **70**, 3531–3535.

Oldfield, E., Norton, R. S., & Allerhand, A. (1975) *J. Biol. Chem.* **250**, 6381–6402.

Othman, S., Le Lirzin, A., & Desbois, A. (1993) *Biochemistry* **32**, 9781–9791.

Otting, G., Liepinsh, E., Farmer, B. T., & Wüthrich, K. (1991) *J. Biomol. NMR 1*, 209–215.

Patel, D. J., & Canuel, L. L. (1976) *Proc. Natl. Acad. Sci. USA* **73**, 1398–1402.

Paterson, Y., Englander, S. W., & Roder, H. (1990) *Science* **249**, 755–759.

Pearce, L. L., Gärtner, A. L., Smith, M., & Mauk, A. G. (1989) *Biochemistry* **28**, 3152–3156.

Pelletier, H., & Kraut, J. (1992) *Science* **258**, 1748–1755.

Perkins, S. J. (1980) *J. Magn. Res.* **38**, 297–312.

Perkins, S. J. (1982) in *Biological Magnetic Resonance* (Berliner, L. J., & Reuben, J., eds.), Vol. 4, pp. 193–336, Plenum, New York.

Pettigrew, G. W., & Moore, G. R. (1987) *Cytochromes* c: *Biological Aspects*, Springer-Verlag, London.

Pfeil, W., & Privalov, P. L. (1976) *Biophysical Chemistry 4*, 41–50.

Phillips, J. N. (1963) *Comprehensive Biochem.* **9**, 34–72.

Pielak, G. J., Mauk, A. G., & Smith, M. (1985) *Nature (London)* **313**, 152–154.

Pielak, G. J., Atkinson, R. A., Boyd, J., & Williams, R. J. P. (1988a) *Eur. J. Biochem.* **177**, 179–185.

Pielak, G. J., Boyd, J., Moore, G. R., & Williams, R. J. P. (1988b) *Eur. J. Biochem.* **177**, 167–177.

Pielak, G. J., Auld, D. S., Beasley, J. R., Befz, S. F., Cohen, D. S., Doyle, D. F., Finger, S. A., Fredericks, Z. L., Hilgen-Willis, S., Saunders, A. J., & Trojak, S. K. (1995) *Biochemistry* **34**, 3268–3276.

Pople, J. A. (1956) *J. Chem. Phys.* **24**, 1111.

Poulos, T. L., & Fenna, R. E. (1994) *Metals in Biological Systems* **30**, 26–75.

Poulos, T. L., & Finzel, B. C. (1984) *Peptide and Protein Reviews 4*, 115–169.

Poulos, T. L., & Kraut, J. (1980) *J. Biol. Chem.* **255**, 10322–10330.

Poulos, T. L., Sheriff, S., & Howard, A. J. (1987) *J. Biol. Chem.* **262**, 13881–13884.

Privalov, P. L., & Khechinashvili, N. N. (1974) *J. Mol. Biol.* **86**, 665–684.

Ptitsyn, O. B. (1992) in *Protein Folding* (Creighton, T. E., ed.), pp. 243–300, Freeman, New York.

Qi, P. X., Urbauer, J. L., Fuentes, E. J., Leopold, M. F., & Wand, A. J. (1994a) *Structural Biology 1*, 378–382.

Qi, P. X., Di Stefano, D. L., & Wand, A. J. (1994b) *Biochemistry* **33**, 6408–6417.

Radford, S. E., Dobson, C. M., & Evans, P. A. (1992) *Nature* **358**, 302–307.

Ragg, E., & Moore, G. R. (1984) *J. Inorg. Biochem.* **21**, 253–261.

Redfield, A. G., & Gupta, R. K. (1971) *Cold Spring Harbor Symp. Quant. Biol.* **36**, 405–411.

Robinson, N. C., & Capaldi, R. A. (1977) *Biochemistry* **16**, 375–381.

Robinson, M. N., Boswell, A. P., Huang, Z.-X., Eley, C. G. S., & Moore, G. R. (1983) *Biochem. J.* **213**, 687–700.

Roder, H., & Elöve, G. A. (1994) *Mechanisms of Protein Folding* (Pain, R. H., ed.), pp. 26–54, Oxford, New York.

Roder, H., Wagner, G., & Wüthrich, K. (1985a) *Biochemistry* **24**, 7396–7407.

Roder, H., Wagner, G., & Wüthrich, K. (1985b) *Biochemistry* **24**, 7407–7411.

Roder, H., Elöve, G. A., & Englander, S. W. (1988) *Nature* **335**, 700–704.

Salemme, F. R. (1976) *J. Mol. Biol.* **102**, 563–568.

Santos, H., & Turner, D. L. (1985) *FEBS Lett.* **184**, 240–244.

Santos, H., & Turner, D. L. (1986) *FEBS Lett.* **194**, 73–77.

Santos, H., & Turner, D. L. (1987) *FEBS Lett.* **226**, 179–185.

Santos, H., & Turner, D. L. (1992) *Eur. J. Biochem.* **206**, 721–728.

Satterlee, J. D. (1986) *Annu. Rep. NMR Spectrosc.* **17**, 79–178.

Satterlee, J. D. (1990) *Concepts in Magnetic Resonance 2*, 69–79.

Satterlee, J. D., Moench, S. J., & Erman, J. E. (1987) *Biochim. Biophys. Acta* **912**, 87–97.

Scarpulla, R. C., & Nye, S. H. (1986) *Proc. Natl. Acad. USA* **83**, 6352–6356.

Schejter, A., Lanir, A., Vig, I., & Cohen, J. S. (1978) *J. Biol. Chem.* **253**, 3768–3770.

Schejter, A., Luntz, T. L., Koshy, T. I., & Margoliash, E. (1992) *Biochemistry* **31**, 8336–8343.

Schowen, K. B. & Schowen, R. L. (1982) *Methods Enzymol.* **87**, 551–606.

Self, S. J., Hunter, C. N., & Leatherbarrow, R. J. (1990) *Biochem. J.* **265**, 599–604.

Senn, H., & Wüthrich, K. (1983) *Biochim. Biophys. Acta* **746**, 48–60.

Senn, H., & Wüthrich, K. (1985) *Quarterly Rev. Biophys.* **18**, 111–134.

Senn, H., Keller, R. M., & Wüthrich, K. (1980) *Biochem. Biophys. Res. Commun.* **92**, 1362–1369.

Senn, H., Eugster, A., & Wüthrich, K. (1983a) *Biochim. Biophys. Acta* **743**, 58–68.

Senn, H., Guerlesquin, F., Bruschi, M., & Wüthrich, K. (1983b) *Biochim. Biophys. Acta* **748**, 194–204.

Senn, H., Billeter, M., & Wüthrich, K. (1984) *Eur. Biophys. J.* **11**, 3–15.

Serrano, L., Bycroft, M., & Fersht, A. R. (1991) *J. Mol. Biol.* **218**, 465–475.

Sherwood, C., & Brayer, G. D. (1985) *J. Mol. Biol.* **185**, 209–210.

Shulman, R. G., Glarum, S. H., & Karplus, M. (1971) *J. Mol. Biol.* **57**, 93–115.

Sibel'dina, L. A., Kayushin, L. P., Kutyshenko, V. P., Lazareva, A. V., & Bronikov, G. E. (1976) *Studia Biophysica* **54**, 15–22.

Smith, G. M. (1979) *Biochemistry* **18**, 1628–1634.

Smith, M., & McLendon, G. (1981) *J. Amer. Chem. Soc.* **103**, 4912–4921.

Smith, M. B., & Millett, F. (1980) *Biochim. Biophys. Acta* **626**, 64–72.

Snel, M. M. E., Kaptein, R., & de Kruijff, B. (1991) *Biochemistry* **30**, 3387–3395.

Sorrell, T. N., Martin, P. K., & Bowden, E. F. (1989) *J. Amer. Chem. Soc.* **111**, 766–767.

Sosnick, T. R., Mayne, L., Hiller, R., & Englander, S. W. (1994) *Structural Biology* **1**, 149–156.

Soussi, B., Bylund-Fellenius, A-C., Scherstén, T., & Ångström, J. (1990) *Biochem. J.* **265**, 227–232.

Spooner, P. J. R., & Watts, A. (1991a) *Biochemistry* **30**, 3871–3879.

Spooner, P. J. R., & Watts, A. (1991b) *Biochemistry* **30**, 3880–3885.

Spooner, P. J. R., & Watts, A. (1992) *Biochemistry* **31**, 10129–10138.

Staudenmayer, N., Smith, M. B., Smith, H. T., Spies, F. K., & Millett, F. (1976) *Biochemistry* **15**, 3198–3205.

Stellwagen, E., & Shulman, R. G. (1973) *J. Mol. Biol.* **75**, 683–695.

Stellwagen, E., Smith, L. M., Cass, R., Ledger, R., & Wilgus, H. (1977) *Biochemistry* **16**, 3672–3679.

Stemp, E. D. A., & Hoffman, B. M. (1993) *Biochemistry* **32**, 10848–10865.

Stockman, B. J., Reily, M. D., Westler, W. M., Ulrich, E. L., & Markley, J. L. (1989) *Biochemistry* **28**, 230–236.

Szilàgyi, L., & Jardetzky, O. (1989) *J. Magn. Res.* **83**, 441–449.

Takano, T., & Dickerson, R. E. (1981a) *J. Mol. Biol.* **153**, 79–94.

Takano, T., & Dickerson, R. E. (1981b) *J. Mol. Biol.* **153**, 95–115.

Theorell, H., & Åkesson, Å. (1941) *J. Amer. Chem. Soc.* **63**, 1812–1818.

Thomas, M. A., Delsuc, M. A., Beloeil, J. C., & Lallemand, J. Y. (1987) *Biochem. Biophys. Res. Commun.* **145**, 1098–1104.

Thurgood, A. G. P., Davies, A. M., Greenwood, C., Mauk, A. G., Smith, M., Guillemette, J. G., & Moore, G. R. (1991a) *Eur. J. Biochem.* **202**, 339–347.

Thurgood, A. G. P., Pielak, G. J., Cutler, R. L., Davies, A. M., Greenwood, C., Mauk, A. G., Smith, M., Williamson, D. J., & Moore, G. R. (1991b) *FEBS Lett.* **284**, 173–177.

Timkovich, R. (1990) *Biochemistry* **29**, 7773–7780.

Timkovich, R. (1991) *Inorg. Chem.* **30**, 37–42.

Timkovich, R., & Cai, M. (1993) *Biochemistry* **32**, 11516–11523.

Timkovich, R., Cai, M. L., & Dixon, D. W. (1988) *Biochem. Biophys. Res. Commun.* **150**, 1044–1050.

Timkovich, R., Walker, L. A., III, & Cai, M. (1992) *Biochim. Biophys. Acta* **1121**, 8–15.

Trewhella, J., Carlson, V. A. P., Curtis, E. H., & Heidorn, D. B. (1988) *Biochemistry* **27**, 1121–1125.

Turner, D. L. (1985) *J. Magn. Res.* **61**, 28–51.

Turner, D. L. (1993) *Eur. J. Biochem.* **211**, 563–568.

Turner, D. L., & Williams, R. J. P. (1993) *Eur. J. Biochem.* **211**, 555–562.

Turner, D. L. (1995) *Eur. J. Biol.* **227**, 829–837.

Tzagoloff, A. (1982) *Mitochondria*, Plenum, New York.

Ulrich, E. L., Markley, J. L., & Kyogoku, Y. (1989) *Protein Sequences and Data Anal.* **2**, 23–37.

Urbanski, G. J., & Margoliash, E. (1977) *J. Immunol.* **118**, 1170–1180.

von Wachenfeldt, C., & Hederstedt, L. (1990) *FEBS Lett.* **270**, 147–151.

Wagner, G., Pardi, A., & Wüthrich, K. (1983) *J. Amer. Chem. Soc.* **105**, 5948–5949.

Wallin, S. T., Stemp, E. D. A., Everest, A. M., Nocek, J. M., Netzel, T. L., & Hoffman, B. M. (1991) *J. Am. Chem. Soc.* **113**, 1842–1844.

Wand, A. J., & Englander, S. W. (1985) *Biochemistry* **24**, 5290–5294.

Wand, A. J., Roder, H., & Englander, S. W. (1986) *Biochemistry* **25**, 1107–1114.

Wand, A. J., Di Stefano, D. L., Feng, Y., Roder, H., & Englander, S. W. (1989) *Biochemistry* **28**, 186–194.

Wendoloski, J. J., Matthew, J. B., Weber, P. C., & Salemme, F. R. (1987) *Science* **238**, 794–797.

Whitford, D., Concar, D. W., Veitch, N. C., & Williams, R. J. P. (1990) *Eur. J. Biochem. 192*, 715–721.

Whitford, D., Gao, Y., Pielak, G. J., Williams, R. J. P., McLendon, G. L., & Sherman, F. (1991a) *Eur. J. Biochem. 200*, 359–367.

Whitford, D., Concar, D. W., & Williams, R. J. P. (1991b) *Eur. J. Biochem. 199*, 561–568.

Wilbur, D. J., & Allerhand, A. (1977) *FEBS Lett. 74*, 272–274.

Williams, G. (1986) *J. Inorg. Biochem. 28*, 373–380.

Williams, G., & Moore, G. R. (1984) *J. Inorg. Biochem. 22*, 1–10.

Williams & Moore (1985) p. 132.

Williams, G., Eley, C. G. S., Moore, G. R., Robinson, M. N., & Williams, R. J. P. (1982) *FEBS Lett. 150*, 293–299.

Williams, G., Clayden, N. J., Moore, G. R., & Williams, R. J. P. (1985a) *J. Mol. Biol. 183*, 447–460.

Williams, G., Moore, G. R., Porteous, R., Robinson, M. N., Soffe, N., & Williams, R. J. P. (1985b) *J. Mol. Biol. 183*, 409–428.

Williams, G., Moore, G. R., & Williams, R. J. P. (1985c) *Comments on Inorganic Chemistry 4*, 55–98.

Williams, R. J. P. (1988) *Z. Phys. Chemie 269*, 387–402.

Williams, R. J. P. (1989) *Eur. J. Biochem. 183*, 479–497.

Williams, R. J. P., & Moore, G. R. (1985) *Trends in Biological Sciences 10*, 96–97.

Williamson, M. P. (1990) *Biopolymers 29*, 1423–1431.

Willie, A., Stayton, P. S., Sligar, S. G., Durham, B., & Millett, F. (1992) *Biochemistry 31*, 7237–7242.

Willie, A., McLean, M., Liu, R.-Q., Hilgen-Willis, S., Saunders, A. J., Pielak, G. J., Sligar, S. G., Durham, B., & Millet, F. (1993) *Biochemistry 32*, 7519–7525.

Wood, L. C., Muthukrishnan, K., White, T. B., Ramdas, L., & Nall, B. T. (1988) *Biochemistry 27*, 8554–8561.

Wooten, J. B., Cohen, J. S., Vig, I., & Schejter, A. (1981) *Biochemistry 20*, 5394–5402.

Wu, J., & Gorenstein, D. G. (1993) *J. Am. Chem. Soc. 115*, 6843–6850.

Wu, L. C., Laub, P. B., Elöve, G. A., Carey, J., & Roder, H. (1993) *Biochemistry 32*, 10271–10276.

Wüthrich, K. (1969) *Proc. Natl. Acad. Sci. USA 63*, 1071–1078.

Wüthrich, K. (1970) *Structure and Bonding 8*, 53–121.

Wüthrich, K. (1976) *NMR in Biological Research: Peptides and Proteins*, North-Holland, New York.

Wüthrich, K. (1986) *NMR of Proteins and Nucleic Acids*, Wiley, New York.

Wüthrich, K., & Wagner, G. (1975) *FEBS Lett. 50*, 265–268.

Yi, Q., Erman, J. E., & Satterlee, J. D. (1992) *J. Am. Chem. Soc. 114*, 7907–7909.

Yi, Q., Erman, J. E., & Satterlee, J. D. (1993) *Biochemistry 32*, 10988–10994.

Yi, Q., Erman, J. E., & Satterlee, J. D. (1994a) *Biochemistry 33*, 12032–12041.

Yi, Q., Erman, J. E., & Satterlee, J. D. (1994b) *J. Am. Chem. Soc. 116*, 1981–1987.

Yu, L. P., & Smith, G. M. (1988) *Biochemistry 27*, 1949–1956.

Yu, L. P., & Smith, G. M. (1990a) *Biochemistry 29*, 2914–2919.

Yu, L. P., & Smith, G. M. (1990b) *Biochemistry 29*, 2920–2925.

Zehfus, M. H., Reily, M. D., Ulrich, E. L., Westler, W. M., & Markley, J. L. (1990) *Arch. Biochem. Biophys. 276*, 369–373.

Zhou, J. S., & Hoffman, B. M. (1994) *Science 265*, 1693–1696.

Zimm, B. H., & Bragg, J. K. (1959) *J. Chem. Phys. 31*, 526–535.

Notes Added in Proof

Contact shifts. Turner (1995) has completely assigned the ^{13}C heme resonances of horse cytochrome *c*. Contact shifts were then estimated by removing the dipolar and diamagnetic contributions and assuming that redox-linked conformation changes are negligible. This study shows how EPR data and contact shifts can be combined to determine the dipolar field of the heme, thereby allowing calculation of pseudo-contact shifts. This process will facilitate complete resonance assignment and preliminary structure determination.

Native tertiary structure in non-native forms. Wu *et al.* (1993) investigated the interaction between the heme-containing fragment comprising residues 1–38 of horse cytochrome *c* and a peptide comprising residues 87–104. Analysis of proton inversion recovery and CD data shows that the interaction induces helicity and that certain residues in the peptide are close to the high-spin heme. The data are consistent with the idea that formation and interaction of the N- and C-terminal helices nucleates folding. Marmorino and Pielak (1995), using *Saccharomyces cerevisiae* iso-1-cytochrome *c* with changes at the interface between the helices, show that this evolutionarily invariant native tertiary interaction is present in the molten globule. This conclusion goes against the concept that native tertiary interactions are absent from molten globules and helps explain the results of Sosnick *et al.* (1994).

The cytochrome c/CCP complex. Using inversion transfer experiments, Yi *et al.* (1994a) show that the lifetime of cytochrome *c*/CN-CCP complex is concentration dependent; the lifetime decreasing with increasing concentration. The data are consistent with either substrate-assisted dissociation or two cytochrome *c* binding sites on CCP. A second, very weak binding site is also observed by excited-state quenching experiments on complexes between Zn^{2+} and Mg^{3+}-substituted CCP and various cytochromes *c* (Stemp and Hoffman, 1993), as well as the complex between native CCP and Zn^{2+}-substituted cytochrome *c* (Zhou and Hoffman, 1994). This site is also observed potentiometrically (Mauk *et al.*, 1994) . The results have an exciting implication: complexes detected by crystallography and NMR may not be catalytically important. Direct characterization of the weak site by NMR will be challenging given its high K_D ($\approx$ mM).

Amide proton exchange within cytochrome *c* in the horse ferricytochrome *c*/CCP complex is reported by Jeng *et al.* (1994). The authors used techniques identical to those employed to study cytochrome *c*/antibody complexes (Paterson *et al.*, 1990; Mayne *et al.*, 1992). The protection factors are exceedingly small (< 6). Amide protons with the largest factors are clustered at the C-terminal end of the N-terminal helix, consistent with all models and structures for cytochrome *c*/redox partner complexes. The size of the protection factors suggests that the complex is dynamic and that, unlike the structures and models, water is not excluded from the interface. Yi *et al.* (1994b) assigned the proton spectrum of wild-type *Saccharomyces cerevisiae* iso-1-ferrocytochrome *c* and then studied amide proton exchange in the iso-1-ferricytochrome *c*/CCP complex. The protection factors (≤ 40) are larger for the homologous complex, but smaller than those for antibody complexes (≤ 340). Many additional amide protons are protected in the homologous complex. This additional protection occurs in positions that are not implicated in the models or crystal structures, specifically the back side of the protein and the C-terminal helix. The results are consistent with the idea that there are major differences between homologous and heterologous complexes.

6

Resonance Raman Spectroscopy of Cytochrome *c*

PETER HILDEBRANDT

Max-Planck-Institut für Strahlenchemie
Mülheim, Germany

I. Introduction

Resonance Raman (RR)[1] spectroscopy is a well-established and powerful tool for elucidating structure–function relationships in biomolecules (Spiro, 1988). The underlying principle of this technique is that the excitation line is tuned to an electronic transition of a chromophore. In this way, the Raman bands of this chromophoric group are strongly enhanced because of the coupling of the Raman transitions with the electronic level under consideration (resonance condition). The intensity gain of the RR bands is sufficiently high that the (nonenhanced) Raman bands of the optically transparent matrix are not detectable, so that such RR spectra exclusively display the vibrational bands of the chromophore. It is this selectivity which makes the technique so attractive for structural studies of large biomolecules. Although RR spectroscopy does not provide information about the three-dimensional architecture of the macromolecule, it can give deep insight into the structure of the chromophore and its immediate environment down to a molecular level which may be beyond the resolution of x-ray crystallography or NMR

---

[1] Abbreviations: RR, resonance Raman; SE(R)R, surface enhanced (resonance) Raman; 6cLS, six-coordinated low-spin; 6cHS, six-coordinated high-spin; 5cHS, five-coordinated high-spin; DMPG, dimyristoylphosphatidyl glycerol; DOPG, dioleoylphosphatidyl glycerol; DOG, dioleoyl glycerol.

spectroscopy. Moreover, the sensitivity of the RR technique is high enough to use low sample concentrations and design experiments which are closely related to physiological conditions.

Following the first experiments in the early 1970s (Spiro and Strekas, 1972a,b; Nafie *et al.*, 1973; Friedman and Hochstrasser, 1973), RR spectroscopy has been applied widely to heme proteins by probing the RR spectrum of the porphyrin chromophore. Such investigations have contributed to a significant extent to the understanding of the biological action of heme proteins.

In this review of the RR spectroscopy of cytochrome *c*, emphasis is placed on those studies which employed the RR technique to elucidate structure–function relationships of cytochrome *c*, while those investigations which were primarily concerned with the scattering mechanism of the heme chromophore are not included. For a discussion of such studies the reader is referred to the literature (Cartling, 1988; Bobinger *et al.*, 1981; Zgierski, 1988; Schomacker and Champion, 1989, and references therein). This review focuses on the results for mammalian cytochrome *c* (horse heart), which was used in most of the RR studies. It also includes work on yeast iso-1-cytochrome *c* because of its high potential for future RR investigations due to the availability of genetically engineered mutants. Other bacterial *c*-type cytochromes are not considered.

II. Experimental Aspects

RR spectra of heme proteins are readily obtained with excitation into the B- and Q-transitions by using Ar^+- and Kr^+-ion laser lines between 406 and 415 nm and 514 and 530 nm in the cw-mode (for details of the experimental setup, see the original papers cited in this review). Depending on the sample arrangement, a cytochrome *c* concentration as low as 10 micromolar can be sufficient to obtain excellent RR spectra. Detection and analysis of the Raman scattered light is possible in two modes. Optical multichannel detection, using a spectrograph equipped with a photodiode array or a CCD camera, offers the advantage of measuring a large frequency region in a short time, but generally lacks the spectral resolution required for a detailed analysis (Yu and Srivastava, 1980; Hildebrandt *et al.*, 1991). Monochromatic detection (double monochromator and photon counting system) is more time-consuming because it requires scanning through the frequency region of interest, but it offers a significantly better resolution and, in addition, can be easily extended to dual-channel Raman difference spectroscopy (Rousseau, 1981; Shelnutt *et al.*, 1981; Hildebrandt and Stockburger, 1989c). This latter technique permits the detection of very small spectral differences between two samples which are measured quasi-simultaneously by depositing the samples in two chambers of a divided rotating cell.

Despite the resonance enhancement, RR scattering of the heme chromophore is still a relatively weak process with a quantum yield on the order of 10^{-5}. Thus, the protein samples must be free of fluorescing impurities

Using a rotating cuvette, even prolonged laser exposure for several hours does not lead to a degradation of cytochrome *c*. However, in some cases photoreduction of ferricytochrome *c* was observed (Hildebrandt *et al.*, 1990b). Addition of potassium ferricyanide as an oxidizing agent is problematic because, particularly at excitation wavelengths below 500 nm, ferricyanide is photodecomposed, and free cyanide can bind to ferricytochrome c by replacing the methionine ligand (Chottard, 1989). Using catalytic amounts of cytochrome c oxidase in an air-saturated solution appears to be a more promising approach to reoxidize ferrocytochrome c (unpublished results).

III. RR Scattering of the Heme Chromophore

Very early in the RR spectroscopic investigations of heme proteins, it was found that the vibrational pattern is intimately related to the symmetry properties of the porphyrin chromophore (Spiro and Strekas, 1972a,b). Assuming a D_{4h} symmetry, the two highest filled molecular orbitals of a porphyrin are very close in energy and belong to the a_{1u} and a_{2u} species, while the lowest unoccupied orbital is degenerate and belongs to the e_g species. Consequently, the two transitions $a_{1u} \rightarrow e_g$ and $a_{2u} \rightarrow e_g$ have E_u symmetry. They are subject to strong configuration interactions yielding an additive and a subtractive combination of the transition dipoles, which correspond to the strong Soret- and the much weaker Q-band, respectively. The latter can borrow intensity from the strongly allowed transition via vibronic coupling leading to a fine structure of the Q-band.

Since the transition dipoles lie in the porphyrin plane, resonance enhancement is only expected for the "gerade" in-plane vibrational modes A_{1g}, B_{1g}, A_{2g}, and B_{2g} (Spiro, 1983; Kitagawa and Ozaki, 1987). However, there is a preferential enhancement of either the totally symmetric (A_{1g}) or the non-totally symmetric (B_{1g}, A_{2g}, B_{2g}) modes, depending on whether excitation occurs in resonance with the strong or weak electronic transition. The A_{1g} modes dominate the RR spectrum upon Soret-band excitation. Since their intensities are proportional to the square of the electronic transition dipole moment (A-term enhancement mechanism; Albrecht, 1961), these bands are relatively weak upon Q-band excitation. On the other hand, the nontotally symmetric modes gain intensity via vibronic coupling (B-term enhancement mechanism) and hence are relatively strong upon excitation in the Q-band region.

The excitation profile is not the only criterion to distinguish between modes of different symmetry. The different types of RR-active modes reveal unique polarization properties which can readily be understood in terms of the classical treatment of the Raman scattering tensor (Spiro and Strekas, 1972b). Only the A_{1g} modes are polarized, while the B_{1g} and B_{2g} modes are depolarized and the A_{2g} modes are anomalously polarized.

While these simple considerations of the RR scattering process, which are based on an adiabatic approximation, are sufficient for a qualitative understanding of the RR spectra, they fail to account for the dispersion of the RR intensities and depolarization ratios in a quantitative manner. In the past 10 years, a number of approaches have been developed which have significantly improved the theoretical description of the RR effect. For a critical discussion of these theories and their applications to the heme chromophore, the reader is referred to the literature (Stallard et al., 1984; Zgierski, 1988; Bobinger et al., 1989, and references therein).

A. Interpretation of the RR Spectra of Cytochrome c

The vibrational assignment is a prerequisite for extracting structural information from the RR spectra of cytochrome c. For model compounds belonging to the D_{4h} symmetry group, normal mode analysis calculations have been carried out which are well supported by a large amount of experimental data (see, for example, Abe et al., 1978; Kitagawa et al., 1978; Li et al., 1989, 1990a, 1990b; Gladkov and Solovyov, 1985a, 1985b, 1986; Lee et al., 1986). All naturally occurring porphyrins including heme c, however, have an effective symmetry lower than D_{4h}, weakening the selection rules and modifying the enhancement pattern and the polarization properties of the RR bands (Spiro, 1983; Kitagawa and Ozaki, 1987; Kubitschek et al., 1986). Nevertheless, the calculated and experimental data for highly symmetric porphyrins can well be used as a starting point for the analysis of the vibrational spectra of the heme group in cytochrome c. However, one must take into account that symmetry lowering can induce RR-activity to those modes which are RR-forbidden within the D_{4h} symmetry group. For example, the asymmetric substitution of the vinyl groups in protoporphyrin IX activates the E_u modes (Choi et al., 1982a,b; Lee et al., 1986). In heme c, the vinyl groups are replaced by thioether bridges and, in fact, the E_u modes v_{37} and v_{38} are also observed in the RR spectra of cytochrome c (Hildebrandt et al., 1992).[2] Symmetry lowering can also be the origin for the observation of out-of-plane modes in the RR spectrum (Choi and Spiro, 1983; Li et al., 1989). A_{2u}-modes can become RR-active because of the loss of the center of symmetry, for example, as the result of an asymmetric axial ligation pattern. A mode of this type has been assigned to the RR band of cytochrome c at ~ 360 cm^{-1} (Hildebrandt et al., 1991). RR-activation of other out-of-plane modes such as A_{1u}, B_{1u}, and B_{2u} requires deformations of the planar structure through ruffling or doming of the porphyrin (Li et al., 1989; Czemuszewicz et al., 1989). Crystal structure data on cytochrome c have revealed a nonplanar geometry of the porphyrin (Takano and Dickerson, 1981a,b; Bushnell et al., 1990; Louie and Brayer, 1990), and indeed, a number of RR bands, in particular in the region between 400 and 800 cm^{-1}, are possible candidates for such out-of-plane modes.

Figure 6.1 displays the RR spectrum of ferrocytochrome c excited at 413 nm. There are two frequency regions which deserve special attention because they include bands which can be correlated with structural parameters of the heme and the heme pocket: the marker band and the fingerprint region.

1. Marker band region

The bands in the high-frequency range (1300–1700 cm^{-1}) originate from modes which include predominantly C–C and C–N stretching vibrations of the porphyrin (Abe et al., 1978). The strongest band in the RR spectrum of ferricytochrome c at 1370.1 cm^{-1} can readily be assigned to the A_{1g} mode v_4, which is a nearly pure C–N stretching vibration (Table 6.1). In ferrocytochrome c, this band shifts down by ~ 10 cm^{-1} to 1360.4 cm^{-1} (Figure 6.1). This can be understood in terms of an in-

---

[2] Mode numbering refers to Abe et al. (1978).

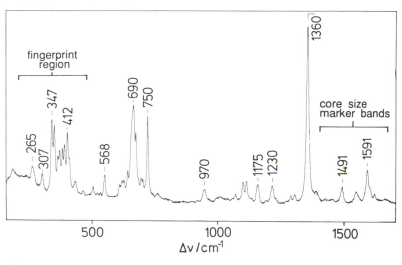

Figure 6.1

RR spectrum of ferrocytochrome c (horse heart; pH 7.0) excited at 413 nm.

creased back donation of electron density into the antibonding π^* orbitals of the porphyrin upon reduction of the heme iron (Kitagawa and Ozaki, 1987; Spiro, 1983), which, in turn, leads to a weakening of the C–N bonds as reflected by the frequency decrease of ν_4. This effect is further enhanced by electron-rich axial ligands such as thiolate (cytochrome P-450), so that this mode can be even below 1350 cm^{-1} (Anzenbacher et al., 1989). Thus, the mode ν_4 can be used as an oxidation state marker band, permitting an unambiguous determination of the redox state of the heme.

Most of the other modes in this frequency range also depend on the redox state. In addition, they are very sensitive to changes of the geometry of the tetrapyrrole macrocycle (Spaulding et al., 1975; Spiro et al., 1979). Based on a large set of experimental data on model compounds for which x-ray structure data are available, it could be shown that the frequencies of most of these bands depend on the core size of the macrocycle, i.e., the pyrrole-nitrogen to core center distance (d_t) (Parthasarathi et al., 1987). Since d_t differs in a characteristic way for high-spin (HS) and low-spin (LS), five- and six-coordinated (5c, 6c), and oxidized and reduced iron porphyrins, these bands can be used to determine the spin, ligation, and oxidation state of the heme iron. For both ferri-and ferrocytochrome c, the assignment of these modes is straightforward (Table 6.1, Figure 6.2; Hildebrandt, 1990; Hildebrandt et al., 1992). For most of these bands there is a good agreement with the expected frequencies for 6cLS-ferric and ferrous hemes. There are only three obvious exceptions. The mode ν_2 of ferrocytochrome c (but not of ferricytochrome c) at 1591 cm^{-1} is significantly higher than that calculated from the d_t–ν relationship. This deviation was attributed to the increased back-donation in the reduced state (Hildebrandt, 1990; Pathasarathi et al., 1987). In contrast, in the oxidized form, the modes ν_{19} at 1583 cm^{-1} (not

Table 6.1
Frequencies and half widths (in cm^{-1}) of the marker bands ν_4 and ν_3 for various species of cytochrome c.

Species[a]			Mode			
			ν_4	$\Delta\nu_4$	ν_3	$\Delta\nu_3$
Reduced	6cLS	diss., pH 7.0	1360.4	9.8	1490.5	12.0
		ads., state I	1358.8	9.9	1489.8	12.3
		ads., state II	1362.5	13.3	1491.1	12.1
Reduced	5cHS	ads., state II	1354.8	14.4	1467.3	14.4
Oxidized	6cLS	diss., pH 7.0	1370.1	15.1	1500.4	11.1
		diss., pH 11.0	1373.3	15.6	1502.1	10.9
		diss., pH 1.8	1373.0	14.6	1502.3	13.8
		ads., state I	1369.0	12.0	1499.5	11.4
		ads., state II	1373.6	13.7	1503.1	11.6
Oxidized	5cHS	diss., pH 1.8	1369.0	12.0	1491.2	14.0
		ads., state II	1367.8	11.8	1488.4	12.5
Oxidized	6cHS	diss., pH 1.8	1365.5	14.6	1480.7	13.9

[a] "diss." and "ads." refer to the cytochrome c species in solution and adsorbed on the Ag electrode. The data for the adsorbed species and for the dissolved cytochrome c at neutral pH were taken from Hildebrandt and Stockburger (1989b); the data for the acid and the alkaline form were taken from Hildebrandt (1990).

observed with Soret-band excitation) and ν_{10} at 1633 cm^{-1} appear at lower frequencies, which, however, can be attributed to the nonplanar structure of the heme in cytochrome c (Hildebrandt, 1990). Also the considerable RR-activity of the E_u-mode ν_{37} ($\sim$1595 cm^{-1} in ferricytochrome c; Figure 6.2) indicate an effective symmetry of the heme lower than D_{4h}.

2. Fingerprint region

The bands below 500 cm^{-1} originate from modes which include considerable contributions from vibrations involving the peripheral substituents of the porphyrin (Abe et al., 1978; Li et al., 1989, 1990a, 1990b; Lee et al., 1986). Hence, these modes should sensitively reflect the specific structure of the heme pocket in cytochrome c

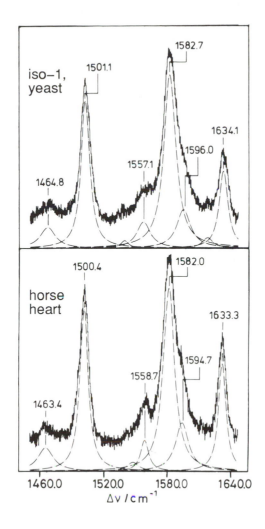

Figure 6.2

RR spectra of ferricytochrome *c* (pH 7.0) from yeast (iso-1) and horse heart, excited at 413 nm (Hildebrandt *et al.*, 1992). The dashed lines represent the fitted Lorentzian bandshapes.

and can be regarded as a fingerprint for heme–protein interactions (Hildebrandt, 1990). In fact, the RR spectra of cytochrome *c* in this region reveal pronounced differences compared to simple iron porphyrin model compounds or *b*-type heme proteins. The number of bands between 300 and 420 cm^{-1} is about twice as high as, for example, in myoglobin (cf. Choi and Spiro, 1983). The unusually rich manifold of sharp but overlapping bands in the RR spectra of ferricytochrome *c* (Figure 6.3a) and ferrocytochrome *c* cannot exclusively be attributed to the effect of symmetry lowering (Valance and Strekas, 1982; Hildebrandt, 1990) or to the involvement of overtones and combination modes (Friedman and Hochstrasser, 1973). It was suggested that some of the low-frequency modes split into two components that

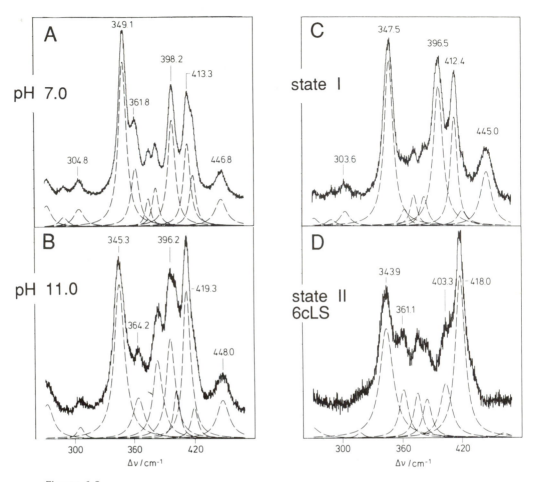

Figure 6.3

RR spectra of ferricytochrome c (horse heart) excited at 413 nm. (A) aqueous solution, pH 7.0; (B) aqueous solution, pH 11.0; (C) state I and (D) state II (6cLS) of ferricytochrome c bound to heteropolytungstates (Hildebrandt, 1990). The dashed lines represent the fitted Lorentzian bandshapes.

correspond to different conformations and orientations of the peripheral substituents (Hildebrandt, 1990; Hildebrandt *et al.*, 1990a). This interpretation requires that the mobility of these substituents be restricted by steric constraints imposed by the protein matrix so that individual conformations of these substituents are trapped. This idea is strongly supported by the extremely narrow band widths (6–9 cm^{-1}), which are about 6 cm^{-1} smaller than those observed in other heme proteins. A considerable increase of the band widths, however, is observed following conformational transitions of cytochrome c which involve a relaxation of the heme pocket (see Section VI.B). Such structural changes are also accompanied by a (partial) removal of the band splitting. It is interesting to note that in ferrocytochrome c the

Table 6.2
RR frequencies (in cm^{-1}) of cytochrome c from horse heart and yeast (iso-1) in the fingerprint region.

Mode[a]	Reduced		Oxidized	
	Horse[b]	Yeast[b]	Horse[c]	Yeast[c]
v_9	262.5	263.2	266.0	265.2
	270.9	272.0	272.0	275.0
v_{52}	—	—	287.6	288.5
v_{51}	306.0	306.8	304.8	305.8
v_8	345.9	345.8	348.9	349.2
pyr[d]	356.9	357.3	361.9	361.2
$2v_{35}$	371.5	370.6	375.8	375.1
	379.6	379.4	381.9	381.9
$v_{34} + v_{35}$	391.1	391.4	397.9	398.1
	400.3	400.3		
$2v_{34}$	412.8	413.4	413.0	413.1
	421.0	421.8	419.0	419.3
prop[d]	445.2	442.6	446.1	442.4
		449.4		449.7

[a] Mode numbering refers to Abe *et al.* (1978).
[b] Unpublished data.
[c] Data taken from Hildebrandt *et al.* (1991, 1992).
[d] Pyr: pyrrole tilting; prop: propionate bending.

band splitting is more pronounced and the band widths are narrower than in ferricytochrome c (unpublished results), suggesting a more rigid fixation of the heme in the protein in the ferrous form. This is in line with the observation that the H/D exchange of the buried Trp-59 is much slower in the reduced than in the oxidized form (Liu *et al.*, 1989).

Based on these considerations, an assignment of the bands in this region has been suggested (Table 6.2; Hildebrandt *et al.*, 1991). Another consequence of the low symmetry of the heme c is that the vibrational modes are not fully delocalized over the entire porphyrin. Lee *et al.* (1986) measured the isotopic shifts of the RR bands of metalloporphyrins which were asymmetrically labeled by deuterated vinyl or methyl substituents. The results clearly show that an asymmetric substitution pattern such as in protoporphyrin IX causes a localization of some of the porphyrin modes in individual pyrrole rings. Such results are of utmost importance for correlating changes in the RR spectra with structural changes in specific regions of the heme pocket.

B. Cytochrome c from Different Species

Shelnutt *et al.* (1981) have carefully examined cytochromes from a large variety of species by employing the Raman difference technique. Even though the secondary and tertiary structures are not substantially different as inferred from the x-ray and NMR data for tuna, horse, and yeast iso-1-cytochrome c (Takano and Dickerson, 1981a,b; Bushnell *et al.*, 1990; Louie and Brayer, 1990; Gao *et al.*, 1990, 1991a; Feng *et al.*, 1989), the RR spectra of these proteins reveal small but distinct frequency shifts which were attributed to three different effects (Shelnutt *et al.*, 1981): (1) replacement of amino acids which have van der Waals contacts with the heme [e.g., Phe-46 (horse)], modifying the vibrational frequencies via a direct interaction with the π^* orbitals of the porphyrin; (2) substitution at position 92, bringing about conformational changes that propagate to the heme and affect the mode v_{19} at ~1584 cm^{-1} (ferrocytochrome c); (3) structural modification near the Cys-14 thioether bridge, influencing the mode v_{21}.

While these experiments were restricted to ferrocytochrome c using Q-band excitation, a recent analysis of Soret-band excited spectra of horse heart and yeast iso-1-cytochrome c provides complementary results (Hildebrandt *et al.*, 1992). In the oxidized form, all the marker bands are upshifted by ~1 cm^{-1} in the yeast protein (Figure 6.2), pointing to a slightly smaller heme core size (Parthasarathi *et al.*, 1987) which could be induced by a conformational change of one of the axial ligands. In the fingerprint region, frequency shifts by up to 1 cm^{-1} were observed in both ferri- and ferrocytochrome c (Table 6.2). The most striking differences of the yeast iso-1 compared to the horse heart protein are the splitting of the propionate bending at ~446 cm^{-1} into two components (442.4 and 449.7 cm^{-1} in the oxidized form) and a band broadening by about 1.5 cm^{-1}. The latter, which was only observed for the oxidized form, can readily be attributed to a more flexible heme pocket structure than in the horse heart protein. Different steric constraints on the heme imposed by the adjacent amino acids may also account for the modified intensity distribution among the bands at 375 and 382 cm^{-1} which were assigned to the same mode of two heme conformers, differing in the orientation of the peripheral substituents (see Section III.A.2). This would imply that the specific structures of the heme pockets stabilize different populations of these conformers. It should be noted that the crystal structures of the proteins do not reveal these differences (Bushnell *et al.*, 1990; Louie and Brayer, 1990), implying that they are beyond the resolution of these x-ray structure data, or accessible to the protein only in solution. The splitting of the 446-cm^{-1} band which was tentatively assigned to a porphyrin mode including a propionate bending vibration was interpreted in terms of a structural change of one of the propionate side chains in iso-1-cytochrome c (Hildebrandt *et al.*, 1992). Crystal structure data point to structural differences between horse heart and yeast cytochrome *c* around the residues 56 and 57, modifying the environment of Trp-59 which is hydrogen-bonded to the propionate of pyrrole ring A (Bushnell *et al.*, 1990; Louie and Brayer, 1990). Hence, the downshifted component of this band at 442.4 cm^{-1} is assigned to the mode involving the propionate of ring A (for the pyrrole ring numbering, see Chapter 3 and Louie and Brayer, 1989).

C. Cytochrome *c* Mutants

The investigations of cytochromes from different species do not give insight into the structural and functional role of individual amino acids. In this respect, however, the recent progress in genetic engineering of iso-1-cytochrome *c* is of utmost importance, because it permits the selective replacement of those phylogenetically invariant amino acids which are thought to be crucial for stabilizing the protein structure and for the electron transfer (Mauk, 1991). Such iso-1-cytochrome *c* mutants substituted at position 82 (Phe) have been characterized by various physicochemical methods (Louie *et al.*, 1988; Piclak *et al.*, 1988; Pearce *et al.*, 1989; Louie and Brayer, 1989; Rafferty *et al.*, 1990; Gao *et al.*, 1991b). The first RR study of these variant proteins analyzed the effect of substitution of Phe-82 by a serine (Hildebrandt *et al.*, 1991). Although the RR spectra were only poorly resolved, spectral changes for both the reduced and the oxidized form were noted which indicated perturbations of the heme–protein interactions, in particular in the buried part of the heme crevice. Reinvestigation of the Phe-82 and Ser-82 mutants (unpublished results) using highly resolved RR spectra confirm the results on the oxidized form, but demonstrate that the spectral changes are much smaller in the reduced state (the previously published spectra presumably reflect contamination by denatured protein). These results are in line with the structural data obtained from NMR spectroscopy (Gao *et al.*, 1991b). The crystal structure of the reduced Ser-82 protein, however, indicates an additional conformational change remote from the mutation site affecting the hydrogen bonding interactions with the propionate group A (Louie *et al.*, 1988). The RR spectra do not reveal any change for the propionate bending modes, and show only small changes for those modes which are regarded to be sensitive to protein structure perturbations in the vicinity of pyrrole rings A and D (unpublished results). On the other hand, a distinct intensification of the 449-cm^{-1} band, attributable to the bending mode of propionate D, is observed in the oxidized form which can readily be understood on the basis of the NMR structural analysis. Besides this localized change, the RR spectra of both oxidation states show intensity redistributions of those band pairs which were assigned to the same mode of different conformers (see Section III.A.2.). In addition, intensity changes and frequency upshifts of the marker bands by about 1 cm^{-1} were observed. It should be noted that such small frequency shifts would correspond to a contraction of the porphyrin core by less than 0.005 Å (Parthasarathi *et al.*, 1987). Evidently, there is a subtle reaccommodation of the porphyrin in the protein matrix following structural changes of the surrounding polypeptide chain due to the substitution of Phe-82 by Ser.

IV. Conformational Changes of Cytochrome *c*

A. Temperature-Induced Conformational Changes

Upon increasing temperature, the native structure of ferricytochrome *c* is destabilized. At $\sim$45 °C (horse heart), a reversible conformational transition takes place which was ascribed to an opening of the heme crevice (Ångström *et al.*, 1982). The loss of the 695-nm absorption band was attributed to a perturbation of the Met-80–iron bond. Employing NMR spectroscopy, it was found that a new 6cLS configuration is formed which coexists with a HS configuration, the latter prevailing above 60 °C. These temperature-induced conformational changes were studied by RR spectroscopy using iso-1-ferricytochrome *c*. The high-temperature spectrum (62 °C) differs remarkably from the room-temperature spectrum. In particular, the characteristic vibrational pattern in the low-frequency region is drastically altered. The mode v_8 is downshifted by $\sim$5 cm^{-1}, and the intensities of the combination mode $v_{34} + v_{35}$ and propionate bending have strongly decreased. These spectral changes can be regarded as sensitive indicators for an open heme crevice, corroborating the conclusions derived from other spectroscopic techniques. In the marker band region, one notes an upshift of the modes v_3, v_2, and v_{10}. These findings confirm the idea that a new 6cLS configuration is formed at elevated temperature. Furthermore, the apparently increased band widths which accompany the frequency shifts also point to an additional contribution from a 5cHS state. Monitoring the intensity of the propionate bending, the RR spectra measured as a function of temperature revealed a transition temperature which compares very well with that obtained by other techniques. Such experiments were extended to the Ser-82 mutant of iso-1-ferricytochrome *c*, which exhibits a significantly lower transition temperature ($\sim$30 °C) and a higher 5cHS content at 45 °C. Apparently, the replacement of Phe-82 by a Ser destabilizes the heme pocket structure so that the dissociation of the Met-80 ligand is facilitated (see also Pearce *et al.*, 1989).

It is well known that the stability of the heme pocket of cytochrome *c* is much greater in the reduced form (Margoliash and Schejter, 1966). Thus, it is not surprising that besides a small band broadening by about 0.5 cm^{-1} (unpublished results), no spectral changes could be detected in the temperature range between 5 and 65 °C (Hildebrandt *et al.*, 1991).

B. pH-Induced Conformational Changes

In the oxidized form, cytochrome *c* undergoes a variety of reversible pH-induced conformational transitions (Theorell and Åkesson, 1941). The first alkaline transition occurs with a pK_a of 9.35 (horse heart) and includes the replacement of the Met-80 ligand, presumably by a lysine residue (Gadsby *et al.*, 1987). The replacement of the axial ligand is reflected by a frequency increase of the marker bands by about 3–4 cm^{-1} in the RR spectrum (see also Table 6.1), and the titration curves based on the band frequencies roughly parallel those obtained from other spectroscopic techniques (Kitagawa *et al.*, 1975, 1977; Myer *et al.*, 1983; Greenwood and Palmer, 1965; Gupta and Koenig, 1971). The RR spectrum in the marker band region of this alkaline form is closely related to the RR spectrum of alkylated ferricytochrome *c* in which Met-80 is replaced by methylamine, supporting the assignment of a lysine residue as the sixth ligand in the alkaline form (Kitagawa *et al.*, 1977). Myer *et al.* (1983) interpreted the frequency upshift of the marker bands in terms of a contraction of the porphyrin core size.

In the fingerprint region, the transition to the alkaline form is accompanied by a moderate increase of the band widths, implying a somewhat greater flexibility of the heme (Figure 6.3b; Hildebrandt, 1990). These findings support the conclusions drawn from the analysis of the depolarization ratios, which suggest much smaller symmetry distortion than in the neutral form (Kubitschek *et al.*, 1986). Apparently, the conformational changes in the heme pocket at high pH remove some of the steric constraints on the heme and allow for a more relaxed geometry. However, the complexity of the vibrational pattern in the fingerprint region is still comparable with that of the neutral form. While the splitting of the $2v_{35}$ overtone (375.8 and 381.9 cm^{-1}) is removed, now the combination mode $v_{34} + v_{35}$ (397.9 cm^{-1}) is split into two components in the alkaline form (Figure 6.3a,b; Table 6.2; Hildebrandt, 1990). This was ascribed to a localized structural change in a part of the heme pocket, following an idea raised by Takano and Dickerson (1981b). These authors suggested that the replacement of Met-80 by Lys-72 or Lys-79 would require a conformational change only in the left part of the heme crevice. This interpretation implicitly rules out a large-scale unfolding of the protein, and indeed, there appears to be only minor disordering on the level of the tertiary structure as concluded from a UV RR experiments monitoring the pH-dependence of the amide bond vibrations (Copeland and Spiro, 1985).

Time-resolved RR studies have shown that the transition to the alkaline state proceeds via an intermediate formed within 20 to 200 milliseconds (Uno *et al.*, 1984). This intermediate is characterized by a band at ~ 1610 cm^{-1} (v_{10}), which indicates a 6cHS configuration. It is speculated that the alkaline transformation is initiated by the replacement of the Met ligand by a hydroxide. Thus, it may be that this 6cHS intermediate corresponds to the hydroxy complex of the heme.

The pK_a value of the alkaline transition of the reduced form has been calculated to be much higher (> 16) than for the oxidized form (Barker and Mauk, 1992), and hence this state is not accessible to spectroscopic studies. However, Valance and Strekas (1982) noted dramatic changes in the RR spectrum of ferrocytochrome c already upon increasing the pH to 13.5. Thus, one may conclude that the underlying structural changes are different from those of the alkaline transition of ferricytochrome c. In the marker band region, the only significant frequency shift is observed for the mode v_{11}, which drops by ~ 10 cm^{-1} to 1535 cm^{-1}. This may be attributed to the replacement of the Met-80 ligand by a hydroxide, in contrast to the nitrogenous ligand of the alkaline form of ferricytochrome c. Since the frequencies of all other marker bands remain largely constant, one can conclude that this ligand exchange is not associated with a change of the spin state. Moreover, the unusually low frequency of v_{11} can be attributed to an increased back-donation, promoted by the hydroxide ligand, to the π^* orbitals of the porphyrin. As pointed out by Parthasarathi *et al.* (1987), this effect can specifically lower the frequency of v_{11}. Furthermore, the splitting of the low-frequency modes between 300 and 450 cm^{-1} is completely removed, and the RR bands are very broad. Valance and Strekas (1982) ascribed these findings to an increase of the effective symmetry of the heme due to a loosening of the heme pocket structure.

The hydroxide ligand can readily be replaced by other strong-field ligands such as cyanide or carbon monoxide. In both cases, the fingerprint region is basically unchanged, indicating that the loose heme pocket structure of this alkaline ferrocytochrome c can accommodate different exogeneous ligands without altering the heme–protein interactions. It is interesting to note that the response of the oxidation marker band v_4 to CO-binding is quite different than in hemoglobin or myoglobin, in which one axial position is also occupied by a histidine. In these proteins, CO-binding to the reduced heme iron causes a frequency upshift of this mode by ~ 10 cm^{-1}, since empty π^* orbitals of the bound CO can effectively compete with the π^* orbitals of the porphyrin for electron density (Spiro, 1983; Kitagawa and Ozaki, 1987). In the alkaline CO-bound ferrocytochrome c, however, the frequency is even further downshifted by 4 cm^{-1} (Valance and Strekas, 1982).

At a pK_a of 2.5, ferricytochrome c is converted to an acidic form (Theorell and Åkesson, 1941) in which the folded protein structure is still largely preserved (Cohen and Hayes, 1974; Copeland and Spiro, 1985). In this state, the Met-80 ligand is also removed, leaving the heme iron in a HS configuration. Although RR spectroscopy can in principle distinguish between a 6cHS and 5cHS state in the oxidized form, the interpretation of the RR data was not unambiguous, because the broad RR bands were not resolved (Kitagawa et al., 1977; Lanir et al., 1979; Myer and Saturno, 1991). A more detailed analysis of the v_3 band region of ferricytochrome c at pH 1.8, however, clearly reveals three components at 1480.7, 1491.2, and 1502.3 cm^{-1}, which correspond to a 6cHS-, a 5cHS-, and a 6cLS-configuration, respectively (Table 6.1; Hildebrandt, 1990). In the 6cHS-state, weak-field ligands such as water must be coordinated to the heme iron, at least at the sixth coordination site. Presumably, the fifth position is still occupied by the (protonated) His-18 (Myer and Saturno, 1991). Binding of water to the heme iron may be very weak, so that this ligand is easily dissociated, leading to the 5cHS form. The 6cLS configuration then may result from the binding of the exogeneous anion of the acidifying agent (chloride). This interpretation is strongly supported by the finding that addition of KCl to the solution leads to a significant increase of the 6cLS content (Lanir et al., 1979; Myer and Saturno, 1991). The overlapping contributions of three configurational states at pH 1.8 indicate that under these conditions ferricytochrome c does not exhibit a uniform heme pocket structure. It may well be that there is already an appreciable contribution from the second acidic state, formed with a pK_a of 1.1, in which the heme exists in a 5cHS configuration (Lanir et al., 1979; Myer et al., 1983; Myer and Saturno, 1991). Myer and Saturno (1991) suggested that in this state the only axial ligand is the protonated or strongly hydrogen-bonded His-18.

Monitoring the titration of ferricytochrome c over a wide pH range by various spectroscopic techniques, including RR spectroscopy, suggests that both the acid and the alkaline transitions are not simple one-step processes but occur via intermediates (Myer et al., 1983; Myer and Saturno, 1991). However, the RR bands which were ascribed to these species are at positions very similar to those in the neutral form. Since these spectra were measured with a low resolution and an unfavorable excitation wavelength, the interpretation in terms of the heme conformation appears to be premature.

V. Redox Transitions

The strong dependence of the RR spectrum of the heme on the oxidation state, in particular the mode v_4, makes this technique most suitable for monitoring redox transitions. Anderson and Kincaid (1978) were the first to combine RR spectroscopy with the potentiometric analysis of cytochrome *c*. Using methylviologen and 1,1′-bis(hydroxymethyl)ferrocene as mediators, rapid electron exchange between a working electrode and the dissolved cytochrome *c* was ensured so that the redox equilibrium of the bulk cytochrome *c* could be measured as a function of the potential. Taking the relative intensities of the v_4 modes as a measure for the relative concentrations of ferri- and ferrocytochrome *c*, a Nernstian fit was obtained with a slope close to the expected value for $n = 1$.

The reduction of ferricytochrome *c* should proceed in at least two steps. Following the electron transfer to the heme, the porphyrin and the surrounding protein environment must relax to the equilibrium conformation of reduced state, which satisfies the altered charge distribution. In an attempt to identify an intermediate redox state, Cartling (1983) studied the radiolytic reduction of ferricytochrome *c* at low temperature in a glycerol/buffer glass. The RR spectra of the radiolytically reduced species reveal some differences compared to the chemically reduced cytochrome *c*, which were ascribed to a reduced heme with a nonrelaxed protein environment. However, at neutral pH, these spectral differences are very small, so they may also be due to a residual contribution of the oxidized form. At pH 10.8, the spectral differences are much more pronounced and cannot be explained by the superposition of the RR spectra of the ferro- and ferricytochrome *c*, but must in fact be attributed to an intermediate state. Since at this high pH the alkaline form is only stable in the oxidized but not in the reduced state, the reduction of ferricytochrome *c* must be followed by a structural rearrangement of the heme pocket back to the conformation of the "neutral" form. It may be that this step is blocked in the glycerol matrix at low temperature.

Forster *et al.* (1982) studied this process at alkaline pH by time-resolved RR spectroscopy at ambient temperature by monitoring the chemical reduction in a continuous mixed-flow experiment. The spectrum obtained 55 ms after mixing with dithionite largely reflects an intermediate state in which the heme is already reduced as indicated by the position of v_4 (1362 cm^{-1}), but which differs from the stable reduced form in the positions of the spin state marker bands. In analogy to the low-temperature experiments, this intermediate may be attributed to a "nonrelaxed" protein conformation.

VI. Cytochrome *c* Bound to Charged Surfaces

It is well known that electrostatic interactions play a key role for the physiological electron transfer reactions of cytochrome *c* (cf. Pettigrew and Moore, 1987, and references therein). The first step of these processes is the formation of a complex between cytochrome *c* and its reaction partner that is stabilized via Coulombic attractions between the positively charged, lysine-rich domain around the heme crevice of cytochrome c and the negatively charged binding domain of the partner protein. It is assumed that these electrostatic interactions ensure an optimal alignment of the electron donor and acceptor groups, facilitating a rapid electron transfer (Koppenol and Margoliash, 1982). However, details of the molecular structure of such complexes are not known.

RR spectroscopic studies of the physiological redox complexes of cytochrome c are hindered by the fact that several chromophoric groups contribute to the spectra, considerably complicating the interpretation of the experimental data. Thus, in a number of investigations a different approach was chosen in which model systems are used to mimic the charged surfaces of the physiological binding domains and thereby avoid contributions to the RR spectrum.

A. Cytochrome c Adsorbed at the Metal/Electrolyte Interface

A particularly attractive model system for studies of this type is the charged silver surface, because it offers the advantage of permitting surface enhanced Raman (SER) spectroscopy (for reviews, see Cotton, 1988; Koglin and Séquaris, 1986). This special Raman spectroscopic technique is based on the enhancement of Raman signals for those molecules which are adsorbed on submicroscopically rough Ag, Au, or Cu surfaces (colloids, electrochemically roughened electrodes). The enhancement factor compared to the dissolved molecules can be up to six orders of magnitude; however, the underlying physical processes are incompletely understood. Most likely, the dominant contribution originates from dipole–dipole interactions between the radiation field and the surface plasmons of the metal. If the excitation line is in resonance with a molecular transition of the adsorbate, the molecular RR and the SER effect can combine (surface enhanced resonance Raman–SERR). In this way, the sensitivity of this technique is further increased so that it is possible to study molecules adsorbed at submonolayer coverages.

Cotton et $al.$ (1980) were the first to demonstrate that SERR spectroscopy is also applicable to cytochrome c although the chromophore is separated from the metal surface by the protein matrix. The spectra show no indications for any SER bands of the protein. Moreover, it was found that the enhancement factor on Ag colloids is extraordinarily high (1.1×10^6 at 457 nm), exceeding even the value for the "naked" iron protoporphyrin (Hildebrandt and Stockburger, 1986). This enhancement was interpreted in terms of a perpendicular orientation of the heme of cytochrome c with respect to the metal surface. However, the early SERR studies revealed distinct differences compared to the RR spectra of the dissolved cytochrome c, stimulating a controversy concerning a possible denaturation of the adsorbed cytochrome c. Smulevich and Spiro (1985) noted spin-state changes (to the 5cHS configuration) of cytochrome c adsorbed on Ag colloids which were attributed to the extraction of the heme from the protein, presumably initiated by the attack of Ag cations on the thioether bridges. Cotton et $al.$ (1989) observed that the transition to the 5cHS configuration was promoted by laser irradiation, suggesting a photodegradation of the adsorbed protein. However, since at low temperature no conversion to the 5cHS state occurred during the RR experiments, it is more likely that the effect of the exciting laser beam is local heating of the sample, which promotes a conformational change. On the other hand, Hildebrandt and

Stockburger (1986, 1989a) demonstrated that these spin-state changes are reversible, i.e., the adsorbed cytochrome *c*–in both the reduced and the oxidized state—exists in a reversible thermal coordination equilibrium between a 5cHS and a 6cLS configuration, with the latter prevailing at low temperature. Therefore, this 5cHS species cannot be ascribed to an irreversibly denatured form of the adsorbed protein. On the other hand, a careful inspection of the SERR spectra of the 6cLS species indicates differences compared to the RR spectrum of the dissolved 6cLS form of cytochrome *c* (Hildebrandt and Stockburger, 1986, 1989a; Cotton *et al.*, 1989).

The conformational changes of cytochrome *c* adsorbed at the Ag/electrolyte interface were analyzed in great detail using a rotating Ag electrode which minimizes local heating and thermally induced desorption due to the laser irradiation (Hildebrandt and Stockburger, 1989b). It was found that adsorption of the heme protein on the Ag electrode at a potential[3] of -0.15 V yields a SERR spectrum which originates from the adsorbed ferrocytochrome *c*. The frequencies, half-widths, and relative intensities of the SERR bands are essentially the same as for the dissolved ferrocytochrome *c*. Increasing the potential to $+0.35$ V leads to a complete oxidation of the adsorbed cytochrome *c*, and again, the SERR spectrum compares very well with the RR spectrum of the dissolved ferricytochrome *c*. However, at this positive potential, time-dependent changes are observed in the SERR spectrum which are complete after about two hours. In the spin-state marker band region, new bands appear which are indicative of a 5cHS configuration, and those bands corresponding to the 6cLS configuration are upshifted by ~ 2 cm^{-1}. These findings imply that at positive potentials, a new conformational state is stabilized which includes both a 5cHS and a 6cLS configuration, the latter being different from the solution form of cytochrome *c*. This state is also immediately obtained if adsorption of the protein is carried out at this positive potential. Once the potential is switched back to negative values, i.e., -0.25 V, this new conformational state, denoted as state II, is completely reduced and yields the characteristic marker bands of the reduced 5cHS and 6cLS configurations. However, at this negative potential the state II is not stable but slowly converts back to the original state (state I), which is closely related to the dissolved species. This implies that the adsorbed cytochrome *c* exists in a potential-dependent equilibrium of the conformational states I and II. The spectral parameters of the marker bands v_3 and v_4 of these species are listed in Table 6.1. All conformational changes below $+0.35$ V and above -0.25 V are reversible. At a potential below the potential of zero charge (~ -0.35 V), rapid and irreversible spectral changes were observed which were ascribed to an unfolding of the protein (denaturation). Based on these findings, an overall reaction scheme of the adsorbed cytochrome *c* was presented by Hildebrandt and Stockburger (1989b). The potential-dependent conformational equilibria of cytochrome *c* on the Ag electrode also provide a reasonable basis for the interpretation of previous SERR experiments of cytochrome *c* on colloidal Ag. Taking into account that small Ag particles behave as microelectrodes with a resting potential of $\sim +0.1$ V (Henglein and Lilie, 1981), the SERR spectra of cytochrome *c* adsorbed on colloidal Ag should originate predominantly from the conformational state II (6cLS and 5cHS).

---

[3] All potentials cited in the text refer to the standard hydrogen electrode.

B. Structural Changes Induced by Electrostatic Interactions

The electrode potential controls not only conformational equilibria between states I and II, but also the coordination equilibria in state II (Hildebrandt and Stockburger, 1989b). The latter also depend on the ionic strength and the kind of anions of the supporting electrolyte (see also Rospendowski et al., 1989). Hence, it was concluded that electrostatic interactions control these equilibria and are the origin for the structural changes in state II (Hildebrandt and Stockburger, 1989b). There is no indication for any specific effect of Ag or Ag cations on the protein. Therefore, it was postulated that similar structural changes should also occur for cytochrome c bound to other types of charged interfaces. Figure 6.4 compares the RR spectra in the region of the marker band v_3 of the free ferro- and ferricytochrome c with the complexes of dimyristoylphosphatidyl glycerol (DMPG). In the latter spectra, the single band of the free cytochrome c is replaced by three bands which were determined by a band-fitting analysis (Hildebrandt et al., 1990b). These components are at nearly the same positions as the v_3 bands of conformational states I and II (6cLS and 5cHS) of the adsorbed cytochrome c (cf. Table 6.1), as demonstrated by the corresponding SERR spectra, which are also included in Figure 6.4. Similar observations were made with a variety of lipid systems (see Section VII.C), inverted micelles and large polyanions (heteropolytungstates) (Hildebrandt and Stockburger, 1989c; Hildebrandt, 1990). In each of these systems, a careful analysis of the RR spectra revealed variable contributions of state I as well as of the 5cHS and 6cLS configurations of state II. In the case of the heteropolytungstates, the distributions among various species strongly depend on the size of the polyanion and, in analogy to the electrode, on the ionic strength and the pH (Hildebrandt, 1990). These results confirm the idea that the structure of the adsorbed cytochrome c is controlled by electrostatic interactions between the positively charged lysine-rich domain around the heme crevice and the negatively charged surfaces of these model systems (Hildebrandt, 1991).

SERR studies on Ag electrodes also offer the advantage of analyzing the redox processes of the adsorbed species. Hildebrandt and Stockburger (1989b) measured the SERR spectra in the region of the marker bands v_4 and v_3 as a function of the potential, choosing conditions which stabilize predominantly state I and state II, respectively (Figure 6.5). Employing a band-fitting analysis, the relative band intensities were determined for each of the adsorbed species and then converted into relative concentrations. In this way Nernstian plots were obtained for the electron transfer processes in states I and II. The slopes of the straight lines were close to the expected values for a one-electron redox couple. In state I, a redox potential of $+0.26$ V was obtained that corresponds to the value for the dissolved cytochrome c ($+0.25$ V; Margalit and Schejter, 1970). This result is not surprising, because the structure must be essentially the same in state I and in the dissolved cytochrome c as demonstrated by the excellent agreement of the RR spectra in the sensitive fingerprint region (cf. Figure 6.3a,c). A careful inspection of the RR(SERR) spectra reveals only frequency shifts in the marker band region of 1 cm^{-1} or less, which were attributed to a Stark effect on the vibrational energy levels of the porphyrin (Hildebrandt and Stockburger, 1989b; Hildebrandt, 1991).

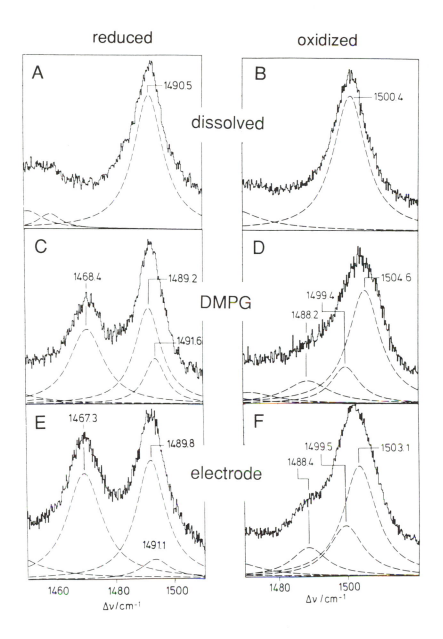

reduced

oxidized

A
1490.5

B
1500.4

dissolved

C
1468.4
1489.2
1491.6

D
1499.4
1488.2
1504.6

DMPG

E
1467.3
1489.8
1491.1

F
1499.5
1488.4
1503.1

electrode

1460   1480   1500
Δv/cm^{-1}

1480   1500
Δv/cm^{-1}

Figure 6.4

RR and SERR spectra of the reduced and oxidized cytochrome c (horse heart) in different states, excited at 407 and 413 nm. (A) and (B) are the RR spectra of the dissolved (free) cytochrome c. (C) and (D) are the RR spectra of the complexes of cytochrome c with DMPG. (E) and (F) are the SERR spectra of mixtures of the states I and II of cytochrome c adsorbed on the Ag electrode. The dashed lines represent the fitted Lorentzian bandshapes. For further experimental details, see Hildebrandt (1991), and references therein.

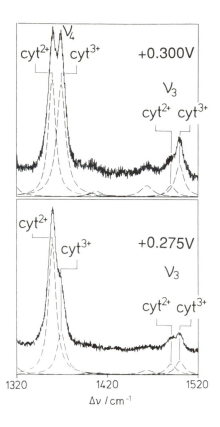

Figure 6.5

SERR spectra of cytochrome *c* (horse heart) in the conformational state I, adsorbed on the Ag electrode at different potentials (vs. standard hydrogen electrode). The excitation wavelength was 413 nm. The bands of the oxidized and reduced cytochrome *c* are denoted by "cyt^{3+}" and cyt^{2+}," respectively. The dashed lines represent the fitted Lorentzian bandshapes (Hildebrandt and Stockburger, 1989b).

The existence of a 5cHS/6cLS-coordination equilibrium in both redox forms of state II implies that one axial bond is significantly weakened. Most likely this is the Met-80 ligand, since the 695-nm absorption band vanishes upon formation of state II (Chottard *et al.*, 1987; Hildebrandt, 1990). This change of the coordination shell is not the result of an extensive rearrangement of the polypeptide chain, because the characteristic pH-dependent conformational changes of ferricytochrome *c* also take place in state II, albeit at slightly different pK_a values (Hildebrandt and Stockburger, 1989b; Hildebrandt, 1990; Vanhecke and Heremans, 1988). Consequently, it is concluded that the structural change associated with the transition to state II is restricted to the heme crevice. Complementary information was provided by recent FT-IR measurements of ferricytochrome *c* bound to phospholipid vesicles under conditions where state II prevails (Muga *et al.*, 1991). These experiments suggest that the secondary structure remains unchanged while the tertiary structure is loosened.

The structural changes of the heme pocket are sensitively reflected in the finger-print region of the RR spectrum, which clearly differs from that of the unbound cytochrome *c* (Figure 6.3; Hildebrandt, 1990, 1991). The ferric 6cLS form of state II reveals pronounced frequency shifts and intensity changes of the bands between 300 and 450 cm^{-1}. In particular, the band of the propionate bending vibrations at $\sim$446 cm^{-1} (cf. Table 6.2) nearly vanishes in state II. Furthermore, the band widths increase by $\sim$3 cm^{-1}. This has been interpreted in terms of inhomogeneous broadening, implying a higher flexibility of the heme in the protein matrix upon binding to the charged surfaces. A loosening of the heme pocket can also be inferred from the convergence of the 413- and 419-cm^{-1} bands to a single band at $\sim$416 cm^{-1}. Such loosening can be attributed to a removal of the steric constraints on the mobility of the porphyrin substituents. As inferred from the further increase of the band widths by $\sim$3 cm^{-1}, the loosening of the heme pocket structure is even more pronounced in the 5cHS form of state II, i.e., upon the dissociation of the Met-80 ligand. In the reduced form of state II, the RR(SERR) spectra reveal a comparable line broadening; other spectral characteristics, such as the intensity decrease of the propionate bending or the convergence of bands in the fingerprint region, can also be detected. These changes imply that the heme crevice is opened in the conformational state II independent of the redox state.

Such a structural change implies high solvent accessibility to the heme that should increase the local dielectric constant and cause a negative shift of the redox potential (Kassner, 1972). This expectation was in fact confirmed by potential-dependent SERR measurements (Hildebrandt and Stockburger, 1989a). Assuming that the electron transfer reactions take place without changing the spin and coordination state, a redox potential of -0.16 V is obtained for the redox couple in the 6cLS configuration. This value is in good agreement with the results obtained from heme *c* model peptides in which the heme is accessible to the solvent (Kassner, 1972). In the 5cHS configuration, the redox potential is -0.06 V, although the opening of the heme pocket is even more pronounced. Apparently, the change of the spin and coordination state has an opposite effect on the redox potential.

Based on these results, previous findings for cytochrome *c* on the Ag electrode can readily be understood. Niki *et al.* (1987) estimated a redox potential of -0.13 V from potential-dependent SERR experiments, indicating that the adsorbed cytochrome *c* was exclusively in state II. The same interpretation may hold for the early SERR results reported by Cotton *et al.* (1980) and Taniguchi *et al.* (1984). In this context it is necessary to emphasize the fact that the SERR spectroscopy exclusively probes the redox processes of the adsorbed species, and hence the results cannot directly be compared with those obtained by electrochemical methods which probe species in solution (Reed and Hawkridge, 1987; Hildebrandt and Stockburger, 1989b).

It is still not completely clear how the electrostatic interaction can cause the structural change in state II. Binding to the negatively charged surfaces (electrodes, polyanions, phospholipid vesicles) must be established via lysine residues around the exposed heme edge. Some of these residues are assumed to be essential for stabilizing the closed heme pocket structure by forming intramolecular salt bridges or hydrogen bonds (Takano and Dickerson, 1981a, 1981b). If those lysines are involved in the binding to anionic surfaces, the heme cleft may be destabilized, leading to a movement of the peptide segment 80–90 away from the heme edge (Michel *et al.*, 1989; Hildebrandt, 1991). This, in turn, would affect not only the Met-80–iron bond, but also the hydrogen bonding interactions in the buried part of the heme pocket, including the propionate side chains of the heme. The latter might be the origin for the intensity decrease of the propionate bending, as well as for the increase of the fluorescence of Trp-59 in state II (Chottard *et al.*, 1987). If binding to the charged surfaces occurs via those lysine residues, which do not have the function of stabilizing the heme pocket structure, no conformational changes are induced, and state I is formed. Following this interpretation, the transition between the states I and II requires, at first, the dissociation of intramolecular ionic bonds of specific lysines, followed by the rebinding via different lysine residues. In these model systems, the dissociation step appears to be rate-determining for the conversion between the states I and II and may be as slow as 2 hours on the electrode or faster than the resolution in the RR experiments in phospholipid systems (i.e., <1 s) (Hildebrandt and Stockburger, 1989b; Hildebrandt, 1991).

It is interesting to note that for ferricytochrome *c* the structural changes in state II are very similar to those induced by increasing the temperature (Hildebrandt *et al.*, 1991). In the reduced form, however, the thermal energy (up to 65 °C) is not sufficient to break the internal hydrogen bonds or salt bridges of the structurally relevant lysine residues, as is possible by intermolecular electrostatic interactions.

The structure of the conformational state II is not exactly the same in the various model systems, as revealed by a careful inspection of the RR/SERR spectra in the fingerprint region. Minor differences of frequencies, band widths, and relative intensities, as well as thermodynamical parameters of state II, were interpreted in terms of a gradual increase of the flexibility of the heme pocket in the order of the model systems "phospholipid vesicles" < "polyanions" < "electrode" (Hildebrandt, 1991).

C. Interactions with Phospholipids

The interactions between cytochrome *c* and model membranes have been studied by a variety of techniques (for a comprehensive review, the reader is referred to Chapter 12 by Jordi and de Kruijff in this volume). Employing RR spectroscopy, it is possible to probe structural changes in the heme pocket which are induced upon binding of cytochrome *c* to negatively charged phospholipids. Based on a band-fitting analysis of the fingerprint and marker band regions, the relative concentrations of conformational states I and II were determined in an attempt to correlate the conformational equilibria with structural properties of the lipid vesicles (Hildebrandt *et al.*, 1990b). Upon binding to dioleoylphosphatidyl glycerol (DOPG), the concentration ratio of state II to state I for ferricytochrome *c* is about 2.6, but it decreases upon adding uncharged dioleoyl glycerol (DOG; Heimburg *et al.*, 1991). These findings provide a reasonable explanation for recent results obtained by FT-IR spectroscopy and differential scanning calorimetry (Muga *et al.*, 1991). However, the RR data do not reveal a linear relationship between the conformational equilibrium of cytochrome *c* and the DOG content. In the range between 0 and 10% DOG, the decrease of the state II is much steeper than in the range between 10 and

30%. On the other hand, there is no significant effect on the binding energy of the cytochrome *c*/lipid complexes, which mainly results from the electrostatic interactions between the lysine residues and the phospholipid headgroups (Heimburg *et al.*, 1991). Hence, it was concluded that in the binary systems, binding of cytochrome *c* causes an asymmetric distribution of the charged and uncharged lipids, so that the number of DOPG molecules per cytochrome *c* binding site remains largely constant even upon dilution by DOG. Consequently, one would expect distortions of the lipid bilayer structure upon cytochrome *c* binding that are in fact observed by ^{31}P NMR spectroscopy. While in the absence of cytochrome *c* the extended bilayer structure prevails up to 30% DOG, the ^{31}P NMR spectra of the cytochrome *c*/lipid complexes are dominated by an isotropic peak which was interpreted in terms of a large number of local curvatures in the bilayer. Such curvatures may be required to adapt the lipid surface to the spherically shaped protein. The state I/state II ratio of ferricytochrome *c* bound to DOPG is independent of the temperature between 15 and 30 °C. On the other hand, the complexes of ferricytochrome *c* with DMPG show a sudden increase of the state II content at about 25 °C that corresponds to the gel-to-fluid transition temperature of the lipid. Apparently, the electrostatic interactions with the bound cytochrome *c* are different in the gel and in the fluid phase. Furthermore, it is remarkable that in both the gel and the fluid phase of DMPG, the state II content is significantly greater than in the fluid phase of DOPG. These findings cannot simply be ascribed to different surface charge densities, but indicate that there are other factors controlling the conformational equilibrium of the bound cytochrome *c*. Most likely, this is the spatial arrangement of the negative charges on the lipid surface, which depends, for example, on the ability of the phospholipid to form local curvatures. These results suggest that the specific arrangement of negative charges in the binding domains of the physiological redox partners of cytochrome *c* (see Section VII) may sensitively control the conformational equilibria of the bound cytochrome *c*. Furthermore, these equilibria may also be affected by the lipid composition of the mitochondrial membrane, pointing to a possible role of charged phospholipids in the electron transport processes in the respiratory chain.

The major anionic phospholipid component in the inner mitochondrial membrane is cardiolipin, which, in fact, is assumed to be functionally relevant for the physiological energy transduction (Spooner and Watts, 1991a,b, and references therein). Hence, the interaction of cytochrome *c* with cardiolipin has been addressed in several studies. A recent detailed NMR investigation (Spooner and Watts, 1991a,b) and an earlier EPR study (Vincent *et al.*, 1987) provide results which can indeed be interpreted in a fashion analogous to that of the DMPG and DOPG systems. However, peculiar results on the cytochrome *c*/cardiolipin complex were reported by Vincent and Levin (1986) employing RR spectroscopy. Although the visible absorption as well as EPR spectra are characteristic for the ferric form, the RR spectra bear a strong resemblance to those of the reduced form, but there are no indications of bands which are characteristic of the conformational state II. At present, there is no satisfactory explanation to reconcile these data with those obtained from other phospholipid systems. It appears to be necessary to reinvestigate the cytochrome *c*/cardiolipin complex by high-resolution RR spectroscopy using Soret excitation.

Finally, it should be mentioned that the structural changes of cytochrome *c* upon complex formation with phospholipids are also of particular interest in the wider context of membrane interactions with peripheral proteins. Taking cytochrome *c* as a representative extrinsic protein, these results can provide an important contribution to the elucidation of the mechanism of enzyme activation by charged lipids (Heimburg *et al.*, 1991; Muga *et al.*, 1991).

VII. Complexes of Cytochrome *c* with Physiological Redox Partners

There is strong evidence that the same lysine residues of cytochrome *c* that interact with the array of negative charges of the model systems are also involved in complex formation with physiological redox partners such as cytochrome b_5, cytochrome *c* oxidase, and cytochrome *c* peroxidase (cf. Hildebrandt and Stockburger, 1989c, and references therein). Thus, in principle, binding to these proteins may cause structural changes in cytochrome *c* similar to those in the case of the model systems.

A. Cytochrome *c* Oxidase

Cytochrome *c* oxidase includes two heme groups (a, a_3), which in the oxidized state exhibit absorption maxima at ~420 nm (Wikström *et al.*, 1981). Hence, using the 407-nm excitation line is favorable for the relative enhancement of the RR bands of ferricytochrome *c* $(\lambda_{max} = 410$ nm) compared to RR bands of the heme *a* chromophores of cytochrome *c* oxidase. In fact, particularly in the fingerprint region, the RR bands of cytochrome *c* oxidase are much weaker than those of cytochrome *c* (Hildebrandt *et al.*, 1990; Figure 6.6). Thus, the low-frequency RR spectrum of the fully oxidized cytochrome *c*–cytochrome *c* oxidase complex is dominated by the RR bands of ferricytochrome *c*. The RR spectrum of the complex is not identical with the sum of the RR spectra of the individual components or the RR spectrum of the physical mixture of both proteins, measured at high ionic strength. This inequivalence is demonstrated by the difference spectrum "complex minus mixture" (Figure 6.6). Since the RR bands of cytochrome *c* oxidase appear at slightly different frequencies than those of cytochrome *c*, the spectral changes can be attributed to the individual components, and it is evident that the most pronounced effects occur to the RR bands of cytochrome *c*. In particular, one notes a significant intensity decrease of the propionate bending at 447 cm^{-1}. As discussed in the previous section, this is a characteristic indicator for conformational state II. Furthermore, the spectral changes in the frequency range between 390 and 410 cm^{-1} can be taken as evidence for a partial conversion of the bound cytochrome *c* to state II. The same conclusion is reached by the analysis of the marker band region. The state II content was estimated to be between 40 and 50%. It should be noted that these results also provide a reasonable explanation for the spectral changes in the MCD and CD spectra of the cytochrome *c*–cytochrome *c* oxidase complex (Weber *et al.*, 1987; Michel *et al.*, 1989).

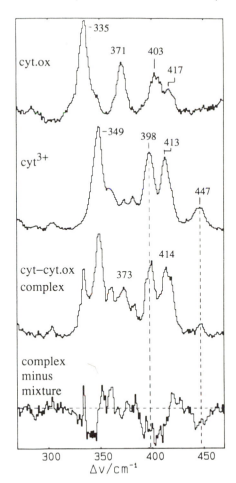

Figure 6.6

RR spectra of the fully oxidized cytochrome *c* oxidase (bovine heart–cyt.ox), oxidized cytochrome *c* (horse heart–cyt^{3+}), and the complex between both proteins (cyt–cyt.ox). The bottom spectrum is the difference between the complex and the mixture of both proteins. The excitation wavelength was 407 nm. For further experimental details, see Hildebrandt *et al.* (1990b).

In addition, the RR spectra of cytochrome *c*–cytochrome *c* oxidase reveal small spectral changes attributable to the heme groups in cytochrome *c* oxidase. These findings indicate that also the heme pockets of both heme *a* and heme a_3 are affected upon complex formation.

B. Cytochrome *c* Peroxidase

The active site of cytochrome *c* peroxidase contains an iron protoporphyrin IX (Yonetani, 1976), which, at physiological pH, exists predominantly in the 5cHS configuration with a histidine as the only axial ligand, and, to a minor extent, in a

6cHS and 6cLS configuration with a water and hydroxide ion, respectively, as the second axial ligand (Hildebrandt *et al.*, 1992). The relative contributions of these configurations can be determined by a band-fitting analysis of the RR spectra in the marker band region. Upon binding of ferricytochrome *c*, such an analysis reveals an increase of the 6cLS content from 9 to 19% at the expense of the 5cHS species, while the 6cHS form remains unchanged. On the other hand, there is no indication for the formation of state II of the bound cytochrome *c*. Identical changes of the spin and coordination equilibrium of cytochrome *c* peroxidase were induced by both horse heart and iso-1-cytochrome *c*; however, there were subtle effects on the propionate bending vibration in iso-1-cytochrome *c* upon complex formation.

According to the structural model of the cytochrome *c*–cytochrome *c* peroxidase complex based on the x-ray data of the individual components, the electrostatic and hydrogen-bonding interactions between the proteins lead to structural changes in cytochrome *c* peroxidase, the most important being a reorientation of His-181 (Poulos and Kraut, 1980b). Since this residue is hydrogen-bonded to propionate-7, which in turn is linked to Arg-48 via a water molecule, cytochrome *c*–induced repositioning of His-181 would perturb hydrogen-bonding interactions in the distal heme pocket of cytochrome *c* peroxidase (Hildebrandt *et al.*, 1992). Considering the crucial role of Arg-48 in the binding of distal ligands (Smulevich *et al.*, 1988; Edwards and Poulos, 1990), alteration of its hydrogen-bonding interactions may increase the amount of 6cLS in cytochrome *c* peroxidase in the complex. In fact, the importance of the distal hydrogen-bonding network is apparent from RR studies on the Gly-181 mutant of cytochrome *c* peroxidase. This mutant, which has lost the hydrogen-bonding capability of His-181, is completely 6cLS at pH 7.0 (Smulevich *et al.*, 1991).

C. Cytochrome b_5

The complexes of cytochrome *c* and cytochrome b_5 were studied in both the fully reduced and the fully oxidized state in the ν_3 band region (Hildebrandt and Stockburger, 1989c). These spectra reveal a downshift of the bands of ferro- and ferricytochrome *c* by 1.5 and 1.1 cm^{-1}, respectively, which is consistent with the formation of state I. There was no evidence for state II. On the other hand, the bands of cytochrome b_5 remain constant within the experimental accuracy. This implies that the electrostatic interactions are not sufficiently strong to perturb the structure of the heme pockets in both proteins. However, it should be noted that in the ν_3 band region, a contribution from state II of less than 15% would be below the limit of detection. Reinvestigation of this system should be extended to the more sensitive low-frequency region.

D. Biological Implications

While the RR spectra provided evidence for conformational state II of ferricytochrome *c* in the fully oxidized complex with cytochrome *c* oxidase, the physiological complex is formed between the reduced cytochrome *c* and the oxidized cytochrome *c* oxidase. However, all studies on model systems have shown that the formation of state II is largely independent of the oxidation state of cytochrome *c* (Hildebrandt, 1991). Thus, it is very likely that this conformational transition also occurs under physiological conditions. The functional significance of this structural change may be the lowering of the activation barrier for the electron transfer due to the opening of the cytochrome *c* heme crevice upon complex formation. This barrier depends on the reorganization energy required for the structural changes of the heme and the protein following the electron transfer step. Crystal structure data suggest that the main conformational differences between the ferrous and the ferric form exist in the vicinity of the Met-80 ligand and in the buried part of the heme crevice (Takano and Dickerson, 1981b; Berghuis and Brayer, 1992). These differences include an increase of the Fe–Met bond angle in the oxidized form, as well as movements of Asn-52 and Tyr-48. These amino acids are part of the hydrogen-bonding network that involves the propionate side chains of the heme. This is just the region of the heme pocket which is subject to conformational changes in state II. In particular, the RR spectra demonstrate a loosening of the protein structure and a weakening of the Fe–Met axial bond. Thus, the higher flexibility of the heme pocket in state II may favor the conformational relaxation subsequent to the electron transfer (Hildebrandt, 1991). Furthermore, the opening of the heme pocket also lowers the redox potential, and hence increases the driving force for the electron transfer.

While the structural changes of cytochrome *c* in the complex with cytochrome *c* oxidase may facilitate electron transfer, the situation is entirely different in the cytochrome *c*–cytochrome *c* peroxidase complex, since no contributions of state II could be detected. The observed structural changes pertain solely to cytochrome *c* peroxidase and indicate a shift of the coordination equilibrium from the 5cHS to the 6cLS configuration. In the latter configuration, the heme iron is in the porphyrin plane, and hence in a position which may favor substrate (peroxide) binding (Hildebrandt *et al.*, 1992). Furthermore, it is reasonable to assume that the formation of the 6cLS configuration of cytochrome *c* peroxidase results from coordination by a hydroxide, provided by a nearby water molecule. This would imply that an adjacent amino acid residue (e.g., His-52) becomes protonated, and that this part of the heme pocket assumes a more polar character that in turn would facilitate the catalytic splitting of the peroxide (Poulos and Kraut, 1980a). On the other hand, the first electron-transfer step involves reduction of compound I by ferricytochrome *c*. Whether or not this state of cytochrome *c* peroxidase is capable of inducing functionally relevant structural changes in the bound cytochrome *c* remains to be investigated.

VIII. Acknowledgment

I thank Professor K. Schaffner for encouragement and continuous support. Dr. G. Heibel is gratefully acknowledged for valuable discussions.

IX. References

Abe, M., Kitagawa, T., & Kyogoku, Y. (1978) *J. Chem. Phys.* **69**, 4526.

Albrecht, A. C. (1961) *J. Chem. Phys.* **34**, 1476.

Anderson, J. L., & Kincaid, J. R. (1978) *Appl. Spectrosc.* **32**, 356.

Ångström, J., Moore, G. R., & Williams, R. J. P. (1982) *Biochim. Biophys. Acta* **703**, 87.

Anzenbacher, P., Dawson, J. H., & Kitagawa, T. (1989) *J. Mol. Struct.* **214**, 149.

Barker, P. D., & Mauk, A. G. (1992) *J. Am. Chem. Soc.* **114**, 3619.

Berghuis, A. M., & Brayer, G. D. (1992) *J. Mol. Biol.* **223**, 959.

Bobinger, U., Schweitzer-Stenner, R., & Dreybrodt, W. (1989) *J. Raman Spectrosc.* **20**, 191.

Bushnell, G. W., Louie, G. V., & Brayer, G. D. (1990) *J. Mol. Biol.* **214**, 585.

Cartling, B. (1983) *Biophys. J.* **43**, 191.

Cartling, B. (1988) in *Biological Applications of Raman Spectroscopy* (Spiro, T. G., ed.), Vol. 3, p. 217, Wiley, New York.

Choi, S., & Spiro, T. G. (1983) *J. Am. Chem. Soc.* **105**, 3683.

Choi, S., Spiro, T. G., Langry, K. C., & Smith, K. M. (1982a) *J. Am. Chem. Soc.* **104**, 4337.

Choi, S., Spiro, T. G., Langry, K. C., Smith, K. M., Budd, D. L., & La Mar, G. N. (1982b) *J. Am. Chem. Soc.* **104**, 4345.

Chottard, G. (1989) *Biochim. Biophys. Acta* **997**, 155.

Chottard, G., Michelon, M., Hervé, M., & Hervé, G. (1987) *Biochim. Biophys. Acta* **916**, 402.

Cohen, J. S., & Hayes, M. B. (1974) *J. Biol. Chem.* **249**, 5472.

Copeland, R. A., & Spiro, T. G. (1985) *Biochemistry* **24**, 4960.

Cotton, T. M. (1988) in *Spectroscopy of Surfaces* (Clark, R. J. H., & Hester, R. E., eds.), Wiley, New York, p. 91.

Cotton, T. M., Schultz, S. G., & Van Duyne, R. P. (1980) *J. Am. Chem. Soc.* **102**, 7960.

Cotton, T. M., Schlegel, V., Holt, R. E., Swanson, B., & Ortiz de Montellano, P. (1989) *SPIE* **1055**, 263.

Czernuszewicz, R. S., Li, X.-Y., & Spiro, T. G. (1989) *J. Am. Chem. Soc.* **111**, 7024.

Edwards, S. L., & Poulos, T. L. (1990) *J. Biol. Chem.* **264**, 2588.

Feng, Y., Roder, H., Englander, S. W., Wand, A. J., & Di Stefano, D. L. (1989) *Biochemistry* **28**, 195.

Forster, M., Hester, R. E., Cartling, B., & Wilbrandt, R. (1982) *Biophys. J.* **38**, 111.

Friedman, J. M., & Hochstrasser, R. M. (1973) *Chem. Phys.* **1**, 457.

Gadsby, P. M. A., Peterson, J., Foote, N., Greenwood, C., & Thomson, A. J. (1987) *Biochem. J.* **246**, 43.

Gao, Y., Boyd, J., Pielak, G. J., & Williams, R. J. P. (1990) *Biochemistry* **29**, 6994.

Gao, Y., Boyd, J., Pielak, G. J., & Williams, R. J. P. (1991a) *Biochemistry* **30**, 1928.

Gao, Y., Boyd, J., Pielak, G. J., & Williams, R. J. P. (1991b) *Biochemistry* **30**, 7033.

Gladkov, L. L., & Solovyov, K. N. (1985a) *Spectrochim. Acta* **41A**, 1437.

Gladkov, L. L., & Solovyov, K. N. (1985b) *Spectrochim. Acta* **41A**, 1443.

Gladkov, L. L., & Solovyov, K. N. (1986) *Spectrochim. Acta* **42A**, 1.

Greenwood, C., & Palmer, G. (1965) *J. Biol. Chem.* **240**, 3660.

Gupta, R. K., & Koenig, S. H. (1971) *Biochem. Biophys. Res. Comm.* **45**, 1134.

Heimburg, T., Hildebrandt, P., & Marsh, D. (1991) *Biochemistry* **30**, 9084.

Henglein, A., & Lilie, J. (1981) *J. Am. Chem. Soc.* **103**, 1059.

Hildebrandt, P. (1990) *Biochim. Biophys. Acta* **1040**, 175.

Hildebrandt, P. (1991) *J. Mol. Struct.* **242**, 379.

Hildebrandt, P., & Stockburger, M. (1986) *J. Phys. Chem.* **90**, 6017.

Hildebrandt, P., & Stockburger, M. (1989a) *Vib. Spectr. Struct.* **17A**, 443.

Hildebrandt, P., & Stockburger, M. (1989b) *Biochemistry* **28**, 6710.

Hildebrandt, P., & Stockburger, M. (1989c) *Biochemistry* **28**, 6722.

Hildebrandt, P., Heimburg, T., Marsh, D., & Powell, G. L. (1990a) *Biochemistry* **29**, 1661.

Hildebrandt, P., Heimburg, T., & Marsh, D. (1990b) *Eur. Biophys. J.* **18**, 193.

Hildebrandt, P., Pielak, G. J., & Williams, R. J. P. (1991) *Eur. J. Biochem.* **201**, 211.

Hildebrandt, P., English, A., & Smulevich, G. (1992) *Biochemistry* **31**, 2384.

Kassner, R. J. (1972) *Proc. Natl. Acad. Sci. USA* **69**, 2263.

Kitagawa, T., & Ozaki, Y. (1987) *Struct. Bonding* **64**, 72.

Kitagawa, T., Kyogoku, Y., Iizuka, T., Ikeda-Saito, M., & Yamanaka, T. (1975) *J. Biochem.* **78**, 719.

Kitagawa, T., Ozaki, Y., Teraoka, J., Kyogoku, Y., & Yamanka, T. (1977) *Biochim. Biophys. Acta* **494**, 100.

Kitagawa, T., Abe, M., & Ogoshi, H. (1978) *J. Chem. Phys.* **69**, 4516.

Koglin, E., & Séquaris, J.-M. (1986) *Topics Curr. Chem.* **134**, 1.

Koppenol, W. H., & Margoliash, E. (1982) *J. Biol. Chem.* **257**, 4426.

Kubitschek, U., Dreybrodt, W., & Schweitzer-Stenner, R. (1986) *Spectrosc. Lett.* **19**, 681.

Lanir, A., Yu, N.-T., & Felton, R. H. (1979) *Biochemistry* **18**, 1656.

Lee, H., Kitagawa, T., Abe, M., Pandey, R., Leung, H.-K., & Smith, K. M. (1986) *J. Mol. Struct.* **146**, 329.

Li, X.-Y., Czernuszewicz, R. S., Kincaid, J. R., & Spiro, T. G. (1989) *J. Am. Chem. Soc.* **111**, 7012.

Li, X.-Y., Czemuszewicz, R. S., Kincaid, J. R., Su, Y. O., & Spiro, T. G. (1990a) *J. Phys. Chem.* **94**, 31.

Li, X.-Y., Czernuszewicz, R. S., Kincaid, J. R., Stein, P., & Spiro, T. G. (1990b) *J. Phys. Chem.* **94**, 47.

Liu, G.-Y., Grygon, C. A., & Spiro, T. G. (1989) *Biochemistry* **28**, 5046.

Louie, G. V., & Brayer, G. D. (1989) *J. Mol. Biol.* **209**, 313.

Louie, G. V., & Brayer, G. D. (1990) *J. Mol. Biol.* **214**, 527.

Louie, G. V., Pielak, G. J., Smith, M., & Brayer, G. D. (1988) *Biochemistry* **27**, 7870.

Margalit, R., & Schejter, A. (1970) *FEBS Lett.* **6**, 278.

Margoliash, E., & Schejter, A. (1966) *Adv. Prot. Chem.* **21**, 113.

Mauk, A. G. (1991) *Struct. Bonding* **75**, 131.

Michel, B., Proudfoot, A. E. I., Wallace, C. J. A., & Bosshard, H. R. (1989) *Biochemistry* **28**, 456.

Muga, A., Mantsch, H. H., & Surewicz, W. K. (1991) *Biochemistry* **30**, 7219.

Myer, Y. P., & Saturno, A. F. (1991) *J. Prot. Chem.* **10**, 481.

Myer, Y. P., Srivastava, R. B., Kumar, S., & Raghavendra, K. (1983) *J. Prot. Chem.* **2**, 13.

Nafie, L. A., Pezolet, M., & Peticolas, W. L. (1973) *Chem. Phys. Lett.* **20**, 563.

Niki, K., Kawasaki, Y., Kimura, Y., Higuchi, Y., & Yasuoka, N. (1987) *Langmuir* **3**, 982.

Parthasarathi, N., Hanson, C., Yamaguchi, S., & Spiro, T. G. (1987) *J. Am. Chem. Soc.* **109**, 3865.

Pearce, L. L., Gärtner, A. L., Smith, M., & Mauk, A. G. (1989) *Biochemistry* **28**, 3152.

Pettigrew, G. W., & Moore, G. R. (1987) *Cytochrome c—Biological Aspects*, Springer Verlag, Berlin.

Pielak, G. J., Atkinson, R. A., Boyd, J., & Williams, R. J. P. (1988) *Eur. J. Biochem.* **177**, 179.

Poulos, T. L., & Kraut, J. (1980a) *J. Biol. Chem.* **255**, 8199.

Poulos, T. L., & Kraut, J. (1980b) *J. Biol. Chem.* **255**, 10322.

Rafferty, S. P., Pearce, L. L., Barker, P. D., Guillemette, J. G., Kay, C. M., Smith, M., & Mauk, A. G. (1990) *Biochemistry* **29**, 9365.

Reed, D. E., & Hawkridge, F. M. (1987) *Anal. Chem.* **59**, 2334.

Rospendowski, B., Schlegel, V. L., Holt, R. E., & Cotton, T. M. (1989) in *Charge and Field Effects in Biosystems* (Allen, M. J., Cleary, S. F., & Hawkridge, F. M., eds.), Vol. 2, Plenum Press, New York, p. 43.

Rousseau, D. L. (1981) *J. Raman Spectrosc.* **10**, 94.

Schomacker, K. T., & Champion, P. M. (1989) *J. Chem. Phys.* **90**, 5982.

Shelnutt, J. A., Rousseau, D. L., Dethmers, J. K., & Margoliash, E. (1981) *Biochemistry* **20**, 6485.

Smulevich, G., & Spiro, T. G. (1985) *J. Phys. Chem.* **89**, 5168.

Smulevich, G., Mauro, J. M., Fishel, L. A., English, A. M., Kraut, J., & Spiro, T. G. (1988) *Biochemistry* **27**, 5477.

Smulevich, G., Miller, M. A., Kraut, J., & Spiro, T. G. (1991) *Biochemislry* **30**, 9546.

Spaulding, L. D., Chang, C. C., Yu, N.-T., & Felton, R. H. (1975) *J. Am. Chem. Soc.* **97**, 2517.

Spiro, T. G. (1983) in *Iron Porphyrins* (Lever, A. B. P., & Gray, H. B., eds.), Part II, p. 89, Addison-Wesley, Reading, Massachusetts.

Spiro, T. G. (1988) *Biological Applications of Raman Spectroscopy*, Vols. 2 and 3, Wiley, New York.

Spiro, T. G., & Strekas, T. C. (1972a) *Biochim. Biophys. Acta* **278**, 188.

Spiro, T. G., & Strekas, T. C. (1972b) *Proc. Natl. Acad. Sci. USA* **69**, 2622.

Spiro, T. G., Stong, J. D., & Stein, P. (1979) *J. Am. Chem. Soc.* **101**, 2648.

Spooner, P. J. R., & Watts, A. (1991a) *Biochemistry* **30**, 3871.

Spooner, P. J. R., & Watts, A. (1991b) *Biochemistry* **30**, 3880.

Stallard, B. R., Callis, P. R., Champion, P. M., & Albrecht, A. C. (1984) *J. Chem. Phys.* **80**, 70.

Takano, T., & Dickerson, R. E. (1981a) *J. Mol. Biol.* **153**, 79.

Takano, T., & Dickerson, R. E. (1981b) *J. Mol. Biol.* **153**, 95.

Taniguchi, I., Iseki, M., Yamaguchi, H., & Yasakouchi, K. (1984) *J. Electroanal. Chem.* **175**, 341.

Theorell, H., & Åkesson, Å. (1941) *J. Am. Chem. Soc.* **63**, 1804.

Uno, T., Nishimura, Y., & Tsuboi, M. (1984) *Biochemistry* **23**, 6802.

Valance, W. G., & Strekas, T. C. (1982) *J. Phys. Chem.* **86**, 1804.

Vanhecke, F., & Heremans, K. (1988) in *The Spectroscopy of Biological Molecules–New Advances* (Schmidt, E. D., Siebert, F., & Schneider, F. W., eds.), Wiley, New York, p. 343.

Vincent, J. S., & Levin, I. W. (1986) *J. Am. Chem. Soc.* **108**, 3551.

Vincent, J. S., Kon, H., & Levin, I. R. (1987) *Biochemistry* **26**, 2312.

Weber, C., Michel, B., & Bosshard, H. R. (1987) *Proc. Natl. Acad. Sci. USA* **84**, 6687.

Wikström, M., Krab, K., & Saraste, M. (1981) *Cytochrome Oxidase–A Synthesis*, Academic Press, New York.

Yonetani, T. (1976) in *The Enzymes* (Boyer, P. D., ed.), Vol. 13, Academic Press, New York, p. 345.

Yu, N.-T., & Srivastava, R. B. (1980) *J. Raman Spectrosc.* **9**, 166.

Zgierski, M. Z. (1988) *J. Raman Spectrosc.* **19**, 23.

PART

IV

Thermodynamic Properties

7

Direct Electrochemistry of Cytochrome *c*

H. ALLEN O. HILL

Inorganic Chemistry Laboratory
Oxford University

L. H. GUO AND GEORGE McLENDON

Department of Chemistry
University of Rochester

I. Introduction

Since the biological function of cytochrome *c* is to carry out oxidation–reduction reactions, it follows that measurements of the oxidation–reduction potentials and associated electron transfer kinetics of cytochrome *c* are central to understanding this protein. These key properties can, in principle, be well addressed by electrochemical techniques.

Two basic types of electrochemical measurements (Bard and Faulkner, 1980) might be considered: potentiometric techniques, where no current flows in the measuring circuit, and voltammetry, which measures current as a function of the potential applied at the working electrode.

Potentiometric techniques have long been used to measure protein redox potentials. Soon after cytochrome *c* was isolated by Keilin, its formal potential was estimated by analytic potentiometric titration using a redox-active dye. In such experiments, the protein generally does not react directly with the electrode. Rather, the electrode and protein each equilibrate with a redox-active small molecule, known as a "mediator." A wide variety of mediators are available, ranging from simple organic dyes to organometallic reagents such as ferrocene. When the dye and protein equilibrate, the Nernstian potential of the dye is sensed at the electrode, while the ratio of Fe(II)cytochrome c/Fe(III)cytochrome c can be determined by an independent spectrophotometric measurement. In an important variant of this approach, the electrode itself may serve as the spectrophotometric "cell." This technique of optically transparent thin layer electrochemistry (OTTLE) played a role in the discovery of direct electrochemistry of cytochrome c, for which mediators are not required to permit equilibration of the protein and electrode, as discussed subsequently.

In a voltammetry experiment, current flows between the electrode and redox-active species. The current is measured as a function of potential (i.e., the potential difference between the working and auxiliary electrodes, measured relative to a reference electrode such as Ag/AgCl). The most popular voltammetric technique is cyclic voltammetry, in which the potential is ramped at a constant rate between preset voltage limits and the voltage waveform reversed (thus "cyclic").

For an ideal, diffusion-controlled electron transfer, this produces a characteristic waveform for the current–voltage plot, as shown in Figure 7.1. Deviations in the shape of this waveform can result if the electron transfer is not diffusion limited, reflecting, for example, slow electron transfer kinetics or reactant adsorption–desorption (*vide infra*). While the cyclic voltammetry experiment works extremely well for small molecules at a "simple" (untreated) electrode surface, early attempts to use these techniques for redox proteins uniformly failed: No currents were observed. This uniform failure resulted in a number of theories of protein electro(in)activity, which mercifully will be left forgotten.

In 1976, one of us (HAOH) reopened investigations of the fundamentals of protein electrochemistry (Eddowes and Hill, 1977). This work was influenced, in part, by the general work on modified electrodes (Murray, 1980) that was gaining currency in the electrochemical community. It was reasoned, based on Taube's work on mediating inner-sphere electron transfer, that pyrazine-type ligands might be able to attach to the electrode and mediate transfer to a protein such as cytochrome c. Consequently, an undergraduate student. Mark Eddowes, took as an undergraduate thesis project the exploration of chemically modified electrodes as an approach to direct protein electrochemistry. Persevering through many unsuccessful attempts, he discovered that 4,4'-bipyridyl would bind to a gold electrode to provide a hydrophilic surface, which in turn would (transiently) bind cytochrome c. This engineered surface then produced a reversible cyclic voltammogram like that shown in Figure 7.1. Clearly, the basic strategy of chemical modification to promote protein electrochemistry was viable.

At about the same time, Kuwana, using optically transparent electrodes, surprisingly discovered (Yeh and Kuwana, 1977) that a mediator was not required to obtain a current between cytochrome c and the transparent electrode (In-doped) SnO_2. As discussed in a subsequent section, SnO_2 also provides essentially a "modified" surface in the form of surface OH functions which can recognize and bind to cytochrome c.

These keynote papers formed the basis for direct protein electrochemistry. Although this review will properly focus only on cytochrome c electrochemistry, we note that it has been possible to develop specific electrode modification protocols

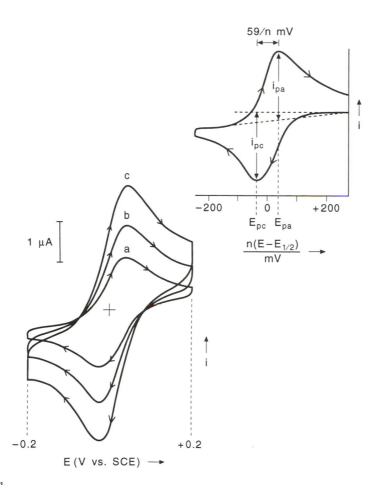

Figure 7.1
Cyclic voltammograms of a diffusion-controlled, reversible redox couple (upper), and horse heart cytochrome c at a gold electrode in the presence of saturated 4,4′-bipyridine (lower). 400 μM cytochrome c in 20 mM phosphate/100 mM perchlorate, pH 7.0. Scan rate: (a) 20 mV/s, (b) 50, (c) 100.

to facilitate the (specific) electrochemistry of a wide variety of redox proteins (Armstrong *et al.*, 1986; Frew and Hill, 1988; Armstrong *et al.*, 1988; Armstrong, 1990), both cationic and anionic. Most recently, it has even been possible to demonstrate the direct electrochemistry of enzymes, (Guo and Hill, 1991) resulting in direct electrochemical enzymatic catalysis, with all the associated implications for applications in (bio)sensor technology.

II. Electrode Modification

Once the direct electrochemistry of cytochrome c had been demonstrated, the next task was to develop an understanding of the modified electrode surface and explore what structural features made possible the electrochemistry response which had pre-

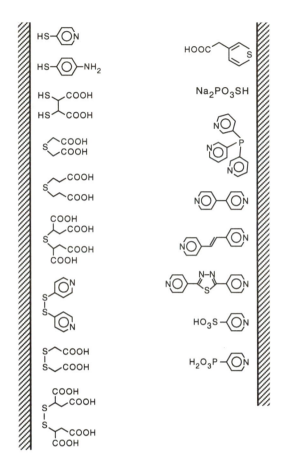

Figure 7.2
Structures and proposed surface conformations of the gold electrode modifiers for the promotion of direct electrochemistry of cytochrome *c*.

viously eluded investigators for nearly 50 years. Thus, over the last few years, major efforts were devoted to understanding the factors which led to direct voltammetric response for proteins so that electrodes could be "designed" for protein electrochemistry. Key studies in this vein were carried out by N. J. Walton, who studied the effects of more than 50 surface modifiers for promoting the reversible electrochemistry to cytochrome *c* at a (modified) gold electrode (Allen *et al.*, 1984) (Figure 7.2). Some of the relevant criteria include the concentration-normalized integrated current, the cathodic–anodic peak separations in the cyclic voltammogram, and long-term electrode stability under repeated scan conditions.

From these studies. the following general conclusions can be drawn:

1. The electrode must possess a hydrophilic surface which is complementary to the binding surface of the protein. For the most successful promoters (e.g., the

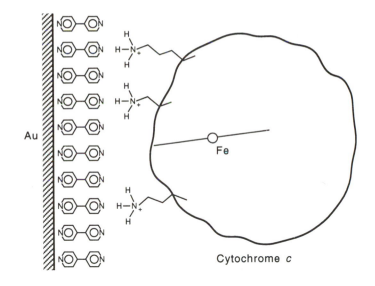

Figure 7.3

Schematic representation of the assumed binding mechanism between the surface lysine amino acid residuals on cytochrome c and the gold electrode modifier, 4,4′-bipyridine.

pyridines), ellipsometry studies have shown that the promoter forms a near close-packed monolayer on gold, with the pyridine function nearly perpendicular to the surface (Elliott *et al.*, 1986).

2. This surface serves to reversibly bind the protein and perhaps to orient it with respect to the metal surface (Figure 7.3). Since electron transfer rates depend exponentially on distance (McLendon, 1988) a small change in orientation could drastically affect the electron transfer rate (Figure 7.4).

3. Finally, a key aspect of the surface modification is to prevent irreversible, adventitious adsorption, including hydrophobic adsorption (perhaps with denaturation) of the protein itself.

We now examine the basis for these conclusions. As detailed elsewhere in this volume, cytochrome c binds to other proteins (e.g., cytochrome oxidase, cytochrome c peroxidase) and to the modified electrodes though a group of lysine residues centered around the heme edge (Lys 13, 27, 72, 79). In order to interact with these lysines, the surface promoter must have a group capable of hydrogen bonding. Thus, 4-thiopyridine acts as an efficient promoter, but isosteric thiophenol does not.

For a gold electrode, the necessary promoters are added to the solution: Generally they contain a group, such as a thiol, which strongly binds to the gold surface. For other types of electrodes, including the aforementioned SnO_2 electrode and the commonly used graphite electrodes. the functionality necessary to recognize and bind cytochrome c may be intrinsic to the surface. For SnO_2, OH functionalities are

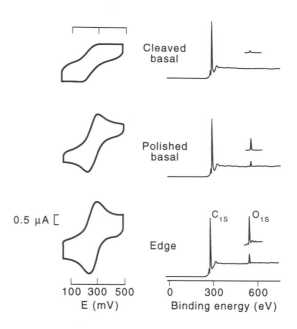

Figure 7.4
Cyclic voltammograms (left) of horse heart cytochrome c at various surfaces of pyrolytic
graphite characterized by x-ray photoelectron spectroscopy (right). Voltammetry was
measured on the fourth scan at 20 mV/s with 150 μM cytochrome c in 5 mM Tricine/
0.1 M NaCl, pH 8.0, 20 °C. The voltammetric potential is referenced against the normal
hydrogen electrode. For XPS, a scale enlargement ($\times 3$) is shown for the O_{1s} peak.

naturally found on the oxide surface. and these can act as the requisite H-bond
acceptors to promote the reversible binding of cytochrome c.

For graphite, the situation is slightly more complex, depending on the way the
electrode has been prepared. In pyrolytic graphite, two quite different surfaces
might be distinguished. Basal plane graphite (BPG) presents a relatively smooth
surface formed by the continuous graphitic structure, while in "edge-plane graphite" (EPG), the graphite rings have been oxidized along (edge-step) termination
sites, thereby providing a highly functionalized surface which includes both alcohol
and carboxylate functions (Figure 7.5). Data from XPS and other techniques have
shown high O/C ratios at EPG but not BPG. These differences result in very
different voltammetric responses for cytochrome c at BPG and EPG (Figure 7.4).
EPG provides high voltammetric currents, with a peak shape consistent with reversible electrochemistry at or near the diffusion-controlled regime. In contrast, BPG,
which is far less functionalized, shows a very different electrochemical response, in
which clear "peaks" in the cyclic voltammogram cannot really be identified. Ini-

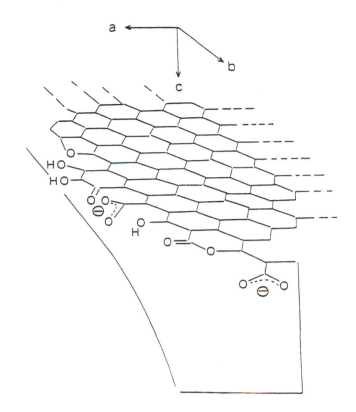

Figure 7.5

The anisotropic structure of pyrolytic graphite: The basal (*a–b*) plane and the edge (*b–c*) plane. Some of the C–O functional groups resulting from polishing in air or chemically/ electrochemically oxidizing the surface are indicated.

tially, it might be assumed that this shape reflects slow (quasi-reversible) electron transfer.

Recent theoretical and experimental work at Oxford suggest a more subtle explanation. Bond suggested (Armstrong *et al.*, 1989; Bond *et al.*, 1992) that if electrochemistry *only* occurs at the functionalized sites, and if the density of functionalized sites were sufficiently low, then the linear diffusion condition which gives the familiar cyclic voltammogram (CV) peak shape will no longer hold. Linear diffusion requires that the active electrode area be much greater than the dimensions of the electroactive molecule. If the site density is low, then the dimension of the "active" electrode surface is made similar to those of the protein, and a "microelectrode" model, characterized by radial diffusion, is appropriate. The peak shape expected for radial diffusion is much different (Figure 7.6).

This insight suggests that the apparent differences in electrochemical response at various graphite electrodes primarily reflect the history of the electrode and specifically the degree of functionalization of the electrode. This model for cyto-

324

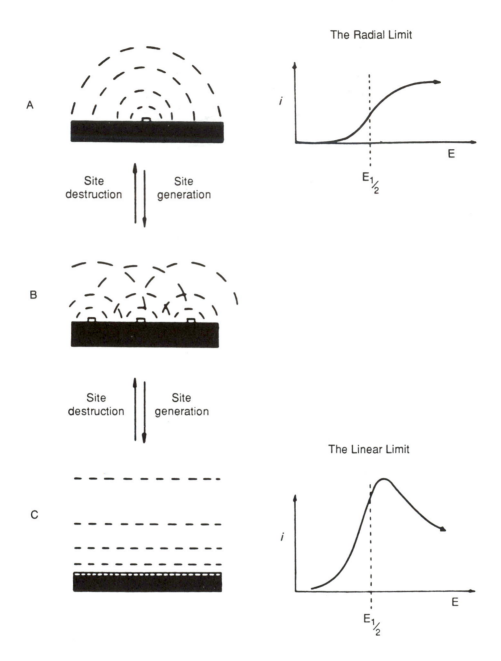

Figure 7.6
Illustration of the conversion of the voltammogram from sigmoidal shape to a peak due to the change of mass transfer from radial to linear diffusion, as the density of specific electroactive surface sites increases (A through C).

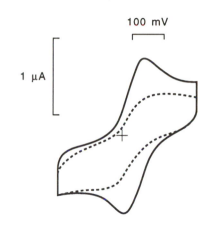

Figure 7.7
Cyclic voltammogram of horse heart cytochrome c at native (solid line) and derivatized (dotted line) edge-plane graphite electrodes. Potential versus SCE. 20 mV/s. The derivatized electrode was silanized by reacting with N-trimethylsilylimidazole.

chrome c has been tested by selectively blocking oxygen functionality of EPG by treatment with silanes (Armstrong and Brown, 1987). Such treatment leads to a smooth transition from linear diffusion to radial diffusion (Figure 7.7). Furthermore, this transformation can be reversed by removing the silyl group with KOH to restore the linear diffusion case. Similarly, if polycations (e.g., Cr^{3+}, Mg^{2+}) are bound to the oxygen surface, this similarly decreases the available site density for electrochemistry of the polycation cytochrome c (Armstrong et al., 1984). [Conversely, such treatment *increases* the site density for anionic proteins such as plastocyanin (Armstrong et al., 1985).]

III. Protein Modification Studies

To understand the factors that promote reversible electroactive protein binding, a complementary approach to modifying the electrode is to modify the protein. This has been done by conventional chemical modification techniques (e.g., lysine acetylation (Hill and Whitford, 1987)), or more recently by the newer genetic engineering techniques. (Burrows et al., 1991). Both approaches give similar results: Modification of any single lysine does not destroy the electrochemical response. Clearly, the protein–electrode recognition is rather plastic: If one lysine is removed, a small rotational reorientation of the protein can allow another stereochemically similar lysine to bind. This facile reorientation ability is important in understanding the electrochemistry of protein complexes (e.g., cytochrome c–cytochrome b_5) as outlined subsequently. Indeed, such facile reorientation probably plays a key role in protein-to-protein electron transfer within protein complexes. If multiple lysines are altered, then electrochemical response can be drastically diminished.

By binding the heme edge lysines, the promoter not only binds cytochrome c, but also orients it so that the heme edge is as close as possible to the surface, thereby minimizing the tunneling distance. However, even single-site protein modification need not be entirely "innocent": If the protein is modified in a way that increases surface hydrophobicity (e.g., Lys 27 → Ile), then the electrochemical response may be significantly diminished. It appears that a key aspect of electrode modification is not only to facilitate productive binding in an optimal orientation for electron transfer, but also to minimize binding in an unproductive mode. For example, an untreated gold electrode surface exposed to ambient laboratory conditions is relatively hydrophobic, and proteins readily adsorb via unfolding to expose hydrophobic sites. When the surface is modified to be hydrophilic (e.g., with thiopyridine), adsorptive blocking of the surface is minimized.

This simple picture predicts that protein adsorption and associated electrochemistry should be quite sensitive to the nature of hydrophilic versus hydrophobic groups at the surface. A simple mutagenesis experiment supports this notion. When the mutant cationic residue at position 13 in cytochrome c (lysine in horse, arginine in yeast) is replaced by a hydrophobic isoleucine, a large change in electrochemical response is observed. The mutant Ile 13 shows the most significant difference in the voltammogram in comparison with wild-type protein: On the first cycle, the peak separation is relatively larger (80 mV); on the following scans both cathodic and anodic currents decrease gradually accompanied by an increase in peak separation leading to a change from peak-shaped to sigmoidal curves (Figure 7.8). A sigmoidal cyclic voltammogram (CV) of cytochrome c can be observed if a macroelectrode is incompletely covered with a promoter. When the size of the local area of adsorbed promoter approaches the molecular dimension, the macroelectrode will in effect behave like a multi-microelectrode with mode of diffusion dominated by radial rather than linear motion. Modification of a gold electrode by dipping in a 1 mM solution of 4,4'-bipyridyl disulfide (SSBP) for two minutes usually results in a peak-shaped CV of cytochrome c. However, if the protein undergoes adsorption at the modified electrode surface with loss of electrochemical activity, the density of active sites on the electrode will be decreased, and a sigmoidal-shaped response is likely to appear.

To investigate further the possible adsorption of cytochrome c mutants on this electrode, two control experiments have been performed. In one of them, the modified electrode is first put into a solution of the Ile 13 cytochrome c, and the corresponding CV is recorded on continuous scans until the curve is sigmoidal in shape. Then the electrode is taken out and rinsed well with water before it is placed in a wild-type protein solution which would give a near-reversible response at freshly modified gold. However, the voltammetric curve is still sigmoidal-shaped on the first scan. On the following scans, the currents grow in with better defined cathodic and anodic peaks, and a near-reversible CV is eventually restored. This experiment suggests that the Ile 13 variant does adsorb on the electrode surface and blocks the active sites. But the adsorption is reversible, and the adsorbates are in equilibrium with the bulk species, because they desorb readily from the electrode when the electrode is placed in a solution containing the wild-type protein.

Examples from studies of mutations at other positions are summarized in Table 7.1. Overall, such studies emphasize how careful a balance is needed for protein recognition at the electrode surface in order to obtain reversible electrochemistry.

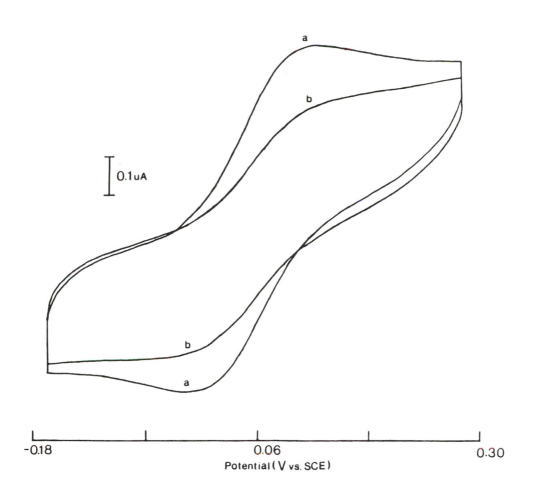

Figure 7.8
Cyclic voltammograms of the yeast cytochrome c mutant, Lys13Ile (180 μM), at a gold electrode modified with 4,4'-bipyridyl disulfide. 20 mV/s, 10 mM HEPES/0.1 M KCl, pH 7.4. (a) Initial scan. (b) Final scan.

IV. Protein–Protein Complexes

As noted earlier, current results are consistent with the hypothesis that specific recognition between cytochrome c and an electrode involves the surface lysine groups of the protein which interact with the functionalities of the electrode surface through hydrogen bonding and/or salt bridging. Thus, an analogy between binding of cytochrome c to an electrode and to another biologicial protein partner is apt.

Most recently this biological analogy has been extended to studies in which direct electrochemistry has been observed within a *ternary* complex formed between the electrode and a protein/protein complex such as cytochrome c/cytochrome b_5. Here the electrode may be viewed not only as an electron source, but also as a separate phase with a unique local chemical potential at which the two redox pro-

Table 7.1
Summary of the cyclic voltammetric results of the cyt c mutants (in 100 mM KCl/ 10 mM HEPES, pH 7.4) at modified gold and EPG electrodes.

	$i_P^o \{A[cm^2(V \ s^{-1})^{1/2} \ mol \ cm^{-3}]^{-1}\}$[a]	ΔE_P (mV)[b]	$E_{1/2}$ (mV)[c]
Wild type	167[d]	71	+39
	301[e]	80	+41
Asn52Ile	211[d]	67	−10
	345[e]	68	−10
Asn52Ala	225[d]	70	+18
	263[e]	152	+12
Lys72Asp	161[d] [e,f]	66	+31
Lys27Gln	86[d] [e,f]	74	+33
Arg13Ile	104[d] [e,f]	68	+62

[a] Normalized peak current $i_P^o = i_P/(Ac_o v^{1/2})$ (at 20 mV s^{-1}).
[b] Peak separation at 100 mVs^{-1}.
[c] Midpoint potential (vs. SCE) measured at 20 mV s^{-1}.
[d] Associated with modified gold electrodes.
[e] Associated with EPG electrodes.
[f] Deterioration of the voltammograms prevented precise measurements.

teins can be specifically recognized and bound. It is thus analogous to the biological membrane which supports redox activity *in vivo*. To form an electroactive ternary complex, the binding domains and (electro)active sites of all three components must be simultaneously optimized. Initial studies by Barker (Barker *et al.*, 1988; Bagby *et al.*, 1990) and subsequent extended studies by Burrows, Guo, and others (Burrows *et al.*, 1991) have in fact shown that it is possible to obtain facile electron transfer within a ternary protein/protein/electrode adduct—e.g., cytochrome c/cytochrome b_5/graphite.

While cytochrome c gives a strong voltammetric response at EPG, cytochrome b_5 is normally not electroactive at the EPG surface. However, when in the presence of cytochrome c, cytochrome b_5 can bind to cytochrome c while the latter is bound to graphite. This ternary complex then simultaneously promotes electrochemistry of both cytochrome c *and* cytochrome b_5 (Figure 7.9). In a key control experiment, Paul Barker replaced Fe cytochrome c by Zn cytochrome c, which cannot transfer electrons below 0.8 V versus SCE. Nonetheless, cytochrome b_5 showed electroactivity when Zn cytochrome c was added to the solution. thereby clearly demonstrating that cytochrome c serves to bind and activate cytochrome b_5. A surprise in these studies is that the *same* lysines which are known to recognize the electrodes are *also*

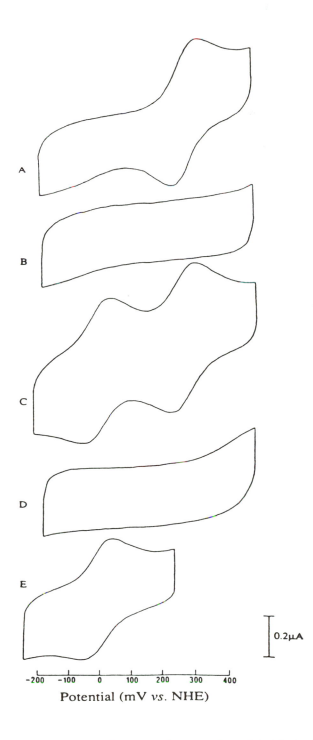

Potential (mV *vs.* NHE)

Figure 7.9

Cyclic voltammograms of (A) horse heart cytochrome *c* alone (95 μM), (B) bovine cytochrome b_5 alone (95 μM), (C) cytochrome *c* and cytochrome b_5 (95 μM each), (D) Zn-substituted cytochrome *c* alone (75 μM), and (E) cytochrome b_5 and Zn–cytochrome *c* (75 μM each) at an edge-plane graphite electrode. 1 mM HEPES/20 mM KCl, pH 7.0, 10 mV/s.

known to recognize b_5. Not surprisingly, these ternary c–b_5–electrode adducts are particularly sensitive to single-site mutations of the critical cytochrome c binding surface (Table 7.2).

The structural and mechanistic implications of a ternary adduct of protein/protein/electrode are of interest not only in bioelectrochemistry and analysis, but also for their biochemical implications. Direct electrochemistry has helped clarify biological issues in the binding and recognition of redox proteins. The ternary protein/protein/electrode complex might be viewed as a model for the important phase equilibria involving cytochromes and the biological membranes in which redox enzymes such as cytochrome oxidase are found. It had been heretofore assumed that the same recognition surface on cytochrome c is involved in binding to other redox proteins, as well as to the electrode surface. This recognition site (or sites) is believed to be primarily electrostatic, focused around lysines 13, 27, 72, and 79. The observation that cytochrome c promotes the electrochemistry of cytochrome b_5 shows that the latter and the electrode cannot compete for the same site. Rather, a variety of (partially overlapping) binding sites may exist on a redox protein such as cytochrome c. These sites may recognize the electrode, or another protein, or both, by a delicate balance of binding interactions. The awareness of such multiple sites is growing in part because of recent careful studies of binding equilibria and dynamics of redox proteins, which are mostly easily interpreted if multiple sites in rapid dy-

Table 7.2

Summary of the cyclic voltammetric results of the complexes between cyt c mutants and cyt b_5 (in 10 mM KCl/1 mM HEPES, pH 7.0). Potential values are quoted with respect to SCE.

	$i_P^\circ \{A[cm^2(V\ s^{-1})^{1/2}\ mol\ cm^{-3}]^{-1}\}$ [a]	ΔE_P (mV) [b]	$E_{1/2}$ (mV) [c]
yeast cyt c/cyt b_5	160[d]	50	+50
	112[e]	85	−220
Ile52/cyt b_5	74[d]	40	−5
	56[e]	210	−210
Ala52/cyt b_5	114[d]	50	+25
	49[e]	185	−235
Gln27/cyt b_5	85[d]	55	+25
	36[e]	205	−285
Asp72/cyt b_5	15[d]		
	22[e]		

[a] Normalized peak current $i_P^\circ = i_P/(Ac_o v^{1/2})$ (at 20 mV s^{-1}).
[b] Peak separation at 20 mVs^{-1}.
[c] Midpoint potential measured at 20 mV s^{-1}.
[d] Associated with cyt c.
[e] Associated with cyt b_5.

namic equilibrium are invoked. In the specific case of the cytochrome *c*/cytochrome b_5 system, it is noteworthy that NMR data suggest that cytochrome *c* can bind two cytochrome b_5 molecules simultaneously (Whitford *et al.*, 1991).

The interactions in the ternary cytochrome *c*/cytochrome b_5/electrode system have observable thermodynamic consequences. The midpoint potential of wild-type cytochrome *c* determined at EPG is +40 mV (vs. SCE), which agrees well with the redox potential obtained by potentiometric determinations. Mediated spectro-electrochemical measurements of the cytochrome *c* (horse heart)/cytochrome b_5 (bovine) mixture show that this complex formation does *not* measurably change the redox potential of either protein [$E^{O'}$(cytochrome *c*) = +20 mV. $E^{O'}$(cytochrome b_5) = −226 mV] (McLendon *et al.*, 1988). However, the midpoint potential of either cytochrome *c* mutant or cytochrome b_5 in the complex measured by direct cyclic voltammetry at EPG differs from the values for the proteins alone (Table 7.2). Clearly, the ternary adduct affords a unique environment which is reflected in the electrochemical potential of cytochrome *c*, but is not found in either of the binary interactions (cytochrome *c*/cytochrome b_5 or cytochrome *c*/electrode). This observation may be related to the unusual electrostatic environments probed by Raman spectroscopy by Hildebrandt *et al.* (Hildebrandt and Stockburger, 1989) (see Chapter 6 by Hildebrandt, this volume). The delicate balance required to produce efficient dynamic binding to both the electrode and cytochrome b_5 suggests that the ternary cytochrome *c*/cytochrome b_5/electrode might be particularly sensitive to small, single-site modifications of the cytochrome *c* binding surface. This is indeed the case.

Given the results that the differences in cytochrome *c*/electrode interactions are enhanced by binding to cytochrome b_5, one might anticipate that different CV responses could be obtained with the same complex at different electrode surfaces. In fact, this has been illustrated by the cyclic voltammetric studies of the cytochrome *c*–plastocyanin complex. At an EPG electrode, Faradaic currents of both proteins were clearly seen, whereas at SSBP-modified gold, heterogeneous reaction was observed only with cytochrome *c* and not with plastocyanin (Barker *et al.*, 1989). This is a further indication of the high sensitivity in the ternary protein/protein/electrode complex.

Overall, the present results support a growing body of data that suggests that the molecular recognition of cytochrome *c* and other redox proteins does not involve "lock and key" epitopes. Instead, flexible binding domains are utilized which can recognize a variety of partners. These different domains may overlap spatially and differ only slightly in free energy, so that different complexes may rapidly interconvert. The flexibility afforded by multiple overlapping domains may allow more than one partner to bind simultaneously, as in the present ternary cytochrome *c*/cytochrome b_5/electrode adduct. The optimum structure for the binary complexes need *not* be optimal in the ternary complex, where reorganization may occur to allow all components to interact simultaneously. Equivalent reorganization may occur in a variety of *in vivo* biological recognitions in which multiple components are involved.

V. Electrocatalysis of Redox Cycles

This approach can be extended to characterize the kinetics of complex respiratory chains in which electroreduced cytochrome c may act as a biologically specific mediator for electron transfer. For example, although carbon monoxide oxidase alone does not show any current, electrocatalytic turnover of carbon monoxide has been demonstrated when cytochrome c is added to a previously inert electrochemical cell containing a graphite electrode and CO oxidase. In like manner, the cytochrome c/cytochrome oxidase system has been investigated by Barker (Barker *et al.*, 1986) and the reaction rates for the individual components obtained.

VI. Acknowledgments

We thank the Science and Engineering Research Council (H.A.O.H) and the National Science Foundation (G.M.) for financial support. G.M. is a fellow of the Guggenheim Foundation (1990), and L.-H.G. is a recipient of the SBFSS Award (1987–1990).

VII. References

Allen, P. M., Hill, H. A. O., & Walton, N. J. (1984) *J. Electroanal. Chem. 178*, 69.

Armstrong, F. A. (1990) *Struct. Bonding 72*, 137.

Armstrong, F. A., & Brown, K. J. (1987) *J. Electroanal. Chem. 219*, 319.

Armstrong, F. A., Hill, H. A. O., Oliver, B. N., & Walton, N. J. (1984) *J. Am. Chem Soc. 106*, 921.

Armstrong, F. A., Hill, H. A. O., Oliver, B. N., & Whitford, D. (1985) *J. Am. Chem. Soc. 107*, 1473.

Armstrong, F. A., Hill, H. A. O., & Walton, N. J. (1986) *Quart. Rev. Biophys. 18*, 261.

Armstrong, F. A., Hill, H. A. O., & Walton, N. J. (1988) *Acc. Chem. Res. 21*, 407.

Armstrong, F. A., Bond, A. M., Hill, H. A. O., Psalti, I. S. M., & Zoski, C. G. (1989) *J. Phys. Chem. 93*, 6485.

Bagby, S., Barker, P. D., Guo, L. H., & Hill, H. A. O. (1990) *Biochemistry 29*, 3213.

Bard, A. J., & Faulkner, L. R. (1980) *Electrochemical Methods. Fundamentals and Applications*, Wiley, New York.

Barker, P. D., Coleman, J. O. D., Hill, H. A. O., Walton, N. J., & Whitford, D. (1986) *Biochem. Soc. Trans. 14*, 130.

Barker, P. D., Guo, L. H., Hill, H. A. O., & Sanghera, G. S. (1988) *Biochem. Soc. Trans. 16*, 957.

Barker, P. D., Hill, H. A. O., & Walton, N. J. (1989) *J. Electroanal. Chem. 260*, 303.

Bond, A. M., Hill, H. A. O., Komorsky-Lovric, S., Lovric, M., McCarthy, M. E., Psalti, I. S. M., & Walton, N. J. (1992) *J. Phys. Chem. 96*, 8100.

Burrows, A. L., Guo, L. H., Hill, H. A. O., McLendon, G., & Sherman, F. (1991) *Eur. J. Biochem. 202*, 543.

Eddowes, M. J., & Hill, H. A. O. (1977) *J. Chem. Soc., Chem. Commun. 771*.

Elliott, D., Hamnett, A., Lettington, O. C., Hill, H. A. O., & Walton, N. J. (1986) *J. Electroanal. Chem. 202*, 303.

Frew, J. E., & Hill, H. A. O. (1988) *Eur. J. Biochem. 172*, 261.

Guo, L. H., & Hill, H. A. O. (1991) in *Advances in Inorganic Chemistry*, Vol. 36 (Sykes, A. G., ed.), p. 341, Academic Press, San Diego.

Hildebrandt, P., & Stockburger, M. (1989) *Biochemistry 28*, 6710.

Hill, H. A. O., & Whitford, D. (1987) *J. Electroanal. Chem. 235*, 153.

McLendon, G. (1988) *Acc. Chem. Res. 21*, 160.

McLendon, G., Magner, E., Zhang, Q. P., Pardue, K., & Simmons, J. (1988) in *Proceeding of the Robert A. Welch Foundation Conference on Chemical Research*, Vol. XXXII, p. 201, Robert A. Welch Foundation, Houston.

Murray, R. W. (1980) *Acc. Chem. Res.* **13**, 135.

Whitford, D., Concar, D., & Williams, R. J. P. (1991) *Eur. J. Biochem.* **199**, 553.

Yeh, P., & Kuwana, T. (1977) *Chem. Lett.*, 1145.

8

Oxidation State–Dependent Properties of Cytochrome *c*

Sackler Institute of Molecular Medicine
Sackler Medical School
Tel-Aviv University

I. Introduction
II. Differences in Hydrodynamic Properties
III. Differences in Ion Binding and Chemical Reactivity
IV. Differences in Spectroscopic Properties
V. Differences in Tertiary Structure and Isotope Exchange
VI. Facts, Explanations, and Implications
VII. References

I. Introduction

Twenty-five years ago, in a review of cytochrome *c*, attention was called to a series of observations of chemical and physicochemical differences between the oxidized and reduced states of the molecule of cytochrome *c* (Margoliash and Schejter, 1966). Some of these differences were apparent in chromatographic behavior, such as the larger affinity of the reduced protein for kaolin columns (Zeile and Reuter, 1933); others, in the easier formation of monomolecular films of the protein in its oxidized state (Jonxis, 1939), or in the higher rate of proteolysis of the latter when compared to that of the ferrous state (Nozaki *et al.*, 1957, 1958; Mizushima *et al.*, 1958; Yamanaka *et al.*, 1959). These facts, as well as the higher stability of the reduced molecule at high temperature (Butt and Keilin, 1962), which had been already noticed by Keilin in his earliest work on the cytochromes (Keilin, 1925, 1930), suggested that the structure of ferrocytochrome *c* was in some way stronger than that of ferricytochrome *c*, and this was attributed to an oxidation state–dependent conformational difference (Margoliash and Schejter, 1966). There was no structural information on cytochrome *c* at that time, but it was known that its ferric and ferrous states crystallized in different forms (Hagihara *et al.*, 1956), a fact that recalled a similar observation on the crystals of oxy- and deoxyhemoglobin (Haurowitz, 1938), which Perutz had shown to be due to the three-dimensional changes caused by ligation (Muirhead and Perutz, 1963).

Comparative studies of cytochrome c have, since then, amply confirmed that its ferric and ferrous forms behave dissimilarly in a wide variety of experimental observations; while crystallography (Takano and Dickerson, 1981a,b; 1982) and, more recently, NMR spectroscopy of cytochrome c in solution, have shown that the structural differences between the two states are very small. The nature of the changes in structure, chemical reactivity, and physicochemical properties that occur when cytochrome c changes its oxidation state, and the possible explanations of this characteristic behavior of the molecule, are the subject of the present article.

II. Differences in Hydrodynamic Properties

A thorough investigation of monomolecular films of oxidized and reduced cytochrome c, while confirming Jonxis' results (Jonxis, 1939), showed that the differences between the oxidized and reduced states were more complicated than it was originally thought, since they were strongly pH-dependent (Chin and Brody, 1975). At pH 7.0, the area per molecule of ferrocytochrome c, measured at a constant surface pressure of 6 dyn/cm, was at a minimum, while that of the ferric molecule was at a maximum; at higher and lower pH, the situation is reversed, and at pH 6.2 and 7.6 they are identical (Chin and Brody, 1975).

The intrinsic viscosity of ferrocytochrome c, measured at neutral pH in unbuffered solutions of 0.05 M ionic strength, was 2.17 mL/g, a significantly lower value than that of 2.53 mL/g found for the oxidized molecule (Fisher et al., 1973). This result implies that the reduced molecule has a more compact structure, a fact borne out by small-angle x-ray scattering measurements, showing that in 5 mM phosphate buffer, pH 7.3, the radius of gyration increases upon oxidation from 13.3–13.6 Å to 14.2 Å (Trewhella et al., 1988). Correspondingly, the maximum linear dimension of the reduced molecule in its low-frequency excursions, 40 ± 2 Å, grows to a value of 44 ± 2 Å in the ferric state (Trewhella et al., 1988). This difference, which does not exist in the crystalline state, is also abolished when the scattering is measured in the same buffer, but with added 0.2 M sodium chloride (Trewhella et al., 1988). Thus, the difference appears again to depend on the salt concentration of the solution.

The reasons underlying these salt effects on ferricytochrome c, which become evident at ionic strengths below 50 mM, or even lower, are probably similar to those causing the appearance of biphasic kinetics in its reduction by ascorbate at low ionic strength (Goldkorn and Schejter, 1977), and the shift of its alkaline pK from 9.4 at $I = 0.5$ M to 8.9 at $I = 10$ mM (Osheroff et al., 1980). Recently, it was reported that in polyethylene glycol, at $I = 45$ mM, ferricytochrome c crystallized in space group $P2_1$ (Walter et al., 1990), while in concentrated ammonium sulfate it crystallizes in the $P4_1$ space group (Takano and Dickerson, 1981a).

If the differences in viscosity and radius of gyration observed at low ionic strength result from larger volume fluctuations of the ferric molecule, then the compressibility of cytochrome c should be larger when its iron is oxidized. This was indeed demonstrated in measurements of sound velocity, carried out in 5 mM phosphate buffer, from which was estimated an increase in the apparent molal adiabatic compressibility of cytochrome *c* upon oxidation equivalent to 30% of the root-mean-square volume fluctuations, from 74 Å^3 in the ferrous state, to 96 Å^3 in the ferric state (Eden *et al.*, 1982). It should be noted that this latter result has been disputed in a later study (Kharakoz and Mkhitaryan, 1986), which concluded that the difference was only 0.2%.

III. Differences in Ion Binding and Chemical Reactivity

With regard to the effects of added salt mentioned earlier, it is of interest that one of the properties in which the ferric and ferrous states of cytochrome *c* differ is that of their affinities for anions and cations. Electrophoretic (Margoliash *et al.*, 1970) and chromatographic (Margalit and Schejter, 1973, 1974) experiments demonstrated that the ferric form of the horse protein binds preferentially anions, such as Cl^-, $H_2PO_4^-$, and ADP, while the reduced form binds preferentially cations, such as Mg^{2+}, Ca^{2+}, Na^+, and K^+. Quantitatively, there are conflicting values of the binding constants in the literature (Stellwagen and Shulman, 1973; Peterman and Morton, 1979; Andersson *et al.*, 1979; Gopal *et al.*, 1988). Furthermore, species differences were found (Barlow and Margoliash, 1966; Moore and Pettigrew, 1990) that were also investigated in more recent studies. For example, while horse ferri-cytochrome *c* binds chloride and phosphate anions more strongly than ferro-cytochrome *c*, for the cow and tuna proteins the inverse situation occurs (Moore and Pettigrew, 1990).

The striking oxidation state dependence of the reactivity of the cytochrome *c* iron towards exogenous ligands was already known in the 1940s from the fact that, while the ferric molecule could form a complex with the typical ferric iron ligands cyanide (Potter, 1941; Horecker and Kornberg, 1946) and azide (Horecker and Stannard, 1948), ferrocytochrome *c* did not bind the Fe^{2+} ligand CO, except under extreme conditions of temperature (Keilin, 1925) or pH (Theorell and Akesson, 1941). It was found that the reactions of ferricytochrome *c* with cyanide (George and Tsou, 1952; George *et al.*, 1967), azide (George *et al.*, 1967), and imidazole (Schejter and Aviram, 1969) were driven by large favorable entropies, which suggested that binding of the metal entailed a conformation change of the protein chain (George and Tsou, 1952; George, 1956).

The dissociation, upon neutralization, of the CN^- and CO complexes of ferro-cytochrome *c* generated at high pH (George and Schejter, 1964) showed that the unusual stability of the reduced "closed crevice" (George and Lyster, 1958) had a thermodynamic origin (George and Schejter, 1964). The NMR observation of the kinetics of crevice closing demonstrated that the rate-determining step was the rebinding of the iron by the native methionine-80 sulfur ligand (Wüthrich, 1969).

The very strength of this bond has made its direct measurement impossible, since no known ligand of Fe can break it in neutral solution. The conformational stability of the whole protein appears to be stronger when the bond is intact in the reduced state, as judged by the higher stability of the latter. Thus, the free energy of unfolding of horse cytochrome *c* is 8.1 kcal/mol in the ferric state (Schejter *et al.*, 1992) and 18.4 in the ferrous state (Jones *et al.*, 1993). Moreover, when all the lysine side chains are maleylated, causing the electrostatic collapse of the ferric protein, the methionine-80–sulfur bond in the ferrous state remains unchanged and unreactive to CO (Aviram and Schejter, 1980).

IV. Differences in Spectroscopic Properties

The optical absorption spectra of ferri- and ferrocytochrome c are typical of the low-spin states of ferric and ferrous iron porphyrins, respectively (Eaton and Hofrichter, 1981). The increased sharpness of the visible bands in the ferrous state can be attributed to the lower degree of mixing of metal and porphyrin orbitals, due to the filled character of the d^6 half-shell (Hochstrasser, 1971). An important characteristic in which the spectrum of ferrocytochrome c is outstanding is that of the enhanced intensity and splitting of its visible bands at liquid-nitrogen temperatures (Keilin and Hartree, 1949; Estabrook, 1957). There may be a connection between this unique property of ferrocytochrome c and the unusual strength of its coordinative linkage to the protein, since the low-temperature effect is lost (Estabrook, 1961) when the protein dimerizes (Margoliash and Lustgarten, 1962), a situation in which the ferrous iron becomes accessible to CO ligation (Margoliash et al., 1959).

Early observations of differences in the optical rotatory dispersion of the reduced and oxidized states had also been interpreted as resulting from conformation changes (Ulmer, 1965; Urry and Doty, 1965), but alternative explanations were given later, when it was found that the circular dichroism in the far ultraviolet region was essentially independent of the oxidation state of the iron (Myer, 1968).

V. Differences in Tertiary Structure and Isotope Exchange

The redox-state insensitivity of circular dichroism below 250 nm is in keeping with the results of x-ray crystallography (Takano and Dickerson, 1981a,b, 1982) and NMR spectroscopy (Feng et al., 1990), which indicate that in aqueous solution the oxidized and reduced structures are very similar, in regard to the folding of the chain and the α-helical and β-sheet structures. There are, however, certain interesting differences, which, both in the crystalline state (Takano and Dickerson, 1982) and in solution (Feng et al., 1990), are concentrated mainly at the bottom of the molecule, especially on its left side. Of particular interest is the effect of reduction on three internal water molecules. One of these is hydrogen-bonded to the side chains of Tyr-67, Asn-52 and Thr-78: in the reduced state, this water moves from 6.1 to 6.6 Å away from the iron, while all its hydrogen-bonding distances to the named side chains increase (Takano and Dickerson, 1981a,b, 1982). Another internal water affected is located on the right side of the molecule and is H-bonded to the peptide chain atoms of residues Lys-25, Lys-27, and Gly-29, the last two being part of a γ-turn with an anomalous internal H-bond configuration (Takano and Dickerson, 1981a,b, 1982). A third internal water sensitive to the oxidation state is that located in the area of residues 38–42 (Takano and Dickerson, 1981a,b, 1982; Feng et al., 1990).

An important ultraviolet resonance Raman study centered on the sensitivity of the only tryptophan residue, Trp-59, to oxidation state and ionic strength changes (Liu *et al.*, 1989). A resonance Raman band assigned to the Trp-59 indole was used to demonstrate that the room-temperature H/D exchange of this –NH group has half-times of 5 and 30 hours in the ferric and ferrous states, respectively, at $I = 5$ mM; at $I = 1.5$ M, the exchange is distinctly slower for both oxidation states, with $t_{1/2}$ of 30 hours for ferricytochrome *c*, and much longer for the ferrous protein. Other observations indicated the weakening or loss of a phenolic side-chain H-bond, tentatively assigned to Tyr-48, upon oxidation (Liu *et al.*, 1989).

The magnitude of the rate of H/D exchange is one of the clearest demonstrations of a difference between the ferric and ferrous states of cytochrome *c*. It was first recognized using infrared spectroscopy, in experiments that indicated a constant difference of about 10% in the rates of rapid, slow, and very slow isotope exchange (Ulmer and Kagi, 1968), the oxidized molecule always exchanging faster. This was later confirmed in NMR studies (Patel and Canuel, 1976), and in recent years a more detailed picture of the exchanging regions of the protein, based on proton NMR assignments, is beginning to take shape (Williams *et al.*, 1985; Wand *et al.*, 1986). While the results of this type of study are eagerly awaited, it is worth noticing that the initial findings implicate as oxidation-state–sensitive areas isotope-exchanging residues quite removed from the lower left side of the molecule previously discussed. For example, while the exchange rates of most of the N-terminal helix amide protons are not affected by the oxidation state of the iron, those of residues Cys-14 and Ala-15, which are closest to the heme group, exchange at a rate 15 times slower in the Fe^{2+} than in the Fe^{3+} state (Wand *et al.*, 1986). This result argues for a higher exposure of the ferric heme environment to the solvent, in agreement with the crystal structure (Takano and Dickerson, 1982) and with measurements of solvent perturbation of the optical absorption spectrum (Schlauder and Kassner, 1979).

VI. Facts, Explanations, and Implications

The preceding sections can be simply summarized by stating that the oxidized and reduced states of mitochondrial cytochrome *c* differ in properties more than they do in structure. The "static" structural changes observed are much less pronounced than those that occur when deoxyhemoglobin binds oxygen; the dynamic changes, which are evident from measurements of hydrodynamic properties and isotope exchange, are real. The use of the adjective "static" should be qualified, since a crystallographic structure is not an absolute "static" measurement: Its precision is limited by thermal disorder in the crystal, reflected in the atomic temperature factors, which show notable differences between the ferric and ferrous states (Takano and Dickerson, 1981a,b, 1982). The expression "differences in conformation" to describe the oxidation state–dependent structural changes of cytochrome *c* is a poor choice, if the mental image that it evokes is restricted to the ligand-dependent structural changes exemplified by hemoglobin. A better alternative might be the phrase "differences in conformational fluctuations"; even better would be to borrow Lindestrom Lang's classic term (Lindestrom-Lang, 1955) and describe the phenomenon as one of "differences in breathing."

This conclusion puts in a different perspective the usual interpretation of the entropy change that accompanies the reduction of cytochrome c. From direct calorimetry (Watt and Sturtevant, 1967; George et al., 1966), ΔS for the reduction of ferricytochrome c by hydrogen has been evaluated as -28 cal-mol^{-1}-K^{-1} (entropy units) at $I = 0.1$ M, while the temperature dependence of the redox potential leads to values of -36 entropy units at $I = 10$ mM, and -13 at $I = 0.23$ M (Margalit and Schejter, 1973). All these values are negative, and this has been commonly taken to imply a more compact conformation of the reduced protein. If this view is now abandoned, and replaced by the "differential breathing" hypothesis, the negative entropy change can be justified on the basis that in the oxidized molecule the dominant conformational state has more substates available than in the reduced one. The effects of ionic strength on the entropy of reduction are particularly revealing in this context, because they are in keeping with the previously described influences of salts: If at very low ionic strengths the breathing of the ferricytochrome c increases more than that of the ferrous molecule, then the entropy change upon observed reduction should be more negative.

The breathing hypothesis explains most of the observations just described, but one result remains unclear: the different reactivities of the Fe^{3+} and Fe^{2+} centers towards added ligands. Binding of an exogenous ligand requires the cleavage of the sulfur–iron bond that closes the cytochrome c crevice. In the ferric protein, this appears to be an S_{N_1} mechanism, as shown in kinetic studies with various ligands, with an opening rate constant of 30 to 60 s^{-1} (Sutin and Yandell, 1972). The equilibrium constant for crevice closing is of the order of 10^2 to 10^3 (Margoliash and Schejter, 1966) and the rate of crevice closing can therefore be estimated as 3×10^3 to 6×10^4 s^{-1}.

In the reduced state, none of these measurements is possible, because at neutral pH, ligands do not bind the iron at all (George and Schejter, 1964): The crevice-closing Met-80–sulfur–iron bond is too strong for the ligands to compete with it. Two questions now arise: What makes this bond so strong, and how is this stability related to the breathing movements of the protein chain?

Attention to the strong affinity of the cytochrome c ferrous iron for the methionine sulfur was first raised when it was shown that a strong complex was formed between methionine and the reduced metal of the 11–21 heme peptide (Harbury et al., 1965), and that the bond between the 1–65 and 66–104 peptides could be reconstructed if the iron was kept in the ferrous state (Corradin and Harbury, 1970, 1971). The suggestion that this strong affinity could be the basis for the rigidity of the reduced protein (Salemme, 1977) received support from measurements of the affinities for several ligands (Schejter and Plotkin, 1988) of ferric and ferrous myoglobin and a derivative of cytochrome c in which the liganding power of its methionine-80 sulfur had been voided by alkylation (Schejter and Aviram, 1970). It was found that reduction of these heme proteins diminished their affinities for imidazole and cyanide, but to a much lesser extent for the modified cytochrome c; at the same time, the affinity of myoglobin for dimethyl sulfide was unaltered by the reduction, while that of the alkylated cytochrome caused its increase by two orders of magnitude (Schejter and Plotkin, 1988).

This striking peculiarity of the mitochondrial cytochrome c iron may be the reason for the prevalence of the methionine–Fe–imidazole coordination shared by the algal photosynthetic c-type cytochromes and by a wide variety of prokaryotic cytochromes, which include the bacterial cytochromes c_2, c-551, c-553 and c_5, and at least one b-type cytochrome, b-562 of *Escherichia coli* (Matthews, 1985). The intriguing nature of this coordinative structure resides in the fact that, while imidazole is a well-known iron ligand (Sundberg and Martin, 1974), methionine is not. Thioether sulfur has negligible affinity for iron (Livingstone, 1965; Murray and Hartley, 1981), thioethers do not bind iron porphyrins (Ellis *et al.*, 1979), and in methionine complexes of free iron in aqueous solution, coordination is through the amino and carboxyl groups of the amino acid (McAuliffe *et al.*, 1966; Halbert and Rogerson, 1972). Nevertheless, the stability of otherwise weak thioether–metal bonds can be enhanced by favorable stereospecific conditions and low water activity (Sigel *et al.*, 1979), two conditions that exist in cytochrome c: The microenvironment of the closed crevice is hydrophobic, and the methionine-80 sulfur is placed in proper coordinating position by the protein fold (Takano and Dickerson, 1981a,b, 1982). But what causes the unequal affinity for the sulfur of the ferric and ferrous iron?

The answer to this question can be found in the ligand qualities of divalent sulfur. The pK's of imidazole and cyanide are 7.0 and 9.4, respectively, and the ratio of their proton affinities is not too different from that of their affinities for the cytochrome c iron (Schejter and Plotkin, 1988). While divalent sulfide is a very strong acid, with a pK of -6.4 (Bonvicini *et al.*, 1972), and consequently a very poor donor, it binds the ferrous cytochrome c iron better than imidazole and only two orders of magnitude less than cyanide (Schejter and Plotkin, 1988). The surprising affinity of divalent sulfur for the cytochrome c iron in the reduced state must stem from its good π-acceptor ability, which is due in turn to the suitability of its empty, low-lying $3d$ orbitals to overlap with the d_{xz} and d_{yz} orbitals of iron (Cotton and Wilkinson, 1980). The d-electrons that populate the t_{2g} orbitals of low-spin cytochrome c can delocalize into the empty $3d$ orbitals of the methionine-80 sulfur, with the consequent strengthening of the bond. Furthermore, upon reduction, the increase of electron density in the iron $3d$ shell increases the extent of back-donation. These electronic properties of the iron–sulfur bond explain its extra strength in the reduced state. For this type of π-accepting behavior, sulfur is frequently characterized as a "b-type" (Ahrland *et al.*, 1958) or "soft" (Pearson, 1963) ligand. Ferric iron is an "a-type" (Ahrland *et al.*, 1958) or "hard" (Pearson, 1963) metal, which prefers binding to hard ligands (Pearson, 1963), but ferrous iron is at the borderline between hardness and softness and is also able to bind soft ligands.

Support for the assumption that metal-to-ligand electron delocalization is responsible for the strength of the iron–sulfur bond in cytochrome c was provided by the spectroscopic effects of replacing the native sulfur ligand by a group of ligands of different donor and acceptor abilities: phosphines, phosphite, thioether, cyanide, isocyanide, and imidazole (Schejter *et al.*, 1991). The complexes with π-acceptor ligands had strikingly similar spectra to that of cytochrome c, especially in the far red. Remarkably, the 695-nm band (Schejter and George, 1964; Eaton and Hochstrasser, 1967), which previously had been found in no ferric heme protein other than cytochrome c, and which had been reconstructed only by adding methionine to the ferric heme undecapeptide (Shechter and Saludjian, 1967), was recovered by addition of dimethyl sulfide to the Met-80–alkylated derivative, from which it is absent; similar bands appeared in the complexes with all the aforementioned ligands, except imidazole. Moreover, bands corresponding to the charge-transfer and $d–d$ near infrared transitions of ferrocytochrome c (Eaton and Charney, 1969) appeared when the π-accepting ligands were added to the ferrous alkylated cytochrome c (Schejter $et\ al.$, 1991). This study also showed that the stabilities of the reduced cytochrome c complexes were correlated, albeit imperfectly, to features of the visible spectrum previously recognized as indicative of the extent of covalency of the sulfur–iron bond (Hill $et\ al.$, 1970).

It is worth noting that various physical properties of the cytochrome c molecule—such as the decrease in the value of the spin–orbit coupling constant, calculated on the basis of the EPR spectrum of ferricytochrome c (Salmeen and Palmer, 1968), and the sizeable quadrupole splitting found in the Mössbauer spectrum of ferrocytochrome c (Cooke and Debrunner, 1968; Lang $et\ al.$, 1968)—have been previously related to possible electron delocalization, or back-bonding. Also, the larger affinity of the derivatized ferrocytochrome c for the π-acceptor ligand CO, relative to other heme proteins, had already been reported (Wilson $et\ al.$, 1973).

A structural explanation of the unusually high back-bonding tendency of the cytochrome c iron, and the consequent higher affinity for π-acceptor ligands, must be sought in the specific characteristics of the heme–protein bonds of this molecule. In myoglobin (Fermi, 1975) and in hemoglobin (Baldwin, 1980), the positions of a carbon atom of the imidazole iron ligand and a pyrrole nitrogen of the porphyrin overlap, which makes the imidazole ring "tilt," pulling the iron out of the porphyrin plane (Gelin and Karplus, 1977), away from the $trans$ ligand, and causing the distribution of the d-electrons in a high-spin configuration. These structural and electronic characteristics of the imidazole–iron bond of the globins are synergic in diminishing the possibility of back-donation to a potential $trans$ axial ligand.

In cytochrome c, the imidazole plane is almost perpendicular to the $N_1–Fe–N_3$ line (Takano and Dickerson, 1981a,b, 1982), eliminating the possibility of imidazole–porphyrin nonbonded contacts. The iron is thus kept in the porphyrin plane, in the low-spin state, favoring back-donation to the $trans$ axial ligand. Moreover, the steric relationship between the imidazole and porphyrin planes minimizes the overlap between the imidazole ring π-orbital and the iron d_{xz}, d_{yz} orbitals (Fergusson $et\ al.$, 1972), but see also (Timkovich, 1979). This, in turn, minimizes the extent of back-donation from the iron to the imidazole, which could compete with back-donation to the other axial ligand. $Trans$ effects of this type across the heme plane, mediated by the iron orbitals, have been extensively discussed in the literature (Caughey $et\ al.$, 1973; Caughey, 1973).

In the globins, the vinyl groups of the protoheme are part of the conjugated porphyrin ring and decrease the basicity of its nitrogens by withdrawing electronic charge. When the vinyl groups are saturated, the electronic charge on the nitrogens increases and is transmitted to the metal, which acquires additional back-bonding power (Caughey $et\ al.$, 1973; Caughey, 1973). This cis effect of the planar ligand has been elegantly demonstrated by following the effect of saturation of the vinyl groups on the stretching frequencies of CO bound to hemes (Caughey, 1973). Clearly in cytochrome *c*, where the formation of the thioether bonds saturates the porphyrin vinyls, the situation will resemble that of an etio-type heme, and the donating power of the iron will be strengthened. Moreover, the lone pairs of the thioether sulfurs of cytochrome *c* may also contribute to the electron density of the porphyrin ring, increasing the *cis* effect; in this regard, it seems worth mentioning that involvement of the thioether sulfur electrons in the conjugated system of the heme through hyperconjugation has been already invoked as a possible avenue for electron transfer in cytochrome *c* (George and Griffith, 1963).

Finally, the hypothesis just described could also provide an explanation of the differences in breathing between the two oxidation states. A stronger sulfur–iron bond in the Fe^{2+} state also implies a diminished rate of its opening, and a lesser frequency of excursions of the Met-80 side chain between its position as iron ligand and that corresponding to the open crevice state. In this respect, it is worth mentioning that the area into which the Met-80 side chain must move during these excursions contains the internal water molecule previously mentioned as hydrogen-bonded to the Tyr-67, Asn-52, and Thr-78 side chains. Movements of this water and the whole surrounding area were singled out among the important oxidation state–dependent structural differences in crystalline cytochrome *c* (Takano and Dickerson, 1982; Dickerson and Timkovich, 1975; also see Chapter 3 by Brayer, this volume), and recent studies with site-directed mutants of the protein have shown that two of these residues, Tyr-67 (Luntz *et al.*, 1989) and Asn-52 (Das *et al.*, 1989; Hickey *et al.*, 1991; Koshy *et al.*, 1994; Schejter *et al.*, 1994) have a significant role in maintaining the stability of the sulfur–iron bond and the stability of the whole protein. These are the first examples of the potential impact of site-directed mutagenesis techniques on the problem of the oxidation-dependent properties of cytochrome *c*, and undoubtedly many more will be soon forthcoming.

VII. References

Ahrland, S., Chatt, J., & Davies, N. R. (1958) *Quart. Rev.* **12**, 265–276.

Andersson, T., Thulin, E., & Forsen, S. (1979) *Biochemistry* **18**, 2487–2493.

Aviram, I., & Schejter, A. (1980) *J. Biol. Chem.* **255**, 3020–3024.

Baldwin, J. M. (1980) *J. Mol. Biol.* **136**, 103–128.

Barlow, E. H., & Margoliash, E. (1966) *J. Biol. Chem.* **241**, 1473–1477.

Bonvicini, P., Levi, A., Lucchini, V., & Scorrano, G. J. (1972) *J. Chem. Soc., Perkin Trans.* 2, 2267–2269.

Butt, W. D., & Keilin, D. (1962) *Proc. Roy. Soc. London, Series B* **156**, 429–458.

Caughey, W. S. (1973) in *Inorganic Biochemistry*, Vol. 2 (Eichhorn, G. L., eds.), pp. 797–831, Elsevier, Amsterdam.

Caughey, W. S., Barlow, C. H., O'Keeffe, D. H., & O'Toole, M. C. (1973) *Ann. N. Y. Acad. Sci.* **206**, 296–309.

Chin, P., & Brody, S. S. (1975) *Biochemistry* **14**, 1190–1193.

Cooke, R., & Debrunner, P. (1968) *J. Chem. Phys.* **48**, 4532–4537.

Corradin, G., & Harbury, H. A. (1970) *Biochim. Biophys. Acta* **221**, 489–496.

Corradin, G., & Harbury, H. A. (1971) *Proc. Natl. Acad. Sci. USA* **68**, 3036–3039.

Cotton, F. A., & Wilkinson, G. (1980) *Advanced Inorganic Chemistry*, John Wiley & Sons, New York.

Das, G., Hickey, D. R., McLendon, D., McLendon, G., & Sherman, F. (1989) *Proc. Natl. Acad. Sci. USA* **86**, 496–499.

Dickerson, R. E., & Timkovich, R. (1975) in *The Enzymes*, Vol. 11 (Boyer, P. D., ed.), pp. 397–547, Academic Press, New York.

Eaton, W. A., & Charney, E. (1969) *J. Chem. Phys.* **50**, 4502–4505.

Eaton, W. A., & Hochstrasser, R. (1967) *J. Chem. Phys.* **46**, 2533–2539.

Eaton, W. A., & Hofrichter, J. (1981) *Methods Enzymol.* **76**, 175–261.

Eden, D., Matthew, J. B., Rosa, J. J., & Richards, F. M. (1982) *Proc. Natl. Acad. Sci. USA* **79**, 815–819.

Ellis, P. E., Jones, R. D., & Basolo, F. (1979) *Proc. Natl. Acad. Sci. USA* **76**, 5418–5420.

Estabrook, R. W. (1957) *J. Biol. Chem.* **223**, 781–794.

Estabrook, R. W. (1961) in *Haematin Enzymes, Part 2* (Falk, J. E., Lemberg, R., & Morton, R. K., eds.), pp. 436–457, Pergamon Press, Oxford.

Feng, Y., Roder, H., & Englander, S. W. (1990) *Biochemistry* **29**, 3494–3504.

Fergusson, J. E., Robinson, W. T., & Rodley, G. A. (1972) *Austral. J. Biol. Sci.* **25**, 1365–1371.

Fermi, G. (1975) *J. Mol. Biol.* **97**, 237–256.

Fisher, W. R., Taniuchi, H., & Anfinsen, C. B. (1973) *J. Biol. Chem.* **248**, 3188–3195.

Gelin, B. R., & Karplus, M. (1977) *Proc. Natl. Acad. Sci. USA* **74**, 801–805.

George, P. (1956) in *Current Topics in Biochemical Research* (Green, D. E., ed.), pp. 338–348, Interscience, New York.

George, P., & Griffith, J. S. (1963) in *The Enzymes*, Vol. 1 (Boyer, P. D., Lardy, H. & Myrback, K., eds.), pp. 397–547, Academic Press, New York.

George, P., & Lyster, R. L. J. (1958) *Proc. Natl. Acad. Sci. USA* **44**, 1013–1029.

George, P., & Schejter, A. (1964) *J. Biol. Chem.* **239**, 1504–1508.

George, P., & Tsou, C. L. (1952) *Biochem. J.* **50**, 440–450.

George, P., Hanania, G. I. H., & Eaton, W. A. (1966) in *Hemes and Hemoproteins* (Chance, B., Estabrook, R. W., & Yonetani, T., eds.), pp. 267–270, Academic Press, New York.

George, P., Glauser, S.C., & Schejter, A. (1967) *J. Biol. Chem.* **242**, 1690–1695.

Goldkorn, T., & Schejter, A. (1977) *FEBS Lett.* **75**, 44–46.

Gopal, D., Wilson, G. S., Earl, R. A., & Cusanovich, M. A. (1988) *J. Biol. Chem.* **263**, 11652–11656.

Hagihara, B., Horio, T., Yamashita, J., Nozaki, M., & Okunuki, K. (1956) *Nature* **178**, 629–632.

Halbert, E. J., & Rogerson, M. J. (1972) *Austral. J. Chem.* **25**, 421–424.

Harbury, H. A., Cronin, J. R., Fanger, M. W., Hettinger, T. P., Murphy, A. J., Myer, Y. P., & Vinogradov, S. N. (1965) *Proc. Natl. Acad. Sci. USA* **54**, 1658–1664.

Haurowitz, F. (1938) *Z. Phyiol. Chem.* **254**, 266–274.

Hickey, D. R., Berghuis, A. U., Lafond, G., Jaeger, J. A., Cardillo, T. S., McLendon, D., Das, G., Sherman, F., Brayer, G. D. and McLendon, G. (1991) *J. Biol. Chem.* **266**, 11686–11694.

Hill, H. A. O., Roder, A., & Williams, R. J. P. (1970) *Struct. Bonding* **8**, 123–151.

Hochstrasser, R. M. (1971) in *Probes of Structure and Function of Macromolecules and Enzymes*, Vol. 1 (Chance, B., Lee, C. P. & Blasie, J. K., eds.), pp. 57–64, Academic Press, New York.

Horecker, M., & Kornberg, A. (1946) *J. Biol. Chem.* **165**, 11–20.

Horecker, M., & Stannard, J. N. (1948) *J. Biol. Chem.* **172**, 589–597.

Jones, C. H., Henry, E. R., Hu, Y., Chan, C-K., Luck, S. D., Bhuyan, A., Roder, H., Hofrichter, J. and Eaton, W. A. (1993), *Proc. Natl. Acad. Sci. USA* **90**, 11860–11864.

Jonxis, J. H. P. (1939) *Biochem. J.* **33**, 1743–1751.

Keilin, D. (1925) *Proc. Roy. Soc. London, Series B* **98**,. 312–339.

Keilin, D. (1930) *Proc. Roy. Soc. London, Series B* **106**, 418–444.

Keilin, D., & Hartree, E. F. (1949) *Nature* **164**, 254–259.

Kharakoz, D. P., & Mkhitaryan, A. G. (1986) *Mol. Biol. (Engl. translation)* **20**, 312–321.

Koshy, T. I., Luntz, T. L., Plotkin, B., Schejter, A. and Margoliash, E. (1994) *Biochem. J.* **299**, 347–350.

Lang, G., Herbert, D., & Yonetani, T. (1968) *J. Chem. Phys.* **49**, 944–950.

Lindestrom-Lang, K. (1955) *Chem. Soc. (London) Spec. Publ.* **2**, 1–20.

Liu, G., Grygon, C. A., & Spiro, T. G. (1989) *Biochemistry* **28**, 5046–5050.

Livingstone, S. E. (1965) *Quart. Rev.* **19**, 386–425.

Luntz, T. L., Schejter, A., Garber, E. A. E., & Margoliash, E. (1989) *Proc. Natl. Acad. Sci. USA* **86**, 3524–3528.

Margalit, R., & Schejter, A. (1973) *Eur. J. Biochem.* **32**, 500–505.

Margalit, R., & Schejter, A. (1974) *Eur. J. Biochem.* **46**, 387–391.

Margoliash, E., & Lustgarten, E. (1962) *J. Biol. Chem.* **237**, 3397–3405.

Margoliash, E., & Schejter, A. (1966) *Adv. Protein Chem.* **21**, 113–283.

Margoliash, E., Frohwirt, N., & Wiener, E. (1959) *Biochem. J.* **71**, 559–572.

Margoliash, E., Barlow, G. H., & Byers, V. (1970) *Nature* **228**, 723–726.

Matthews, F. S. (1985) *Prog. Biophys. Molec. Biol.* **45**, 1–56.

McAuliffe, C. A., Quagliano, J. V., & Vallarino, L. M. (1966) *Inorg. Chem.* **5**, 1996–2003.

Mizushima, H., Nozaki, M., Horio, T., & Okunuki, K. (1958) *J. Biochem. (Tokyo)* **45**, 845–846.

Moore, G. R., & Pettigrew, G. W. (1990) *Cytochromes c*, Springer Verlag, Berlin.

Muirhead, H., & Perutz, M. F. (1963) *Nature* **199**, 633–638.

Murray, S. G., & Hartley, F. R. (1981) *Chem. Rev.* **81**, 365–414.

Myer, Y. P. (1968) *J. Biol. Chem.* **243**, 2115–2122.

Nozaki, M., Yamanaka, T., Horio, T., & Okunuki, K. (1957) *J. Biochem. (Tokyo)* **44**, 453–464.

Nozaki, M., Mizushima, H., Horio, T., & Okunuki, K. (1958) *J. Biochem. (Tokyo)* **45**, 815–823.

Osheroff, N., Borden, D., Koppenol, W. H., & Margoliash, E. (1980) *J. Biol. Chem.* **255**, 1689–1697.

Patel, D. J., & Canuel, L. L. (1976) *Proc. Natl. Acad. Sci. USA* **73**, 1398–1402.

Pearson, R. G. (1963) *J. Am. Chem. Soc.* **85**, 3533–3539.

Peterman, B. F., & Morton, R. A. (1979) *Can. J. Biochem.* **57**, 372–377.

Potter, V. R. (1941) *J. Biol. Chem.* **137**, 13–20.

Salemme, F. R. (1977) *Annu. Rev. Biochem.* **46**, 299–329.

Salmeen, I., & Palmer, G. (1968) *J. Chem. Phys.* **48**, 2049–2052.

Schejter, A., & Aviram, I. (1969) *Biochemistry* **8**, 149–153.

Schejter, A., & Aviram, I. (1970) *J. Biol. Chem.* **295**, 1552–1557.

Schejter, A., & George, P. (1964) *Biochemistry* **3**, 1045–1049.

Schejter, A., & Plotkin, B. (1988) *Biochem. J.* **255**, 353–356.

Schejter, A., Luntz, T. L., Koshy, T. I. and Margoliash, E. (1992) *Biochemistry* **31**, 8336–8343.

Schejter, A., Plotkin, B., & Vig, I. (1991) *FEBS Lett.* **280**, 199–201.

Schejter, A., Koshy, T. I., Luntz, T. L., Sanishvili, R., Vig, I. and Margoliash, E. (1994) *Biochem. J.* **302**, 95–101.

Schlauder, J., & Kassner, R. J. (1979) *J. Biol. Chem.* **254**, 4110–4113.

Shechter, E., & Saludjian, P. (1967) *Biopolymers* **5**, 788–790.

Sigel, H., Rheinsberger, V. M., & Fischer, B. E. (1979) *Inorg. Chem.* **18**, 3334–3339.

Stellwagen, E., & Shulman, R. G. (1973) *J. Mol. Biol.* **75**, 683–695.

Sundberg, R. J., & Martin, R. B. (1974) *Chem. Rev.* **74**, 471–517.

Sutin, N., & Yandell, J. K. (1972) *J. Biol. Chem.* **247**, 6932–6936.

Takano, T., & Dickerson, R. E. (1981a) *J. Mol. Biol.* **153**, 79–94.

Takano, T., & Dickerson, R. E. (1981b) *J. Mol. Biol.* **153**, 95–115.

Takano, T., & Dickerson, R. E. (1982) in *Electron Transport and Oxygen Utilization*, Vol. 2 (Ho, C., ed.), pp. 18–26, Elsevier North Holland, Amsterdam.

Theorell, H., & Akesson, A. (1941) *J. Am. Chem. Soc.* **63**, 1804–1821.

Timkovich, R. (1979) in *The Porphyrins*, Vol. 7 (Dolphin, D., ed.), pp. 261–294, Academic Press, New York.

Trewhella, J., Carlson, V. A. P., Curtis, E. H., & Heidorn, D. B. (1988) *Biochemistry* **27**, 1121–1125.

Ulmer, D. D. (1965) *Biochemistry* **4**, 902–907.

Ulmer, D. D., & Kagi, J. H. R. (1968) *Biochemistry* **7**, 2710–2717.

Urry, D. W., & Doty, P. (1965) *J. Am. Chem. Soc.* **87**, 2756–2758.

Walter, M. H., Westbrook, E. M., Tykodi, S., Uhm, A. M., & Margoliash, E. (1990) *J. Biol. Chem.* **265**, 4177–4180.

Wand, A. J., Roder, H., & Englander, S. W. (1986) *Biochemistry* **25**, 1107–1114.

Watt, G. D., & Sturtevant, J. M. (1967) *Biochemistry* **8**, 4567–4571.

Williams, G., Clayden, N. J., Moore, G. R., & Williams, R. J. P. (1985) *J. Mol. Biol.* **183**, 447–460.

Wilson, M. T., Brunori, M., Rotilio, G., & Antonini, E. (1973) *J. Biol. Chem.* **248**, 8162–8169.

Wüthrich, K. (1969) *Proc. Natl. Acad. Sci. USA* **63**, 1071–1075.

Yamanaka, T., Mizushima, H., Nozaki, M., Horio, T., & Okunuki, K (1959) *J. Biochem. (Tokyo)* **46**, 121–132.

Zeile, K., & Reuter, F. (1933) *Z. Physiol. Chem.* **221**, 101–116.

9

Electrostatic Analysis of Cytochrome *c*

M. R. GUNNER

Department of Physics
City College of New York

BARRY HONIG

Department of Biochemistry and Molecular Biophysics
Columbia University

I. Introduction

The *c*-type cytochromes are important and much-studied electron transfer proteins. They all use heme as a redox active cofactor. Most are small, soluble proteins with a single heme, although larger multiheme proteins and a variety of transmembrane proteins are also included in this family. All cytochromes use iron–protoporphyrin IX with propionic acids as peripheral substituents (meso-heme). In the *c*-type cytochromes, the heme is covalently attached to the protein via peripheral thioether links to two cysteine groups. The protein also supplies the heme with two axial ligands, either a histidine and a methionine or two histidines. However, despite the uniformity of the cofactor and its ligands, the electrochemical midpoint potentials of different cytochromes range over more than 600 mV (Clark, 1960; Meyer and Kamen, 1982).

There have been various suggestions as to how proteins modulate cofactor midpoint potentials. Clark (1960) pointed out that the difference in behavior is an

347

expression of the differential affinity of oxidized and reduced heme for each protein. Kassner (1972, 1973) recognized that burying the heme in the low-dielectric protein environment would destabilize the charged, oxidized heme and raise the midpoint potential relative to a heme in solution. It was also realized that short-range interactions with the electron-donating axial ligands (Moore and Williams, 1977) and longer-range interactions with charges on the heme propionic acids (Moore, 1983; Reid *et al.*, 1984; Rogers *et al.*, 1985) or in the rest of the protein (Rees, 1985) would modulate electrochemical behavior. All of these factors are likely to contribute to the *in situ* midpoints (Moore *et al.*, 1986). However, until fairly recently it has not been possible to analyze them in a self-consistent manner to compare their relative importance. Recent advances in the application of classical electrostatics allow each of these contributions to the free energy of oxidation to be calculated for proteins of known structure (Warwicker and Watson, 1982; Warshel and Russell, 1984; Klapper *et al.*, 1986; Warwicker, 1986; Gilson and Honig, 1988a,b) (for reviews, see also Matthew, 1985; Matthew *et al.*, 1985; Honig *et al.*, 1986; Rogers, 1986; Harvey, 1989; Karshikov *et al.*, 1989; Sharp and Honig, 1990b).

The central problem addressed in this review will be how proteins achieve the large variation in the electrochemistry of the bound heme. The primary goal will be to calculate the energy of removing an electron from a heme given the distribution of charges in the protein and the polarizability of the protein and its surroundings. We will consider how to describe the reorganization of electrons and dipoles in and around the protein in response to changes in charge, motions which affect both the thermodynamics and the kinetics of heme oxidation. The first section will provide an overview of the underlying physics (Feynman *et al.*, 1963; Bockris and Reddy, 1973; Purcell, 1985). Solution of the relevant equations for proteins is deferred to Section III, while the information about the numerical values that are put into the equations will be found in Section IV. Finally, Section V will describe the results of electrostatic analysis of the behavior of several cytochromes.

II. Electrostatic Contributions to the Free Energy of Heme Oxidation

In the discussion presented here, the free energy of heme oxidation (ΔG_{ox}) will be divided as follows:

$$\Delta G_{ox} = \Delta G_{crg} + \Delta G_{solv} + \Delta G_{other(protein)} + \Delta G_{other(heme)} \qquad (9.1)$$

The first two terms are the electrostatic contributions, where ΔG_{crg} contains the interactions with charges in the protein and ΔG_{solv} describes the solvation of the heme charge by its surroundings. The challenge is to include the modulation of these interactions by the polarization of the protein and solvent. The third and fourth terms are contributions that lie beyond a purely classical electrostatic description. $\Delta G_{other(protein)}$ may include conformational changes induced by the redox reaction or binding of protons or ions to specific sites on the protein. The division between motions that can be treated as a statistically averaged dielectric response (i.e., in term 1 and 2) and those that must be treated explicitly in term 3 is an active question. The fourth term contains the quantum-mechanical effects, such as the free energy of oxidizing the heme in some reference state and perturbations of the heme charge distribution by the environment.

A. Contribution of Charges in the Protein

In an electrostatic analysis, the modification of the free energy of heme oxidation by charges within the protein is given by

$$\Delta G_{\text{crg}} = \sum_j \Phi_j \Delta q_j, \qquad (9.2)$$

where Δq_j is the change in electron density at heme atom j on oxidation, and Φ_j is the electrostatic potential at j due to other charges in the system. The partial charges within neutral, dipolar groups (e.g., carbonyls, amides, hydroxyls), as well as the real charges on ionized residues, will contribute to Φ (see Section IV.B). If the charges were immobile, then the electrostatic potential throughout space would be simply

$$\Phi_j = C \sum_{i \neq j} \frac{q_i}{r_{ij}}, \qquad (9.3)$$

where q_i is the charge on atom i, and r_{ij} is the distance between i and j. The sum runs over all charges in the protein and heme not including the atoms in the heme that change electron density on oxidation. C is equal to 14.6 eV/esu (331 kcal/esu) if the distance is in angstroms and q_i is in multiples of the charge on an electron (esu). Thus, with the position of all charges fixed, the electrostatic contribution to the free energy of heme oxidation could be evaluated analytically by Coulomb's law:

$$\Delta G_{\text{crg}} = C \sum_j \sum_{i \neq j} \frac{\Delta q_j \times q_i}{r_{ij}}. \qquad (9.4)$$

However, adding or removing a charge causes small movements of other charges in the protein and solvent, modifying Φ_j. These motions will be considered in two categories: electronic and nuclear. The separation can be justified in a variety of ways. One is that the rearrangement of the electrons is much faster ($\tau < 10^{-15}$ s) than that of the nuclei (10^{-14} s $< \tau <$ seconds). While this is important when considering electron transfer rates that take place on time scales faster than nuclear rearrangement, it is less important when considering equilibrium properties such as reaction free energies. Perhaps a better justification is that the motion of electrons in response to a charge is similar for all organic molecules, so the electronic polarization is relatively uniform. In contrast, there are a variety of nuclear rearrangements of different magnitudes and time scales that need to be considered independently.

1. Electronic polarization

The polarization of electrons in matter in response to a charge diminishes the interaction between charges. In nonconductors the electrons are tightly associated with individual atoms, so they can move only a small amount. The resultant electrostatic potential in a nonpolar, organic medium such as hexane is half what would be found for the same charge in vacuum; that is, the dielectric constant (ε) is 2 (Lide, 1990). If the electrons move freely, as they do in a conductor, ε approaches infinity.

The Poisson equation,

$$\nabla \cdot (\varepsilon(r)\nabla\Phi(r)) = -4\pi\rho(r), \tag{9.5}$$

can be used to account for the effect of the medium on the electrostatic potential. This relates the dielectric constant $\varepsilon(r)$ and the charge distribution $\rho(r)$ to the electrostatic potential. There is no requirement that the dielectric constant be uniform, and this fundamental expression holds even if ε varies over atomic dimensions. The degree of detail can also be controlled by viewing the charges as a distribution of point charges (q_i) or as a smoothly varying charge density [$\rho(r)$]. Solution of the Poisson equation with distributions of material with different dielectric response is described in sections III.A and III.B. When the dielectric constant is the same throughout space, Equation (9.5) simplifies to

$$\varepsilon\nabla^2\Phi(r) = -4\pi\rho(r), \tag{9.6}$$

which, for point charges, reduces to

$$\Phi_j = \frac{C}{\varepsilon} \sum_{i \neq j} \frac{q_i}{r_{ij}}, \tag{9.7}$$

reinstating Coulomb's law, so that

$$\Delta G_{\mathrm{crg}} = \frac{C}{\varepsilon} \sum_j \sum_{i \neq j} \frac{\Delta q_j \times q_i}{r_{ij}}. \tag{9.8}$$

A second method of treating electronic polarization is to place an inducible point dipole (μ_i) on each atom (Warshel and Russell, 1984):

$$\mu_i = -\alpha\mathbf{E}_i, \tag{9.9}$$

where α is the polarizability of each atom, and $\mathbf{E}_i$ is the electric field from all other charges, dipoles, and induced dipoles, given by

$$\mathbf{E}_i = \sum_{j \neq i} \left(\frac{q_i\hat{\mathbf{r}}_{ij}}{r_{ij}^2} + \left[\frac{\mu_j - 3(\mathbf{r} \cdot \mu_j)\hat{\mathbf{r}}_{ij}}{r_{ij}^3} \right] \right). \tag{9.10}$$

μ, $\mathbf{E}$, and $\mathbf{r}$ are vector quantities (denoted by bold type). $\hat{\mathbf{r}}$ is the unit vector in the direction of $\mathbf{r}$. The electrostatic potential is related to the field by

$$\nabla\Phi = -\mathbf{E}(r). \qquad (9.11)$$

The application of this method is described in Section III.C.

These two methods of describing the effects of electronic polarization on Φ contain much of the same information. The Poisson–Boltzmann equation views the effects of electronic polarization as being distributed over the entire atom or averaged over a larger region of space, while the protein dipole model reduces it to an induced dipole at the atom center. A more detailed comparison is given in Section III.C.2.

2. Influence of water on the electrostatic potential

Nuclear motions also modify the electrostatic potential resulting from a charge distribution. Water molecules with their large dipole moments have the largest influence. If a uniform dielectric is assumed, a unit charge (equivalent to one electron) would shift an E_m by 60 meV (or a pK_a by 1 pH unit) if it is 120 Å away in a medium where ε is 2, while the charge has to be within 3 Å of the site for the same effect when ε is 80. And yet, the motions of water producing this response are still very small-scale. The dielectric of 80 requires that only one water in a million rotates by 180°, or an average rotation of 10^{-4} degree, in response to a field of 1 volt/cm (Debye, 1929).

The effects of the reorientation of dipoles can be separated from the polarization of electrons by observing the dielectric response to an oscillating electric field (Pethig, 1979). The dielectric constant of water is 80 when the field changes at frequencies of less than 1 GHz. Above 200 GHz, ε is reduced to 4. Thus, the motions of water that make the most significant contribution to the dielectric response need about 10 ps to come to equilibrium with the field. At optical frequencies (>1000 GHz), ε is 1.93, similar to that found for nonpolar molecules at any frequency field (Lide, 1990).

There are a variety of ways to model the dielectric response of water. One is to treat it as a dielectric continuum where ε is 80. While sacrificing all information about water structure, this allows the statistically averaged effects of water to be incorporated into the Poisson equation (Section III.B) (Sharp and Honig, 1990b). Another tactic is to simulate the reorientation of water dipoles by point dipoles, fixed on a grid, whose orientation and dipole strength changes in response to a field (Section III.C) (Warshel and Russell, 1984). The influence of water can also be included by using an atomic model of water in molecular simulations (Section III.D).

3. The influence of dipolar groups in proteins

Polar groups in proteins influence the reduction potential in two ways. Their charges contribute explicitly to generating Φ. In addition, reorientation of dipoles in response to a charge modulates the potential. Since proteins are close-packed, these motions are highly constrained. If the motions result in a relatively uniform polarizability throughout the protein, are of small amplitude, and vary linearly with the field, they can be modeled by an averaged dielectric response. This assumption would allow their effects to be included in the Poisson equation (see Section IV.B). Free energy simulations provide another method of modeling the response of dipolar groups to changing charges (see Beveridge and DiCapua, 1989, for a review).

4. Ionic strength effects

If there is salt in the medium, the redistribution of ions around charges affects the electrostatic potential in and around a protein. The simplest method of incorporating this effect into the analysis is to treat the ion distribution statistically. The charges on the protein generate a concentration gradient governed by Boltzmann's equation:

$$n_i(r) = n, \exp(-z_i \varepsilon(r) \Phi(r)/k_b T), \tag{9.12}$$

where $n_i(r)$ is the concentration of ion i at r, n is its bulk concentration, z_i is its charge, k_b is Boltzmann's constant, and T is the absolute temperature. Combining this relationship with the Poisson equation leads to the Poisson–Boltzmann equation:

$$\nabla \cdot (\varepsilon(r) \nabla \Phi(r)) = -4\pi\rho(r) + \varepsilon\kappa^2 \sinh[\Phi(r)], \tag{9.13}$$

where κ is given by

$$\kappa^2 = \frac{8\pi N z^2 I}{1000 \varepsilon k_b T}, \tag{9.14}$$

where N is the number of ions per unit volume and I is the ionic strength. κ has the units of length^{-1} and provides a measure of how salt attenuates the effect of a charge with distance. Most applications use the linearized Poisson–Boltzmann equation:

$$\nabla \cdot (\varepsilon(r) \nabla \Phi(r) \kappa^{-2} \Phi(r) = 4\pi\rho(r), \tag{9.15}$$

an approximation that is valid when $\Phi \ll kT$, although the nonlinear Poisson–Boltzmann equation has also been used (Sharp and Honig, 1990a). When the dielectric constant is uniform, Equation (9.15) simplifies to the Debye–Hückel equation:

$$\Phi_j = \frac{C}{\varepsilon} \sum_{i \neq j} \frac{q_i e^{-\kappa r}}{r_{ij}}. \tag{9.16}$$

Ions can also be included in the analysis by adding them explicitly to molecular simulations (see Sharp and Honig, 1990b, for a review).

B. The Solvation of the Heme Charge

It is well known that it is easier to put a charge in hexane than in vacuum, and it is still easier to put it in water. This dielectric response of the medium to a charge decreases the free energy of heme oxidation. The idea that cytochromes could modulate their electrochemistry by varying how deeply the heme is buried away from water in the low-dielectric protein was proposed initially by Kassner (1972, 1973) and explored by a variety of workers (Stellwagen, 1978; Tollin *et al.*, 1986). A microscopic view of the interaction between the charge and its environment is that the field from the positive charge on the heme polarizes the electrons and dipoles in its vicinity to induce a small opposing or reaction field. There is a favorable interaction between the reaction field and the heme charge that lowers the energy of the system by an amount designated ΔG_{solv}. Several approaches to calculations of ΔG_{solv} are described in sections III.A.2, III.B.2, and III.C.

III. Calculating the Electrostatic Contribution to the Free Energy of Oxidation

If protein and water responded identically to electric fields, Coulomb's law could be used to calculate the electrostatic contribution to the free energy of oxidizing a cytochrome. However, the differences in polarizability of protein and surrounding solvent means the simple expression will not hold. Several methods have been developed to deal with this fundamental difficulty in analyzing electrostatics in proteins. We will focus on those that use the Poisson–Boltzmann equation, but we will also discuss techniques that use atomic dipoles for electronic polarizability and Langevin dipoles for water, as well as the contributions of more detailed molecular simulations.

A. Using the Poisson–Boltzmann Equation

1. Analytic solutions of the Poisson–Boltzmann equation

There are analytic solutions to the Poisson–Boltzmann equation [Equation (9.13)] only for a few simple cases. The most important is, of course, when ε is uniform. Here, Coulomb's law applies. The Poisson–Boltzmann equation can also be solved analytically for systems with spherical or cylindrical symmetry (Böttcher, 1973). Early attempts to address the problems of placing charges in organic molecules focused on understanding the solubility of zwitterions (Kirkwood, 1934) and the pK_a's of diacids (Kirkwood and Westheimer, 1938) and proteins (Tanford, 1957a,b; Tanford and Kirkwood, 1957; Tanford and Roxby, 1972). Calculations of pK_a's and E_m's are essentially equivalent since, to first order, both problems deal with the stability of charges within proteins (see Section IV.C.2). Tanford and Kirkwood rationalized the altered ionization of amino acids in terms of electrostatic interactions between charges within the protein, which was itself viewed as a low-dielectric sphere within the high-dielectric water (Tanford and Kirkwood,

1957; Tanford and Roxby, 1972). Charges were either smeared over the surface (Linderstrom-Lang, 1924) or placed as points near the surface (Tanford, 1957a). The assumption of a spherical protein allows the Poisson–Boltzmann equation to be solved analytically and was reasonable at the time, since there was no information about protein structure. More recent attempts have been made to incorporate atomic resolution information into the Tanford-Kirkwood formalism by maintaining a spherical picture of the protein but mapping charges from their true positions onto the surface of the sphere (Shire et al., 1974; Matthew et al., 1985; States and Karplus, 1987). While the Tanford–Kirkwood analysis has been useful, there are now a variety of numerical methods to solve the Poisson–Boltzmann equation that do not rely on assumptions about the shape of the regions of differing dielectric constant. Thus, in future work it should be possible to use all the information about protein shapes contained in x-ray or NMR structures.

2. The solvation energy of spherical ions

For simple ions, the assumption of spherical geometry is reasonable. Their solvation free energy is largely due to electrostatic interactions, with only small contributions from hydrophobic forces. Thus, measured ion solvation free energies provide a good test of methods that treat the solvent as a dielectric continuum. The free energy of transferring a spherical ion with charge q from one medium of dielectric constant ε_i to another of dielectric ε_f is given by the Born equation (Born, 1920):

$$\Delta G_{transfer} = \Delta \Delta G_{solv} = \frac{Cq^2}{2r}\left[\frac{1}{\varepsilon_f} - \frac{1}{\varepsilon_i}\right], \tag{9.17}$$

where r is the radius of the cavity formed by the ion in the solvent.

Measurements of vacuum-to-water transfer free energy (see Marcus, 1985) or of the enthalpy of solution for salts in water (see Bockris and Reddy, 1973) can be used to test the Born equation. The only unknown in Equation (9.17) is the radius of the water cavity formed by the ion. Use of an ionic radius determined from x-ray structures of ionic crystals provides values for ΔG_{solv} that are reasonably accurate for anions. However, for cations, calculated values for ΔG_{solv} are too large (Bockris and Reddy, 1973). Rashin and Honig (1985) argued that the cation radius in covalent molecules provides a better measure of how close water can actually come to the ion. These radii yield excellent agreement with experimental solvation free energies for a variety of ions.

The success of this extremely simple theory motivated a series of Monte Carlo simulations to try to understand why water structure around the ion can be treated with a continuum model (Jayaram et al., 1989). Simple dielectric theory assumes that the orientation of dipoles by a charge is small, so that the reorientation can increase linearly with the charge. If dipoles become fully oriented around an ion, the effective polarization saturates and no longer increases with increasing charge. This so-called "dielectric saturation" has been thought to produce a low-dielectric water region around an ion. While this is found in the simulations, it appears that its effect is balanced by electrostriction, where waters are pulled towards the ion to increase the effective polarization. These two effects tend to cancel, allowing the simple model to account surprisingly well for the observed phenomena.

B. Numerical Solutions of the Poisson–Boltzmann Equation (FDPB)

Although the Poisson–Boltzmann equation cannot generally be solved analytically, there are numerical techniques for rapidly solving this differential equation (Warwicker and Watson, 1982; Klapper et al., 1986; Gilson et al., 1987; Davis and McCammon, 1990; Nicholls and Honig, 1991). This makes it possible to obtain electrostatic potentials from an atomic-resolution description of the distribution of charges and polarizability throughout a protein. Most methods use a finite difference algorithm and are designated FDPB (finite difference Poisson–Boltzmann).

In finite difference methods, space is divided with a grid. The protein is placed within the grid, and charges on atoms distributed to the eight nearest grid points (Gilson et al., 1987; Nicholls and Honig, 1991). The midpoint between each grid element is assigned a dielectric constant, determined by whether it is inside or outside of an atom. Generally the problem is simplified by assigning the same dielectric constant to all regions within the protein (Sharp and Honig, 1990b). A probe with the radius of water (e.g., 1.4 Å) identifies areas that are accessible to water, which can include cavities within the protein (Lee and Richards, 1971; Connolly, 1983). If salt is included, a layer of water around the protein that excludes salt can also be defined. The Stern layer is approximately 2 Å wide (Bockris and Reddy, 1973). The electrostatic potential at each grid point (Φ_0) is determined from the charge assigned to that site (q_0), the electrostatic potential of its six neighbors (Φ_x), and the dielectric constant of the six grid midpoints (ε_x). Thus, for each grid point:

$$\Phi_0 = \frac{\sum \varepsilon_x \Phi_x + 4\pi q_0}{\sum \varepsilon_x + \kappa^2 b^2}, \tag{9.18}$$

where κ is the reciprocal of the Debye length [Equation (9.14)] and b the grid spacing (Klapper et al., 1986; Gilson et al., 1987). The values of Φ are iterated until the potentials at all grid points form a self-consistant set of values. The electrostatic potentials at specific atoms are obtained by interpolation from the Φ's at the real grid points. There are several methods to obtain convergence of the coupled equations, and new techniques are being developed to increase speed and accuracy (Davis and McCammon, 1990; Nicholls and Honig, 1991).

One problem with finite difference methods is the finite dimensions of the grid. The protein is represented best when grid points are closely spaced relative to atomic dimensions. It is especially important to accurately define the boundary between regions with different dielectric constants, such as between protein and water. However, computation time increases with the number of grid points. Generally, FDPB methods use between 65^3 and 100^3 grid points. A second problem is that the values of Φ at the grid boundaries cannot be obtained from Equation (9.18), since all neighboring points are not included. The boundary values are, therefore, fixed at the start of the calculation. Estimates obtained from the protein charges, the external dielectric constant, and the Debye–Hückel approximation give only small errors if the protein is far from the boundary (Klapper *et al.*, 1986).

A method designated "focusing" can produce an accurate, high-resolution picture of the electrostatic potential (Gilson *et al.*, 1987). A series of calculations are run in which the protein initially fills a small portion of the grid. Boundary potentials are estimated with the Debye–Hückel equation (or Coulomb's law if there is no salt). Successive calculations double the protein's size relative to the grid. The distribution of charges and dielectric in the grid is redefined at each scale, and the boundary values are taken from the previous, lower-resolution calculation. The final grid dimensions, centered at a point of interest, can be arbitrarily small. Much of the protein will lie out of the grid, but the influence of the charges and dielectric in these regions is contained at somewhat lower resolution in the boundary values.

1. Calculations of the electrostatic potential due to other charges

ΔG_{crg} for heme oxidation can be obtained with Equation (9.2) using Φ at the positions of atoms in the heme calculated with the FDPB method. The total ΔG_{crg} is a simple sum of the interactions of heme atoms with individual charges. Thus, the influence of a subset of atoms, such as the protein backbone or an ionized residue, can be obtained by including only these charges in the calculations of Φ.

2. Calculations of the solvation free energy

The solvation free energy (ΔG_{solv}) measures the decrease in free energy of the system due to the polarization of the environment by the heme charge. It can be obtained with the FDPB method by calculating the free energy of transferring the heme from a reference state to its position in the protein surrounded by water (Gilson and Honig, 1988b). The polarization that generates the reaction field requires a dielectric boundary (Onsager, 1936). Therefore, an appropriate reference state is one where the medium (ε_{out}) has the same dielectric constant as the heme (ε_{in}), so ΔG_{solv} is zero.

The total energy of a charged system is given by

$$\Delta G_{tot} = 0.5 \sum_j q_j \Phi_j. \tag{9.19}$$

For this calculation, the sum runs over the charges in the group whose solvation energy is desired, such as the heme or an ionizable amino acid, while all other charges are zero. Φ_j contains the energy of assembling the point charges themselves, and the energy of interaction of the charges with each other and with the external medium. The last two terms provide ΔG_{solv}. If the Poisson–Boltzmann equation were solved analytically, ΔG_{tot} would be infinite since it involves evaluating Equation (9.3) as $r_{ij} \to 0$; in the numerical approximation the result is large, but finite. However, the problematic term, which is the ΔG of charge assembly, depends only on ε_{in} and not on ε_{out}. Therefore, ΔG_{solv} can be obtained from

$$\Delta G_{solv} = \Delta G_{tot}(\varepsilon_{out} = 80) - \Delta G_{tot}(\varepsilon_{out} = \varepsilon_{in}), \tag{9.20}$$

where in both calculations of ΔG_{tot}, ε_{in} is a dielectric constant appropriate to the heme itself.

C. Protein Dipole, Langevin Dipole (PDLD) Method

A second method for calculating electrostatic free energies using atomic detail information about a protein has been developed by Warshel and colleagues and is called the PDLD method (protein dipole Langevin dipole) (Warshel and Russell, 1984; Russell and Warshel, 1985; Churg and Warshel, 1986). The polarization of the electrons in the protein is approximated by an induced point dipole (μ_i) at the atom centers [Equation (9.9)]. Each dipole is dependent on the electric field produced by the charges and other dipoles in the protein ($\mathbf{E}_i$) [Equation (9.10)]. The reorganization of water in response to the field is modeled by discrete dipoles (μ_L) placed on a grid spaced to maintain the density of water. These point in the direction of the field created by the charges in the protein ($\hat{\mathbf{e}}_L$) [the first term in Equation (9.10)]. The effective dipole strength, which is the projection of the dipole in the direction of the field, increases with the total field ($\mathbf{E}_L$). $\mathbf{E}_L$ includes the contribution of the charges on the protein, protein-induced dipoles, and water dipoles. A Langevin function describes the relationship between $\mathbf{E}_L$ and μ_L:

$$\mu_L = \hat{\mathbf{e}}_L \mu_0 (\coth(X_i) - X_i^{-1}), \tag{9.21}$$

where $\hat{\mathbf{e}}_L$ is the unit vector in the direction of the field produced by the charges in the protein, μ_0 is the upper limit of the dipole at infinite field strength, and X_i is

$$X_i = C' \mu'(\mathbf{E}_L') \cdot \hat{\mathbf{e}}_L / kT. \tag{9.22}$$

C' is a constant, and $\mathbf{E}_L'$ is the total field at the dipole not including the field due to the nearest neighbors. The values of C' and μ' are derived semi-empirically from calculations that reproduce measured solvation energies.

The polarization of electrons in the protein [Equation (9.9)] and the effective water dipoles [Equations (9.21) and (9.22)] are connected via the electric field [Equation (9.10)]. In practice, Equations (9.9) and (9.10) are evaluated until a self-consistent set of values of μ_i and $\mathbf{E}_i$ for the protein is derived in the absence of the water dipoles. This field is used as an input for determining μ_L and $\mathbf{E}_L$ again by an iterative procedure. If necessary, the protein atomic dipoles can be reevaluated given the contribution of the fields induced by the oriented water dipoles (Russell and Warshel, 1985). With the coarse grid spacing used to maintain the appropriate density of Langevin dipoles, the results are dependent on the position of protein relative to the grid. The calculations are performed several times, moving the protein in relation to the grid. The position that provides the lowest total energy is used (Russell and Warshel, 1985).

1. Caculations of the electrostatic free energy of heme oxidation

The electrostatic free energy of heme oxidation is evaluated with the PDLD method by determining a consistent set of μ_i, μ_L, and fields in the presence of the charges on the reduced cytochrome and again with the charge distribution of the oxidized protein (Russell and Warshel, 1985; Churg and Warshel, 1986; Warshel, 1987). The free energy of interaction with charges in the protein is given by Equation (9.4), implying a dielectric constant of 1. The modulation of the interaction by the electronic polarization is given by

$$\Delta G_{\mathrm{PD}} = \frac{C}{2}\left[\left(\sum_i \mathbf{E}_i \cdot \mu_i\right)_{\mathrm{ox}} - \left(\sum_i \mathbf{E}_i \cdot \mu_i\right)_{\mathrm{red}}\right]. \tag{9.23}$$

The first term is calculated with an oxidized heme, and the second is calculated with the reduced heme. The sum runs over heme atoms as well as atoms in the rest of the protein, thereby counting each interaction twice. This is dealt with by the prefactor of one-half. The effect of the polarization of Langevin dipoles is obtained from

$$\Delta G_{LD} = \frac{C}{2}\left[\left(\sum_L \mathbf{E}_L \cdot \mu_L\right)_{\mathrm{ox}} - \left(\sum_L \mathbf{E}_L \cdot \mu_L\right)_{\mathrm{red}}\right]. \tag{9.24}$$

The region of water modeled by the Langevin dipoles generally extends about 20 Å from the protein center. The solvation of the cytochrome by water farther away is obtained by considering a spherical region of Langevin dipoles and protein immersed in a continuum solvent with a dielectric of 80. The solvation of the protein by the continuum water is accounted for by the Born equation:

$$\Delta G_{\mathrm{bulk}} = \frac{C}{2r}\left[\frac{1}{80} - \frac{1}{1}\right][z_{\mathrm{ox}}^2 - z_{\mathrm{red}}^2], \tag{9.25}$$

where r is the radius of the sphere including the protein and the Langevin dipoles, and z is the total charge on the oxidized or reduced cytochrome. This bulk water is not included in the calculations of the induced dipoles.

2. Comparison of free energies derived by the PDLD and FDPB methods

The FDPB and PDLD methods include the electrostatic contributors to the free energy of heme oxidation differently (e.g., Russell and Warshel, 1985; Churg and Warshel, 1986; Warshel, 1987; and Gilson and Honig, 1987; Gilson et al., 1987). The PDLD method treats the charge–charge interactions as occurring in a vacuum. The polarization of the electrons induced by all charges in the protein and the heme is grouped together in ΔG_{PD}. The polarization of water by all charges is included in ΔG_{LD} and ΔG_{bulk}. In contrast, the FDPB method accounts for polarization of the protein and water by the constant charges in the protein as ε in Equations (9.5) or (9.17). The dielectric constant modifies Φ and thus ΔG_{crg}. The polarization of the medium by the changing heme charge on oxidation is defined as ΔG_{solv}. Comparison of FDPB and PDLD calculations for polar organic molecules shows similar values of ΔG_{solv} can be obtained by the two methods (Jean-Charles et al., 1991).

A second distinction between the methods is that the **FDPB** method can treat small motions of dipolar groups in a protein as part of the dielectric response (Gilson and Honig, 1986) (see Section IV.B). The PDLD method includes only the electronic polarization of the protein. The contribution of movements of dipoles in proteins that are either seen explicitly in the different x-ray structures (Churg *et al.*, 1983; Warshel and Churg, 1983) or obtained with molecular simulations can be included in either method (see Section III.E).

D. Molecular Mechanics and Electrostatics

Eventually all theoretical questions may be amenable to some form of theoretical treatment where the motion of proteins and the surrounding solvent are reliably simulated on a computer. If the forces were accurately described, the changes in molecular structure over time, which provide kinetic properties and the statistical averaging that provides thermodynamic properties, could be fully understood at a molecular level. However, despite much progress there are still problems that remain to be solved in simulation methodology (see McCammon and Harvey, 1987; Beveridge and DiCapua, 1989; Karplus and Petsko, 1990, for reviews).

One important problem in such simulations is that it is difficult to correctly include electrostatic forces in calculations where there are regions of varying dielectric constant (see Harvey, 1989). This review is primarily concerned with methods for correctly solving this problem for a single static structure. The biggest problem is modeling the response of the solvent water. Simulations that contain explicit water molecules around the protein should provide correct values for the electrostatic interactions, although at increased cost in computer time for the calculation. In addition, the motions of water that yield the dielectric response are on the picosecond time scale. Thus, for simulation of very fast time scale properties, models using a continuum water dielectric constant of 80 will overestimate the electrostatic damping. However, if longer time scale or equilibrium properties are desired, simulations must be run long enough to average the water response statistically. In addition, if mobile ions are to be included, the time scale needed for them to diffuse into an equilibrium distribution is > 200 ps (Bockris and Reddy, 1973).

Simulations are often performed without any solvent molecules present. A variety of *ad hoc* methods are used to account for the effect of water on the electrostatic forces. These methods generally assume that forces can be obtained with Coulomb's law, but they use a dielectric constant that increases with increasing distance from the charge. The rationale is that the influence of water will be larger for interactions between charges that are far apart than for nearby charges. Detailed comparison of distance-dependent dielectric constants with the results of solving the Poisson equation (Bashford *et al.*, 1988; Gilson and Honig, 1988b) or adding discrete waters to a molecular simulation (Wendoloski and Matthew, 1989) show the distance-dependent dielectric model is inadequate. In addition, a vacuum simulation cannot include the interaction of the charge with the solvent reaction field that provides a force pulling the charge into the water (Sharp, 1991). Attempts are being made to develop methods that either speed up solution of the Poisson–Boltzmann equation so that it can be incorporated into dynamical algorithms (Sharp, 1991) or to provide other fast, yet accurate approximations of the effective electrostatic forces (Gilson and Honig, 1991).

E. Electrostatic Contributions to the Reorganization Energy for Electron Transfer

The motions of nuclei in response to a charge affect electron transfer rates. The reorganization energy, which is the energy necessary to move the nuclei from the equilibrium position of the reduced system to that of the oxidized state without transferring the electron, is often used in electron transfer rate expressions (see Chance *et al.*, 1979; Devault, 1980; Sutin, 1982; Marcus and Sutin, 1985; Closs *et al.*, 1986; Mayo *et al.*, 1986, for reviews). Changes in the nuclear conformation in the heme, protein, and surrounding solvent all contribute to this energy. These motions have been traditionally separated into outer-sphere (λ_{out}) and inner-sphere contributions (λ_{in}). λ_{out} is the free energy required to change the low-frequency motions of the heme surroundings, while λ_{in} is the energy required to change the higher-frequency motions of the intramolecular bonds in the heme and ligands.

λ_{out} has been modeled by Marcus and others as the dielectric response of the dipoles, primarily water, to the reaction (Devault, 1980). It has been calculated by the Born equation, assuming a spherical geometry for the redox sites. Both the FDPB and PDLD formalisms should allow the calculation of the solvent reorganization energy with more realistic geometry.

If there is information available about the differences in structure of the oxidized and reduced cytochrome, then electrostatic contributions to λ_{in} can also be calculated. Information about the change in structure can come from x-ray crystal structures (Takano and Dickerson, 1981a,b), NMR (Feng *et al.*, 1989; Wand *et al.*, 1989), other experimental methods (Moss *et al.*, 1990), or from molecular simulations (Northrup *et al.*, 1980, 1981; Wendoloski *et al.*, 1987). The contribution of changes in ΔG_{crg} to λ_{in} will be

$$\Delta \Delta G_{crg} = (\Phi_{ox} - \Phi_{red})\Delta q_j, \qquad (9.26)$$

where Φ_{ox} is the electrostatic potential at the heme calculated with the oxidized equilibrium protein structure, while Φ_{red} is calculated with the reduced structure (Churg *et al.*, 1983; Warshel and Churg, 1983).

IV. The Parameters Used in the Calculations

To obtain the electrostatic contribution to the free energy of cytochrome oxidation, one requires not only a tractable computation scheme describing the protein and solvent, but also accurate values for the required physical parameters. Both the FDPB and PDLD formalisms need the position of atoms and the charge associated with each atom. The FDPB formalism then uses the dielectric constant of different regions, while the PDLD formalism needs values of atomic polarizability. As will be described in this section, the appropriate values for these parameters are not fully established. Thus, each successful calculation of some experimental observation helps refine the methods and the input parameters, in addition to providing increased understanding of how proteins work.

A. Protein Structures

The positions of charged atoms and regions of different polarizability are provided by an atomic-resolution protein structure and are generally derived from x-ray crystallography. These structures give the positions of the heavy atoms, but usually not of the hydrogens. Polar hydrogens are important contributors to local charge–charge interactions. Several molecular mechanics programs are available to add protons to satisfy the valence of the heavy atoms in proteins. These include DISCOVER (Biosym) (Hagler and Moult, 1978) and XPLOR (Brünger *et al.*, 1987). Given the heavy atom positions, hydroxyl groups and lysine ε-amino groups have the only polar hydrogens with rotational degrees of freedom. Energy minimization routines in these packages allow these protons to explore allowed positions to maximize favorable local electrostatic interactions while minimizing steric overlaps.

A water molecule that is bound with a fixed geometry in the protein needs to be treated as an explicit, polar group distinct from the solvent. Such groups can strongly influence local electrostatic potentials. Hydrogen bonds from water to buried charges or dipoles will play a significant role in stabilizing these charges within the protein (Rashin and Honig, 1984; Rashin *et al.*, 1986; Luntz *et al.*, 1989). A serious problem is to decide which waters to include. While some waters are localized within an x-ray structure, only in the highest-resolution pictures is it likely that all are seen. Experimental and theoretical evidence suggests that when there are available hydrogen bonds, cavities in proteins will be filled with water (see, e.g., Rashin and Honig, 1984; Kossiakoff, 1985; Otting and Wüthrich, 1989; Cheng and Schoenborn, 1990).

A final uncertainty in the information provided by crystal structures is the presence of bound ions. It is generally not known if any of the electron density assigned to neutral waters might actually be ions such as ammonium. Also, some of the ions localized in a structure under the high-salt conditions of crystallization will not be present under normal solution conditions. Buried ions will significantly affect the electrostatic potential at long distances from the charge.

B. The Dielectric Constant of Proteins

The dielectric constant of a protein is a concept subject to much controversy and confusion. A soluble protein is a structure of low polarizability surrounded by water. The effect of one charge on another will be exquisitely dependent on the distance between the charges and their positions relative to the dielectric boundary (see Gilson *et al.*, 1985; Harvey, 1989). Therefore, knowledge of the interaction between two charges in the protein provides no knowledge of how other charges will interact with each other. However, this does not imply that the protein itself does not have a relatively uniform polarizability; rather, it is a result of the profound influence of the surrounding water.

There are several approaches to obtaining a dielectric constant for protein. The bulk dielectric constant of dry or partially hydrated protein has been measured, providing values for ε of 2 to 4 (Takashima and Schwan, 1965; Pennock and Schwan, 1969; Harvey and Hoekstra, 1972). Theoretical studies of the normal modes of random α-helices packed with the density found in protein also yield values for ε between 2 and 4 (Gilson and Honig, 1986; Nakamura et al., 1988). Calculation of experimental properties with the FDPB method can provide another check of the value of ε. The best tests will involve successful calculation of several properties within a protein with a single value of ε. Bashford and Karplus (1990) calculated the pK_a's of 21 ionizable residues in lysozme and obtained reasonable agreement with the measured values using a dielectric constant of 4. Gunner and Honig (1991) calculated the E_m's for the four hemes in the cytochrome subunit of the reaction center from *Rhodopseudomonas viridis* with ε equal to 2, 4, or 6. The best agreement with experiment was obtained with a dielectric constant of 4 (see Section V.A).

The dielectric model for protein is computationally attractive, because it uses a statistical average of the protein response rather than a detailed molecular picture. This model will fail when there are local conformational changes that are coupled to a reaction (Wendoloski and Matthew, 1989), so it should, therefore, be applied on a case-by-case basis. However, systems where reaction involves little rearrangement, such as many electron transfer proteins, are good candidates for analysis in this manner.

C. Charges

1. Distribution of charge within the protein

The charge on a residue is not uniformly distributed. Thus, polar side chains and the peptide backbone, though neutral overall, are treated as being made up of atoms with positive or negative partial charges. The charges on ionized residues are also distributed among several atoms in the side chain. The separation of charge in polar groups is significant. For example, the carbon of a backbone carbonyl may be assigned a positive charge of $+0.5$, while the attached oxygen has -0.5 charge (Brooks et al., 1983). Though the influence of dipoles falls off with distance, dipoles produce large effects on the local electrostatic potential. In fact, in most modern molecular simulations of proteins, the attractive forces of hydrogen bonds are obtained solely from the electrostatic attraction of appropriately oriented dipoles in close contact.

The amino acid partial charge sets are generally obtained from a variety of theoretical considerations. These include *ab initio* molecular orbital calculations (Momany et al., 1975; Hagler and Moult, 1978; Jorgensen and Tirado-Rives, 1988), rules based on the relative electronegativity of bonded atoms (No et al., 1990a,b), or electron density measurements of electron diffraction patterns from very high resolution x-ray structure determinations (Coppens, 1989; Destro et al., 1989). Several charge distributions have been proposed for the amino acids in proteins (Momany et al., 1975; Hagler and Moult, 1978; Brooks et al., 1983; Weiner et al., 1986; Jorgensen and Tirado-Rives, 1988). These charges are derived as part of force fields used for molecular mechanics simulations. The force fields include information about intramolecular potentials for bond lengths and angles and the van der Waals and electrostatic forces that control intermolecular interactions. The parameter values are tuned to fit experimental observables such as atom positions in known structures, rotational energy barriers, heats of vaporization, vibrational frequencies, and dipole moments. These properties are functions of all force field parameters, not just the distribution of partial charges. In addition, the simulations run to test the parameters may not correctly incorporate electrostatic forces (see Section III.D). Even dipole moments, which depend most simply on the charge distribution, can be obtained with many different distributions of charge within a fixed molecular geometry. Thus, all charge sets are highly underdetermined. This does not invalidate the work that relies on them, but it should make clear the need to test how a calculation depends on the assigned distribution of charges in the protein. In addition, the comparison of measured effects of specific amino acids on E_m's or pK_a's can occasionally provide tests of the charges.

2. Are acidic and basic residues ionized?

A major uncertainty in the calculations of electrostatic potential within a protein is the charge state of ionizable residues. In isolation from the protein, lysines, arginines, and aspartic, glutamic, and propionic acids would be charged at the pH's of most experiments. Histidines, tyrosines, and cysteines also have pK_a's near physiological pH. However, the free energy of protonation is influenced by the same electrostatic forces that affect the free energy of heme oxidation (Russell and Fersht, 1987; Soman *et al.*, 1989; Bashford and Karplus, 1990; Sharp and Honig, 1990b). Calculation of changes in pK_a's in proteins with site-directed mutations provided an early test of electrostatic calculations (Gilson and Honig, 1987; Sternberg *et. al.*, 1987).

The factors that affect ionization within a protein are the loss of solvation energy as groups are shielded from water and the charge–charge interaction with ionized residues and nearby polar groups. The loss of solvation energy always favors the neutral form of a residue, but other charges or properly oriented dipoles can stabilize the ionized form. It appears to be possible to obtain reasonable estimates of the ionization state of acidic and basic residues with electrostatic calculations (Bashford and Karplus, 1990; Yang *et al.*, 1993). Using the division of free energy of ionization described for the FDPB method, the pK_a of site m in the protein is given by (Bashford and Karplus, 1990; Sharp and Honig, 1990b)

$$pK_{(protein)}(m) = pK_{(soln)}(m) - 2.303 \, kTc_m \left[\Delta\Delta G_{(desolv)}(m) \right.$$

$$\left. + \Delta G_{(dipolar)}(m) + \sum_{\substack{n \neq m \\ \text{ionizable sites}}} f_n \Phi(\Delta q_n, m) \Delta q_m \right], \qquad (9.27)$$

where $pK_{(soln)}$ is the pK_a of the amino acid in solution; c_m is $+1$ for a base and -1 for an acid; $\Delta\Delta G_{(desolv)}$ is the difference between the solvation energy of m in water and within the protein; $\Delta G_{(dipolar)}$ is obtained with Equation (9.2), where Φ_m

is calculated with only partial charges on non-ionizable side chains and backbone atoms; and Δq_m is the change in charge on ionization of m. The last term is the influence of other ionizable sites (n), where $\Phi(\Delta q_n, m)$ is the electrostatic potential at m due to the changing charge on ionization of n. The fraction of n ionized is f_n. A self-consistent set of pK_a's must be obtained for all ionizable groups, because the pK_a of each residue is dependent on the the ionization state of all other residues.

One method to estimate pK_a's is to use the standard expression

$$\log(f_n/[1 - f_n]) = pK(n) - pH \tag{9.28}$$

to get f_n (Tanford and Roxby, 1972). Equations (9.26) and (9.27) are then iterated to obtain a consistent set of $pK_a(m)$'s and f_m's for all residues (Tanford and Roxby, 1972; Bashford and Karplus, 1990). This method can be rapidly solved, even for large numbers of residues. However, it can yield significant errors. A system with two nearby acidic residues (A and B) with the same pK_a in the absence of the other site provides an example. The iterative algorithm results in the site which is ionized first raising the pK_a of its neighbor. Thus, site A might be fully ionized, while site B is completely neutral (or vice versa). In reality there will be a range of pH's where 50% of the molecules have A ionized and 50% have B ionized. Thus, the correct picture will be a Boltzmann distribution of all possible ionization states. The free energy of each state is the term in brackets in Equation (9.27), where f_n is now 0 or 1.

The problem with methods that require consideration of all possible states is that for n ionizable residues, there are 2^n states. For proteins with more than 20 ionizable residues, this number becomes extremely large. It may then become necessary to restrict the calculation to sites near a region of interest (Bashford and Karplus, 1991; Yang *et al.*, 1992), or to use procedures such as Monte Carlo methods to sample possible ionization states to find the lowest-energy conformations (Beroza *et al.*, 1991; Gunner and Honig, 1993).

The ability to determine the ionization state of a protein should allow the pH dependence of cytochrome E_m's to be predicted from calculations. In addition, in systems where conformational changes are small, the stoichiometry and pH dependence of proton release on cytochrome oxidation should be amenable to analysis. Much information of this kind is available (e.g., Pettigrew *et al.*, 1975, 1978; Moore *et al.*, 1980, 1984; Barakat and Strekas, 1982; Moore, 1983; Altman *et al.*, 1989; Cutler *et al.*, 1989; Gao *et al.*, 1990; Barker *et al.*, 1991), which should provide good tests of the methods outlined here.

3. The change in electron density on heme oxidation

The electrostatic free energy of heme oxidation described in this review is simply the change in the interaction energy of the heme with the protein and solvent when the heme loses an electron. Therefore, some information must be available about the oxidation state–linked change in distribution of charges on the heme (Δq_{ox}). Various quantum-mechanical calculations have been carried out on iron porphyrins (Warshel and Lappicirella, 1981; Rawlings *et al.*, 1985a,b,c; Rohmer, 1985; Tollin *et al.*, 1986). There is also a great deal known about the electrochemistry of isolated hemes with different ligands (Kadish, 1986). These include studies of small cytochrome *c* fragments (Harbury and Loach, 1960; Harbury *et al.*, 1965; Warme and Hager, 1970; Wilson, 1974), as well as hemes from other sources (Mashiko *et al.*, 1979, 1981).

It seems reasonable to see if the dependence of the reduction potential of isolated hemes on axial and peripheral ligands and solvent can be reproduced by electrostatic calculations. Of particular relevance to cytochromes is the observation that changing the axial ligands from bis-histidine to histidine–methionine raises the heme E_m by 150 meV (Harbury *et al.*, 1965; Wilson, 1974; Mashiko *et al.*, 1979). This result is at most weakly dependent on the solvent (Kadish, 1986). Also, hemes with axial histidine ligands are easier to oxidize if there is a proximal hydrogen bond acceptor (Doeff *et al.*, 1983; Valentine *et al.*, 1979).

Gunner and Honig (1991) analyzed the oxidation of isolated bis-histidine and histidine–methionine hemes with the FDPB method to try to reproduce the 150 meV difference in E_m's. This assumed that axial ligands modify midpoint potentials by local electrostatic interactions with the heme. Sensitivity of the calculations to the delocalization of the charge on the heme was tested by comparing the results with a Δq_{ox} of $+1.0$ at the heme iron or $+0.25$ at each of the heme nitrogens. Various charge distributions on the amino acid ligands were tested by assuming values, taken from a number of different charge sets. The difference between the bis-histidine and histidine–methionine hemes with $\Delta q_{ox} = 1.0$ at the iron ($\Delta q_{ox} = 0.25$ at each heme nitrogen in parentheses) was 158 (74) meV with CHARMM charges (Brooks *et al.*, 1983), 207 (84) meV with DISCOVER charges (Hagler and Moult, 1978), and -20 (6) meV with OPLS (Jorgensen and Tirado-Rives, 1988). In all cases, the dependence of the E_m on the dielectric constant of the solvent was small.

Thus, the electrostatic calculations can reproduce the variation in heme E_m's with changing axial ligands if an appropriate charge set is used. It appears that the best results are obtained if the loss of electron density on oxidation is localized at the iron. OPLS incorrectly orders the midpoints because its methionine sulfur has a negative partial charge similar to that of the histidine nitrogen; this assumption is unlikely to be correct, given the relative electronegativity of these atoms. DISCOVER overestimates the ligand dependence primarily because its sulfur has a small positive charge (0.12). CHARMM, which places a small negative charge on the sulfur (-0.12), shows the best match with experiment.

Modeling the contribution of axial ligands to heme oxidation as the simple effect of a nearby static charge ignores any stronger coupling of the ligand and heme electronic states. This approximation is supported by studies showing that ring substituents change a pyridine's pK_a and E_m of a heme bound to the pyridine by the same amount (Kadish and Bottomley, 1980). Pyridine substituents change the pK_a of the pyridine by modifying the charge density at the nitrogen, thereby affecting the electrostatic potential felt by the proton. The midpoint potential of heme appear to be affected through the same mechanism. When the loss of electron density on oxidation is distributed throughout the macrocycle, as for the oxidation of Fe(I) hemes, substituent effects on E_m's are smaller than on pK_a's (Kadish and Bottomley, 1980). This behavior is mirrored in the smaller calculated dependence of the E_m on the ligand when Δq_{ox} is at the heme nitrogens.

4. Comparing the results to experiment: The relationship of the calculations to the standard hydrogen electrode

The implicit reference state for the calculations described here is that of heme oxidation in the absence of any external charges in a medium with the same polarizability as the heme. In contrast, the measured E_m's use the standard hydrogen electrode as a reference. Calculations on isolated hemes with well-defined ligands have been used to find an additive constant to connect the results of experiment and theory (Churg and Warshel, 1986; Gunner and Honig, 1991). This factor will also include the other contributions to heme oxidation that are assumed to be independent of the environment [i.e., $\Delta G_{other(heme)}$ in Equation (9.1)].

V. Electrostatic Calculations on Cytochromes

A. Calculating Cytochrome Reduction Potentials

Churg and Warshel (1986) calculated the midpoint potential of tuna cytochrome c using the structure of Takano *et al.* (1981a,b) by the PDLD method. Calculations compared the contribution of dipolar charges to ΔG_{crg}, ΔG_{PD}, ΔG_{LD}, and G_{bulk} for the cytochrome and an octapeptide fragment containing the heme and its ligands. The E_m of the cytochrome is 300 meV higher than that of the heme–octapeptide (Harbury *et al.*, 1965). The calculated difference between these factors in the protein and octapeptide is approximately 400 meV. If the propionic acids were included in this analysis, the calculated difference in midpoints was closer to that measured. The contributions of no other ionizable residues were considered in this study.

Gunner and Honig (1991) used the FDPB method to analyze the four-heme cytochrome bound to the reaction center from *Rhodopseudomonas viridis* (Deisenhofer *et al.*, 1984, 1985; Michel and Deisenhofer, 1988; Deisenhofer and Michel, 1989). The hemes in this protein are found in a roughly linear arrangement, and their potentials range over more than 450 meV. As one moves from the membrane out into the periplasm, the heme midpoints are ordered: the highest-potential site (380 meV), a low-potential site (20 meV), a high-potential heme (310 meV), the lowest-potential heme (−70 meV) (Dracheva *et al.*, 1988; Fritzsch *et al.*, 1989; Nitschke and Rutherford, 1989; Alegria and Dutton, 1991) (see Gunner, 1991, for a review). It was found that the FDPB method could reasonably reproduce the observed E_m's of all four sites. Both DISCOVER and CHARMM protein partial charge sets were tested. Both sets reproduced the pattern of *in situ* E_m's, but DISCOVER produced somewhat better quantitative agreement with the experimental measurements. The average error with the DISCOVER charges was less than 60 meV when the calculations used the solution E_m of histidine–methionine hemes as a standard (Warme and Hager, 1970; Wilson, 1974; Mashiko *et al.*, 1979, 1981). The error in the relative midpoints of the four hemes was less than 30 meV.

It is interesting to consider how the heme midpoints are controlled in the reaction center. A number of factors contribute. The loss in solvation energy burying the hemes in the protein raises the E_m's by 300 ± 40 meV. Thus, this factor is an important determinant of the absolute midpoint, but it provides little discrimination between sites. The contribution of ionized residues to ΔG_{crg} varies over 220 mV for the four hemes. These groups provide much of the difference between the highest-potential heme at one end of the cytochrome and the lowest-potential site at the other. Three of the hemes have histidine–methionine ligands. The second heme has bis-histidine ligands. The second histidine ligand of this latter heme ensures that this is a low-potential site. The position of the third heme, bracketed by two low-potential sites, is one of the reasons for its high midpoint potential. The positive charges on the adjacent hemes, which will be oxidized first, raise the E_m of the centrally located heme by more than 100 meV.

B. Calculating the pH Dependence of Cytochrome E_m's

An early use of the finite difference method to solve Poisson's equation was to calculate the interaction energy between the propionic acids and the heme in cytochrome c_{551} using the structure from *Pseudomonas aeruginosa* (Matsuura *et al.*, 1982). Moore (1983) had noted that in cytochromes where one propionic acid has a pK_a in a pH range that is accessible for titration, the ionized propionic acid lowers the E_m of the heme by 65 meV. Appropriately, the oxidized heme raises the pK_a of the acid by 1.1 pH units. Rogers and colleagues (1985) calculated the interaction energy between these two charges using an FDPB method and uniform or distance-dependent dielectric models. The FDPB method gave an interaction energy of 90 meV, in reasonable agreement with the experiment. The other calculations gave interaction energies of greater than 200 meV, demonstrating that these approximations for electrostatic interactions can be very unreliable. A similar study of the variation in pK_a of several histidines in azurin with oxidation state has also been carried out (Bashford *et al.*, 1988).

C. Changes in Electrochemistry on Mutation

Arginine-38 of yeast cytochrome c was mutated to several other amino acids (Cutler *et al.*, 1989). These mutations were found to lower the heme E_m by about 50 meV. The variation in the charge at this site did not bring the pK_a of the heme propionic acids into a region that is experimentally accessible. This result was unexpected since the propionic acids in cytochromes without an arginine at this position have pK_a's between 6 and 8 (Moore *et al.*, 1984). Small changes in the pK_a's of other residues were measured. The change in midpoint was calculated by the free energy perturbation method and found to be in excellent agreement with experiment (Cutler *et al.*, 1989).

VI. What Has Been Learned from Electrostatic Calculations on Cytochromes?

A. How Proteins Modify Midpoint Potentials

One goal of the electrostatic analysis of cytochromes is to understand qualitatively how proteins might modify heme E_m's. The electrostatic contributors to ΔG_{ox} can be compared in water, in a high-dielectric, polarizable solvent, and in the rigid, low-dielectric environment provided by protein (Gunner and Honig, 1991). The solvation of a charge is always smaller in protein, raising the midpoint potential (Kassner, 1972, 1973). However, long-range, pairwise interactions between charges can play a greater role in a protein environment than in water. Thus, the propionic acids or other negatively charged groups can lower the midpoint potential to a greater extent, thus compensating for the loss of solvation energy in the protein. The interactions between charges in van der Waals contact such as the axial ligands are relatively insensitive to their surroundings.

The loss of electron density in the oxidation of ferric to ferrous heme (Δq_{ox}) appears to be primarily localized to the iron. This has consequences for how cytochromes can modulate heme midpoint potentials. Burying the heme in the protein appears to raise the E_m of the heme by 300 ± 50 meV. The hemes in large cytochromes (Gunner and Honig, 1991) and smaller soluble cytochromes (Gunner and Honig, unpublished observations) all experience solvation penalties in this relatively narrow range. This is a consequence of the limited opportunity to change the approach of water to iron in a six-coordinate heme. Therefore, the large variation in cytochrome E_m's cannot be due to small changes in heme edge accessibility (Stellwagen, 1978), although exposure of the heme edge may still influence electron transfer rates (Tollin *et al.*, 1986).

The pairwise interactions of the heme iron with ionized groups in the protein can be important at distances of greater than 10 Å. However, the magnitude of effect depends on how deeply the charges are buried in the protein, as well as on their separation. This is illustrated by the stabilization of the oxidized heme by propionic acids in cytochrome c_{551} (Rogers *et al.*, 1985) or the cytochrome subunit of *Rs. viridis* reaction centers (Gunner and Honig, 1991). These negatively charged groups lower the E_m by 60 to 200 meV. The variation is not a consequence of distance, as the propionic acid oxygens are all approximately the same distance from the iron (7–9 Å). Rather, the interaction energy is correlated with the propionic acid desolvation penalty. Pairwise interactions with well-solvated propionic acids provide the least stabilization of the oxidized heme, while those acids that are more deeply buried have a more significant influence (Gunner and Honig, 1991).

With Δq_{ox} localized on the iron, the axial ligands are the only part of the protein within 5 Å of the oxidation site. This distance dependence limits the influence of polar side-chain and backbone partial charges, because interactions between dipoles and charges fall off rapidly at distances greater than the separation between the charges in the dipole.

B. How Proteins Modify pK_a's

The pK_a's of ionizable residues are modified by electrostatic factors in proteins in the same way as the electrochemistry of the heme. However, there are differences in how a protein can tune residue pK_a's and heme E_m's, because the change in charge on protonation is generally at the end of a side chain instead of buried as in a fully coordinated heme. The rules governing long-range interaction with other ionized residues are similar for hemes and protonatable residues. However, the differences in solvation penalty between residues at different positions can be large, so local environment can make a substantial contribution to the variation in pK_a's (Bashford and Karplus, 1990; Gunner and Honig, 1991). Also, dipolar groups can come close enough to the ionized side chain to provide significant stabilization of a buried charge (Bashford and Karplus, 1990; Gunner and Honig, 1991).

VII. Conclusions

The cytochrome family of proteins provides an excellent test system for evaluation of strategies for modeling protein electrostatic properties. Furthermore, the function of these proteins can in turn be illuminated by such analysis. Probably the central challenge is to elucidate the interactions between protein and heme that produce the wide spread of E_m's found in different cytochromes. In addition, there is a wide variation in the pH dependence of cytochrome redox potentials that should be a consequence of electrostatic factors. The conclusions of the analysis of E_m's or pK_a's can be tested with site-directed mutations. Aspects of the outer- and inner-sphere reorganization energy that influence the rates of cytochrome electron transfers should also be dependent on electrostatic factors.

We find that careful calculations can reproduce experimental midpoint potentials surprisingly well. However, each of the small set of cytochromes that has been studied achieves the observed midpoint potentials by different combinations of interactions. Thus, future analysis of cytochromes should help provide both the specific and the general rules for how proteins exploit electrostatic interactions.

VIII. Acknowledgments

We thank Anthony Nicholls, Kim Sharp, An-Suei Yang, S. Shridaran, Rosemary Sampogna, and Arald Jean-Charles for stimulating discussions. The financial support of the National Institute of Health GM12897 (M.R.G.) and National Science Foundation DMB-89-03484 are gratefully acknowledged.

IX. References

Alegria, G., & Dutton, P. L. (1991) *Biochem. Biophys. Acta* **1057**, 258–272.
Altman, J., Lipka, J. J., Kuntz, I., & Waskell, L. (1989) *Biochemistry* **28**, 7516–7523.
Barakat, R., & Strekas, T. C. (1982) *Biochem. Biophys. Acta* **679**, 393–399.
Barker, P. D. Mauk, M. R., & Mauk, A. G. (1991) *Biochemistry* **30**, 2377–2383.
Bashford, D., & Karplus, M. (1990) *Biochemistry* **29**, 10219–10225.
Bashford, D., & Karplus, M. (1991) *J. Phys. Chem.* **95**, 9556–9561.
Bashford, D., Karplus, M., & Canters, G. W. (1988) *J. Mol. Biol.* **203**, 507–510.
Beroza, P., Fredkin, D. R., & Okamura, M. Y. (1991) *Proc. Natl. Acad. Sci. USA* **88**, 5804–5808.
Beveridge, D. L., & DiCapua, F. M. (1989) *Annu. Rev. Biophys. Biophys. Chem.* **18**, 432–492.
Bockris, J. O., & Reddy, A. K. N. (1973) *Modern Electrochemistry*, Plenum, New York.
Born, M. (1920) *Z. Phys.* **1**, 45.
Böttcher, C. J. F. (1973) *Theory of Electric Polarization*, Elsevier, Amsterdam.
Brooks, B. R., Bruccoleri, R. E., Olafson, B. D., States, D. J., Swaminathan, S., & Karplus, M. (1983) *J. Comput. Chem.* **4**, 187–217.
Brünger, A. T., Kuriyan, J., & Karplus, M. (1987) *Science* **235**, 458–460.
Chance, B., Devault, D., Frauenfelder, H., Marcus, R. A., Schrieffer, J. R., & Sutin, N. (1979) *Tunneling in Biological Systems*, Academic Press, New York.
Cheng, X., & Schoenborn, B. P. (1990) *Acta Cryst.* **B46**, 195–208.
Churg, A. K., & Warshel, A. (1986) *Biochemistry* **25**, 1675–1681.
Churg, A. K., Weiss, R. M., Warshel, A., & Takano, T. (1983) *J. Phys. Chem.* **87**, 1683–1694.
Clark, W. M. (1960) *Oxidation–Reduction Potentials on Organic Systems*, Waverly Press, Baltimore.
Closs, G. L., Calcaterra, L. T., Green, N. J., Penfield, K. W., & Miller, J. R. (1986) *J. Phys. Chem.* **90**, 3673–3683.
Connolly, M. L. (1983) *Science* **221**, 709–713.
Coppens, P. (1989) *J. Phys. Chem.* **93**, 7979–7984.
Cutler, R. L., Davies, A.M., Creighton, S., Warshel, A., Moore, G. R., Smith, M., & Mauk, A. G. (1989) *Biochemistry* **28**, 3188–3197.
Davis, M. E., & McCammon, J. A. (1990) *J. Comp. Chem.* **10**, 386.
Debye, P. (1929) *Polar Molecules*, Chemical Catalog, New York.
Deisenhofer, J., & Michel, H. (1989) *The EMBO Journal* **8**, 2149–2170.
Deisenhofer, J., Epp, O., Miki, K., Huber, R., & Michel, H. (1984) *J. Mol. Biol.* **385**, 385.
Deisenhofer, J., Epp, O., Miki, R., & Michel, H. (1985) *Nature* **318**, 618–624.
Destro, R., Bianchi, R., & Morosi, G. (1989) *J. Phys. Chem.* **93**, 4447–4457.
Devault, D. (1980) *Q. Rev. Biophys.* **13**, 387–564.
Doeff, M. M., Sweigart, D. A., & O'Brien, P. (1983) *Inorg. Chem.* **22**, 851–852.
Dracheva, S. M., Drachev, L. A., Konstantinov, A. A., Semenov, A. Y., Skulachev, V. P., Arutjunjan, A. M., Shuvalov, V. A., & Zaberezhnaya, S. M. (1988) *Eur. J. Biochem.* **171**, 253–264.
Feng, Y., Roder, H., Englander, S. W., Wand, A. J., & Di Stefano, D. L. (1989) *Biochemistry* **28**, 195–203.
Feynman, R. P., Leighton, R. B., & Sands, M. (1963) *The Feynman Lectures in Physics*, Addison-Wesley, Reading, Massachusetts.
Fritzsch, G., Buchanan, S., & Michel, H. (1989) *Biochim. Biophys. Acta* **977**, 157–162.
Gao, J.-L, Shopes, R. J., & Wraight, C. A. (1990) *Biochim. Biophys. Acta* **1015**, 96–108.
Gilson, M. K., & Honig, B. (1986) *Biopolymers* **25**, 2097–2199.
Gilson, M. K., & Honig, B. (1987) *Nature* **330**, 84–86.
Gilson, M. K., & Honig, B. (1988a) *Proteins* **4**, 7–18.
Gilson, M. K., & Honig, B. (1988b) *Proteins* **3**, 32–52.
Gilson, M. K., & Honig, B. (1991) *J. Computer-Aided Molec. Design* **5**, 5–20.
Gilson, M. K., Rashin, A., Fine, R., & Honig, B. (1985) *J. Mol. Biol.* **183**, 503–516.
Gilson, M. K., Sharp, K. A., & Honig, B. H. (1987) *J. Comp. Chem.* **9**, 327–335.
Gunner, M. R. (1991) *Current Topics in Bioenergetics* **16**, 319–367.
Gunner, M. R., & Honig, B. (1991) *Proc. Natl. Acad. Sci. USA* **88**, 9151–9155.
Gunner, M. R., & Honig, B. (1993) in *The Photosynthetic Bacterial Reaction Center: Structure, Spectroscopy and Dynamics II* (Breton, J., and Vermeglio, A., eds.), Plenum, New York, 403–410.
Hagler, A. T., & Moult, J. (1978) *Nature* **272**, 222–226.

Harbury, H. A., & Loach, P. A. (1960) *J. Biol. Chem.* **255**, 3640–3645.

Harbury, H. A., Cronin, J. R., Fanger, M. W., Hettinger, T. P., Murphy, A. J., Meyer, Y. P., & Vinogradov, S. N. (1965) *Proc. Natl. Acad. Sci. USA* **54**, 1658–1664.

Harvey, S. (1989) *Proteins* **5**, 78–92.

Harvey, S. C., & Hoekstra, P. (1972) *J. Phys. Chem.* **76**, 2987–2994.

Honig, B., Hubbell, W., & Flewelling, R. (1986) *Annu. Rev. Biophys. Biophys. Chem.* **15**, 163–193.

Jayaram, B., Fine, R., Sharp, K., & Honig, B. (1989) *J. Phys. Chem.* **93**, 4320–4327.

Jean-Charles, A., Nicholls, A., Sharp, K., Honig, B., Tempczyk, A., Hendrickson, T. F., & Still, W. C. (1991) *J. Am. Chem. Soc.* **113**, 1454–1455.

Jorgensen, W. L., & Tirado-Rives, J. (1988) *J. Am. Chem. Soc.* **110**, 1657–1666.

Kadish, K. M. (1986) *Prog. Inorg. Chem.* **34**, 435–605.

Kadish, K. M., & Bottomley, L. A. (1980) *Inorg. Chem.* **19**, 832–836.

Karplus, M., & Petsko, G. A. (1990) *Nature* **347**, 631–639.

Karshikov, A. D., Engh, R., Bode, W., & Atanasov, B. P. (1989) *Eur. Biophys. J.* **17**, 287–297.

Kassner, R. J. (1972) *Proc. Natl. Acad. Sci. USA* **69**, 2263–2267.

Kassner, R. J. (1973) *J. Am. Chem. Soc.* **95**, 2674–2676.

Kirkwood, J. G. (1934) *J. Chem. Phys.* **2**, 351–361.

Kirkwood, J. G., & Westheimer, F. H. (1938) *J. Chem. Phys.* **6**, 506–517.

Klapper, I., Hagstrom, R., Fine, R., Sharp, K., & Honig, B. (1986) *Proteins* **1**, 47–59.

Kossiakoff, A. A. (1985) *Ann. Rev. Biochem.* **54**, 1195–1227.

Lee, B., & Richards, F. M. (1971) *J. Mol. Biol.* **55**, 379–400.

Lide, D. R., ed. (1990) *CRC Handbook of Chemistry and Physics*, CRC Press, Boca Ratom, Florida.

Linderstrom-Lang, K. (1924) *C.R. Trav. Lab. Carlsberg* **15**, 1–29.

Luntz, T. L., Schejter, A., Garber, E. A. E., & Margoliash, E. (1989) *Proc. Natl. Acad. Sci. USA* **86**, 3524–3528.

Marcus, R. A., & Sutin, N. (1985) *Biochim. Biophys. Acta* **811**, 265–322.

Marcus, Y. (1985) *Ion Solvation*, Wiley, Chichester.

Mashiko, T., Marchon, J.-C., Musser, D. T., Reed, C. A., Kastner, M. E., & Scheidt, W. R. (1979) *J. Am. Chem. Soc.* **101**, 3653–3655.

Mashiko, T., Reed, C. A., Haller, K. J., Kastner, M. E., & Scheidt, W. R. (1981) *J. Am. Chem. Soc.* **103**, 5758–5767.

Matsuura, Y., Takano, T., & Dickerson, R. E. (1982) *J. Mol. Biol.* **156**, 389–409.

Matthew, J. B. (1985) *Ann. Rev. Biophys. Chem.* **14**, 387–417.

Matthew, J. B., Gurd, F. R. N., Garcia-Moreno, B., Flanagan, M. A., March, K. L., & Shire, S. J. (1985) *CRC Critical Reviews in Biochemistry* **18**, 91–197.

Mayo, S. L., Ellis, W. R., Crutchley, R. J., & Gray, H. B. (1986) *Science* **233**, 948–952.

McCammon, J. A., & Harvey, S. C. (1987) *Dynamics of Proteins and Nucleic Acids*, Cambridge University Press, Cambridge, U.K.

Meyer, T. E. .& Kamen, M. D. (1982) *Adv. Prot. Chem.* **35**, 105–212.

Michel, H., & Deisenhofer, J. (1988) *Biochemistry* **27**, 1–7.

Momany, F. A., McGuire, R. F., Burgess, A. W., & Scheraga, H. A. (1975) *J. Phys. Chem.* **79**, 2361–2380.

Moore, G. R. (1983) *FEBS Letts.* **161**, 171–175.

Moore, G. R., & Williams, R. J. P. (1977) *FEBS Letts.* **79**, 229–232.

Moore, G. R., Pettigrew, G. W., Pitt, R. C., & Williams, R. J. P. (1980) *Biochim. Biophys. Acta* **590**, 261–271.

Moore, G. R., Harris, D. E., Leitch, F. A., & Pettigrew, G. W. (1984) *Biochem. Biophys. Acta* **764**, 331–342.

Moore, G. R., Pettigrew, G. W., & Rogers, N. K. (1986) *Proc. Natl. Acad. Sci. USA* **83**, 4998–4999.

Moss, D., Nabedryk, E., Breton, J., & Mantele, W. (1990) *Eur J Biochem.* **187**, 565–572.

Nakamura, H., Sakamoto, T., & Wada, A. (1988) *Protein Engineering* **2**, 177–183.

Nicholls, A., & Honig, B. (1991) *J. Comp. Chem.* **12**, 435–445.

Nitschke, W., & Rutherford, A. W. (1989) *Biochemistry* **28**, 3161–3168.

No, K. T., Grant, J. A., & Scheraga, H. A. (1990a) *J. Am. Chem. Soc.* **94**, 4712–4739.

No, K. T., Grant, J. A., & Scheraga, H. A. (1990b) *J. Am. Chem. Soc.* **94**, 4740–4746.

Northrup, S. H., Pear, M. R., McCammon, J. A., Karplus, M., & Takano, T. (1980) *Nature* **287**, 659–660.

Northrup, S. H., Pear, M. R., Morgan, J. D., McCammon, J. A., & Karplus, M. (1981) *J. Mol. Biol.* **153**, 1087–1109.

Onsager, L. (1936) *J. Am. Chem. Soc.* **58**, 1486–1493.

Otting, G., & Wüthrich, K. (1989) *J. Am. Chem. Soc.* **111**, 1871–1875.

Pennock, B. D., & Schwan, H. P. (1969) *J. Phys. Chem.* **73**, 2600–2610.

Pethig, R. (1979) *Dielectric and Electronic Properties of Biological Materials*, John Wiley, New York.

Pettigrew, G. W., Meyer, T. E., Bartsch, R. G., & Kamen, M. D. (1975) *Biochim. Biophys. Acta* **503**, 509–523.

Pettigrew, G. W., Bartsch, R. G., Meyer, T. E., & Kamen, M. D. (1978) *Biochim. Biophys. Acta* **503**, 509–523.

Purcell, E. M. (1985) *Electricity and Magnetism*, McGraw-Hill, New York.

Rashin, A. A., & Honig, B. (1985) *J. Phys. Chem.* **89**, 5588–5593.

Rashin, A. A., Iofin, M., & Honig, B. (1986) *Biochem.* **25**, 3619–3625.

Rawlings, D. C., Gouterman, M., Davidson, E. R., & Feller, D. (1985a) *Int. J. Quantum Chem.* **28**, 773–796.

Rawlings, D. C., Gouterman, M., Davidson, E. R., & Feller, D. (1985b) *Int. J. Quantum Chem.* **28**, 797–822.

Rawlings, D. C., Gouterman, M., Davidson, E. R., & Feller, D. (1985c) *Int. J. Quantum Chem.* **823**, 823–842.

Rees, D. C. (1985) *Proc. Natl. Acad. Sci. USA* **82**, 3082–3085.

Reid, L. S., Mauk, M. R., & Mauk, A. G. (1984) *J. Am. Chem. Soc.* **106**, 2182–2185.

Rogers, N. K. (1986) *Prog. Biophys. Molec. Biol.* **48**, 37–66.

Rogers, N. K., Moore, G. R., & Sternberg, M. J. E. (1985) *J. Mol. Biol* **182**, 613–616.

Rohmer, M.-M. (1985) *Chem. Phys. Lett.* **116**, 44–49.

Russell, A. J., & Fersht, A. R. (1987) *Nature* **328**, 496–500.

Russell, S. T., & Warshel, A. (1985) *J. Mol. Biol.* **185**, 389–404.

Sharp, K. (1991) *J. Comp. Chem.* **12**, 454–468.

Sharp, K. A., & Honig, B. (1990a) *J. Phys. Chem.* **94**, 7684–7692.

Sharp, K. A., & Honig, B. (1990b) *Annu. Rev. Biophys. Biophys. Chem.* **19**, 301–332.

Shire, S. J., Hanania, G. I. H., & Gurd, F. R. N. (1974) *Biochemistry* **13**, 2967–2974.

Soman, K., Yang, A.-S., Honig, B., & Fletterick, R. (1989) *Biochemistry* **28**, 9918–9926.

States, D. J., & Karplus, M. (1987) *J. Mol. Biol* **197**, 122–130.

Stellwagen, E. (1978) *Nature* **275**, 73–74.

Sternberg, M. J. E., Hays, F. R. F., Russell, A. J., Thomas, P. G., & Fersht, A. R. (1987) *Nature* **330**, 86–88.

Sutin, N. (1982) *Acc. Chem. Res.* **15**, 275–282.

Takano, T., & Dickerson, R. E. (1981a) *J. Mol. Biol.* **153**, 79–94.

Takano, T., & Dickerson, R. E. (1981b) *J. Mol. Biol.* **153**, 95–115.

Takashima, S., & Schwan, H. P. (1965) *J. Phys. Chem.* **69**, 4176–4182.

Tanford, C. (1957a) *J. Am. Chem. Soc.* **79**, 5348–5352.

Tanford, C. (1957b) *J. Am. Chem. Soc.* **79**, 5340–5347.

Tanford, C., & Kirkwood, J. G. (1957) *J. Am. Chem. Soc.* **79**, 5333–5339.

Tanford, C., & Roxby, R. (1972) *Biochemistry* **11**, 2192.

Tollin, G., Hanson, L. K., Caffrey, M., Meyer, T. E., & Cusanovich, M. A. (1986) *Proc. Natl. Acad. Sci. USA* **83**, 3693–3697.

Valentine, J. S. Sheridan, R. P., Allen, L. C., & Kahn, P. C. (1979) *Proc. Natl. Acad. Sci. USA* **76**, 1009–1013.

Wand, J. A., Di Stefano, D. L., Feng, Y., Roder, H., & Englander, S. W. (1989) *Biochemistry* **28**, 186–194.

Warme, P. K., & Hager, L. P. (1970) *Biochemistry* **9**, 1606–1614.

Warshel, A. (1987) *Nature* **330**, 15–16.

Warshel, A., & Churg, A. K. (1983) *J. Mol. Biol.* **168**, 687–694.

Warshel, A., & Lappicirella, A. (1981) *J. Am. Chem. Soc.* **103**, 4664–4673.

Warshel, A., & Russell, S. T. (1984) *Q. Rev. Biophys.* **17**, 283–422.

Warwicker, J. (1986) *J. Theor. Biol.* **121**, 199–210.

Warwicker, J., & Watson, H. C. (1982) *J. Mol. Biol.* **157**, 671–679.

Weiner, S. J., Kollman, P. A., Nguyen, D. T., & Case, D. A. (1986) *J. Comp. Chem.* **7**, 230–252.

Wendoloski, J. J., & Matthew, J. B. (1989) *Proteins* **5**, 313–321.

Wendoloski, J. J., Mathew, J. B., Weber, P. C., & Salamme, F. R. (1987) *Science* **238**, 794–797.

Wilson, G. S. (1974) *Bioelectrochem. and Bioenerg.* **1**, 172–179.

Yang, A.-S., Gunner, M. R., Sampogna, R., Sharp, K., & Honig, B. (1993) *Proteins* **15**, 252–265.

10

Differential Protection Techniques in the Analysis of Cytochrome *c* Interaction with Electron Transfer Proteins

HANS RUDOLF BOSSHARD

Biochemisches Institut der Universität Zürich
Zürich, Switzerland

I. Introduction

Cytochrome *c* forms complexes with many different electron-donating and -accepting proteins. These complexes are important to the biological function of both eukaryotic and prokaryotic cytochrome *c*. Many of the complexes can be isolated and studied in solution. This chapter deals with a protein-chemical approach that is useful for investigating the intermolecular interface of the complexes.

Work on the binding of cytochrome *c* to enzymes, proteins, and membranes has been reviewed by Nicholls (1974), Dickerson and Timkovich (1975) and Pettigrew and Moore (1987). Proteins binding to cytochrome *c* can be divided into three categories: (i) true physiological redox partners; (ii) proteins that are not physiological redox partners but are able to exchange electrons with cytochrome *c in vitro*; (iii) proteins that bind cytochrome *c* but cannot accept or donate electrons. Members of the first category include the mitochondrial inner membrane enzymes cytochrome *c* oxidase and ubiquinol:cytochrome *c* oxidoreductase (cytochrome bc_1-complex) and the enzymes sulfite oxidase and cytochrome *c* peroxidase resident in the intermembrane space of mitochondria. Group (ii) contains ferredoxin, plastocyanin, NADH-cytochrome P450 reductase, and bacterial photosynthetic reaction centers. These proteins are efficient redox partners for cytochrome *c* in the test tube, but they do not encounter cytochrome *c in vivo*. The acidic egg protein phosvitin is a member of category (iii). Besides binding to proteins, cytochrome *c* can bind to acidic polar headgroups of natural and artificial membranes (see, for example, Nicholls, 1974; Brown and Wüthrich, 1977; Teissie, 1981; Demel *et al.*, 1989).

It has been known for a long time that cytochrome *c* complexes are stabilized by ionic bonds between lysine residues of cytochrome *c* and acidic residues of the complexed protein [review of early literature by Nicholls (1974) and Dickerson and Timkovich (1975)]. Thus, nearly all these complexes dissociate at high ionic strength. Washing of mitochondria with 0.15 M NaCl removes bound cytochrome *c*, presumably by weakening this ionic binding, and abolishes respiratory activity (Jacobs and Sanadi, 1960). At low ionic strength, usually 10 mM or less, the dissociation constant of the complexes is sufficiently low to enable their isolation, for example by gel filtration.

Many of the complexes have been analyzed by a variety of techniques in many different laboratories. Central questions asked were: What surface area of cytochrome *c* is recognized by the bound protein? Do different redox partners bind to unique sites, or is there a common site of reaction for all redox partners? Work on isolated cytochrome *c* complexes can only help to answer these two related questions if the complexes studied *in vitro* are valid models of electron transfer complexes as they occur during cytochrome *c*–mediated electron transfer reactions. The validity of this relationship has not been proven unambiguously. At best, an isolated cytochrome *c* complex is a virtual model of a real electron transfer complex. By this it is meant that the complex may differ only by some subtle detail that is not crucial to the electron transfer function. There is no general solution to this problem.

II. Principle of Differential Protection Techniques

The rationale of the technique, also known as "differential chemical modification" (Bosshard, 1979), is to compare the relative rates of chemical modification of free and complexed cytochrome *c* and to deduce the location of the binding site on cytochrome *c* from the differential chemical reactivity of amino acid side chains. Residues at the intermolecular interface are expected to be protected in the complex. The actual degree of protection may vary. The technique is equally applicable to both partners of the complex. So far, more data have been collected on protected residues of cytochrome *c* than on protected residues of its redox partners.

Chemical modification of a protein molecule in different functional states is well known. However, in conventional chemical modification of enzymes in the presence

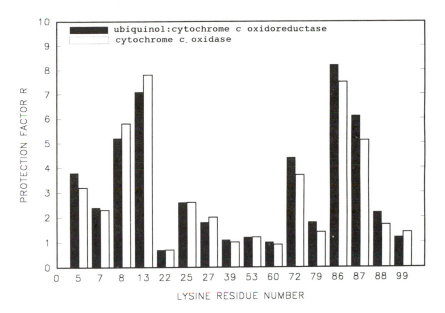

Figure 10.1

Differential protection of lysine residues of horse cytochrome c by ubiquinol:cytochrome c oxidoreductase (filled bars) and cytochrome c oxidase (open bars). The protection factor indicates the degree to which a residue is less labeled by acetic anhydride or formaldehyde/borohydride when the molecule is bound to the reductase or the oxidase in a 1:1 complex (Rieder and Bosshard, 1980; Margoliash and Bosshard, 1983). Protection factors shown for residues 53, 72, and 99 are averages over residues 53 + 55, 72 + 73, and 99 + 100, respectively. Later it was shown that residues 55 and 100 are unprotected in both complexes and that residue 73 is much less protected than residue 72 (H.R.B. and R. Rieder, unpublished experiments).

and absence of low-molecular-weight substrates, coenzymes, inhibitors, etc., the information collected is qualitative, or at best semiquantitative: Residues are found to be "buried" or "exposed" or "half-buried." If we are to map an intermolecular interface, this kind of information is insufficient. We need a quantitative measure of the degree of protection of residues at an intermolecular interface. A useful quantitative measure is the rate by which functional amino acid side chains react with a modifying reagent. Absolute rates can be measured, but the procedure is cumbersome (Kaplan *et al.*, 1971). It is easier to measure relative rates, namely the ratio of the rate of reaction of a side chain in free and in complexed cytochrome c. This ratio is called the *protection factor* **R**. As will be shown later, **R** can be determined from the degree of modification of a side chain in free and complexed cytochrome c if the modification reaction is performed with a trace of radioactively labeled reagent under pseudo–first order conditions with the modifiable group in excess of the modifying reagent.

As a first example, Figure 10.1 shows the protection factors for the 19 lysine residues of horse cytochrome c when this molecule is bound to cytochrome c oxidase or ubiquinol:cytochrome c oxidoreductase from beef heart. The two enzymes protect some but not all lysine residues from reaction with acetic anhydride. In this particular experiment, the most protected residues were eight times less reactive in complexed than in free cytochrome c.

III. Theoretical Considerations

Before discussing the data of Figure 10.1 and their relationship to the electron transfer function of cytochrome c, we must deal with the general background of the technique. If we consider a single class of functional groups, each member of the class displays a somewhat different reactivity toward the same chemical reagent. Because lysine residues were long known to be important in binding of redox partners, their protection in cytochrome c complexes was examined. We restrict the following general discussion to the modification of ε-amino groups of lysines. However, many of the considerations are equally applicable to the modification of carboxyl, sulfhydryl, or imidazole groups. Differential protection of carboxyl groups is discussed in Section X.

The reactivity of a particular amino group is governed by its pK value and nucleophilicity, which in turn are influenced by the microenvironment of the group. In the present context, the microenvironment encompasses neighboring atoms of cytochrome c, as well as of the cytochrome c-bound redox partner. We may formally discern two modes by which the microenvironment can change the chemical reactivity of an amino group: by altering the pK of the group, or by steric factors. Proximity effects of the microenvironment can increase or decrease pK's. The formation of an ionic bond between the protonated lysine side chain and a neighboring carboxylate will increase the pK. Creation of a less polar microenvironment around the lysine side chain in a cytochrome c complex lowers the pK. If the microenvironment only affects the pK, the reactivity change will be the same irrespective of the modifying reagent (as long as chemically similar modification reactions are compared). Steric effects, on the other hand, may prevent modification of a side chain by a bulky reagent even if the pK has not changed. In practice, the microenvironment almost always affects both the pK and the steric accessibility of the side chain.

Functional groups amenable to chemical modification usually contain a dissociable proton. The amino group reacts in the deprotonated form with reagents such as acetic anhydride. The rate of chemical modification is therefore related to the concentration of the unprotonated form [A], which depends on pH in the manner indicated by Equation (10.1):

$$[A] = [A_t]K_A/(K_A + [H^+]) = \alpha_A[A_t]. \qquad (10.1)$$

$[A_t]$ is the total concentration of the amino group, K_A is the acid dissociation constant, and α_A denotes the fraction of the ε-amino group in the deprotonated form. The amino group reacts with acetic anhydride (RX) according to $A + RX \rightarrow AR + X$, where AR is the N^ε-acetyl function and X is acetate. The rate of the reaction is

$$d[AR]/dt = [A_t][RX]k_A K_A/(K_A + [H^+]) = k_A[RX]\alpha_A[A_t], \qquad (10.2)$$

where k_A is the pH-independent second-order rate constant of the acetylation reaction. From (10.2) it follows that the rate of acetylation decreases if K_A decreases—that is, when the pK of the ε-amino group rises.[1] This will occur, as mentioned earlier, when the lysine side chain enters an electrostatic bond with an acidic group of the redox partner. However, the situation is more complicated because k_A also depends on K_A. For groups behaving "normally," the dependency is given by the Brønsted relationship $\log k_A = \beta \cdot pK + \gamma$, where β and γ are constants. The relationship predicts that if the steric accessibility of the ε-amino group remains unchanged, which may be the case for small reagents such as acetic anhydride, $\log k_A$ rises when the pK increases. This means that, although a pK increase of the ε-amino group reduces the overall rate of acetylation according to Equation (10.2), the Brønsted equation predicts that at the same time the pH-independent second-order rate constant k_A increases. In other words, part of the rate decrease consequent to the formation of an ionic bond at the intermolecular interface (pK increase) is lost through a rise of k_A. With a bulkier reagent, the effect of a pK increase on the overall rate of modification should be larger because the reagent's access to the ε-amino group will be more hindered. Bulkier reagents "lower" k_A, which, translated into the formalism of the Brønsted equation, means a displacement from the straight line of the Brønsted equation. It was shown that surface accessible ε-amino groups behave as predicted by the Brønsted equation (Bosshard, 1981), whereas buried groups fall below the Brønsted line (Kaplan $et\ al.$, 1971; Kaplan, 1972).

IV. Determination of Protection Factors

It follows from the foregoing considerations that the protection factor **R** is a valid measure for the chemical reactivity of ε-amino groups only if **R** is proportional to the ratio of the rates of acetylation in free and complex-bound cytochrome c:

$$(d[AR]/dt)_{free}/(d[AR]/dt)_{bound} = \mathbf{R}. \qquad (10.3)$$

Equation (10.3) shows that protection by a redox partner will increase **R** above unity. $\mathbf{R} = 1$ if the redox partner has no effect on the rate of modification, and $\mathbf{R} < 1$ if the redox partner increases the rate of modification.

The easiest experimental way to measure **R** is to use radioactively labeled acetic anhydride. In two parallel experiments, free and bound cytochrome c are treated first with a small amount of [³H]acetic anhydride, and then with an excess of non-radioactive acetic anhydride under denaturing conditions. In this way one obtains a chemically homogeneous but isotopically heterogeneous derivative of cytochrome c. The two cytochrome c derivatives, one labeled in presence and the other in absence of the redox partner, can then be digested by suitable proteases, the peptides purified by HPLC, and the specific radioactivity of each peptide determined. If peptides contain several acetylated lysines, stepwise degradation of peptides by the

---

[1] The acetylation of ε-amino groups is usually performed near pH 7 to 8, hence $[H^+] \gg K_A$; therefore $(K_A + [H^+])$ of Equation (10.2) changes very little when K_A decreases.

Edman procedure will yield radioactive thiohydantoin derivatives of the acetylated lysines. In this way one obtains the data necessary to calculate **R** according to

$$\frac{\text{specific radioactivity of AR in free cytochrome } c}{\text{specific radioactivity of AR in bound cytochrome } c} = \mathbf{R}. \qquad (10.4)$$

It must be emphasized that Equation (10.4) is valid only if the reaction with the radioactive anhydride is performed under conditions where, on average, less than one lysine per cytochrome c molecule is acetylated (Bosshard, 1979). This is very important because acetylation of one lysine may influence the reactivity of a neighboring lysine of cytochrome c. Consequently, changes in the rate of acetylation could reflect events other than binding of cytochrome c to the redox partner. As a precaution, the number of cytochrome c molecules with two or more [³H]acetyl groups must be kept negligibly small. In practice, the degree of radioactive acetylation is kept at 0.1 to 0.5 mol [³H]acetyl per mol of cytochrome c.

Determining **R** by use of Equation (10.4), though conceptually simple, is technically demanding. The reason is that to calculate the specific radioactivities one must know the absolute amount of the labeled peptide or thiohydantoin derivative. Amino acid analysis can be used to quantify peptides, and UV absorbance can be used to quantify the thiohydantoin derivative of N^ε-acetyl lysine. In our hands, the reproducibility of these measurements is not satisfactory. A way out of this problem is to use two different radioactive isotopes, namely [³H]- and [¹⁴C]acetic anhydride. Two experimental protocols were devised. The first is explained in Scheme I (Bosshard, 1979; Rieder and Bosshard, 1980; Bechtold and Bosshard, 1985). In a first step, free (left branch in Scheme 10.1) and complexed cytochrome c (right branch) are acetylated with a trace amount of [³H]acetic anhydride of high specific radioactivity (>1 Ci/mmol). Trace-labeled cytochrome c is separated from the redox partner by gel filtration in high ionic strength buffer. In step 2, equimolar amounts of ³H-trace-labeled cytochrome c are mixed with equal amounts of a cytochrome c derivative that had been acetylated with [¹⁴C]acetic anhydride. This derivative is prepared in a separate experiment.[2] Unlabeled carrier cytochrome c may be added in step 2 to facilitate the separation and sequence analysis of peptides in subsequent steps.

In step 3, the mixture of ³H- and ¹⁴C-labeled proteins, which still contain some free ε-amino groups, are completely acetylated under denaturing conditions [6 M guanidine hydrochloride or saturated sodium acetate solution (Riordan and Vallee, 1972)]. The ensuing material is now chemically homogeneous—that is, all lysine residues are acetylated, but the material is heterogeneous with regard to the isotopic label. Finally, in step 4, the derivatives are digested and the peptides purified by HPLC and degraded by the Edman procedure. For each labeled thiohydantoin

---

[2] The derivative need not be completely acetylated at all its lysine residues if the same batch of ¹⁴C-labeled material is used for mixing with the two samples of ³H-trace-labeled cytochrome c. Actually, the cost of [¹⁴C]acetic anhydride is prohibitively high to allow for complete labeling of milligram quantities of cytochrome c.

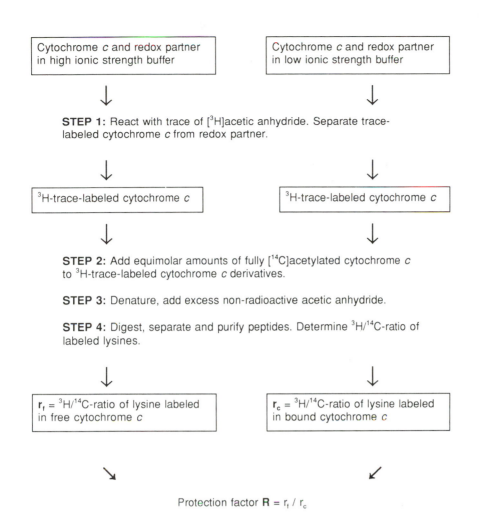

Cytochrome *c* and redox partner in high ionic strength buffer

Cytochrome *c* and redox partner in low ionic strength buffer

STEP 1: React with trace of [³H]acetic anhydride. Separate trace-labeled cytochrome *c* from redox partner.

³H-trace-labeled cytochrome *c*

³H-trace-labeled cytochrome *c*

STEP 2: Add equimolar amounts of fully [¹⁴C]acetylated cytochrome *c* to ³H-trace-labeled cytochrome *c* derivatives.

STEP 3: Denature, add excess non-radioactive acetic anhydride.

STEP 4: Digest, separate and purify peptides. Determine ³H/¹⁴C-ratio of labeled lysines.

r_f = ³H/¹⁴C-ratio of lysine labeled in free cytochrome *c*

r_c = ³H/¹⁴C-ratio of lysine labeled in bound cytochrome *c*

Protection factor $\mathbf{R} = r_f / r_c$

SCHEME I

derivative, the ³H/¹⁴C ratio is determined, and **R** is calculated as

$$\frac{{}^3\mathrm{H}/{}^{14}\mathrm{C} \text{ of AR in free cytochrome } c}{{}^3\mathrm{H}/{}^{14}\mathrm{C} \text{ of AR in bound cytochrome } c} = \mathbf{R}. \qquad (10.5)$$

The second experimental protocol for double-labeling of ε-amino groups is simpler (Bosshard *et al.*, 1987b). Trace-labeling of free cytochrome *c* is performed with one isotope, and labeling of complexed cytochrome *c* is performed with the other. Both trace-labeled derivatives are then mixed together and completely acetylated under denaturing conditions with nonradioactive reagent, as described for the first protocol. The protection factors are now obtained directly from

$$\frac{{}^3\mathrm{H} \text{ of AR in free cytochrome } c}{{}^{14}\mathrm{C} \text{ of AR in bound cytochrome } c} = \mathbf{R}. \qquad (10.6)$$

Simple as it is, this protocol demands that [^{3}H]- and [^{14}C]acetic anhydride be of exactly the same specific radioactivity and added at exactly the same concentration to free and complexed cytochrome c, which in turn must both be present at identical concentrations. Adjustment of specific radioactivities must be done very carefully, and great care is necessary to keep experimental conditions such as protein concentrations and the mode of addition of acetic anhydride, stirring of the reaction solution, and temperature control identical in both experiments. This is particularly important because the reaction is fast. The concentration of acetic anhydride should be kept low enough to render the reaction pseudo–first-order.

The determination of ^{3}H/^{14}C ratios requires further comment. In earlier work, the ^{3}H/^{14}C-ratio of labeled peptides known to contain only a single lysine was taken to represent the ^{3}H/^{14}C-ratio of that particular lysine (Rieder and Bosshard, 1978a, 1980). However, we now find that more accurate and more reproducible ^{3}H/^{14}C-ratios are measured at the level of the thiohydantoin derivative of individual lysine residues. These are obtained by stepwise Edman degradation of labeled peptides. If a labeled peptide has two or more lysines, the ^{3}H/^{14}C ratios for single residues can be obtained only by stepwise Edman degradation. The yield of repetitive Edman degradation steps is critical in the case of sequential lysines, e.g., the sequence ^{86}Lys–Lys–Lys^{88} in horse cytochrome c. Automatic sequencing with the help of a gas-phase sequenator is best, and long peptides can be handled easily. The radioactivity of thiohydantoin derivatives of residues other than lysine must also be checked to control the cleavage yield at each step (Bosshard et $al.$, 1987b). Unfortunately, most cytochromes c have a blocked N-terminus and cannot be sequenced directly. In addition, lysine-modifying reagents block free N-termini. Proteolysis and peptide separation is, therefore, always necessary before ^{3}H/^{14}C ratios can be determined by Edman degradation.

V. Choice of Reaction Conditions in the Trace-Labeling Step

A complex of cytochrome c with a redox partner is in rapid equilibrium with its free components. Because of this, even a residue which is completely unreactive in the complex will be modified to a small extent, which means that **R** values are never very large. The highest values observed for cytochrome c labeled in the presence and absence of redox partners were around 10. For cytochrome c bound to high-affinity monoclonal antibodies, **R** values above 20 have been observed (Saad et $al.$, 1988; Oertle et $al.$, 1989). In any case, the protection factor **R** is always a lower estimate of the actual degree of protection of a lysine residue. In practice, the concentrations of cytochrome c and its redox partner should be at least one order of magnitude greater than the value of the dissociation constant of the complex. The ratio of the initial rates of modification of a residue in bound and free cytochrome c is proportional to $(K/[\text{cyt } c_{\text{tot}}])^{1/2}$, where K is the dissociation constant of the complex. This relationship holds for the case in which equimolar amounts of cytochrome c and redox partner are present at a concentration well above the dissociation constant K of the complex (Bosshard, 1979).

Another difficulty arises from the fact that the redox partner also contains modifiable amino groups. If trace labeling is performed in the presence and absence of the redox partner, the concentration of reactive amino groups differs in the two parallel experiments. This may lead to erroneous **R** values. To counter this difficulty, the redox partner can be added to both trace-labeling experiments, and the ionic strength can be varied so that binding of cytochrome *c* to the redox partner takes place in only one of the two experiments. However, in this case the reaction conditions are not identical because the ionic strengths differ (e.g., 10 mM and 250 mM). To test whether the ionic strength affects the rate of acetylation of ε-amino groups, one may perform a control experiment in which only the ionic strength changes (Bosshard *et al.*, 1987b). The measured **R** values are then corrected by the **R** values due to the change of ionic strength. The corrected values are $\mathbf{R} = \mathbf{R}_{measured}/\mathbf{R}_{ionic\ strength}$.[3]

Another way to test for differences in reaction conditions is to include an internal standard nucleophile in both trace-labeling reactions (Kaplan *et al.*, 1971). This test assumes that the standard nucleophile experiences the same change of reaction conditions as cytochrome *c*. The $^{3}H/^{14}C$-ratio calculated for the standard nucleophile is then used to correct for differences in reaction conditions (Rieder and Bosshard, 1978a; Bosshard, 1979). The α-amino group of phenylalanine has been used as a standard nucleophile. The problem with this kind of correction is that the ionic strength may affect the reactivity of the internal standard to a different extent than that of individual amino groups. In our experience, a control experiment in which the degree of labeling is measured at low and high ionic strength provides more reliable results than a control using an internal standard.

If for some reason labeling of the free protein has to be done in the absence of the binding reaction partner, the amount of the labeling reagent may be adjusted to account for the varying amount of potentially modifiable amino groups. For example, in epitope mapping with a monoclonal antibody, the antigen–antibody complex cannot be dissociated by high ionic strength. In trace labeling of free and monoclonal antibody–bound cytochrome *c*, the amount of modifying reagent is increased in proportion to the extra amino groups introduced by the antibody (Oertle *et al.*, 1989; Saad and Bosshard, 1990; Tinner *et al.*, 1990).

VI. Binding of Horse Cytochrome *c* to Ubiquinol:cytochrome *c* Oxidoreductase and Cytochrome *c* Oxidase

When the first differential protection experiments with cytochrome *c* were undertaken more than 15 years ago, the opinion prevailed that cytochrome *c* oxidase and ubiquinol:cytochrome *c* oxidoreductase bind to separate binding sites on cytochrome *c* (Margoliash *et al.*, 1973; Smith *et al.*, 1973; Aviram and Schejter, 1973;

---

[3] The correction of **R** published earlier (Rieder *et al.*, 1985; Bosshard *et al.*, 1987b) is wrong, though the published values differ little from those obtained when the proper correction is made.

Dickerson and Timkovich, 1975). Cytochrome c was envisaged as a static electron-conducting molecule which could bind simultaneously to cytochrome c oxidase and ubiquinol:cytochrome c oxidoreductase, accepting an electron from the reductase and donating it to the oxidase. In a dynamic model, cytochrome c oscillates between ubiquinol:cytochrome c oxidoreductase and cytochrome c oxidase, i.e., two-dimensional diffusion occurs within the plane of the inner mitochondrial membrane or between the membrane and the intermembrane space (three-dimensional diffusion). Differential protection of cytochrome c by cytochrome c oxidase and ubiquinol:cytochrome c oxidoreductase provided the first clear evidence for a single interaction domain on cytochrome c for both its physiological electron donor and its acceptor. Soon after the first protection experiments with cytochrome c oxidase, it was found that ubiquinol:cytochrome c oxidoreductase and cytochrome c_1, the immediate electron donor to cytochrome c, protect the very same residues on cytochrome c that are protected by the oxidase (Rieder and Bosshard, 1978b; Bosshard et al., 1979; Rieder and Bosshard, 1980). If these data can be extrapolated to the in vivo situation, they mean that in its respiratory function cytochrome c must be mobile, as both large enzyme complexes could not occupy the same surface area of such a small molecule at the same time. The mobility of cytochrome c in mitochondria was also demonstrated later by other independent experiments (Gupte and Hackenbrock, 1988, and references therein) and has become textbook knowledge (Voet and Voet, 1990).

The protection factors reported in this early work are summarized in Figure 10.1. Lysine residues 5, 8, 13, 72 + 73, 86, and 87 are consistently protected by both physiological redox partners. The remaining residues are about equally or only slightly less reactive in bound cytochrome c. The protected residues are grouped above and to the left of the solvent-exposed heme edge on the "front side" of cytochrome c. The approximate spatial position of the lysine residues of mitochondrial cytochrome c is depicted in the top stereo pair of Figure 10.2. Lysines 8, 27, 72, 79, 86, and 87 are located to the left and to the right of the exposed heme edge. Lys-13 is just above the heme cleft.

Could a residue of the intermolecular interface have escaped detection because its reactivity did not change? From the preceding discussion, it follows that this may indeed happen if, for example, the pK value of the residue decreases upon binding of cytochrome c to the redox partner, yet the small reagent molecule could still easily penetrate into the intermolecular interface. Another, although less likely, possibility is that a reagent molecule binds preferentially near a residue at the interface, masking the protective effect. Hence, equal reactivity of a residue need not necessarily indicate that it is outside the interaction interface.

What about those residues that exhibit protection? A lysine side chain may become less reactive in bound cytochrome c through the influence of a conformational change in cytochrome c induced by the redox partner. Such a residue need not be located at the intermolecular interface and could be erroneously assigned to the binding site of the redox partner.

There is yet a third problem with data such as those presented in Figure 10.1 that was stated briefly in the Introduction. Namely, the differential protection experiments can very often only be done with isolated complexes in which cyto-

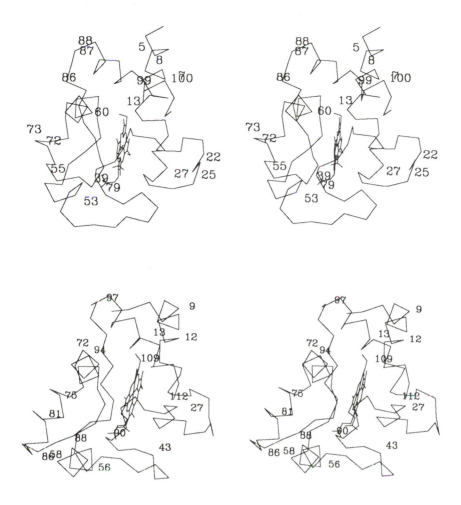

Figure 10.2

Spatial distribution of lysine residues in mitochondrial cytochrome c and bacterial cytochrome c_2. Molecules oriented with the "front side" facing the viewer. Upper stereo pair: C_α-backbone structure of tuna cytochrome c (Takano and Dickerson, 1981). Numbers indicate position of lysine residues of horse cytochrome c whose polypetide backbone folding is nearly identical to that of tuna cytochrome c. Lower stereo pair: C_α-backbone structure of cytochrome c_2 from *Rhodospirillum rubrum* (Salemme et al., 1973). Numbers indicate position of lysine residues.

chrome c and redox partner are at the same redox state, usually fully oxidized. Electron transfer, however, occurs from reduced cytochrome c to oxidized heme a of cytochrome c oxidase, or from reduced cytochrome c_1 to oxidized cytochrome c. Moreover, the ionic strength of the medium surrounding cytochrome c in the inner mitochondrial membrane is not known. In the labeling experiment, the ionic strength must be kept at a value that is very low and quite different from the ionic strength at the inner membrane. In conclusion, our interpretation of the differential protection experiments is based on the following crucial assumptions: (i) The protected residues are at the intermolecular interface and are not "pseudo-protected" through a conformational change resulting from the binding of cytochrome c to its redox partner. (ii) The complexes composed of cytochrome c and its redox partner are valid models for transitory electron transfer complexes.

In complementary work addressing the same questions, Millett and co-workers (Smith *et al.*, 1977; Ahmed *et al.*, 1978; Smith *et al.*, 1980) and Margoliash and co-workers (Brautigan *et al.*, 1978a,b; Ferguson-Miller *et al.*, 1978; Speck *et al.*, 1979; Osheroff *et al.*, 1980) independently prepared lysine-modified cytochrome c derivatives. These cytochrome c derivatives modified at single lysine residues were used as substrates for the reaction with ubiquinol:cytochrome c oxidoreductase and cytochrome c oxidase. Comparison of the electron transfer properties of these derivatives revealed that, in general, modification of the same lysine residues led to reduced rates of electron transfer with both cytochrome c oxidase and ubiquinol: cytochrome c oxidoreductase (Ahmed *et al.*, 1978; Speck *et al.*, 1979). By this entirely independent experimental approach, our conclusion (Rieder and Bosshard, 1978b, 1980) that reductase and oxidase bind to the very same surface area of cytochrome c was supported: Modification of the same residues that were protected in bound cytochrome c led to reduced electron transfer rates or to increased K_m values.

In Figure 10.3, a comparison is made between the data from differential protection of cytochrome c complexes and kinetic measurements with cytochrome c derivatives modified at single lysine residues. A quantitative comparison of these unrelated sets of data is difficult. Figure 10.3 is meant only to give an idea of the agreement achieved by the two approaches. To this end, each protection factor is expressed as a percentage of the maximum protection factor observed for each cytochrome c complex. Similarly, the electron transfer rate observed for each single lysine-modified derivative is presented as a percentage of the maximum inhibition observed in a particular set of experiments (see legend to Figure 10.3 for details).

Both approaches provide evidence for the importance of lysines 13, 86, and 87 in the interactions between cytochrome c and its redox partners. Furthermore, both approaches indicate that lysines 22, 39, 53, 55, 60, 99, and 100 are not involved in the interactions. These latter residues are mainly on the back side of the molecule (see the upper stereo pair of Figure 10.2). The results for residues 25 and 27, to the right of the heme edge (Figure 10.2, upper stereo pair), are conflicting. In the isolated complex, these residues are protected little if at all (Figure 10.1). Modification of the same residues to the N^ε-(4-carboxy-2,6-dinitrophenyl) derivative (CDNP) (Margoliash and co-workers) inhibits the reaction with both oxidase and reductase. Modification of residue 25 to the N^ε-trifluoroacetyl (TFA) or the N^ε-(4-trifluoromethylphenylcarbamoyl) derivative (TFC) (Millett and co-workers) inhibits the reaction catalyzed by the reductase (Figure 10.3, top panel) but not that catalyzed by the oxidase. Conversely, the same type of modification of lysine 27 inhibits the reaction with oxidase, but not that with reductase (Figure 10.3, bottom panel). There could be several reasons for these discrepancies. As discussed earlier, the **R** value determined from differential protection data may stay close to unity even if a resi-

385

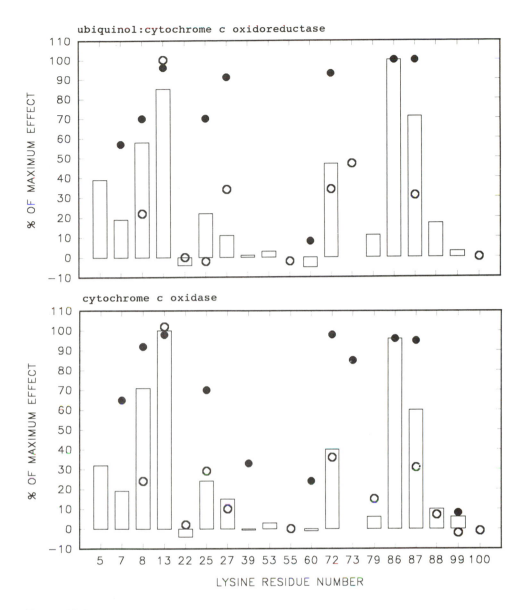

Figure 10.3

Comparison of differential protection of cytochrome c complexes and kinetic analysis of the reaction of lysine-substituted cytochrome c derivatives with ubiquinol:cytochrome c oxidoreductase (upper panel) and cytochrome c oxidase (lower panel). Open circles are for data published by the group of Millett, closed circles for data published by the group of Margoliash. In the case of differential protection, which is indicated by open bars, "% of maximum effect" indicates the percentage of the maximum protection observed, calculated as $100(R-1)/(R_{max}-1)$. For the kinetic data, the open circles refer to the data of Table II of Smith $et\ al.$ (1980), which have been transformed to $(V/K_m)_{native}/(V/K_m)_{derivative}$ as a percentage of the highest value observed for $(V/K_m)_{native}/(V/K_m)_{derivative}$ (i.e., the derivative which exhibits the lowest activity). Closed circles refer to data of Figures 1 and 2 of Koppenol $et\ al.$ (1982), which have been transformed to $100[(TN/S)-(TN/S)_{min}]/[(TN/S)_{max}-(TN/S)_{min}]$, where $(TN/S)_{max}$ stands for native cytochrome c and $(TN/S)_{min}$ for the derivative exhibiting the lowest reactivity.

due is located at the intermolecular interface. In the case of the lysine-substituted derivatives, the substitution may affect the kinetics of electron transfer by direct interference with binding of cytochrome c to the redox partner; by interfering with the intermolecular electron transfer reaction proper; or through a modification-induced conformational change of cytochrome c, which, in turn, influences the electron transfer reaction. This last possibility haunts all the chemical modification experiments as well as the more recent site-specific mutagenesis experiments, and it can never be excluded except with the help of independent data from unrelated experiments. In the differential protection technique, however, modification-induced conformational changes cannot perturb the analysis because the degree of labeling is kept very low during trace-labeling: Even if chemical modification of residue A gives rise to a conformational change which alters the chemical reactivity of residue B, the changed reactivity of residue B will not perturb the **R** value of B because, on average, less than one mole of lysine per mole of cytochrome c is modified during the trace-labeling reaction.

In general, modification of cytochrome c with the negatively charged CDNP group affects the electron transfer reactions of the protein to a greater extent than observed following modification with TFA and TFC. Interestingly, almost all CDNP-modified derivatives reacted more slowly with both cytochrome c oxidase and ubiquinol:cytochrome c oxidoreductase, even if the substituent clearly was at a residue on the back side of the protein, far removed from the solvent-accessible heme area (Figure 10.2). For example, CDNP–Lys-39 and CDNP–Lys-60 derivatives were significantly less reactive with the oxidase than was native horse cytochrome c (Figure 10.3, bottom panel). A possible interpretation of these results is that, depending on the location of the modification, the CDNP substituent can alter the magnitude and the orientation of the molecular dipole of cytochrome c in a functionally detectable manner. As a consequence, the orientation of the cytochrome c derivative in the electrostatic field of the redox partner may change and give rise to less reactive or unproductive complexes (Koppenol and Margoliash, 1982; Koppenol *et al.*, 1982; Margoliash and Bosshard, 1983). This interpretation is valid only if product dissociation is rate-determining. It is not clear if this is the case (Brzezinski and Malmstrøm, 1986).

In summary, the impression obtained from Figure 10.3 is that the different approaches are supportive of a common interaction domain on cytochrome c for both reaction partners in the inner mitochondrial membrane. Small differences between the sites may exist, as also indicated by data not mentioned here (Pande *et al.*, 1987). Most importantly, the possible shortcomings of the two approaches—analysis of an isolated complex versus modification-induced conformational changes—are quite different. This distinction adds further strength to our conclusions.

VII. Binding of Mitochondrial Cytochrome c to Nonphysiological Redox Partners

Mitochondrial cytochrome c functions as an almost universal electron donor or acceptor in *in vitro* systems. This lack of specificity may be explained by the high evolutionary stability of mitochondrial c-type cytochromes. In particular, the

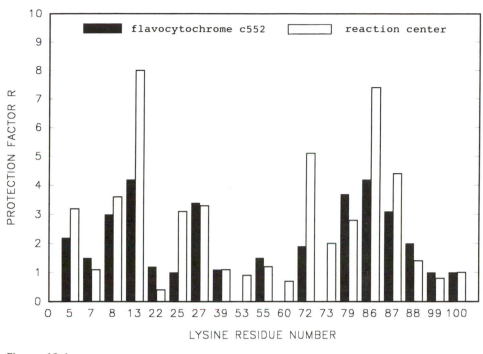

Figure 10.4

Differential protection of lysine residues of horse cytochrome *c* by flavocytochrome c_{552} from *Chromatium vinosum* (filled bars; Bosshard *et al.*, 1986) and by the photosynthetic reaction center from *Rhodospirillum rubrum* (open bars; Bosshard *et al.*, 1987a). The protection factor indicates the degree to which a lysine residue is less labeled when cytochrome *c* is bound to flavocytochrome c_{552} and the reaction center, respectively.

pattern of lysine residues surrounding the solvent-accessible heme edge has been strongly preserved. We believe this to be the major reason for the observation that horse cytochrome *c* binds to many nonphysiological redox partners in much the same way as it binds to the physiological partners ubiquinol:cytochrome *c* oxido-reductase and cytochrome *c* oxidase. The bacterial photosynthetic reaction center of *Rhodospirillum rubrum* and flavocytochrome c_{552} from *Chromatium vinosum* are proteins that function as electron transfer partners of mitochondrial cytochrome *c* *in vitro* but not *in vivo* (Rickle and Cusanovich, 1979; Gray and Knaff, 1982). As shown in Figure 10.4, lysines 5, 8, 13, 72, 86, and 87 are significantly shielded by both the reaction center and the flavoprotein. Both these proteins form stable 1:1 complexes with cytochrome *c* (Bosshard *et al.*, 1986, 1987a). Lysines 13 and 86 are most shielded, as was observed with the physiological complexes (compare Figures 10.1 and 10.3). Some differences do occur, however. Lys-25 is not shielded by the reaction center, but it is somewhat shielded by the mitochondrial redox partners. Lys-79 is more shielded by the nonphysiological than by the physiological partners. The variation of the degree of protection probably reflects variations in chemistry at the intermolecular interface. In general, however, there is no doubt that the two nonphysiological redox partners are bound to the same surface domain on the front side of horse cytochrome c. The protection pattern with the reaction center has been confirmed by kinetic analysis of lysine-substituted cytochrome c derivatives (van der Wal *et al.*, 1987). Parenthetically, it should be noted that the structures of the four binding partners, ubiquinol:cytochrome c oxidoreductase, cytochrome c oxidase, flavocytochrome c_{552}, and reaction center, are not related to each other.

VIII. Binding of Bacterial Cytochrome c_2 to Photosynthetic Reaction Center and Cytochrome bc_1 Complex

Cytochrome c_2 is an electron carrier involved in cyclic electron transport in purple bacteria. It provides an electron to the reaction center subsequent to the primary photochemical process. This function corresponds to the electron transfer from reduced mitochondrial cytochrome c to cytochrome c oxidase. The bacterial cytochrome bc_1 complex catalyzes the electron flow from ubiquinol to cytochrome c_2 and has a function analogous to that of mitochondrial ubiquinol:cytochrome c oxidoreductase (Wynn *et al.*, 1986).[4] *In vitro*, cytochrome c_2 can be replaced by cytochrome c in both the reaction with the cytochrome bc_1 complex and that with the photosynthetic reaction center. The folding of the polypeptide backbone of the two cytochromes is very similar (Figure 10.2). As expected, both the reaction center and the cytochrome bc_1 complex form stable 1:1 complexes with cytochrome c_2. These complexes were analyzed by the differential protection technique.

The *R. rubrum* cytochrome bc_1 complex protects lysines 12, 13, and 97 of *R. rubrum* cytochrome c_2 (Bosshard *et al.*, 1987b). Lysines 13 and 97 are sequence-homologous to lysines 13 and 87 of horse cytochrome c, both of which are strongly protected by the mitochondrial ubiquinol:cytochrome c oxidoreductase. Additional residues protected by the mitochondrial enzyme are lysines 5, 8, 72, and 86 (Figure 10.1). Lysines 8 and 72 are sequence-homologous to lysines 9 and 75 of cytochrome c_2. However, residues 9 and 75 are not protected by the bacterial cytochrome bc_1 complex as one might have expected on the basis of the results obtained in the mitochondrial system. There are no lysines of cytochrome c_2 which are sequence-homologous to cytochrome c lysines 5 and 86. Comparing the spatial location of front surface lysines in cytochrome c and cytochrome c_2 and taking into account the protection pattern for mitochondrial cytochrome c, one would predict that lysines 9, 13, 12, 27, 90, 75, 94, and 97 (viewed clockwise around the exposed heme edge) are likely to be protected by redox partners of cytochrome c_2 (compare the two stereo pairs of Figure 10.2).

Derivatives of *R. rubrum* cytochrome c_2 substituted at single lysine residues are not available. Mixtures of CDNP derivatives which contain either front surface-protected or back surface-protected lysines were tested as electron acceptors in the

---

[4] To distinguish, within the context of the present chapter, between the bacterial and the mitochondrial enzyme, the name "cytochrome bc_1 complex" is used only for the bacterial enzyme. In general, "cytochrome bc_1 complex" is used interchangeably for both the bacterial and the mitochondrial enzyme.

steady-state oxidation of ubiquinol catalyzed by *R. rubrum* cytochrome bc_1 complex. The results indicate that, as expected, front surface-modification reduces the rate of electron transfer up to 20-fold, whereas back surface-modification has almost no effect (Hall *et al.*, 1987a). Since no derivative with residue 97 modified has been available for kinetic tests, the importance of this residue, indicated by the data from differential protection, could not be corroborated. It is not clear if there is a difference in the binding and reaction domains of cytochrome c_2 for the cytochrome bc_1 complex in the *R. rubrum* system.

The situation is even less clear for the binding of *R. rubrum* cytochrome c_2 to the *R. rubrum* photosynthetic reaction center. Modification of the cytochrome c_2– reaction center complex has indicated a binding site on the back surface of cytochrome c_2. Lysines 109 and 112 and the α-amino group of Glu-1 are protected in the reaction center–cytochrome c_2 complex (see lower stereo pair of Figure 10.2; Rieder *et al.*, 1985). The kinetic analysis of mixtures of derivatives modified at single lysine residues, however, indicates a reaction domain on the front surface (Hall *et al.*, 1987b). The biological relevance of a binding site on the back surface of cytochrome c_2 is not known.

IX. Binding of Cytochrome c to Other Proteins

Its high isoelectric point and large molecular dipole moment make cytochrome c bind to almost any macromolecule with a negative electrostatic surface potential (Nicholls, 1974).

A. Binding to Phosvitin

The phosphoprotein phosvitin binds up to 22 molecules of cytochrome c (experiments with cytochrome c from *Candida crusei* by Yoshimura *et al.*, 1985). Distinct binding modes of high and low affinity are observed. Because of the acidic nature of phosvitin, one would expect that it binds to the surface area of cytochrome c where the positive end of the molecular dipole is centered at the β-carbon of phenylalanine 82, close to the solvent-accessible heme edge (Koppenol and Margoliash, 1982). We have tested the protection of cytochrome c lysines by phosvitin from egg yolk. The complex investigated had four molecules of cytochrome c bound per molecule of phosvitin (H.R.B. and L. Braun, unpublished). Binding of the four cytochrome molecules is thermodynamically stable (average $K_d = 2 \times 10^{-7}$ M, from spectrophotometric titration; C. Weber and H.R.B., unpublished). Protection factors are shown in Table 10.1. The pattern is unequivocal: strong protection of lysines around the solvent-accessible heme edge, little or no protection of those lysines which are removed from the heme edge (Figure 10.2). Protection is more uniform than in the case of the specific, physiologically relevant complexes. **R** values for lysines 7, 8, 13, 22 + 25 + 27, 72 + 73, 79, and 86–88 are between 6.8 and 10.4. Lysines 5, 39, 60, and 99 + 100 are probably unprotected, though their **R** values are not unity. This latter finding could be attributable to different reaction conditions in the presence and absence of phosvitin, for which no correction was made.

Table 10.1

Protection of horse cytochrome c by phosvitin.[a] A differential protection experiment was performed with cytochrome c (32 μM) in the presence and absence of phosvitin (8 μM). Some protection factors are average values for 2–3 lysine residues.

Lysine residue(s)	Protection factor
5	5.6
7	6.8
8	8.6
13	8.3
22 + 25 + 27	10.4
39	3.8
53 + 55	4.2
60	2.9
72 + 73	9.4
79	7.8
86 + 87 + 88	7.4
99 + 100	3.9

[a] H. R. Bosshard and L. Braun, unpublished.

B. Binding to Cytochrome c–Specific Monoclonal Antibodies

The differential protection technique was used to map antigenic determinants (epitopes) recognized by monoclonal antibodies (Burnens *et al.*, 1987). The antigenic structure of cytochrome c is described by Paterson in Chapter 11 of this volume. Some areas of cytochrome c seem to be immunodominant, but in general the entire surface of cytochrome c is antigenic (Benjamin *et al.*, 1984). Moreover, many antigenic determinants are conformation-dependent and composed primarily of residues brought together by the folding of the polypeptide chain. Detection of such determinants by conventional mapping techniques is difficult. Mapping techniques based on synthetic peptides may lead to erroneous results (Schwab *et al.*, 1993). Differential protection is a new way of gaining insight into the antigenic structure of cytochrome c (Burnens *et al.*, 1987; Saad *et al.*, 1988; Oertle *et al.*, 1989; Saad and Bosshard, 1990).

X. Protection of Carboxyl Groups of Cytochrome c Peroxidase by Cytochrome c

A variety of experimental methods have been used in an attempt to elucidate the structure of the complex composed of yeast cytochrome c peroxidase and cytochrome c: computer-aided model building (Poulos and Finzel, 1984), chemical cross-linking of the two hemoproteins (Bisson and Capaldi, 1981; Pettigrew and Seilman, 1982; Waldmeyer and Bosshard, 1985), kinetics with lysine-substituted cytochrome c derivatives (Kang et al., 1978; Smith and Millett, 1980), and differential protection techniques (Pettigrew, 1978; Bechtold and Bosshard, 1985). At present, this complex may be one of the most thoroughly studied protein electron transfer complexes (review by Bosshard et al., 1990). A fairly coherent picture of the main intermolecular interface of the complex has emerged. Some of the data are summarized in Table 10.2.

Table 10.2
Residues proposed to be located at the intermolecular interface of the complex composed of cytochrome c and cytochrome c peroxidase.

Residues of cyto-chrome c peroxidase	Residues of cyto-chrome c	Prediction from model building[a]
Asp-33, Asp-34	Lys-8, Gln-12	
Asp-37, Leu-82	Lys-13, Gln-16	
Gln-86, Asn-87	Lys-72, Ile-81	
His-181, Asp-217	Phe-82, Lys-87	
Asp-33, Asp-34		differential protection of COOH groups in free and cytochrome c–bound cytochrome c peroxidase[b]
Asp-37, Asp-221		
Asp-224, Glu-290		
Glu-291, Leu-294		
	Lys-8, Lys-13	Rate of oxidation of lysine-substituted cytochrome c[c]
	Lys-72, Lys-86	
	Lys-87	
Peptide 32–37	Lys-13, Lys-86	Complex cross-linked by carbodiimide[d]

[a] Poulos and Finzel (1984).
[b] Bechtold and Bosshard (1985).
[c] Kang et al. (1978).
[d] Waldmeyer and Bosshard (1985).

As expected, the binding site for cytochrome *c* peroxidase on cytochrome *c* is almost indistinguishable from that for the mitochondrial enzymes ubiquinol:cytochrome *c* oxidoreductase and cytochrome *c* oxidase, as indicated by the rate of electron transfer with different lysine-substituted derivatives (Kang *et al.*, 1978; Smith and Millett, 1980). In the covalent complex, formed by treating equimolar amounts of the two hemoproteins with 1-ethyl-3-(3-dimethylaminopropyl)carbodiimide-hydrochloride (EDC), cross-linking is to lysines 13 and 86 (Waldmeyer and Bosshard, 1985). These residues are shielded in the noncovalent complex (Pettigrew, 1978).

Model building for this complex starts from the known crystal structure of the two hemoproteins and attempts to optimize distances and geometries of ionic and hydrogen bonds that could form at the protein–protein interface. This strategy takes into account the well-known observation that the complex is stabilized through ionic bonds to which the peroxidase contributes a series of acidic residues (Table 10.2). The contribution of these acidic residues to cytochrome *c* binding has been analyzed experimentally by the differential protection technique. The carboxyl group activating reagent EDC, in combination with aminoethane sulfonic acid (taurine), has been used to modify carboxyl groups (Bechtold and Bosshard, 1985). Taurine with a ^{3}H and a ^{35}S isotopic label is used in a double-labeling procedure as described earlier for acetic anhydride. Side reactions such as intermolecular and intramolecular cross-linking of proteins may complicate the modification of carboxyl groups by a carbodiimide (Mauk and Mauk, 1989).

Reaction of cytochrome *c* peroxidase with EDC/taurine in the presence of cytochrome *c* has revealed that most of the residues that were predicted to juxtapose with lysines of cytochrome *c* are indeed protected (Table 10.2). Model prediction and chemical data are, however, not completely congruent. Asp-217 juxtaposes with Lys-72 of cytochrome *c* in the model, but is not among the protected residues. Nearby Asp-221 and Asp-224 are, however, protected. In addition, Glu-290, Glu-291 and possibly the carboxy-terminal residue Leu-294 are protected. The carboxy-terminal end of the peroxidase is not part of the cytochrome *c* binding site predicted by model building. It should be noted that a single molecule of cytochrome *c* is too small to protect at the same time all the acidic residues which were found to be protected and which are listed in Table 10.2.

The discrepancy between the predictions from model building and the results from differential protection highlights a conceptual difficulty. Model-building is based on the tacit assumption of a unique and specific electron transfer complex as a prerequisite for efficient electron transfer. This assumption may not be correct. For a 1:1 complex, there may exist more than one orientation of the two proteins that can permit efficient electron transfer. Results from differential protection of the complex may pertain to an average over several structures, which are in rapid equilibrium with one another. Indeed, calculations of the electrostatic effects on the dynamics of complex formation indicate an ensemble of electrostatically stable encounter complexes rather than a single dominant complex. Simulation of the diffusional association of cytochrome *c* with cytochrome *c* peroxidase predicts an ionic bond between Glu-290 and lysines of cytochrome *c* in some, though not all, of several electrostatically stabilized encounter complexes (Northrup *et al.*, 1987, 1988). Glu-290 is protected by cytochrome *c* (Table 10.2). The possibility of more than a single electron transfer complex must be seriously considered.[5] The faster electron transfer observed at higher ionic strength where the complex is only transitory supports a similar line of thinking (Hazzard *et al.*, 1988). Still, the overall agreement between the data summarized in Table 10.2 is very good, and there is no compelling reason to doubt the existence of a dominant, though possibly not singular, orientation of the two hemoproteins in the electrostatically stabilized complex.

Recently, the crystal structure of the 1:1 complex of cytochrome *c* and cytochrome *c* peroxidase has been solved (Pelletier and Kraut, 1992). Surprisingly, the structure differs significantly from the previous model predictions based on electrostatic considerations and characterized by direct charge–charge interactions (Poulos and Finzel, 1984). To see if electrostatic attraction contributes to the crystalline complex, we have calculated the surface potential distribution of cytochrome *c* peroxidase and yeast cytochrome *c* (unpublished experiments by I. Jelesarov and H.R.B.; calculations based on items 1CCP and 1YCC of Brookhaven Data Bank, calculations performed by program DelPhi, Biosym Technologies, San Diego). On cytochrome c peroxidase, two adjacent domains of strong negative potential are seen at a contour level of 10 kJ/mol (293 K, 150 mM ionic strength). The larger domain is almost identical with the binding site for cytochrome *c* in the published crystal structure. Acidic residues Asp-33, Asp-34, Glu-35, and Glu-290 contribute to this domain. The same residues were protected from chemical modification (Bechtold and Bosshard, 1985; Table 10.2). In the crystal of the complex, Asp-34 and Glu-290 are close to Lys-87 and Lys-73 near the heme cleft of cytochrome *c*, but distances are too large for charge-mediated hydrogen bonds (Pelletier and Kraut, 1992). Where are the residues located that, in view of the crystal structure of the complex, seem to have been wrongly assigned to the cytochrome c binding site of the peroxidase? Asp-217 (Poulos and Finzel, 1984) and the pair Asp-221/Asp-224 (Bechtold and Bosshard, 1985) are in a smaller negative potential domain of the peroxidase. Leu-82, Gln-86, Asn-87, His-181, and notably the heme edge (Poulos and Finzel, 1984) are in a neutral to weakly positive potential domain. Thus, while the previous model predictions allowed for several direct charge–charge interactions at the intermolecular interface of the cytochrome *c*/cytochrome *c* peroxidase complex, the crystal structure and our unpublished surface potential calculations point to a far better electrostatic attraction without the need of formation of charge-mediated hydrogen bonds.

XI. Conclusions

The differential protection technique applied to cytochrome *c* was instrumental in delineating the intermolecular interface of protein–protein complexes composed of cytochrome *c* and several of its redox partners. The beauty of the method lies in the

---

[5] Apparent protection of residues close to the C-terminus may also be explained by a conformational change induced by binding of cytochrome *c* and leading to reduced rates of modification of residues near to the C-terminus. This explanation was originally favored (Bechtold and Bosshard, 1985), but now seems less likely in view of the molecular dynamics calculations (Northrup *et al.*, 1987, 1988).

ability to assess in a single experiment the contribution of all of the lysines to the formation of electron transfer complexes. In addition, differential protection of surface carboxyl groups can be used to map the complementary binding sites on reaction partners of cytochrome c, as exemplified by differential protection of cytochrome c peroxidase by cytochrome c. Differential chemical modification, as the method is also called, has its pitfalls like any other method. The assignment of a residue to the binding site may sometimes be equivocal. In addition, the technique can be applied only to isolated cytochrome c complexes that are stable at low ionic strength. Such isolated complexes may differ somewhat from functional electron transfer complexes. However, there is no compelling evidence for gross differences between the complexes studied in solution and the transitory, functional electron transfer complexes.

A common surface area of cytochrome c is certainly important for complex formation. However, there is growing evidence that cytochrome c does not form a single unique electron transfer complex with a given redox partner. Rather, there may be an ensemble of similar complexes in which electron transfer can proceed. Both cytochrome c and its redox partner may be free to move against each other about a specific surface domain. Such a "rolling ball" mechanism (Burch et al., 1990; Roberts et al., 1991) is also indicated by the differential protection of cytochrome c peroxidase by cytochrome c.

XII. Acknowledgments

I am grateful to the many collaborators who over the years have contributed to developing and improving the differential protection technique. Work from the author's own laboratory was supported in part by the Swiss National Science Foundation and the Kanton of Zürich.

References

Ahmed, A. I., Smith, H. T., Smith, M. B., & Millett, F. (1978) *Biochemistry* **17**, 2479–2483.

Aviram, I., & Schejter, A. (1973) *FEBS Lett.* **36**, 174–176.

Bechtold, R., & Bosshard, H. R. (1985) *J. Biol. Chem.* **260**, 5191–5200.

Benjamin, D. C., Berzofsky, J. A., East, I. J., Gurd, F. R. N., Hannum, C., Leach, S. J., Margoliash, E., Michael, J. G., Miller, A., Prager, E. M., Reichlin, M., Sercarz, E., Smith-Gill, S. J., Todd, P. E., & Wilson, A. C. (1984) *Ann. Rev. Immunol.* **2**, 67.

Bisson, R., & Capaldi, R. A. (1981) *J. Biol. Chem.* **256**, 4362–4367.

Bosshard, H. R. (1979) *Meth. Biochem. Anal.* **25**, 273–301.

Bosshard, H. R. (1981) *J. Mol. Biol.* **153**, 1125–1149.

Bosshard, H. R., Zuerrer, M., Schaegger, H., & Von Jagow, G. (1979) *Biochem. Biophys. Res. Commun.* **89**, 250–258.

Bosshard, H. R., Davidson, M. W., Knaff, D. B., & Millett, F. (1986) *J. Biol. Chem.* **261**, 190–193.

Bosshard, H. R., Snozzi, M., & Bachofen, R. (1987a) *J. Bioenerg. Biomembr.* **19**, 375–382.

Bosshard, H. R., Wynn, R. M., & Knaff, D. B. (1987b) *Biochemistry* **26**, 7688–7693.

Bosshard, H. R., Anni, H., & Yonetani, T. (1990) in *Peroxidases in Chemistry and Biology* (Everse, J., & Grisham, M. B., eds.), CRC Press, Boca Raton, Florida, pp. 51–84.

Brautigan, D. L., Ferguson Miller, S., & Margoliash, E. (1978a) *J. Biol. Chem.* **253**, 130–139.

Brautigan, D. L., Ferguson Miller, S., Tarr, G. E., & Margoliash, E. (1978b) *J. Biol. Chem.* **253**, 140–148.

Brown, L. R., & Wüthrich, K. (1977) *Biochim. Biophys. Acta* **468**, 389–410.

Brzezinski, P., & Malmstrom, B. G. (1986) *Proc. Natl. Acad. Sci. USA* **83**, 4282–4286.

Burch, A. M., Rigby, S. E. J., Funk, W. D., MacGillivray, R. T. A., Mauk, M. R., Mauk, A. G., & Moore, G. R. (1990) *Science* **247**, 831–833.

Burch *et al.* (1990) p. 31.

Burnens, A., Demotz, S., Corradin, G., Binz, H., & Bosshard, H. R. (1987) *Science* **235**, 780–783.

Demel, R. A., Jordi, W., Lambrechts, H., van Damme, H., Hovius, R., & de Kruijff, B. (1989) *J. Biol. Chem.* **264**, 3988–3997.

Dickerson, R. E., & Timkovich, R. (1975) in *The Enzymes* (Boyer, P. D., ed.), pp. 397–547, Academic Press, New York.

Ferguson-Miller, S., Brautigan, D. L., & Margoliash, E. (1978) *J. Biol. Chem.* **253**, 149–159.

Gray, G. O., & Knaff, D. B. (1982) *Biochim. Biophys. Acta* **680**, 290–296.

Gupte, S. S., & Hackenbrock, C. R. (1988) *J. Biol. Chem.* **263**, 5248–5253.

Hall, J., Kriaucionas, A., Knaff, D., & Millett, F. (1987a) *J. Biol. Chem.* **262**, 14005–14009.

Hall, J., Ayres, M., Zha, X. H., O'Brien, P., Durham, B., Knaff, D., & Millett, F. (1987b) *J. Biol. Chem.* **262**, 11046–11051.

Hazzard, J. T., McLendon, G., Cusanovich, M. A., & Tollin, G. (1988) *Biochem. Biophys. Res. Commun.* **151**, 429–434.

Jacobs, E. E., & Sanadi, D. R. (1960) *J. Biol. Chem.* **235**, 531–534.

Kang, C. H., Brautigan, D. L., Osheroff, N., & Margoliash, E. (1978) *J. Biol. Chem.* **253**, 6502–6510.

Kaplan, H. (1972) *J. Mol. Biol.* **72**, 153–162.

Kaplan, H., Stevenson, K. J., & Hartley, B. S. (1971) *Biochem. J.* **124**, 289–299.

Koppenol, W. H., & Margoliash, E. (1982) *J. Biol. Chem.* **257**, 4426–4437.

Koppenol, W. H., Ferguson-Miller, S., Osheroff, N., Speck, S. H., & Margoliash, E. (1982) in *Oxidases and Related Redox Systems. Proceedings of the Third International Symposium* (King, T. E., Mason, H. S., & Morrison, M., eds.), pp. 1037–1053, Pergamon Press, Oxford.

Margoliash, E., & Bosshard, H. R. (1983) *Trends Biochem. Sci.* **8**, 316–320.

Margoliash, E., Ferguson-Miller, S., Tulloss, J., Kang, C. H., Feinberg, B. A., Brautigan, D. L., & Morrison, M. (1973) *Proc. Natl. Acad. Sci. USA* **70**, 3245–3249.

Mauk, M. R., & Mauk, A. G. (1989) *Eur. J. Biochem.* **186**, 473–486.

Nicholls, P. (1974) *Biochim. Biophys. Acta* **346**, 261–310.

Northrup, S. H., Boles, J. O., & Reynolds, C. L. (1987) *J. Phys. Chem.* **91**, 5991–5998.

Northrup, S. H., Boles, J. O., & Reynolds, J. L. (1988) *Science* **241**, 67–70.

Oertle, M., Immergluck, K., Paterson, Y., & Bosshard, H. R. (1989) *Eur. J. Biochem.* **182**, 699–704.

Osheroff, N., Brautigan, D. L., & Margoliash, E. (1980) *J. Biol. Chem.* **255**, 8245–8251.

Pande, J., Kinnally, K., Thallum, K. K., Verma, B. C., Myer, Y. P., Rechsteiner, L., & Bosshard, H. R. (1987) *J. Protein Chem.* **6**, 295–319.

Pelletier, H., & Kraut, J. (1992) *Science* **258**, 1748–1755.

Pettigrew, G. W. (1978) *FEBS Lett.* **86**, 14–16.

Pettigrew, G. W., & Seilman, S. (1982) *Biochem. J.* **201**, 9–18.

Pettigrew, G. W., & Moore, G. R. (1987) *Cytochromes c, Biological Aspects*, Springer, Berlin.

Poulos, T. L., & Finzel, B. C. (1984) *Pept. Protein Rev.* **4**, 115–171.

Rickle, G. K., & Cusanovich, M. A. (1979) *Arch. Biochem. Biophys.* **197**, 589–598.

Rieder, R., & Bosshard, H. R. (1978a) *J. Biol. Chem.* **253**, 6045–6053.

Rieder, R., & Bosshard, H. R. (1978b) *FEBS Lett.* **92**, 223–226.

Rieder, R., & Bosshard, H. R. (1980) *J. Biol. Chem.* **255**, 4732–4739.

Rieder, R., Wiemken, V., Bachofen, R., & Bosshard, H. R. (1985) *Biochem. Biophys. Res. Commun.* **128**, 120–126.

Riordan, J. F., & Vallee, B. L. (1972) *Meth. Enzymol.* **25**, 494–499.

Roberts, V. A., Freeman, H. C., Olson, A. J., Tainer, J. A., & Getzoff, E. D. (1991) *J. Biol. Chem.* **266**, 13431–13441.

Saad, B., & Bosshard, H. R. (1990) *Eur. J. Biochem.* **187**, 425–430.

Saad, B., Corradin, G., & Bosshard, H. R. (1988) *Eur. J. Biochem.* **178**, 219–224.

Salemme, F. R., Freer, S. T., Xuong, N. H., Alden, R. A., & Kraut, J. (1973) *J. Biol. Chem.* **248**, 3910–3921.

Schwab, C., Twardek, A., Lo, T. P., Brayer, G. D., & Bosshard, H. R. (1993) *Protein Science* **2**, 175–182.

Smith, H. T., Staudenmayer, N., & Millett, F. (1977) *Biochemistry* **16**, 4971–4974.

Smith, L., Davies, H. C., Reichlin, M., & Margoliash, E. (1973) *J. Biol. Chem.* **248**, 237–243.

Smith, M. B., & Millett, F. (1980) *Biochim. Biophys. Acta* **626**, 64–72.

Smith, M. B., Stonehuerner, J., Ahmed, A. J., Staudenmayer, N., & Millett, F. (1980) *Biochim. Biophys. Acta* **592**, 303–313.

Speck, S. H., Ferguson-Miller, S., Osheroff, N., & Margoliash, E. (1979) *Proc. Natl. Acad. Sci. USA* **76**, 155–159.

Takano, N., & Dickerson, R. E. (1981) *J. Mol. Biol.* **153**, 95–115.

Tanaka, N., Yamane, T., Tsukihara, T., Ashida, T., & Kakudo, M. (1975) *J. Biochem. (Tokyo)* **77**, 147–162.

Teissie, J. (1981) *Biochemistry* **20**, 1554–1560.

Tinner, R., Oertle, M., Heizmann, C. W., & Bosshard, H. R. (1990) *Cell Calcium* **11**, 19–23.

van der Wal, H. N., van Grondelle, R., Millett, F., & Knaff, D. B. (1987) *Biochim. Biophys. Acta* **893**, 490–498.

Voet, D., & Voet, J. G. (1990) *Biochemistry*. John Wiley & Sons, Inc., New York.

Waldmeyer, B., & Bosshard, H. R. (1985) *J. Biol. Chem.* **260**, 5184–5190.

Wynn, R. M., Gaul, D. F., Choi, W. K., Shaw, R. W., & Knaff, D. B. (1986) *Photosynth. Res.* **9**, 181–195.

Yoshimura, T., Matsushima, A., & Aki, K. (1985) *Arch. Biochem. Biophys.* **241**, 50–57.

11

The Immunology of Cytochrome c

YVONNE PATERSON

Department of Microbiology
University of Pennsylvania

I. Introduction

The immune system is capable of responding to almost any molecular form from small organic haptens to complex macromolecules such as nucleic acids, polysaccharides, and proteins, but the recognition of protein antigens has held the greatest fascination for immunologists for several reasons. For example, although antibody production to a variety of antigens is important in the host defense system against infectious and parasitic organisms, protein antigens are the main site of attack for the cellular immune response. Furthermore, the ability of the organism to distinguish the $\sim 10^5$ proteins that constitute its own self from those that are of foreign origin is the main constituent of immunological tolerance, the breakdown of which is responsible for auto-immune disease. Finally, the recognition of closely related cell surface protein antigens of the Major Histocompatibility Complex (MHC) is the basis of the rejection of transplanted tissues. Thus, studies on the perception of proteins as antigenic molecules by cells of the immune system go beyond an academic interest in immune recognition to providing the knowledge base for curing diverse human diseases. Nevertheless, a fundamental knowledge of how the cellular receptors of the immune system, which are themselves protein molecules, interact with protein antigens also provides important model systems for studying protein/protein interactions in general. This is particularly true for studies on the structurally well-characterized model antigens, of which cytochrome c is one.

The ability of proteins to elicit an immune response that results in the production of serum antibodies (immunogenicity) and the factors involved in the interaction of the antigen with antibody or with the T-cell receptors (TCR) of the cellular immune system (antigenicity) have traditionally been studied using model protein antigens such as myoglobin, lysozyme, insulin, and cytochrome c (Benjamin et al., 1984; Tainer et al., 1985; Schroer et al., 1983), which have proved invaluable in studying the role of protein structure and conformation in the immune recognition of proteins. Although these model antigens are, in general, not related to the disease process, they have been used because, unlike most pathogenic proteins, their primary and tertiary structures are known. In addition, they all belong to evolutionarily related families of proteins that have strong homology with respect to three-dimensional structure and amino acid sequence. This characteristic allows them be used in fine-specificity analyses to determine key residues in the protein antigen that interact with antibodies (Benjamin et al., 1984; Tainer et al., 1985; Schroer et al., 1983) or are involved in T-cell receptor recognition (Berkower et al., 1982; Schwartz, 1985; Shastri et al., 1985).

Cytochrome c has been one of the most intensely and successfully studied model protein antigens. Earlier immunological studies on cytochrome c were facilitated by the ease of preparation of the protein from a wide variety of sources (Urbanski and Margoliash, 1977). The pioneering work of Margoliash and co-workers resulted in sequence determination of more than 100 cytochromes (Borden and Margoliash, 1976; Carlson et al., 1977)—knowledge which is essential for the interpretation of fine-specificity analyses. In addition, the crystal structure of a number of species of eukaryotic cytochromes c, in both oxidation states, had been solved by 1980 and a number of others have been solved since then (see Chapter 3 by Brayer in this volume). These structures show a strong conservation of the polypeptide backbone fold of the molecule with expected changes in side-chain conformations where sequence differences occur between the species. Thus, differences in affinity of antibodies for various species of molecules can be attributed to local sequence variation rather than global changes in structure.

Many of the early studies on the immunological properties of cytochrome c have been reviewed with respect to recognition by antibodies (Benjamin et al., 1984; Urbanski and Margoliash, 1977) and helper T cells (Benjamin et al., 1984; Schwartz, 1985). In this chapter we will only briefly review this work for historical purposes and will concentrate on summarizing the many studies performed since then. These recent studies have emphasized the use of cytochrome c in gaining insights into the structural constraints of the molecular events involved in the interaction of protein antigenic determinants with antibodies and T-cell receptors.

II. The Recognition of Cytochrome *c* by B Cells and Their Antibody Products

A. Serum Antibody Responses

The production of serum antibodies by an animal when challenged with a foreign protein is brought about by a complex series of cellular events, a very simplified view of which is shown in Figure 11.1. Antigen-dependent activation of B cells into antibody-secreting cells (plasmocytes) requires not only the engagement of the immunoglobulin receptor of the B cell, but the presence of T cells of the helper phenotype. The immunogenicity of a protein is thus dependent not only on the presence of B cells bearing a suitable immunoglobulin receptor which can interact with the antibody, but also on its ability to stimulate T helper cells by a mechanism which we will discuss in the section of this chapter on the T-cell recognition of cytochrome *c*.

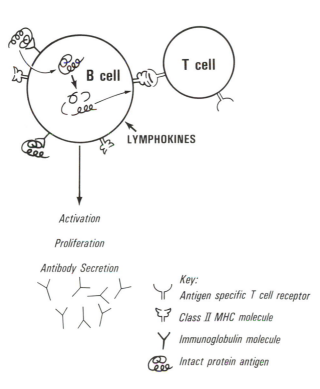

Figure 11.1
Cellular interactions in the clonal selection of resting B cells and their activation to antibody-secreting cells by protein antigens.

Because cytochrome *c* is highly conserved within the mammalian species (see Figure 11.2), it is poorly immunogenic in mammals due to immunological tolerance. This mechanism, which ensures that the organism does not respond to its own tissue proteins, results in the elimination of cells from the B cell and T cell repertoire which bear receptors that are specific for self proteins (reviewed by Nossal, 1989).

```
                                          10                    20
MOUSE/RAT                      * G D V E K G K K I F   V Q K C A Q C H T V
HORSE                          * - - - - - - - - - -   - - - - - - - - - -
COW                            * - - - - - - - - - -   - - - - - - - - - -
RABBIT                         * - - - - - - - - - -   - - - - - - - - - -
DOG                            * - - - - - - - - - -   - - - - - - - - - -
HUMAN                          * - - - - - - - - - -   I M - - S - - - - -
PIGEON                         * - - I - - - - - - -   - - - - S - - - - -
CHICKEN                        * - - I - - - - - - -   - - - - S - - - - -
MOTH                         G V P A - N A D N - - - - - - R - - - - - - -

                                          30                    40                    50
MOUSE/RAT               E K G G K H K T G P   N L H G L F G R K T   G Q A A G F S Y T D
HORSE                   - - - - - - - - - -   - - - - - - - - - -   - - - P - - T - - -
COW                     - - - - - - - - - -   - - - - - - - - - -   - - - P - - - - - -
RABBIT                  - - - - - - - - - -   - - - - - - - - - -   - - - V - - - - - -
DOG                     - - - - - - - - - -   - - - - - - - - - -   - - - P - - S - - -
HUMAN                   - - - - - - - - - -   - - - - - - - - - -   - - - P - S Y - - A
PIGEON                  - - - - - - - - - -   - - - - - - - - - -   - - - E - - - - - -
CHICKEN                 - - - - - - - - - -   - - - - - - - - - -   - - - E - - - - - -
MOTH                    - A - - - - - V - -   - - - - F - - - - -   - - - P - - - - S N

                                          60                    70                    80
MOUSE/RAT               A N K N K G I T W G   E D T L M E Y L E N   P K K Y I P G T K M
HORSE                   - - - - - - - - - K   - E - - - - - - - -   - - - - - - - - - -
COW                     - - - - - - - - - -   - E - - - - - - - -   - - - - - - - - - -
RABBIT                  - - - - - - - - - -   - E - - - - - - - -   - - - - - - - - - -
DOG                     - - - - - - - - - -   - E - - - - - - - -   - - - - - - - - - -
HUMAN                   - - - - - - - I - -   - - - - - - - - - -   - - - - - - - - - -
PIGEON                  - - - - - - - - - -   - - - - - - - - - -   - - - - - - - - - -
CHICKEN                 - - - - - - - - - -   - - - - - - - - - -   - - - - - - - - - -
MOTH                    - - - A - - - - - Q   D - - - F - - - - -   - - - - - - - - - -

                                          90                    100             104
MOUSE/RAT               I F A G I K K K G E   R A D L I A Y L K K   A T N E
HORSE                   - - - - - - - - T -   - E - - - - - - - -   - - - -
COW                     - - - - - - - - - -   - E - - - - - - - -   - - - -
RABBIT                  - - - - - - - - D -   - - - - - - - - - -   - - - -
DOG                     - - - - - - T - - -   - - - - - - - - - -   - - K -
HUMAN                   - - V - - - - E -     - - - - - - - - - -   - - - -
PIGEON                  - - - - - - - A -     - - - - - - - - Q     - - A K
CHICKEN                 - - - - - - - S -     - V - - - - - - D     - - S K
MOTH                    V - - - L - - A N E   - - - - - - - - Q     - - K
```

* ACETYLATED AMINO TERMINI

Figure 11.2
Amino acid sequences of some eukaryotic cytochromes *c*.

The greatest effect is on the T cell repertoire, but T cell tolerance can often be bypassed either by polymerizing a self protein or by covalently coupling it to a foreign protein. These molecular forms of the antigen can stimulate the helper T cell response required for serum antibody responses to protein antigens, whereas the monomeric self-like protein may not. In this way, B cell responses can often be achieved for foreign proteins which have high homology with the self variant. However, the antibodies produced are almost always specific for regions of the molecule that contain at least one amino-acid residue difference between the immunizing and the host species. Cytochrome *c* is no exception in this regard. Early studies (reviewed by Urbanski and Margoliash, 1977; Benjamin *et al.*, 1984; and Tainer *et al.*, 1985), which characterized the serum antibody response to horse cytochrome *c* in mouse and to mouse, horse, and guanaco cytochromes *c* in rabbit, identified three regions of the molecule as immunodominant. These regions were found where sequence variations occurred between these species, i.e. residues 44 and 47, 60 and 62, and 89 and 92 (see Figure 11.2). In addition the rabbit and mouse antibody responses to pigeon cytochrome *c* confirm this "rule," because antibodies were found to be directed against four sites on the pigeon molecule which included all seven of the residues which differ (Figure 11.2) between rabbit and pigeon (Hannum and Margoliash, 1985) and mouse and pigeon (Hannum *et al.*, 1985).

The studies just described imply that the immune response to proteins with homology to self is confined to regions of the molecule which contain non-homologous residues. As we will discuss in the next section, conserved residues must also be part of these sites. Where such antigenic sites do not exist, e.g., in rabbits immunized with rabbit cytochrome *c*, then no immune response should occur. However, breakdown in tolerance can often be induced by immunizing with self proteins, and such is the case for mouse cytochrome *c* in some strains of mice (Cooper and Paterson, unpublished observations) and for rabbit cytochrome *c* in rabbits (Nisonoff *et al.*, 1967; Jemmerson and Margoliash, 1979; Jemmerson *et al.*, 1985; Cooper *et al.*, 1989). In addition, human anti–cytochrome *c* antibodies have been known to arise in auto-immune disease (Mamula *et al.*, 1990).

It has been suggested, largely from studies on the immunogenicity of cytochrome *c*, that even in those instances when breakdown in tolerance to a self protein has occurred, the response is still directed to regions that are sites of evolutionary variability for the protein (Jemmerson and Margoliash, 1979; Kieber-Emmons and Kohler, 1986). Thus, it has been postulated that the immune system possesses some kind of evolutionary memory, perhaps brought about by the deletion of those genetic elements from the immunoglobulin gene pool of mammalian species, that could generate anti-self responses (Jemmerson and Margoliash, 1979). However, more recent studies have shown that B cell tolerance is affinity-dependent (Riley and Klinman, 1986), so those immature B cells which bear immunoglobulin receptors with affinities for self antigens which are too low to induce a functional deletion may remain in the mature repertoire. The antibody products of such cells, when stimulated with self antigens, will of course be of low affinity and undetectable in the solution-phase assays which were the analytical tool of the earlier studies (Jemmerson and Margoliash, 1979). Later studies, using more sensitive assays, have revealed that although a small population of antibodies to the carboxy-terminal helical region of the molecule, which contains the evolutionary residues 89 and 92, could be detected in the response to native rabbit cytochrome *c* in rabbits, the majority of the antibodies were to the evolutionarily conserved amino-terminal region of the molecule (Jemmerson *et al.*, 1985; Cooper *et al.*, 1989). Thus, B cells do exist in the mature repertoire that can respond to evolutionarily conserved regions of self molecules. Rabbit anti–rabbit cytochrome *c* antibodies were found to be at least 100-fold lower in affinity compared with rabbit anti–horse cytochrome *c* antibodies (Cooper *et al.*, 1989), indicating that those B cells that escape tolerance induction bear lower-affinity immunoglobulin receptors.

The regions on cytochrome *c* which are immunogenic are clearly influenced by self tolerance and are restricted to a few regions of the molecule. It should be noted, however, that this is not generally the case for protein antigens. Although a number of factors, e.g., surface mobility (Moore and Williams, 1980; Tainer *et al.*, 1985), have been correlated with protein antigenicity, the most widely accepted view of the antigenicity of proteins is that most of the water-accessible surface area of a protein molecule is potentially immunogenic. Thus, although the surface regions of cytochromes *c* which have one or more side-chain differences from that of the host will dominate the immune response to this molecule, the responses to proteins which have little homology to self proteins are not so limited. Monoclonal antibodies to lysozyme, for example, have been prepared which overlap most of the surface structure of the protein (Smith-Gill *et al.*, 1984). However, there are restrictions in the response by individual animals, even to foreign proteins, which are imposed by the factors that govern the helper T-cell response, by immunoregulatory phenomena such as suppressor T cells or anti-idiotypic networks, and by the immune status of the host animal. Thus, the specificities of the secondary serum antibody response to lysozyme are not simply a combination of all of the known monoclonal antibody specificities, reflecting a bias imposed by the *in vivo* environment in which the precursors of the plasmocytes developed (Metzger *et al.*, 1984). In addition the composition of the serum antibody population may vary with respect to individual specificities from animal to animal and within a species, as has been shown for the protein myohemerythrin (Geysen *et al.*, 1987); it may also vary within the same animal with subsequent exposure to antigen. For example, the specificity of antibodies in the primary response to lysozyme is quite different to those in the secondary response to that protein in the same animal (Miller *et al.*, 1989). For these reasons it is unwise to draw conclusions about the overall immunogenicity of a protein from the specificities represented in a few examples of antisera.

B. Probing the Antigenicity of Cytochrome *c* Using Monoclonal Antibodies

Studies on the recognition of protein antigens by B cells have been facilitated in the past decade by the availability of epitope-specific monoclonal antibodies produced by hybridoma technology. Thus, it is now possible to study the specificities of individual B cell clones (Paterson, 1989; Benjamin *et al.*, 1984; Berzofsky, 1985). Monoclonal antibody–secreting hybridomas have been raised against a wide range of species-variant cytochromes *c*, including yeast iso-1-cytochrome *c* (Silvestri and

Taniuchi, 1988), *Paracoccus denitrificans* (Kuo *et al.*, 1984, 1985; Kuo and Davies, 1983), cow (Kuo and Davies, 1983; Kuo *et al.*, 1984), horse (Carbone and Paterson, 1985; Burnens *et al.*, 1987; Kim *et al.*, 1987; Goshorn *et al.*, 1991; Jemmerson and Johnson, 1991), rat (Kim and Nolla, 1986; Goshorn *et al.*, 1991; Jemmerson and Johnson, 1991) and human (Kuo *et al.*, 1986). One or two residues in the epitopes recognized by the antibodies to the mammalian cytochromes have been identified by fine-specificity analyses with panels of homologous molecules such as those shown in Figure 11.2. For example, the monoclonal antibody to human cytochrome *c* binds to a region around Ile^{58}, since it does not bind to cytochrome *c* from *Macacca mulatta* which differs from the human antigen only by the replacement of residue 58 with threonine (Kuo *et al.*, 1986). However, some of these monoclonal antibodies are to species of cytochrome *c* which have multiple sequence changes within the family of evolutionarily variant cytochromes, and thus individual residues on the surface of cytochrome *c* bound by the antibody cannot be determined using this panel. The constraint of evolutionary variability in fine-specificity analyses with homologous proteins can be overcome by using mutant molecules produced by site-directed mutagenesis. Although this approach has been used with other model protein antigens such as lysozyme (Kam-Morgan *et al.*, 1993) and staphylococcal nuclease (Smith *et al.*, 1991; Smith and Benjamin, 1991), it has not yet been applied to epitope mapping on cytochrome *c*. The horse cytochrome *c*–specific monoclonal antibodies produced in our laboratory (Carbone and Paterson, 1985) are perhaps the best characterized because we have used them extensively as tools for three-dimensional epitope mapping over the last few years (Paterson, 1989; Paterson *et al.*, 1990; Mayne *et al.*, 1992). These studies will be described extensively in the next section.

Monoclonal antibodies can be generated by techniques other than by the hybridoma technology pioneered by Kohler and Milstein (1975, 1976). Prior to the development of the cell fusion procedure, homogeneous antigen-binding monoclonal antibodies were produced by clonal B-cell culturing. Limiting dilution cloning of antigenic B cells in micro–tissue culture, in the presence of antigen, filler cells, and growth factors, allows for the collection of small quantities of homogeneous antibody before the cells die (Quintáns and Lefkovits, 1973; Pike, 1975). More robust and longer-lasting B-cell clones can be produced by the splenic fragment assay technique (Klinman, 1969, 1972). In this procedure, B cells are isolated and purified from antigen-immunized mice and adoptively transferred in limiting numbers into recipient mice. The recipients are generally lethally irradiated to minimize the background contribution of the endogenous B cells; they are also primed in advance with carrier protein to provide T helper cells (which are more radio-resistant than B cells) to stimulate the clonal outgrowth of the transferred B cells. This rather complex culturing procedure has the advantage over hybridoma technology in that it better represents the entire repertoire of specificities produced by B cells to a single antigen, and it can be used to study changes in the B cell repertoire during ontogeny from the pre-B cell to the secondary B cell (Klinman and Press, 1975; Klinman and Linton, 1988; Johnson and Jemmerson, 1991), during aging from the neonatal to the old mouse (Klinman and Press, 1975; Klinman and Linton, 1988) and during tolerance induction (Riley and Klinman, 1986; Johnson and Jemmerson, 1992). In addition, a very large number of clones can be produced, yielding large numbers of monoclonal antibodies, albeit in very small quantities.

Using the splenic fragment assay procedure, the secondary B cell murine response to horse cytochrome c and to the self molecule rat cytochrome c [mouse and rat cytochrome c have an identical sequence (Carlson et al., 1977)] have been extensively analyzed by Jemmerson and co-workers (Jemmerson, 1987a, 1987b; Jemmerson and Blankenfeld, 1988; Johnson and Jemmerson, 1991). More than 100 monoclonal specificities to horse cytochrome c were examined which displayed 22 reactivity patterns when subjected to fine-specificity analyses with the mammalian cytochromes shown in Figure 11.2 and with cyanogen bromide–cleaved fragments of horse cytochrome c, 1–65, 66–80, 81–104. As was expected, all of the monoclonal antibodies were directed to the three immunodominant regions of the molecule where sequence differences between horse and mouse cytochrome c occur, i.e., 44 and 47, 60 and 62, 89 and 92 (Jemmerson, 1987a). However, the recognition of these regions differed in subtly different ways, indicating that the three antigenic regions consist of multiple overlapping epitopes. This has also been shown for monoclonal antibodies to horse cytochrome c produced by cell fusion (Carbone and Paterson, 1985; Paterson, 1989; Mayne et al., 1992; Jemmerson and Johnson, 1991; Goshorn et al., 1991). Only a small percentage of the monoclonal antibodies examined by Jemmerson and colleagues could recognize the cyanogen bromide–cleaved fragments of cytochrome c, thereby showing the strong preference for the native conformation by most of the specificities produced; this preference is also displayed by other monoclonal antibodies to native proteins (Benjamin et al., 1984; Berzofsky, 1985; Jemmerson and Paterson, 1986b) and is discussed later.

In a subsequent study, Jemmerson and Blankenfeld (1988) showed that unlike the varied response seen to horse cytochrome c, 50% of the 556 antibody products of B lymphocyte clones produced in splenic fragments against rat (self) cytochrome c were sensitive to an Asp to Glu change at position 62 in the sequence. In the case of the horse cytochrome c specific clones, the highest frequency measured for the ~100 clonotypes was 26%. Of course, the identification of a single residue in an epitope only locates the face of the protein to which the antibody binds. Given that x-ray crystallographic analyses of immune complexes have shown that as many as 17 residues on a protein surface may interact with an antibody (Davies et al., 1988), clearly a wide number of specificities could include any one residue on a protein surface. Nevertheless, it is of interest that self-responses to cytochrome c appear to be more restricted in specificity than those to heterologous molecules in both mice (Jemmerson and Blankenfeld, 1988) and rabbits (Cooper et al., 1989).

C. Antibody/Antigen Interactions at the Molecular Level

A detailed molecular analysis of the interface between two proteins can only be achieved currently by x-ray crystallography. At present, such analyses have been performed for three protein antigens in six different complexes with Fab fragments of their specific monoclonal antibodies (Amit et al., 1986; Fischmann et al., 1991; Sheriff et al., 1987; Padlan et al., 1989; Colman et al., 1987; Tulip et al., 1989); for none of these was the antigen cytochrome c. The general features of these structures will be described here, however, because of the information they provide on anti-body–antigen recognition at the stereochemical level. Two of the complexes studied involved the large viral protein neuraminidase (Colman *et al.*, 1987; Tulip *et al.*, 1989, 1992), but three of the monoclonal antibodies crystallized are with lysozyme (Amit *et al.*, 1986; Fischmann *et al.*, 1991; Sheriff *et al.*, 1987; Padlan *et al.*, 1989), a small globular protein of only slightly greater molecular mass than cytochrome c (129 amino acid residues, M_r 14,200–14,600). These three structures revealed that an extensive surface of this small protein antigen, about 750 $Å^2$, interacts with the antibody (Davies *et al.*, 1988). Each of these surfaces is made up of 14 to 16 amino acid residues located in two to four discontiguous stretches of polypeptide back-bone. The three epitopes together occupy about 50% of the water-accessible surface of the molecule, but actually represent only a small number of the total number of epitopes determined for lysozyme (Lavoie *et al.*, 1989).

Several questions emerge from the ability to study antibody/antigen complexes at the atomic level. The first is the nature of the interaction and the degree of conformational change that takes place in either the antibody or the antigen on complex formation. Since both neuraminidase (Varghese *et al.*, 1983) and lysozyme (Blake *et al.*, 1965) have been crystallized in the free form, direct comparisons can be made between the antibody-bound and -free forms of these antigens. Differences in the degree of complementarity have been observed in these existing antibody/antigen structures. Thus, no significant conformational change in the structure of lysozyme on binding to D1.3 was observed (Amit *et al.*, 1986), whereas movements of 1.0–2.0 Å in the C_α backbone of the antigen have been observed in the binding of the antibody HyHEL 5 to lysozyme (Sheriff *et al.*, 1987) and in the anti-neuraminidase structure (Colman *et al.*, 1987). Rotations about side-chain bonds in residues both in and remote from the epitope also occur (Davies *et al.*, 1988). In addition, side-chain rotations of residues remote from the epitope are also observed for antigen in the HyHel 5 complex (Sheriff *et al.*, 1987), whereas only minor local changes are observed for lysozyme in the antibody–lysozyme complex of HyHEL 10. These findings, together with our own investigations on antibody/protein complexes using H–D exchange and 2D NMR which will be discussed in detail later, indicate that in some cases small changes can take place in the three-dimensional fold of the antigen that improve the complementarity between the interaction sites. Strong complementarity, however, may be a prerequisite for the binding of monoclonal antibodies to protein antigens, as we have found for some of the mono-clonal antibodies to horse cytochrome c which we have prepared in our laboratory (Carbone and Paterson, 1985). One of these, E8, has been crystallized in complex with horse cytochrome c and is currently being analyzed by x-ray crystallography (Mylvaganum *et al.*, 1991). As the crystal structure of horse cytochrome c was recently solved to high resolution (Bushnell *et al.*, 1990), a direct comparison of the complexed and uncomplexed antigen can be made to address the role of complementarity in this high-affinity interaction.

The six crystal structures thus far solved provide no concrete evidence on whether conformational changes can also take place in the antibody on interaction with a protein antigen, since the free antibody structure has not yet been solved for any of them. Two crystal structures of antibody fragments to peptides, both free and complexed with their short peptidic antigens, have recently been published (Stanfield *et al.*, 1990; Rini *et al.*, 1992). In these cases, only a much-reduced area of contiguous peptide chain of a few residues contacts the antibody-combining site, compared to the topographic surface involved when a protein is the antigen. Nevertheless, small but significant (1 to 2 Å) side-chain and main-chain rearrangements do take place in the complementarity-determining loops of the antibody combining site on binding the peptide antigen. In addition, one of the anti-lysozyme antibodies has been crystallized complexed to an anti-idiotypic antibody Fab fragment (Bentley *et al.*, 1990). The structure of the idiotype bound to lysozyme compared to the immunoglobulin anti-idiotype indicates that a large number of the residues in the combining site are found to be interacting with the epitopes of both complexes, and that side-chain conformations of these residues differ in the two complexes. Nevertheless, the conformation of the main chain of the idiotype is essentially the same in both structures, which provides circumstantial evidence that this is also the case for the uncomplexed form. Combined experimental and computational evidence also indicates that sequence changes in antibodies may result in local side-chain rearrangements to position side chains in the combining site and alter binding activity (Chien *et al.*, 1989). We are currently collecting x-ray crystallographic data on Fab fragments of monoclonal antibody E8 both complexed to the antigen molecule and free (Mylvaganam *et al.*, 1991). In addition, the antibody to lysozyme for which crystal structures in two complexed forms has been solved (Amit *et al.*, 1986; Bentley *et al.*, 1990), has been crystallized, and its structure is currently under investigation (Bentley *et al.*, 1989). These studies will address this important issue directly.

There has been a suggestion that paratope rearrangement may take place on antigen engagement in the case of one of the antibody/neuraminidase structures, since the orientation of the variable domains of the light and heavy chains of the Fab fragment of this antibody is different from that in other Fab fragments of uncomplexed antibodies (Colman *et al.*, 1987). However, a further refinement of this structure (Tulip *et al.*, 1992) and a comparison to another antibody/ neuraminidase complex (Colman *et al.*, 1989) leads one to question this conclusion. This reassessment emphasizes the need to study the same antibody Fab in free and antigen-bound form to establish whether such domain realignment or shifting takes place on antigen binding. Thus, whether such conformational changes in the antibody do indeed take place awaits confirmation from a structural analysis of the uncomplexed molecule. This issue has important implications for B-cell triggering in the response to protein antigens.

Receptor signaling in the activation of resting B lymphocytes results in clonal expansion of antigen-specific B cells and their subsequent differentiation to antibody-secreting plasma cells. Immunoglobulin receptor engagement resulting in signal transduction is known to play an important role in stimulating growth and differentiation in the response of B cells to multivalent antigens, to mitogens such as LPS, and to polyvalent anti-immunoglobulin stimulation (DeFranco *et al.*, 1987; Cambier and Ransom, 1987). However, the molecular mechanism by which the receptor communicates the presence of these stimuli to the cell, with subsequent activation of the phospho-inositide pathway, is still largely unknown. What remains even more controversial is the role of surface immunoglobulin engagement in the activation of T-dependent, monovalent protein antigens where receptor cross-linkage by the antigen cannot take place. It is possible that B-cell activation in such cases is induced solely by T-cell stimulation; the purpose of the surface-bound immunoglobulin receptor may be only to convey antigen specificity via passive receptor uptake and processing. However, if immunoglobulin receptor engagement plays a more active role in B-cell triggering, either directly via signal transduction or indirectly by up-regulating receptor-mediated endocytosis of antigen, then in the absence of receptor linkage, the communication of receptor engagement may be through a conformational change invoked in the immunoglobulin on antigen binding. Such a mechanism, involving changes in the elbow bend on antigen binding, has been suggested (Huber *et al.*, 1976). However, different crystal forms of antibody HyHEL 5 complexed to lysozyme show quite different elbow-bend angles, which suggests that this is simply an inherently flexible region of the antibody molecule (Sheriff *et al.*, 1987). In addition, the elbow bends of an anti-peptide antibody, free and complexed, were similar and fell in the midrange of angles observed for other Fab structures (Stanfield *et al.*, 1990).

A further question that arises in considering the data derived from the crystal structure analysis of antibody/antigen complexes is the role of specific types of inter-residue interactions to the free energy of association. A comparison of the three Fab/lysozyme complexes indicates that a similar number of residues for the lysozyme and antibody are involved in each interface, despite differences of two orders of magnitude among the binding affinities of the complexes. The interface of all three complexes comprise several hydrogen bonds and aromatic residue contacts. In addition, the high affinity (2×10^{-9} M) of HyHEL 5-lysozyme (Sheriff *et al.*, 1987) is further stabilized by two salt bridges between arginine residues in the epitope of the lysozyme and glutamic residues in the V_h region of the Fab. For the second high-affinity antibody structure (HyHEL 10-lysozyme = 10^{-9} M), a much weaker ion pair at the interface was observed (Padlan *et al.*, 1989). In the lower-affinity structure (D1.3-lysozyme = 10^{-7} M), interactions between charged residues did not appear to make any contribution to the stability of the complex (Amit *et al.*, 1986). Novotny and co-workers (1989) have calculated the free energy of interaction of these structurally defined immune complexes both from the experimentally measured affinities of the interactions and using energy calculations, and they have concluded that only a small number of the amino acid residues occluded by the antibody combining site contribute to the binding energy. Thus, it appears that not every residue identified in the interface of antibody/antigen complexes is important to the stability of the complex and involved in determining the specificity of the antibody for its antigen. Clearly, although x-ray crystallography is the definitive method for antigenic analysis at the molecular level, it also requires the support of biochemical techniques for generating antigenic variants with which to measure the contributions of individual residues in the protein/protein interface to the affinity of the antibody/antigen complex. In addition, it provides a static view of antibody/antigen interactions which may not prevail in solution, and it is limited by the difficulty in obtaining crystals of adequate quality for x-ray diffraction.

D. Structural Analyses of Antibody/Antigen Interactions Using Biochemical Techniques

The technical limitations in studying a large number of different types of antigenic sites on proteins by x-ray crystallography provides the impetus for developing methods for three-dimensional epitope mapping of protein antigen/antibody combining sites which may be more generally applicable to other protein/protein interactions. In one approach, overlapping synthetic peptides that together represent the complete polypeptide chain of the antigen are screened for binding to antibodies specific for the intact protein. When binding occurs, this procedure can provide information on contiguous stretches of a determinant that are part of the complete epitope (Geysen et al., 1987; Getzoff et al., 1988). Unfortunately, negative results with this technique are difficult to interpret because the failure of a peptide region to bind an antibody does not imply that it is not part of an epitope (Savoca et al., 1991). Instead, this failure may merely reflect the strong complementarity between the paratope of the antibody and the epitope of the protein antigen. Thus, in our experience with monoclonal antibodies to cytochrome c, linear peptide fragments that represent parts of the epitopes, defined by three-dimensional mapping procedures, were not effective at binding the relevant antibodies (Carbone and Paterson, 1985; Jemmerson and Paterson, 1986a). This limitation can sometimes be overcome by using conformationally constrained peptides which more closely mimic the native state of the protein than short linear peptides (Jemmerson and Hutchinson, 1990).

For these reasons our main approach to mapping those antigenic sites on proteins that interact with antibodies has been a biochemical strategy, in which we attempt to maintain the antigen in native conformation. Thus, epitopes which contain many residues that are discontiguous in sequence and are required to be in native conformation can be studied. We will illustrate the use of these techniques by describing our studies on four monoclonal antibodies, C3, C7, E3, and E8, raised against horse cytochrome c to which they bind with high affinity ($K_{ass} \simeq 10^8$). In this work we have defined residues on the surface of the cytochrome c molecule that are important in stabilizing the antigen/antibody complex, and we have also determined the boundaries of the epitopes.

We have already described the application of fine-specificity analyses to characterizing polyclonal antiserum specificities. This classical approach measures the ability of antibodies to bind to panels of evolutionarily variant molecules, which differ in sequence in only one or two residues and have essentially identical tertiary structure. This strategy is even more useful in determining key residues in the epitopes of proteins recognized by monoclonal antibodies of single specificity. Applying fine-specificity analyses to monoclonal antibodies E3, C3, C7, and E8 to cytochrome c identified a single residue in the epitopes of three of the monoclonal antibodies and two residues, either of which must be in the epitope of the fourth antibody, E8 (Table 11.1). As the panel of homologous molecules is often quite limited, even for model protein antigens, it is often useful to extend the number of variants available for fine-specificity analyses by using chemical techniques to modify residues on the intact protein antigen. In some cases, these can be specifically directed to single

Table 11.1
Residues in the epitopes of monoclonal antibodies to horse cytochrome *c* identified using biochemical techniques.

| | Residues Identified for Monoclonal Antibodies | | |
C3	C7	E3	E8
Using fine-specificity analyses with evolutionarily variant cytochromes (Carbone and Paterson, 1985)			
Pro^{44}	Pro^{44}	Thr^{47}	Lys^{60} or Thr^{89}
Using chemically modified or protein engineered cytochromes (Cooper *et al.*, 1987; Collawn *et al.*, 1988)			
N.I.[a]	His^{26}	Thr^{40}[b]	Glu^{66}
Using immunoprotective techniques:			
(a) By tryptic proteolysis (Jemmerson and Paterson, 1986a; Cooper *et al.*, 1987)			
His^{26} & Pro^{44}	His^{26} & Pro^{44}	N.D.[c]	Lys^{60}, Glu^{66}
(b) By acetylation of nonshielded lysines (Oertle *et al.*, 1989)			
Lys^{53}, Lys^{79}	Lys^{53}	Lys^{53}, Lys^{79}	Lys^{60}, Lys^{99}

[a] No residues identified in this way for this monoclonal antibody.
[b] J. M. Collawn, C. J. A. Wallace, and Y. Paterson, unpublished observations.
[c] Not examined by this procedure.

residues in the molecule (Brautigan *et al.*, 1978a,b; Cooper *et al.*, 1987; Collawn *et al.*, 1988). Thus, we identified a single histidine in the epitope of monoclonal antibody C7 (Cooper *et al.*, 1987) and a glutamic acid residue in the epitope of E8 (Collawn *et al.*, 1988) by chemical manipulation of the sequence of horse cytochrome *c* (see Table 11.1). Clearly, the utility of site-directed modifications of cytochrome *c* in epitope mapping will ultimately be extended by the generation of mutant antigens with single-residue substitutions by site-directed mutagenesis. This approach has already been successful with other antigens (Benjamin, 1991; Smith *et al.*, 1991; Smith and Benjamin, 1991; Kam-Morgan *et al.*, 1993). It must be borne in mind, however, that the chemical or genetic modification of phylogenetically conserved residues can result in conformational changes in the protein. This observation has been made with *N*-formyl-tryptophan^{59} horse cytochrome *c* (Cooper *et al.*, 1987) and with the substitution of norvaline for glutamic acid^{66} in horse cytochrome *c* (Wallace and Corthésy, 1986).

Fine-specificity analysis with natural or engineered variants of the antigen can be used to identify residues in the epitope where the substitutions have a detectable effect on the affinity of the antibody/antigen interaction. Other residues in the interface of the antibody/antigen complex can be identified by immunoprotective techniques. In this approach, an immune complex of the antibody or Fab fragment and

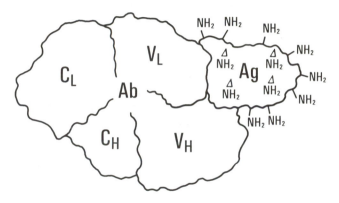

Figure 11.3
The protection of lysine residues on the surface of a protein antigen by binding to a monoclonal antibody.

the antigen is made, which protects residues in the epitope from enzymatic cleavage (Jemmerson and Paterson, 1986a) or chemical modification (Burnens *et al.*, 1987; Oertle *et al.*, 1989). In the case of horse cytochrome *c*, which has 19 lysine residues evenly distributed over the surface of the molecule (see Figure 11.3), we have used protection from trypsin proteolysis to identify two discontiguous regions of the polypeptide chain in the epitopes of each of C3, C7, and E8 (see Table 11.1) (Jemmerson and Paterson, 1986a; Cooper *et al.*, 1987). We have also examined the protective effect of these monoclonal antibodies on the acetylation of lysine residues on the surface of cytochrome *c* (Oertle *et al.*, 1989). This method, which was developed by Bosshard, has been applied to the antibody/antigen system as a special case of the general use of differential chemical modification in the study of protein/protein interactions, reviewed by Bosshard (1979) (also see Chapter 10 by Bosshard in this volume). This method is very useful when the residues of interest may be modified under conditions which do not disrupt the antibody/antigen complex.

Although these methods, even in combination, can only provide partial information on the residues of the antigen molecule involved in contact with antibody, they can be most effective in defining the surfaces of the antigen on which the epitope lies and in determining some of the residues that contribute to the overall affinity of the interaction. This is particularly the case for monoclonal antibodies such as C3, C7, E3, and E8 that have high affinity for the purified, soluble, native protein and that will not bind to peptide fragments of the antigen. Table 11.1 summarizes our findings using this combination of approaches. At least three residues have been identified in each epitope, thus defining the surface of the cytochrome molecule to which the antibody binds. The surfaces of the antigen to which three of the monoclonal antibodies (C3, E3, and C7) bind overlap, but the fourth (E8) binds to a region of the antigen on the opposite face of the molecule. These findings are consistent with blocking studies with these monoclonal antibodies (Carbone and

Paterson, 1985) which demonstrated that no two pairs of E3, C3, and C7 could bind the antigen simultaneously. However, the antigen could accommodate E8 with any one of the other three monoclonal antibodies. In addition to determining the lysine residues recognized by the monoclonal antibodies, the differential acetylation of lysine residues in the immune complex also defines the lysine residues on the antigen that are not protected by complex formation with the antibody. This result provides an estimate of the perimeter of the epitope.

E. Structural Analyses of Antibody/Antigen Interactions Using 2-D NMR

We have recently been exploring the use of high-resolution NMR to examine antibody/antigen interactions to gain a more complete view of the epitope than that provided by the immunoprotective techniques described above. Recent progress in 2-D NMR spectroscopy has made it possible to obtain complete proton resonance assignments for small proteins and to determine their structures in solution (reviewed by Wüthrich, 1986, and Markley, 1989). However, previous NMR studies on antibody/antigen interactions have been confined to examining the interaction of small molecules with antibodies. For example, phosphorus and fluorine NMR have been used to study the kinetics of hapten/antibody interactions (Kooistra and Richards, 1978; Goetze and Richards, 1978) and magnetization transfer ^{1}H NMR and 2-D transferred NOE difference spectroscopy have been used by Anglister and colleagues to study anti-hapten (Anglister *et al.*, 1987) and anti-peptide antibody complexes (Anglister *et al.*, 1988; 1989). The latter studies have concentrated on the contribution of individual aromatic residues to the interaction with ligand, in the combining site of the antibody. A more recent study by Cheetham and colleagues (1991) measured the changes in line widths and chemical shifts of the $\alpha\beta$ and $\beta\gamma$ cross peaks of residues in the 2-D COSY spectrum of a 28-residue peptide from lysozyme brought about by binding to an anti-peptide antibody Fab. They showed that a limited number of the residues are tightly bound to the antibody combining site, while the rest retain considerable mobility.

In general, ^{1}H NMR spectroscopy has not been applied directly to antibody/antigen interactions where the antigen is a protein molecule because of the large size of such complexes. We have recently solved this problem by immobilizing the antibody on a solid support to allow rapid dissociation of the antigen and thereby permit us to perform 2-D NMR spectroscopy on the protein antigen, in isolation from the antibody (Figure 11.4) (Paterson *et al.*, 1990; Paterson, 1992; Mayne *et al.*, 1992). We have measured the rate of deuterium–hydrogen (D–H) exchange with the amide protons of antigen/antibody complexes of horse cytochrome *c* and the three monoclonal antibodies E3, C3, and E8, and have used 2-D COSY NMR to identify those protons protected by complex formation. For monoclonal antibody E8, we found that the H-exchange rate of residues in three discontiguous regions of the cytochrome *c* polypeptide backbone is slowed by 7-to 340-fold in the antibody–antigen complex compared with free cytochrome *c* (see Table 11.2). The protected residues, 36–38, 59–67, and 100–102, and their H-bond acceptors, are brought together in the three-dimensional structure to form a contiguous, largely exposed pro-

412

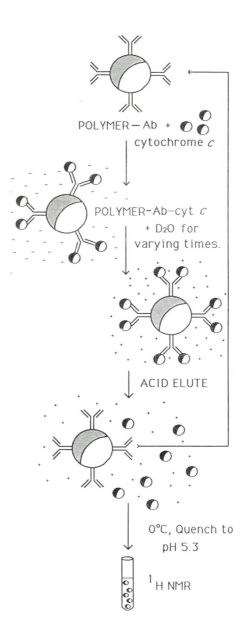

POLYMER — Ab + cytochrome *c*

POLYMER-Ab⋯cyt *c*
+ D₂O for
varying times.

ACID ELUTE

0°C, Quench to
pH 5.3

¹H NMR

Figure 11.4
Procedure for hydrogen exchange on antibody-bound cytochrome *c*. Affinity-purified
monoclonal antibody is coupled to a solid support. The polymer-bound antibody is
incubated with oxidized horse cytochrome *c* and then transferred to a column. Unbound
cytochrome *c* is removed by washing copiously, and exchange is initiated by washing the
column with D_2O. The antibody/antigen complex is incubated at 20°C in D_2O for time
intervals ranging from 1 hour to 11 days, and the cytochrome *c* is then eluted in a minimal
volume of 0.2 M acetic acid. The fractions are pooled and concentrated for NMR analysis.

Table 11.2
Effects on ferricytochrome *c* H-exchange induced by binding of the E8 monoclonal antibody[a].

Residue[b]	H-bond acceptor[c]	k_{free} (h^{-1})	k_{bound} (h^{-1})	Protection factor ($k_{free/bound}$)
Phe 36	H_2O	0.074	<0.001	>75
Gly 37	Trp 59	0.33	0.0028	120
Arg 38	Leu 35	1.44	0.0042	340
Trp 59	Arg 38	0.065	<0.0005	>130
Lys 60	Thr 63	0.018	<0.0003	>60
Leu 64	Lys 60	0.014	<0.0003	>50
Met 65	Glu 61	0.004	<0.0005	>10
Glu 66	Glu 62	0.62	0.073	8
Tyr 67	Thr 63	0.04	0.0058	7
Lys 100	Ala 96	0.27	0.0011	250
Ala 101	Tyr 97	0.41	0.0018	230

[a] Taken from Paterson *et al.*, 1990.

[b] Residues with protection factors greater than 3 are listed, except for Gln[12], which was originally thought to have a protection factor of 20. We now believe (Mayne *et al.*, 1992) this high protection factor is due to spectral overlap with Thr[102]. Residues with measurable exchange rates in both the free and the bound form, and with k_{free}/k_{bound} of 0.5 to 3, are as follows: Lys[7] (k_{free}/k_{bound} = 2.6), Lys[8] (1), Ile[9] (0.5), Val[11] (0.5), Lys[13] (07) Cys[14] (0.9), Ala[15] (0.6), His[18] (1.0), Thr[19] (0.9), Gly[29] (1.1), Leu[32] (0.9), His[33] (0.9), Gln[42] (∼3), Glu[69] (3.0), Asn[70] (1), Tyr[74] (1.8), Ile[75] (2.2), Ile[85] (1), Arg[91] (0.9), Glu[92] (0.7), and Asp[93] (1.1). The effect of antibody binding could not be determined for Phe[10], Leu[68], and Leu[94] through Lys[99] because their exchange rates were too slow to measure even in free cytochrome *c*. For the remaining 60 nonproline residues, exchange is too fast for 2-D NMR analysis under the conditions used in both free ferricytochrome *c* and the E8 complex.

[c] H-bond acceptors are the main-chain carbonyl O atoms except for Lys[60], the acceptor for which is the Oγ atom of the Thr side chain (Bushnell *et al.*, 1990).

tein surface. The interaction site defined in this way is consistent with prior epitope mapping studies on E8, and the size of the apparent epitopic surface ($\cong 750$ Å^2) is comparable to that determined by x-ray crystallographic analysis of other monoclonal antibodies complexed with small protein antigens. Figure 11.5 is a representation of the antigenic site recognized by E8 that was defined in this way (Paterson *et al.*, 1990).

More recent studies on two other anti–cytochrome *c* antibodies, E3 and C3, indicate somewhat different effects on the kinetics of H–D exchange on complex formation (Mayne *et al.*, 1992). They show that the binding of E3 and C3 antibodies to horse cytochrome *c* affects the hydrogen exchange behavior of regions not

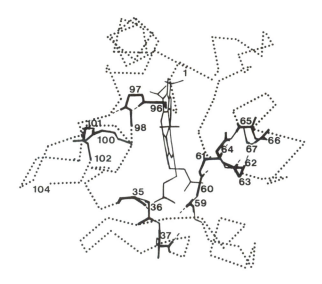

Figure 11.5
Computer graphics representation of the E8 epitope on horse cytochrome c determined by
H-exchange labeling (Paterson *et al.*, 1990). The alpha carbon backbone of horse cyt-c is
shown as a dotted line with the first and last (104) residues numbered. The main chain of
the residues identified and their H acceptors, which are also numbered, is shown as a solid
line.

only within the epitope, but also beyond the immediate binding site, apparently by
restricting local conformational flexibility. E3 and C3 appear to bind to a closely
overlapping surface on cytochrome c centered on the 37–59 Ω loop with some
contact with residues around 29–32 and 79–80. Some effects of binding appear to
spread beyond this area of direct contact to residues 18, 19, 60, and 64 which appear
to be slowed by restricting local unfolding motions. The two antibodies had been
shown previously to bind to a similar surface on cytochrome c, and this was
confirmed by the H–D exchange data in that a similar set of residues were slowed
by each antibody but by different rates (see Table 11.3). We have previously shown
that the largest protection factors that are measured in these experiments are
governed by the affinity of the interaction and occur when exchange can only take
place when the antigen is off the antibody (Paterson *et al.*, 1990). Although the
buried surface on cytochrome c is similar for both E3 and C3, the set of residues
with the maximum measurable protection factors within this area differ for the two
antibodies (Table 11.3). Thus, residues within the same surface can differ considerably in their ability to contribute to the "strength" of the binding in that they differ
in their ability to restrict motions that allow hydrogen exchange. This finding is
consistent with the notion of Novotny and co-workers (1989) of a "functional
epitope" within the buried surface which consists of a relatively small number of
residues within the interface that make the largest contributions to the stability of
the complex. Our findings also suggest that certain buried residues have sufficient

Table 11.3

Ratios of the D–H exchange rate constant for free, oxidized horse cytochrome *c* (free) to C3 and E3 bound cytochrome *c* (bound) for significantly protected[a] residues.

Residue	H-bond acceptor	$k_{free/bound}$ for C3	$k_{free/bound}$ for E3	E3/C3[b]
His 18	Ala 15	2.6	0.62	0.2
Thr 19	H_2O	30	3.2	0.1
Gly 29	Cys 17	55	68	1.2
Leu 32	Thr 19	14	50	3.6
His 33	Asn 31 (Nγ)	1.8	3.3	1.8
Arg 38	Leu 35	15	42	2.8
Gln 42	H_2O	11	18	1.6
Asn 52	Thr 49 (Oγ)	<2	11	>5
Trp 59	Arg 38	2.8	45	16
Lys 60	Thr 63 (Oγ)	2.9	14	4.8
Leu 64	Lys 60	1.9	32	17
Tyr 74	Asn 70	1.9	3.2	1.7
Ile 75	Pro 71	1.8	6.2	3.4
Lys 79	Heme O1δ	3	14	4.7
Met 80	Thr 78 (Oγ)	<6	>30	>5

[a] Significantly protected residues are considered to be those where $k_{free/bound} \geq 3$ or <1 for either antibody epitope. Taken from Mayne *et al.*, 1992.

[b] Values in this column represent the ratio of protection factors (i.e., $k_{free/bound}$) for each antibody.

access to solvent that exchange takes place even when antibody is bound to the antigen. This interesting finding indicates that even for high-affinity antibody/antigen complexes in solution, solvent may diffuse into the antigen/antibody interface.

In addition to providing evidence for local conformational flexibility in the cytochrome *c* molecule, these studies also address the issue of whether complex formation between antibody and antigen can result in conformational changes in either molecule. In the case of E8, which binds to a surface on cytochrome *c* which is relatively immobile, there appears to be little conformational reorientation of the antigen on antibody binding (Paterson *et al.*, 1990). E3 and C3, however, bind to a region, the 37–59 Ω loop, which has high segmental flexibility (Tainer *et al.*, 1985), and in so doing stabilize a network of hydrogen bonds which restricts the conformational flexibility of regions remote from the epitope. The differing effects of these antibodies on HX exchange kinetics can be accounted for by considering the local segmental flexibility of the surfaces of cytochrome c to which they bind.

To investigate how the differences in the binding of E3 and E8 to horse cytochrome c, determined using H–D exchange, influence the thermodynamics of the interactions, we determined the difference in configurational entropic costs on binding of E3 versus E8 to cytochrome c using isothermal titration calorimetry (ITC) (Murphy et $al.$, 1995). We show in this study that the cost in terms of configurational entropy for binding of cytochrome c to E3 is considerably greater than for E8, confirming that the anti–horse cytochrome c antibodies E8 and E3 have different configurational effects on the antigen on complex formation, as predicted by our NMR studies. We also determined the solvation entropy for the two interactions and made structural energetic calculations of the total burial area on complex formation for the antibodies E3 and E8. We found that about 1500 Å² of protein surface is buried on E8 binding to cytochrome c, and about 2330 Å² is buried for E3. The H–D exchange experiments indicate that E8 interacts with about 750 Å² of the cytochrome c surface. If the antibody surface is approximately the same size, then the total area buried is 1500 Å², as estimated from the thermodynamics of binding. Based on both the H–D exchange and other epitope mapping data (Paterson, 1989), we have estimated the size of the E3 epitope on cytochrome c to be about 1200 Å² (Mayne et $al.$, 1992), which is consistent with the thermodynamic calculations presented here, but is about half the total area in which H–D exchange is slowed by antibody binding.

Taken together, the findings of our NMR studies and thermodynamic measurements indicate that the area protected from H–D exchange on cytochrome c by E8 is that occluded directly by antibody binding, whereas binding of E3 to horse cytochrome c stabilizes the conformation of the molecule outside the immediate epitope buried by the antibody combining site.

F. Limitations and Assumptions in the Application of H–D Exchange to Antibody/Antigen Interactions

Deuterium exchange with the amide protons of a protein is measured as the reduction of peak height for a 1-D spectrum or a reduction of cross peak intensities in the J-correlated NH-C$_\alpha$H region of a 2-D spectrum. In addition to its use in examining antibody/antigen interactions, this technique has also been used with great success to define those amide protons which become protected against exchange during the refolding of horse cytochrome c (Wand et $al.$, 1986; Roder et $al.$, 1988). There are a number of limitations to applying this technique to studying antibody/antigen interactions. Clearly, for H-exchange labeling studies of refolding or antigen/antibody interactions to be undertaken, the complete assignment of the ^{1}H NMR resonances of the protein antigen must be measurable. This currently limits the technique to small proteins of molecular mass less than 20,000. For horse cytochrome c, MW $\cong$ 12,500, complete assignments had been made for both oxidation states (Wand and Englander, 1986; Wand et $al.$, 1989; Feng et $al.$, 1989), and the D–H exchange rates of the amide protons of native horse cytochrome c had also been measured (Wand et al., 1986). An increasing number of small proteins are being examined by 2-D NMR spectroscopy. Currently more than 100 small proteins have assigned NMR spectra, and the solution structure has been determine for about 50 of them (Wüthrich, 1986; Markley, 1989).

A further limitation in applying H–D exchange to epitope mapping is that not all amide protons in a protein have measurable H–D exchange rates. Many peptide hydrogens exchange too quickly to be measured in a COSY experiment, and those that are buried in the polypeptide fold are too slow. Those amide protons that are surface-exposed and, therefore, likely to be the target of antibody binding will tend to have measurable exchange rates only if they are involved in a hydrogen bond, which must transiently open for exchange to take place (Englander and Kallenbach, 1984; Englander and Mayne, 1992). In the case of cytochrome c, about two-thirds of the amide protons have measurable exchange rates in the free form of the molecule (Wand et al., 1989), and 12 to 15 residues showed slowed exchange rates on antibody binding (Paterson et al., 1990; Mayne et al., 1992). This situation may not always prevail for each antibody/antigen complex. For example, Benjamin and co-workers (1992) have examined a hen egg lysozyme/antibody complex of known x-ray structure (Sheriff et al., 1987). Only a total of five of the 23 residues buried by antibody binding in the crystal structure had measurable H–D exchange rates under the conditions used in their studies (Benjamin et al., 1992). Although the authors made the interesting observation—also made by ourselves for E3 and C3, vide supra—that some residues outside the boundary of the epitope show slowed amide proton exchange when bound to antibody, very little information about the lysozyme/antibody interaction per se was forthcoming. Under such circumstances, the exchange conditions may need to be adjusted with respect to pH and temperature to maximize the number of surface amide protons to be measured and to render measurable some of the faster or slower exchanging amide protons of interest.

For H–D exchange epitope mapping to be successful, clearly both antigen and antibody must remain conformationally intact during the exchange process, during disruption of the complex, and in the conditions under which the antigen is concentrated and the COSY spectrum is measured. The conditions which were successful for the cytochrome c complexes may require some adjustments for protein antigens which are very unstable at low temperatures or acid pH.

Interpretation of H–D exchange data may also require the support of other epitope mapping techniques to determine which protected residues are directly interacting with the antibody combining site and which are influenced by cooperative effects (Mayne et al., 1992). If the affinity constant of the interaction is known, it is possible to calculate the maximum protection factor for the interaction, which is when exchange can take place only when the antigen is off the antibody (Paterson et al., 1990). If the purpose for mapping the epitope is to identify residues for mutagenesis, to improve the affinity of the interaction, then the most promising targets will be those residues with peptide protons that have the largest protection factors.

Despite these limitations, our studies using the H-exchange labeling approach have demonstrated that it can be used to map binding sites on small proteins in antibody–antigen complexes (Paterson et al., 1990) and to explore the effect of antibody binding on local structural flexibility of the protein antigen (Mayne *et al.*, 1992). We anticipate that this approach will find increasing use in studies on the interaction of small proteins with antibody and to protein–protein, protein–ligand, and protein–DNA interactions in general.

G. Conformational Requirements for Antibody/Antigen Interactions

Antibody binding sites on proteins have long been categorized into two different types, originally designated sequential (also called linear or continuous) and conformational (also called topographic) (Sela *et al.*, 1967) depending on the ability of the antibodies to bind peptide fragments of the protein antigen. It was on this premise that the use of synthetic peptides to probe the antigenicity of a protein was first conceived more than 20 years ago (Fujio *et al.*, 1968; Arnon *et al.*, 1971). Until recently, it was assumed that most antibodies that bind to peptides interact with sequential antigenic sites on proteins that consist of residues located on a contiguous region of the polypeptide chain. Conformational determinants were thought to be composed of regions of the protein that are remote in sequence but topographically close, and that could be detected only by methods such as fine-specificity analyses, which require that the antigen remain in its native conformation.

The structural analyses of antigen–monoclonal antibody interactions we discussed earlier have confirmed the wide acceptance in the last few years that most epitopes on protein antigens are of the conformational or discontiguous variety (Benjamin, 1991; Jemmerson and Paterson, 1986b; Paterson, 1991). Why, however, do some anti-protein antibodies appear to interact with peptides, whereas others do not? Findings from our laboratory with cytochrome *c* (Cooper and Paterson, 1987) and theoretical considerations (Barlow *et al.*, 1986) suggest that the distinction between contiguous and discontiguous determinants is artificial, and that even so-called contiguous determinants, detected by peptide reactivity, are only part of a larger, discontiguous protein epitope. Taking the surface area of a protein antigen that interacts with antibody to be about the size of the buried surface of one of the crystallographic structures of a lysozyme/antibody complex (Amit *et al.*, 1986), Barlow and co-workers (1986) showed that no protein epitope could consist of a single segment of the polypeptide backbone.

If all epitopes are to some extent discontiguous, then the peptide fragments of a protein which can cross-react with anti-protein antibodies must represent only part of a larger surface determinant. We have confirmed that this is exactly the case for a low-affinity population of anti-self cytochrome *c* antibodies that bound two peptide fragments of cytochrome *c*, one from the carboxy-terminal region of cytochrome *c* (sequence 81–104) and one from the amino-terminal region of the molecule. These two regions of cytochrome *c* are α-helical and are folded in space to form a single contiguous region of the molecular surface (see Figure 11.6). It seemed possible, therefore, that the antibodies were interacting with two peptide fragments from the same epitope. We isolated the population of antibodies that bound the 81–104 fragment by affinity chromatography, using this peptide covalently coupled to a solid support, and showed that these antibodies could also bind the amino-terminal peptide. This observation demonstrated that the definition of protein epitopes as

Figure 11.6

Computer graphics representation of a discontiguous determinant on rabbit cytochrome *c* determined by peptide binding (Cooper *et al.*, 1989). The full backbone of the peptide sequences 81–104 and 1–9 is shown; the rest of the molecule is represented only by the α-carbon backbone.

sequential or conformational is largely operational and not structural. The question still remains, however, as to why some antibodies will cross-react with partial epitopes represented by peptide fragments, and some will not.

In the case just described, peptide sequences which were part of the antigenic site on cytochrome *c* were able to bind the antibodies in isolation from the native protein. On the other hand, we and many other workers have experienced difficulty in mapping the binding sites of some antibodies using synthetic peptides. Thus, in many studies, peptides fashioned to mimic antigenic sites, previously delineated by other methods, have failed to bind both monoclonal antibodies and polyclonal antisera (Benjamin *et al.*, 1984; Berzofsky, 1985; Jemmerson and Paterson, 1986b; Savoca *et al.*, 1991). This may be because the antibodies used in these studies possessed high affinities for the native antigen. For example, for the high-affinity monoclonal antibodies to horse cytochrome *c*, E3, E8, C3, and C7, described earlier, we can detect no binding to a range of peptide fragments that includes the residues shown in Table 11.1, even when using methods that detect interactions of very low affinity (Carbone and Paterson, 1985; Jemmerson and Paterson, 1986b).

Studies in which peptide reactivities have been successful in mapping antibody binding sites on proteins have often used solid-phase assays where either the peptide antigen is attached to a micro-titer plate (Jemmerson *et al.*, 1985; Cooper and Paterson, 1987) or the peptide is both synthesized and assayed on the same solid support (Leach, 1983; Geysen *et al.*, 1984, 1987; Paterson, 1985; Savoca *et al.*, 1991). Antibodies which have an affinity for an antigen too low to be detected in solution assays will frequently bind antigen in solid-phase assays, where affinity limitations are overcome by increases in avidity due to multisite attachment onto a surface. Our studies with self responses to cytochrome *c* (Cooper and Paterson, 1987; Cooper *et al.*, 1989) indicate that, at least in some instances, antibodies with the ability to bind peptides have low affinity for the protein to which they were raised, whereas antibodies that will not bind peptide fragments are of high affinity. Clearly, high-affinity interactions involve many more and closer residue contacts between antibody and antigen than low-affinity interactions. Such complementarity is only likely to be met by the tertiary fold of the intact sequence. Low-affinity interactions, however, may involve fewer contact residues and may be more able to tolerate the conformational flexibility of a peptide antigen.

The majority of studies with high-affinity monoclonal antibodies to a variety of model protein antigens (Lavoie *et al.*, 1989; Berzofsky, 1985; Paterson, 1989) indicate a stringent requirement for native conformation of the antigen for antibody binding. The inability of peptide fragments to provide this high degree of complementarity between epitope and paratope required for binding is unremarkable when compared to the lack of reactivity of E8, E3, C3, and C7 with three protein-engineered antigens (Collawn *et al.*, 1988) (Table 11.4) which were synthesized by Wallace and co-workers (Wallace, 1984; Proudfoot *et al.*, 1986) by methods described in Chapter 21 of this volume. The analogs 1–37:38–104 and 1–38:39–104 represent intact cytochrome *c* with the peptide bond missing on either side of the arginine residue at position 38. 1–39–56–104 is a covalent complex with the 40–55 loop region deleted. CD spectra of these structures indicated few differences from that of native, unmodified, cytochrome *c* (see Table 11.4). However, recognition of the analogs was almost totally abrogated for all of the monoclonal antibodies, presumably as a result of subtle conformational disturbances invoked by these modifications over a wide range of the antigenic surface. In addition, simply breaking one bond in the polypeptide chain of the antigen at 37:38 or 38:39 is sufficient to drastically reduce recognition by all the monoclonal antibodies. These results explain why peptide probes with presumably less native structure than these noncovalent complexes of intact cytochrome *c* are often ineffective in defining the antigenic sites for high-affinity monoclonal antibodies of this type.

The analog 1–39–56–104 was prepared to examine the importance of the Ω-loop 40–55 in the folding of cytochrome *c*. Ω-loops have been suggested to be independent modules of folding (Rose *et al.*, 1985). In the cytochrome *c* molecule, the 40–55 loop is associated with the wider loop region containing residue 26 and with the heme group by a network of hydrogen bonds. Although the elimination of the sequence 40–55 gives an analog that has retained some functional activity and a CD spectrum similar to that of the native protein (see Table 11.4), none of the monoclonal antibodies binds to this analog. This result would be anticipated for antibodies E3, C3, and C7, which are known to directly engage this region of cytochrome *c* (Mayne *et al.*, 1992), but is unexpected for E8. These studies indicate that monoclonal antibodies can be more sensitive probes of conformational changes than spectroscopic techniques such as circular dichroism.

Table 11.4
Conformational changes in protein engineered cytochrome *c* detected by monoclonal antibodies to the native protein[a].

	Cytochrome *c* analog		
	1–37:38–104	1–38:39–104	1–39–56–104
Monoclonal antibody binding to analog as % of native cytochrome *c*:			
C3	-0.7 ± 5.9	3.3 ± 7.6	1.0 ± 7.1
C7	12.0 ± 5.9	13.7 ± 4.9	14.7 ± 6.1
E3	6.3 ± 8.3	-5.7 ± 8.7	5.7 ± 9.5
E8	-1.0 ± 2.1	-1.0 ± 2.1	-2.7 ± 2.1
Secondary structure content[b] of analog as % of native cytochrome *c*:			
α-helix	81	100	89
β-bends	113	70	78
Other	110	113	123
Biological activity[c] as % of native cytochrome *c*:			
	55–70	20–30	13–17

[a] Data taken from Collawn *et al.*, (1988).
[b] Calculated from circular dichroic spectra as described in Collawn *et al.* (1988).
[c] O_2 uptake by a succinate oxidase system using each of these analogs to promote electron transfer was measured in a depleted mitochondrian assay. These data were taken from Wallace and Corthésy (1986) and Proudfoot *et al.* (1986).

III. Cellular Immune Responses to Protein Antigens

Differences between the determinants on proteins that are recognized by T and B lymphocytes are directly related to differences in the recognition event. Whereas the immunoglobulin receptor on B cells interacts with the intact protein molecule through a binary mode of recognition involving the docking of protein surfaces, T cells recognize peptide fragments from protein antigens in association with gene products of the major histocompatability complex (MHC) on the surface of an antigen-presenting cell (APC) (Figure 11.1). These "presenting" molecules are hetero-dimeric, membrane-bound glycoproteins, are members of the immunoglobulin supergene family, and fall into two categories. The class I molecules are involved in cytotoxic T-cell recognition and consist of one membrane-bound polypeptide chain (the α chain) that has three domains (M_r 45 kDa). The domains in the α chain are numbered from the amino-terminal, membrane distal end of the polypeptide chain.

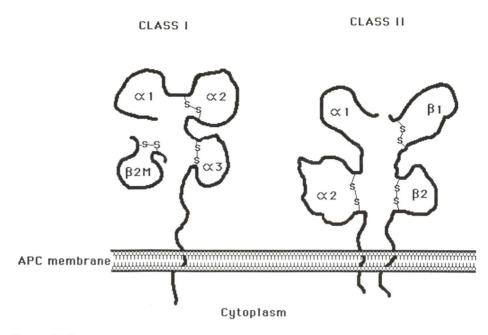

Figure 11.7
A schematic representation of the structure of the two classes of heterodimeric, membrane-bound restriction elements of the MHC.

The membrane proximal α_3 domain is associated with the non–membrane-bound, single-chain protein β_2 microglobulin (M_r 12 kDa). The class II molecules consist of two polypeptide chains (the α and β chains), both of which extend through the membrane. The α chain is heavier (M_r 35 kDa) than the β chain (M_r 28 kDa) because it has extra glycosylation sites, but both chains consist of two domains; α_1 and β_1 are membrane-distal, and α_2 and β_2 are membrane-proximal. A schematic diagram of their structure is shown in Figure 11.7. The class II molecules are the presenting molecules of antigenic peptides for helper T-cell recognition (Figure 11.1).

The MHC complex encodes a large number of closely related alleles of both the class I and class II molecules. Thus, any particular outbred animal will express a number of these highly polymorphic gene products. The polymorphic regions of these polypeptides are concentrated in the membrane-distal domains of the protein chains, in the α_1 and β_1 domains for class II molecules and the α_1 and α_2 domains for class I molecules. Only a restricted number of MHC molecules is capable of presenting any particular antigen. This limit imposes a genetic restriction on the T-cell response to protein antigens in inbred animals which express only one or two of these molecules (Schwartz, 1985). However, to ensure effective cellular immunity to any protein antigen, MHC molecules have evolved so that each polymorphic variant can associate with peptide fragments from protein antigens from widely different sources. Clearly, the immune competence of any species which exhibits a wide polymorphism is improved.

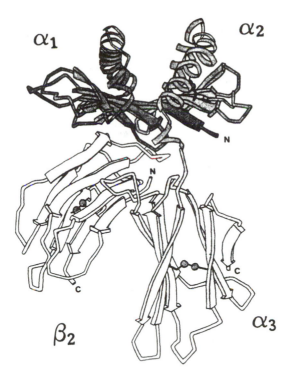

Figure 11.8
A ribbon model of the MHC Class I molecule HLA-A2. The polymorphic α_1 and α_2 domains fold to form a cleft into which peptides derived by antigen processing are believed to bind for presentation to cytotoxic T cells. From Bjorkman *et al.* (1987a), with permission.

The three-dimensional structures of three MHC class I molecules, and one MHC class II molecule, are known. Two of these are of closely related products of the human class I locus (Figure 11.8) (Bjorkman *et al.*, 1987a; Garrett *et al.*, 1989). A large groove (25 Å long, 10 Å wide and 11 Å deep) at the top of the molecule comprising both the α_1 and α_2 domains of the Class I molecule was identified as the binding site for processed peptides (Bjorkman *et al.*, 1987a,b). By comparing these two closely related MHC Class I alleles, six peptide-binding pockets designated A to F were located within the antigen-binding cleft (Garrett *et al.*, 1989). Differences in the size and depth of these pockets allows for diversity in the peptide residues which can be accommodated by various class I MHC molecules. More recently, the x-ray crystal structures of several Class I MHC/peptide complexes with a homogeneous peptide bound in the cleft (Fremont *et al.*, 1992; Matsumura *et al.*, 1992; Zhang *et al.*, 1992; Madden *et al.*, 1991, 1992) have provided a clearer understanding of how the peptide backbone and side chains interact with class I MHC residues within the antigen-binding cleft. Peptides are bound to the MHC cleft in an extended configuration and appear to be "tethered" within the cleft by interactions between the amino-terminus of the peptide and pocket A and by binding of the carboxyl-terminus to pocket F of the peptide-binding cleft, allowing the central residues to bulge out of the cleft where they are exposed to solvent or TCR.

With very few exceptions (Lee *et al.*, 1988), it is well established that the protein antigen is degraded into peptide fragments by intracellular proteolysis within the APC prior to the T-cell recognition event (Shimonkevitz *et al.*, 1983; Chesnut *et al.*, 1982; Ziegler and Unanue, 1982; Unanue, 1984). The cellular compartments in which antigen degradation and association with the Class I or Class II molecules take place are thought to differ between the two classes of MHC molecules (Braciale *et al.*, 1987); these differences have important consequences for cytotoxic versus helper T-cell recognition of protein antigens. In both cases, however, the signaling event involves the formation of a ternary complex among the MHC molecule, the T-cell receptor, and a peptide fragment derived from the intact protein antigen.

A major focus of studies on the T-cell recognition of protein antigens is to understand the molecular features underlying the dual T-cell specificity for peptide and the MHC restriction element. Studies on the molecular features of this ternary complex are limited because, unlike the B-cell receptor, which is available in secreted form as antibody, two of the interacting species are membrane-bound molecules. The situation is further complicated by the involvement of other cell surface molecules on the T cell which serve to stabilize the T cell/APC complex and increase the functional affinity of the T cell/peptide/MHC interaction. These nonpolymorphic cell-adhesion molecules are also members of the immunoglobulin supergene family, and a number of them may be important in the activation of T cells (reviewed by Shevach, 1989). However, two of these accessory molecules help distinguish between the helper and cytotoxic T-cell subsets (Cantor and Boyse, 1975a,b) and are known to be directly involved in stabilizing the association of the TCR with the peptide/MHC ligand by direct interaction with either the class I or the class II molecule. Thus, helper T cells which are class II restricted express a molecule called L3T4 in mice and CD4 in humans, and cytotoxic class I restricted T cells use a molecule called Lyt-2/3 in mice and CD8 in humans (Swain, 1983).

Despite the difficulties in elucidating the molecular features of T-cell recognition of antigen imposed by the cellular nature of the interaction, T-cell studies are facilitated by the fact that the T-cell epitope is almost always a small peptide. This fact has allowed the application of the powerful techniques of peptide chemistry and conformational analysis to the assessment of the structure of T-cell epitopes. This approach has been used extensively for class II restricted T helper cell recognition for a number of years. In addition, the T helper cell recognition of many model protein antigens, such as myoglobin (Berzofsky, 1985; Berkower *et al.*, 1986; Kurata and Berzofsky, 1990), lysozyme (Manca *et al.*, 1984; Gammon *et al.*, 1987; Allen *et al.*, 1984, 1987a), and pigeon cytochrome *c* (Bhayani *et al.*, 1988; Carbone *et al.*, 1987a,b; Fox *et al.*, 1987, 1988; Hansburg *et al.*, 1981, 1983; Hedrick *et al.*, 1982; Heber-Katz *et al.*, 1982; Schwartz *et al.*, 1985; Sorger *et al.*, 1987) has been well characterized in terms of the genetic restriction imposed by the MHC molecule and the primary sequence of the minimal determinant.

A. Helper T-Cell Response to Cytochrome *c*

The study of the immunological recognition of the cytochromes *c*, particularly pigeon cytochrome *c*, has played a critical role in elucidating the specificity of the T-cell receptor for protein antigens (reviewed by Hedrick, 1988) and the genetic restrictions imposed by the proteins encoded by the MHC class II gene complex on recognition by helper T cells (reviewed by Schwartz, 1985). This review will concentrate on recent studies on the T-cell recognition of cytochrome *c* to determine the physical and chemical nature of T-cell determinants derived from protein antigens (Baumhüter *et al.*, 1987; Bhayani and Paterson, 1989; Carbone *et al.*, 1987a,b; Collawn *et al.*, 1989; Lakey *et al.*, 1986; Paterson, 1989; Sorger *et al.*, 1987, 1990; Vita *et al.*, 1990). The dominant T-cell epitopes on the protein backbone of several species of cytochrome *c* have been identified, including cow (Corradin and Chiller, 1979, 1981; Corradin *et al.*, 1981), horse (Baumhüter *et al.*, 1987), and pigeon and moth (Hansburg *et al.*, 1983; Solinger *et al.*, 1979; Ultee *et al.*, 1980).

Corradin and co-workers have shown that distinct I-A^b restricted T-cell epitopes lie within the regions 1–39 and 39–53 of apo-horse cytochrome *c* (Baumhüter *et al.*, 1987). A great deal of heterogeneity was displayed by the T-cell clones, which recognized multiple antigenic determinants within these regions even though the peptide sequence was as short as 48–53. Multiple sites within a short peptide sequence that are recognized by diverse T cell clones have also been observed for other protein antigens. For example, the I-A^k restricted determinants 24–45 (Lambert and Unanue, 1989) and 74–86 (Shastri *et al.*, 1985) from hen egg lysozyme; the I-A^b restricted determinant 81–96, also from hen egg lysozyme (Shastri *et al.*, 1985); and the myoglobin peptide 102–118, which contains multiple T-cell determinants for I-A^d, I-A^k, and I-A^s restricted T-cell clones (Cease *et al.*, 1986; Brett *et al.*, 1989) have been found. We discuss later, with reference to the recognition of pigeon cytochrome *c*, how such a short peptide might create multiple antigenic determinants when bound to a single MHC class II molecule.

In the case of pigeon and moth cytochromes *c*, the strongest T-cell response is seen in mice, which express the MHC class II heterodimer, designated I-E^k (Schwartz, 1985). The entire response was shown in early studies to be directed to a cyanogen bromide fragment from the carboxy-terminus of the molecule that is created by cleaving the polypeptide chain at Met^{80}. Within this sequence, there are four amino acid residue differences between pigeon and mouse (self) cytochrome *c*, and there is strong homology between the pigeon sequence, p81–104, and the moth sequence, m81–103 (see Figure 11.2). In addition to this I-E^k restricted immunodominant T-cell epitope, a much weaker response to pigeon cytochrome *c* is seen in mice that express the I-A^b class II molecule. This response is directed to residues 43–58 (Suzuki and Schwartz, 1986; Ogasawara *et al.*, 1989).

The I-E^k restricted pigeon cytochrome *c* immune response results in the generation of four phenotypic T helper cell specificities (Table 11.5) that were distinguished on the basis of their ability to recognize p81–104 and/or m81–103 in the context of three closely related I-E^k haplotypes which have the same α chain but different β

Table 11.5

The definition of functional phenotypes for pigeon cytochrome c–specific T-cell hybridomas.

Functional phenotypes	Cytochromes c that can be presented to them by MHC class II molecules		
	$E_\alpha^k E_\beta^k$	$E_\alpha^k E_\beta^b$	$E_\alpha^k E_\beta^s$
I	P, M[a]	M	None
II[b]	P, M	M	None
III	P, M	None	P
IV	P	P	None

[a] P = pigeon cytochrome c, M = moth cytochrome c.

[b] T-cell hybridomas with this phenotype also recognize the class II MHC molecule $A_\alpha^s A_\beta^s$ in the absence of any antigen.

chains. The sequence differences among these β chains are shown in Figure 11.9; it can be seen that there are few differences among the three polypeptides E_β^k, E_β^b, and E_β^s in the domain β_1, which is thought to be involved in peptide binding (Brown *et al.*, 1988, 1993). However, E_β^d, which cannot present p88–104 or m88–103, has a wide number of sequence differences. Type I and Type II specificities recognize p81–104 in the context of the $E_\alpha^k E_\beta^k$ only and m81–103 on both $E_\alpha^k E_\beta^k$ and $E_\alpha^k E_\beta^b$. The Type II clones also show autoreactivity to $A_\alpha^s A_\beta^s$ molecules in the absence of antigen. The Type III T-cell specificity can recognize p81–104 in the context of $E_\alpha^k E_\beta^s$ in addition to the $E_\alpha^k E_\beta^k$ molecule, and m81–103 only on the $E_\alpha^k E_\beta^k$ molecule. The Type IV specificity can recognize p81–104 on APCs expressing either the $E_\alpha^k E_\beta^b$ or the $E_\alpha^k E_\beta^k$ molecule and does not recognize m81–103 at all (Carbone *et al.*, 1987b; Sorger *et al.*, 1987). In general, the major phenotype expressed is of the Type I variety, which accounts for 80–85% of the T-cell specificities observed in the I-E^k restricted response to pigeon cytochrome c (Fink *et al.*, 1986). Slight differences in the four major patterns of response may be observed from T-cell clone to T-cell clone, particularly with regard to the antigen concentration required to achieve maximal stimulation in the context of each class II molecule (Fink *et al.*, 1986; Carbone *et al.*, 1987b; Sorger *et al.*, 1987). These differences are probably due to differences in the functional affinity of the T-cell receptor for the peptide/MHC ligand. Such differences have been demonstrated using anti-L3T4 and anti–class II monoclonal antibodies for a Type I pigeon cytochrome c–specific T-cell hybridoma, which had a much stronger response to m81–103 than to p81–104 (Lakey *et al.*, 1986). Thus, although both anti–I-E^k and anti-L3T4 antibodies blocked the T-cell response of this hybridoma to p81–104, the anti-L3T4 antibody did not block the response to m81–103, and complete blocking of the recognition

```
        6            13                                    29    32
Eᵏ      P W F L E Y C K S E C H F Y N G T Q R V R L L V R Y F

Eᵇ      L - - - - - - - - - - - - - - - - - - - - - - E - - -

Eˢ      ? - - - - - S T - - - - - - - - - - - - - - E - - -

Eᵈ      - R - - - - V T - - - - - - - - - H - - F - E - F I


        69       72           79              87          93
Eᵏ      E F L E Q K R A E V D T V C R H N Y E I F D N F L V P

Eᵇ      - - - - - - - - - - - - - - - - - - - - S - K - - - R

Eˢ      - - - - R - - A - - - Y - - - - - - - - L - K - - - -

Eᵈ      - I - - D A - - S - - - Y - - - - - - - S - L - - - R
```

Figure 11.9
Amino acid sequences of hypervariable regions of the first domain of the Class II MHC beta chains, Eᵏ, Eᵇ, Eˢ, and Eᵈ.

of this peptide by anti–I-Eᵏ antibodies could not be demonstrated at the concentrations of antibodies used. These studies implied that the T-cell hybridoma's requirement for the interaction of L3T4 with the class II molecule was less stringent for the more potent antigen and demonstrate the difficulties in determining the affinities of cell-cell interactions.

The ability of antigens to be recognized by T cells is usually determined in most studies of this type by the level of lymphokine released (most commonly the interleukin, IL-2) by the T cell at different antigen concentrations. The amount of IL-2 released is measured by uptake of ^{3}H-thymidine by a T-cell tumor line which is dependent on this interleukin for growth (Gillis and Smith, 1977; Gillis et al., 1978). Figure 11.10 shows the antigen-dependent release of IL-2 for Type I, III, and IV T-cell hybridomas in response to pigeon and moth cytochrome c peptides in association with I-$E_\alpha^k E_\beta^k$, I-$E_\alpha^k E_\beta^b$, and I-$E_\alpha^k E_\beta^s$.

In later studies using synthetic peptides, it was discovered that the minimal determinant which contained all the residues necessary for T-cell recognition within the cyanogen bromide fragment 81–104 was the sequence 95–104 (Schwartz et al., 1985; Carbone et al., 1987a; Bhayani et al., 1988). This sequence differs from the moth sequence by only the absence of an alanine residue at position 103 (see Figure 11.2). We then took advantage of the ability of three of the different T-cell phenotypes specific for p81–104 to recognize the same peptide on closely related class II molecules to distinguish residues in the sequence 95–104 that interact with the class II MHC molecule from those which interact with the TCR (Bhayani and Paterson, 1989). We compared the ability of the singly substituted p95–104 analogs (Figure

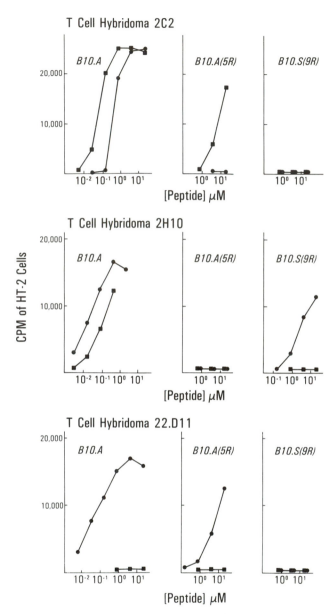

Figure 11.10
Antigen specificity and MHC presentation for T-cell hybridomas 2C2 (phenotype I), 2H10 (phenotype III) and 22D11 (phenotype IV). The hybridomas were co-cultured with varying concentrations of p81–104 (●), or m81–103 (■), and splenic APC bearing I-$E_\alpha^k E_\beta^k$ (from B10.A mice), I-$E_\alpha^k E_\beta^b$ (from B10.A[5R] mice), or I-$E_\alpha^k E_\beta^s$ (from B10.S[9R] mice). The IL-2 released in the culture supernatant was measured as the uptake of ^{3}H-thymidine by the IL-2 dependent cell line HT-2.

```
88              95              104        88              95              104

K A E R A D L I A Y L K Q A T A K          K A E R A D L I A Y L K Q A T A K

- - - - - - - - - - - - - - - - R          - - - - - - - - - - - A - - - - K

- - - - - - - - - - - - - - - A A          - - - - - - - - - - R - - - - K

- - - - - - - - - - - - - - K A            - - - - - - - - - - A - - - - - K

- - - - - - - - - - - - - K K              - - - - - - - - - F - - - - - K

- - - - - - - - - - - - - - I K            - - - - - - - - - A - - - - - - K

- - - - - - - - - - - - - A - K            - - - - - - - - - I - - - - - - K

- - - - - - - - - - - - S - K              - - - - - - - - Y - - - - - - - K

- - - - - - - - - - - I - - K              - - - - - - - - I - - - - - - - K

- - - - - - - - - - - Y - - K              - - - - - - - A - - - - - - - - K

- - - - - - - - - - A - - - K              - - - - - - Y - - - - - - - - - K

- - - - - - - - - - N - - - K
```

Figure 11.11

Single-residue peptide analogs of peptide 88–104 from pigeon cytochrome *c* used to determine the residues in the pigeon cytochrome *c* determinant 95–104 which are important for T cell recognition. From Bhayani and Paterson (1989), with permission. Similar analogs of the moth peptide 88–103 in which Ala^{103} is deleted were also used.

11.11) and a similar set of m95–103 analogs to stimulate Type I, III, and IV clones on APCs bearing one of the three closely related I-E^k haplotypes. We reasoned that differences in the ability of the peptides to be recognized on two different MHC implied that the substituted residue was important for binding to the MHC molecule. Similar experiments have been performed by Schwartz and colleagues to establish that the heteroclitic response to moth cytochrome *c* shown by Type I specificities is due to a change at residue 103 which enhances the ability of the peptide to bind to I-E^k (Matis *et al.*, 1983).

We also examined which residues within the sequence 95–104 were important for direct recognition by the T-cell receptor. Any peptide analog which did not stimulate the pigeon cytochrome *c*–specific T-cell clone was used in blocking assays to determine whether it could prevent the recognition of the native sequence. If it did, we reasoned that the amino acid substitution in the analog was not hindering the ability of the peptide to bind to the class II presenting molecule, and that the lack of potency of the peptide was due to the direct recognition of the residue substitution by the T-cell receptor.

A compilation of our identifications for the three specificities is shown in Figure 11.12. It can be seen not only that almost every residue is involved in some way in the recognition process for all specificities, but that differing sets of residues are used in stabilizing the peptide/MHC complex in each case. On examining the segmental distributions of residues used by each hybridoma to interact with the MHC molecule versus the T-cell receptor (Figure 11.12), it is difficult to imagine how any single conformation of the peptide could generate them. We believe, therefore, that p88–104 can interact with the MHC in different orientations, thus creating multiple antigenic determinants, and that those T-cell specificities are selected which recognize the particular array of residues exposed. That the Type IV specificity is rare in the T-cell response to pigeon cytochrome c (Carbone et $al.$, 1987b; Sorger et $al.$, 1987) ($<1\%$ of the total response) may simply reflect the distribution of the particular peptide conformation that generates it within the ensemble of conformations that p88–104 can adopt when it interacts with I-E^k. Thus, a single class II MHC molecule appears to be able to present the same peptide in different configurations depending upon the TCR that makes up the ternary complex. Our results suggest that a single peptide may generate diversity in the T-cell response by virtue of its conformational flexibility within the TCR–MHC–antigen complex (Bhayani and Paterson, 1989). This conclusion has been extended to a wider variety of T-cell clones to pigeon cytochrome c (Sorger et $al.$, 1990) and to the helper T-cell recognition of two other peptide/class II MHC antigenic systems (Kurata and Berzofsky, 1990; Boyer et $al.$, 1990).

B. Cytochrome c Peptide/MHC Binding

The functional competition assays we have described demonstrate direct peptide/class II binding in the presence of a T-cell receptor. However, competition assays of this type are highly influenced by the relative affinities of the stimulatory and blocking antigens. In studies by other laboratories using different antigenic systems (Babbit et $al.$, 1986; Guillet et $al.$, 1987), it was necessary to use weakly stimulating analogs of the native antigen to demonstrate blocking by nonstimulatory peptides. Thus, for antigenic peptide analogs of low affinity for class II, it may be difficult to observe blocking of presentation to peptide-specific T cells.

A direct approach to examine the ability of antigens to bind to class II molecules that has been pioneered by the laboratories of Unanue, Grey, McConnell, and other workers is to examine directly the binding of peptide antigens to isolated class II molecules in $vitro$ either in detergent solution (Buus et $al.$, 1986a,b, 1987; Allen et $al.$, 1984, 1987a) or reconstituted in planar membranes (Watts et $al.$, 1986; Watts and McConnell, 1987; Sadegh-Nasseri and McConnell, 1989). Applying these techniques to the pigeon cytochrome c peptide/I-E^k system, direct binding of p81–104 and m81–103 has been demonstrated in cold competition assays where the capacity of peptides to inhibit the binding of ^{125}I-labeled peptide to detergent-solubilized I-E^k is measured (Buus et $al.$, 1987). In addition, peptides from other protein antigens known to bind I-E^k will inhibit the binding of p88–104 to I-E^k (Guillet et $al.$, 1987; Buus et $al.$, 1987). In a wide variety of antigenic systems, a clear correlation exists between the ability of a T-cell antigenic peptide to bind to a member of the

Pigeon cyt c as recognized by 22D11.

9 5	9 6	9 7	9 8	9 9	1 0 0	1 0 1	1 0 2	1 0 3	1 0 4
U	U	E	A	E	A	E	E / A	E	E

Pigeon cyt c as recognized by 2H10.

9 5	9 6	9 7	9 8	9 9	1 0 0	1 0 1	1 0 2	1 0 3	1 0 4
A	U	E/A	A	U	U	E/A	U	E	U

Moth cyt c as recognized by 2H10.

9 6	9 7	9 8	9 9	1 0 0	1 0 1	1 0 2	1 0 3
U	E	U	U	U	U	E	U

Pigeon cyt c as recognized by 2B4.

9 5	9 6	9 7	9 8	9 9	1 0 0	1 0 1	1 0 2	1 0 3	1 0 4
U	U	E	-	E	U	E	U	-	E

Moth cyt c as recognized by 2B4.

9 6	9 7	9 8	9 9	1 0 0	1 0 1	1 0 2	1 0 3
-	U	A	U	-	E	U	E / A

Figure 11.12

A summary of TCR and MHC contact residues used by T-cell hybridomas 22.D11, 2H10 and 2B4 (phenotype I) to interact with the pigeon cytochrome *c* sequence 95–104 and the moth cytochrome *c* sequence 95–103 bound to I-E^k. E = TCR contact residue, A = MHC contact residue, U = insufficient data to assign a function to this residue, and - = no effect on T-cell recognition. From Bhayani and Paterson (1989), with permission.

class II polymorphic family *in vitro* and the ability of mice expressing that particular MHC class II molecule to respond to protein antigens containing that peptide.

Direct binding studies by equilibrium dialysis, using detergent-solubilized MHC class II molecules, have indicated small association constants (≤ 1 μM) for complex formation between MHC and peptides (Babbitt *et al.*, 1985; Buus *et al.*, 1986a). However, there is also evidence that the complexes have slow association rates and, once formed, are long-lived (Buus *et al.*, 1986b; Watts and McConnell, 1986). Indeed, such complexes may be isolated and inserted into artificial planar membranes which will stimulate T-cell hybridomas (Watts and McConnell, 1986). These *in vitro* data, however, contrast with *in vivo* observations using flow cytometry to measure the increase of intracellular calcium concentrations in T cells triggered by contact with APC pulsed with an antigenic peptide from tetanus toxoid (Roosneck *et al.*, 1988). In these studies the formation of MHC–peptide complexes was too rapid to be observable within the three minutes required to perform the experiment.

In vitro determinations of peptide/MHC kinetics ignore the collision limit that governs the bimolecular reaction kinetics between ligands and membrane-bound receptors (Abbot and Nelsestuen, 1988; Sargent and Schwyzer, 1986). Thus, the first step involved in the interaction between a peptide and receptor involves the diffusion necessary to bring the reactants into close proximity and into the proper orientation for association to take place. In the solution phase, which pertains when a membrane receptor is solubilized by detergent, this limit is determined by Brownian motion. However, for cell membrane-bound receptors which are often at high receptor density, a membrane surface may become uniformly reactive so that nearly every collision results in binding. An explanation for the divergent observations *in vitro* and *in vivo* is provided by an elegant kinetic study using the p81–104 peptide and I-E^k molecules reconstituted in planar membranes (Sadegh-Nasseri and McConnell, 1989). The kinetics of association between peptide and I-E^k in this system indicated the rapid formation of an intermediate complex which is short-lived, followed by the formation of a long-lived terminal complex. Thus, the conflict between *in vivo* and *in vitro* measurements might be resolved by supposing that both the intermediate and the long-lived complex may be recognized by T cells. The presence of the TCR *in vivo* may catalyze the formation of the long-lived complex and shorten the transition time between this and the intermediate complex. Our finding (Bhayani and Paterson, 1989) that the same peptide, p88–104, may bind an MHC class II molecule in different conformations for recognition by different T-cell clones is consistent with a requirement for the TCR in stabilizing the complex *in vivo* as suggested by others (Watts *et al.*, 1986; Ashwell and Schwartz, 1986). The study by Sadegh-Nasseri and McConnell (1989) in the same antigenic system supports this notion.

C. Structural Analyses of T-cell Determinants of Cytochrome *c*

An analysis of the peptide sequences which had been identified as T-cell epitopes in proteins indicated that they had a propensity to adopt an amphipathic α-helical configuration (DeLisi and Berzofsky, 1985; Margalit *et al.*, 1987; Berzofsky *et al.*, 1987). Given that x-ray crystallographic analysis of the MHC class II molecule has shown that the bound peptides must adopt largely extended conformations (Brown *et al.*, 1993), this propensity is clearly not related to the configuration the peptide adopts when actually bound to the MHC. It could, however, be required for other processing or transport events prior to MHC binding.

We have shown that the sequence 95–104 contains the residues necessary for T-cell recognition of the molecule, but that residues to the amino-terminus of this peptide are required for complete stimulation (Carbone *et al.*, 1987a; Bhayani *et al.*,

1988). The ability of these residues in the native sequence to enhance T-cell recognition had been attributed to the helix-promoting properties of this sequence (Pincus *et al.*, 1983; Schwartz *et al.*, 1985; Vasquez *et al.*, 1987). In a series of studies, however, we have shown that while there is some correlation with helical content of synthetic peptides, as measured in nonaqueous solvents, factors other than conformational stability are probably responsible for their full potency in stimulating T-cell hybridomas (Carbone *et al.*, 1987a; Bhayani *et al.*, 1988). We have examined the physical properties of pigeon cytochrome *c* peptides containing native and nonnative sequences to the amino-terminus of residue 95 in an attempt to determine how these residues modulate the recognition of the core determinant 95–104. We have not been able to demonstrate a correlation between the T-cell stimulatory ability of different nonnative amino terminal sequences and the propensity of these residues to stabilize an α-helix as measured by circular dichroism in the helix-promoting solvent trifluoroethanol (Bhayani *et al.*, 1988; Collawn *et al.*, 1989) or in aqueous or lipid environments (Collawn *et al.*, 1989). A comparison of the ability of some of these peptides to stimulate two pigeon cytochrome *c*–specific T-cell hybridomas with their α-helical content in a variety of solvents is shown in Table 11.6. Two amphipathic analogs K K L L K K L-95–104 and E E L L E E L-95–104 show strong α-helical forming properties in a membrane environment or an aqueous environment, respectively, but are not the most stimulatory of the analogs with respect to T-cell recognition. The other analogs do not have structure in these solvents, but show varying degrees of α-helicity in the helix-promoting solvent trifluoroethanol (TFE). Table 11.6 shows that there is no correlation between T-cell recognition and α-helical structure in this solvent.

These findings are in contrast with the proposal that the α-helix may be a general requirement for most, if not all, T-cell determinants (DeLisi and Berzofsky, 1985; Margalit *et al.*, 1987; Berzofsky *et al.*, 1987). Indeed, the modulation of T-cell recognition of the pigeon cytochrome *c* determinant 95–104 by adding nonnative leader sequences (Table 11.6) is clearly not related to their ability either to stabilize the determinant in an α-helical conformation or to form amphipathic structures. This lack of correlation is consistent with our findings (*vide supra*) that the residues in the epitope 95–104 show no specific conformational pattern. When plotted as a helical wheel (Figure 11.13), the residues identified as contacting TCR or MHC clearly do not segregate on distinct sides of an α-helix. This has also been shown to be the case for the minor pigeon cytochrome *c* determinant 43–58 (Ogasawara *et al.*, 1989). An α-helical conformation is thought to be adopted by the immunodominant peptide in the I-A^k restricted response to hen egg lysozyme (Allen *et al.*, 1987b), but α-helical propensity in a peptide from the glycoprotein D of the herpes simplex virus appeared to reduce T cell recognition (Heber-Katz *et al.*, 1985). Vita and co-workers (1990) also failed to show a correlation between structural properties and T-cell recognition of two peptide analogs of the horse cytochrome *c* sequence 39–53. Thus, the ability to adopt an α-helical conformation is clearly not a general requirement for T-cell recognition of antigenic peptides.

Another possible role that we investigated for nonnative leader sequences is that they enhance the ability of the determinant to interact directly with the lipid bilayer itself. This model has been suggested by various authors for membrane-

Table 11.6

Physical properties of peptide analogs[a] of the pigeon cytochrome c determinant 88–104 compared to their T-cell stimulatory capacity for two pigeon cytochrome c-specific hybridomas[b].

Peptide	Charge	Membrane binding[c]	T-cell response[d]	α-helical content[e]		
				H_2O	TFE	Lipid
K A E R A D L	+3	1.6 ± 0.2	1/1	0	62	0
K A K A K A K	+7	3.1 ± 0.3	0.2/0.1	0	64	0
K K L L K K L[f]	+7	82.0 ± 0.4	0.3/0.8	0	71	69
R R P P R R P	+7	3.2 ± 0.4	0.4/0.3	0	31	0
P G P G P G P	+3	1.0 ± 0.3	0.5/0.2	0	46	0
R T R T R T R	+7	3.6 ± 0.2	2.8/1.8	0	60	0
R P R P R P R	+7	1.4 ± 0.2	3.1/0.5	0	5	0
E E L L E E L[f]	−1	4.2 ± 0.5	7.1/16.8	24	60	10
D D P P D D P	−1	1.8 ± 0.1	14.3/0.9	0	44	0
D T D T D T D	−1	0.3 ± 0.0	>100/11.6	0	55	0
D P D P D P D	−1	0.0 ± 0.0	>200/5.9	4	48	0
E A E A E A E	−1	0.1 ± 0.0	>200/42.0	2	63	0

[a] Only the leader sequences 88–94 are given, in the single-letter code, since the native sequence, 95–104, I A Y L K Q A T A K, is common to all of the analogs.

[b] The data shown in this table were taken from Collawn et al. (1989).

[c] μM peptide bound to liposomes composed of phosphatidylcholine : phosphatidylserine, 9 : 1.

[d] Relative stimulatory capacity compared to 88–104 for two T-cell hybridomas of phenotypes III/IV (see Table 11.4).

[e] Calculated from circular dichroic measurements. The aqueous environment used was tris buffered saline, TFE = trifluoroethanol, and the lipid environment was phosphatidylcholine : phosphatidylserine, 9 : 1.

[f] These two peptides were designed to be amphipathic α-helices of opposite charge; their conformational properties are described in detail in Collawn and Paterson (1990).

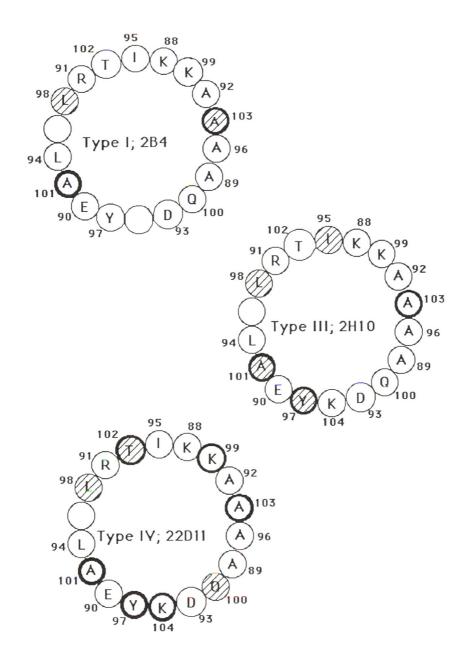

Figure 11.13

Residue assignments summarized in Figure 11.12 plotted as a helical wheel to illustrate the position of MHC contact residues and TCR contact residues in the peptide if it adopted an α-helical conformation. Hatched circles represent MHC contact residues and bold circles represent TCR contact residues. The assignments for p88–104 are shown for 22.D11 and 2H10. MHC contact residue assignments for 2B4 can be made for m88–103 only. From Bhayani and Paterson (1989), with permission.

mediated peptide–hormone–receptor interactions (Deber and Benham, 1984; Kaiser and Kezdy, 1984). Generally, in these proposed systems, the peptide accumulates in the vicinity of the receptor as a consequence of lipid association, and this results in an apparent enhancement in receptor association. The association with lipid is mediated both by electrostatic interactions between the negatively charged membrane and positively charged residues and by hydrophobic interactions between nonpolar residues and the lipid surface (Deber and Benham, 1984). Studies by Falo and colleagues (1986, 1987a,b) have shown that several peptide antigens appear to be stably associated with the APC plasma membrane, and that phospholipase-treated APCs can no longer present insulin to I-A^d-insulin–specific T cells, even though recognition by T cells allospecific for the I-A^d molecule was not affected. These findings support a role for antigen–lipid association in T-cell recognition.

We have measured the ability of the peptide analogs with leader sequences to bind to anionic phospholipid vesicles (see Table 11.6) and find that there is not a good correlation between lipid binding and T-cell recognition for the analogs tested (Collawn et al., 1989). For example, the amphipathic analog, K K L L K K L-95–104 interacts better with phospholipid membranes than does any other peptide we have tested (Carbone et al., 1987a; Collawn et al., 1989), but is not the best-recognized of the analogs.

When one examines the peptide properties of the peptides shown in Table 11.6, the most striking correlation with T-cell recognition appears to be with the charge of the peptide. Thus, those peptides bearing positively charged side chains are significantly more potent than peptides that are negatively charged. This putative relationship led us to examine the influence of electrostatic complementarity in enhancing the interaction of positively charged T-cell determinants with a negatively charged class II binding site. The role of long-range electrostatic forces in guiding the diffusion of ligand into binding sites has been recognized for a wide range of biological intermolecular interactions, including substrate-enzyme binding (Getzoff et al., 1983), drug–receptor recognition (Dean, 1981) hapten–antibody interactions (Karush, 1962), and other protein–protein interactions (see Chapter 16 by Northrup in this volume), and it seemed possible that the leader sequences were playing a similar role. We examined the charges of the 37 residues that are a part of the hypothetical class II foreign antigen binding site (Brown et al., 1988) using the I-E^k sequence. This model is based on the three-dimensional structure of the foreign antigen binding site of a class I histocompatability antigen described by Bjorkman and colleagues (1987a,b) (Figure 11.8). Our inspection indicated that the binding site has 26 neutral residues, 8 negatively charged residues and 3 positively charged residues, with a net charge of −5. If electrostatic complementarity is important, then other determinants that are restricted to I-E^k should also be positively charged. We found that this is indeed the case (Collawn et al., 1989). An argument against this hypothesis would be that many T-cell determinants have a net positive charge, and it may simply be a coincidence that the I-E^k binding site has a net negative charge. Upon examining class II binding sites (Collawn et al., 1989), we found other I-E molecules have a net negative charge, e.g., I-E$_\alpha^k$E$_\beta^b$ = −6, I-E$_\alpha^k$E$_\beta^s$ = −5, and I-E^d = −5. The I-A class II molecules, however, have a lower net charge in their binding sites, e.g., I-A^b = −3, I-A^d = −3, and I-A^k = 0. This analysis suggests that electrostatic interactions might be more important with I-E molecules than with I-A class II molecules. In independent studies, Sette and co-workers (1989) have also concluded that basic residues are important for peptide binding to the I-E class II molecule and that hydrophobic residues are important for I-A binding.

IV. The Immune Response to Heme

Until recently, there has been no investigation into the immunological properties of the heme moiety of cytochrome *c*. As it is not a protein or peptide and bears no amino acid residues (see Figure 11.14), one would predict on the basis of current models of T-cell epitopes (Schwartz, 1985; Buus *et al.*, 1987; Berzofsky *et al.*, 1987) that such a small molecule would be poorly or nonimmunogenic. Nevertheless, heme is a potent T-cell stimulant in both mice (Cooper *et al.*, 1988) and humans (Novogorodsky *et al.*, 1989, 1991; Lander *et al.*, 1992), which is an unusual finding in several respects. In the first case, it is unusual for structural reasons. Heme, the iron derivative of protoporphyrin IX (Figure 11.14), is a rigid planar molecule comprising four pyrrole groups linked by methene bridges to form a tetrapyrrole ring. When the heme group forms the prosthetic group of cytochrome *c*, it is

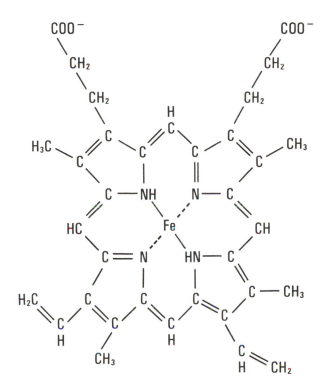

Figure 11.14
The structure of protoheme IX.

covalently bound to the polypeptide backbone of the molecule through thioether linkages at the vinyl side chains to cysteine residues at position 14 and 17 in the polypeptide chain. The overall dimensions of the MHC class II molecule are similar to those of the class I molecule (25 Å long, 10 Å wide, and 11 Å deep) (Brown *et al.*, 1993), the major difference in the structure of the two molecules being that the class II binding cleft is more open so that longer peptides can be accommodated within it by extending out of either end of the groove. Thus, the heme moiety (9 Å × 9 Å) could fit into the MHC binding site for presentation to the TCR, perhaps even in the presence of a short peptide. The propionyl groups of the heme would be available to make hydrogen-bonded or ionic interactions with adjacent side chains within the groove or with the T-cell receptor. Hydrophobic interactions are also possible between substituent groups of the heme and amino acid side chains bearing ring structures (i.e., tryptophan, histidine, and tyrosine). In addition, the fifth and sixth coordination sites of the central iron atom could also interact with appropriate side chains. Nevertheless, structurally rigid, small haptenic molecules of the size of heme are very rarely capable of stimulating a T-cell response. The only other reported example, to our knowledge, is the class II restricted cytotoxic T-cell response to tyrosine–azobenzene–arsonate (Spragg and Goodman, 1987), which is derived by antigen priming in the conventional way.

A second unusual feature of the T-cell response to heme is that the molecule induces a vigorous and specific proliferative response in lymph node (LN) T-cell populations obtained from naive mice. Using limiting dilution analyses, we have examined the frequency of heme-responsive T cells in the spleens of several strains of mice, some of which share matched genetic backgrounds, and found that the response varies according to the MHC molecules expressed by each strain. The frequency of responding T cells is highest in B10.S (I-A^s) and SJL (I-A^s) mice. B10.Br (I-A^k) and Balb/k (I-A^k) are intermediate responders, whereas B10.D2 (I-A^d) and Balb/c (I-A^d), have the lowest responses (Sutherland *et al.*, 1995). However, even the lowest responses, which we measured for I-A^d bearing mice, are much greater than the frequency of antigen-reactive T cells after priming with foreign antigen (Bradley *et al.*, 1993; Sutherland *et al.*, 1995).

The precursor frequency for single specificities in the T-cell repertoire is usually undetectable in the proliferative response in naive T-cell populations. One exception to this rule is the repertoire of T cells specific for foreign MHC products, i.e., the alloreactive response (Wilson *et al.*, 1968), which is responsible for the rejection of foreign tissue grafts. The other two exceptions belong to the group of molecules designated "superantigens" by Marrack and Kappler (1990) because of their ability to stimulate the proliferation of up to 10% of the T cells in the mouse T-cell repertoire. They are the response to the determinants known as minor lymphocyte-stimulating antigens which is observed between certain sets of H-2 congenic mice with different background genes (Festenstein, H., 1974), i.e., the MLS system, and the strong proliferative response seen to staphylococcal enterotoxins and a group of related proteins (reviewed by Marrack and Kappler, 1990). The strong proliferative response by naive T cells to the heme molecule provides another example of a high precursor frequency in certain H-2 congenic mice. However, unlike the "super-antigens," the reactivity to the heme moiety by responder T cells falls under strict

MHC restriction, thus behaving as a "classical" T-cell response to a foreign antigen. In addition, it has been shown that T cells responsive to superantigens have highly restricted variable gene usage (Marrack and Kappler, 1990).

The specificity of TCR, like antibody specificity, is determined by a combination of different variable components of genes (V, D, J). However, it has been shown that, unlike antibodies, T cells often use restricted sets of V genes for given antigen specificities. For example, the autoreactive, myelin basic protein-specific T cells in experimental allergic encephalitis (Urban *et al.*, 1988) often use $V_\beta 8.2$. There is a preferential gene usage of $V_\beta 8.1$, $V_\beta 6$, and $V_\beta 3$ in superantigen, MLS responses (Kappler *et al.*, 1988; Abe *et al.*, 1988) and of $V_\beta 17a$ and 11 in I-E reactive T cells (Kappler *et al.*, 1987b; Bill *et al.*, 1989). In addition, the predominant usage of $V_\alpha 3$ by tyrosine–azobenzene–arsonate–specific T cells (Tan *et al.*, 1988) and of $V_\alpha 11$ by pigeon cytochrome *c*-specific T cells (Fink *et al.*, 1986) has been reported. Characterization of V gene usage in these instances has provided a better understanding of the molecular mechanism of tolerance induction and TCR recognition of antigen/ MHC complex. We have determined the V genes used by three heme-reactive T-cell specificities selected at random from the peripheral T-cell repertoire and have shown that both the V_β and V_α genes are different for each specificity (Sutherland *et al.*, 1995). It seems unlikely, therefore, that the presence of large numbers of heme-reactive T cells in the peripheral lymphoid organs of responder mice is related to superantigen-like responses.

However, the high precursor frequency to heme in the peripheral T-cell repertoire is, in many ways, similar to alloreactive responses in that it is highly polyclonal, and the frequencies of reactive T cells (about 1 in 1000 to 1 in 100) are also quite similar (Miller and Stutman, 1982). The ability of foreign MHC to stimulate naive T cells is thought to be due to differences between the ability of individual MHC haplotypes to bind self peptides (Rötzschke *et al.*, 1991). The peripheral lymphoid system of an individual animal will not contain T cells that can be activated by self peptides bound to self MHC molecules because the development of such cells will have been aborted during ontogeny in the thymus. However, recent studies on the nature of the naturally processed peptides found bound to MHC Class II molecules have shown that different MHC molecules bind different sets of self peptides (Rudensky *et al.*, 1992; Hunt *et al.*, 1992). Thus "foreign" tissues bear MHC molecules that have peptides associated with them for which the recipient animal is not tolerant. It has been estimated that about 1000 different peptides can associate with a single MHC haplotype (Hunt *et al.*, 1992). These peptide/MHC complexes would provide a very diverse source of antigen for which there will be many specificities in the naive peripheral T-cell repertoire. We believe that heme may exert its stimulatory effect by changing the set of peptides associated with the MHC molecule, particularly the high responder haplotype, I-A^s. There are a number of ways by which this might be achieved. One possibility is that the heme molecule interacts directly with the MHC molecule and changes its ability to bind certain peptides. Another is that soluble heme competes with the MHC molecule for binding to self peptides, and thus gives peptides which have a lower affinity for the MHC molecule the opportunity to bind. In support of the second explanation is the finding that the most abundant naturally processed peptides found associated with the high-responder MHC haplotype, I-A^s, all contain a histidine residue (Rudensky *et al.*, 1992), a residue which is known to preferentially ligand with heme (Moore and Pettigrew, 1990). The occurrence of heme as the prosthetic group for a wide number of respiratory proteins makes it an ubiquitous self molecule, and T cells responsive to it should be eliminated in the organism by tolerance. An explanation that the stimulatory effects of heme are due, not to recognition of heme itself, but to its influence on the peptides bound to MHC is, therefore, also consistent with heme as a tolerance-inducing molecule.

V. Class I Restricted T-Cell Recognition of Cytochrome *c*

Until quite recently, it was thought that, like B cells, class I restricted cytotoxic T cells recognized membrane-bound, intact antigenic proteins associated in some way with the class I MHC (Zinkernagel and Doherty, 1979). With the cloning and sequencing of the T-cell receptor, it became clear that there are many similarities between class I and class II restricted recognition of antigens by TCR; indeed, the same pool of V_α, J_α and V_β, D_β, and J_β gene segments is used to generate class I and class II restricted TCR (reviewed by Hedrick, 1988). Thus, it seemed likely that the ligands for these receptors are essentially similar. This notion is supported by the fact that the two chains of the class II MHC have a similar polypeptide backbone fold to that of the single chain of the class I molecule (Brown *et al.*, 1993). In the last five years, extensive evidence has shown that the stimulation of both class I and class II restricted T cells can indeed be achieved by incubating APCs with peptide fragments (reviewed by Carbone, 1990) which in some instances may be identical (Perkins *et al.*, 1989).

The cellular consequences of MHC class I and MHC class II restricted recognition by T cells are almost always, however, quite different. In the case of class II restricted recognition, the result of T-cell activation is the secretion of lymphokines which stimulate the growth and differentiation of cells of the immune system, whereas class I restricted recognition results in the targeting of the APC for lysis by the killing mechanism of the cytotoxic T cell. The functions of these two types of T-cell subsets, therefore, are quite dissimilar, and they have evolved in the immune system to perform quite different tasks in protecting the host organism from invasion by pathogens. Class II restricted T cells have evolved to help other lymphocytes differentiate and mature and can only be stimulated by class II expressing cells (see Figure 11.1). Only a very limited number of cell types in the organism (e.g., those which are related to the immune system, such as macrophages and B cells) express class II molecules. The role of class I restricted cytotoxic T cells is to survey the host organism's cells for the expression of foreign antigens, the result usually of infection, and to destroy them. Class I MHC molecules are expressed by almost all nucleated cells of the mammalian organism.

These functional differences appear to have resulted in the evolution of different pathways in the APC where the antigenic peptides are generated and where they associate with the relevant MHC molecule (Braciale *et al.*, 1987). In the case of class I restricted recognition, the antigenic peptides are derived from intracellular processing of newly synthesized proteins within the cell cytoplasm, the so-called endogenous pathway (Moore *et al.*, 1988; Yewdell *et al.*, 1988). Peptides derived from this pathway have been shown to associate with class I molecules within the endoplasmic reticulum (Yewdell and Bennink, 1989; Nuchtern *et al.*, 1989) and in so doing, to catalyze its folding and egress from this compartment (Townsend *et al.*, 1989; Lie *et al.*, 1990). The class I restricted T cells which recognize the peptides derived in this way are cytotoxic and are directed to endogenous foreign proteins such as viral antigens. Class I processing is, therefore, quite different from the exogenous pathway of antigen processing whereby external soluble proteins enter an endosomal compartment of the cell in which they are processed.

Clearly, the differences between the processing pathways and their respective requirements for antigen localization determine whether presentation will be restricted to class I or class II MHC molecules (Braciale *et al.*, 1987). Thus, the T-cell response to priming by soluble protein antigens such as cytochrome *c* always results in class II restricted T-cell recognition. It is possible, however, to supply the requirement for class I presentation (that the protein enter the endogenous pathway) by several artificial mechanisms, and thus to raise a cytotoxic T-cell response to soluble protein antigens. One such way is to introduce the soluble protein into the cytoplasm of cells by osmotic loading or by transfecting cells with the gene of the protein, and then using these cells to immunize syngeneic animals (Moore *et al.*, 1988; Carbone and Bevan, 1990). Another approach is to bypass the intracellular processing requirement entirely by taking the spleen cells of unprimed animals and immunizing them, *in vitro*, by repeatedly stimulating them with APC and peptides from the soluble protein (Carbone *et al.*, 1988). In a few instances (Carbone and Bevan, 1989; Ishioka *et al.*, 1989; Aichele *et al.*, 1990), this can also be achieved *in vivo*, and peptide priming in this case will also result in a class I restricted cytotoxic T-cell response.

In collaboration with the laboratory of Sheil, we have been using *in vitro* peptide priming to examine the ability of cytochrome *c* to interact with the MHC class I molecules of the murine H-2^b locus and to stimulate the peripheral repertoire of cytotoxic T cells. By priming spleen cells from H-2^b mice with tryptic digests of horse cytochrome *c*, we have generated class I restricted T-cell clones which recognize both horse and bovine cytochrome *c* digests, but not pigeon or chicken cytochrome *c* digests (Sheil *et al.*, 1994). The assay system for class I restricted T-cell recognition involves measuring the ability of the T cells to lyse APC in the presence of peptide antigens; the degree of lysis is measured by the release of ^{51}Cr which is used to label the target cells prior to the assay.

We have isolated the tryptic peptide that stimulates the T-cell response by HPLC purification and have identified it as residues 40–53. Using a series of truncated synthetic peptides, we found that the optimal sequence required to give the maximal T-cell response was peptide 41–49. This peptide contains two sequence differences among the avian, cow, horse, and mouse (self) cytochromes (see Figure 11.2). At position 44, the horse and cow sequences have Pro, the mouse sequence has Ala, and the avian sequences have Glu. At position 47, the horse sequence has Thr, and the avian, mouse, and cow sequences have Ser. Thus, it appears that the amino acid residue at position 44 is crucial for correct T-cell recognition, and we have now confirmed this using synthetic peptide analogs with residue 44 substituted by either Ala or Gly; these peptides could not target APC for killing by the class I restricted horse cytochrome *c*-specific T-cell clones.

These findings further extend our knowledge of T-cell tolerance of self peptides to class I restricted recognition. It is of interest that the Pro for Ala substitution at residue 44 in horse versus mouse cytochrome *c* renders this region of horse cytochrome *c* immunogenic for B cells (Urbanski and Margoliash, 1977; Carbone and Paterson, 1985), T helper cells (Baumhüter *et al.*, 1987), and, as we have described here, cytotoxic T cells. Our findings also provide further evidence that the immense repertoire of T-cell receptors in the peripheral lymphoid organs of mice (estimated to be about 10^{15} by Davis and Bjorkman, 1988) contains specificities which can distinguish single-residue differences in the context of class I MHC presentation, in addition to the much more extensively studied MHC class II restricted recognition of peptides by T cells.

VI. Concluding Remarks

In the introduction to this chapter, the importance of protein antigens to cellular immunity and to the immune phenomena of tolerance and MHC restriction was discussed. Cytochrome *c* is not a foreign antigen that is likely to be introduced to the immune system under normal circumstances. Nevertheless, because of the wealth of structural information about this molecule and the availability of proteins of mammalian origin, cytochrome *c* has played an important role in clarifying the complex mechanisms by which an immune response to protein antigens is mounted in the mammalian organism. In this chapter I have attempted to review this role. In addition, I have described the way in which the interaction of cytochrome *c* with monoclonal antibodies can be used as a model system to extend theories and techniques for studying protein/protein interactions in general.

VII. Acknowledgments

I would like to thank Dr. Elizabeth Getzoff, and Dr. John Tainer, Scripps, for continued help in preparing computer graphics representations of the cytochrome *c* molecule which display the various results of our studies. I am supported by a Faculty Research Award (FRA-372) from the American Cancer Society.

VIII. References

Abbott, A. J., & Nelsestuen, G. L. (1988) *FASEB J.* **2**, 2858–2866.
Abe, R., Vacchio, M. S., Fox, B., & Hodes, R. J. (1988) *Nature (London)* **335**, 827–830.
Aichele, P., Hengartner, H., Zinkernagel, R. M., & Schulz, M. (1990) *J. Exp. Med.* **171**, 1815–1820.
Allen, P. M., Strydom, D. J., & Unanue, E. R. (1984) *Proc. Natl. Acad Sci.* **81**, 2489–2493.
Allen, P. M., Babbit, B. P., & Unanue, E. R. (1987a) *Imm. Revs.* **98**, 171–187.

Allen, P. M., Matsueda, G. R., Evans, R. J., Dunbar, J. B., Marshall, G. R., & Unanue, E. R. (1987b) *Nature (London)* **327**, 713–715.

Amit, A. G., Mariuzza, R. A., Phillips, S. E. V., & Poljak, R. J. (1986) *Science* **233**, 747–753.

Anglister, J., Bond, M. W., Frey, T., Leahy, D., Levitt, M., McConnell, H. M., Rule, G. S., Tomasello, J., & Whittaker, M. (1987) *Biochemistry* **26**, 6058–6064.

Anglister, J., Jacob, C., Assulin, O., Ast, G., Pinker, R., & Arnon, R. (1988) *Biochemistry* **27**, 717–724.

Anglister, J., Levy, R., & Scherf, T. (1989) *Biochemistry* **28**, 3360–3365.

Arnon, R., Maron, E., Sela, M., & Anfinsen, C. B. (1971) *Proc. Natl. Acad. Sci. USA* **68**, 1450–1455.

Ashwell, J. D., & Schwartz, R. H. (1986) *Nature (London)* **320**, 176–179.

Babbitt, B. P., Allen, P. M., Matsueda, G., Haber, E., & Unanue, E. R. (1985) *Nature (London)* **317**, 359–361.

Babbitt, B. P., Matsueda, G., Haber, E., Unanue, E. R., & Allen, P. M. (1986) *Proc. Natl. Acad. Sci. USA* **83**, 4509–4513.

Barlow, D. J., Edwards, M. S., & Thornton, J. M. (1986) *Nature (London)* **322**, 747–748.

Baumhüter, S., Wallace, C. J. A., Proudfood, A. E. I., Bron, C., & Corradin, G. (1987) *Eur. J. Immunol.* **17**, 651–656.

Benjamin, D. C. (1991) *Int. Revs. of Immunol.* **7**, 149–164.

Benjamin, D. C., Berzofsky, J. A., East, I. J., Gurd, F. R. N., Hannum, C., Leach, S. J., Margoliash, E., Michael, J. G., Miller, A., Prager, E. M., Reichlin, M., Sercarz, E. E., Smith-Gill, S. J., Todd, P. E., & Wilson, A. C. (1984) *Ann. Rev. Immunol.* **2**, 67–101.

Benjamin, D. C., Williams, D. C. Jr., Smith-Gill, S. J., & Rule, G. S. (1992) *Biochemistry* **31**, 9539–9545.

Bentley, G. A., Bhat, T. N., Boulot, G., Fischmann, T., Navaza, J., Poljak, R. J., Riottot, M.-M., & Tello, D. (1989) *Cold Spring Harbor Symposium on Quantitative Biology* **54 Pt. 1**, 239–245.

Bentley, G. A., Boulot G., Riottot, M.-M., & Poljak, R. J. (1990) *Nature (London)* **348**, 254–257.

Berkower, I., Buckenmeyer, G. K., Gurd, F. R. N., & Berzofsky, J. A. (1982) *Proc. Natl. Acad. Sci. USA* **79**, 4723–4727.

Berkower, I., Buckenmeyer, G. K., & Berzofsky, J. A. (1986) *J. Immunol.* **136**, 2498–2503.

Berzofsky, J. A. (1985) *Science* **229**, 932–940.

Berzofsky, J. A., Cease, K. B., Cornette, J. L., Spouge, J. L., Margalit, H., Berkower, I. J., Good, M. F., Miller, L. H., & DeLisi, C. (1987) *Imm. Rev.* **98**, 9–52.

Bhayani, H., & Paterson, Y. (1989) *J. Exp. Med.* **170**, 1609–1625.

Bhayani, H., Carbone, F. R., & Paterson, Y. (1988) *J. Immunol.* **141**, 377–382.

Bill, J., Kanagawa, O., Woodland, D. K., & Palmer, E. (1989) *J. Exp. Med.* **169**, 1405–1419.

Bjorkman, P. J., Saper, M. A., Samraoui, B., Bennett, W. S., Strominger, J. L., & Wiley, D. C. (1987a) *Nature (London)* **329**, 506–512.

Bjorkman, P. J., Saper, M. A., Samraoui, B., Bennett, W. S., Strominger, J. L., & Wiley, D. C. (1987b) *Nature (London)* **329**, 512–518.

Blake, C. C. F., Koenig, D. F., Mair, G. A., North, A. C. T., Phillips, D. C., & Sarma, V. R. (1965) *Nature (London)* **206**, 757–763.

Borden, D., & Margoliash, E. (1976) in *Handbook of Biochemistry and Molecular Biology, Proteins* Vol. III (Fasman, G. D., ed.), pp. 268–279, The Chemical Rubber Co., Cleveland, Ohio.

Bosshard, H. R. (1979) *Methods Biochem. Anal.* **25**, 273–301.

Boyer, M., Novak, Z., Fraga, E., Oikawa, K., Kay, C. M., Fotedar, A., & Singh, B. (1990) *Int. Immunol.* **2**, 1221–1233.

Braciale, T. J., Morrison, L. A., Sweetser, M. T., Sambrook, J., Gething, M. J., & Braciale, V. L. (1987) *Immunol. Rev.* **98**, 95–114.

Bradley, L. M., Duncan, D. D., Yoshimoto, K., & Swain, S. L. (1993) *J. Immunol.* **150**, 3119–3130.

Brautigan, D. L., Ferguson-Miller, S., & Margoliash, E. (1978a) *J. Biol. Chem.* **253**, 140–148.

Brautigan, D. L., Ferguson-Miller, S., & Margoliash, E. (1978b) *Methods in Enzymology* **53**, 128–164.

Brett, S. J., Cease, K. B., Ouyang, C-A. S., & Berzofsky, J. S. (1989) *J. Immunol.* **143**, 771–779.

Brown, J. H., Jardetzky, T. S., Saper, M. A., Samraoui, B., Bjorkman, P. J., & Wiley, D. C. (1988) *Nature (London)* **332**, 845–850.

Brown, J. H., Jardetzky, T. S., Gorga, J. C., Stern, L. J., Urban, R. G., Strominger, J. L., & Wiley, D. C. (1993) *Nature (London)* **364**, 33–39.

Burnens, A., Demotz, S., Corradin, G., & Bosshard, H. R. (1987) *Science* **235**, 780–783.

Bushnell, G. W., Louie, G. V., & Brayer, G. D. (1990) *J. Mol. Biol.* **214**, 585–595.

Buus, S., Colon, S., Smith, C., Freed, J. H., Miles, C., & Grey, H. M. (1986a) *Proc. Natl. Acad. Sci.* **83**, 3968–3971.

Buus, S., Sette, A., Colon, S., Jenis, D. M., & Grey, H. M. (1986b) *Cell* **47**, 1071–1077.

Buus, S., Sette, A., & Grey, H. M. (1987) *Immunol. Rev.* **98**, 115–141.

Cambier, J. C., & Ransom, J. T. (1987) *Ann. Rev. Immunol.* **5**, 175–199.

Cantor, H., & Boyse, E. A. (1975a) *J. Exp. Med.* **141**, 1376–1389.

Cantor, H., & Boyse, E. A. (1975b) *J. Exp. Med.* **141**, 1390–1399.

Carbone, F. R. (1990) *Int. Revs. of Immunol.* **7**, 129–138.

Carbone, F. R., & Paterson, Y. (1985) *J. Immunol.* **135**, 2609–2616.

Carbone, F. R., & Bevan, M. J. (1989) *J. Exp. Med.* **169**, 603–612.

Carbone, F. R., & Bevan, M. J. (1990) *J. Exp. Med.* **171**, 377–387.

Carbone, F. R., Fox, B. S., Schwartz, R. H., & Paterson, Y. (1987a) *J. Immunol.* **138**, 1838–1844.

Carbone, F. R., Staerz, U. D., & Paterson, Y. (1987b) *Eur. J. Immunol.* **17**, 897–899.

Carbone, F. R., Moore, M. W., Sheil, J. M., & Bevan, M. J. (1988) *J. Exp. Med.* **167**, 1767–1779.

Carlson, S. S., Mross, G. A., Wilson, A. C., Mead, R. T., Wolin, L. D., Bowers, S. F., Foley, N. T., Muijers, A. O., & Margoliash, E. (1977) *Biochemistry* **16**, 1437–1442.

Cease, K. B., Berkower, I., York-Jolley, J., & Berzofsky, J. A. (1986) *J. Exp. Med.* **164**, 1779–1784.

Cheetham, J. C., Raleigh, D. P., Griest, R. E., Redfield, C., Dobson, C. M., & Rees, A. R. (1991) *Proc. Natl. Acad. Sci. USA* **88**, 7968–7972.

Chesnut, R. W., Colon, S. M., & Grey, H. M. (1982) *J. Immunol.* **129**, 2382–2388.

Chien, N. C., Roberts, V. A., Guisti, A., Scharff, M. D.., & Getzoff, E. D. (1989) *Proc. Natl. Acad. Sci. USA* **86**, 5532–5536.

Collawn, J. F., & Paterson, Y. (1990) *Biopolymers* **29**, 1289–1296.

Collawn, J. F., Wallace, C. J. A., Proudfoot, A. E. I., & Paterson, Y. (1988) *J. Biol. Chem.* **263**, 8625–8634.

Collawn, J. F., Bhayani, H., & Paterson, Y. (1989) *Mol. Immunol.* **26**, 1069–1079.

Colman, P. M., Laver, W. G., Varghese, J. N., Baker, A., Tulloch, P., Air, G. M., & Webster, R. (1987) *Nature (London)* **326**, 358–363.

Colman, P. M., Tulip, W. R., Varghese, J. N., Tulloch, P., Baker, A. T., Laver, W. G., Air, G. M., & Webster, R. G. (1989) *Phil. Trans. Roy. Soc. Lond. Ser. B* **323**, 511–518.

Cooper, H. M., & Paterson, Y. (1987) *Protides of the Biological Fluids* **35**, 523–526.

Cooper, H. M., Jemmerson, R., Hunt, D. F., Griffin, P. R., Yates, J. R., Shabanowitz, J., Zhu, N.-Z., & Paterson, Y. (1987) *J. Biol. Chem.* **262**, 11591–11597.

Cooper, H. M., Corradin, C., & Paterson, Y. (1988) *J. Exp. Med.* **168**, 1127–1143.

Cooper, H. M., Klinman, N. R., & Paterson, Y. (1989) *Eur. J. Immunol.* **19**, 315–322.

Corradin, G., & Chiller, J. M. (1979) *J. Exp. Med.* **149**, 436–447.

Corradin, G., & Chiller, J. M. (1981) *Eur. J. Immunol.* **11**, 115–119.

Corradin, G., Zubler, R. H., & Engers, H. D. (1981) *J. Immunol.* **127**, 2442–2446.

Davies, D. R., Sheriff, S., & Padlan, E. A. (1988) *J. Biol. Chem.* **263**, 10541–10544.

Davis, M. M., & Bjorkman, P. J. (1988) *Nature (London)* **334**, 395–402.

Dean, P. M. (1981) *Br. J. Pharmac.* **74**, 39–46.

Deber, C. M., & Behnam, B. A. (1984) *Proc. Natl. Acad. Sci. USA* **81**, 61–65.

DeFranco, A. L., Gold, M. R., & Jakway, J. P. (1987) *Immunol. Revs.* **95**, 161–176.

DeLisi, C., & Berzofsky, J. A. (1985) *Proc. Natl. Acad. Sci. USA* **82**, 7048–7052.

Englander, S. W., & Kallenbach, N. R. (1984) *Q. Rev. Biophys.* **16**, 521–655.

Englander, S. W., & Mayne, L. (1992) *Annu. Rev. Biophys. Biomol. Struct.* **21**, 243–265.

Falo, L. D., Jr., Benacerraf, B., & Rock, K. L. (1986) *Proc. Natl. Acad. Sci. USA* **83**, 6994–6997.

Falo, L. D., Jr., Benacerraf, B., Rothstein, L., & Rock, K. L. (1987a) *J. Immunol.* **139**, 3918–3923.

Falo, L. D., Jr., Haber, S. I., Herrmann, S., Benacerraf, B., & Rock, K. L. (1987b) *Proc. Natl. Acad. Sci. USA* **84**, 522–526.

Feng, Y., Roder, H., Englander, S. W., Wand, A. J., & DiStefano, D. L. (1989) *Biochemistry* **28**, 195–203.

Festenstein, H. (1974) *Transplantation* **18**, 555–557.

Fink, P. J., Matis, L. A., McElligott, D. L., Bookman, M., & Hedrick, S. M. (1986) *Nature (London)* **321**, 219–226.

Fischmann, T. O., Bentley, G. A., Bhat, T. N., Boulot, G., Mariuzza, R. A., Phillips, S. E., Tills, D., & Poljak, R. J. (1991) *J. Biol. Chem.* **266**, 12915–12920.

Fox, B. S., Chen, C., Fraga, E., French, C. A., Singh, B., & Schwartz, R. H. (1987) *J. Immunol.* **139**, 1578–1588.

Fox, B. S., Carbone, F. R., Germain, R. N., Paterson, Y., & Schwartz, R. H. (1988) *Nature (London)* **331**, 538–540.

Fremont, D. H., Matsumura, M., Stura, E. A., Peterson, P. A., & Wilson, I. A. (1992) *Science* **257**, 919–926.

Fujio, H., Imanishi, M., Nishioka, K., & Amano, T. (1968) *Biken J.* **11**, 207–218.

Gammon, G., Shastri, N., Cogswell, J., Wilbur, S., Sadegh-Nasseri, S., Kryzych, U., Miller, A., & Sercarz, E. E. (1987) *Imm. Revs.* **98**, 53–74.

Garrett, T. P. J., Saper, M. A., Bjorkman, P. J., Strominger, J. L., & Wiley, D. C. (1989) *Nature (London)* **342**, 692–696.

Getzoff, E. D., Tainer, J. A., Weiner, P. K., Kollman, P. A., Richardson, J. S., & Richardson, D. C. (1983) *Nature (London)* **306**, 287–290.

Getzoff, E. D., Tainer, J. A., Lerner, R. A., & Geysen, H. M. (1988) *Adv. in Immunol.* **43**, 1–84.

Geysen, H. M., Meloen, R. H., & Barteling, S. J. (1984) *Proc. Natl. Acad. Sci. USA* **81**, 3998–4002.

Geysen, H. M., Tainer, J. A., Rodda, S. J., Mason, T. J., Alexander, H., Getzoff, E. D., & Lerner, R. A. (1987) *Science* **235**, 1184–1190.

Gillis, S. M. M., & Smith, K. A. (1977) *Nature (London)* **268**, 154–156.

Gillis, S. M. M., Ferm, W., Ou, W., & Smith, K. A. (1978) *J. Immunol.* **120**, 2027–2032.

Goetze, A. M., & Richards, J. H. (1978) *Biochemistry* **17**, 1733–1739.

Goshorn, S. C., Retzel, E., & Jemmerson, R. (1991) *J. Biol. Chem.* **266**, 2134–2142.

Guillet, J. C., Lai, M. Z., Briner, T. J., Buus, S., Sette, A., Grey, H. M., Smith, J. A., & Gefter, M. L. (1987) *Science* **235**, 865–870.

Hannum, C. H., & Margoliash, E. (1985) *J. Immunol.* **135**, 3303–3313.

Hannum, C. H., Matis, L. A., Schwartz, R. H., & Margoliash, E. (1985) *J. Immunol.* **135**, 3314–3322.

Hansburg, D., Hannum, C., Inman, J. K., Appella, E., Margoliash, E., & Schwartz, R. H. (1981) *J. Immunol.* **127**, 1844–1851.

Hansburg, D., Fairwell, T., Schwartz, R. H., & Appella, E. (1983) *J. Immunol.* **131**, 319–324.

Heber-Katz, E., Schwartz, R. H., Matis, L. A., Hannum, C., Fairwell, T., Apella, E., & Hansburg, D. (1982) *J. Exp. Med.* **155**, 1086–1099.

Heber-Katz, E., Hollosi, M., Dietzschold, B., Hudecz, F., & Fasman, G. D. (1985) *J. Immunol.* **135**, 1385–1390.

Hedrick, S. M. (1988) *Adv. in Immunol.* **43**, 193–234.

Hedrick, S. M., Matis, L. A., Hecht, T. T., Samelson, L. E., Longo, D. L., Heber-Katz, E., & Schwartz, R. H. (1982) *Cell* **30**, 141–152.

Huber, R., Deisenhofer, J., Colman, P. M., Matsushima, M., & Palm, W. (1976) *Nature (London)* **264**, 415–420.

Hunt, D. F., Michel, H., Dickinson, T. A., Shabanowitz, J., Cox, A. L., Sakaguchi, K., Appella, E., Grey, H. M., & Sette, A. (1992) *Science* **256**, 1817–1820.

Ishioka, G. Y., Colon, S., Miles, C., Grey, H. M., & Chesnut, R. W. (1989) *J. Immunol.* **143**, 1094–1100.

Jemmerson, R. (1987a) *J. Immunol.* **139**, 1939–1945.

Jemmerson, R. (1987b) *J. Immunol.* **138**, 213–219.

Jemmerson, R., & Blankenfeld, R. (1988) *J. Immunol.* **140**, 1762–1769.

Jemmerson, R., & Hutchinson, R. M. (1990) *Eur. J. Immunol.* **20**, 579–585.

Jemmerson, R., & Johnson, J. G. (1991) *Proc. Natl. Acad. Sci. USA* **88**, 4428–4432.

Jemmerson, R., & Margoliash, E. (1979) *Nature (London)* **282**, 468–471.

Jemmerson, R., & Paterson, Y. (1986a) *Science* **232**, 1001–1004.

Jemmerson, R., & Paterson, Y. (1986b) *BioTechniques* **4**, 18–31.

Jemmerson, R., Morrow, P. R., Klinman, N. R., & Paterson, Y. (1985) *Proc. Natl. Acad. Sci. USA* **82**, 1508–1512.

Johnson, J. G., & Jemmerson, R. (1991) *Eur. J. Immunol.* **21**, 951–958.

Johnson, J. G., & Jemmerson, R. (1992) *J. Immunol.* **148**, 2682–2689.

Kaiser, E. T., & Kezdy, F. J. (1984) *Science* **223**, 249–255.

Kam-Morgan, L. N. W., Smith-Gill, S. J., Taylor, M. G., Zhang, L., Wilson, A. C., & Kirsch, J. F. (1993) *Proc. Natl. Acad. Sci. USA* **90**, 3958–3962.

Kappler, J. W., Roehm, N., & Marrack, P. (1987a) *Cell* **49**, 273–280.

Kappler, J. W., Wade, T., White, J., Kushnir, E., Blackman, M., Bill, J., Roehm, N., & Marrack, P. (1987b) *Cell* **49**, 263–271.

Kappler, J. W., Staerz, U., White, J., & Marrack, P. C. (1988) *Nature (London)* **332**, 35–40.

Karush, F. (1962) *Adv. Immun.* **2**, 1–40.

Kieber-Emmons, T., & Köhler, H. (1986) *Proc. Natl. Acad. Sci. USA* **83**, 2521–2525.

Kim, I. C., & Nolla, H. (1986) *Biochem. Cell. Biol.* **64**, 1211–1217.

Kim, I. C., Nolla, H., & Priola, S. (1987) *Biochem. Cell. Biol.* **65**, 783–789.

Klinman, N. R. (1969) *Immunochemistry* **6**, 757–762.

Klinman, N. R. (1972) *J. Exp. Med.* **136**, 241–260.

Klinman, N. R., & Linton, P.-J. (1988) *Adv. in Immunol.* **42**, 1–93.

Klinman, N. R., & Press, J. L. (1975) *Transplant Rev.* **24**, 41–83.

Kohler, G., & Milstein, C. (1975) *Nature (London)* **256**, 495–497.

Kohler, G., & Milstein, C. (1976) *Eur. J. Immunol.* **6**, 511–519.

Kooistra, D. A., & Richards, J. H. (1978) *Biochemistry* **17**, 345–351.

Kuo, L.-M., & Davies, H. C. (1983) *Mol. Immunol.* **20**, 827–838.

Kuo, L.-M., Davies, H. C., & Smith, L. (1984) *Biochim. Biophys. Acta* **766**, 472–482.

Kuo, L.-M., Davies, H. C., & Smith, L. (1985) *Biochim. Biophys. Acta* **809**, 388–395.

Kuo, L.-M., Davies, H. C., & Smith, L. (1986) *Biochim. Biophys. Acta* **848**, 247–255.

Kurata, A., & Berzofsky, J. A. (1990) *J. Immunol.* **144**, 4526–4535.

Lakey, E. K., Margoliash, E., Fitch, F. W., & Pierce, S. K. (1986) *J. Immunol.* **136**, 3933–3938.

Lambert, L. E., & Unanue, E. R. (1989) *J. Immunol.* **143**, 802–807.

Lander, H. M., Levine, D. M., & Novogrodsky, A. (1992) *Febs Letters* **303**, 242–246.

Lavoie, T. B., Kam-Morgan, L. N. W., Hartman, A. B., Mallett, C. P., Sheriff, S., Saroff, D. A., Mainhart, C. R., Hamel, P. A., Kirsch, J. F., Wilson, A. C., & Smith-Gill, S. J. (1989) in *The immune response to strucutrally defined proteins: The Lysozyme Model* (Smith-Gill, S., & Sercarz, E., eds.), pp. 151–168, Adenine Press, Guilderland, New York.

Leach, S. J. (1983) *Biopolymers* **22**, 425–440.

Lee, P., Matsueda, G. R, & Allen, P. M. (1988) *J. Immunol.* **140**, 1063–1068.

Lie, W.-R., Myers, N. B., Gonka, J., Rubocki, R. J., Connolly, J. M., & Hansen, T. H. (1990) *Nature (London)* **344**, 439–441.

Madden, D. R., Gorga, J. C., Strominger, J. L, & Wiley, D. C. (1991) *Nature (London)* **353**, 321–325.

Madden, D. R., Gorga, J. C., Strominger, J. L., & Wiley, D. C. (1992) *Cell* **70**, 1035–1048.

Mamula, M. J., Jemmerson, R., & Hardin, J. A. (1990) *J. Immunol.* **144**, 1835–1840.

Manca, F., Clarke, J. A., Miller, A., Sercarz, E. E., & Shastri, N. (1984) *J. Immunol.* **133**, 2075–2078.

Margalit, H., Spouge, J. L., Cornette, J. L., Cease, K. B., DeLisi, C., & Berzofsky, J. A. (1987) *J. Immunol.* **138**, 2213–2229.

Markley, J. L. (1989) *Meth. Enzymol.* **176**, 12–64.

Marrack, P., & Kappler, J. (1990) *Science* **248**, 705–711.

Matis, L., Longo, G. L., Hedrick, S. M., Hannum, C., Margoliash, E., & Schwartz, R. H. (1983) *J. Immunol.* **130**, 1527–1535.

Matsumura, M., Fremont, D. H., Peterson, P. A., & Wilson, I. A. (1992) *Science* **257**, 927–934.

Mayne, L., Paterson, Y., Cerasoli, D., & Englander, S. W. (1992) *Biochemistry* **31**, 10678–10685.

Metzger, D. W., Ch'ng, L.-K., Miller, A., & Sercarz, E. E. (1984) *Eur. J. Immunol.* **14**, 87–93.

Michaeleck, M. T., Benacerraf, B., & Rock, K. L. (1989) *Proc. Natl. Acad. Sci. USA* **86**, 3316–3320.

Miller, R. A., & Stutman, O. (1982) *J. Immunol.* **128**, 2258–2264.

Miller, A., Hsu. D.-H., Benjamin, C., Ch'ng, L.-K., Harvey, M. A., Kaplan, M. A., Kawahara, D., Keller, M. A., Metzger, D. W., Wicker, L. S., & Sercarz, E. E. (1989) in *The Immune Response to Structurally Defined Proteins: The Lysozyme Model* (Smith-Gill, S., & Sercarz, E., eds.), pp. 341–351, Adenine Press, Guilderland, New York.

Moore, G. R., & Pettigrew, G. W. (1990) in *Cytochromes* c: *Evolutionary, Structural and Physicochemical Aspects*, Springer-Verlag, Berlin.

Moore, G. R., & Williams, R. J. P. (1980) *Eur. J. Biochem.* **103**, 543–550.

Moore, M. W., Carbone F. R., & Bevan, M. J. (1988) *Cell* **54**, 777–785.

Murphy, K. P., Freire, E., & Paterson, Y. (1995) *Prot. Struct. Funct. Gen.* **21**, 83–90.

Mylvaganum, S. E., Paterson, Y., Kaiser, K., Bowdish, K., Tainer, J. A., & Getzoff, E. D. (1991) *J. Mol. Biol.* **221**, 455–462.

Nisonoff, A., Margoliash, E., & Reichlin, M. (1967) *Science* **155**, 1273–1275.

Nossal, G. J. V. (1989) in *Fundamental Immunology, Second Edition* (W. E. Paul, ed.), pp. 571–586, Raven Press, New York.

Novogorodsky, A., Suthanthiran, M., & Stenzel, K. H. (1989) *J. Immunol. 143*, 3981–3987.

Novogorodsky, A., Suthanthiran, M., & Stenzel, K. H. (1991) *Cell. Immunol. 133*, 295–305.

Novotny, J., Bruccoleri, R. E., & Saul, F. A. (1989) *Biochemistry 28*, 4735–4749.

Nuchtern, J. G., Bonifacino, J. S., Biddison, W. E., & Klausner, R. D. (1989) *Nature (London) 339*, 223–226.

Oertle, M., Immergluck, K., Paterson, Y., & Bosshard, H. (1989) *Eur. J. Biochem. 182*, 699–704.

Ogasawara, K., Maloy, W. L., Beverly, B., & Schwartz, R. H. (1989) *J. Immunol. 142*, 1448–1456.

Padlan, E. A., Silverton, E. W., Sheriff, S., Cohen, G. H., Smith-Gill, S. J., & Davies, D. R. (1989) *Proc. Natl. Acad. Sci. USA 86*, 5938–5942.

Paterson, Y. (1985) *Biochemistry 24*, 1048–1054.

Paterson, Y. (1989) in *The Immune Response to Structurally Defined Proteins: The Lysozyme Model* (Smith-Gill, S., & Sercarz, E. E., eds.), pp. 177–189, Adenine Press, Guilderland, New York.

Paterson, Y. (1991). *Int. Revs. of Immunol. 7*, 121–128.

Paterson, Y. (1992) *Nature (London) 356*, 456–457.

Paterson, Y., Englander, S. W., & Roder, H. (1990) *Science 249*, 755–759.

Perkins, D. L., Lai, M.-Z., Smith, J. A., & Gefter, M. L. (1989) *J. Exp. Med. 170*, 279–289.

Pike, B. L. (1975) *J. Immunol. Meth. 9*, 85–104.

Pincus, M. R., Gerewitz, F., Schwartz, R. H., & Scheraga, H. A. (1983) *Proc. Natl. Acad. Sci. USA 80*, 3297–3300.

Proudfoot, A. E. I., Wallace, C. J. A., Harris, D. E., & Offord, R. E. (1986) *Biochem. J. 239*, 333–337.

Quintáns, J., & Lefkovits, I. (1973) *Eur. J. Immunol. 3*, 392–397.

Riley, R. L., & Klinman, N. R. (1986) *J. Immunol. 136*, 3147–3154.

Rini, James M., Schulze-Gahmen, U., & Wilson, I. A. (1992) *Science 255*, 959–965.

Roder, H., Elöve, G. A., & Englander, S. W. (1988) *Nature (London) 335*, 700–704.

Roosneck, E., Demotz, S., Corradin, G., & Lanzavecchia, A. (1988) *J. Immunol. 140*, 4079–4082.

Rose, G. D., Gierasch, L. M., & Smith, J. A. (1985) *Adv. in Protein Chemistry 37*, 1–109.

Rötzschke, O., Falk, K., Faath, S., and Rammensee, H.-G. (1991) *J. Exp. Med. 172*, 1059–1071.

Rudensky, A. Y., Preston-Hurlburt, P., Al-Ramadi, B. K., Rothbard, J., & Janeway, C. A. (1992) *Nature (London) 359*, 429–431.

Sadegh-Nasseri, S., & McConnell, H. M. (1989) *Nature (London) 337*, 274–276.

Sargent, D. F., & Schwyzer, R. (1986) *Proc. Natl. Acad. Sci. USA 83*, 5774–5778.

Savoca, R., Schwab, C., & Bosshard, H. R. (1991) *J. Immunol. Methods 141*, 245–252.

Schroer, J. A., Bender, T., Feldmann, R. J., & Kim, K. J. (1983) *Eur. J. Immunol. 13*, 693–700.

Schwartz, R. H. (1985) *Ann. Rev. Immunol. 3*, 237–261.

Schwartz, R. H., Fox, R. S., Fraga, E., Chen, C., & Singh, B. (1985) *J. Immunol. 135*, 2598–2608.

Sela, M., Schechter, B., Schechter, I., & Borek, F. (1967) *Cold Spring Harbor Symp. Quant. Biol. 32*, 537–545.

Sette, A., Adorini, L., Appela, E., Colon, S., Miles, C., Tanaka, S., Erhardt, C., Doria, G., Nagy, Z. A., Buus, S., & Grey, H. M. (1989) *J Immunol. 143*, 3289–3294.

Shastri, N., Oki, A., Miller, A., & Sercarz, E. E. (1985) *J. Exp. Med. 162*, 332–345.

Sheil, J. M., Schell T. D., Shepherd, S. E., Klimo, G. F., Kioschos, J. M., & Paterson, Y. (1994) *Eur. J. Immunol. 24*, 2141–2149.

Sheriff, S., Silverton, E. W., Padlan, E. A., Cohen, G. H., Smith-Gill, S. J., Finzell, B. C., & Davies, D. R. (1987) *Proc. Natl. Acacl. Sci. USA 84*, 8075–8079.

Shevach, E. M. (1989) "Accessory Molecules" in *Fundamental Immunology*, 2nd Ed. (Paul, W. E., ed.), pp. 413–441, Raven Press, New York.

Shimonkevitz, R., Kappler, J., Marrack, P., & Grey, H. M. (1983) *J. Exp. Med. 158*, 303–316.

Silvestri, I., & Taniuchi, H. (1988) *J. Biol. Chem. 263*, 18702–18715.

Smith, A. M., & Benjamin, D. C. (1991) *J. Immunol. 146*, 1259–1264.

Smith, A. M., Woodward M. P., Hershey C. W., Hershey, E. D., & Benjamin, D. C. (1991) *J. Immunol. 146*, 1254–1258.

Smith-Gill, S. J., Lavoie, T. B., & Mainhart, C. R. (1984) *J. Immunol. 133*, 384–393.

Solinger, A. M., Ultee, M. E., Margoliash, E., & Schwartz, R. H. (1979) *J. Exp. Med. 150*, 830–848.

Sorger, S. B., Hedrick, S. M., Fink, P. J., Bookman, M. A., & Matis, L. A. (1987) *J. Exp. Med. 165*, 279–301.

Sorger, S. B., Paterson, Y., Fink, P. J., & Hedrick, S. M. (1990) *J. Immunol. 144*, 1127–1135.

Spragg, J. H., & Goodman, J. W. (1987) *J. Immunol. 138*, 1169–1177.

Stanfield, R. L., Fieser, T. M., Lerner, R. A., & Wilson, I. A. (1990) *Science 248*, 712–719.

Sutherland, R. M., Pan Z.-K., Caton, A. J., Tang, X., Cerasoli, D., & Paterson, Y. (1995) *Int. Immunol. 7*, 771–783.

Suzuki, G., & Schwartz, R. H. (1986) *J. Immunol. 136*, 230–239.

Swain, S. L. (1983) *Immunol. Rev. 74*, 129–141.

Tainer, J. A., Getzoff, E. D., Paterson, Y., Olson, A. J., & Lerner, R. A. (1985) *Ann. Rev. Immunol. 3*, 501–535.

Tan, K.-N., Datlof, B. M., Gilmore, J. A., Kronman, A. C., Lee, J. H., Mascom, A. M., & Rao, A. (1988) *Cell 54*, 247–261.

Townsend, A., Öhlén, C., Bastin, J., Ljunggren, H.-G., Foster, L., & Kärre, K. (1989) *Nature (London) 340*, 443–448.

Tulip, W. R., Varghese, J. N., Webster, R. G., Air, G. M., Laver, W. G., & Colman, P. M. (1989) *Cold Spring Harbor Symp. Quant. Biol. 54*, 257–263.

Tulip, W. R., Varghese, J. N., Laver, W. G., Webster, R., & Colman, P. M. (1992) *J. Mol. Biol. 227*, 122–148.

Ultee, M. E., Margoliash, E., Lipkowski, A., Flouret, G., Solinger, A. M., Lebwohl, D., Matis, L. A., Chen, C., & Schwartz, R. H. (1980) *Mol. Immunol. 17*, 809–822.

Unanue, E. R. (1984) *Ann. Rev. Immun. 2*, 395–428.

Urban, J. L., Kumar, V., Kono, D., Gomez, C., Horvath, S. J., Clayton, J., Lando, D. G., Sercarz, E. E., & Hood, L. (1988) *Cell 54*, 577–592.

Urbanski, G. J., & Margoliash, E. (1977) in *The Immunochemistry of Enzymes and Their Antibodies* (Salton, M. R. H., ed.), pp. 203–225 John Wiley & Sons Inc., New York.

Varghese, J. N., Laver, W. G., & Colman, P. M. (1983) *Nature (London) 303*, 35–40.

Vasquez, M., Pincus, M. R., & Scheraga, H. A. (1987) *Biopolymers 26*, 373–386.

Vita, C., Baumhüter, S., & Corradin, G. (1990) *Mol. Immunol. 27*, 291–295.

Wallace, C. J. A. (1984) *Biochem. J. 217*, 589–594.

Wallace, C. J. A., & Corthésy, B. E. (1986) *Protein Engineering 1*, 589–594.

Wand, A. J., & Englander, S. W. (1986) *Biochemistry 25*, 1100–1106.

Wand, A. J., Roder, H., & Englander, S. W. (1986) *Biochemistry 25*, 1107–1114.

Wand, A. J., DiStefano, D. L., Feng, Y., Roder, H., & Englander, S. W. (1989) *Biochemistry 28*, 186–194.

Watts, T. H., & McConnell, H. M. (1986) *Proc. Natl. Acad. Sci. USA 83*, 9660–9664.

Watts, T. H., & McConnell, H. M. (1987) *Ann. Rev. Immunol. 5*, 461–475.

Watts, T. H., Gaub, H. E., & McConnell, H. M. (1986) *Nature (London) 320*, 179–181.

Wilson, D. B., Blyth, J. L., & Nowell, P. C. (1968) *J. Exp. Med. 128*, 1157–1181.

Wüthrich, K. (1986) *NMR of Proteins and Nucleic Acids*, John Wiley & Sons, New York.

Yewdell, J. W., & Bennink, J. R. (1989) *Science 244*, 1072–1075.

Yewdell, J. W., Bennink, J. R., & Hosaka, Y. (1988) *Science 239*, 637–640.

Zhang, W., Young, A. C., Imarai, M., Nathenson, S. G., & Sacchettini, J. C. (1992) *Proc. Natl. Acad. Sci. USA 89*, 8403–8407.

Ziegler, H. K., & Unanue, E. R. (1982) *Proc. Natl. Acad. Sci. USA 79*, 175–178.

Zinkernagel, R. M., & Doherty, P. C. (1979) *Adv. Immunol. 27*, 51–177.

12

Apo- and Holocytochrome c—Membrane Interactions

WILCO JORDI* AND BEN DE KRUIJFF

Centre for Biomembranes and Lipid Enzymology
University of Utrecht

I. Introduction

Mitochondrial cytochrome c molecules are 12-kDa proteins for which the net positive charge has been conserved during evolution (Mathews, 1985) and which are localized in the mitochondrial intermembranous space (Gupte and Hackenbrock, 1988a,b). Cytochrome c functions in the respiratory chain and interacts with its redox partners, cytochrome bc_1 and cytochrome oxidase, on the outside of the inner mitochondrial membrane. The protein is synthesized as a precursor, apocytochrome c, without an amino-terminal cleavable presequence (Stuart and Neupert, 1990; Zimmerman et al., 1979; Stewart et al., 1971; Zitomer and Hall, 1976; Smith et al., 1979; Matsuura et al., 1981; Stuart et al., 1987) and is thus not proteolytically processed. Only the amino-terminal methionine is removed from the primary translation product (Zimmerman et al., 1979; Matsuura et al., 1981), and in many mammals the amino-terminal glycine is acetylated (Mathews, 1985).

Cytochrome c molecules have a heme group which is generally covalently attached to two cysteine residues in the amino-terminal region of the protein by the enzyme cytochrome c heme lyase (Nicholson et al., 1987; Drygas et al., 1989;

---

* Present address: CABO, Bornsesteeg 65, 6708 PD Wageningen, The Netherlands

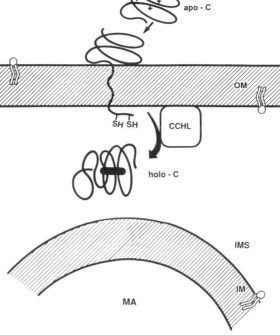

Figure 12.1
Model for the location of apo- and holocytochrome c with respect to the four subcompart-
ments of the mitochondrion outer membrane (OM), intermembranous space (IMS), inner
membrane (IM), and matrix (MA). After synthesis in the cytoplasm, the positively charged
apocytochrome c molecule inserts into the outer membrane and interacts with cytochrome c
heme lyase (CCHL), which covalently attaches the heme (dark bar) to two amino-terminal
cysteine residues. The covalent attachment of heme results in folding of the polypeptide,
which is released as the functional holoprotein in the intermembranous space.

Dumont *et al.*, 1987). Figure 12.1 illustrates the location of the apo- and holoprotein
in the mitochondrion. Since the amino acid sequences of apo- and holocytochrome
c are identical, except for the amino-terminal methionine, removal of the heme
yields a polypeptide with the primary structure of the precursor protein apocytochrome c (Fisher *et al.*, 1973). The availability of chemical amounts of both the
precursor and mature protein has stimulated biophysical studies on the interaction
of the polypeptides with model membranes. The aim of this chapter is twofold: first,
to review these studies, and second, to discuss the consequences of the polypeptide–
lipid interactions for import of apocytochrome c into mitochondria.

II. Lipid Composition of Mitochondrial Membranes

The main constituents of outer and inner mitochondrial membrane are proteins and
phospholipids. The phospholipid composition of outer and inner mitochondrial
membranes of various organisms has been described (for review, see Daum, 1985).

Table 12.1

Data are represented as means ±SD. Phospholipid compositions are expressed as molar percentages of total lipid phosphorus.

	Outer membrane	Inner membrane
nmol phospholipid/mg protein	1500 ± 150[a]	450 ± 90
Phosphatidylcholine	48 ± 5	33 ± 3
Phosphatidylethanolamine	31 ± 1.6	41 ± 3
Cardiolipin	9 ± 2	24 ± 4
Phosphatidylinositol	9.9 ± 1.7	1.0 ± 0.2
Phosphatidylserine	1.0 ± 0.3	0 ± 0

[a] Corrected for contaminating membranes (Hovius *et al.*, 1990).

Table 12.1 shows the lipid composition of outer and inner mitochondrial membrane fractions of highly purified rat liver mitochondria (Hovius *et al.*, 1990). This composition is similar to that of mitochondria from other organisms.

Mitochondrial membranes have certain distinctive characteristics. The phospholipid/protein ratio of the outer mitochondrial membrane is high compared to other cellular membranes such as the inner membrane (Daum, 1985). Furthermore, both mitochondrial membranes contain substantial amounts of negatively charged phospholipids. Phosphatidylinositol (PI) and phosphatidylserine (PS) are preferentially localized in the outer mitochondrial membrane. Cardiolipin is present in both membranes, but is most abundant in the inner membrane (Table 12.1). Eukaryotic cardiolipin is unique for a number of reasons: (i) It is exclusively localized in the mitochondria (Daum, 1985); (ii) its synthesis only occurs in the mitochondrial inner membrane (Daum, 1985); (iii) in several tissues, cardiolipin contains an extraordinarily high percentage of unsaturated fatty acids (Daum, 1985); (iv) several inner mitochondrial membrane enzymes, e.g., cytochrome *c* oxidase, require cardiolipin for optimal activity (Fry and Green, 1980); and (v) cardiolipin has unique physical properties which enable the phospholipid to adopt nonbilayer lipid structures, e.g., as a consequence of protein–lipid interactions (De Kruijff and Cullis, 1980).

The exact asymmetry of phospholipids of the inner mitochondrial membrane is yet to be established. However, several studies indicate that the major portion of cardiolipin is located on the matrix side of the mitochondrial inner membrane (Harb *et al.*, 1981; Krebs *et al.*, 1979; Nilsson and Dallner, 1977). PI and phosphatidylcholine (PC) are evenly distributed between both sides of the outer membrane of *Saccharomyces cerevisiae*, whereas the majority of phosphatidylethanolamine (PE) is oriented towards the intermembranous space (Sperka-Gottlieb *et al.*, 1988). The sidedness of cardiolipin and PS in the outer membrane is unknown.

III. Apo- and Holocytochrome *c*

Mitochondrial cytochrome *c* molecules are water-soluble proteins with a net positive charge (Mathews, 1985). In general, a heme group is covalently attached to two
cysteine residues in the amino-terminal region of the protein (Mathews, 1985). The
protein is folded around the heme group to produce a globular structure with α-
helical stretches oriented so that the side chains of the hydrophobic amino acid
residues interact with the heme in the interior of the protein (Bushnell *et al.*, 1990).

Apocytochrome *c* can be readily prepared from cytochrome *c* by chemical removal of the heme (Fisher *et al.*, 1973). Circular dichroism measurements of this
chemically prepared polypeptide revealed that the protein exhibits a largely random
coil structure (e.g., Fisher *et al.*, 1973), and the extreme protease sensitivity compared to that of cytochrome *c* suggests a loosely folded conformation (Jordi *et al.*,
1989a). Recently, small amounts of apocytochrome *c* molecules have been purified
from *in vitro* transcription–translation systems (Hakvoort *et al.*, 1990). These
polypeptides exhibit a higher affinity for binding to mitochondria than chemically
prepared apocytochrome *c* molecules (Hakvoort *et al.*, 1990).

IV. Nature and Specificity of the Interaction of Horse Heart Apo- and Holocytochrome *c* with Lipids in Model Membranes

A. Electrostatics

A number of experimental observations emphasize the importance of the electrostatic interaction between the positively charged cytochrome *c* molecule and phospholipids organized in bilayers. The interaction of cytochrome *c* with model membranes is strongly reduced in the presence of a high ionic strength (Nichols, 1974;
Quin and Dawson, 1969a,b; Kimelberg and Papahadjopoulos, 1971a; Görrissen *et
al.*, 1986; Demel *et al.*, 1989; Teissie and Baudras, 1977; Mustonen *et al.*, 1987).
Furthermore, efficient interaction requires negatively charged phospholipids (Quin
and Dawson, 1969b; Kimelberg and Papajadjopoulos, 1971a; Teissie and Baudras,
1977; Mustonen *et al.*, 1987; Rietveld *et al.*, 1983; Van der Kooij *et al.*, 1973;
Teissie, 1981; Faucon *et al.*, 1976; Brown and Wüthrich, 1977). The importance of
the electrostatic interaction for the binding to liposomes can be illustrated by
experiments in which free protein and bound protein are separated by centrifugation. Figure 12.2 shows this for binding of holocytochrome *c* and its heme-free precursor apocytochrome *c* to liposomes of different lipid composition. For both proteins, efficient binding is only observed to the negatively charged PS liposomes and
not to the zwitterionic PC liposomes (Rietveld *et al.*, 1983, 1986a). Both the affinity
of the protein for the PS liposomes and the amount of protein bound to the
liposomes are greater for apocytochrome *c* than for the holoprotein. Binding of
apocytochrome *c* is also strongly inhibited at high ionic strength (Demel *et al.*, 1989;
Pilon *et al.*, 1987). Recent binding experiments with *in vitro*–synthesized *Neurospora
crassa* apocytochrome *c* demonstrate that the *in vitro*–synthesized apocytochrome

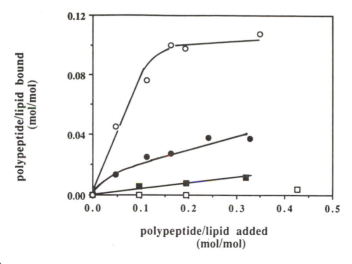

Figure 12.2

Binding of cytochrome c to liposomes of PC (■) and PS (●); binding of apocytochrome c to liposomes of PC (□) and PS (○). Increasing amounts of the polypeptides were added to a constant amount of liposomal phospholipid. Data were calculated from Rietveld et al. (1983). See that reference for experimental details.

c also has a strong preferential interaction with negatively charged phospholipids (Jordi et al., 1992).

The importance of various regions of apo- and holocytochrome c for interaction with model systems has been studied using chemically and enzymatically prepared peptide fragments of horse heart apo- and holocytochrome c (Jordi et al., 1989a,b; Li-Xin et al., 1988; Jordi et al., 1990a; Snel et al., 1991). Figure 12.3 shows the peptides studied and their net positive charge. Binding experiments (Figure 12.4) clearly demonstrated that the carboxy-terminal part of fragment 60–104 with a net charge of +2 has a lower affinity for PS vesicles compared to the amino-terminal peptide 1–59 (net charge +7). The small central fragment 66–80 with a net charge of +1 has no detectable affinity for the PS vesicles. The correlation between the efficiency of binding and the net positive charge of the peptides emphasizes the importance of the electrostatic interaction for the binding of the peptides to a lipid bilayer.

Additional evidence for the electrostatic component in the binding of apocytochrome c to model membranes is obtained by studying in a comparative manner apocytochrome c molecules with a different net positive charge. Figure 12.5 shows that the binding of apocytochrome c from horse (net charge +9) to vesicles containing negatively charged lipids is much stronger than binding of apocytochrome c from Candida krusei (net charge +4) to liposomes. Moreover, for proteins the efficiency of binding decreases when the negatively charged phospholipid content decreases.

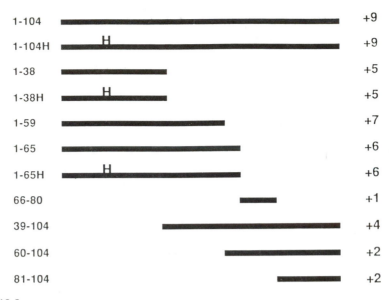

Figure 12.3
Location of the various peptides derived from apocytochrome c (1–104) and cytochrome c (1–104H). The net charges of the various proteins and peptides at pH 7.0 as calculated from the side chains of the amino acids (Arg and Lys positive, Asp and Glu negative) are given at the right of the figure, and the location of the heme group is indicated by the letter H.

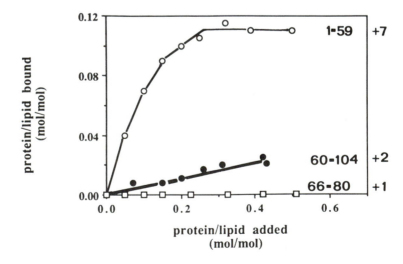

Figure 12.4
Binding of apocytochrome c–derived peptides to large unilamellar PS vesicles. Increasing amounts of the various polypeptides were added to a constant amount of PS (from beef brain) vesicles. Amino-terminal apopeptide 1–59 (o), central fragment 66–80 (□), and carboxy-terminal peptide 60–104 (●). Data from Jordi *et al.* (1989a). See that reference for experimental details.

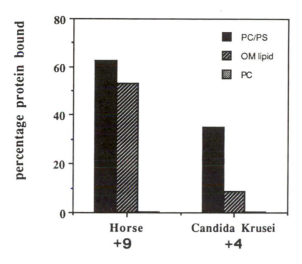

Figure 12.5

Binding of apocytochrome c from horse heart and *Candida krusei* to hand-shaken liposomes with three different lipid compositions: PC/PS (molar ratio 1:1); a lipid mixture resembling the lipid composition of the outer mitochondrial membrane: dioleoyl-PC/dioleoyl-PE/dioleoyl-PS/cardiolipin/cholesterol in a molar ratio of 50:30:10:5:7 (OM lipid); and pure dioleoyl-PC. The average of three independent experiments is shown. The liposomes (800 nmol phosphorus) were incubated for 30 min at 30 °C with 35 μg apocytochrome c in 200 μL buffer that contained 10 mM Pipes, pH 7.0, 50 mM NaCl. Subsequently, bound and nonbound protein were separated by centrifugation for 25 min at 37,000 × g, 4 °C, and the protein and phosphorus content of the supernatant was determined.

B. Membrane Penetration

Apo- and holocytochrome c are both water-soluble proteins according to sequence-prediction programs (Eisenberg et al., 1984), but there are striking differences in the efficiency with which they penetrate model membranes. A readily measured parameter to study membrane penetration is the change in surface pressure of mono-molecular lipid layers at a constant surface area. These surface pressure changes are interpreted as a result of the insertion of the protein in between the lipids. Figure 12.6 compares the interaction of apo- and holocytochrome c with monolayers composed of the total lipid fraction of the outer mitochondrial membrane of rat liver at different initial surface pressures. For both apo- and holocytochrome c, there is a decline in protein penetration with increasing initial surface pressure (Figure 12.6), that is, increasing molecular packing of the lipid layer. Extrapolation to high surface pressures would let the penetration of apocytochrome c cease at a pressure of 42 mN/m, whereas penetration of cytochrome c already ceases at initial surface pressures of 30 mN/m. The differences in the pressure range of 30–35 mN/m are particularly interesting because these conditions are considered to be most physio-

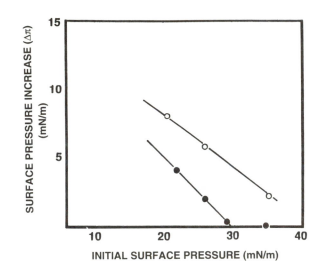

Figure 12.6
Effect of apocytochrome c (o) and holocytochrome c (●) on the surface pressure of monomolecular layers composed of outer mitochondrial membrane lipids at different initial surface pressures. The subphase contained 10 mM Pipes, pH 7.0, 50 mM NaCl. Data from Rietveld *et al.* (1986b). See that reference for experimental details.

logically relevant (Demel *et al.*, 1975). Therefore, these data indicate that in strong contrast to the apoprotein, cytochrome c does not efficiently penetrate the mono-layer under physiological conditions. More efficient penetration of cytochrome c into monolayers has been observed using a low-ionic-strength subphase (Nichols, 1974; Quin and Dawson, 1969b; Demel *et al.*, 1989; Pilon *et al.*, 1987) or using pure negatively charged phospholipids (Nichols, 1974; Quin and Dawson, 1969a,b; Demel *et al.*, 1989; Pilon *et al.*, 1987; Kimelberg and Papahadjopoulos, 1971b; Papahadjopoulos *et al.*, 1973).

Information on the type of interaction of proteins with model membranes can be obtained by monitoring the effect of proteins on the energy content of the gel-to-liquid-crystalline phase transition of lipid dispersions. Figure 12.7 shows the effect of apo- and holocytochrome c on the energy content of the phase transition in dielaidoylphosphatidylserine (DEPS) dispersions. DEPS undergoes an endothermic transition at the convenient temperature of 25 °C with ΔH of 5.7 kcal/mol. Binding of increasing amounts of apocytochrome c to DEPS dispersion causes a linear de-crease in the ΔH of the gel-to-liquid-crystalline transition (Figure 12.7) without a large effect on the transition temperature (Li-Xin *et al.*, 1988). Such behavior, which was also observed for dimyristoyl-PS and dimyristoyl-phosphatidylglycerol (Rietveld *et al.*, 1983), can most simply be interpreted as the result of perturbation of gel-state packing by membrane-penetrated polypeptide (McElhaney, 1986). The cooperative transition is completely removed by 40 nmol apocytochrome c bound per micromole DEPS. However, in strong contrast, binding of increasing amounts

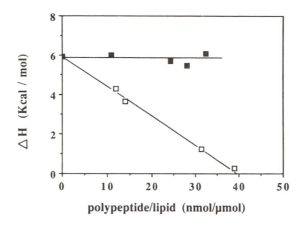

Figure 12.7

Influence of increasing amounts of bound apocytochrome c (□) and cytochrome c (■) on the energy content of the gel-to-liquid-crystalline phase transition of DEPS. Data from Li-Xin et al. (1988).

of cytochrome c has no effect on both the transition temperature (Li-Xin et al., 1988) and ΔH of the transition of DEPS (Figure 12.7), indicating that cytochrome c penetrates less efficiently than apocytochrome c, in agreement with the monolayer measurements. Cytochrome c is able to influence to some extent the energy content of the gel-to-liquid-crystalline transition of lipid dispersions with a looser molecular packing (Papahadjopoulos et al., 1975; Chapman et al., 1974; Chapman and Urbina, 1971) compared to the packing of DEPS liposomes (Demel et al., 1987).

Further evidence for the difference in the efficiency of penetration of apo- and holocytochrome c into model membranes comes from electron spin resonance (ESR) measurements with phospholipids spin labeled in the acyl chain. Figure 12.8a shows the ESR spectrum of a phosphatidylglycerol (PG) molecule spin-labeled at the 12-position of the acyl chain in a pure PS dispersion. Such a spectrum is typical of phospholipids experiencing the motions present in fluid bilayers. On binding of apocytochrome c, a second, broader spectral component is observed (indicated with an arrow in Figure 12.8b) (Görrissen et al., 1986; Jordi et al., 1989b). This demonstrates that apocytochrome c restricts the motion of a fraction of the PG spin label at the 12-position of the acyl chain. The ability to perturb the mobility of a spin label attached to the acyl chains of the phospholipids can be interpreted as a result of penetration into the hydrophobic part of the membrane. In strong contrast, no motionally restricted component is observed in the presence of holocytochrome c (Figure 12.8c), demonstrating that the protein has no effect on the mobility of the spin label attached at the C12 position of the acyl chain, and, therefore, the protein is not penetrating deeply into the hydrophobic core of the membrane.

These experiments demonstrate that after the initial electrostatic interaction, apocytochrome c penetrates more efficiently than cytochrome c into model membranes. The interpretation that apocytochrome c penetrates the hydrophobic core

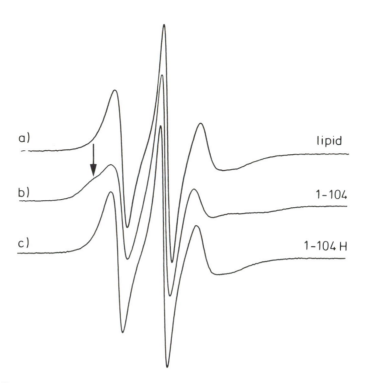

Figure 12.8
ESR spectra of the 12-PGSL spin label in PS dispersions in 10 mM Hepes, 10 mM NaCl, 0.1 mM EDTA, pH 7.0 (containing 0.01% mercaptoethanol), in the absence (a), and presence of saturating amounts (5 mg peptide or protein/mg lipid) of apocytochrome c (b), and cytochrome c (c). Total scan width = 100 G; $t = 30$ °C. Data from Jordi *et al.* (1989b). See that reference for experimental details.

of the bilayer is further corroborated by the fact that part of apocytochrome c is protected against trypsin digestion in the protein–lipid complex, and that this protected fragment could be labeled from the hydrophobic interior of the bilayer by dansylchloride (Dumont and Richards, 1984; Lee and Kim, 1989). Furthermore, it was demonstrated that the carboxy-terminus of apocytochrome c is entirely protected against carboxypeptidase Y digestion after interaction with lipids (Lee and Kim, 1989). The efficiency of labeling cytochrome c from deeply inside the hydrophobic core of the bilayer with a photoactivatable phospholipid is much less than that of the labeling of apocytochrome c (Lee and Kim, 1989). Additionally, the induction of release of vesicle content upon binding of cytochrome c (Kimelberg and Papajadjopoulos, 1971a; Rietveld *et al.*, 1983; Li-Xin *et al.*, 1988; Kimelberg and Papahadjopoulos, 1971b; Papahadjopoulos *et al.*, 1973) is much less efficient compared to the effect of apocytochrome c (Rietveld *et al.*, 1983; Li-Xin *et al.*, 1988). For some reason, cytochrome c exhibits a less efficient membrane penetration. The major difference between both proteins is the covalently attached heme. We will now focus on the role of the heme for the protein–lipid interaction.

C. Role of the Heme

The role of the heme in the protein–lipid interaction has been studied by comparing the penetration of heme-containing and heme-free peptides into model membranes (Jordi *et al.*, 1989a; Demel *et al.*, 1989; Jordi *et al.*, 1989b; Li-Xin *et al.*, 1988; Jordi *et al.*, 1990a; Snel *et al.*, 1991). Figure 12.9 shows the effect of covalent attachment of the heme to apocytochrome *c* (1–104) and to the amino-terminal peptide residue numbers 1–65 on the ability to induce surface pressure increases for a pure dioleoyl-phosphatidylserine (DOPS) monolayer at an initial surface pressure of 20 mN/m. The covalent coupling of the heme moiety to apocytochrome *c* (1–104) strongly reduces the ability of the protein to cause a pressure increase for a DOPS mono-layer, as described earlier. In contrast, the amino-terminal heme-containing peptide (residue numbers 1–65) is even slightly more efficient in causing a pressure increase than the corresponding apopeptide, strongly suggesting that the heme itself does not cause the decrease in membrane penetration of cytochrome *c*. This is further corroborated by vesicle leakage experiments (Li-Xin *et al.*, 1988), differential scan-ning calorimetry measurements (Li-Xin *et al.*, 1988), ESR measurements with spin-labeled lipids (Jordi *et al.*, 1989b) and ^{2}H NMR measurements with deuterated phospholipids (Jordi *et al.*, 1990a) investigating the interaction of apo- and heme-containing peptides with model membrane systems. These measurements also dem-onstrated that differences in polypeptide–lipid interaction cannot be explained by a different amount of protein bound to the liposomes, but rather reflect a different type of interaction with the lipids.

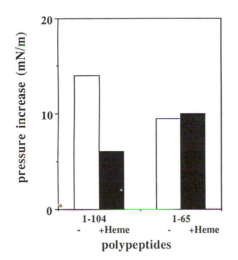

Figure 12.9
Interaction of apocytochrome *c* (1–104), holocytochrome *c* (1–104 +heme), the amino-terminal apopeptide residue numbers 1–65, and the corresponding heme-containing peptide (1–65 +heme) with DOPS monolayers at an initial surface pressure of 20 mN/m. Average experimental error 1 mN/m. Data from Jordi *et al.* (1989a). See that reference for experimental details.

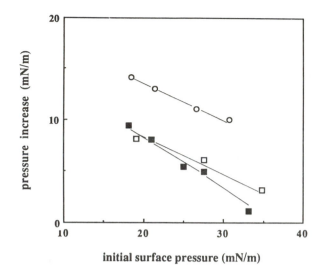

Figure 12.10

Effect of unfolding of cytochrome c by urea on the surface pressure increase of DOPS monomolecular layers at different initial surface pressures: untreated cytochrome c (■), cytochrome c incubated with 8 M urea (○), cytochrome c incubated with 8 M urea and subsequently dialyzed for 16 h (□). Data from Görrissen *et al.* (1986). See that reference for experimental details.

To investigate the conformational dependence of cytochrome c penetration of model membranes, the penetration of cytochrome c unfolded by heat treatment or incubation in 8 M urea (Myer, 1968) into model membranes was analyzed by monolayer measurements. Unfolded cytochrome c exhibits an enormous increase in surface pressure relative to that of the native cytochrome c. In particular, at initial pressures where the native cytochrome c no longer responds (> 34 mN/m), the heated protein yields pressure increases nearly identical to those of apocytochrome c (Figure 12.10). Refolding of the polypeptide by dialysis removal of urea leads again to a reduced affinity for PS (Figure 12.10). The monolayer measurements indicate that the inability of cytochrome c to penetrate is due to its compact globular conformation.

V. Structural Consequences of the Protein–Lipid Interaction

A. Protein Folding

A number of experiments indicate that the conformation of the protein changes as a consequence of the interaction of the protein with acidic phospholipids. Resonance Raman spectroscopy data (Vincent and Levin, 1986; Hildebrandt and Stockburger, 1989; Hildebrandt *et al.*, 1990) and magnetic resonance data (Vincent *et al.*, 1987;

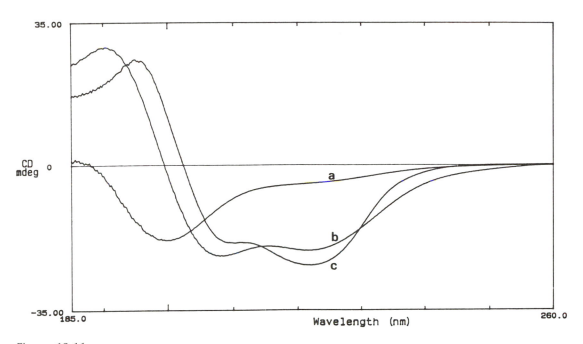

Figure 12.11
Circular dichroism spectra of apocytochrome c in buffer (a), apocytochrome c in the presence of DOPS large unilamellar vesicles (b) and cytochrome c in buffer (c). Data from De Jongh and De Kruijff (1990). See that reference for experimental details.

Soussi et al., 1990) indicate that both the porphyrin ring conformation and the heme coordination are changed upon binding of the protein to acidic phospholipids. In addition, the tertiary structure of cytochrome c is destabilized upon binding to anionic lipids (Muga et al., 1991b; Spooner and Watts, 1991a,b). Association of cytochrome c with detergents that mimic anionic phospholipids results in a highly dynamical, partially α-helical conformation of the protein, as shown by the hydrogen and deuterium exchange of protein backbone amide groups (De Jongh et al., 1992).

Apocytochrome c in solution has hydrodynamic properties (Stellwagen and Rysavy, 1972) and a featureless circular dichroism (CD) spectrum, indicative of a largely random coil structure (Figure 12.11a) (Fisher et al., 1973; Jordi et al., 1989a; Rietveld et al., 1985; Toniolo et al., 1975; Walter et al., 1986; De Jongh and De Kruijff, 1990; Stellwagen and Rysavy, 1972). In contrast, cytochrome c shows extreme ellipticity at 209 and 222 nm typical of α-helical structures (Figure 12.11c) in agreement with the x-ray crystal structure of the protein (Bushnell et al., 1990). Interaction of apocytochrome c with negatively charged detergents (Jordi et al., 1989a; Rietveld et al., 1985) or vesicles composed of negatively charged phospholipids (Walter et al., 1986; De Jongh and De Kruijff, 1990) induces a partial α-helical conformation in the protein in both cases (Figure 12.11b). Because the

Table 12.2
α-Helix formation in apocytochrome c and derived
peptides in 10 mM phosphate buffer or 0.5% SDS.
Data from Jordi *et al.* (1989a).

Peptide or protein	α-Helix (%)	
	buffer	SDS
Apocytochrome c	9	21
1–38	9	16
1–59	2	10
1–65	6	19
66–80	6	no fit obtained
39–104	9	20
60–104	8	26
81–104	13	31

turbidity contribution to the spectrum is less in the case of detergents, these systems are more suitable to quantification of the spectral changes. To investigate the localization of the lipid-induced α-helices in apocytochrome c, CD measurements were performed with the various peptides (for schematic representation of the various peptides used, see Figure 12.3). All peptides have CD spectra typical of polypeptides with a largely random coil structure (Jordi *et al.*, 1989a; Toniolo *et al.*, 1975; De Jongh and De Kruijff, 1990; Babul *et al.*, 1972). However, in the presence of SDS (Jordi *et al.*, 1989a) or PS vesicles (De Jongh and De Kruijff, 1990), the polypeptides, except fragment 66–80, convert into in a partly α-helical conformation. Fragment 66–80, on the other hand, has a spectrum indicative of a large amount of β-turn (De Jongh and De Kruijff, 1990). From the CD spectra, the secondary structure of the proteins and peptides was determined by mathematical curve-fitting using the method and reference spectra of Chen *et al.* (1974). The spectral analysis shows that apocytochrome c and the various amino-terminal and carboxy-terminal peptides in buffer have only a small amount of α-helix (Table 12.2).

For all fragments, SDS induces a strong increase in the calculated amount of α-helical content. In general, the carboxy-terminal fragments have a higher α-helical content than the amino-terminal fragments in the presence of SDS. It can be suggested that the tendency to form an α-helix is expressed by the presence of negatively charged lipids and detergents or after covalent coupling of the heme to the apoprotein. There are strong correlations between the α-helical content of the various peptides in the presence of SDS and the α-helical content of these parts of the cytochrome c molecule judged from the x-ray structure (Jordi *et al.*, 1989a; Toniolo *et al.*, 1975; De Jongh and De Kruijff, 1990). This suggests that the helices, which are induced in apocytochrome *c* by interaction with lipids, are localized at positions similar to those in the mature protein. Fourier-transform infrared (FTIR) spectroscopy measurements also demonstrated transitions from random coil conformation to an α-helical structure at high lipid-to-protein ratios (>40:1) (Muga *et al.*, 1991a). The induced folding of apocytochrome *c* at water–lipid interfaces is highly dynamical, judging from the hydrogen and deuterium exchange of protein backbone amide groups monitored by NMR (De Jongh *et al.*, 1992).

B. Lipid Packing and Organization

FTIR, ^{2}H, and ^{31}P NMR measurements have been used to study the effect of the binding of apo- and holocytochrome *c* on the lipid packing and organization of model membranes. The interaction of cytochrome *c* with PS dispersion shows no indication of changes in the overall bilayer configuration (Jordi *et al.*, 1990a; Smith *et al.*, 1983; Devaux *et al.*, 1986). Furthermore, only small changes in the orientation of the PS headgroup and no modification in the order of the acyl chains of the PS molecules have been reported on binding of cytochrome *c* (Jordi *et al.*, 1990a; Devaux *et al.*, 1986). In mixed PC/PS systems, no evidence for protein-induced lateral phase segregation was found (Devaux *et al.*, 1986). X-ray and electron microscopy data show that negatively charged phospholipids spontaneously form unilamellar vesicles in excess buffer (e.g., Hauser, 1984). Addition of cytochrome *c* converts these vesicles into a multilamellar structure caused by stacking of the negatively charged bilayers by the positively charged protein (Jordi *et al.*, 1990a; Papahadjopoulos and Miller, 1967; Gulik-Krzywicki *et al.*, 1969; Shipley *et al.*, 1969; Kimelberg *et al.*, 1970).

The consequences of the interaction of apocytochrome *c* with negatively charged phospholipid dispersions for lipid packing and organization have been extensively studied by ^{2}H, ^{31}P NMR, and Fourier-transform infrared techniques. FTIR measurements demonstrated that binding of the protein to model membranes composed of DMPG resulted in a drastic perturbation of phospholipid structure, at the level of both the acyl chains and the carbonyl groups. Binding of the protein to dioleoyl-PS liposomes results in ^{31}P NMR spectra typical of phospholipids in liquid-crystalline bilayers, with a reduced residual chemical shift anisotropy and an increased line width. ^{2}H NMR spectra on headgroup-deuterated dioleoyl-PS dispersions demonstrated a decrease in quadrupolar splitting and a broadening of the signal on interaction with apocytochrome *c*. Both results are consistent with an increased motional averaging within a bilayer orientation (Jordi *et al.*, 1990a). Figure 12.12 shows the deuterium spectrum of hydrated dioleoyl-PS deuterated at the 11-position of the acyl chain. The distance between the two peaks (the quadrupolar splitting) is a measure of the order of the carbon–deuterium bond in the acyl chain. Addition of apocytochrome *c* to the acyl chain deuterated dioleoyl-PS dispersions results in the gradual appearance of a second component in the spectrum with a 44% reduced quadrupolar splitting (Jordi *et al.*, 1990a). At a lipid:protein molar ratio of 10:1, the original splitting disappears, and the resulting ^{2}H NMR spectrum indicates a strong disorder of the phospholipid acyl chains on binding of apocytochrome *c*

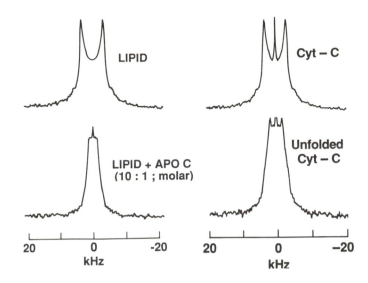

Figure 12.12
^{2}H NMR spectra of [11,11-^{2}H$_2$]-DOPS dispersions in the absence or presence of 10 : 1 lipid : protein (molar) of apocytochrome c, cytochrome c, and heat-denatured (unfolded) cytochrome c at 30 °C. Data from Jordi *et al.* (1990a).

(Figure 12.12). In contrast, addition of similar amounts of cytochrome c has no effect on the quadrupolar splitting of the spectrum other than the induction of a small isotropic component due to deuterons present in the buffer. Unfolding of cytochrome c by heat treatment restores the ability of the protein to influence the acyl chain order (Figure 12.12). The interaction of apocytochrome c with mixtures of negatively charged and zwitterionic phospholipids does not result in macroscopic phase segregation between the negatively charged and zwitterionic lipid component (Rietveld *et al.*, 1986b). However, fluorescence spectroscopy (Berkhout *et al.*, 1987), ESR measurements (Rietveld *et al.*, 1986b) and NMR measurements (Jordi *et al.*, 1990a) demonstrate that there is a small but significant preferential interaction with the negatively charged lipid component. The interaction of apocytochrome c with negatively charged liposomes results in stacking of bilayers, as evidenced by small-angle x-ray measurements on the protein–lipid complex (Jordi *et al.*, 1990a), and its effect is comparable to that of cytochrome c on the lipid dispersions. The ability of the apoprotein to cause close contact between model membranes is much greater than for the holoprotein, as evidenced by the ability of the apo- and not the holo-protein to cause contacts between anionic lipids containing vesicles and a lipid monolayer (Demel *et al.*, 1989).

Several lines of experimental evidence suggest a special interaction between cytochrome c and the mitochondrial phospholipid cardiolipin. Cytochrome c exhibits a higher affinity for cardiolipin model membranes than for other negatively charged phospholipids (Demel *et al.*, 1989; Teissie and Baudras, 1977; Mustonen *et*

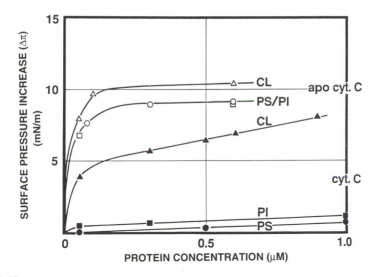

Figure 12.13

Surface pressure increase of cardiolipin (CL), dioleoyl-PS, and PI monomolecular layers as a function of apocytochrome c and cytochrome c concentration at an initial surface pressure of 34 mN/m. Apocytochrome c: $\triangle$, CL; $\circ$, PS; $\square$, PI. Cytochrome c: $\blacktriangle$, CL; $\bullet$, PS; $\blacksquare$, PI. The subphase contained 10 mM Pipes, pH 7.0, 50 mM NaCl. Data from Demel *et al.* (1989). See that reference for experimental details.

al., 1987; Rietveld *et al.*, 1983). This preferential interaction of cytochrome c with cardiolipin is illustrated in Figure 12.13. Apocytochrome c induces comparable pressure increases for monolayers prepared from three chemically different negatively charged phospholipids. In strong contrast, cytochrome c causes virtually no increase in surface pressure for the PI and PS monolayer, but it efficiently induces increases in the surface pressure for cardiolipin monolayers (Demel *et al.*, 1989). In contrast, the interaction of unfolded cytochrome c has a phospholipid specificity similar to that of apocytochrome c. This demonstrates that upon folding into its functional conformation, cytochrome c acquires the specific affinity for cardiolipin. Brown and Wüthrich (1977) showed by ^{1}H and ^{13}C NMR spectroscopy that in mixtures of PC and cardiolipin, PC is segregated out on binding of cytochrome c to the cardiolipin component. Cardiolipin model membranes can, as a consequence of the interaction with cytochrome c, partly convert from bilayer into nonbilayer states, as demonstrated by changes in the ^{31}P NMR spectra (De Kruijff and Cullis, 1980; Rietveld *et al.*, 1983). That these are of an inverted nature is suggested by monolayer measurements (Demel *et al.*, 1989) and by the efficient extraction of cytochrome c with negatively charged phospholipids into an organic phase (for a review, see Nichols, 1974). Waltham *et al.* (1986) and Spooner and Watts (1991a,b) have reported that the average orientation of the cardiolipin headgroup is not greatly affected on binding of cytochrome c, but there is a pronounced reduction in T_1 values of the cardiolipin phosphorus. Waltham *et al.* interpret these results

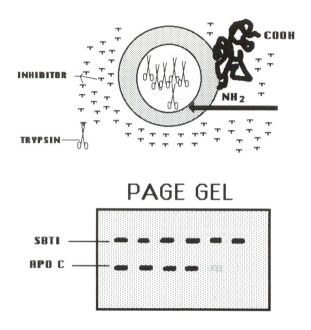

Figure 12.14
Schematic illustration of the principle of the translocation assay. Apocytochrome c is added to large unilamellar vesicles containing trypsin. Residual non-enclosed trypsin is blocked by soybean trypsin inhibitor (SBTI). At different times after the addition of apocytochrome c to the vesicles, samples are drawn and processed for polyacrylamide gel electrophoresis (PAGE). The lower part of the figure shows a schematic representation of a gel of translocation experiments with six time points. A decrease in the amount of apocytochrome c relative to the amount of SBTI in the incubation mixture is interpreted as (partial) translocation of the protein to the vesicle interior.

as a reduction of the headgroup motion, whereas Spooner and Watts describe the results in terms of a conformational change of cytochrome c which allows efficient paramagnetic interaction of the protein with the cardiolipin headgroup.

In summary, apocytochrome c efficiently penetrates into negatively charged lipid dispersions and exhibits strong disordering effects on the headgroup and acyl chains of the phospholipids, whereas the overall bilayer configuration is maintained. In contrast, cytochrome c penetrates efficiently only into cardiolipin dispersions and can convert the cardiolipin bilayer into inverted nonlamellar lipid structures.

C. Membrane Insertion/Translocation

To determine if membrane penetration of apocytochrome c results in exposure of (part of) the protein to the opposite interface of the bilayer, a liposomal protein "translocation" assay has been described (Jordi *et al.*, 1989a; Rietveld *et al.*, 1986a; Dumont and Richards, 1984; Rietveld and De Kruijff, 1984). The principle of the assay is illustrated in Figure 12.14. Apocytochrome c is added to large unilamellar

Table 12.3
Trypsin-containing vesicles composed of PS/egg-PC
(molar ratio 1:1) incubated with the various polypeptides
for 2 h at 30 °C.[a]

Peptide or protein	Peptide or protein present after 2 h incubation (% of $t = 0$)
1–38H	112 ± 14
1–65H	107 ± 6
Apocytochrome *c*	53 ± 7
1–38	73 ± 7
1–59	54 ± 11
1–65	59 ± 10
39–104	100 ± 13
60–104	95 ± 12
81–104	102 ± 6

[a] The digestion is presented as a percentage of the amount initially
bound to the vesicles. The error represents the standard deviation; $N = 4$.
Data are from Jordi *et al.* (1989a). See that reference for details.

vesicles containing trypsin. The bulk of non-enclosed trypsin is removed by centrifugation of the vesicle suspension, and residual non-enclosed trypsin is blocked by
a large molar excess of the soybean trypsin inhibitor (SBTI). A decrease in the
amount of apocytochrome *c* relative to the amount of SBTI in the incubation
mixture as visualized by SDS PAGE can, with the appropriate control experiments
as described (Rietveld *et al.*, 1986a; Rietveld and De Kruijff, 1984), be interpreted as
(partial) translocation of the protein to the vesicle interior, where it is digested by
the protease. The "translocation assay" strongly discriminates between various polypeptides. Geller and Wickner (1985) have described that the precursor protein for
M13 spontaneously translocates across the lipid bilayer in a very similar assay. In
strong contrast, cytochrome *c* does not spontaneously translocate across pure lipid
bilayers (Geller and Wickner, 1985), in agreement with the inefficient insertion
of the protein into the hydrophobic core of the bilayer. We have determined the
"translocation" of various cytochrome *c*- and apocytochrome *c*–derived peptides
across a lipid bilayer composed of egg PC/bovine brain PS (molar ratio 1:1).
Table 12.3 shows the percentage of peptides or proteins that were digested in 2 h by
the enclosed trypsin. Control experiments carried out as described (Rietveld and De
Kruijff, 1984) showed that the lipid barrier of the vesicles remained intact during the
translocation assay, with respect to both trypsin and a small chromogenic substrate
of trypsin. Since it was also demonstrated that digestion of the peptides could only

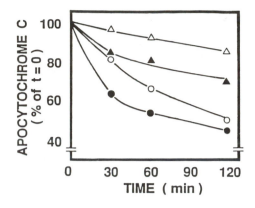

Figure 12.15

Time dependence of apocytochrome c translocation across model membranes composed of equimolar DOPS/DOPC. The trypsin-containing vesicles were incubated at 0 °C ($\triangle$), 5 °C ($\blacktriangle$), 30 °C (o), or 37 °C (●). Data from Li-Xin *et al.* (1988). See that reference for experimental details.

occur inside the vesicles, it can be concluded that apocytochrome c and the amino-terminal fragments 1–38, 1–59, and 1–65 are able to reach the opposite interface of the bilayer at least partially. The other peptides were not digested to a significant extent during the incubation.

For membrane translocation, there is an absolute requirement for negatively charged phospholipids (Rietveld *et al.*, 1986a), since only bound protein potentially can penetrate and (partially) translocate across the bilayer. The molecular packing of the lipids plays an additional role in translocation of the protein across the bilayer. This role can be illustrated by considering the effect of a membrane-active compound on the translocation process. Phenethylalcohol (PEA) is such a molecule, in that it causes a large disordering effect on the acyl chains of phospholipids (Jordi *et al.*, 1990b). Inclusion of 1% (v/v) PEA in the translocation assay facilitates the digestion of apocytochrome c by trypsin enclosed in large unilamellar vesicles, indicating that the protein "translocates" faster. For DOPS:DOPC (molar ratio 1:1) vesicles, the half life, $t_{1/2}$, of apocytochrome c translocation decreases from 20 in the absence of 1% (v/v) PEA to 6 min in its presence (Jordi *et al.*, 1990b). Additional information on the influence of the physical state of lipids on apocytochrome c translocation can be derived from the temperature dependency of apocytochrome c translocation. The bilayer of DOPS/DOPC large unilamellar vesicles is in a liquid-crystalline state under the experimental conditions used for the translocation experiments. Figure 12.15 shows that the amount of digested apocytochrome c is strongly temperature-dependent and that translocation is virtually blocked at 0 °C, emphasizing the importance of a loosely packed bilayer for efficient protein translocation (Li-Xin *et al.*, 1988).

VI. Consequences for Apocytochrome c Import into Mitochondria

The characteristics of the import pathway of apocytochrome c into mitochondria are different compared to those for other mitochondrial precursor proteins (for a review on the import pathway of apocytochrome c, see Stuart and Neupert, 1990). The protein is synthesized without a cleavable amino terminal presequence, and the import is not dependent on nucleoside triphosphates or the presence of a membrane potential across the inner mitochondrial membrane. We have suggested that the special protein–lipid interactions of the precursor and the holoprotein are essential for the unidirectional translocation across the outer mitochondrial membrane (Jordi et al., 1989a). The fact that apocytochrome c exhibits a high-affinity binding for negatively charged phospholipids and spontaneously translocates across a lipid bilayer may enable apocytochrome c to circumvent early steps in the general import pathway into mitochondria. We will first summarize the data which indicate that apocytochrome c initially interacts with the negatively charged lipids of the outer mitochondrial membrane: (i) No protease-sensitive components exist on the mito- chondrial surface which mediate binding and import of apocytochrome c into mitochondria (Nicholson et al., 1988); (ii) two identified proteinaceous receptors for mitochondrial precursor proteins MOM19 and MOM72 are not involved in the import of apocytochrome (Söllner et al., 1989, 1990); (iii) apocytochrome c binds to lipid domains in arrays of mitochondrial outer membrane channels (Manella et al., 1987); (iv) the apparent K_D for binding of apocytochrome c to liposomes containing negatively charged phospholipids (e.g., Rietveld et al., 1986a) and mitochondria (e.g., Hennig et al., 1984) is of the same order of magnitude; and (v) binding of in vitro–synthesized Neurospora crassa apocytochrome c to negatively charged liposomes can compete with mitochondria for import of apocytochrome c into the organelle (Jordi et al., 1992).

After binding of apocytochrome c, it is proposed that insertion of the initially random coil protein into the lipid part of the outer membrane is accompanied by the formation of three α-helices in apocytochrome c (Figure 12.16). These α-helices are localized in regions of the apocytochrome c molecule that will remain α-helical in the mature holocytochrome c, indicating that the polypeptide–lipid interaction induces the basic secondary structure elements of mature holocytochrome c. During the insertion of the protein, there is a preferential interaction with negatively charged phospholipids. In contrast to the compact structure of holocytochrome c, we think that at this stage in the import pathway the tertiary structure of apocyto- chrome c is still loosely folded to allow efficient attachment of the heme to the two cysteine residues at a later stage in the import route. The amino-terminal α-helix (amino acid residue numbers 4–17) is particularly interesting, since this part of the polypeptide contains the two cysteine residues (numbers 14 and 17 in horse apocytochrome c) to which the heme is covalently attached by cytochrome c heme lyase in the intermembranous space. The translocation experiments with an amino- terminal peptide fragment (amino acid residue numbers 1–38) demonstrated that

470

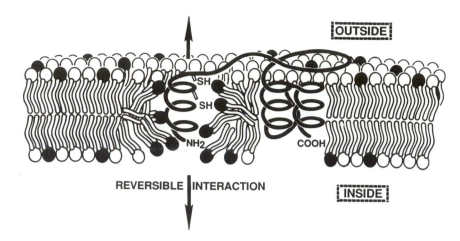

Figure 12.16

Model for the penetration into and translocation of apocytochrome c across a bilayer composed of mixed lipids. The lipids with the dark headgroups represent negatively charged phospholipids, which are thought to be enriched especially around the amino-terminal region of the protein. The amino-terminal region of the protein is supposed to exhibit a reversible translocation across the bilayer, causing a disorder in the phospholipids. Three α-helices are indicated. The indicated orientation of the α-helices is speculative. The approximate location of the thiol groups of the cysteine residues is shown.

this part of the polypeptide is able to (partially) translocate across a lipid bilayer. We think that the initial penetration of the α-helix disorders the phospholipid molecules and allows deeper insertion of the amino-terminal region of the protein into the outer mitochondrial membrane in such a way that the relatively hydrophobic heme can be attached to the two cysteine residues by cytochrome c heme lyase, which has its active site exposed to the mitochondrial intermembranous space (Nicholson *et al.*, 1988). This results in a further refolding of the protein in such a way that the hydrophobic residues of the protein are facing the heme moiety in the interior of the protein. As a consequence of refolding of the protein, the affinity for the outer mitochondrial membrane is strongly reduced, especially at physiologically relevant ionic strength. This results in release of the protein from the outer mitochondrial membrane into the intermembranous space.

VII. Acknowledgments

The authors thank the various scientists who contributed to the research described in this chapter. Drs. P. J. R. Spooner and A. Watts are thanked for communicating studies on cytochrome c dynamics prior to publication. This work was carried out under the auspices of the Netherlands Foundation of Chemical Research (SON) and with the financial aid from the Netherlands Organization for Scientific Research (NWO).

VIII. References

Babul, J., McGowan, E. B., & Stellwagen, E. (1972) *Arch. Biochem. Biophys.* *148*, 141–147.
Berkhout, T. A., Rietveld, A., & De Kruijff, B. (1987) *Biochim. Biophys. Acta* *897*, 1–4.
Brown, L. R., & Wüthrich, K. (1977) *Biochim. Biophys. Acta* *468*, 389–395.
Bushnell, G. W., Louie, G. V., & Brayer, G. D. (1990) *J. Mol Biol.* *214*, 585–595.
Chapman, D., & Urbina, J. (1971) *FEBS Lett.* *12*, 169–172.
Chapman, D., Urbina, J., & Keough, K. M. (1974) *J. Biol. Chem.* *249*, 2512–2521.
Chen, Y. H., Yang, J. T., & Chau, K. H. (1974) *Biochemistry* *13*, 3350–3359.
Daum, G. (1985) *Biochim. Biophys. Acta* *822*, 1–42.
De Jongh, H. H. J., & De Kruijff, B. (1990) *Biochim. Biophys. Acta* *1029*, 105–112.
De Jongh, H. H. J., Killian, J. A., & De Kruijff, B. (1992) *Biochemistry* *31*, 1636–1643.
De Kruijff, B., & Cullis, P. R. (1980) *Biochim. Biophys. Acta* *602*, 477–490.
Demel, R. A., Geurts van Kessel, W. S. M., Zwaal, R. F. A., Roelofsen, B., & Van Deenen, L. L. M. (1975) *Biochim. Biophys. Acta* *406*, 97–107.
Demel, R. A., Paltauf, F., & Hauser, H. (1987) *Biochemistry* *26*, 8659–8665.
Demel, R. A., Jordi, W., Lambrechts, H., Van Damme, H., Hovius, R., & De Kruijff, B. (1989) *J. Biol. Chem.* *264*, 3988–3997.
Devaux, P. F., Hoatson, G. L., Favre, E., Fellmann, P., Farren, B., MacKay, A. L., & Bloom, M. (1986) *Biochemistry* *25*, 3804–3812.
Drygas, M. E., Lambowitz, A. M., & Nargang, F. E. (1989) *J. Biol. Chem.* *264*, 17897–17906.
Dumont, M. E., & Richards, F. M. (1984) *J. Biol. Chem.* *259*, 4147–4156.
Dumont, M. E., Ernst, J. F., Hampsey, D. M., & Sherman, F. (1987) *EMBO J.* *6*, 235–241.
Eisenberg, D., Schwarz, E., Komaromy, M., & Wall, R. (1984) *J. Mol. Biol.* *179*, 125–142.
Faucon, J. F., Dufourcq, J., Lussan, C., & Bernon, R. (1976) *Biochim. Biophys. Acta* *435*, 283–294.
Fisher, W. R., Taniuchi, H., & Anfinsin, C. B. (1973) *J. Biol. Chem.* *248*, 3188–3195.
Fry, M., & Green, D. E. (1980) *Biochem. Biophys. Res. Commun.* *93*, 1238–1246.
Geller, B. L., & Wickner, W. (1985) *J. B ol. Chem.* *260*, 13281–13285.
Görrissen, H., Marsh, D., Rietveld, A., & De Kruijff, B. (1986) *Biochemistry* *25*, 2904–2910.
Gulik-Krzywicki, T., Schechter, E., Luzzati, V., & Faure, M. (1969) *Nature* *223*, 1116.
Gupte, S. S., & Hackenbrock, C. R. (1988a) *J. Biol. Chem.* *263*, 5241–5247.
Gupte, S. S., & Hackenbrock, C. R. (1988b) *J. Biol Chem.* *263*, 5248–5253.
Hakvoort, T. B. M., Sprinkle, J. R., & Margolis, E. (1990) *Proc. Natl. Acad. Sci. USA* *87*, 4996–5000.
Harb, J. S., Comte, J., & Gautheron, D. C. (1981) *Arch. Biochim. Biophys.* *208*, 305–518.
Hauser, H. (1984) *Biochim. Biophys. Acta* *988*, 1–45.
Hennig, B., Koehler, H., & Neupert, W. (1984) *Proc. Natl. Acad. Sci. USA* *80*, 4963–4967.
Hildebrandt, P., & Stockburger, M. (1989) *Biochemistry* *28*, 6722–6728.
Hildebrandt, P., Heimburg, T., & Marsh, D. (1990) *Eur. Biophys. J.* *18*, 193–201.
Hovius, R., Lambrechts, H., Nicolay, K., & De Kruijff, B. (1990) *Biochim. Biophys. Acta* *1021*, 217–226.
Jordi, W., Li-Xin, Z., Pilon, M., Demel, R. A., & De Kruijff, B. (1989a) *J. Biol. Chem.* *264*, 2292–2301.
Jordi, W., De Kruijff, B., & Marsh, D. (1989b) *Biochemistry* *28*, 8998–9005.
Jordi, W., De Kroon, A. I. P. M., Killian, J. A., & De Kruijff, B. (1990a) *Biochemistry* *29*, 2312–2321.
Jordi, W., Nibbeling, R., & De Kruijff, B. (1990b) *FEBS Lett.* *261*, 55–58.
Jordi, W., Hergersberg, C., & De Kruijff, B. (1992) *Euro J. Biochem.* *204*, 841–846.
Kimelberg, H. K., & Papahadjopoulos, D. (1971a) *J. Biol. Chem.* *246*, 1142–1148.
Kimelberg, H. K., & Papahadjopoulos, D. (1971b) *Biochim. Biophys. Acta* *233*, 805–809.
Kimelberg, H. K., Lee, C. P., Claude, A., & Mrena, E. (1970) *J. Membr. Biol.* *2*, 235–251.
Krebs, J. J. R., Hauser, H., & Carafoli, E. (1979) *J. Biol. Chem.* *254*, 5308–5316.
Lee, S., & Kim, H. (1989) *Arch. Biochem. Biophys.* *271*, 188–199.
Li-Xin, Z., Jordi, W., & De Kruijff, B. (1988) *Biochim. Biophys. Acta* *942*, 115–124.
Manella, C. A., Ribeiro, A. J., & Frank, J. (1987) *Biophys. J.* *51*, 221–226.
Mathews, F. S. (1985) *Progr. Biophys. Mol. Biol.* *45*, 1–56.
Matsuura, S., Arpin, M., Hannum, C., Margoliash, E., Sabatini, D. D., & Morimoto, T. (1981) *Proc. Natl. Acad. Sci. USA* *78*, 4368–4372.

McElhaney, R. N. (1986) *Biochim. Biophys. Acta* **864**, 361–421.

Muga, A., Mantsch, H. H., & Surewicz, W. K. (1991a) *Biochemistry* **30**, 2629–2635.

Muga, A., Mantsch, H. H., & Surewicz, W. K. (1991b) *Biochemistry* **30**, 7219–7224.

Mustonen, P., Virtanen, J. A., Somerharju, P. J., & Kinnuunen, P. K. J. (1987) *Biochemistry* **26**, 2991–2997.

Myer, Y. P. (1968) *Biochemistry* **7**, 765–776.

Nichols, P. (1974) *Biochim. Biophys. Acta* **346**, 261–310.

Nicholson, D. W., Köhler, H., & Neupert, W. (1987) *Eur. J. Biochem.* **164**, 147–157.

Nicholson, D. W., Hergersberg, C., & Neupert, W. (1988) *J. Biol. Chem.* **263**, 19034–19042.

Nilsson, O. S., & Dallner, G. (1977) *Biochim. Biophys. Acta* **464**, 453–458.

Papahadjopoulos, D., & Miller, N. (1967) *Biochim. Biophys. Acta* **135**, 624.

Papahadjopoulos, D., Cowden, H., & Kimelberg, H. (1973) *Biochim. Biophys. Acta* **330**, 8–26.

Papahadjopoulos, D., Moscarello, M., Eylar, E. H., & Isac, T. (1975) *Biochim. Biophys. Acta* **401**, 317–335.

Pilon, M., Jordi, W., De Kruijff, B., & Demel, R. A. (1987) *Biochim. Biophys. Acta* **466**, 10–22.

Quin, P. J., & Dawson, R. M. C. (1969a) *Biochem. J.* **113**, 791–803.

Quin, P. J., & Dawson, R. M. C. (1969b) *Biochem. J.* **115**, 65–75.

Rietveld, A., & De Kruijff, B. (1984) *J. Biol. Chem.* **259**, 6704–6704.

Rietveld, A., Sijens, P., Verkleij, A. J., & De Kruijff, B. (1983) *EMBO J.* **2**, 907–913.

Rietveld, A., Ponjee, G. A. E., Schiffers, P., Jordi, W., Van de Coolwijk, P. J. F. M., Demel, R. A., Marsh, D., & De Kruijff, B. (1985) *Biochim. Biophys. Acta* **818**, 398–409.

Rietveld, A., Jordi, W., & De Kruijff, B. (1986a) *J. Biol. Chem.* **261**, 3846–3856.

Rietveld, A., Berkhout, T. A., Roenhorst, A., Marsh, D., & De Kruijff, B. (1986b) *Biochim. Biophys. Acta* **858**, 38–46.

Shipley, G. G., Leslie, R. B., & Chapman, D. (1969) *Nature* **222**, 561.

Smith, M., Leung, D. W., Gillam, S., Astell, C. R., Montgomery, D. L., & Hall, B. D. (1979) *Cell* **16**, 753–761.

Smith, R., Cornell, B. A., Keniry, M. A., & Separovic, F. (1983) *Biochim. Biophys. Acta* **732**, 492–498.

Snel, M. M. E., Kaptein, R., & De Kruijff, B. (1991) *Biochemistry* **30**, 3387–3395.

Söllner, T., Griffith, G., Pfaller, R., Pfanner, N., & Neupert, W. (1989) *Cell* **59**, 1061–1070.

Söllner, T., Pfaller, R., Griffith, G., Pfanner, N., & Neupert, W. (1990) *Cell* **62**, 107–115.

Soussi, B., Bylund-Fellenuis, A. N., Schersten, T., & Angstrom, J. (1990) *Biochem. J.* **265**, 227–232.

Sperka-Gottlieb, C. D. M., Hermetter, A., Paltauf, F., & Daum G. (1988) *Biochim. Biophys. Acta* **946**, 227–234.

Spooner, P. J. R., & Watts, A. (1991a) *Biochemistry* **30**, 3871–3879.

Spooner, P. J. R., & Watts, A. (1991b) *Biochemistry* **30**, 3880–3885.

Stellwagen, E., & Rysavy, R. (1972) *J. Biol. Chem.* **247**, 8074–8077.

Stewart, J. W., Sherman, F., Shipman, N. A., & Jackson, M. (1971) *J. Biol. Chem.* **246**, 7429–7445.

Stuart, R. A., & Neupert, W. (1990) *Biochimie* **72**, 115–121.

Stuart, R. A., Neupert, W., & Tropschug, M. (1987) *EMBO J.* **6**, 2131–2137.

Teissie, J. (1981) *Biochemistry* **20**, 1554–1560.

Teissie, J., & Baudras, A. (1977) *Biochimie* **59**, 693–703.

Toniolo, C., Fontana, A., & Scoffone, E. (1975) *Eur. J. Biochem.* **50**, 367–374.

Van der Kooij, J., Erecinska, M., & Change, B. (1973) *Arch. Biochem. Biophys.* **154**, 219–229.

Vincent, J. S., & Levin, I. W. (1986) *J. Am. Chem. Soc.* **108**, 3551–3554.

Vincent, J. S., Kon, H., & Levin, I. W. (1987) *Biochemistry* **26**, 2312–2314.

Walter, A., Margolis, D., Mohan, R., & Blumenthal, R. (1986) *Membr. Biochem.* **6**, 217–237.

Waltham, M. C., Cornell, B. A., & Smith, R. (1986) *Biochim. Biophys. Acta* **862**, 451–456.

Zimmermann, R., Paluch, U., & Neupert, W. (1979) *FEBS Lett.* **108**, 141–146.

Zitomer, R. S., & Hall, B. D. (1976) *J. Biol. Chem.* **251**, 6320–6326.

Electron Transfer Kinetics

13

Steady-State Electron Transfer Reactions Involving Cytochrome *c*

FRANCIS MILLETT

Department of Chemistry and Biochemistry
University of Arkansas

I. Introduction

The electron transfer reactions between cytochrome *c* and its physiological redox partners have most commonly been measured under steady-state conditions, so that the rate of the reaction is slow enough to be detected by standard spectrophotometers. This requires that the concentration of cytochrome *c* (the substrate) be large compared to that of the redox partner (the enzyme). The steady-state reactions of cytochrome *c* generally display saturation kinetics at sufficiently high cytochrome *c* concentrations, and often obey Michaelis–Menten kinetics with a Michaelis constant, K_m and a maximum velocity, V_{max}. In many cases, the kinetics display several different phases, each characterized by a V_{max} and K_m value. While the description of the kinetics is relatively straightforward, the interpretation of the kinetic parameters is extremely complex, and it is often not possible to assign them to discrete steps in the overall reaction. In general, the reaction will involve binding cytochrome *c* to the redox partner to form a substrate complex, ES; electron transfer within the complex to yield a product complex, EP; and dissociation of the product complex to release free cytochrome *c*. The rate-limiting step in the reaction could involve any one of these events, or an internal event in the electron transfer partner not directly related to the cytochrome *c* reaction. The Michaelis constant K_m is often not a direct measure of the dissociation constant of the ES complex, K_d, and the maximum velocity, V_{max}, is usually not a measure of the actual electron transfer step between cytochrome *c* and the initial acceptor. In spite of these complexities, the steady-state reactions of cytochrome c have been extensively characterized, and under appropriate experimental conditions the kinetic parameters can be related to individual molecular events in the overall reaction.

II. Reaction between Cytochrome c and Cytochrome c Oxidase

A. General Properties of Cytochrome c Oxidase

Cytochrome c oxidase, the terminal member of the mitochondrial electron transport chain, is a redox-linked proton pump that oxidizes four molecules of ferrocytochrome c and uses the electrons to reduce molecular oxygen to two water molecules:

$$4\,c^{2+} + O_2 + 8\,H^{+}_{in} \rightarrow 4\,c^{3+} + 2\,H_2O + 4\,H^{+}_{out}.$$

It is generally accepted that four protons are pumped across the membrane for every four electrons that are transferred to oxygen (Wikström et al., 1981). Cytochrome oxidase isolated from beef heart mitochondria is normally a dimer of molecular weight 300–400 kDa, and consists of three large subunits synthesized inside the mitochondria, and 10 smaller subunits synthesized in the cytoplasm (Mihara and Blobel, 1980). Each monomer unit contains two heme groups, cytochrome a and cytochrome a_3, and two copper atoms, Cu_A and Cu_B, which are redox-active. Einarsdottir and Caughey (1985) have found that the total metal content is 2.5 Cu, 2 Fe, 1 Zn, and 1 Mg per monomer unit. Available evidence suggests that cyt a, cyt a_3 and Cu_B are located in subunit I (Shapleigh et al., 1992), while Cu_A is liganded to Cys 196 and Cys 200 in subunit II (Winter et al., 1980; Martin et al., 1988; Hall et al., 1988). Cyt a_3 and Cu_B are directly involved in the reduction of oxygen to water, while cyt a and Cu_A transfer electrons from cytochrome c to the cyt a_3 and Cu_B.

B. Spectrophotometric Assay of Steady-State Reaction

The steady-state reaction between ferrocytochrome c and cytochrome oxidase can be measured by following either the oxidation of ferrocytochrome c spectrophotometrically, or the consumption of oxygen polarographically. There are significant differences in the kinetics observed for the two assays, and they will be discussed separately. In the spectrophotometric assay, the oxidation of ferrocytochrome c is usually detected at 550 nm, and the enzyme is either a cytochrome c–depleted Keilin–Hartree heart muscle preparation, or purified cytochrome oxidase. Keilin (1930) first observed that the initial velocity of the reaction had a hyperbolic dependence on the concentration of cytochrome c. Extending these observations, Slater (1949) determined that the reaction obeyed Michaelis–Menten kinetics under many different conditions, suggesting the formation of an enzyme–substrate complex. A puzzling aspect of the kinetics, however, was that a first-order time course was observed at all cytochrome c concentrations, whereas a simple Michaelis–Menten mechanism would predict zero-order kinetics at saturating cytochrome c concentrations (Albaum et al., 1946). Since the first-order rate constant decreased with increasing cytochrome c concentrations, Smith and Conrad (1956) suggested that cytochrome c was itself an inhibitor of the reaction. The most elegant explanation of these results was provided by Minnaert (1961), who demonstrated that both hyperbolic kinetics and a first-order time course would occur if ferrocytochrome c formed an active substrate complex with cytochrome oxidase, and the product ferricytochrome c acted as a competitive inhibitor with the same binding affinity:

$$E + S \underset{k_2}{\overset{k_1}{\rightleftharpoons}} ES \overset{k_3}{\rightarrow} EP \underset{k_6}{\overset{k_5}{\rightleftharpoons}} E + P.$$

The steady-state solution for this mechanism is

$$v = \frac{V_{max}S}{K_{mS} + S + (K_{mS}/K_{mP})P},$$

where

$$V_{max} = \frac{k_3 k_5 e_0}{k_3 + k_5},$$

$$K_{mS} = \frac{k_2 k_5 + k_3 k_5}{k_1(k_3 + k_5)},$$

$$K_{mP} = \frac{k_2 k_5 + k_3 k_5}{k_6(k_2 + k_3)},$$

$$e_0 = \text{total enzyme concentration}.$$

If $K_{mS} = K_{mP}$, then this reduces to

$$v = \frac{V_{max}S}{K_{mS} + (S + P)} = k_{obs}S.$$

This equation describes a first-order reaction with a rate constant, k_{obs}, that decreases as a function of $(S + P)$, the total concentration of ferro- and ferricytochrome c. Support for this hypothesis was provided by Yonetani and Ray (1965), who found that the K_i for ferricytochrome c was the same as the K_m for ferrocytochrome c under conditions where first-order kinetics were obeyed.

An additional complexity in the spectrophotometric kinetics is that two phases are observed at low ionic strength: a high-affinity phase with low turnover, and a low-affinity phase with high turnover (Errede et al., 1976; Mochan and Nicholls, 1972; Smith et al., 1979; Rosevear et al., 1980; Sinjorgo et al., 1986). In 25 mM Tris-Mops buffer, pH 7.8, the high-affinity phase has $K_m = 0.04$ μM and $TN_{max} = 33$ s^{-1}, and the low-affinity phase has $K_m = 40$ μM and $TN_{max} = 255$ s^{-1} (Figure 13.1; Sinjorgo et al., 1986). The K_m value for the high-affinity phase is nearly the same as the equilibrium dissociation constant of the 1:1 cytochrome c–cytochrome oxidase complex (Garber and Margoliash, 1990). Stopped-flow studies have shown that the rate of dissociation of ferricytochrome c from the $1:1$ complex with cytochrome oxidase is the same as the V_{max} for the high-affinity phase of the reaction, indicating that product dissociation is the rate-limiting step in this phase of the reaction (Wilms et al., 1981). The K_m for the low-affinity phase is about the same as the equilibrium dissociation constant, K_d, for the $2:1$ complex of cytochrome c with cytochrome oxidase, suggesting that this phase somehow involves binding a second molecule of cytochrome c. A number of different models have been proposed to explain the biphasic steady-state kinetics of the reaction, as recently reviewed by Cooper (1990). Nicholls (1965), Errede et al. (1976), and Ferguson-Miller et al. (1976) proposed that there were two separate catalytic binding sites for cytochrome c with different kinetic parameters. However, models involving two catalytic sites are inconsistent with several studies showing that covalently cross-linking a single molecule of cytochrome c to the high-affinity site completely inhibited electron transfer from free cytochrome c in solution (Birchmeier et al., 1976; Bisson et al., 1978, 1980; Moreland and Dockter, 1981; Fuller et al., 1981). Since the rate-limiting step in the high-affinity phase of the reaction is product dissociation, a number of workers have suggested that the binding of a second molecule of cytochrome c to the low-affinity site increases the rate of dissociation of ferricytochrome c from the high-affinity catalytic site, thus accounting for the low-affinity phase of the reaction (Capaldi et al., 1982; Speck et al., 1984; Sinjorgo et al., 1984, 1986; Garber and Margoliash, 1990). This effect could be caused by direct electrostatic repulsion between the cytochrome c molecules occupying the two sites, or a change in protein conformation accompanying the binding of cytochrome c to the low-affinity site. Stopped-flow studies have demonstrated that a second molecule of ferrocytochrome c can react rapidly with the $1:1$ ferricytochrome c–cytochrome oxidase complex, indicating that cytochrome c can rapidly exchange between the low-affinity and high-affinity sites (Veerman et al., 1980). In another type of model it was proposed that during the electron transfer process, cytochrome oxidase oscillates between two different conformations that are compulsory intermediates in vectorial proton translocation (Brzezinski and Malmstrom, 1986; Michel and Bosshard, 1989). In this model, biphasic kinetics is an intrinsic property of redox-linked proton translocation if the two different conformations have different affinities for cytochrome c.

As the ionic strength is increased, the K_m and V_{max} of the high-affinity phase both increase, while the extent of the low-affinity phase decreases (Figure 13.1; Sinjorgo et al., 1986). At ionic strengths above 150 mM, only a single kinetic phase is observed, which is equivalent to the high-affinity phase of the reaction at low ionic strength. The effect of specific cytochrome c lysine modifications on the steady-state kinetics is dramatically different at low and high ionic strength (see Chapter 17 by Millett and Durham, this volume). At low ionic strength, modification of the heme crevice lysines increased the V_{max} of the high-affinity phase of the reaction, but had relatively little effect on the K_m (Ferguson-Miller et al., 1978; Smith et al., 1982). This is in marked contrast to equilibrium binding studies, which showed that these same modifications caused large increases in K_d. At high ionic strength, modification of heme crevice lysines caused large increases in the K_m of the single phase observed under these conditions, with little effect on V_{max} (Ferguson-Miller et al., 1978; Smith et al., 1982). The steady-state kinetics are much simpler at high

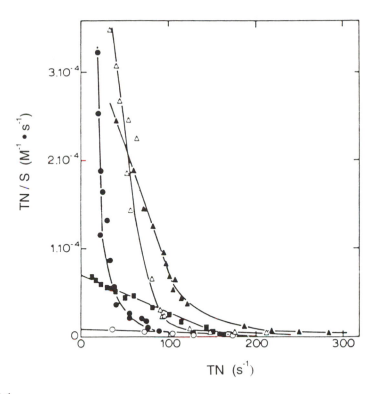

Figure 13.1
Steady-state oxidation of horse heart ferrocytochrome *c* by purified bovine heart cytochrome oxidase, measured spectrophotometrically in Tris–Mops buffers at pH 7.8 (reprinted from Sinjorgo *et al.*, 1986). The ionic strength was 25 mM (●); 35 mM (△); 50 mM (▲); 100 mM (■); 200 mM (○). The lines are an aid to the eye.

ionic strength, and K_m provides a more reliable indication of the binding interaction between ferrocytochrome *c* and the high-affinity reaction site. These conditions are therefore preferable for characterizing the reactions of derivatives or variant forms of cytochrome *c* with cytochrome oxidase.

C. Polarographic Assay of Steady-State Reaction

Multiphase kinetics are also observed in the polarographic steady-state reaction between cytochrome *c* and cytochrome oxidase, but the kinetic parameters are quite different than in the spectrophotometric assay (Ferguson-Miller *et al.*, 1976, 1978; Smith *et al.*, 1979, 1982). This difference is attributed to the use of the artificial reducing agents TMPD and ascorbate in the polarographic assay to maintain cytochrome *c* in the reduced form (for a recent review, see Cooper, 1990).

The K_m value of the high-affinity phase of the polarographic assay is the same as the K_m for the high-affinity phase of the spectrophotometric assay, but the V_{max} is

10-fold greater (Ferguson-Miller *et al.*, 1976, 1978). A direct correspondence has been established between the V_{max} of the high-affinity phase of the polarographic assay and the rate of reduction of bound ferricytochrome by TMPD (Osheroff *et al.*, 1983). Therefore, the larger V_{max} observed in the polarographic assay is attributed to the direct reduction of ferricytochrome c bound to cytochrome oxidase without the necessity for a slow dissociation step:

$$E + S \rightleftarrows ES \rightarrow EP \rightleftarrows E + P.$$

$$TMPD \quad TMPD^+$$

The K_m of the high-affinity phase is also the same as the equilibrium dissociation constant K_d of the 1 : 1 cytochrome c–cytochrome oxidase complex, indicating that K_m is a direct measure of binding energy. Specific cytochrome c lysine modifications have essentially the same effect on K_m and K_d, providing further evidence for this assumption (Ferguson-Miller, 1978; Osheroff *et al.*, 1980). This is in marked contrast to the spectrophotometric assay at low ionic strength, where specific lysine modifications have a relatively small effect on the high-affinity K_m value (Ferguson-Miller *et al.*, 1978; Smith *et al.*, 1982). Two different low-affinity phases are observed in the polarographic assay, with K_m values of 0.84 μM and 17 μM, respectively (Garber and Margoliash, 1990). The K_m for the first low-affinity phase is 10-fold smaller than the K_d value for binding a second molecule of cytochrome c, indicating that the transition between the high-affinity phase and the first low-affinity phase involves only a 1 : 1 complex. Garber and Margoliash have recently proposed a hysteresis model to account for this phase. In this model, the enzyme is primed for rapid turnover following electron transfer from ferrocytochrome c, and if the solution concentration of cytochrome c is sufficiently high, the next molecule of ferrocytochrome c can react rapidly. At lower solution concentrations of ferrocytochrome c, the enzyme would relax back to the less active form observed in the high-affinity phase. This mechanism was supported by studies of the effect of viscosity on the low-affinity phases (Garber and Margoliash, 1990). The K_m for the second low-affinity phase is comparable to the K_d value for binding a second molecule of cytochrome c, and so this phase can be identified with the 2 : 1 complex. The polarographic assay is useful for studying the reaction of variants of cytochrome c with cytochrome oxidase, but it is necessary to characterize the dependence of the kinetics on TMPD.

D. Properties of the Interaction Domain

A number of studies have been carried out to identify the location of the high- and low-affinity binding sites on cytochrome oxidase. Arylazido-lysine 13 cytochrome c was found to specifically cross-link to His 161 of subunit II on cytochrome oxidase, and block the high-affinity site (Bisson *et al.*, 1980, 1982). Millett *et al.* (1983) found that the water-soluble carbodiimide EDC specifically modified five carboxylate groups on subunit II and inhibited the high-affinity reaction with cytochrome c. Binding one molecule of cytochrome c to cytochrome oxidase protected Asp 112, Glu 114, Asp 115, and Glu 198 from modification by EDC and prevented the loss in electron transfer activity. These negatively charged residues therefore appear to be involved in binding cytochrome c to the high-affinity site. The involvement of Glu 198 is particularly interesting, since it is located between the conserved cysteines 196 and 200 that ligate Cu_A (Hall $et\ al.$, 1988). Cu_A might therefore be the initial electron acceptor in cytochrome oxidase. Asp 112, Glu 114, and Asp 115 are located in a highly conserved sequence (104–115) containing alternating acidic and aromatic residues. These aromatic residues could serve as an electron transfer pathway from cytochrome c to cytochrome a or Cu_A. Yeast cytochrome c modified at Cys 107 on the backside of the protein was cross-linked to subunit III of dimeric cytochrome oxidase from both yeast (Moreland and Dockter, 1981), and beef heart (Fuller $et\ al.$, 1981). On the basis of these results, Capaldi $et\ al.$ (1982) proposed that the high-affinity site is located at a cleft between the two monomers of cytochrome oxidase. The ring of lysines surrounding the heme crevice at the front of cytochrome c interact with the carboxylates on subunit II of one monomer, while the backside of cytochrome c is close to subunit III on the other monomer. This model has been confirmed by fluorescence energy transfer studies carried out by Hall $et\ al.$ (1988). Subunit II is the major site of interaction, since modification of residues on the backside of cytochrome c have relatively little effect on binding (Ferguson-Miller $et\ al.$, 1978; Smith $et\ al.$, 1977). The low-affinity site appears to involve the tightly bound cardiolipin associated with cytochrome oxidase, since extraction of the cardiolipin abolished the low-affinity phase of the reaction (Vik $et\ al.$, 1981). The high-affinity phase was retained, but the V_{max} value was decreased to 25% relative to the native enzyme.

III. Reaction between Cytochrome c and the Cytochrome bc_1 Complex

A. Steady-State Kinetics

The cytochrome bc_1 complex, perhaps the most universal electron transfer complex, is found in eukaryotic mitochondria as well as many prokaryotes (Crofts, 1986). The complex contains two b cytochromes, b-566 and b-562, located in a single polypeptide, the Rieske iron–sulfur protein, and cytochrome c_1. The complex purified from photosynthetic bacteria such as $Rb.\ sphaeroides$ contains only four polypetides (Yu $et\ al.$, 1984), while the beef heart mitochondrial complex contains 8–10 (Capaldi, 1982). It is generally accepted that electron transfer through the complex involves a Q-cycle mechanism in which four protons are translocated to the cytoplasmic side of the membrane per two electrons transferred to cytochrome c (Crofts, 1986). The steady-state reaction between ferricytochrome c and the cytochrome bc_1 complex is generally detected spectrophotometrically by following the reduction of cytochrome c at 550 nm. In kinetic studies of the purified cytochrome bc_1 complex a short-chain ubiquinone analog such as 2,3-dimethoxy-5-methyl-6-decylhydroquinone is used as the source of electrons (Speck *et al.*, 1979). Succinate is generally used as the source of electrons in studies of purified succinate–cytochrome *c* reductase and Keilin–Hartree particles. Takemori and King (1964) observed that the steady-state reduction of ferricytochrome *c* by succinate–cytochrome *c* reductase obeyed single-phase Michaelis–Menten kinetics with an apparent K_m of 4.2 μM in 0.1 M phosphate, pH 7.4. Smith *et al.* (1974) found that the time course of the reaction was mixed zero-, first-order, and that the first-order rate constant decreased with increasing cytochrome *c* concentration. Ahmed *et al.* (1978) confirmed these observations, and in addition found that at very low cytochrome *c* concentrations, <0.5 μM, the reaction was purely first-order, indicating that the rate-limiting step involved cytochrome *c*. The complex effect of ionic strength on the steady-state kinetics was studied by Ahmed *et al.* (1978) and Speck *et al.* (1979). Increasing the ionic strength from 25 mM to 100 mM had relatively little effect on the kinetics, increasing both V_{max} and K_m by about 50%, but leaving V_{max}/K_m unchanged (Figure 13.2). Further increases in ionic strength led to large increases in K_m, and essentially no change in V_{max}. The effect of specific cytochrome *c* lysine modifications on the kinetics was dramatically different at low and high ionic strengths (Ahmed *et al.*, 1978; Speck *et al.*, 1979; see Chapter 17 by Millett and Durham, this volume). At low ionic strength, modification of heme-crevice lysine amino groups had little or no effect, while at high ionic strength it caused large increases in K_m. The insensitivity of K_m to both changes in ionic strength and specific lysine modifications at low ionic strength suggests that it is not a function of cytochrome *c* binding. Indeed, K_m is 0.49 μM at 25 mM ionic strength, which is over an order of magnitude larger than the equilibrium dissociation constant, K_d, which is <0.02 μM (Speck and Margoliash, 1984). K_d is very sensitive to both ionic strength and specific lysine modifications under these conditions. Furthermore, the binding of an additional molecule of cytochrome *c* to the cytochrome bc_1 complex was observed in the same concentration range as K_m. These observations led Speck and Margoliash (1984) to propose a mechanism in which the binding of a second molecule of cytochrome *c* to a noncatalytic site can affect the kinetic parameters of the catalytic site. Thus, the kinetics observed at low ionic strength actually monitor the reaction of solution cytochrome *c* with a complex already containing a bound cytochrome *c*, accounting for the insensitivity of the kinetics to specific lysine modifications and changes in ionic strength. At ionic strengths above 0.1 M, K_m becomes equal to K_d and monitors the interaction of cytochrome *c* with the free enzyme. Therefore, the most reliable information on the interaction between cytochrome *c* and the cytochrome bc_1 complex is obtained at ionic strengths above about 0.2 M.

B. Properties of the Interaction Domain

A number of studies have focused on the location of the cytochrome *c* binding site on the cytochrome bc_1 complex. Stonehuerner *et al.* (1985) have found that the water-soluble carbodiimide EDC selectively labeled carboxylate groups in the sequence 63–81 of cytochrome c_1, and cytochrome *c* binding protected these residues from labeling. This highly acidic sequence contains 7 Asp and Glu residues that could potentially be involved in binding cytochrome *c*. Broger *et al.* (1983) have

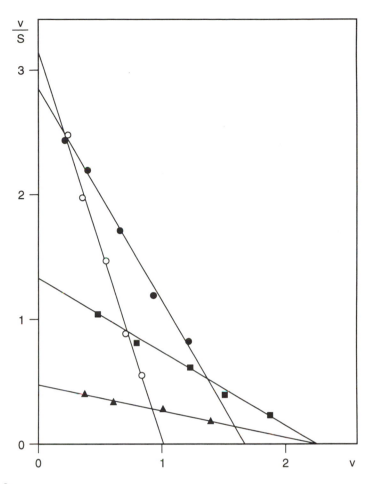

Figure 13.2
Steady-state reduction of horse heart ferricytochrome c by succinate–cytochrome c reductase (Ahmed *et al.*, 1978). The initial rate of reduction is in μM cytochrome c reduced per minute, and the concentration of cytochrome c, S, is in μM. The ionic strength of the Tris-HCl buffer, pH 7.5, was 0.02 M (o); 0.1 M (●); 0.2 M (■); 0.3 M (▲). The lines are the best fits to the data with V_{max} values of 1.0, 1.7, 2.2, and 2.2 μM/min, and V_{max}/K_m values of 3.3, 2.9, 1.3, and 0.4 min^{-1} for ionic strengths of 0.02, 0.1, 0.2, and 0.3 M, respectively.

shown that arylazido-lysine 13 cytochrome c is cross-linked to cytochrome c_1 somewhere in the highly acidic sequence 165–174, suggesting that carboxylate groups in this region might also be involved in cytochrome c binding. These studies suggest that at least two different regions of the cytochrome c_1 sequence are folded together to form the cytochrome c binding site. The highly acidic "hinge" peptide in the cytochrome bc_1 complex has also been implicated in cytochrome c binding by chemical modification and cross-linking studies (Kim and King, 1983; Stonehuerner *et al.*, 1985).

IV. Reaction between Cytochrome *c* and Cytochrome *c* Peroxidase

A. Steady-State Kinetics

Cytochrome *c* peroxidase has become the most thoroughly characterized redox partner of cytochrome *c*, and detailed studies of the reactions between these two proteins have provided considerable insight into the nature of biological electron transfer reactions in general. Cytochrome *c* peroxidase isolated from yeast mitochondria is a heme protein containing 293 amino acids with a MW of 35,235 (Takio and Yonetani, 1980). The x-ray crystal structure has been determined for a number of different forms of the enzyme (Finzel *et al.*, 1984; Edwards *et al.*, 1987), and molecular modeling techniques have been used to develop a hypothetical structure for the 1 : 1 complex between cytochrome *c* and cytochrome *c* peroxidase (Poulos and Kraut, 1980). Cytochrome *c* peroxidase catalyzes the following reaction:

$$2\,c^{2+} + H_2O_2 + 2H^+ \rightarrow 2\,c^{3+} + 2\,H_2O.$$

Although the mechanism is still not completely understood, three individual steps have been identified (Kim *et al.*, 1990):

$$CcP + H_2O_2 \rightarrow CcP\text{-}I + H_2O, \tag{13.1}$$

$$CcP\text{-}I + c^{2+} + H^+ \rightarrow CcP\text{-}II + c^{3+}, \tag{13.2}$$

$$CcP\text{-}II + c^{2+} + H^+ \rightarrow CcP + c^{3+} + H_2O. \tag{13.3}$$

CcP-I (Compound I) is oxidized two equivalents above the resting enzyme, Ccp, and contains an oxyferryl heme Fe(IV) and a free radical located on Trp 191 (Mauro *et al.*, 1988; Sivaraja *et al.*, 1989). The single oxidized site on CcP-II (Compound II) can be either the oxyferryl heme Fe(IV) or the free radical, depending on conditions (Coulson *et al.*, 1971; Ho *et al.*, 1983). The first detailed steady-state kinetic study was carried out by Yonetani and Ray (1966), who observed that the initial velocity obeyed simple Michaelis–Menten kinetics for both the cytochrome *c* and H_2O_2 substrates in acetate buffer at pH 6. They interpreted their results using a compulsory ordered two-substrate mechanism. Nicholls and Mochan (1971) also obtained simple Michaelis–Menten kinetics in phosphate buffers at pH 7, but proposed a mechanism involving random addition of the two substrates rather than ordered addition. In contrast, Kang *et al.* (1977) reported biphasic kinetics as a function of cytochrome *c* concentration for a variety of different conditions with both horse cytochrome *c* and baker's yeast iso-1 cytochrome *c*. They also carried out equilibrium binding studies indicating that more than one cytochrome *c* binding site was present on cytochrome *c* peroxidase at low ionic strength. However, the K_m value of the high-affinity phase of the steady-state kinetics was over two orders of magnitude larger than the high-affinity equilibrium dissociation constant, K_d, at low ionic strength. Furthermore, the K_d increased rapidly as the ionic strength increased, while the K_m for the high-affinity phase was insensitive to ionic strength over the range 10 mM to 200 mM. The V_{max} of the high-affinity phase was much more sensitive to ionic strength, displaying a maximum at 50 mM Tris acetate, pH 6.0, and decreasing at both lower and higher ionic strength. In phosphate buffers, the maximum V_{max} occurred at much lower ionic strength, 30 mM. The effect of specific cytochrome c lysine modifications on the kinetics was also dramatically dependent on ionic strength (Kang *et al.*, 1978; Smith and Millett, 1980). At low ionic strength, 10 mM, modification of lysines surrounding the heme crevice increased the V_{max} of the high-affinity phase, with relatively minor effect on K_m, while at high ionic strength, the V_{max} was decreased. Kang *et al.* (1978) interpreted these results to mean that at low ionic strength the V_{max} of the high-affinity phase was rate-limited by product dissociation, and specific lysine modifications or increasing ionic strength increased the rate of this step. The interpretation of K_m in these assays is not clear, but it is obviously not sensitive to the interaction between cytochrome c and cytochrome c peroxidase. Although Kang *et al.* (1978) have interpreted the biphasic kinetics in terms of a model involving the binding of two molecules of cytochrome c to cytochrome c peroxidase, Kang and Erman (1982), Summers and Erman (1988), and Kim *et al.* (1990) have developed mechanisms involving only a single catalytic site. Summers and Erman (1988) reported that in stopped-flow experiments at low ionic strength, ferrocytochrome c preferentially reduces the oxyferryl heme Fe(IV) site in CcP-I rather than the free radical site. They propose that the free radical form CcP-II$_R$, then slowly equilibrates to the more stable oxyferryl form, CcP-II$_F$, which can then react with an additional molecule of ferrocytochrome c. Summers and Erman (1988) and Kim *et al.* (1990) attribute the biphasic kinetics observed in the steady-state assay to an equilibrium between two different forms of CcP-I: the stable form which contains the free radical, CcP-I$_B$, and a short-lived transient form, CcP-I$_A$, which does not contain the free radical. At low cytochrome c concentrations corresponding to the high-affinity phase, the normal CcP-I$_B$ is formed, which requires the enzyme to react through the CcP-II$_R$ intermediate containing the free radical site, and the overall rate is limited by the transition between CcP-II$_R$ and CcP-II$_F$. At high cytochrome c concentrations, CcP-I$_A$ is reduced directly to CcP-II$_F$, thus bypassing the unreactive intermediates containing the free radical. The key assumption in the preceding mechanism is that cytochrome c is unreactive with the radical, both in CcP-I$_B$ and CcP-II$_R$. However, Geren *et al.* (1991) and Hahm *et al.* (1992) found that a number of different ruthenium-modified cytochrome c derivatives reacted much more rapidly with the radical in CcP-I$_B$ than with the oxyferryl heme Fe(IV) under all ionic strength conditions. Thus, the mechanism of the reaction between cytochrome c and cytochrome c peroxidase is still being intensively studied.

B. Properties of the Interaction Domain

Pelletier and Kraut (1992) have recently determined the X-ray crystal structure of a 1:1 complex between yeast cytochrome c and cytochrome c peroxidase, which has a distinctly different binding domain than that proposed by Poulos and Kraut (1980). The binding domain includes hydrophobic interactions as well as electrostatic interactions between cytochrome c lysine amino groups and cytochrome c peroxidase carboxylate groups. An efficient electron transfer pathway was identified which extends from the cytochrome *c* heme methyl group through cytochrome *c* peroxidase residues Ala-194, Ala-193, and Gly-192 to the indolyl radical on Trp-191, which is in van der Waals contact with the heme group. This pathway is thus consistent with rapid electron transfer from cytochrome *c* to the Trp-191 indolyl radical cation in CcP-I$_B$, as found by Geren *et al.* (1991) and Hahm *et al.* (1992).

V. Acknowledgment

This work was supported in part by NIH Grant GM 20488.

VI. References

Ahmed, A. J., Smith, H. T., Smith, M. B., & Millett, F. (1978) *Biochemistry 17*, 2479.

Albaum, H. G., Tepperman, J., & Bodansky, L. (1946) *J. Biol. Chem. 163*, 641–647.

Birchmeier, W., Kohler, C. E., & Schatz, G. (1976) *Proc. Natl. Acad. Sci. USA 73*, 4334–4338.

Bisson, R., Azzi, A., Gutweniger, H., Colonna, R., Montecucco, C., & Zanotti, A. (1978) *J. Biol. Chem. 253*, 1874–1880.

Bisson, R., Jacobs, B., & Capaldi, R. A. (1980) *Biochemistry 19*, 4173.

Bisson, R., Steffens, G. M., Capaldi, R. A., & Buse, G. (1982) *FEBS Lett. 144*, 359.

Broger, C., Slardi, S., & Azzi, A. (1983) *Eur. J. Biochem. 131*, 349–352.

Brzezinski, P., & Malmstrom, B. G. (1986) *Proc. Natl. Acad. Sci. USA 83*, 4282–4286.

Capaldi, R. A. (1982) *Biochim. Biophys. Acta 694*, 291–306.

Capaldi, R. A., Darley-Usmar, V., Fuller, S., & Millett, F. (1982) *FEBS Lett. 138*, 1–7.

Cooper, C. E. (1990) *Biochim. Biphys. Acta 1017*, 187–203.

Coulson, A. F. W., Erman, J. E., & Yonetani, T. (1971) *J. Biol. Chem. 246*, 917–924.

Crofts, A. R. (1986) *Journal of Bioenergetics and Biomembranes 18*, 437.

Edwards, S. L., Xuong, N. H., Hamlin, R. C., & Kraut, J. (1987) *Biochemistry 26*, 1503–1511.

Einarsdottir, O., & Caughey, W. S. (1985) *Biochem. Biophys. Res. Comm. 125*, 840–847.

Errede, B., Haight, G. P., & Kamen, M. D. (1976) *Proc. Natl. Acad. Sci. USA 73*, 113–117.

Ferguson-Miller, S., Brautigan, D. L., & Margoliash, E. (1976) *J. Biol. Chem. 251*, 1104–1115.

Ferguson-Miller, S., Brautigan, P. L., & Margoliash, E. (1978) *J. Biol. Chem. 253*, 149–159.

Finzel, B. C., Poulos, T. L., & Kraut, J. (1984) *J. Biol. Chem. 259*, 13027–13036.

Fuller, S. D., Darley-Usmar, V. M., & Capaldi, R. A. (1981) Biochemistry *20*, 7046–7053.

Garber, E. A. E., & Margoliash, E. (1990) *Biochim. Biophys. Acta 1015*, 279–287.

Geren, L., Hahm, S., Durham, B., & Millett, F. (1991) *Biochemistry 30*, 9450–9457.

Hahm, S., Durham, B., & Millett, F. (1992) *Biochemistry 31*, 3472–3477.

Hall, J., Moubarak, A., O'Brien, P., Pan, L. P., Cho, I., & Millett, F. (1988) *J. Biol. Chem. 263*, 8142–8149.

Ho, P. S., Hoffman, B. M., Kang, C. H., & Margoliash, E. (1983) *J. Biol. Chem. 258*, 4356–4363.

Kang, C. H., Ferguson-Miller, S., & Margoliash, E. (1977) *J. Biol. Chem. 252*, 919–926.

Kang, C. H., Brautigan, D. L., Osheroff, N., & Margoliash, E. (1978) *J. Biol. Chem. 253*, 6502–6510.

Kang, D. S., & Erman, J. E. (1982) *J. Biol. Chem. 257*, 12775–12779.

Keilin, D. (1930) *Proc. R. Soc. London Ser. B 106*, 418–444.

Kim, C. H., & King, R. E. (1983) *J. Biol. Chem. 258*, 13543–13551.

Kim, K. L., Kang, D. S., Vitello, L. B., & Erman, J. E. (1990) *Biochemistry 29*, 9150–9159.

Martin, C. T., Scholes, C. P., & Chan, S. I. (1988) *J. Biol. Chem. 263*, 8420–8429.

Mauro, J. M., Fishel, L. A., Hazzard, J. T., Meyer, T. E., Tollin, G., Cusanovich, M. A., & Kraut, J. (1988) *Biochemistry 27*, 6243–6256.

Michel, B., & Bosshard, H. R. (1989) *Biochemistry 28*, 244–252.

Mihara, K., & Blobel, G. (1980) *Proc. Natl. Acad. Sci. USA 77*, 4160–4164.

Millett, F., deJong, C., Paulson, L., & Capaldi, R. A. (1983) *Biochemistry 22*, 546.

Minnaert, K. (1961) *Biochim. Biophys. Acta* **50**, 23–34.

Mochan, E., & Nicholls, P. (1972) *Biochim. Biophys. Acta* **267**, 309–319.

Moreland, R. B., & Dockter, M. E. (1981) *Biochem. Biophys. Res. Comm.* **99**, 339–346.

Nicholls, P. (1965) in *Oxidases and Related Redox Systems* (Kint, T. E., Mason, H. S., & Morrison, M., eds.), Wiley, New York, pp. 764–777.

Nicholls, P., & Mochan, E. (1971) *Biochem. J.* **121**, 55–67.

Osheroff, N., Brautigan, D. L., & Margoliash, E. (1980) *J. Biol. Chem.* **255**, 8245–8251.

Osheroff, N., Speck, S., Margoliash, E., Veerman, E. C. I., Wilms, J. Konig, B. W., & Muijsers, A. O. (1983) *J. Biol. Chem.* **258**, 5731–5738.

Pelletier, H., & Kraut, J. (1992) *Science* **258**, 1748–1755.

Poulos, T. L., & Kraut, J. (1980) *J. Biol. Chem.* **255**, 10322–10330.

Rosevear, P., van Aken, T., Baxter, J., & Ferguson-Miller, S. (1980) *Biochemistry* **19**, 4108–4115.

Shapleigh, J. P., Hosler, J. P., Tecklenburg, M. M. J., Younkyoo, K., Babcock, G. T., Gennis, R. B., & Ferguson-Miller, S. (1992) *Proc. Natl. Acad. Sci. USA* **89**, 4786.

Sinjorgo, K. M. C., Meijling, J. H., & Muijsers, A. O. (1984) *Biochim. Biophys. Acta* **767**, 48–56.

Sinjorgo, K. M. C., Steinebach, O. M., Dekker, H. L., & Muijsers, A. O. (1986) *Biochim. Biophys. Acta* **850**, 108–115.

Sivaraja, M., Goodin, D. B., Smith, M., & Hoffman, B. M. (1989) *Science* **245**, 738–740.

Slater, E. C. (1949) *Biochem. J.* **44**, 305–318.

Smith, H. T., Staudenmeyer, N., & Millett, F. (1977) *Biochemistry* **16**, 4971.

Smith, L., & Conrad, H. (1956) *Arch. Biochem. Biophys.* **63**, 403–413.

Smith, L., Davies, H. C., & Nava, M. (1974) *J. Biol. Chem.* **249**, 2904–2910.

Smith, L., Davies, H. C., & Nava, M. E. (1979) *Biochemistry* **18**, 3140–3146.

Smith, L., Davies, H. C., Nava, M. E., Smith, H. T., & Millett, F. (1982) *Biochim. Biophys. Acta* **700**, 184–191.

Smith, M. B., & Millett, F. (1980) *Biochim. Biophys. Acta* **626**, 64.

Speck, S. H., & Margoliash, E. (1984) *J. Biol. Chem.* **259**, 1064–1072.

Speck, S. H., Ferguson-Miller, S., Osheroff, N., & Margoliash, E. (1979) *Proc. Nat. Acad. Sci. USA* **76**, 155–160.

Speck, S. H., Dye, D., & Margoliash, E. (1984) *Proc. Natl. Acad. Sci. USA* **81**, 347–351.

Stonehuerner, J., O'Brien, P., Geren, L., Millett, F., Steidl, J., Yu, L., & Yu, C.-A. (1985) *J. Biol. Chem.* **260**, 5392–5398.

Summers, F. E., & Erman, J. E. (1988) *J. Biol. Chem.* **263**, 14267–14275.

Takemori, S., & King, T. E. (1964) *J. Biol. Chem.* **239**, 3546–3558.

Takio, K., & Yonetani, T. (1980) *Arch. Biochem. Biophs.* **203**. 605–614.

Veerman, E. C. I., Wilms, J., Casteleijn, G., & van Gelder, B. F. (1980) *Biochim. Biophys. Acta* **590**, 117–127.

Vik, S. B., Georgevich, G., & Capaldi, R. A. (1981) *Proc. Natl. Acad. Sci. USA* **78**, 1456–1460.

Wikström, M., Krab, K., & Saraste, M. (1981) *Cytochrome Oxidase, A Synthesis*, Academic Press, New York.

Wilms, J., Veerman, E. C. I., Konig, B. W., Dekker, H. L., & Van Gelder, B. F. (1981) *Biochim. Biophys. Acta* **635**, 13–24.

Winter, D. B., Bruyninckx, W. J., Foulke, F. G., Grinich, N. P., & Mason, H. S. (1980) *J. Biol. Chem.* **255**, 11408–11414.

Yonetani, T., & Ray, G. S. (1965) *J. Biol. Chem.* **240**, 3392–3398.

Yonetani, T., & Ray, G. S. (1966) *J. Biol. Chem.* **241**, 700–706.

Yu, L., Mei, W. C., & Yu, C.-A. (1984) *J. Biol. Chem.* **259**, 5752–5760.

14

Kinetics of Electron Transfer of *c*-Type Cytochromes with Small Reagents

MICHAEL A. CUSANOVICH
GORDON TOLLIN

Department of Biochemistry
University of Arizona

I. Introduction

The redox kinetics of cytochrome *c* have been extensively studied with small inorganic and organic reactants. Thus, compounds such as the iron hexacyanides, flavin semiquinones, and iron EDTA have played an important role in our understanding of the mechanism of electron transfer by *c*-type cytochromes. To a large extent, the early focus on small reactants was driven by their ready availability and by the fact that they generally yielded interpretable results. This was in sharp contrast to systems involving protein–protein interactions, such as cytochrome *c*-cytochrome *c* oxidase, where both reactants are macromolecules, where the structures of both reactants are generally not known, and where the kinetics is complex. Thus, in simplest terms, the use of small reactants allows one to focus on the interaction domain of the protein reactant, i.e., cytochrome *c*, since small redox compounds have well-defined structures and chemical properties. Through the use of small reactants, researchers were able to establish that cytochrome *c* undergoes outer-sphere electron transfer and, importantly, has kinetic properties describable in terms of Marcus theory (Marcus and Sutin, 1985). This latter point is crucial, because Marcus theory is well-developed and allows one to obtain insights into reaction mechanisms in terms of the various parameters controlling electron transfer kinetics. Thus, in many systems the contribution of driving force (the difference in redox potential between the reactants), electrostatics (ion–ion interactions), and sterics (surface topography of the reactants) can be quantified (Tollin *et al.*, 1986a). Less directly, information on distance between redox centers, dynamics in the transition state complex, and the relative orientation of redox centers is sometimes accessible.

As noted earlier, the use of small reactants as oxidants and reductants for *c*-type cytochromes has provided researchers the opportunity to define interaction domains on the cytochrome *c* surface. Thus, depending on the reactant used, a reasonably well-defined region of the molecular surface of cytochrome *c* can be described which has properties dictated by specific amino acid side chains present. Importantly, it is now clear that the interaction domains for small reactants are to a first approximation the same as those determined for protein–protein interactions, although somewhat smaller in terms of the molecular surface involved. Thus, small molecule reactants have turned out to be excellent models for physiological systems (Cusanovich *et al.*, 1987).

In what follows we will attempt to extract from a very large literature the principal features of our knowledge of the redox kinetics of cytochrome *c* with small reactants. This will not be an exhaustive analysis, and the reader is referred to a number of recent reviews that address the issues and results discussed here (Moore *et al.*, 1984; Marcus and Sutin, 1985; Tollin *et al.*, 1986a; Cusanovich *et al.*, 1987; Gray and Malmström, 1989).

II. Biological Diversity

Although the principal features of the properties of cytochromes *c* are discussed elsewhere in this volume, it is important to define some parameters that strongly influence the study of the redox kinetics of the cytochromes *c* and their relationship to function. This discussion will focus on the Class-I *c*-type cytochromes (Ambler, 1977), which are defined by the presence of protoheme IX bound near the N-terminus of the protein with the characteristic sequence Cys–x–x–Cys–His. Moreover, for the Class-I *c*-type cytochromes, the heme out-of-plane ligands are Met and His. The Class-I *c*-type cytochromes, although showing wide variation in amino acid sequence, have the same basic structure (see Chapter 3 by Brayer, this volume). Thus, it is characteristic of this family that, when viewed from what is defined as the front (Figure 14.1), the heme is positioned so that its edge is exposed to the solvent, but the bulk of the heme, including the iron, is buried in the interior of the protein. This so-called "exposed heme edge" will be referred to throughout the discussion and is apparently critical in defining a portion of the interaction domain mediating electron transfer. Class-I *c*-type cytochromes typically contain approximately 50% helix, with the N- and C-terminal helices most prominent. Moreover, it is the amino acid side chains surrounding the exposed heme edge that appear to be most important in influencing the electron transfer kinetics and that define the interaction domain.

Some discussion will be presented of the Class-II *c*-type cytochromes. This structural family is distinct from the Class-I cytochromes in that heme attachment is near the C-terminus, and the proteins have a secondary structure which is largely

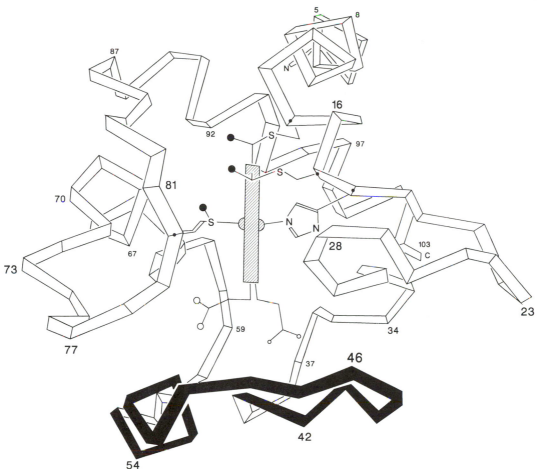

Figure 14.1
Ribbon drawing of tuna cytochrome *c*. The shaded portion, consisting of 16 residues, is missing in the smaller cytochromes *c*, for example *Ps. aeruginosa* cytochrome *c*-551. In the larger cytochromes, insertions are found in the bottom left, the upper right, and, in some cases, the C-terminus.

helical. Moreover, the heme out-of-plane ligands are Met/His in the low-spin forms and vacant/His in the more common high-spin form (cytochrome *c'*). In the case of the high-spin *c*-type cytochromes, the heme lies at the bottom of a large crevice with a very substantial fraction of the heme accessible to solvent (approximately twice as much as with Class-I cytochromes) and to small molecule reactants (Tollin *et al.*, 1986b).

The Class-I cytochromes *c* are widely distributed in nature and are found in most, but not all, bacteria and in all eukaryotic organisms. Thus, nature has preserved the structural motif of the Class-I *c*-type cytochromes throughout evolution (see Chapter 2 by Meyer, this volume). Nevertheless, in a functional sense, the

Class-I c-type cytochromes show a large variation in physiological partners. Although thought of primarily in terms of respiration, where cytochrome c mediates electron transfer between cytochrome b/c_1 and cytochrome oxidase, the Class-I c-type cytochromes also can participate in many other reactions. For example, in some photosynthetic bacteria, cytochrome c_2 (a subclass of Class-I c-type cytochromes) also can mediate electron transfer between cytochrome b/c_1 and the photosynthetic reaction center, as well as participate in reactions involving alternate electron donors and acceptors (Meyer and Cusanovich, 1989). Thus, Class-I c-type cytochromes are important in both sulfur and nitrogen metabolism, as well as in respiration and photosynthesis.

It is important to recognize the wide range of functional properties of the Class-I c-type cytochromes, because this apparently results in quite different physicochemical properties. Although horse heart cytochrome c and eukaryotic cytochromes c in general are considered prototypic of Class-I c-type cytochromes, they are in many respects atypical. Eukaryotic cytochrome c is a very basic protein ($pI \sim 10$) with a redox potential of 260 ± 5 mV (vs. NHE). This is in sharp contrast to prokaryotic Class-I c-type cytochromes, which have redox potentials ranging from 20 to 450 mV and isoelectric points from 4 to 10, yet are structurally homologous to eukaryotic cytochrome c. Table 14.1 summarizes the redox potentials and net charges of a number of Class-I c-type cytochromes.

From a functional standpoint, the large variation in properties found among the eukaryotic Class-I c-type cytochromes results from the need to adapt to a variety of environmental conditions, as well as participation in a wide variety of electron transfer pathways. In this context, it is important to note that in bacteria

Table 14.1
Electrostatic analysis of reactions of Class-I and Class-II c-type cytochromes with FMN semiquinone.[a]

Cytochrome	$E_{m,7}$ (mV, vs. NHE)	Net charge	V_{ii}[b] (kcal/mol)	$k_\infty \times 10^{-7}$[c] (M^{-1} s^{-1})
Class I				
R. tenue 3761 c-553	400	4+	−1.6	2.4
R. tenue 2761 c-553	390	5+	−1.5	2.4
Euglena sp. c-552	370	8−	+1.8	5.2
R. spheroides c_2	370	2−	−2.0	2.5
R. vaniellii c_2	354	2+	−1.5	2.3
R. capsulatus c_2	350	1+	−2.1	2.0
R. purpureus c-553	340	5+	−1.7	1.9
R. rubrum c_2	324	0	−2.2	2.5

Table 14.1 (*continued*)

Cytochrome	$E_{m,7}$ (mV, vs. NHE)	Net charge	V_{ii}^{b} (kcal/mol)	$k_{\infty} \times 10^{-7c}$ ($M^{-1}\,s^{-1}$)
R. salexigens c_2	314	5−	−1.3	2.2
P. aeruginosa c-551	270	2−	+1.1	3.0
C. krusei c	265	5+	−3.0	1.6
Tuna *c*	260	7+	−2.9	1.4
Horse *c*	260	7+	−3.0	1.3
P. denitrificans c_2	250	7−	−1.8	2.2
C. thiosulfatophilum c-555	150	6+	−2.2	1.3
Paracoccus sp. c-554 (548)	114	7−	+1.7	1.5
E. halophila c-551	58	10−	+1.7	2.2
R. gelatinosa c-550	28	5+	−1.2	0.8
Class II				
R. palustris c-556	230	0	−1.0	3.4
R. spheroides c-554	203	(10−)	+2.4	7.0
Alcaligenes sp. c'	130	3+	−2.5	2.7
R. salexigens c'	95	(0)	+2.6	6.4
R. palustris c'	94	3+	−2.4	2.0
R. gelatinosa c'	60	2+	−1.2	2.4
R. capsulatus c'	51	7−	+0.8	2.7
R. tenue 3761 c'	45	2+	−2.7	1.4
R. spheroides c'	30	5−	0	3.4
C. vinosum c'	18	5−	+1.0	4.4
R. molischianum c'	14	1+	−1.6	1.9
R. photometricum c'	14	1−	−1.1	2.4
R. rubrum c'	3	1−	−0.8	2.3
R. purpureus c'	3	2+	−2.0	1.0

[a] Data from Meyer *et al.* (1984, 1986). FMN has a net charge of −1.9 under these conditions. A positive value of V_{ii} indicates an electrostatically repulsive interaction.

[b] V_{ii} is the electrostatic interaction energy due to ion–ion interactions.

[c] k_{∞} is the rate constant at infinite ionic strength where ion–ion interactions are fully screened.

Class-I c-type cytochromes are periplasmic (Wood, 1983), and hence exposed to the media in which the bacteria live. Thus, Class-I c-type cytochromes are found in bacteria which live in environments with pH values ranging from 2 to 10 and salt (NaCl) concentrations ranging from a few millimolar to 4 M. It follows, then, that the electrostatic properties, surface topography, and driving force in bacterial cytochromes can modulate the kinetics of electron transfer in very different ways, and thus influence biological specificity. In recent years it has been found that electrostatics, sterics, and driving force can each contribute up to a factor of 10^3 to the kinetics of electron transfer (Tollin *et al.*, 1986a). It follows then that nature, by making the appropriate amino acid substitutions in the interaction domain, can alter electrostatics or topography, and by changing the heme environment, alter the oxidation–reduction potential, and hence the driving force, to provide a kinetic modulation of 10^9. It is this ability to modulate the kinetics of electron transfer which accounts for biological specificity in reactions involving c-type cytochromes. This is of critical importance, particularly in bacterial systems, inasmuch as thermo-dynamically there is no means to control electron transfer pathways. That is, a large number of donors and acceptors are present which could short-circuit pathways. No doubt kinetic control as described earlier minimizes nonproductive pathways and maximizes the physiologically relevant paths. In what follows, it will be shown that the use of small inorganic and organic redox reagents has played a crucial role in elucidating the contribution of electrostatics, sterics, and driving force in cyto-chrome electron transfer.

III. Application of Marcus Electron Transfer Theory

The transfer of an electron between a small molecule and a redox protein can con-veniently be thought of as occurring within a transiently formed complex, via an outer-sphere mechanism which involves relatively weak through-space orbital inter-actions. This can be expressed in terms of the following *minimal* kinetic mechanism (it should be noted that more than one complex can in principle exist along the reaction pathway; however, this complication will be ignored in what follows):

$$A_{ox} + B_{red} \underset{k_{-1}}{\overset{k_1}{\rightleftarrows}} [A_{ox} B_{red}] \overset{k_{et}}{\rightarrow} A_{red} + B_{ox}. \tag{14.1}$$

In this equation, k_{et} does not necessarily refer to the actual electron transfer event, but rather includes all intracomplex processes which occur prior to product forma-tion. Thus, for example, conformational transitions or solvent rearrangements may contribute to the observed value of this constant. The magnitudes of the complex formation constant ($K = k_1/k_{-1}$) and the lifetime of the complex will determine whether or not this intermediate can be experimentally detected. Complex forma-tion is most often indicated by the existence of a nonlinear dependence of the observed pseudo–first-order rate constant for product formation (k_{obs}) on the con-centration of the reactant which is present in excess. If such is the case, k_{et} can be directly determined either by extrapolating k_{obs} to infinite concentration, or by a nonlinear least squares fit to the analytical solution of the exact differential equations for the preceding mechanism (Ahmad *et al.*, 1981). When the intermediate complex cannot be detected in this way, two limiting cases are generally recognized:

(a) $k_{et} \gg k_{-1}$: under these conditions, $k_{obs} = k_1[A_{ox}]$, and thus the calculated second-order rate constant will reflect *only* the complex formation step.

(b) $k_{et} \ll k_{-1}$: in this case (which is often referred to as the rapid pre-equilibrium condition), $k_{obs} = Kk_{et}[A_{ox}]$, and the calculated second-order rate constant contains *both* the complex formation and the electron transfer steps, and also corresponds to a minimum value for k_1.

For the reactions between c-type cytochromes and small molecules to be discussed here, intermediate complex formation has generally not been directly observed. However, as will be shown, the evidence clearly indicates that condition (b) applies, i.e., that observed rate constants reflect the processes involved in both complex formation and electron transfer. If we assume this to be the case, we can then relate the calculated second-order rate constant [which we will refer to as k_2, and which will be numerically equal to Kk_{et} in the limit of condition (b)] to the theoretical principles contained within the Marcus formulation of electron transfer processes (*cf.* Marcus and Sutin, 1985).

An electron transfer reaction occurring within a complex of donor and acceptor molecules has a rate constant which can be written as $k_{et} = (2\pi/\hbar)|V_{ab}|^2 \cdot \text{FCWD}$. The first term in this equation is the barrier crossing rate as defined by transition state theory. The second term, V_{ab}, usually referred to as the tunneling matrix element, describes the mixing of donor and acceptor wave functions (determined mainly by orbital overlap and symmetry considerations) which allows the system to cross over from the potential energy surface characteristic of the reactants to that characteristic of the products (*cf.* Figure 14.2). In a simple uniform dielectric model, the magnitude of this term depends exponentially on the distance (R) between the donor and the acceptor [$|V_{ab}| = \text{const. } \exp(-\alpha R)$, where α will generally approximate unity, although values from 0.3 to 3.3 have been found in some systems]. Orientation effects on V_{ab} are also expected, especially for π-electron systems, because of the nonspherical nature of the electronic orbitals, and for molecules having nonspherical shapes (such as heme). Chapter 15 by Scott in this volume discusses this in more detail. It should be kept in mind that for bimolecular reactions in fluid media, electron transfer will occur from a variety of distances and mutual orientations, and thus the observed rate constants will reflect an average over these parameters. The coupling between the donor and acceptor orbitals can also be influenced by the nature of the intervening medium, via through-bond coupling mechanisms such as superexchange (*cf.* Closs and Miller, 1988, for description). The third term in the equation, the so-called Franck–Condon weighted density of states (FCWD), describes the nuclear motions along the reaction coordinate that accompany the electron transfer event (*cf.* Marcus and Sutin, 1985). These time-dependent structural events can involve processes such as bond lengthening due to electron transfer into an antibonding orbital [in the case of tuna cytochrome *c*, iron–ligand bond distances do not change appreciably upon reduction (Takano and Dickerson, 1981)], and solvent and protein dipole reorientation in response to changes in

496

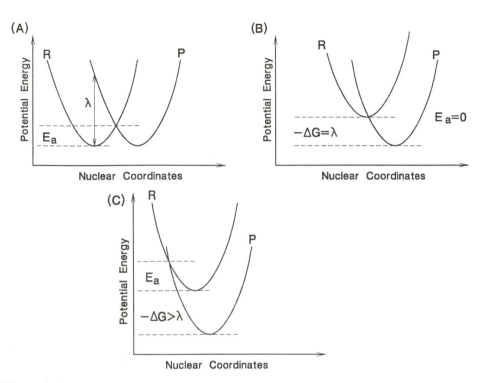

Figure 14.2
Potential energy curves for electron transfer systems as a function of thermodynamic free energy difference (ΔE) between reactants (R) and products (P). E_a (activation energy) corresponds to the energy required to reach the crossover point between curves R and P. Reorganization energy (λ) is illustrated in panel (A); panels (B) and (C) illustrate the Marcus inverted region.

electrostatic charge distributions. The quantitative importance of these latter processes in cytochrome reactions is difficult to assess, although attempts to do this have been made (*cf*. Marcus and Sutin, 1985). According to the Franck–Condon principle, such nuclear movements must take place *prior* to the actual physical transfer of the electron, which is usually considered to occur when the energy of the system becomes independent of whether the electron resides on the donor or on the acceptor, i.e., at the crossing point between the potential energy surfaces of the reactants and products (Figure 14.2). Thermal fluctuations of the nuclear coordinates from those which exist near the bottom of the potential surface defining the reactants are required for the system to reach this intersection region and thus cross over onto the product surface (this "classical" view of nuclear motion can be extended to include nuclear tunneling through the barrier between the potential surfaces). It should also be noted that these potential surfaces are a function of *all* the nuclear coordinates of the system (including solvent) that change as a consequence of the electron transfer reaction, and thus are highly complex and multidimensional, rather than being parabolic and unidimensional as in Figure 14.2.

It is clear from the preceding considerations that the calculation from first principles of rate constants for electron transfer reactions involving species as structurally complex as redox proteins is beyond our capabilities at the present time. However, despite this, the Marcus formulation of electron transfer theory allows some useful insights to be obtained. Thus, the activation energy for the electron transfer reaction (E_a), and hence the rate constant k_{et}, can be related to the free energy difference between reactants and products (ΔG°), and a quantity called the "reorganization energy" (λ), which corresponds to the energy required to accomplish a Franck–Condon transition (i.e., one occurring without nuclear movement) between the reactant and product potential surfaces when $\Delta G^\circ = 0$ (cf. Figure 14.2a):

$$E_a = (\Delta G^\circ - \lambda)^2/4. \tag{14.2}$$

Whereas ΔG° reflects the overall thermodynamics of the reaction (i.e., the driving force), the reorganization energy represents the sum of contributions from each nuclear coordinate undergoing a displacement during the reaction, and thus is a measure of structural differences between reactants and products (including any solvent contributions). Most significantly, the latter term is an experimentally accessible quantity.

Expressed in terms of rate constants and the redox potential difference between reactants and products (ΔE_m), Equation (14.2) becomes:

$$\ln k_{et} = \ln v_{et} - (\lambda - \Delta E_m)^2/4k_B\lambda T, \tag{14.3}$$

where v_{et} is the limiting rate constant as E_a approaches zero (i.e., when $\lambda = \Delta E_m$; cf. Figure 14.2b); k_B is the Boltzmann constant; and T is the absolute temperature. Note that this equation predicts a decrease in the rate constant for $\Delta E_m > \lambda$, the so-called Marcus inverted region (Figure 14.2c). This is not normally observed for bimolecular reactions in fluid solutions, in some cases because diffusion becomes rate-limiting at high exothermicities (Closs et al., 1986), or possibly because of conformational effects which result in a change in the rate-limiting reaction in the inverted region (Brunschwig and Sutin, 1989). As a consequence of this, other semiempirical equations have been developed to describe the free energy dependence of electron transfer rate constants, such as the Rehm–Weller equation (Rehm and Weller, 1969) and the Marcus exponential equation (Marcus, 1968; Agmon and Levine, 1977). Inasmuch as we will make use of the latter equation below, it will be presented here:

$$\ln k_{et} = \ln v_{et} + [\Delta E_m - (\lambda/2.8) \ln(1 + \exp(-2.8\Delta E_m/\lambda))]/k_B T. \tag{14.4}$$

Although the meaning of the other parameters in this equation remain the same as in Equation (14.3), the v_{et} term now represents an extrapolation to infinite thermodynamic driving force and can be thought of as defining an "intrinsic reactivity" (see further discussion).

The most important conclusion from the preceding analysis for our present considerations is that the rate constants of electron transfer reactions involving a series of structurally homologous redox proteins and a given small molecule should depend on the midpoint potential difference between the donors and the acceptors. This, of course, requires that other factors, such as the steric and electrostatic properties of the electron transfer site, and any electronic effects involving specific amino acid side chains, remain the same within the series. As will be documented later, in fact, kinetic measurements can be used to probe such parameters. Several examples of the application of Marcus theory to electron transfer reactions involving cytochrome c can be found in the review by Marcus and Sutin (1985).

IV. Reactions with Organic Reagents

A. Quinols

The reduction of oxidized cytochrome c by hydroquinone at neutral pH has been shown to be an autocatalytic process, mediated by the semiquinone anion radical, $Q^{\bar{}}$ (Williams, 1963; Yamazaki and Ohnishi, 1966). Second order rate constants for the reactions of benzosemiquinone ($BQ^{\bar{}}$) and menasemiquinone ($MQ^{\bar{}}$) with horse heart cytochrome c have been determined to be 2.5×10^6 $M^{-1}s^{-1}$ and 2×10^8 $M^{-1}s^{-1}$, respectively (Yamazaki and Ohnishi, 1966; Ohnishi et al., 1969). This order of rate constants is in accord with the respective reduction potentials for these two quinols (i.e., $+81$ mV for $BQ/BQ^{\bar{}}$ and -160 mV for $MQ/MQ^{\bar{}}$; cf. Rich and Bendall, 1980), as predicted from Marcus theory. At lower pH values, reduction of the cytochrome occurs via the anionic quinol, $Q^{\bar{}}$ and the fully protonated quinol (Rich and Bendall, 1980). For these species (as derived from benzhydroquinone), rate constants of 1.8×10^3 $M^{-1}s^{-1}$ and 1.3×10^{-2} $M^{-1}s^{-1}$ have been measured (Rich and Bendall, 1980). For the anionic quinol reaction, an extensive series of reductants has been investigated, and a clear Marcus-type relationship between rate constant and reduction potential has been obtained over a range of approximately 500 mV (Rich and Bendall, 1980).

B. Flavins

Figure 14.3 shows free energy plots of the second-order rate constants for the one-electron transfer reactions of oxidized Class I and Class II c-type cytochromes with the semiquinone forms of lumiflavin and riboflavin generated by laser flash photolysis (Tollin et al., 1986a; Cusanovich et al., 1987). The solid lines correspond to plots of the Marcus exponential relation [Equation (14.4)]. Several points can be noted from these plots.

(1) For a given reductant, all of the Class I cytochromes fall approximately on the same theoretical curve. This constitutes strong evidence that these second-order rate constants at least partly reflect the electron transfer event. The monotonic nature of the free energy dependence over a wide range of biological sources implies, as pointed out earlier, that any differences in steric and intervening media effects are not dominant in these reactions (electrostatic effects are not involved here inasmuch as the reductants are electrically neutral; however, see later discussion). This is prob-

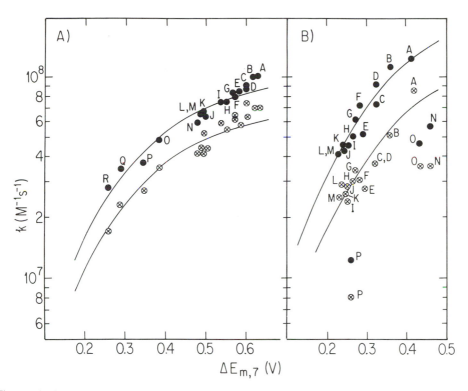

Figure 14.3

Plot of second-order rate constant for reduction of c-type cytochromes by flavin semiquinones vs. difference in redox potential. Solid lines are theoretical curves using Equation (14.4); see text for parameters used in fitting data. (●) lumiflavin; (⊗) riboflavin. Data from Meyer *et al.* (1983, 1984, 1986). (A) Class I cytochromes: A, *R. tenue* 3761 *c*-553; B, *R. tenue* 2761 *c*-553; C, *Euglena* sp. *c*-552; D, *R. spheroides* c_2; E, *R. vanniellii* c_2; F, *R. capsulatus* c_2; G, *R. purpureus* *c*-553; H, *R. rubrum* c_2; I, *R. salexigens* c_2; J, *P. aeruginosa* *c*-551; K, *C. krusei* *c*; L, tuna *c*; M, horse *c*; N, *P. denitrificans* c_2; O, *C. thiosulfatophilum* *c*-555; P, *Paracoccus* sp. *c*-554(548); Q, *E. halophila* *c*-551; R, *R. gelatinosa* *c*-550. (B) Class II cytochromes: A, *Agrobacterium* *c*-556; B, *Alcaligenes* sp. *c'*; C, *R. salexigens* *c'*; D, *R. palustris* *c'*; E, *R. gelatinosa* *c'*; F, *R. capsulatus* *c'*; G, *R. tenue* 3761 *c'*; H, *R. spheroides* *c'*; I, *C. vinosum* *c'*; J, *R. molischianum* *c'*; K, *R. photometricum* *c'*; L, *R. rubrum* *c'*; M, *R. purpureus* *c'*; N, *R. palustris* *c'*; O, *R. spheroides* *c*-554; P, *E. halophila* *c'*.

ably a consequence of the relatively simple nature of the reductants, and it is consistent with what is known from x-ray structural studies of this group of cytochromes. In the case of the Class II cytochromes, although the majority of the proteins cluster about a single line for both lumiflavin and riboflavin, there are clearly several proteins which are less reactive. Interestingly, this is true for both of the reductants. We must conclude from this that these proteins belong to a different group with respect to steric and/or electronic properties (although only three proteins are involved, it is possible to draw curves through the data points which are parallel to those for the majority of the proteins). Although this remains to be confirmed by direct structural studies, these results point up the value of this approach in classifying groups of homologous proteins with respect to structural properties (kinetic taxonomy).

(2) In all cases, the v_{et} values for riboflavin semiquinone are smaller than those for the lumiflavin species (the values for the Class I cytochromes are 0.7×10^8 $M^{-1}s^{-1}$ and 1.0×10^8 $M^{-1}s^{-1}$, respectively; those for the Class II cytochromes are 1.5×10^8 $M^{-1}s^{-1}$ and 2.8×10^8 $M^{-1}s^{-1}$, respectively). This is interpreted as reflecting steric effects of the ribityl side chain $[-CH_2(CHOH)_3CH_2OH]$. It is also important to note that these values lie considerably below the diffusion-controlled limit. Thus, physicochemical factors must be controlling these limiting rate constants, which validates the contention that they represent an "intrinsic" reactivity. This is further demonstrated by the fact that second-order rate constants obtained using 5-deazariboflavin semiquinone as a reductant of this same group of Class I cytochromes (Meyer *et al.*, 1983) are in the range of 10^9 $M^{-1}s^{-1}$, indicating that these latter reactions are indeed diffusion-limited. The deazaflavin species has a considerably lower reduction potential than does the flavin semiquinone (-650 mV vs. -230 mV, respectively), and is evidently a more reactive species towards Class I cytochromes. The so-called "inverted region" (Marcus and Sutin, 1985) has not been observed for small molecule–cytochrome interactions. This most likely results from the fact that with large driving force—for example, deazariboflavin semiquinone—it is diffusion-controlled collision that is rate-limiting, not electron transfer (Meyer *et al.*, 1983).

(3) For both reductants, the intrinsic reactivities of the Class II cytochromes (of the majority type) are significantly larger than for the Class I cytochromes. Although only a single x-ray structure presently exists for a Class II cytochrome (Weber *et al.*, 1981), this shows a clear difference in the extent of exposure of the heme at the protein surface, with the Class II cytochrome exposure being larger than that of all of the Class I cytochromes for which x-ray structures are known (*cf.* Meyer *et al.*, 1986). This suggests that, in this case, steric effects are the dominant factor in controlling reactivity, although one cannot exclude the possibility that electronic effects are also contributing.

(4) The values of the reorganizational energies are moderate, being 11.0 kcal/mol (0.48 eV) for the Class I proteins and 12.0 kcal/mol (0.52 eV) for the Class II cytochromes for both lumiflavin and riboflavin. The fact that they are independent of the flavin reductant (see later discussion) indicates that they reflect mainly protein structural parameters, although we cannot specify precisely what features are involved. It is also of interest that the values are closely similar for the two classes of cytochrome. Again, the structural implications of this are unclear, although it seems reasonable to conclude that the main contribution reflects processes occurring in the vicinity of the heme cofactor, rather than more global events. It is interesting that these values compare favorably with a recent theoretical estimate of 11.9 kcal/mol for the reorganizational energy upon electron transfer into tuna cytochrome *c* (Marcus and Sutin, 1985). Inasmuch as the model used was relatively crude, it is not clear whether or not this is merely coincidental. Thus, Warshel and co-workers have calculated a value which is approximately twice as large using a different model (Warshel and Churg, 1983; Churg *et al.*, 1983).

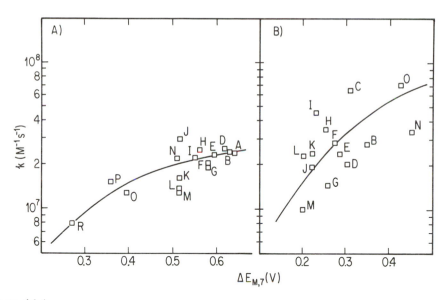

Figure 14.4

Plot of extrapolated second-order rate constants (k_∞) for reduction of Class I (A) and Class II (B) c-type cytochromes by FMN semiquinone vs. difference in redox potential. Labels are defined in caption to Figure 14.3. The solid lines are theoretical curves using Equation (14.4); see text for parameter values. Data from Meyer *et al.* (1984, 1986).

Very little information is currently available with regard to Class III c-type cytochromes. The single example that has been determined, *D. vulgaris* cytochrome c_3, is approximately 16 times more reactive towards lumiflavin semiquinone relative to Class I cytochromes than would be predicted based upon its redox potential (Meyer *et al.*, 1983). This is consistent with the presence of four hemes in this protein, each of which is substantially more solvent-exposed than is the case for the Class I proteins.

Figure 14.4 shows free energy plots for FMN semiquinone reactions with Class I and II cytochromes. Inasmuch as the reductant in this case carries a net negative electrostatic charge, the rate constant values in this plot were obtained by extrapolating ionic strength dependence data (*cf.* Figure 14.5 for examples) to infinite ionic strength using a formalism developed by Watkins (1986) and described previously (*cf.* Cusanovich *et al.*, 1987; Watkins *et al.*, 1994). Note that the ionic strength dependence gives information concerning the sign and magnitude of the electrostatic charge on the cytochrome at the electron transfer site, i.e., the electrostatic free energy of the interaction between donor and acceptor (V_{ii}). Values for these quantities obtained from the theoretical fit are given in Table 14.1. Although the absolute values are model-dependent, the relative values are not, and thus comparisons among the various proteins are valid. In general, the results are consistent with what is known from structural data about the region surrounding the exposed heme, suggesting that *local* charge effects dominate in most cases. This is especially

502

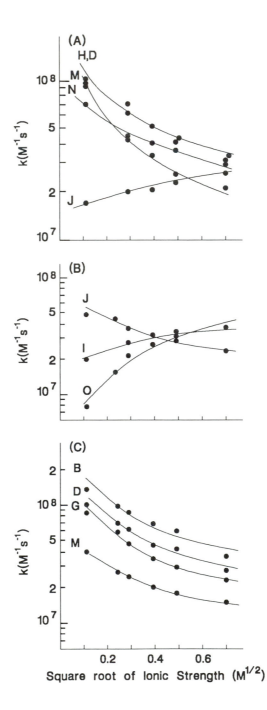

Figure 14.5
Ionic strength dependence of second-order rate constants for reduction of Class I (A) and Class II (B, C) c-type cytochromes by FMN semiquinone. Labels are defined in caption to Figure 14.3. Solid lines are theoretical fits using electrostatic formalism of Watkins (1986). Data from Meyer *et al.* (1984, 1986).

apparent for the *c*-type cytochromes (e.g., *R. spheroides*), in which the charge distribution is quite asymmetric, with the positive charge localized close to the exposed heme edge (Weber and Tollin, 1985). For those proteins with little or no charge near the exposed heme surface (e.g., *Pseudomonas c*-551), more global charge distributions become important.

From the plots of Figure 14.4, it is evident that a considerably larger degree of scattering exists for the FMN reactions than was the case for lumiflavin or riboflavin. This is attributed to the larger size of the ribityl phosphate side chain, which therefore samples a larger fraction of the protein surface during electron transfer and is thus more sensitive to differences in steric properties (some of this scattering may also be due to the requirement for extrapolation of ionic strength data, which introduces an additional source of error). The larger steric effect is also indicated by the smaller v_{et} values (0.27×10^8 M^{-1}s^{-1} for Class I cytochromes and 1.2×10^8 M^{-1}s^{-1} for Class II cytochromes). Interestingly, the λ values are the same as were found for lumiflavin and riboflavin, again suggesting that these quantities reflect protein properties.

One further piece of information can be obtained from the FMN studies. If one plots the values of ln k_∞ for a group of Class I cytochromes having approximately the same redox potential, and for which structural information is available, against a distance calculated from the methyl group of the substituted vinyl group of the heme to a plane defined by the positions of the terminal atoms of Gln-16, Val-28 and Ile-81 in horse cytochrome *c*, or the analogous sequence positions in the other cytochromes, one obtains a straight line (Figure 14.6) with a slope of 0.44 (this would correspond to a value for α in the distance dependence of the tunneling matrix element; see earlier discussion). Thus, the expected exponential distance depen-

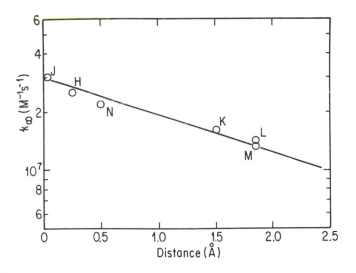

Figure 14.6

Plot of k_∞ for FMN semiquinone reduction of some Class I *c*-type cytochromes vs. the distance of closest approach (as described in text and in Cusanovich *et al.*, 1987). Labels are defined in caption to Figure 14.3.

dence relationship is apparently obeyed, which is consistent with the interpretation that the main determinant of the deviation of the FMN rate constants from the theoretical Marcus curve is the steric interference of the protein with accessibility to the heme edge.

Flavin analogs have also been used as photo-oxidants of reduced cytochromes (Roncel *et al.*, 1990). Comparisons of the kinetics of cytochrome oxidation with quenching of the flavin triplet state have demonstrated that the latter species is the oxidant. This is a very efficient reaction, with second-order rate constants in the range of $1.5–4.0 \times 10^9$ M^{-1}s^{-1} for lumiflavin and riboflavin with horse cytochrome *c* and a high potential algal cytochrome (*c*-552), approximately two orders of magnitude larger than the corresponding rate constants for the reduction reactions with the semiquinone species. This is attributed to the excited state nature of the triplet state and its corresponding high energy content (i.e., to an effect of thermodynamic driving force). Steric and electrostatic effects similar to those previously observed for the photo-reductions were found for the photo-oxidation reactions. This is consistent with the same (or closely adjacent) sites on the cytochromes being used for both types of reactions, which is in agreement with earlier studies of cytochrome oxidation and reduction using other reactants (Post *et al.*, 1977; Moore *et al.*, 1982, 1984).

V. Reactions with Inorganic Reagents

A large body of data has been collected using inorganic oxidants and reductants. These data have played a critical role in developing our current understanding of biological electron transfer. Most importantly, it was studies with inorganic reactants which led to the first convincing kinetic evidence that electron transfer takes place at the exposed heme edge in Class-I *c*-type cytochromes (Hodges *et al.*, 1974; Cusanovich and Miller, 1974). Nevertheless, studies with inorganic reactants have been fraught with difficulties and resulted in substantial controversy and confusion. As will be demonstrated later, this results primarily from the fact that inorganic reactants are, in general, highly charged, and their reaction kinetics with *c*-type cytochromes are quite ionic strength–dependent. In order to discuss kinetic studies with inorganic reactants, Table 14.2 summarizes rate constants for the oxidation and reduction of several different cytochromes with a variety of reactants. For the purposes of discussion, the rate constants can be viewed as Kk_{et} in terms of Equation (14.1). This list is by no means exhaustive, but represents a reasonable sampling of published rate constants. What is most notable is that there is no obvious correlation between the measured rate constants and driving force (ΔE_m). This observation is contrary to the expectation of Marcus theory and quite different from the results obtained with quinols and flavin semiquinones discussed in the previous section. It is the factors that contribute to this apparent lack of correlation between driving force and rate constants that will be the focus of the discussion that follows.

Prior to further analysis, it is important to note some limitations with the data given in Table 14.2. These data were compiled from a number of references where the experimental conditions were not identical and, as a result, are not precisely comparable in all cases. For example, the ionic strength is on the order of 100 mM,

Table 14.2
Cytochrome reaction kinetics.

Cytochrome	Reactant	$\Delta E_m{}^a$ (mV)	$k_{red} \times 10^{-4}$ $M^{-1} s^{-1}$	$k_{ox} \times 10^{-4}$ $M^{-1} s^{-1}$
Horse c^d	$Fe(CN)_6{}^{4-/3-}$	160	3.5	920
Horse c^b	$Fe(CN)_6{}^{4-/3-}$	160	5.1	860
C. krusei c^b	$Fe(CN)_6{}^{4-/3-}$	160	22.0	2100
Ps. aeruginosa c-551^c	$Fe(CN)_6{}^{4-/3-}$	150	—	6.0
Horse c^e	$Co(dipic)_2{}^-$	140	—	1.1
Ps. aeruginosa c-551^e	$Co(dipic)_2{}^-$	130	—	0.06
Horse c^c	$Co(phen)_3{}^{3+}$	110	—	0.3
Ps. aeruginosa c-551^c	$Co(phen)_3{}^{3+}$	100	—	8.0
Horse c^d	$Fe(CN)_5(imid)^{3-}$	56	42.0	—
Horse c^e	$Fe(dipic)_2{}^{2-}$	18	1.5	—
Ps. aeruginosa c-551^e	$Fe(dipic)_2{}^{2-}$	8	3.2	—
Horse c^f	$Fe(EDTA)^{2-}$	−140	2.6	—
Ps. aeruginosa c-551^g	$Fe(EDTA)^{2-}$	−145	0.4	—

[a] $\Delta E_m = E_m$ (reactant) $- E_m$ (cytochrome).
[b] Eley et al. (1984); $\mu = 120$ mM, pH 7.4, 25°.
[c] Cheddar et al. (1989); $\mu \sim 100$ mM, pH 7.2, 23.5°.
[d] Butler et al. (1981); $\mu = 100$ mM, pH 7.2, 25°.
[e] Bis(dipicolinato) cobaltate (III), bis(dipicolinato) ferrate (II), Mauk et al. (1979); $\mu = 200$ mM, pH 7.0, 25°.
[f] Hodges et al. (1974); $\mu = 100$ mM, pH 7.0, 25°.
[g] Coyle and Gray (1976); $\mu = 100$ mM, pH 7.0, 25°.

the temperature varies between 23.5° and 25°, and the pH is 7.2 to 7.4. The iron hexacyanide data given in Table 14.2 require special note. For the cytochromes listed, the reaction with $Fe(CN)_6{}^{4-}$ does not go to completion even at very high concentrations. Thus, the observed rate constant for reduction by $Fe(CN)_6{}^{4-}$ does not reflect the true rate constant, and the data (absorbance versus time) have been treated for equilibration kinetics to obtain the appropriate rate constant (Butler et al., 1981; Ohno and Cusanovich, 1981). It is also notable that the iron hexacyanide data for both stopped-flow measurements (Butler et al., 1981) and NMR experiments (Eley et al., 1984) are in excellent agreement (Table 14.2).

The apparent lack of correlation between driving force and measured rate constants results from a number of factors. At low ionic strengths (< 500 mM), electrostatic interactions strongly influence the reaction kinetics. Hence, depending upon the protein charge and the reactant charge, the effect of ionic strength can have quite dramatic effects on the reaction kinetics (*cf.* Figure 14.5). Plus–plus or minus–minus interactions result in an increase in the measured rate constant with increasing ionic strength. Similarly, plus–minus interactions have rate constants which decrease with increasing ionic strength. However, it is important to note that the net protein charge is not a good predictor of the ionic strength effects. For example, the acidic cytochromes c_2 behave as cations, consistent with a strong positive charge in the region of the exposed heme edge in spite of a net negative protein charge, with the iron hexacyanides (Post *et al.*, 1977) as well as the flavin semiquinones (Meyer *et al.*, 1984; Tollin *et al.*, 1984; see Table 14.1). These observations represent definitive evidence for well-defined interaction domains on the cytochrome molecular surface. Thus, to compare reactants in order to quantitate the role of driving force, rate constants at infinite ionic strength are required. Unfortunately, very little data are available as a function of ionic strength, and thus detailed analysis is precluded in most cases.

Recently, an effort has been made to resolve ionic strength effects with two inorganic oxidants (Cheddar *et al.*, 1989). In this study, the oxidation of *Ps. aeruginosa* cytochrome *c*-551, horse cytochrome *c*, *Paracoccus denitrificans* cytochrome c_2, and *Azotobacter vinelandii* cytochrome c_5 ($E_{m,7} = 315$ mV) by $Fe(CN)_6^{3-}$ and $Co(phen)_3^{3+}$ was studied as a function of ionic strength. These cytochromes were chosen because they have quite distinct electrostatic surface potentials: that for horse cytochrome *c* is generally positive; *Paracoccus* cytochrome c_2 has a strong positive surface potential in the vicinity of the exposed heme edge and a strong negative field on the backside; and *Ps. aeruginosa* cytochrome *c*-551 and *Azotobacter* cytochrome c_5 have no significant electrostatic potential near the heme edge, but have a generally negative field elsewhere. Rate constants were extrapolated to infinite ionic strength after making measurements over the ionic strength range 10 mM to 800 mM. Table 14.3 presents the driving force (ΔE_m), the rate constant at infinite ionic strength (k_∞), and the apparent charge on the cytochrome (Z_1). Values of Z_1 were obtained by fitting the data with the electrostatic model developed by Watkins (1986), as noted earlier. In general, smooth monotonic curves were obtained when the log of the observed second-order rate constant was plotted against the square root of the ionic strength. However, in the case of the reaction of *Paracoccus* cytochrome c_2 with $Co(phen)_3^{3+}$, the rate constant initially decreased with increasing ionic strength, and then increased with increasing ionic strength (above 100 mM). Both regions were fitted, and the parameters obtained are given in Table 14.3. It is important to note that extrapolation from the high ionic strengths used here to infinite ionic strength is not particularly model-dependent, because there is in general very little change in rate constant above 500 mM ionic strength. Thus, although models for analyzing the effect of ionic strength on cytochrome reaction kinetics are somewhat controversial, the values of k_∞ presented here are quite reasonable estimates independent of the model used.

Table 14.3
Effect of ionic strength on cytochrome reaction kinetics.

Cytochrome	Co(phen)$_3$$^{3+}$			Fe(CN)$_6$$^{3-}$		
	ΔE_m[b] (mV)	$k_\infty \times 10^{-4}$ (M^{-1} s^{-1})	Z_1[c]	ΔE_m[b] (mV)	$k_\infty \times 10^{-4}$ (M^{-1} s^{-1})	Z_1[c]
Paracoccus c_2	120	3.5/.006[a]	+3.6/−6.8[a]	170	2.0	+2.1
Horse c	110	0.7	+1.9	160	1.2	+4.0
Ps. aeruginosa c-551	100	3.2	−1.7	150	0.3	−2.4
Azotobacter c_5	55	4.9	−1.0	105	0.2	−1.5

[a] High μ/low μ.
[b] $\Delta E_m = E_m$ (reactant) − E_m (cytochrome).
[c] Apparent charge on cytochrome (see text).

A number of important points can be derived from these studies. First, for *Paracoccus* cytochrome c_2 it is clear that the ferricyanide reaction occurs at a positively charged region with a rate constant similar to that for horse cytochrome c, as expected from the similar oxidation–reduction potentials. However, at low ionic strength, Co(phen)$_3$$^{3+}$ reacts at a site with a negative electrostatic field and appears to switch to a region with a positive electrostatic field at higher ionic strengths (Table 14.3). These results suggest that at low ionic strength, Co(phen)$_3$$^{3+}$ interacts at a site which is distant from the heme (10^2–10^3-fold less reactive), presumably the backside, and switches to a more reactive site when the repulsive interactions are suppressed by increased ionic strength, yielding a k_∞ for the high-ionic-strength site which is comparable to those of the other cytochromes. Second, ferricyanide is 4–10 times more reactive (k_∞) with the horse and *Paracoccus* cytochromes than with cytochrome c-551 and c_5, while Co(phen)$_3$$^{3+}$ is substantially less reactive with horse cytochrome c and *Paracoccus* cytochrome c_2 (low ionic strength). The simplest interpretation is that ferricyanide reacts at the exposed heme edge with the *Paracoccus* and horse cytochromes because of the strong positive fields. However, cytochromes c_5 and c-551 have relatively uncharged nonpolar environments in the region of the exposed heme edge, which apparently facilitates reaction of the ferricyanide at more distant sites. What is interesting is that ferricyanide would apparently rather react at an electrostatically repulsive site on cytochrome c_5 and c-551 than at a closer more hydrophobic site. As an inverse of this situation, it appears that Co(phen)$_3$$^{3+}$ has a greater intrinsic reactivity with cytochromes c-551 and c_5, possibly because of the nonpolar character of the cobalt ligands, which would allow a closer approach to the exposed heme edge where that region is more hydrophobic in nature.

To summarize to this point, it is clear that electrostatics play a critical role in interpreting redox kinetics using inorganic reactants. Moreover, it is apparent that the c-type cytochromes are capable of using more than one site of electron transfer (also see later discussion) depending on the electrostatics (and, thus, ionic strength), as well as the chemistry of the inorganic complex ligands. In view of this, it is clear that drawing definitive conclusions concerning redox kinetics from the type of data presented in Table 14.2 is difficult, if not impossible, in many cases, and considerable effort is required to fully exploit studies with polyionic reactants.

As complex as the situation is with the inorganic reactants, at least with eukaryotic cytochromes c, analysis is even more difficult than suggested earlier. This results from the facts that the basic cytochromes c readily bind anions (Barlow and Margoliash, 1966; Margoliash et al., 1970), and that multiple anion binding sites exist (Moore et al., 1984). In recent work, the anion binding properties of three eukaryotic cytochromes have been characterized (Gopal et al., 1988). These cytochromes were from horse, bovine, and tuna, and the ions studied were chloride, phosphate, and the supposedly nonbinding buffer, Tris-cacodylate. Binding constants and, in parentheses, the approximate number of binding sites are summarized in Table 14.4. As can be seen, the binding constants and number of binding sites are quite sensitive to the redox state of a particular cytochrome and quite different for different cytochromes c. Interestingly, even the supposed nonbinding buffer Tris-cacodylate shows weak but measurable binding. The data presented established that anion binding by eukaryotic cytochromes c can be measured quantitatively, and that ion binding is quite sensitive to experimental conditions. Importantly, binding is also sensitive to the amino acid sequence of the cytochrome c, because horse, bovine, and tuna cytochrome c show different affinities and stoichiometries for anion binding.

Moore et al. (1984) have summarized data characterizing ion binding by horse cytochrome c and have identified three possible binding sites. Site I, at the top left of the exposed heme edge (lysines 5, 86, 87, and 88); Site II, at the front right heme edge (lysines 7, 23, and 27); and Site III, at the front left (lysines 13, 72, and 86). All of these sites are in the vicinity of the exposed heme edge, and the presence of bound anions can be expected to substantially alter the electrostatic surface potential. It follows then that the ion binding differences between different cytochromes c must result from structural changes which alter electrostatic fields and, thus, ion binding. Moreover, the difference in ion binding between redox states of a particular cytochrome suggests that structural and/or dynamic changes take place on reduction which are different for the three cytochromes studied. It is important to realize that, because of the ion binding properties of the cytochromes c, very different kinetics of electron transfer can be expected with the same cytochrome, depending on the ionic strength and ionic composition, which will control both the number of sites occupied and the electrostatics of the site.

It has been known for some time that the iron hexacyanides tightly bind to eukaryotic cytochrome c (Stellwagen and Cass, 1975). Indeed, at low ionic strength, it requires electrodialysis to completely remove ferricyanide from ferricytochrome c (Vorkink and Cusanovich, 1974; Peterman and Morton, 1977). In addition, a variety of other inorganic complexes, including $Fe(CN)_6^{4+}$, $Co(CN)_6^{3-}$, $Cr(C_2O_4)_3^{3-}$,

Table 14.4
Ion binding by eukaryotic cytochromes c.

Cytochrome c	K (M^{-1})		
	Chloride	Phosphate	Tris-cacodylate
Horse (Fe^{3+})	2224 (2)[a]	1556 (2)	74[b]
(Fe^{2+})	295 (3)	476 (1–2)	64
Bovine (Fe^{3+})	239 (2)	293 (3)	3
(Fe^{2+})	847 (3)	1330 (3)	25
Tuna (Fe^{2+})	389 (2–3)	16 (3–4)	20
(Fe^{2+})	5170 (1–2)	2089 (3)	8

[a] Value in parentheses is approximate number of ions bound from modified Scatchard plots.
[b] Because of the weak binding, reliable estimates of the number of Tris-cacodylate binding sites could not be made.

and $Fe(EDTA)_2^{2-}$, have been shown to bind at multiple sites (2–3) with moderate affinities (100–400 M^{-1}) at ionic strengths on the order of 100 mM (see Moore *et al.*, 1984, for a review). The affinity of these reactants is ionic strength–dependent, increasing with decreasing ionic strength. In view of the foregoing, it is apparent that the two-step mechanism given in Equation (14.1) is incomplete, at least in the case of the iron hexacyanides, and a three-step mechanism [Equation (14.5)] is required:

$$\text{cyto. (Fe}^{+3}) + \text{Fe(CN)}_6^{4-} \underset{k_{21}}{\overset{k_{12}}{\rightleftharpoons}} C_1 \underset{k_{32}}{\overset{k_{23}}{\rightleftharpoons}} C_2 \underset{k_{43}}{\overset{k_{34}}{\rightleftharpoons}} \text{cyto. (Fe}^{+2}) + \text{Fe(CN)}_6^{3-}, \quad (14.5)$$

where C_1 is a ferrocyanide-ferricytochrome c complex (precursor complex), and C_2 a ferricyanide–ferrocytochrome c complex (successor complex).

For the binding of $Fe(CN)_6^{4-}$ to ferricytochrome c, there appears to be general agreement between kinetic and NMR measurements that the binding constant is 100–150 M^{-1} at pH 7–7.4, 20–25°, and at moderate ionic strength (100–120 mM) (Ragg and Moore, 1984; Ohno and Cusanovich, 1981). However, there are differences between the NMR and kinetic analysis in terms of the binding constant for $Fe(CN)_6^{3-}$ to ferrocytochrome c. Kinetic analysis suggests a binding constant for ferricyanide on the order of 8×10^4 M^{-1} (Ohno and Cusanovich, 1981), while NMR experiments suggest a value of 130 M^{-1} (Eley *et al.*, 1984) at 100 to 120 mM ionic strength. It is important to note that the $Fe(CN)_6^{3-}$ binding to ferrocytochrome c was not measured directly in the NMR studies; rather, it was measured by a competitive method using $Co(CN)_6^{3-}$, which is redox-inactive. Similarly, the kinetic analysis is indirect, with the binding constant derived from mathematical analysis of absorbance versus time data. At present, it is not possible to resolve the apparent differences in binding constants obtained by the two methods.

Direct or indirect measurements of k_{et} [k_{23} or k_{32} in Equation (14.5)] for the reaction of c-type cytochromes with the iron hexacyanides are scarce. Eley *et al.* (1984) determined a value of approximately 780 s^{-1} for k_{et} for the reduction of horse and *C. krusei* cytochrome c by ferrocyanide. In addition, they estimated k_{et} for oxidation by ferricyanide to be 6.6×10^4 s^{-1}, assuming that the binding constants for ferricyanide to ferrocytochrome c and for ferrocyanide to ferricytochrome c were equivalent. These values yield an overall equilibrium constant (K_{eq}) of 1.2×10^{-2} for Equation (14.5). Unfortunately, this is not in good agreement with the value of K_{eq} expected from redox potentials ($\sim 1.8 \times 10^{-3}$). Using a different approach, Ohno and Cusanovich (1988) determined k_{23} and k_{32} [Equation (14.5)] for a series of Class I c-type cytochromes at 100 mM ionic strength (20°). This analysis used numerical integration to solve the simultaneous differential equations which describe Equation (14.5) and utilized data for both oxidation and reduction of c-type cytochromes by ferri- and ferrocyanide, as well as the overall equilibrium constant. In this analysis, all six rate constants defined in Equation (14.5) were determined. It was found that k_{12}, k_{21}, k_{34}, and k_{43} were severely constrained. However, only minimum values of k_{23} and k_{32} could be obtained. That is, any value above the minimum would yield an equally good fit as long as the ratio of k_{23}/k_{32} was constant. Interestingly, this approach yielded a ratio of k_{23}/k_{32} of approximately 1 (0.57 to 1.06) for the 11 cytochromes analyzed. Minimum values for k_{23} and k_{32} ranged from 800 s^{-1} to 5600 s^{-1}, depending on the cytochrome. No correlation between k_{et} (k_{23} or k_{32}) and redox potential was found, as can be inferred from the fact that the ratio of k_{23}/k_{32} was approximately 1.

What became apparent was that the difference in oxidation–reduction potential was related to the ratio k_{12}/k_{43} in all cases analyzed, contrary to the expectation of Marcus theory. This analysis yielded an equilibrium constant of 8×10^4 M^{-1} for the binding of ferricyanide to ferrocytochrome c, as described earlier. The discrepancy between k_{et} and the fact that the redox potential appeared to be reflected in k_{12} and k_{43} led Ohno and Cusanovich (1981) to propose a five-step mechanism. In this mechanism, reversible first-order steps are inserted prior to and following the electron transfer step in Equation (14.5). These first-order steps were viewed as structural changes (reorganizations) within the transient iron hexacyanide–cytochrome complex which were required to reach the transition state so that electron transfer could take place. This approach yields a physically reasonable model which, in terms of Marcus theory, makes the reorganization energy [λ, Equation (14.2)] the controlling factor. Nevertheless, this analysis is not completely satisfying, and further study is required before definitive conclusions can be drawn. In particular, data at infinite ionic strength are required to fully understand the iron hexacyanide–cytochrome c interaction.

Analysis of the kinetics of interaction of the iron hexacyanides with eukaryotic cytochrome c is further complicated by the existence of multiple binding sites for these reactants (Ragg and Moore, 1984). Thus, it is not clear at this time if all binding sites are redox-active. However, if they are, they must have similar kinetics (within a factor of 3) since the observed kinetics are monophasic, consistent with a single kinetic process. This expectation is not unreasonable, since the binding sites identified to date (Moore *et al.*, 1984) are in the general region of the exposed heme edge, and hence roughly at equal distance from the heme. However, it is possible that under the appropriate experimental conditions, one binding site is preferentially reactive. Ragg and Moore (1984) have reported that the two binding sites for $Co(CN)_6^{3-}$ have different affinities (approximately 10-fold) at 70 mM ionic strength but are similar at 120 mM. Given the complexities described here, it is not surprising that comparing the redox kinetics for different cytochromes with different inorganic reactants (Table 14.2) is difficult. The key issue in terms of Marcus theory is the value of k_{et} [k_{23} for reduction, k_{32} for oxidation, equation (14.5)], which should relate driving force to kinetics. Unfortunately, because of the strong effects of ionic strength and complex formation between cytochrome c and anionic reactants, analysis of k_{et} or, if a rapid equilibrium exists, Kk_{et} for oxidation and reduction is precluded for most of the systems described in Table 14.2, since the effect of ionic strength and specific ions has not generally been characterized. Moreover, as discussed earlier, the nature of the metal ligands is also important.

Although future studies with inorganic reactants could be carried out to resolve the complexities (for example, Cheddar et al., 1989), these would seem to be peripheral to the main issue. Thus, as described in Sections III and IV, the use of quinols and flavin semiquinones appears to provide the means to apply Marcus theory and to elucidate the principal features of electron transfer by c-type cytochromes. Moreover, as the study of biological electron transfer has matured, the major focus is now on protein–protein and bound small molecule–protein interactions, not on protein–free small molecule interactions. Thus, although inorganic and organic reactants have served us well in providing an initial understanding of electron transfer by c-type cytochromes, as described elsewhere in this volume the major emphasis in the future will undoubtedly be on other systems (see chapters by Millett, Scott, and Millett and Durham, this volume).

VI. Summary

It is apparent from the discussion presented here that Marcus theory is applicable to biological electron transfer. Moreover, by the judicious choice of experimental conditions, the impact of electrostatic interactions, complex formation, and surface topography of interacting molecules can be described and, in some cases, quantified. This is important in terms of Equations (14.2)–(14.4) because it means that one can focus on λ (the reorganization energy) and v_{et} (intrinsic reactivity). It is these terms that contain information on the contributions of the distance between redox centers, the relative orientation of chromophores, molecular dynamics in the transition-state complex, and the intervening media between redox centers to electron transfer kinetics. In simplest terms, biological electron transfer can be divided into two steps: complex formation and subsequent electron transfer. The work using small-molecule reactants has provided the basis for understanding the role of electrostatics and surface topography in complex formation and has provided some information on the contribution of driving force to k_{et}. The role of the intervening medium is more difficult to determine and will probably depend heavily in the future on the use of site-directed mutagenesis approaches.

Central to our understanding of biological electron transfer is elucidating the structural and dynamic factors that control k_{et} in any specific system, as described earlier. In this context, a complication exists which currently limits our ability to fully exploit available kinetic studies. This is the fact that the measurement of electron transfer (k_{et}) is constrained to the rate-limiting first-order step. Thus, in the context of Equation (14.5), this would be k_{23} for reduction or k_{32} for oxidation. However, if electron transfer is preceded by a dynamic process (i.e., a structural change required to reach the transition state) which is rate-limiting, then analysis in terms of Marcus theory will not be possible. This concept, termed "conformational gating" (Hoffman and Ratner, 1987), has been observed in several systems—for example, in studies to measure intracomplex electron transfer in the cytochrome b_2–cytochrome c complex (McLendon *et al.*, 1987)—and cannot be excluded *a priori* in other systems. In this situation, a minimum five-step mechanism would be required, with reversible conformational changes preceding electron transfer in either direction (Ohno and Cusanovich, 1981). This complication is a particular challenge because there is no direct way to distinguish between first-order structural changes and first-order electron transfer in kinetic experiments.

Current studies in many laboratories now focus on the characterization of intramolecular electron transfer in biologically relevant protein–protein complexes. As will be discussed in part elsewhere in this volume, intracomplex and/or intramolecular electron transfer can be characterized in a wide variety of systems involving such complexes—for example, cytochrome c–cytochrome b_5; cytochrome c–flavodoxin; cytochrome c–cytochrome c peroxidase; cytochrome c–cytochrome oxidase; cytochrome c–reaction centers; and cytochrome c–cytochrome b/c_1.

In view of the progress in recent years and, importantly, with the development of new technologies such as NMR to measure dynamic properties and site-directed mutagenesis to alter specific amino acid side chains, it is easy to be optimistic that substantial progress will be made in the future towards the goal of understanding biological electron transfer.

VII. References

Agmon, N., & Levine, R. D. (1977) *Chem. Phys. Lett.* **52**, 197–201.
Ahmad, N., Cusanovich, M. A., & Tollin, G. (1981) *Proc. Natl. Acad. Sci. USA* **78**, 6724–6728.
Ambler, R. P. (1977) in *The Evolution of Metalloenzymes, Metalloproteins and Related Materials* (Leigh, G. J., ed.), pp. 100–118, Symposium Press, London.
Barlow, G. H., & Margoliash, E. (1966) *J. Biol. Chem.* **241**, 1423–1477.
Brunschwig, B. S., & Sutin, N. (1989) *J. Am. Chem. Soc.* **111**, 7454–7465.
Butler, J., Davies, D. M., & Sykes, A. G. (1981) *J. Inorg. Biochem.* **15**, 41–53.
Cheddar, G., Meyer, T. E., Cusanovich, M. A., Stout, C. D., & Tollin, G. (1989) *Biochem.* **28**, 6318–6322.
Churg, A. K., Weiss, R. M., Warshel, A., & Takano, T. (1983) *J. Phys. Chem.* **87**, 1683–1694.
Closs, G. L., & Miller, J. R. (1988) *Science* **240**, 440–447.
Closs, G. L., Calcaterra, L. T., Green, N. J., Penfield, K. W., & Miller, J. R. (1986) *J. Phys. Chem.* **90**, 3673.
Coyle, C. L., & Gray, H. B. (1976) *Biochem. Biophys. Res. Comm.* **73**, 1122–1127.
Cusanovich, M. A., & Miller, W. G. (1974) *Bioelectrochem. and Bioenergetics* **1**, 448–458.
Cusanovich, M. A., Meyer, T. E., & Tollin, G. (1987) *Adv. Inorg. Chem.* **7**, 37–91.
Eley, C. G. S., Ragg, E., & Moore, G. R. (1984) *J. Inorg. Biochem.* **21**, 295–310.

Gopal, D., Wilson, G. S., Earl, R. A., & Cusanovich, M. A. (1988) *J. Biol. Chem.* **263**, 11652–11656.

Gray, H. B., & Malmstrom, B. G. (1989) *Biochem.* **28**, 7499–7505.

Hodges, H. L., Holwerda, R. A., & Gray, H. B. (1974) *J. Am. Chem. Soc.* **96**, 3132–3137.

Hoffman, B. M., & Ratner, M. A. (1987) *J. Am. Chem. Soc.* **109**, 6237–6243.

Marcus, R. A. (1968) *J. Phys. Chem.* **72**, 891–899.

Marcus, R. A., & Sutin, N. (1985) *Biochim. Biophys. Acta* **811**, 265–322.

Margoliash, E., Barlow, G. H., & Byers, V. (1970) *Nature* **228**, 723–726.

Mauk, A. G., Coyle, C. L., Bordignon, E., & Gray, H. B. (1979) *J. Am. Chem. Soc.* **101**, 5054–5056.

McLendon, G., Pardue, K., & Bak, P. (1987) *J. Am. Chem. Soc.* **109**, 7540–7541.

Meyer, T. E, & Cusanovich, M. A. (1989) *Biochim. Biophys. Acta* **975**, 1–28.

Meyer, T. E., Przsysiecki, C. T., Watkins, J. A., Bhattacharyya, A., Simondsen, R. P., Cusanovich, M. A., & Tollin, G. (1983) *Proc. Natl. Acad. Sci. USA* **80**, 6740–6744.

Meyer, T. E., Watkins, J. A., Przysiecki, C. T., Tollin, G., & Cusanovich, M. A. (1984) *Biochemistry* **23**, 4761–4767.

Meyer, T. E., Cheddar, G., Bartsch, R. G., Getzoff, E. D., Cusanovich, M. A., & Tollin, G. (1986) *Biochemistry* **25**, 1383–1390.

Moore, G. R., Huang, Z. X., Eley, C. G. S., Barker, H. A., Williams, G., Robinson, M. N., & Williams, R. J. P. (1982) *Faraday Disc. Chem. Soc.* **74**, 311–329.

Moore, G. R., Eley, C. G. S., & Williams, G. (1984) *Adv. Inorg. Bioinorg. Mech.* **3**, 1–96.

Ohnishi, T., Yamazaki, H., Iyanagi, T., Nakamura, T., & Yamazaki, I. (1969) *Biochim. Biophys. Acta* **172**, 357–369.

Ohno, N., & Cusanovich, M. A. (1981) *Biophys. J.* **36**, 589–605.

Peterman, B. F., & Morton, R. A. (1977) *Can. J. Biochem.* **55**, 796–803.

Post, C. B., Wood, F. E., & Cusanovich, M. A. (1977) *Arch. Biochem. Biophys.* **184**, 586–595.

Ragg, E., & Moore, G. R. (1984) *J. Inorganic Biochem.* **21**, 253–261.

Rehm, D., & Weller, A. (1969) *Ber. Bunsenges. Phys. Chem.* **73**, 834–839.

Rich, P. R., & Bendall, D. K. (1980) *Biochim. Biophys. Acta* **592**, 506–518.

Roncel, M., Hervas, M., Navarro, J. A., De la Rosa, M., & Tollin, G. (1990) *Eur. J. Biochem.* **191**, 531–536.

Stellwagen, E., & Cass, R. D. (1975) *J. Biol. Chem.* **250**, 2095–2098.

Takano, T., & Dickerson, R. E. (1981) *J. Mol. Biol.* **153**, 79–115.

Tollin, G., Cheddar, G. Watkins, J. A., Meyer, T. E., & Cusanovich, M. A. (1984) *Biochem.* **23**, 6345–6349.

Tollin, G., Meyer, T. E., & Cusanovich, M. A. (1986a) *Biochim. Biophys. Acta* **853**, 29–41.

Tollin, G., Hanson, L. K., Caffrey, M., Meyer, T. E., & Cusanovich, M. A. (1986b) *Proc. Natl. Acad. Sci. USA* **83**, 3693–3697.

Vorkink, W. P., & Cusanovich, M. A. (1974) *Photochem. Photobiol.* **19**, 205–215.

Warshel, A., & Churg, A. K. (1983) J. Mol. Biol. **168**, 693–697.

Watkins, J. A. (1986) Ph. D. Thesis, University of Arizona, Tucson, Arizona.

Watkins, J. A., Cusanouich, M. A., Meyer, T. E., & Tollin, G. (1994) *Prot. Sci.* **3**, 2104–2114.

Weber, P. C., & Tollin, G. (1985) *J. Biol. Chem.* **260**, 5568–5573.

Weber, P. C., Howard, A., Xuong, N. H., & Salemme, F. R. (1981) *J. Mol. Biol.* **153**, 399–424.

Williams, G. R. (1963) *Can. J. Biochem. Physiol.* **41**, 231–237.

Wood, P. M. (1983) *FEBS Letts.* **164**, 223–226.

Yamazaki, I., & Ohnishi, T. (1966) *Biochim. Biophys. Acta* **112**, 469–481.

15

Long-Range Intramolecular Electron Transfer Reactions in Cytochrome c

ROBERT A. SCOTT

Center for Metalloenzyme Studies
University of Georgia

I. Introduction

Long-range biological electron transfer is at the heart of many important metabolic processes. The coupling of respiration to energy conservation in the mitochondrial electron transport chain and photosynthetic charge separation take advantage of the evolutionary development of donor/acceptor systems that can carry out efficient and selective electron transfer reactions. A molecular-level understanding of the "design principles" involved in these systems has been the goal of considerable research during the past decade or so, some of which will be reviewed in this chapter. Of particular interest has been identification of the molecular and electronic structural characteristics of donor/acceptor sites and of the intervening medium that might be used to control the rate and specificity of electron transfer. Such donor/acceptor characteristics as driving force and reorganizational energy (inherent reactivity), as well as aspects of the arrangement of the donor and acceptor, including their distance of separation, relative orientation, and nature of the intervening medium, are expected to be important controlling factors (for recent reviews, see Bowler *et al.*, 1990; Gray and Malmström, 1989; Isied, 1991b; Marcus and Sutin, 1985; McLendon, 1988; Onuchic *et al.*, 1992; Winkler and Gray, 1992).

Not surprisingly, cytochrome c has been one of the main subjects for such investigations. It is a soluble component of the mitochondrial respiratory chain, shuttling electrons from the cytochrome bc_1 complex to cytochrome aa_3 (complex IV). Homologous high-potential cytochromes c are ubiquitous throughout evolution. As a paradigm for studying electron transfer on a molecular level, it is very well studied (witness the existence of this book), it is structurally characterized both in crystals and solution, it is amenable to genetic manipulation, and it is commercially available. Bimolecular electron-transfer reactions of cytochrome c with small-molecule redox agents and with other proteins abound in the literature. Cytochrome c is also the most popular protein partner for formation of electrostatic protein–protein complexes within which intracomplex electron-transfer reactions can be studied. Another approach for which cytochrome c is a prototype has involved surface labeling of proteins with redox-active small molecules to generate "synthetic bimetallic proteins." Studies of intramolecular electron transfer rates within such hybrid proteins have improved our understanding of the factors affecting long-range biological electron transfer, and it is this approach with which this chapter will be concerned.

II. Theoretical Background

The analysis of long-range electron transfer reactions has taken advantage of a solid theoretical base developed for small-molecule electron transfer reactions (Marcus and Sutin, 1985). The rate constant (k_{et}) for electron transfer from a donor site (D) to an acceptor site (A) can be simply represented as the product of three factors (Equation (15.1), Sutin, 1982),

$$k_{et} = \nu_n \kappa_{el} \kappa_n, \tag{15.1}$$

ν_n is a nuclear frequency (often assumed to be 10^{13} s^{-1}), representing the rate at which the reactants (D, A) approach the potential barrier for conversion to products (D$^+$, A$^-$); κ_{el} is an electronic transmission coefficient that incorporates the interaction between the donor and acceptor electronic wavefunctions; and κ_n is a nuclear factor that describes the activation of the system achieved by nuclear reorganization. A classical description of this nuclear reorganization using simple harmonic potentials for (D, A) and (D$^+$, A$^-$) results in an expression for κ_n as a function of the driving force ($-\Delta G°$) and reorganization energy (λ):

$$\kappa_n = \exp[-(\lambda + \Delta G°)^2/4\lambda k_B T] \tag{15.2}$$

(Marcus and Sutin, 1985). This predicts a parabolic dependence of $\ln k_{et}$ on $-\Delta G°$, with $\ln k_{et}$ first increasing with increasing driving force, then leveling off at $-\Delta G° = \lambda$, and finally *decreasing* with increasing driving force in what is referred to as the Marcus inverted region. (Quantum mechanical treatments of the nuclear factor retain this behavior; *vide infra*.) The reorganization energy is made up of two components [Equation (15.3)]: λ_{in}, describing the bond-stretching and -bending energy required to distort D and A to structures more like D$^+$ and A$^-$; and λ_{out}, describing the reorientation of solvent (or other medium) to accommodate the charge movement when (D, A) $\rightarrow$ (D$^+$, A$^-$):

$$\lambda = \lambda_{in} + \lambda_{out}(d), \tag{15.3}$$

Equation (15.3) reflects the dependence of λ_{out} on the distance between A and D (d) resulting from the more extensive charge movement at longer d (Isied *et al.*, 1988).

One simplification describes the electronic transmission coefficient (κ_{el}) as if it resulted from the overlap of Gaussian wavefunctions on D and A through a constant dielectric intervening medium. This results in a simple exponential distance dependence (Marcus and Sutin, 1985; McLendon, 1988):

$$\kappa_{el} = \kappa_{el}^{\circ} \exp[-\beta (d - d_0)]. \tag{15.4}$$

Using the dielectric continuum model, the distance dependence of λ_{out} can also be described as approximately exponential, so that the β of Equation (15.4) incorporates both electronic- and nuclear-based distance dependences, as explicitly described by

$$\beta = \beta_e + \beta_n. \tag{15.5}$$

(Bowler *et al.*, 1990; Isied *et al.*, 1988). The distance, d, in Equation (15.4) is usually considered to be the minimum edge-to-edge separation between A and D, assuming major lobes of the donor HOMO and acceptor LUMO are centered on edge atoms. d_0 is the edge-to-edge distance at van der Waals separation and is often assumed to be 3 Å (Axup *et al.*, 1988). In the case of strong A–D electronic coupling at van der Waals contact ($d = d_0$), the electron-transfer reaction becomes adiabatic ($\kappa_{el} = \kappa_{el}^{\circ} = 1$). For the long-range ($d > d_0$) electron transfers that will be discussed in this chapter, nonadiabaticity is assumed. Some evidence exists (*vide infra*) that peptide media offer only weak electronic coupling resulting in $\kappa_{el}^{\circ} \ll 1$.

An analogous quantum-mechanical description of electron transfer gives rise to a k_{et} expression:

$$k_{et} = (4\pi^2/h)(H_{RP})^2(\text{FC}). \tag{15.6}$$

In this expression, the nuclear term is replaced by a Franck–Condon overlap factor (FC) describing the Boltzmann-weighted sum of overlaps of vibrational wavefunctions for reactants (D, A) and products (D^+, A^-) (Bertrand, 1991; Marcus and Sutin, 1985). The electronic term is replaced by $(H_{RP})^2$, the square of the matrix element describing the electronic coupling between reactants (R) and products (P). This approach incorporates the possibility of a tunneling contribution to nonadiabatic electron transfer, since nonzero Franck–Condon overlaps exist for vibrational levels that lie below the classical activation energy (DeVault, 1984). The classical [Equation (15.1)] and quantum-mechanical [Equation (15.6)] descriptions can be compared by using the classical version of the Franck–Condon factor:

$$\text{FC} = (4\pi\lambda k_B T)^{-1/2} \exp[-(\lambda + \Delta G^{\circ})^2/4\lambda k_B T] \tag{15.7}$$

(for the quantum-mechanical version, see Bertrand, 1991; Brunschwig and Sutin, 1987; Jortner, 1976; Marcus and Sutin, 1985; Siders and Marcus, 1981), and a simple exponential distance-dependent description of the electronic factor (Jacobs *et al.*, 1991):

$$(H_{RP})^2 = (H_{RP}^{\circ})^2 \chi_{DA}^2; \tag{15.8}$$

$$\chi_{DA}^2 = \exp[-\beta(d - d_0)]. \tag{15.9}$$

For systems with strong electronic coupling that become adiabatic at van der Waals separation, $(H_{RP}{}^\circ)^2$ is given by

$$(H_{RP}{}^\circ)^2 = h\nu_n(\lambda k_B T/4\pi^3)^{1/2}. \tag{15.10}$$

For typical values of $\lambda = 1-2$ eV, $H_{RP}{}^\circ$ is then expected to be $\sim 200-250$ cm^{-1} at 300 K ($\nu_n = 10^{13}$ s^{-1}). $H_{RP}{}^\circ$ may be expected to decrease for systems that have weak electronic coupling even at close approach.

Equations (15.4) and (15.9) derive from the assumption that the material between D and A can be treated as a constant dielectric medium. More recently, a number of approaches have been devised to account for the electronic structure of the intervening (polypeptide) medium in the computation of χ_{DA} [Equation (15.9)] (Beratan *et al.*, 1987; Broo and Larsson, 1991; Christensen *et al.*, 1990, 1992; Goldman, 1991; Kuki, 1991; Kuki and Wolynes, 1987; Larsson, 1983; Onuchic and Beratan, 1990; Onuchic *et al.*, 1992; Siddarth and Marcus, 1990, 1992). These approaches use various levels of sophistication to calculate the electronic coupling of A and D orbitals to "bridge" (polypeptide) orbitals and the decay of orbital overlap through the bridge medium. Experimental effort is currently being focused on uncovering evidence for the necessity of this increased theoretical sophistication.

III. Synthetic and Kinetic Methods

The selection of redox active groups to be used in metalloprotein attachment for the study of intramolecular electron transfer must follow certain criteria: (a) The attached group must be chemically robust (for transition metal complexes, substitution-inert) in both oxidation states. Lability in one oxidation state results in loss of the attached acceptor (donor) during the electron transfer reaction. Ruthenium is one of the few transition metals that is substitution-inert in both common oxidation states [Ru(II) and Ru(III)], which explains the predominance of ruthenium complexes used in these studies. (b) The attached group must not significantly alter the tertiary structure of the host protein. This effectively limits the modifications to the surface of the metalloprotein. (c) The modification reaction must be selective enough so that relatively few products are formed that can be separated and characterized. The separation usually requires that the modified protein have a different charge than the native protein.

The first work in this area used the imidazole side chain of surface histidine residues as a ligand in a substitution reaction with $A_5Ru(H_2O)^{2+}$ ($A \equiv NH_3$), generating a Ru–N$_{imid}$ bond (see structure **1** in Figure 15.1). This reaction was first employed to label ribonuclease A (Matthews *et al.*, 1978, 1980, 1981; Recchia *et al.*, 1982), but its use in the study of electron transfer was pioneered by the groups of Gray and Isied (Isied *et al.*, 1982; Winkler *et al.*, 1982; Yocom *et al.*, 1982). It remains the most popular strategy (by far) for labeling a variety of proteins for electron-transfer studies (see Table 15.1 and references therein). One advantage to this "ruthenation" (Gray, 1986) reaction is the ability to tune the reduction potential and reorganization energy contribution of the ruthenium probe by replacing one

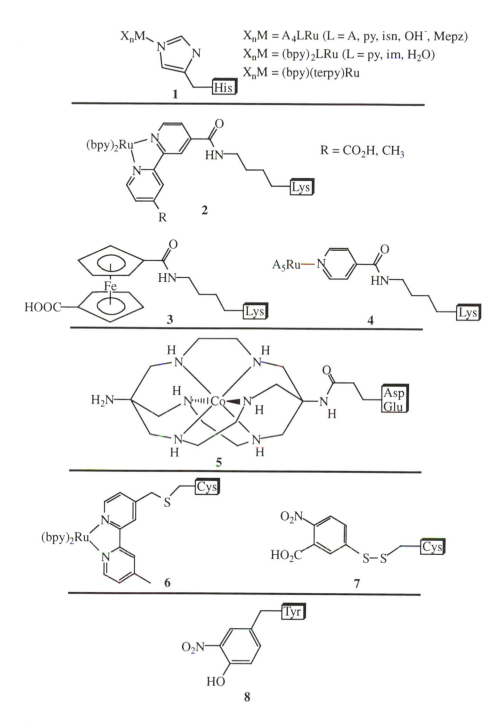

Figure 15.1

The redox-active groups that have been attached to metalloproteins for the study of intramolecular electron transfer. The abbreviations are the same as those used in Table 15.1.

Table 15.1
Rate constants, thermodynamic, and kinetic parameters for protein D → A electron transfer reactions.[a]

D ——[d]—— A (Å)	$-\Delta G°$ (eV)	$k_{et} \sim 25C$ (s⁻¹)	$\Delta H^{\ddagger}$ (kcal/mol)	$\Delta S^{\ddagger}$ (eu)	References[b]
Ru(II)A$_5$(H33)[11.1] cyt c Fe(III)P	0.18	30(5)	2.0(5)	−43(5)	a, b, c, d, e, at, au
Ru(II)A$_5$(H33)[11.1] cyt c Fe(III)P	0.13	53(2)	3.5(2)	−39(1)	f, g
Ru(II)A$_5$(H33)[11.1] cyt c Fe(III)P	0.13	39(1)	—	—	av
c-Ru(II)A$_4$OH(H33)[11.1] cyt c Fe(III)P	0.27	5 × 10²	—	—	h, i, j
c-Ru(II)A$_4$py(H33)[11.1] cyt c Fe(III)P	−0.10	2.0	—	—	h
t-Ru(II)A$_4$py(H33)[11.1] cyt c Fe(III)P	−0.11	1.5	—	—	h
Ru(II)A$_5$(H33)[11.1] cyt c Fe(III)P-im	—	1.2	—	—	k
Ru(II)A$_5$(H62)[14.8] Sc cyt c Fe(III)P	0.20	1.7(1)	—	—	e, l, at
Ru(II)A$_5$(H39)[12.3] Ck cyt c Fe(III)P	0.13	87(2)	—	—	av
Ru(II)A$_5$(H58)[13.4] Sc cyt c Fe(III)P	0.20	43	—	—	e
c-Ru(I)A$_4$Mepz(H33)[11.1] cyt c Fe(III)P	0.28	6 × 10²	—	—	h
Ru(I)(bpy)$_2$py(H33)[11.1] cyt c Fe(III)P	1.56	2.8 × 10⁵	—	—	h, i, j
Ru(I)(bpy)$_2$im(H33)[11.1] cyt c Fe(III)P	1.56	2.0 × 10⁵	—	—	h, i, j
Co(II)(diAMsar-D2)[20(1)] cyt c Fe(III)P	0.64	2.9(3)	—	—	m, n
Co(II)(diAMsar-E4)[19(1)] cyt c Fe(III)P	0.64	3.2(2)	—	—	m, n
Co(II)(diAMsar-E21)[18(1)] cyt c Fe(III)P	0.65	1.0(2)	—	—	m, n
Co(II)(diAMsar-E61)[17(1)] cyt c Fe(III)P	0.62	2.0(3)	—	—	m, n
Co(II)(diAMsar-E62)[15(1)] cyt c Fe(III)P	0.62	1.9(3)	—	—	m, n
Co(II)(diAMsar-E66)[14(1)] cyt c Fe(III)P	0.64	2.8(2)	—	—	m, n

		1./(1)			m, n
Co(III)(ГаГлэа LLO4)[10(1)] cyt c Fe(III)F	0.64	—	—	—	
cyt c Fe(II)P[11.1](H33)Ru(III)(bpy)$_2$im	0.74	$2.6(3) \times 10^6$	—	—	o, p, at, aw
Ck cyt c Fe(II)P[12.3](H39)Ru(III)(bpy)$_2$im	0.74	$3.2(4) \times 10^6$	—	—	p, aw
Sc cyt c Fe(II)P[14.8](H62)Ru(III)(bpy)$_2$im	0.74	$1.0(2) \times 10^4$	—	—	p, aw
cyt c Fe(II)P[8.4](H72)Ru(III)(bpy)$_2$im	0.74	$9.0(3) \times 10^5$	—	—	p, aw
cyt c Fe(II)P[4.5](H79)Ru(III)(bpy)$_2$im	0.74	$>10^8$	—	—	aw
cyt c Fe(II)P[11.1](H33)t-Ru(III)A$_4$isn	0.18	$<10^{-2}$	—	—	q
cyt c Fe(II)P[11.1](H33)c-Ru(III)A$_4$isn	0.18	$<10^{-2}$	—	—	h
cyt c Fe(II)P[11.1](H33)c-Ru(III)A$_4$Mepz	0.46	~ 1.5	—	—	h
cyt c Fe(II)P[11.1](H33)Ru(III)(bpy)$_2$py	0.66	40	—	—	h, i, j
cyt c Fe(II)P[11.1](H33)Ru(III)(bpy)$_2$im	0.53	55	—	—	h, i, j
cyt c Fe(II)P[11.1](H33)Ru(III)(bpy)(terpy)	0.48	40	—	—	i, j
cyt c Fe(II)P[11.1](H33)Ru(III)(bpy)$_2$OH$_2$	0.39	40	—	—	i, j
cyt c Fe(II)P[6–12](K27-dcbpy)Ru(III)(bpy)$_2$	1.12	$3.0(5) \times 10^7$	—	—	r, s, t
cyt c Fe(II)P[6–10](K13-dcbpy)Ru(III)(bpy)$_2$	1.12	$2.6(5) \times 10^7$	—	—	r, s, t
cyt c Fe(II)P[8–16](K72-dcbpy)Ru(III)(bpy)$_2$	1.12	$2.4(5) \times 10^7$	—	—	r, s, t
cyt c Fe(II)P[9–16](K25-dcbpy)Ru(III)(bpy)$_2$	1.12	$1.5(3) \times 10^6$	—	—	r, s, t
cyt c Fe(II)P[9–16](K7-dcbpy)Ru(III)(bpy)$_2$	1.12	$6(2) \times 10^5$	—	—	r, s, t
Ru(II)*(bpy)$_2$im(H33)[11.1] cyt c Fe(III)P	~ 1	$2(1) \times 10^5$	—	—	p, aw
Ru(II)*(bpy)$_2$im(H39)[12.3] Ck cyt c Fe(III)P	~ 1	$1.4(5) \times 10^6$	—	—	p, aw
Ru(II)*(bpy)$_2$im(H62)[14.8] Sc cyt c Fe(III)P	~ 1	$1.1(2) \times 10^5$	—	—	p, aw
Ru(II)*(bpy)$_2$im(H72)[8.4] cyt c Fe(III)P	~ 1	$3.4(7) \times 10^5$	—	—	p, aw
Ru(II)*(bpy)$_2$im(H79)[4.5] cyt c Fe(III)P	~ 1	$>5 \times 10^7$	—	—	aw

(continued)

Table 15.1 (continued)

D ——[d]—— A (Å)	$-\Delta G^\circ$ (eV)	$k_{et} \sim 25C$ (s^{-1})	$\Delta H^\ddagger$ (kcal/mol)	$\Delta S^\ddagger$ (eu)	References[b]
Ru(II)*(bpy)₂(dcbpy-K27)[6–12] cyt c Fe(III)P	0.88	$2.0(3) \times 10^7$	—	—	r, s, t
Ru(II)*(bpy)₂(dcbpy-K13)[6–10] cyt c Fe(III)P	0.88	$1.6(3) \times 10^7$	—	—	r, s, t
Ru(II)*(bpy)₂(dcbpy-K72)[8–16] cyt c Fe(III)P	0.88	$1.4(3) \times 10^7$	—	—	r, s, t
Ru(II)*(bpy)₂(dcbpy-K25)[9–16] cyt c Fe(III)P	0.88	$1.0(3) \times 10^6$	—	—	r, s, t
Ru(II)*(bpy)₂(dcbpy-K7)[9–16] cyt c Fe(III)P	0.88	$3(1) \times 10^5$	—	—	r, s, t
cyt c ³ZnP*[11.1](H33)Ru(III)A₅	0.70	$7.7(8) \times 10^5$	1.7(4)	−27(5)	d, e, u, v, w, x, at, au
cyt c ³ZnP*[11.1](H33)Ru(III)A₅	0.70	—	1.6 (100–150K) 5.6 (>150K)	—	y
cyt c ³ZnP*[11.1](H33)Ru(III)A₄isn	1.05	$2.9(3) \times 10^6$	<0.5	−30(5)	d, e, u, v, w, at, au
cyt c ³ZnP*[11.1](H33)Ru(III)A₄py	0.97	$3.3(3) \times 10^6$	2.2(4)	−22(5)	d, e, u, v, w, at, au
Ck cyt c ³ZnP*[13.0](H39)Ru(III)A₅	0.70	$1.5(2) \times 10^6$	1.3(3)	−27(5)	e, w, z, at, au
Ck cyt c ³ZnP*[13.0](H39)Ru(III)A₄py	0.97	$8.9(9) \times 10^6$	0.2(2)	−27(5)	e, w, z, at, au
Ck cyt c ³ZnP*[13.0](H39)Ru(III)A₄isn	1.05	$1.0(1) \times 10^7$	0.2(2)	−27(5)	e, w, z, at, au
Sc cyt c ³ZnP*[15.5](H62)Ru(III)A₅	0.70	$6.5(7) \times 10^3$	1.4(3)	−37(5)	w, at, au
Sc cyt c ³ZnP*[15.5](H62)Ru(III)A₄py	0.97	$3.6(4) \times 10^4$	—	—	w, at, au
Sc cyt c ³ZnP*[15.5](H62)Ru(III)A₄py	0.97	2.7×10^4	—	—	e
Ru(II)A₅(H33)[11.1] cyt c ZnP•⁺	1.01	$1.6(4) \times 10^6$	—	—	d, e, u, v, w, x, at, au
Ru(II)A₄py(H33)[11.1] cyt c ZnP•⁺	0.74	$3.5(4) \times 10^5$	<0.5	−34(5)	d, e, u, v, w, at, au
Ru(II)A₄isn(H33)[11.1] cyt c ZnP•⁺	0.66	$2.0(2) \times 10^5$	<0.5	−35(5)	d, e, u, v, w, at, au
Ru(II)A₄isn(H39)[13.0] Ck cyt c ZnP•⁺	0.66	$6.5(7) \times 10^5$	−1.7(4)	−39(5)	e, w, z, at, au

Ru(II)A$_4$py(H39)[13.0] Ck cyt c ZnP$^{\bullet+}$	0.74	$1.5(2) \times 10^6$	$-37(5)$	e, w, z, at, au
Ru(II)A$_5$(H39)[13.0] Ck cyt c ZnP$^{\bullet+}$	1.01	$5.7(6) \times 10^6$	$-29(5)$	e, w, z, at, au
Ru(II)A$_4$py(H62)[15.5] Sc cyt c ZnP$^{\bullet+}$	0.74	$8.1(8) \times 10^3$	—	w, at, au
Ru(II)A$_4$py(H62)[15.5] Sc cyt c ZnP$^{\bullet+}$	0.74	2.6×10^3	—	e
Ru(II)A$_5$(H62)[15.5] Sc cyt c ZnP$^{\bullet+}$	1.01	$2.0(2) \times 10^4$	$-37(5)$	w, at, au
Ru(II)A$_5$(H33)[11.1] cyt c ^{3}ZnP*	0.36	2.4×10^2	-41	u, x
Ru(II)A$_5$(H47)[7.9] $P s.s.$ cyt c_{551} Fe(III)**P**	0.15	13(2)	—	aa
cyt Tb_5 Fe(II)**P**[12.1](H26)Ru(III)A$_5$	0.08	1.4(1)	—	ab, at, au
cyt LMb_5 Fe(II)**P**[12.0](H26)Ru(III)A$_5$	0.10	5.9(5)	—	ab, at, au
cyt Tb_5 Fe(II)**P**$_d$[12.9](H26)Ru(III)A$_5$	0.13	0.2(1)	—	ab, at, au
Ru(II)A$_5$(H48)[12.7] Mb Fe(III)**P**	-0.02	0.019(2)	7.4(5)	d, e, ac, ad, ae
Ru(II)A$_5$(H48)[12.7] Mb Mg**P**$^{\bullet+}$	0.91	$4.9(5) \times 10^4$	—	at, au
Ru(II)A$_5$(H48)[12.7] Mb Cd**P**$^{\bullet+}$	0.92	$2.1(2) \times 10^5$	—	at, au
Ru(II)A$_5$(H48)[12.7] Mb Zn**P**$^{\bullet+}$	0.96	$1.5(5) \times 10^5$	—	at, au
Ru(II)A$_4$isn(H48)[12.7] Mb Zn**P**$^{\bullet+}$	0.61	$1.4(5) \times 10^4$	—	at, au
Ru(II)A$_4$py(H48)[12.7] Mb Zn**P**$^{\bullet+}$	0.69	$2.0(5) \times 10^4$	—	at, au
Mb Fe(II)**P**[12.7](H48)Ru(III)A$_5$	0.02	0.040(5)	19.5(5)	d, ac, ad, ae, af, at, au
Mb Fe(II)**P**[12.7](H48)Ru(III)A$_4$py	0.28	2.5(5)	—	af, at, au
Mb Fe(II)**P**[12.7](H48)Ru(III)A$_4$isn	0.35	3.0(4)	—	at, au
Mb Fe(II)**P**(CNBr)[12.7](H48)Ru(III)A$_4$isn	0.26	5.5(5)	—	at, au
Ru(II)A$_5$(H48)[12.7] Mb Fe(III)**P**(CNBr)	0.09	2.0(5)	—	at, au
Mb ^{3}Zn**P***[12.7](H48)Ru(III)A$_5$	0.82	$7.0(7) \times 10^4$	2.2(5)	d, e, ag, ah, ai, at, au
Mb ^{3}Zn**P***[12.7](H48)Ru(III)A$_4$py	1.09	$2.0(2) \times 10^5$	—	at, au

(continued)

Table 15.1 (continued)

D $\overline{\qquad[d]\qquad}$ A (Å)	$-\Delta G°$ (eV)	$k_{et} {}^{\sim 25 C}$ (s^{-1})	$\Delta H^{\ddagger}$ (kcal/mol)	$\Delta S^{\ddagger}$ (eu)	References[b]
Mb ^{3}ZnP*[12.7](H48)Ru(III)A$_4$isn	1.17	$2.9(3) \times 10^5$	—	—	at, au
Mb ^{3}ZnP*[19.3](H81)Ru(III)A$_5$	0.82	86(10)	5.6(25)	—	d, ag, ah, at
Mb ^{3}ZnP*[20.1](H116)Ru(III)A$_5$	0.82	89(3)	5.4(4)	—	d, ag, ah, at
Mb ^{3}ZnP*[22.0](H12)Ru(III)A$_5$	0.82	$1.0(1) \times 10^2$	4.7(9)	—	d, ag, ah, at
Mb ^{3}MgP*[12.7](H48)Ru(III)A$_5$	0.81	$9.5(9) \times 10^4$	—	—	at, au
Mb ^{3}MgP$_{da}$*[12.7](H48)Ru(III)A$_5$	0.87	$5.7(4) \times 10^4$	2.2(3)	$-11(7)$	d, e, ai, aj, ak
Mb ^{3}MgP$_{da}$*[19.3](H81)Ru(III)A$_5$	0.87	82(12)	5.8(5)	1(6)	d, ai, aj, ak
Mb ^{3}MgP$_{da}$*[20.1](H116)Ru(III)A$_5$	0.87	69(3)	5.2(5)	3(7)	d, ai, aj, ak
Mb ^{3}MgP$_{da}$*[22.0](H12)Ru(III)A$_5$	0.87	67(11)	5.2(5)	$-1(6)$	d, ai, aj, ak
Mb ^{3}MgP$_{de}$*[12.7](H48)Ru(III)A$_5$	<0.87	$3.2(5) \times 10^4$	—	—	d, ai, aj, ak
Mb ^{3}MgP$_{de}$*[19.3](H81)Ru(III)A$_5$	<0.87	48(1)	—	—	d, ai, aj, ak
Mb ^{3}MgP$_{de}$*[20.1](H116)Ru(III)A$_5$	<0.87	49(8)	—	—	d, ai, aj, ak
Mb ^{3}MgP$_{de}$*[22.0](H12)Ru(III)A$_5$	<0.87	39(9)	—	—	d, ai, aj, ak
Mb ^{3}PdP*[12.7](H48)Ru(III)A$_5$	0.64	$9.1(9) \times 10^3$	—	—	d, e, ai, al, at, au
Mb ^{3}PdP*[12.7](H48)Ru(III)A$_4$py	0.91	$9.0(9) \times 10^4$	—	—	al, at, au
Mb ^{3}PtP$_{da}$*[12.7](H48)Ru(III)A$_5$	0.73	$1.2(1) \times 10^4$	—	—	d, e, ai, ak
Mb ^{3}CdP$_{da}$*[12.7](H48)Ru(III)A$_5$	0.85	$6.3(5) \times 10^4$	2.8	—	d, e, ai, aj, ak
Mb ^{3}CdP*[12.7](H48)Ru(III)A$_5$	0.79	$4.5(5) \times 10^4$	—	—	at, au
Mb ^{3}H$_2$P$_{da}$*[12.7](H48)Ru(III)A$_5$	0.53	$7.6(6) \times 10^2$	—	—	d, e, ai, ak
Mb ^{3}H$_2$P*[12.7](H48)Ru(III)A$_5$	0.39	$7.6(8) \times 10^2$	—	—	at, au

RSS$^{\bullet-}$(C104)[~12] Hb-α Fe(III)	0.44	$1.9(2) \times 10^2$	—	—	am
(C3)SS$^{\bullet-}$(C26)[~26] *Pa* Az Cu(II)	0.71	44(7)	11.4(10)	−14(2)	ax, ay
(C3)SS$^{\bullet-}$(C26)[~26] *As* Az Cu(II)	~0.7	8.5(15)	4.0(4)	−41(4)	ax
(C3)SS$^{\bullet-}$(C26)[~26] *Pf* Az Cu(II)	0.76	22(3)	8.7(3)	−23(1)	ay
(C3)SS$^{\bullet-}$(C26)[~26] *Af* Az Cu(II)	0.68	11(2)	13.0(3)	−10(2)	ay
Ru(II)A$_5$(H83)[11.8] *Pa* Az Cu(II)	0.27	1.9(4)	<0.8	—	d, an, ao
Ru(II)A$_5$(H83)[11.8] *Pa* Az Cu(II)	0.24	2.4(8)	—	—	ap
Ru(II)A$_5$(H59)[11.9] *Av* Pc Cu(II)	0.26	<0.08	—	—	ap
O$_2$N$^-$(Y83)[10.8] *Sp.o.* Pc Cu(II)	0.77	>10^7	—	—	aq, az
O$_2$N$^-$(Y83)[10.8] *Pe.s.* Pc Cu(II)	0.77	>10^7	—	—	aq, az
O$_2$N$^-$(Y62)[15.9] *Pe.s.* Pc Cu(II)	0.77	>10^7	—	—	az
Ru(II)A$_5$(H59)[~11] *Sc.o.* Pc Cu(II)	0.29	<0.26	—	—	ap
Ru(II)A$_5$(H32/100)[—] *Rv* St Cu(II)	0.10	0.05	—	—	ar
Ru(II)A$_5$(H42)[7.9] *Cv* HiPIP [Fe$_4$S$_4$]$^{3+}$	0.26	18(2)	—	—	as

a Abbreviations: C, cysteine; D, aspartic acid; E, glutamic acid; H, histidine; K, lysine; Y, tyrosine; A, NH$_3$; bpy, 2,2'-bipyridine; dcbpy, 4,4'-dicarboxy-2,2'-bipyridine; diAMsar, 1,8-diamino-3,6,10,13,16,19-hexaazabicyclo-(6.6.6)icosane; im, imidazole; isn, isonicotinic acid (4-carboxypyridine); Mb Fe(II)**P**(CNBr), myoglobin ferrous heme active site modified at H64 by CNBr (Kamiya et al., 1991); Mepz, methylpyrazine; py, pyridine; terpy, 2,2':2',2''-terpyridine; **P**, protoporphyrin-IX; **P**$_d$, deuteroporphyrin-IX; **P**$_{da}$, mesoporphyrin-IX diacid; **P**$_{de}$, mesoporphyrin-IX diethyl ester; *As*, *Alcaligenes* spp.; *Av*, *Anabaena variabilis*; *Ck*, *Candida krusei*; *Cv*, *Chromatium vinosum*; *Pa*, *Pseudomonas aeruginosa*; *Pe.s.*, *Petroselinum sativum*; *Pf*, *Pseudomonas fluorescens*; *Ps.s.*, *Pseudomonas stuzeri*; *Rv*, *Rhus vernicifera*; *Sc*, *Saccharomyces cerevisiae*; *Sc.o.*, *Scenedesmus obliquus*; *Sp.o.*, *Spinacea oleracea*; Az, azurin; cyt Tb$_5$, trypsin-solubilized cytochrome b_5; cyt LMb$_5$, lipase-solubilized mutant cytochrome b_5 (Jacobs et al., 1991); cyt c, cytochrome c (horse heart unless otherwise specified); cyt c_{551}, cytochrome c_{551}; Hb-α, hemoglobin, α subunit; HiPIP, high-potential iron protein; Mb, myoglobin; Pc, plastocyanin; St, stellacyanin.

b References (from the literature through 1992): a, Winkler et al., 1982; b, Yocom et al., 1983; c, Nocera et al., 1984; d, Bowler et al., 1990; e, Therien et al., 1991b; f, Isied et al., 1982; g, Isied et al., 1984; h, Isied, 1990; i, Isied, 1991a; j, Isied, 1991b; k, Bechtold et al., 1986a; l, Bowler et al., 1989; m, Scott et al., 1991; n, Conrad et al., 1992; o, Chang et al., 1991; p, Wuttke et al., 1992b; q, Bechtold et al., 1986b; r, Durham et al., 1989; s, Durham et al., 1990; t, Millett and Durham, 1991; u, Meade et al., 1989; v, Gray and Malmström, 1989; w, Therien et al., 1991a; x, Elias et al., 1988; y, Zang and Maki, 1990; z, Therien et al., 1990; aa, Osvath et al., 1988; ab, Jacobs et al., 1991; ac, Crutchley et al., 1985; ad, Crutchley et al., 1986; ae, Mayo et al., 1986; af, Lieber et al., 1987a; ag, Axup et al., 1988; ah, Lieber et al., 1987b; ai, Cowan and Gray, 1989; aj, Cowan and Gray, 1988; ak, Cowan and Gray, 1989; al, Karas et al., 1988; am, Faraggi and Klapper, 1988; an, Kostic et al., 1983; ao, Margalit et al., 1984; ap, Jackman et al., 1988b; aq, Govindaraju et al., 1990; ar, Farver and Pecht, 1989b; as, Jackman et al., 1988a; at, Winkler and Gray, 1992; au, Onuchic et al., 1992; av, Wishart et al., 1992; aw, Wuttke et al., 1992a, ax, Farver and Pecht, 1989a; ay, Farver and Pecht, 1992; az, Govindaraju et al., 1993.

or more of the NH_3 ligands with other ligands (Figure 15.1). Also, compared to other transition metal labeling reactions, this results in the shortest "tether" connecting the metal to the protein backbone (the Ru is five bonds removed from the C_α, of the histidine).

All other electron-transfer probe attachment strategies (summarized in Figure 15.1) involve organic modification reactions to form covalent connections between the probe molecule and a specific amino acid residue side chain. When the probe molecule is a transition metal complex, the attachment is through a pendant functional group on one of the ligands. The ε-amino groups of lysine side chains are relatively easily modified with pendant carboxylate functional groups on a probe ligand. A carbodiimide-activated carboxylate can be used directly, as in the attachment of 1,1'-dicarboxyferrocene (3) (Gorren et al., 1992; Scott et al., 1991), or a more weakly activated ester (based on N-hydroxysuccinimide) can be used as an intermediate. The latter strategy has been used by Millett and co-workers to attach (bpy)$_2$Ru(dcbpy) (bpy = 2,2'-bipyridine; dcbpy = 4,4'-dicarboxy-2,2'-bipyridine) to lysine residues (2) (Durham et al., 1989) and by Scott and co-workers to attach A$_5$Ru(isn) (isn = isonicotinic acid, 4-carboxypyridine) (4) (Scott et al., 1991) or (bpy)$_2$Ru(cmbpy) (cmbpy = 4-carboxy-4'-methyl-2,2'-bipyridine) (2) (Scott et al., 1991) to lysine residues. The construction of the (bpy)$_2$Ru(dcbpy)–lysine derivatives used a two-step procedure in which the dcbpy ligand was attached first, the modified derivatives separated, then (bpy)$_2$Ru(CO$_3$) reacted with the individual derivatives to form the final modified proteins (Pan et al., 1988). This aided in the separation of the various lysine-modified derivatives because the (bpy)$_2$Ru(dcbpy)–lysine group has the same charge as the original lysine, and cation-exchange separation of the final ruthenium-labeled products would have been difficult. In the pyridine-based lysine-attached probes, the metal is 11 bonds removed from the lysine C_α. The carbodiimide-assisted amide bond formation strategy has also been used to attach the amine functionality of Co(diAMsar) to the carboxylates of glutamic or aspartic acid residue side chains (5) (Conrad and Scott, 1989; Conrad et al., 1992). All of these amide-bond formation attachment strategies have emphasized cytochrome c as the protein of choice.

More recently, a surface cysteine residue on yeast iso-1-cytochrome c was modified by reaction with (4-bromomethyl-4'-methyl-2,2'-bipyridine)Ru(bpy)$_2$ (6), and the modified cytochrome was used in a photoinduced intracomplex electron-transfer reaction with cytochrome c peroxidase (Geren et al., 1991). In another approach, generation of a heterodisulfide (7) from reaction of 5,5'-dithiobis(2-nitrobenzoic acid) with Cys-104 of the human hemoglobin α subunit allowed pulse radiolytic formation of the disulfide anion radical which in turn reduced the Fe(III) heme by intramolecular electron transfer (Faraggi and Klapper, 1988). Farver and Pecht took advantage of a native disulfide in azurins for pulse-radiolytic generation of an anion radical donor to the Cu(II) site (Farver and Pecht, 1989a, 1992). Another organic-radical probe was generated by pulse radiolytic reduction of nitro-tyrosine-83 (8, generated by chemical modification of Tyr-83) of spinach (*Spinacea oleracea*) plastocyanin (Govindaraju et al., 1990, 1993).

First-order rate constants for intramolecular electron transfer in these modified protein systems have been measured (by spectrophotometric monitoring) over the range $\sim 10^{-2}$ to $\sim 10^{7}$ s^{-1} (see Table 15.1). This usually requires a kinetic technique that is sufficiently fast to allow rapid reactions to be observed; either pulse radiolysis or (conventional or laser) flash photolysis have been used in virtually all the measurements. When observing electron transfer from the ground state of the donor to the ground state of the acceptor, the experiment must be designed to "inject" an electron into the oxidized donor of the fully oxidized system ($D^+ \sim A$, Scheme I) or "extract" an electron from the reduced acceptor of the fully reduced system ($D \sim A^-$, Scheme II) in a bimolecular reaction that is rapid enough to reach completion before the intramolecular electron transfer (k_{et}, Schemes I, II) takes place. [Note

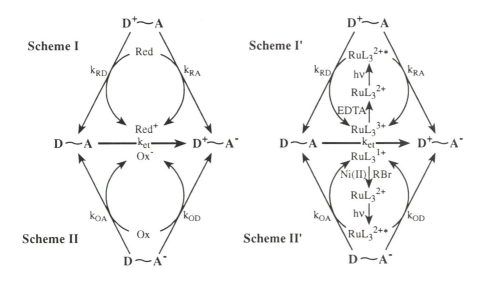

that, by definition, the donor is the site with the lower reduction potential, although in a few cases in which the driving forces are small (Crutchley *et al.*, 1985; Isied, 1990), "uphill" intramolecular electron transfer reactions have been observed.] Ideally, the reductant used for the electron injection would react selectively with the donor site ($k_{RA} \ll k_{RD}$ in Scheme I) or the oxidant used for electron extraction would react selectively with the acceptor site ($k_{OD} \ll k_{OA}$ in Scheme II). This is not usually observed in practice, nor is it required, because the fraction of $D^+ \sim A^-$ molecules formed directly by the k_{RA} or k_{OD} pathways does not interfere with the measurement of k_{et}. Intramolecular rate constants have been measured with as little as $\sim 20\%$ of the molecules going through the k_{RD}, k_{et} sequence (Scheme I).

Stopped-flow or other rapid-mixing techniques do not usually suffice to generate $D \sim A$ fast enough by bimolecular electron transfer, so *in situ* techniques are required. Pulse radiolysis has found application in this regard because generation of appropriate reducing radicals by selected scavenging of the primary e_{aq}^-, OH$^•$ and H$^•$ on the submicrosecond timescale is possible. These secondary radicals can then play the part of the external reductant (Red) in Scheme I. For example, the formate radical ($CO_2^{•-}$) is predominantly produced when an aqueous solution containing N_2O and formate (HCO_2^-) is subjected to a short pulse of high-energy electrons

(Faraggi and Klapper, 1988; Isied *et al.*, 1984):

$$H^+ + e_{aq}^- + N_2O \rightarrow N_2 + OH^\cdot,$$

$$OH^\cdot + HCO_2^- \rightarrow CO_2^{\cdot-} + H_2O,$$

$$H^\cdot + HCO_2^- \rightarrow CO_2^{\cdot-} + H_2.$$

Starting with the earliest work of Gray and co-workers (Winkler *et al.*, 1982), a scheme using the flash-photolytically generated triplet charge-transfer excited state of $Ru(bpy)_3^{2+}$ (RuL_3^{2+} in Scheme I') as a strong reductant $\{E^{\circ\prime}[Ru(bpy)_3^{2+*}/Ru(bpy)_3^{3+}] = -0.7$ V (Whitten, 1980)$\}$ has been employed extensively for these measurements. $Ru(bpy)_3^{2+*}$ reduces either the donor site (k_{RD}, Scheme I') or the acceptor site (k_{RA}, Scheme I') when photolytically generated in the presence of the fully oxidized molecule ($D^+ \sim A$), producing a proportion of $D \sim A$ molecules that decay by intramolecular electron transfer (k_{et}) to $D^+ \sim A^-$ and oxidized ground-state $Ru(bpy)_3^{3+}$ (RuL_3^{3+} in Scheme I'). This latter product must be reductively scavenged to avoid rapid bimolecular back-electron transfer, $Ru(bpy)_3^{3+} + D \sim A \rightarrow Ru(bpy)_3^{2+} + D^+ \sim A$, before k_{et} can occur. Sacrificial reductive scavengers that have been used successfully include ethylenediaminetetraacetic acid (EDTA) (Winkler *et al.*, 1982), (dimethylamino)benzoate (Pan *et al.*, 1990), and aniline (Pan *et al.*, 1990), the last being particularly useful when working at low ionic strength. Scheme II' shows that, when photolytically generated in the presence of the fully reduced $D \sim A^-$ molecule, $Ru(bpy)_3^{2+*}$ can also act as a strong oxidant $\{E^{\circ\prime}[Ru(bpy)_3^+/Ru(bpy)_3^{2+*}] = +1.1$ V (Whitten, 1980)$\}$, generating the same $D \sim A$ and $D^+ \sim A^-$ products as well as ground-state $Ru(bpy)_3^+$, which must be *oxidatively* scavenged to avoid back-electron transfer. Lieber *et al.* (1987a) first described the use of a Ni(II) complex, Ni(II)hexamethyltetraazacyclodecane, and an alkyl bromide, 3-bromopropionic acid. $Ru(bpy)_3^+$ reduces the Ni(II) complex to Ni(I), which then reacts irreversibly with 3-bromopropionic acid, most likely by abstraction of the alkyl radical (Becker *et al.*, 1981). The ability of $Ru(bpy)_3^{2+*}$ to act as either an oxidant or reductant can be used to advantage. For example, if $Ru(bpy)_3^{2+*}$ reduces the A site in preference to the D site in the photoreductive scheme ($k_{RA} \gg k_{RD}$, Scheme I'), producing the thermodynamically stable "mixed-valence" molecule, $D^+ \sim A^-$, then this preferential reactivity for A over D can be used to advantage in the photooxidative scheme, since it implies that $k_{OA} \gg k_{OD}$, and thus *oxidation* by $Ru(bpy)_3^{2+*}$ would produce predominantly the *unstable* mixed-valence molecule $D \sim A$ (Scheme II').

Considering only electron transfer between ground-state D and A, it is often difficult to design systems with high driving forces for the estimation of reorganizational energies (*vide infra*). To access this high-driving-force regime, a number of schemes have been developed to utilize an electronically excited state of D or A, allowing a flash photolysis–based initiation of the electron transfer. The most common application of this strategy is generalized in Scheme III, in which a donor excited state (D*) decays by electron transfer quenching to yield $D^+ \sim A^-$ (k_{et}^{D*}, Scheme III). This product is unstable to back electron transfer to form the starting

Scheme III

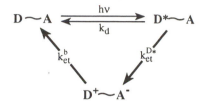

material $D \sim A$ (k_{et}^b, Scheme III) and in many cases, both forward ($D^* \rightarrow A$) and reverse ($A^- \rightarrow D^+$) rates can be measured (see Table 15.1). D^* must have a lifetime that is long enough for electron transfer quenching (k_{et}^{D*}, Scheme III) to compete with other decay processes (k_d, Scheme III). In heme proteins, a simple way of providing a long-lived excited state (D^*) is by using a variety of metal-substituted porphyrins (**MP**) or the metal-free porphyrin in place of the native iron porphyrin; Zn, Mg, Pd, Pt, and Cd all have a long-lived triplet excited state (^{3}**MP**) and have all been used successfully (Table 15.1). Iron-protoporphyrin IX in myoglobin (Mb) is easily substituted with **MP**, and the first use of this scheme involved ruthenated Mb (Axup *et al.*, 1988). The covalently attached heme c in cytochrome c makes metal substitution somewhat more difficult, but Zn-substituted cytochrome c has been used in such measurements as well (Elias *et al.*, 1988; Meade *et al.*, 1989; Therien *et al.*, 1990, 1991a; Zang and Maki, 1990). The oxidized product (D^+) is the metalloporphyrin π-cation radical, **MP$^{\bullet+}$**, a strong oxidant that drives the back electron transfer. The ^{3}**MP** excited state is also susceptible to reduction to the π-anion radical, **MP$^{\bullet-}$**, making the metalloporphyrin excited state the acceptor (Scheme IV)

Scheme IV

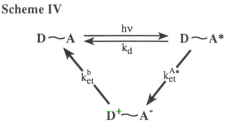

(Elias *et al.*, 1988). Excited states of the attached probe can be used in the same fashion, the main example being Ru(bpy)$_3$ analogs attached covalently to lysine side chains (Pan *et al.*, 1988). In this case, Ru(bpy)$_3^{2+}$ assumes the role of D in Scheme III, being photolyzed to the triplet charge-transfer excited state which decays by intramolecular electron transfer to heme c Fe(III) (k_{et}^{D*}, Scheme III). k_{et}^b then involves rereduction of ground-state Ru(bpy)$_3^{3+}$ by heme c Fe(II). [Millett and co-workers have also shown that Ru(bpy)$_3^{2+*}$ can act as an excited-state oxidant in the presence of heme c Fe(II) (Pan *et al.*, 1988); this is described by Scheme IV with A = Ru(bpy)$_3^{2+}$, D = heme c Fe(II).] Scheme V illustrates a relatively recent strategy (Chang *et al.*, 1991) to access moderately high driving forces in a ground-state intramolecular electron-transfer reaction. The probe is Ru(bpy)$_2$(imid)—attached

Scheme V

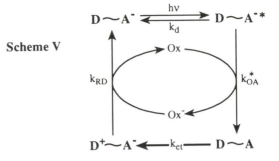

to His-33 of cytochrome c, which has photophysical properties very similar to those of Ru(bpy)$_3$. In Scheme V, D is heme c Fe(II) and A$^-$ is the Ru(II) form of this probe, which is excited by flash photolysis to the triplet charge-transfer excited state (A$^-$*). Final preparation of the "reactant" D $\sim$ A is accomplished by rapid bimolecular oxidative quenching of this excited state with an oxidant (Ox) in solution (RuA$_6{}^{3+}$). Intramolecular electron transfer (k_{et}, Scheme V) then occurs to form D$^+$ $\sim$ A$^-$, which is converted back to the original D $\sim$ A$^-$ starting material by slow (seconds) bimolecular reduction of D$^+$ by RuA$_6{}^{2+}$ in solution. Since ground-state A (Ru(bpy)$_3{}^{3+}$) is a strong oxidant, this gives a substantial driving force for electron transfer, without restricting the timescale for measurable k_{et}. In Schemes III and IV, $k_{et}{}^{D*}$ and $k_{et}{}^{A*}$ have to be faster than k_d to be measurable (otherwise, D* and A* will simply decay by other pathways before electron transfer occurs). Since k_{et} in Scheme V is a ground-state reaction, the only limitation is that k_{et} is faster than the bimolecular reduction of A by Ox$^-$. Schemes III, IV, and V have the advantage (over Schemes I and II) that the flash-photolytically triggered processes are all reversible, allowing multiple flash-induced kinetic traces to be collected and signal-averaged.

IV. Interpretation of Intraprotein Electron Transfer Rates

Table 15.1 summarizes intramolecular electron transfer rates (k_{et}) for metalloproteins with attached redox-active probe molecules. This work involves several electron-transfer metalloproteins: cytochromes (c, c_{551}, and b_5), blue-copper proteins [azurin (Az), plastocyanin (Pc), and stellacyanin (St)], and iron–sulfur proteins [high-potential iron protein (HiPIP)], as well as some non-electron transfer metalloproteins: myoglobin (Mb) and hemoglobin (Hb). The majority of the rate constants ($\sim 65\%$) were measured on cytochrome c–based systems. Also given in Table 15.1 are the driving force ($-\Delta G°$) and $\Delta H^{\ddagger}$ and $\Delta S^{\ddagger}$, the latter two derived from the temperature dependence of k_{et}.

A. Reorganizational Energy

The major current goal of these studies is to understand the effect of the intervening medium on electron transfer rates, i.e., to evaluate in each system the electronic factor $(H_{RP})^2$ [Equation (15.6)] and to correlate this electronic coupling with D–A distance or the molecular details of the D–A "bridge." The evaluation of $(H_{RP})^2$ requires an independent estimate of the nuclear Franck–Condon factor (FC) [Equation (15.6)], which depends on the driving force $(-\Delta G^{\circ})$ and the reorganizational energy (λ) of the system. In the high-temperature limit (where low-frequency protein and solvent vibrational modes can be treated classically), Equation (15.7) gives the temperature dependence of FC, which allows an estimate of λ from $\Delta H^{\ddagger}$ (or the activation energy, E_a) (Bertrand, 1991; Marcus and Sutin, 1985). (This procedure also requires the temperature dependence of $-\Delta G^{\circ}$.) Thus, measurement of the temperature dependence of k_{et} (and the component redox potentials) for a single system can be used to estimate λ and FC, allowing calculation of $(H_{RP})^2$ from k_{et} at a given temperature. The danger of this approach rests in the assumption that room temperature (around which the rates are generally measured) is high enough that treating the higher-frequency vibrational modes quantum-mechanically can be successfully avoided (Bertrand, 1991; Marcus and Sutin, 1985). An alternative approach to evaluating λ is to assume the classical expression for the Franck–Condon factor (FC) [Equation (15.7)] (or make assumptions about the high-frequency vibrational modes and use the quantum-mechanical expression) and use the dependence of k_{et} on driving force $(-\Delta G^{\circ})$ at constant temperature to estimate λ. Variation of the driving force usually involves chemical changes at the donor or acceptor site to alter redox potentials. The danger in this approach is that these chemical changes may affect other parameters as well, in particular the electronic coupling $(H_{RP})^2$ or the reorganizational energy itself. Minimization of these other effects is required for accurate application of this approach, as is a wide enough range of $-\Delta G^{\circ}$ to see curvature in the $\ln k_{et}$ vs. $-\Delta G^{\circ}$ plot as the inverted region is approached. Since λ is a combination of contributions from D and A, any independent knowledge of λ_{in} for isolated D or A electron self-exchange can be useful in estimating λ. Also, dielectric continuum models have been developed for the estimation of λ_{out} for such systems (Brunschwig et al., 1986). One useful application of these independent estimates is the extraction of the reorganizational energy component from the metalloprotein active site, a measure of its intrinsic ability to transfer electrons.

Early temperature-dependent measurements of the rate of Ru(II) → Fe(III) electron transfer in RuA$_5$(H33)cyt c yielded low values of $\Delta H^{\ddagger} = 1.1(4)$ (Nocera et al., 1984) or 3.5(2) kcal-mol^{-1} (Isied et al., 1984). With an estimate of the reorganizational energy of the probe analog RuA$_5$(py) and ΔH° from the temperature dependence of the reduction potentials, Nocera et al. (1984) were able to use their value of $\Delta H^{\ddagger}$ to estimate that the enthalpic component of the heme c reorganizational energy was <8 kcal-mol^{-1} (0.35 eV). Similar calculations on RuA$_5$(H83)Az yielded a maximum reorganization enthalpy for the blue copper site of 7.1 kcal-mol^{-1} (0.31 eV) (Kostic et al., 1983; Margalit et al., 1984). These reorganizational enthalpies are quite low and suggest that minimal inner-sphere reorganization is required for electron transfer at low-spin heme and blue copper sites. By contrast, the high-spin heme active site of myoglobin in RuA$_5$(H48)Mb exhibits a reorganizational enthalpy of about 20 kcal-mol^{-1} (0.87 eV) (Crutchley et al., 1985, 1986), which was suggested to arise from the transfer of a $d\sigma$ electron and the coupling of H$_2$O dissociation from Fe during redox (Gray, 1986). Extension of the (H48)Mb measurements to include the $RuA_4(py)-$ probe and Zn-, Pd-, and Mg-mesoporphyrin substitutions gave a set of rate constants at driving forces ($-\Delta G°$) ranging from 0.0 to 0.9 V (Cowan and Gray, 1988; Karas et al., 1988; Lieber et al., 1988). A plot of $\ln k_{et}$ vs. $-\Delta G°$ allowed fits based on Equations (15.1) and (15.2) that gave a value of λ between 1.9 and 2.4 eV (Cowan and Gray, 1988; Karas et al., 1988). The fact that all the rates could be fitted with a single curve, even though λ_{in} for the metal-substituted porphyrins (acting as ^3MP donors) is expected to be much smaller than that for the high-spin Fe porphyrin, suggested that the reorganizational energy was dominated by λ_{out} contributions, mainly from the solvent environment of the RuA_5- probe. Additional rate constants from Pt-, Cd-, and H_2-mesoporphyrin substitution into myoglobin allowed a more detailed analysis of the driving force dependence using a quantum-mechanical expression for the Franck–Condon factor derived by Jortner (1976). From this analysis, λ was calculated to be ~ 1.3 eV, which agrees well with the value [1.5(3) eV] calculated from the activated portion of the 77–300 K temperature dependence for the RuA_5(H48)Mb derivative (Cowan et al., 1989). Figure 15.2a shows the driving-force dependence of photoexcited $^3MP^* \to (H48)Ru(III)A_4L$ (M = Zn, Mg, Pd, Pt, Cd, H_2; L = A, py, isn) electron-transfer rates (and back electron-transfer rates) compared to the prediction of classical theory [Equations (15.6) and (15.7)] for $\lambda = 1.26$ eV (Winkler and Gray, 1992). With an estimated contribution from the RuA_5- probe of 0.6 eV, the remaining MP and solvent contribution to λ was estimated to be about 0.7 eV (Cowan et al., 1989).

In cytochrome c, studies of λ have used RuA_4L(H33)[Zn]cyt c derivatives (L = A, py, isn), in which both photoinduced charge separation [$^3ZnP^* \to Ru(III)$] and charge recombination [$Ru(II) \to ZnP^{\cdot+}$] electron transfer reactions can be measured (Elias et al., 1988; Meade et al., 1989). The dependence of $\ln k_{et}$ on $-\Delta G°$ is consistent with $\lambda = 1.1$ eV for charge separation and 1.2 eV for charge recombination (Meade et al., 1989). Extension of this strategy to RuA_4L(H39)-[Zn]cyt c (the *Candida krusei* protein) (Therien et al., 1990) and RuA_4L(H62)[Zn]cyt c (the N62H variant of the yeast iso-1 protein) (Bowler et al., 1989; Therien et al., 1991a) yields $\lambda = 1.2$ eV for each derivative when both charge separation and recombination are considered collectively (Figure 15.2b) (Gray and Malmström, 1989; Therien et al., 1990, 1991a,b; Winkler and Gray, 1992). An estimate of λ_{in} for the D and A sites can be made by comparing redox-induced structural changes for Zn-porphyrin and $RuA_5(py)$, yielding $\lambda_{in}[Zn] \leq 0.15$ eV and $\lambda_{in}[Ru] \approx 0.05$ eV, leaving ~ 1 eV of λ_{out} (from solvent and peptide reorganization) for these systems (Meade et al., 1989). The solvent-based λ_{out} is expected to contribute the majority (~ 0.6 eV) of this reorganizational energy (Onuchic et al., 1992; Winkler and Gray, 1992). These λ_{in} estimates are probably low, because a simple dielectric continuum calculation of solvent-based λ_{out} suggests that it varies almost linearly from 0.38 to 0.63 eV as a spherically modeled RuA_5 probe rolls around the surface of a spherical cytochrome c from an edge-to-edge distance of 10.2 to 21.8 Å (Therien et al., 1991a,b).

There appears to be a convergence of opinion that RuA_4L/cyt c systems exhibit reorganization energies of $\lambda \approx 1.2$ eV for edge-to-edge distances in the 10–20 Å range (Onuchic et al., 1992), while RuA_4L/Mb systems have slightly higher $\lambda \approx$

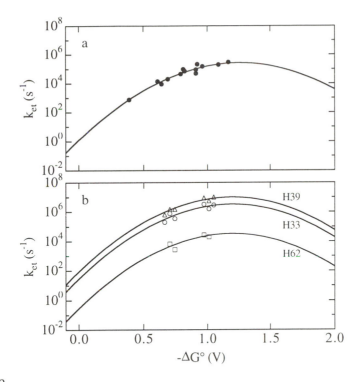

Figure 15.2
Driving-force dependences of $\log_{10} k_{et}$ (from Table 15.1) for (a) RuA_4L(His-48) derivatives of M-substituted myoglobins (M = Zn, Mg, Pd, Cd, H_2 for L = A; M = Zn, Pd for L = py; and M = Zn for L = isn) and (b) RuA_4L(H33)[Zn]cyt c (L = A, py, isn), RuA_4L(H39)[Zn] Ck cyt c (L = A, py, isn), and RuA_4L(H62)[Zn]Sc cyt c (L = A, py). The theoretical lines are calculated using Equations (15.6) and (15.7) with (a) $\lambda = 1.26$ eV, $H_{RP} = 0.034$ cm^{-1}, and (b) $\lambda = 1.2$ eV, $H_{RP} = 0.21$ cm^{-1} (H39), $\lambda = 1.2$ eV, $H_{RP} = 0.12$ cm^{-1} (H33), $\lambda = 1.2$ eV, $H_{RP} = 0.012$ cm^{-1} (H62).

1.3 eV for similar D–A distances. The dielectric continuum calculations imply that $\lambda(RuA_4L/$cyt $c)$ may decrease to ~1.0 eV at shorter distances (≤ 10 Å). For Ru-based probes structurally related to Ru(bpy)$_3$, the reorganizational energy is expected to be lower because Ru(bpy)$_3$ has a much faster self-exchange electron transfer rate than RuA_5(py) $\{\lambda[Ru(bpy)_3] = 0.57$ eV; $\lambda[RuA_5(py)] = 1.20$ eV (Brown and Sutin, 1979; Marcus and Sutin, 1985)$\}$, and more recent Ru(bpy)$_2$(im)/cyt c electron transfer reactions have been analyzed assuming $\lambda = 0.89$ eV in these systems (Chang *et al.*, 1991; Millett and Durham, 1991; Wuttke *et al.*, 1992b).

B. Electronic Factor

The electronic coupling of donor and acceptor through the intervening medium described by H_{RP} [Equation (15.6)] controls how the D → A electron-transfer rate decays with the D–A separation or with "loss of connectivity" through the bridge. In the simplest theory, exponential wave-function decay through a constant-dielectric medium gives rise to an exponential distance dependence defined by the decay factor, β [Equation (15.9)]. The dependence of $\ln k_{et}$ on d should define β (from the slope), which describes the ability of a protein medium to support electron transfer. The intercept (at $d = d_0 = 3$ Å, the vdW contact distance) yields a rate that can be used to calculate $H_{RP}°$ [Equations (15.8) and (15.9)] (Mosc et al., 1992). In principle, measurement of k_{et} for one derivative and independent knowledge of $-\Delta G°$, λ, and d should allow calculation of β by use of Equations (15.6)–(15.9), given the assumption that the electron transfer would be adiabatic at vdW contact [Equation (15.10)]. This assumption may not be valid for the protein systems being considered (*vide infra*).

The first attempt to correlate k_{et} with D–A distance based on the types of modified proteins described in Table 15.1 was made using $RuA_5(Hn)[Zn]Mb$ ($n = 12, 48, 81, 116$) $^3ZnP^* \rightarrow Ru(III)$ charge-separation electron-transfer rates (Axup et al., 1988; Lieber et al., 1987b, 1988). The ground-state $Ru(II) \rightarrow Fe(III)$ electron-transfer rate measured for $RuA_5(H33)cyt\ c$ was also included in the analysis, and a β of 0.87–0.99 Å^{-1} was derived from the $\ln k_{et}$ vs. d plot. The H12 derivative was suspected to be anomalous because of the presence of a tryptophan (W14) between the RuA_5- probe and the heme, and leaving the H12 rate out of the calculation raised the value of β to 0.99–1.12 Å^{-1} (Lieber et al., 1987b, 1988). A similar study with [Mg]Mb derivatives (Cowan and Gray, 1988) gave a similar distance dependence from which a β value of 0.78–0.88 Å^{-1} was derived (Cowan et al., 1989). Inclusion of the $^3ZnP^* \rightarrow Ru(III)$ charge-separation electron-transfer rate measured in $RuA_5(H33)[Zn]cyt\ c$ (Elias et al., 1988) gave a best-fit β value of 0.91 Å^{-1} (Gray and Malmström, 1989). Figure 15.3 displays the distance dependence of the $^3MP^* \rightarrow (Hn)Ru(III)A_5$ ($n = 48, 81, 116, 12$) electron-transfer rates for M = Zn, Mg, compared to the theoretical prediction for $\beta = 0.91$ Å^{-1}. $Ru(II) \rightarrow Fe(III)$ electron-transfer rate constants in a series of cytochrome c derivatives containing a $Ru(bpy)_3$ analog covalently attached to different lysine residues (Pan et al., 1988) also can be explained using a β value of 0.9 Å^{-1} (although the through-space distances in these derivatives are less well defined) (Durham et al., 1989, 1990). A recent study that combined rate constants for electron transfer between cofactors in photosynthetic reaction centers as well as a few of the derivatives in Table 15.1 derived a value of β of 1.4–1.7 Å^{-1} for rate constants that spanned ~ 12 orders of magnitude in systems with edge-to-edge D–A distances of 5–24 Å (Moser and Dutton, 1992; Moser et al., 1992), suggesting that polypeptide can be treated generally as a homogeneous dielectric medium. Wuttke et al. (1992b) have suggested the use of this correlation as a reference line, with rate constants that fall below this line demonstrating the importance of considering intervening medium inhomogeneities in these cases. The data analyzed by Moser et al. (Moser and Dutton, 1992; Moser et al., 1992) behave ideally in the sense that the $\ln k_{et}$ vs. d plot has an intercept of k_{et} (3.6 Å) $= 1 \times 10^{13}$ s^{-1}, the assumed value of ν_n [Equation (15.1)] (i.e., these electron transfer reactions would become adiabatic at 3.6 Å). This behavior is definitely not observed for the $RuA_5(Hn)[M]Mb$ derivatives ($n = 12, 48, 81, 116$; M = Zn, Mg) because the $d = 3$ Å intercept gives effective ν_n values ranging from $\sim 10^6$ to

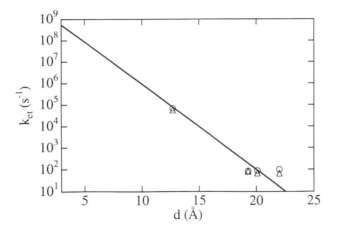

Figure 15.3

Distance dependence of $\log_{10} k_{et}$ (from Table 15.1) for $^3MP^* \to Ru(III)$ electron transfer in RuA$_5$(Hn)[M]Mb derivatives [n = 48, 81, 116, 12; M = Zn (○), Mg (△)]. The theoretical line is calculated using Equations (15.6)–(15.9) with $\lambda = 1.3$ eV, $H_{RP}^\circ = 3.0$ cm^{-1}, $\beta = 0.91$ Å^{-1}, and has an intercept of k_{et} ($d = 3$ Å) = 5.4×10^8 s^{-1}.

10^{10} s^{-1} (Cowan *et al.*, 1989; Gray and Malmström, 1989; Meade *et al.*, 1989; Winkler and Gray, 1992). Recently, Co(II) → Fe(III) electron-transfer rates were measured in a series of cytochrome *c* derivatives with various glutamate or aspartate residues labeled by Co(diAMsar) (Conrad *et al.*, 1992). In seven derivatives with apparent through-space distances ranging from 14 to 20 Å, the rate constants were all nearly identical (k_{et} = 1–3 s^{-1})!

A variety of polypeptide structural features, including the main chain, H-bonds, hydrophobic channels, side chains, and π orbitals on aromatic residues, may be involved in facilitating protein-based electron transfer reactions (Isied, 1984, 1991b). In the last few years, attention has been given to developing theoretical and computational approaches to evaluating the effect of this molecular structure on D → A electron-transfer rates (Beratan *et al.*, 1987; Broo and Larsson, 1991; Christensen *et al.*, 1990, 1992; Goldman, 1991; Kuki, 1991; Kuki and Wolynes, 1987; Larsson, 1983; Onuchic and Beratan, 1990; Onuchic *et al.*, 1992; Siddarth and Marcus, 1990, 1992). Perhaps the most straightforward approach is the Pathways algorithm (Beratan *et al.*, 1987; Onuchic and Beratan, 1990; Onuchic *et al.*, 1992), which evaluates the electronic coupling decay through covalent bonds, H-bonds, and through-space jumps in searching for dominant physical coupling pathways from donor to acceptor. This approach has been developed into a predictive tool that purports to identify relatively "hot" and "cold" surface regions on a metalloprotein that should be relatively well or poorly coupled, respectively, to the active site metal (Beratan *et al.*, 1991). The early suggestion that an intervening tryptophan might be responsible for the unexpectedly high $^3ZnP^* \to Ru(III)$ rate in RuA$_5$(H12)[Zn]Mb (Axup *et al.*, 1988) led to a test of the possible coupling enhancement due to the presence of a tryptophan (W59) ring and a methionine (M64) sulfur in a pathway from the heme of yeast iso-1-cytochrome c to a surface site near residue 62 (Bowler *et al.*, 1989). RuA$_5$–labeling of the N62H variant of this protein and subsequent measurement of the Ru(II) → Fe(III) electron-transfer rate gave a value of 1.7(1) s^{-1}, which agreed reasonably well with predicted rates from an exponential distance analysis (0.4–2.0 s^{-1}) or from a Pathways analysis (0.4 s^{-1}) (Bowler *et al.*, 1989). To date, no unambiguous evidence exists for the facilitation of electron transfer by aromatic side-chains in proteins. With Zn substitution for measurement of charge separation and recombination rates, the use of RuA$_4$L– probes (L = A, py, isn) to estimate λ from the driving-force dependence, and the addition of the *Candida krusei* (H39)cyt c (Therien *et al.*, 1990, 1991a), a comparison of through-space and through-bonds predictions with experiment can be made. For edge-to-edge distances of 11.1 (H33), 12.3 (H39), and 14.8 (H62) Å (Wuttke *et al.*, 1992b), a value of $\beta = 0.9$ Å^{-1} would predict *relative* rates of 1 (H33) > 0.34 (H39) $\gg 0.04$ (H62), whereas the measured rates give 1 (H33) < 3.1 (H39) $\gg 0.007$ (H62) (Therien *et al.*, 1991b). Calculations based on the Pathways algorithm give 1 (H33) < 3.8 (H39) $\gg 0.006$ (H62) (Beratan *et al.*, 1990), in much better agreement with the observed rates. The "good" pathway in the H39 derivative involves a hydrogen bond from G41 to the heme (Beratan *et al.*, 1990; Therien *et al.*, 1990). In another test of the Pathways approach, Fe(II) → Ru(III) electron-transfer rates were measured in RuA$_5$(H26)cyt b_5 derivatives in which either the native protoporphyrin or a substituted deuteroporphyrin formed the heme active site (Jacobs *et al.*, 1991). The absence of a vinyl side chain on the deuteroporphyrin and a corresponding increase in the distance of a through-space jump connecting the heme to the dominant pathway to H26 were suggested as the explanation for the significantly slower rate observed in the deuteroporphyrin-substituted derivative (Jacobs *et al.*, 1991).

The electronic couplings ($\Pi\varepsilon$) calculated by Pathways for the dominant paths can be converted into "effective σ-bond distances" $\{\sigma l = 1.4[\ln(\Pi\varepsilon)/\ln(0.6)]$ Å$\}$ allowing plots of $\ln k_{et}$ vs. either through-space (d) or σ-bond distances (σl) to be compared. The appropriateness of each type of "distance" measurement can then be judged by the linearity of such plots. Such a comparison has been done recently for Fe(II) → Ru(III) electron-transfer rates in Ru(bpy)$_2$(im)(Hn)cyt c derivatives (n = 33, 39, 62, 72, 79, the latter two from semisynthetic K72H and K79H variants) (Beratan *et al.*, 1992; Wuttke *et al.*, 1992a,b). This study actually compared k_{max} values [k_{max} was defined as k_{et} at $-\Delta G° = \lambda$, assuming $\lambda = 0.8$ eV (Chang *et al.*, 1991) and $-\Delta G° = 0.74$ V for these systems] and found that σl was a much better predictor of the rate constants than d. [For example, k_{max}(H72) = 9.4×10^5 s^{-1} and k_{max}(H33) = 2.7×10^6 s^{-1}, even though the through-space distances are 8.4 (H72) and 11.1 (H33) Å; the σ-bond distances are 24.6 (H72) and 19.5 (H33) Å because of a 3.88-Å through-space jump from P71 to the M80 heme ligand in the H72 derivative.] An intriguing feature of this analysis is that the $\ln k_{max}$ vs. σ-bond distance line exhibited a slope of 0.71 Å^{-1} [similar to that seen in covalently coupled D–A systems (Closs and Miller, 1988)], and extrapolation to the one-bond (1.4-Å) limit yielded $k_{max} = 3 \times 10^{12}$ s^{-1}, which is close to the expected ideal value for an adiabatic electron transfer ($v_n = 10^{13}$ s^{-1}) (Beratan *et al.*, 1992; Wuttke *et al.*, 1992a,b). Inclusion of rate constants from the cytochrome b_5 experiments discussed earlier yields similar values for the slope and k_{max} (Onuchic *et al.*, 1992). In contrast to these results, the distance-independent Co(II) → Fe(III) rates in seven Co(diAMsar)–cytochrome *c* derivatives were analyzed using time-averaged Pathways electronic couplings from 100-ps molecular dynamics simulations; the predicted relative rates based on this approach were slightly worse at explaining the constant rates observed than a simple through-space distance approach (neither was satisfactory) (Conrad *et al.*, 1992).

C. Conformational Dynamics

Given the suspected sensitivity of D–A electronic coupling to the molecular structure of intervening polypeptide medium and the sensitivity of reorganizational energies to structural details at the D and A sites, conformational changes of the modified proteins being discussed herein might be expected to have a significant influence on electron transfer rates. A situation can be imagined in which the protein must undergo a conformational transition to get to a structure in which rapid electron transfer is allowed (i.e., a structure with good electronic coupling or low λ). In such a case, the rate measured by monitoring formation of the D^+–A^- product would actually be the rate of the prerequisite conformational transition; i.e., the electron transfer would be "conformationally gated" (Hoffman and Ratner, 1987). Although indistinguishable from the case in which the formation of products is controlled by the factors of Equation (15.1), the rates of conformationally gated electron transfer would not be expected to show the usual dependences on driving force, distance, etc. For example, Conrad *et al.* (1992) have discussed a possible explanation based on a conformational gating hypothesis for the distance-independent Co(II) → Fe(III) rates in a series of Co(diAMsar)–cytochrome *c* derivatives.

An extension of these ideas allows the possibility that a stable electron transfer-inactive conformational state could produce an intramolecular electron transfer that appears "directional" (Brunschwig and Sutin, 1989). Bechtold *et al.* (1986b) failed to observe Fe(II) → Ru(III) electron transfer in *trans*-RuA$_4$(isn)(H33)cyt *c* (k_{et} < 10^{-2} s^{-1}), which has a 0.18-V driving force for this direction of electron transfer, even though the Ru(II) → Fe(III) electron-transfer rate in RuA$_5$(H33)cyt *c* (k_{et} = 30–50 s^{-1}) was about four orders of magnitude higher for virtually the same driving force (Isied *et al.*, 1984; Nocera *et al.*, 1984). The proposed explanation involved a rapid conformational change of the radical-produced Ru(III)/Fe(II) species to [Ru(III)/Fe(II)]*, a conformer that is inactive toward intramolecular electron transfer to form the thermodynamically stable Ru(II)/Fe(III) form (Bechtold *et al.*, 1986b). This suggests that the reduced Fe(II) form of cytochrome *c* has a stable conformation in which the heme is electronically uncoupled from the H33 surface site, and that the oxidized Fe(III) protein does not occupy this conformation [since Ru(II) → Fe(III) electron transfer is relatively rapid]. This sort of directionality has not been observed in other systems: Ru(II) → Fe(III) and Fe(II) → Ru(III) electron-transfer rates in RuA$_5$(H48)Mb are very similar (Crutchley *et al.*, 1985; Lieber *et al.*, 1987a); Fe(II) → Ru(III) electron-transfer rates in Ru(bpy)$_2$(dcbpy-K*n*)cyt *c* derivatives (*n* = 7, 13, 25, 27, 72) appear to be predictably rapid when compared to Ru(II)* → Fe(III) rates in the same derivatives (Durham *et al.*, 1989); and Fe(II) →

Ru(III) electron-transfer rates in Ru(bpy)$_2$(im)(Hn)cyt c derivatives ($n = 33$, 39 in *Candida krusei*, 62 in *Saccharomyces cerevisiae* N62H, 72 in semisynthetic K72H) are also predictable when compared to the Ru(II)* → Fe(III) rates (Wuttke *et al.*, 1992b). The last example suggests that there is nothing special about the H33 surface site coupling to the heme in Fe(II) cytochrome c compared to the Fe(III) form (Chang *et al.*, 1991; Winkler and Gray, 1992).

V. Future Directions

The rate and specificity of biological electron transfer is critical to photosynthesis and respiration, and the development of these complex electron-transport systems likely involved evolutionary molecular structural changes that resulted in fine tuning of the parameters that control electron transfer [as described by the factors in Equation (15.1)]. For example, the evolution of a metal active site with a structure midway between the preferred structures for the oxidized and reduced metal complex would produce a metalloprotein with a low reorganizational energy (λ) that would support rapid electron transfer (Gray, 1986; Gray and Malmström, 1989). Gray has coined the name "long-range electron transferases" to describe such proteins (Mayo *et al.*, 1986). It has been long recognized that another characteristic that can be used to fine-tune the nuclear factor [Equations (15.2) and (15.7)] is the driving force for the reaction (i.e., the relative reduction potentials of the donor and acceptor sites). In recent years, emphasis has shifted to exploring the electronic factor [Equations (15.4) and (15.8)] and how evolution of the biological matrix that imbeds donor and acceptor sites can affect the electronic coupling between donor and acceptor. Cytochrome c has played (and will continue to play) a crucial role in these studies, not so much as a physiological example, but as a prototypical polypeptide framework upon which to test ideas about how the nature and extent of intervening polypeptide affects donor–acceptor electronic coupling. There is now experimental evidence suggesting the importance of the detailed molecular structure of the intervening medium (Beratan *et al.*, 1990; Jacobs *et al.*, 1991; Wuttke *et al.*, 1992b), although in some systems treatment of the polypeptide as a homogeneous dielectric milieu suffices (Moser and Dutton, 1992; Moser *et al.*, 1992), and in others the electron-transfer rates are curiously insensitive to the nature or extent of the intervening medium (Conrad *et al.*, 1992).

Future experimental work will be directed toward obtaining more evidence for (or against) the importance of the detailed structure of the intervening polypeptide medium. Protein engineering (by genetic or semisynthetic strategies) has contributed so far by providing variant proteins with new surface sites to attach probe molecules (Bowler *et al.*, 1989; Wuttke *et al.*, 1992b) or by limiting the number of possible probe attachment sites (Jacobs *et al.*, 1991). The next generation of protein engineering experiments will target the intervening medium itself, altering the type of residue side chain involved in weak links of coupling paths to change the hydrogen bonding or through-space van der Waals contacts, and measuring the effects of such modifications on the rate of electron transfer. Theoretical and computational effort will increase the sophistication of treatment of the electronically coupled intervening molecular structure and, it is to be hoped, will continue to address the dynamics of the intervening medium and how protein vibrational motion affects the electronic factor in addition to the known effects on the nuclear factor. The ultimate application of our increased understanding of these electron-transfer processes to the design and construction of functional molecular electronic devices or new redox enzymes is a very attractive prospect, and efforts directed toward this goal will certainly gain emphasis as the field moves into the next century.

VI. References

Axup, A. W., Albin, M., Mayo, S. L., Crutchley, R. J., & Gray, H. B. (1988) *J. Am. Chem. Soc.* *110*, 435–439.

Bechtold, R., Gardineer, M. B., Kazmi, A., van Hemelryck, B., & Isied, S. S. (1986a) *J. Phys. Chem.* *90*, 3800–3804.

Bechtold, R., Kuehn, C., Lepre, C., & Isied, S. S. (1986b) *Nature* *322*, 286–288.

Becker, J. Y., Kerr, J. B., Pletcher, D., & Rosas, R. (1981) *J. Electroanal. Chem.* *117*, 87–99.

Beratan, D. N., Onuchic, J. N., & Hopfield, J. J. (1987) *J. Chem. Phys.* *86*, 4488–4498.

Beratan, D. N., Onuchic, J. N., Betts, J. N., Bowler, B. E., & Gray, H. B. (1990) *J. Am. Chem. Soc.* *112*, 7915–7921.

Beratan, D. N., Betts, J. N., & Onuchic, J. N. (1991) *Science* *252*, 1285–1288.

Beratan, D. N., Onuchic, J. N., Winkler, J. R., & Gray, H. B. (1992) *Science* *258*, 1740–1741.

Bertrand, P. (1991) *Structure and Bonding* *75*, 1–47.

Bowler, B. E., Meade, T. J., Mayo, S. L., Richards, J. H., & Gray, H. B. (1989) *J. Am. Chem. Soc.* *111*, 8757–8759.

Bowler, B. E., Raphael, A. L., & Gray, H. B. (1990) in *Progress in Inorganic Chemistry*, Vol. 38 (Lippard, S. J., ed.), pp. 259–322, Wiley & Sons, New York.

Broo, A., & Larsson, S. (1991) *J Phys. Chem.* *95*, 4925–4928.

Brown, G. M., & Sutin, N. (1979) *J. Am. Chem. Soc.* *101*, 883–892.

Brunschwig, B. S., & Sutin, N. (1987) *Comments Inorg. Chem.* *6*, 209–235.

Brunschwig, B. S., & Sutin, N. (1989) *J. Am. Chem. Soc.* *111*, 7454–7465.

Brunschwig, B. S., Ehrenson, S., & Sutin, N. (1986) *J. Phys. Chem.* *90*, 3657–3668.

Chang, I.-J., Gray, H. B., & Winkler, J. R. (1991) *J. Am. Chem. Soc.* *113*, 7056–7057.

Christensen, H. E. M., Conrad, L. S., Mikkelsen, K. V., Nielsen, M. K., & Ulstrup, J. (1990) *Inorg. Chem.* *29*, 2808–2816.

Christensen, H. E. M., Conrad, L. S., Hammerstad-Pedersen, J. M., & Ulstrup, J. (1992) *FEBS Lett.* *296*, 141–144.

Closs, G. L., & Miller, J. R. (1988) *Science* *240*, 440–447.

Conrad, D. W., & Scott, R. A. (1989) *J. Am. Chem. Soc.* *111*, 3461–3463.

Conrad, D. W., Zhang, H., Stewart, D. E., & Scott, R. A. (1992) *J. Am. Chem. Soc.* *114*, 9909–9915.

Cowan, J. A., & Gray, H. B. (1988) *Chem Scr.* *28A*, 21–26.

Cowan, J. A., & Gray, H. B. (1989) *Inorg. Chem.* *28*, 2074–2078.

Cowan, J. A., Upmacis, R. K., Beratan, D. N., Onuchic, J. N., & Gray, H. B. (1989) *Ann. N. Y. Acad. Sci.* *550*, 68–84.

Crutchley, R. J., Ellis, W. R., Jr., & Gray, H. B. (1985) *J. Am. Chem. Soc.* *107*, 5002–5004.

Crutchley, R. J., Ellis, W. R., Jr., & Gray, H. B. (1986) in *Frontiers in Bioinorganic Chemistry* (Xavier, A. V., ed.), pp. 679–693, VCH, Weinheim.

DeVault, D. (1984) *Quantum-Mechanical Tunnelling in Biological Systems*, Cambridge University Press, Cambridge, U.K.

Durham, B., Pan, L. P., Long, J. E., & Millett, F. (1989) *Biochemistry* *28*, 8659–8665.

Durham, B., Pan, L. P., Hahm, S., Long, J., & Millett, F. (1990) in *Advances in Chemistry Series. Electron Transfer in Biology and the Solid State*, Vol. 226 (Johnson, M. K., King, R. B., Kurtz, D. M., Jr., Kutal, C., Norton, M. L. & Scott, R. A., eds.), pp. 181–193, ACS, Washington, D.C.

Elias, H., Chou, M. H., & Winkler, J. R. (1988) *J. Am. Chem. Soc.* *110*, 429–434.

Faraggi, M., & Klapper, M. H. (1988) *J. Am. Chem. Soc.* **110**, 5753–5756.

Farver, O., & Pecht, I. (1989a) *Proc. Natl. Acad. Sci. USA* **86**, 6968–6972.

Farver, O., & Pecht, I. (1989b) *FEBS Lett.* **244**, 379–382.

Farver, O., & Pecht, I. (1992) *J. Am. Chem. Soc.* **114**, 5764–5767.

Geren, L., Hahm, S., Durham, B., & Millett, F. (1991) *Biochemistry* **30**, 9450–9457.

Goldman, C. (1991) *Phys. Rev. A* **43**, 4500–4509.

Gorren, A. C. F., Chan, M. L., & Scott, R. A. (1992) *Bioconj. Chem.* **3**, 291–294.

Govindaraju, K., Salmon, G. A., Tomkinson, N. P., & Sykes, A. G. (1990) *J. Chem. Soc., Chem. Commun.* 1003–1004.

Govindaraju, K., Christensen, H. E. M., Lloyd, E., Olsen, M., Salmon, G. A., Tomkinson, N. P., & Sykes, A. G. (1993) *Inorg. Chem.* **32**, 40–46.

Gray, H. B. (1986) *Chem. Soc. Rev.* **15**, 17–30.

Gray, H. B., & Malmström, B. G. (1989) *Biochemistry* **28**, 7499–7505.

Hoffman, B. M., & Ratner, M. A. (1987) *J. Am. Chem. Soc.* **109**, 6237–6243.

Isied, S. S. (1984) in *Progress in Inorganic Chemistry*, Vol. 32 (Lippard, S. J., ed.), pp. 443–517, Wiley & Sons, New York.

Isied, S. S. (1990) in *Advances in Chemistry Series. Electron Transfer in Biology and the Solid State*, Vol. 226 (Johnson, M. K., King, R. B., Kurtz, D. M., Jr., Kutal, C., Norton, M. L., & Scott, R. A., eds.), pp. 91–100, ACS, Washington, D.C.

Isied, S. S. (1991a) in *Advances in Chemistry Series. Electron Transfer in Inorganic, Organic, and Biological Systems*, Vol. 228 (Bolton, J. R., Mataga, N., & McLendon, G., eds.), pp. 229–245, ACS, Washington, D.C.

Isied, S. S. (1991b) in *Metal Ions in Biological Systems*, Vol. 27 (Sigel, H., & Sigel, A., eds.), pp. 1–56, Marcel Dekker, Inc., New York.

Isied, S. S., Worosila, G., & Atherton, S. J. (1982) *J. Am. Chem. Soc.* **104**, 7659–7661.

Isied, S. S., Kuehn, C., & Worosila, G. (1984) *J. Am. Chem. Soc.* **106**, 1722–1726.

Isied, S. S., Vassilian, A., Wishart, J. F., Creutz, C., Schwarz, H. A., & Sutin, N. (1988) *J. Am. Chem. Soc.* **110**, 635–637.

Jackman, M. P., Lim, M.-C., Sykes, A. G., & Salmon, G. A. (1988a) *J. Chem. Soc., Dalton Trans.* 2843–2850.

Jackman, M. P., McGinnis, J., Powls, R., Salmon, G. A., & Sykes, A. G. (1988b) *J. Am. Chem. Soc.* **110**, 5880–5887.

Jacobs, B. A., Mauk, M. R., Funk. W. D., MacGillivray, R. T. A., Mauk, A. G.. & Gray. H. B. (1991) *J. Am. Chem. Soc.* **113**, 4390–4394.

Jortner, J. (1976) *J. Chem. Phy.* **64**, 4860–4867.

Kamiya, N., Shiro, Y., Iwata, T., Iizuka, T., & Iwasaki, H. (1991) *J. Am. Chem. Soc.* **113**, 1826–1829.

Karas, J. L., Lieber, C. M., & Gray, H. B. (1988) *J. Am. Chem. Soc.* **110**, 599–600.

Kostic, N. M., Margalit, R., Che, C.-M., & Gray, H. B. (1983) *J. Am. Chem. Soc.* **105**, 7765–7767.

Kuki, A. (1991) *Structure and Bonding* **75**, 49–83.

Kuki, A., & Wolynes, P. G. (1987) *Science* **236**, 1647–1652.

Larsson, S. (1983) *J. Chem. Soc. Faraday Trans.* **279**, 1375–1388.

Lieber, C. M., Karas, J. L., & Gray, H. B. (1987a) *J. Am. Chem. Soc.* **109**, 3778–3779.

Lieber, C. M., Karas, J. L., Mayo, S. L., Albin, M., & Gray, H. B. (1987b) "Long-Range Electron Transfer in Proteins," The Robert A. Welch Foundation Conference on Chemical Research XXXI, pp. 9–26.

Lieber, C. M., Karas, J. L., Mayo, S. L., Axup, A. W., Albin, M., Crutchley, R. J., Ellis, W. R., Jr., & Gray, H. B. (1988) in *Trace Elements in Man and Animals*, Vol. 6 (Hurley, L. S., Keen, C. L., Lönnerdal, B. & Rucker, R. B., eds.), pp. 23–27, Plenum Press, New York.

Marcus, R. A., & Sutin, N. (1985) *Biochim. Biophys. Acta* **811**, 265–322.

Margalit, R., Kostic, N. M., Che, C.-M., Blair, D. F., Chiang, H.-J., Pecht, I., Shelton, J. B, Shelton, J. R., Schroeder, W. A., & Gray. H. B. (1984) *Proc. Natl. Acad. Sci. USA* **81**, 6554–6558.

Matthews, C. R., Erickson, P. M., Van Vliet, D. L., & Petersheim, M. (1978) *J. Am. Chem. Soc.* **100**, 2260–2262.

Matthews, C. R., Erickson, P. M., & Froebe, C. L. (1980) *Biochim. Biophys. Acta* **624**, 499–510.

Matthews, C. R., Recchia, J., & Froebe, C. L. (1981) *Anal. Biochem.* **112**, 329–337.

Mayo, S. L., Ellis, W. R., Crutchley, R. J., & Gray, H. B. (1986) *Science* **233**, 948–952.

McLendon, G. (1988) *Acc. Chem. Res.* **21**, 160–167.

Meade, T. J., Gray, H. B., & Winkler, J. R. (1989) *J. Am. Chem. Soc. 111*, 4353–4356.

Millett, F., & Durham, B. (1991) in *Metal Ions in Biological Systems*, Vol. 27 (Sigel. H., & Sigel, A., eds.), pp. 223–264, Marcel Dekker, Inc., New York.

Moser, C. C., & Dutton, P. L. (1992) *Biochim. Biophys. Acta 1101*, 171–176.

Moser, C. C., Keske. J. M., Warncke, K., Farid, R. S.. & Dutton, P. L. (1992) *Nature 355*, 796–802.

Nocera, D. G., Winkler, J. R., Yocom, K. M., Bordignon, E., & Gray, H. B. (1984) *J. Am. Chem. Soc. 106*, 5145–5150.

Onuchic, J. N., & Beratan, D. N. (1990) *J. Chem. Phys. 92*, 722–733.

Onuchic, J. N., Beratan, D. N., Winkler, J. R., & Gray, H. B. (1992) *Annu. Rev. Biophys. Biomol. Struct. 21*, 349–377.

Osvath, P., Salmon. G. A., & Sykes. A. G. (1988) *J. Am. Chem. Soc. 110*, 7114–7118.

Pan, L. P., Durham, B., Wolinska, J., & Millett, F. (1988) *Biochemistry 27*, 7180–7184.

Pan, L. P., Frame, M., Durham, B., Davis, D., & Millett, F. (1990) *Biochemistry 29*, 3231–3236.

Recchia, J., Matthews, C. R., Rhee, M.-J., & Horrocks, W. D., Jr. (1982) *Biochim. Biophys. Acta 702*, 105–111.

Scott, R. A., Conrad, D. W., Eidsness, M. K., Gorren, A. C., & Wallin, S. A. (1991) in *Metal Ions in Biological Systems*, Vol. 27 (Sigel, H. & Sigel, A., eds.), pp. 199–222, Marcel Dekker, New York.

Siddarth, P., & Marcus, R. A. (1990) *J. Phys. Chem. 94*, 8430–8434.

Siddarth, P., & Marcus, R. A. (1992) *J. Phys. Chem. 96*, 3213–3217.

Siders, P., & Marcus, R. A. (1981) *J. Am. Chem. Soc. 103*, 741–747.

Sutin, N. (1982) *Acc. Chem. Res. 15*, 275–282.

Therien, M. J., Selman, M., Gray, H. B., Chang, I. J., & Winkler, J. R. (1990) *J. Am. Chem. Soc. 112*, 2420–2422.

Therien, M. J., Bowler, B. E., Selman, M. A., Gray, H. B., Chang, I.-J., & Winkler, J. R. (1991a) in *Advances in Chemistry Series. Electron Transfer in Inorganic, Organic, and Biological Systems*, Vol. 228 (Bolton, J. R., Mataga, N., & McLendon, G.. eds.), pp. 191–199, ACS, Washington, D.C.

Therien, M. J., Chang, J., Raphael, A. L., Bowler, B. E., & Gray. H. B. (1991b) *Structure and Bonding 75*, 109–129.

Whitten, D. G. (1980) *Acc. Chem. Res. 13*, 83–90.

Winkler, J. R., & Gray, H. B. (1992) *Chem. Rev. 92*, 369–379.

Winkler, J. R., Nocera, D. G., Yocom, K. M., Bordignon, E., & Gray, H. B. (1982) *J. Am. Chem. Soc. 104*, 5798–5800.

Wishart, J. F., van Eldik, R., Sun, J., Su, C., & Isied, S. S. (1992) *Inorg. Chem. 31*, 3986–3989.

Wuttke, D. S., Bjerrum, M. J., Chang, I.-J., Winkler, J. R., & Gray, H. B. (1992a) *Biochim. Biophys. Acta 1101*, 168–170.

Wuttke, D. S., Bjerrum, M. J., Winkler, J. R., & Gray, H. B. (1992b) *Science 256*, 1007–1009.

Yocom, K. M., Shelton, J. B., Shelton, J. R., Schroeder, W. A., Worosila, G., Isied, S. S., Bordignon, E., & Gray, H. B. (1982) *Proc. Natl. Acad. Sci. USA 79*, 7052–7055.

Yocom, K. M., Winkler, J. R., Nocera, D. G., Bordignon, E, & Gray. H. B. (1983) *Chem. Scr. 21*, 29–33.

Zang, L.-H., & Maki, A. H. (1990) *J. Am. Chem. Soc. 112*, 4346–4351.

16

Theoretical Simulation of Protein–Protein Interactions

SCOTT H. NORTHRUP

Department of Chemistry
Tennessee Technological University

I. Introduction

A. Early Static Docking Studies

The detailed electrostatic charge distribution on the cytochromes is thought to play a major role in facilitating the electron transfer (ET) between these proteins with the electrostatic forces selectively promoting favorable docking arrangements for ET. A number of varied and powerful computational methods have been used to simulate the mechanisms of interaction involved in complex formation between cytochrome c and other ET proteins. Early studies of Salemme (1976) suggested a hypothetical structural complex of cytochrome ($cytc$) and cytochrome b_5 ($cytb5$), making use of crystallographic coordinates for each protein, and using a least-squares fitting procedure to establish charge complementarity between highly conserved charged groups surrounding the heme crevices. Subsequent predictions along similar lines facilitated by computer graphics display systems have been made of complex structures of cytochrome c–cytochrome c peroxidase (Poulos and Kraut, 1980), and cytochrome c–flavodoxin (Simondsen et al., 1982). The latter complex was used subsequently (Matthew et al., 1983) to assess the role of complementary electrostatics in the preorientation of molecules enhancing the reaction kinetics. These studies have been useful in establishing the role of the conserved positively charged amino acids (lysines 13, 27, 72, 86, and 87) surrounding the heme crevice in the interactions between $cytc$ and other ET species.

543

B. Ignored Dynamical Features of Complexation

The studies just mentioned are *static* in nature and have focused on the construction and analysis of a single putative ET complex, essentially ignoring three important features which are emerging as important considerations in the mechanism of inter-action of the cytochromes, and which require more robust simulation methodology. First is the influence of protein flexibility and internal dynamics on an association complex. Molecular dynamics simulations of a complex of cytochrome c and cyto-chrome b_5 (Wendoloski *et al.*, 1987) indicate a flexible association complex that is able to sample alternative interheme geometries. For example, through dynamical relaxation the complex was able to achieve a more intimate conformation in which the inter-iron distance decreased by as much as 2 Å after less than 5 picoseconds of simulation. Dramatic conformational changes occurred in which Phe 82 of cyto-chrome c moved from its crystallographic location, where the phenyl side chain is packed near the heme, to a position where it could conceivably bridge the two heme group π-orbital systems. The flexibility observed within the complex could explain the relative lack of recognition specificity between ET proteins and explain why some nonphysiological partners such as cytochrome c/cytochrome b_5 react at physiological rates.

A second consideration is the protein structural influence on dynamical pro-cesses leading up to formation of a productive ET complex (Northrup *et al.*, 1988). Bimolecular ET reactions require the initial transport of reactants through space by a diffusional mechanism, and in some cases the diffusional encounter of species limits the overall rate (Berg and von Hippel, 1985). Thus, a knowledge of the dynamics of such encounters is of fundamental importance in the understanding of ET. Since the ET region of cytochromes is small relative to their overall size, strict orientational criteria for their reaction exist. The cytochromes have overcome this formidable obstacle by exploiting the so-called "reduction in dimensionality" princi-ple (Adam and Delbruck, 1968). The species diffuse in three dimensions and initially associate in unreactive configurations by ubiquitous nonspecific forces of attraction, predominantly electrostatic. This association is followed by a rotational diffusional search on a lower dimensional configurational surface of associated molecules, until ultimate production of a properly oriented pair. The magnitude of forces promoting the initial nonspecific association must be finely tuned to a range allowing particles to remain in juxtaposition for time scales required for the rotational diffusive search, but not so strong that encountered particles are locked into unproductive orientations. Additionally, for highly asymmetrically charged species, a "steering" effect may operate which selectively preorients the diffusing particles into productive configurations even prior to initial encounter.

A third important consideration in the mechanism of interaction of the cyto-chromes is the possible existence of an ensemble of near-optimal docking geometries rather than a single complex through which ET might occur. This is a feature apart from the internal and surface group flexibility available within a given complex, as described in the first consideration earlier. Recent theoretical studies (Weber and Tollin, 1985; Mauk *et al.*, 1986) of the electrostatic interactions of three different c-type cytochromes with flavodoxin reveal that variations in the computed electro-static stabilization reflects differences in the distribution of all charged surface groups, and not simply those localized at the site of intermolecular contact. On the basis of electrostatic stability alone, their calculations do not predict a dominant protein–protein interaction, but rather a mixture of complexes whose relative populations shift with pH and ionic strength.

C. The Brownian and Molecular Dynamics Method

For the consideration of these important factors, a robust theoretical simulation approach has been devised based on Brownian dynamics (BD) algorithms (Ermak and McCammon, 1978; Northrup *et al.*, 1984; Allison *et al.*, 1985a,b; Northrup *et al.*, 1986a,b; Ganti *et al.*, 1985). This method is able to compute a representative set of dynamical trajectories of bimolecular diffusion and association in the complicated electrostatic force and torque field arising from the actual charge distribution of the cytochromes. The atomic-resolution irregular surface topography of the molecules is taken into account in computing excluded volume interactions, and the mutual geometric disposition of the heme groups is monitored and incorporated into various ET reaction criteria. Extensive analysis is possible of the electrostatic contacts involved in a large number of generated stable complexes which meet the ET geometric criteria. Brownian simulation of a series of species having modified charged residues has allowed a quantitative assessment of the influence of amino acid composition on rate of attainment of productive complexes. The BD method is complementary to the molecular dynamics method in that the latter is able to simulate in detail a single given intimate complex on a short time scale of tens of picoseconds, while BD is capable of generating a large representative set of docked complexes for molecular dynamical simulation. The BD method has been applied to the study of the association and ET between horse ferrocytochrome *c* [*cytc(r)*] and yeast cytochrome *c* peroxidase (*cyp*) (Northrup *et al.*, 1987a, 1987b, 1988); horse ferricytochrome *c* [*cytc(o)*] and bovine ferrocytochrome b_5 (*cytb5*) (Eltis *et al.*, 1991); yeast iso-1-ferricytochrome *c* (*ycytc*) and *cytb5* (Northrup *et al.*, 1993); self-exchange of *P. aeruginosa* cytochrome c_{551} (*c551*) (Herbert and Northrup, 1989); and self-exchange in *cytc* and *cytb5* (Andrew *et al.*, 1993). In this chapter an overview of the methodology is given, followed by a brief description of some simple applications of the method using a simple dipolar sphere model of these cytochrome reactions, and finally some major results of rigorous simulations.

II. Brownian Dynamics Methodology

A. Treatment of Dynamics

In BD the Brownian motion of interacting macromolecules in a solvent is simulated stochastically by a series of small displacements chosen from a distribution that is equivalent to the short-time solution of the Smoluchowski diffusion equation with forces. The basic Ermak–McCammon (1978) algorithm for free displacements Δr in the relative separation vector r of reactant centers of mass in time step Δt is

$$\Delta r = DF(k_B T)^{-1}\Delta t + S. \tag{16.1}$$

Here D is the translational diffusion coefficient for relative motion and is assumed to be spatially isotropic. The vector F is the systematic interparticle force, which is ordinarily of a Coulombic nature; $k_B T$ is the Boltzmann constant times absolute temperature; and S is the stochastic component of the displacement arising from collisions of particles with solvent molecules and is generated by taking normally distributed random numbers obeying the average relationship $\langle S^2 \rangle = 2D\Delta t$. An equation similar to Equation (16.1) governs the independent rotational Brownian motion of each particle, where force is replaced by torque, and D is replaced by an isotropic rotational diffusion coefficient D_{ri} for each particle i. This feature has been added in subsequent work (Northrup et al., 1986c) to simulate the interaction between two whole proteins, where rotation of one or both molecules is important. Extension to a more generalized treatment which includes hydrodynamic interactions should be straightforward (Dickinson, 1985). In all previously reported work, the particles have been treated as rigid bodies with no internal dynamics. Initial tests are underway of extensions which include flexibility of the most mobile side chains of proteins.

While the Ermak–McCammon algorithm is quite general in allowing one to compute trajectories of reactants of arbitrary shape and complexity, it is limited in that it treats reactant motion in the absence of absorbing and reflecting boundaries. Thus, one is forced to take quite small diffusive time steps near such boundaries, affecting the efficiency of the algorithm. In simple cases where smooth spherical boundaries exist, such as the reflection of two spheres, or the truncation of trajectories at the outer spherical boundary (Northrup et al., 1986a), a 1-D algorithm (Lamm and Schulten, 1983; Lamm, 1984) has been adapted to explicitly include reflection and absorption effects. This dramatically improves the efficiency of the Brownian step algorithm in cases where geometric conditions are not too complicated.

B. Bimolecular Rate Constants

The important connection of BD results with experimental data is made by a strategy which enables one to extract bimolecular rate constants from trajectory fate statistics (Northrup et al., 1984). Trajectories of diffusing species are begun at random orientations from a separation $r = b$ outside the region of asymmetric Coulombic forces, and are truncated at an outer spherical surface $r = c$ (typically 200 Å). Values chosen for b range from 60 Å in c551 self-exchange simulations to 65 Å in the other simulations, in which stronger electric fields are exerted. This essentially separates the problem into the easy one of centrosymmetric diffusion to a starting surface b from outside, followed by diffusion in a complicated force field inside b, where BD simulation provides the description. A large number of trajectories are monitored to obtain the probability p of association of pairs in favorable geometries for reaction prior to ultimate separation to distance $r = c$. The diffusion-controlled bimolecular reaction rate constant k may then be extracted from p by applying the formula

$$k = k_D(b)\frac{p}{1 - (1 - p)k_D(b)/k_D(c)} \qquad (16.2)$$

Here, diffusive rate constants $k_D(s)$ for first arrival at spherical surface s is given by the Smoluchowski–Debye expression when centrosymmetric forces apply in the region $r > s$:

$$k_D(s) = \frac{4\pi}{\displaystyle\int_s^\infty dr [D(r)r^2]^{-1} e^{u(r)/k_B T}}, \tag{16.3}$$

where s is a starting or truncation surface radius, $D(r)$ is a spatially dependent relative translational diffusion coefficient, and $u(r)$ is a centrosymmetric potential energy of interaction between the diffusing pair.

In actual application (Northrup *et al.*, 1986a), trajectories are not terminated when reactive criteria are met, but are continued until the truncation surface c is attained. With each trajectory a parallel set of survival probabilities w_i are monitored which correspond to a parallel set of reaction criteria i. For example, one can study the influence of a whole range of different reaction boundary conditions simultaneously in a single simulation, monitoring the survival of the trajectory with respect to each of these criteria simultaneously. Thus, trajectories are simulated in a reference system having unreactive (e.g., reflecting) inner boundary conditions. The reaction probability p_i for reactive criterion i is then $p_i = 1 - w_i$.

Increased efficiency of the basic strategy of obtaining rate constants has been obtained by breaking up the trajectory simulation calculations into stages (Allison *et al.*, 1985b). This has been useful in problems that involve complicated short-range interactions between reactant and target, where the probability for a reactant to diffuse to a surface relatively close to the target is calculated, and then the probability of diffusing from this intermediate surface to the reaction site is computed (Allison *et al.*, 1988).

C. Modeling of Molecular Structural Details

The starting points for simulational modeling of proteins by BD are the high-resolution x-ray crystallographic structures. From these structures, models are constructed of the protein shapes and placement of charges on the reactants at varying levels of complexity, ranging from simple spherical models where each sphere has an embedded monopole and dipolar charges, or perhaps a more complicated array of charges, to rigorous studies in which every atom is modeled as a partial charge in its crystallographic position and the irregular surface topography is treated.

The protonation state of each titratable amino acid residue is estimated by performing a Tanford–Kirkwood calculation with static-accessibility modification (Matthew, 1985; Tanford and Kirkwood, 1957; Tanford and Roxby, 1972; Shire *et al.*, 1974). Thus, each residue is assigned a net charge based on its protein environment, pH, ionic strength, and temperature. At or near neutral pH, as in these studies, lysines and arginines are fully protonated and carboxylates are fully dissociated. The more ambiguous assignments of charge of amino termini and histidines depend on these environmental parameters and are estimated by the Tanford–Kirkwood calculation. The fractional net charge calculated for histidines are incremented onto the charge at either the $N_{\delta 1}$ or $N_{\varepsilon 2}$ atom, whichever leads to greater overall stability of the protein.

The effect of site-specific mutations and species variations on the rates of bimolecular reactions can be estimated in BD simulations by first performing "computer mutagenesis" on the initial crystallographic structures (Northrup et al., 1987a). This technique was also employed to estimate the positions of 68 atoms of yeast *cyp* missing from the crystallographic structure because of motional disorder in the crystal. Substituted side chains are placed into the coordinate set using standard residue coordinates, and steepest descents energy minimization is performed essentially to relax steric conflicts. More rigorous minimization schemes are superfluous considering the level of approximations made in the BD simulations.

III. BD Using Simple Spherical Models

A. Implementation

A study of a particular reaction is typically begun using a simple dipolar sphere model (DSM) (Northrup et al., 1986c; Herbert et al., 1989; Eltis et al., 1991) before implementing the rigorous full description. Examination of simple models is instructive in that they can more quickly generate rate constants over a wide range of parameters and environmental conditions than the more rigorous simulations, and give one important qualitative insights. The DSM is illustrated schematically in Figure 16.1. Each protein is modeled as a sphere having a centrally embedded monopolar charge q carrying the net protein charge and two dipolar charges placed along the dipole moment axis μ, with magnitude q_d computed to produce the appropriate dipole moment calculated (Koppenol and Margoliash, 1982) from the x-ray coordinate set (or model-built set). The dipole moments of the proteins are computed using the set of partial charges on every atom after the adjustment of net residue charges estimated by the Tanford–Kirkwood theory. The dipolar charge pair is embedded 1.0 Å inside the protein surface (1.5 Å for studies of the *cytc/cytb5* reaction), providing for an ion exclusion radius around the dipolar surface charges and reflecting the fact that most charged groups on proteins are near the surface. Table 16.1 summarizes the parameters characterizing the proteins treated thus far by the DSM.

Since the proteins are somewhat elliptical in *shape* rather than spherical, the estimation of effective radius R_i is somewhat arbitrary. Choice of radii can be made by several considerations: (i) the hydrodynamic radii; (ii) the fall-off profile of the atom number density versus distance to the center of mass; (iii) radii which allow embedded reaction centers (reference points used to prescribe the reaction criteria discussed later) to achieve realistic values at particle juxtaposition.

In the DSM the electrostatic *potential energy* of interaction W between proteins (and the corresponding forces and torques) are computed at each Brownian step by an empirically modified Coulomb/Debye expression (Herbert et al., 1989; Eltis et al., 1991),

$$W = \sum_i \sum_j q_i q_j \frac{\exp(-\kappa(r_{ij} - B_{ij}))}{\varepsilon r_{ij}(1 + \kappa B_{ij})} \qquad (16.4)$$

Table 16.1
Parameters for the dipolar sphere model.[a]

Parameter	Horse cytochrome c (cytc)	Yeast cytochrome c peroxidase (cyp)	Bovine cytochrome b_5 (cytb5)	P. a. cytochrome c_{551} (c551)
R_i, protein hard sphere radius (Å)	16.6	21.0	15.9	13.0
D (cm²/s)	1.5×10^{-6}	1.2×10^{-6}	1.5×10^{-6}	2.7×10^{-6}
D_{ri} (s⁻¹)	4.0×10^7	2.0×10^7	4.6×10^7	1.2×10^8
γ	30.5° (ox) 34.5° (red)	137°	142°	43.7° (ox) 40.3° (red)
q, net monopole	+7.1e (ox) +7.6e (red)	−9.7e	−9.1e	−3e (ox) −4e (red)
$\pm q_d$, dipolar charges	1.96e (ox) 1.49e (red)	3.85e	4.6e	2.47e (ox) 2.29e (red)
μ, dipole moment (Debye)	284 (ox) 224 (red)	739	638	268 (ox) 264 (red)

[a] c551 at pH = 7, I = 0.1, 313 K; cytc (red) at pH = 6, I = 0.1, 298 K; cytc (ox) and cytb5 at pH = 7, I = 0.094, 298 K; cyp at pH = 6, I = 0.1, 298 K.

where summations i and j are over charges on proteins 1 and 2, q_i is electrostatic charge of charge i, r_{ij} is the separation distance between charges i and j, $\kappa = 0.328 I^{1/2}$ is the Debye–Huckel screening parameter in units Å⁻¹, and I is ionic strength. Term B_{ij} is the sum of the embedding depth of charges i and j and is a distance shifting factor for charge i, j interaction included to account for the exclusion of diffusible ion atmosphere from penetrating to the various charge locations. This approximate finite ion size correction is found to be extremely important in simulations of large molecules (Northrup et al., 1986c; Herbert et al., 1989).

In the DSM, several models have been constructed to embody the chemical *reaction criterion*. The *reactive patch model* includes an axially symmetric reactive patch on the surface of each protein prescribed by angle θ, as illustrated in Figure 16.1. For ET reactions between heme proteins, this patch axis is taken to pass through the Fe reaction center, or the most exposed heme-edge atom (in the case of *cyp*). The rate of reaction is thus sensitive to patch sizes and to the angle γ between the patch axis and the dipole vector. Reaction is said to occur upon collision of these patch regions. Patch size qualitatively reflects heme surface exposure. A second model which has been used in our rigorous studies is the *cutoff distance model* (Northrup et al., 1987a,b; Herbert et al., 1989), where reaction occurs when a pre-

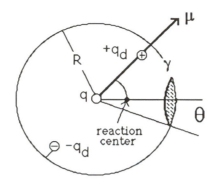

Figure 16.1
A schematic representation of the dipolar sphere model (DSM).

scribed distance is achieved between reaction centers placed inside each sphere (e.g., the inter-iron distance r_{Fe} between two heme proteins, or the heme-edge distance d_{edge}). A third model studied is the *exponential reactivity model* employed in our study of self-exchange in cytochrome *c551* (Herbert *et al.*, 1989) and in the *cytc/cytb5* reaction (Eltis *et al.*, 1991). This ideally provides a more physically realistic model of the ET event in which the intrinsic spatially dependent ET rate constant $k_{et}(r_{Fe})$ is given as a function of inter-iron distance by

$$k_{et}(r_{Fe}) = k_{et}{}^{\circ} \exp[-\beta(r_{Fe} - r_o)]. \tag{16.5}$$

The exponential Equation (16.5) has the flavor of the Siders, Cave, and Marcus theory (1984) for orientation and distance dependence of ET between porphyrins. Here $k_{et}{}^{\circ}$ is the ET rate constant when porphyrins are in direct contact edge-on, which was assumed for the *cytc/cytb5* reaction and *c551* self-exchange study to be 10^{11} s^{-1}. In the *ycytc/cytb5* mutant study (Section IV.G) and in the *cytc* and *cytb5* self-exchange studies (Section IV.C), this parameter was adjusted to give agreement with experiment for the native species and then applied to mutants. The minimum inter-iron distance of two porphyrins in an edge-on arrangement is $r_o = 11.7$ Å. Even faster ET rates would occur for porphyrins in a sandwich orientation at contact (Siders *et al.*, 1984), but such orientations are excluded in cytochromes since the heme groups are generally perpendicular to the protein surface. The ET distance-dependent factor β (see also Chapter 15 by Scott, this volume) is a quantity of intense interest (McLendon, 1988) and is an adjustable parameter which can be varied to fit the experimental data. Thus an estimate of this important factor can be obtained from BD. The exponential factor k_{et} is incorporated into BD as follows. The spatially dependent probability P that the reactant pair will survive a given Brownian step Δt without ET is $P = \exp(-k_{et}\Delta t)$. This probability is multiplicative throughout the trajectory, finally giving the escape probability for that trajectory.

B. Simple Model Results

Figures 16.2 and 16.3 and Table 16.2 summarize the results obtained (Northrup and Herbert, 1990; Eltis *et al.*, 1991) from simulations of the DSM for three ET

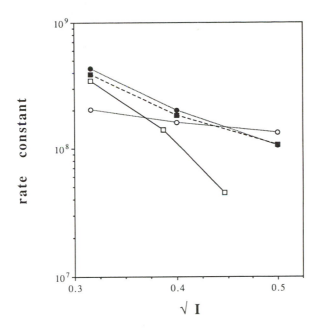

Figure 16.2
Second-order ET rate constants vs. $\sqrt{I}$ for the reaction $cytc(r)/cyp$ at pH = 6.0 and 298 K. BD results for the dipolar sphere model with patch reactivity (20°, 20°; computed dipole) (o); (5°, 5°; aligned dipole) (●), and also the rigorous BD treatment with d_{edge} = 20 Å and ψ = 40° (■) are compared with the experiment of Kang *et al.* (1978) (□).

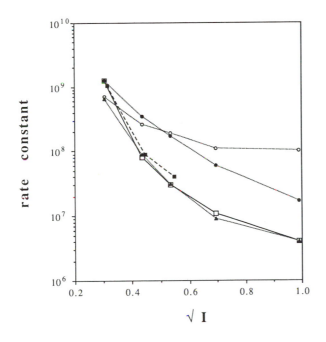

Figure 16.3
Second-order ET rate constants vs. $\sqrt{I}$ for the reactions $cytc(o)/cytb5$ at pH = 7.0 and 298 K. BD results for the dipolar sphere model with patch reactivity (20°, 20°; computed dipole (o); (5°, 5°; aligned dipole) (●); the exponential reactivity criterion using β = 1.2 Å^{-1} (▲); and also the rigorous BD treatment with d_{edge} = 8 Å and ψ = 30° (■) are compared with the experiment of Eltis *et al.* (1991) (□).

Table 16.2
Second-order rate constants in units $M^{-1}s^{-1}$ for the self-exchange reaction of *c551*
at pH = 7.0 and T = 313 K at two ionic strengths.

	$I = 0.1$	$I = 0.6$
Experiment (Timkovich *et al.*, 1988)	1.2×10^7	2.0×10^7
Simple Smoluchowski–Debye theory	3.2×10^9	4.7×10^9
BD DSM with (10°, 10°) reactive patches	1.2×10^7	1.6×10^7
Rigorous BD, distance cutoff model with $d_{edge} = 8$ Å and $\psi = 40°$	1.3×10^7	1.6×10^7

reactions, *cytc(r)/cyp*, *cytc(o)/cytb5*, and *c551(o)/c551(r)*, respectively. In each reaction comparison is made with experiments (Kang *et al.* 1978; Eltis *et al.*, 1991; Timkovich *et al.*, 1988) under the appropriate conditions of pH, T and I. The parameters of the DSM used in each of these systems are summarized in Table 16.1.

Let us first consider the gross features. The steep negative slope of the rate constants versus ionic strength for the *cytc(r)/cyp* and *cytc(o)/cytb5* reactions is as expected based on the complementarity of electrostatic charges. In both cases the reaction rate is enhanced not only by the monopole/monopole interaction, but also by the alignment of the appropriate end of the dipoles with the region of the exposed heme. Furthermore, the end of each dipole is attracted to the monopole of the other protein. On the contrary, the self-exchange of *c551* involves bringing together two proteins of small net negative charge, a rate-retarding influence by itself. There is also a slight enhancement by the interaction of the positive end of the each dipole, which is loosely correlated with the ET region, with the oppositely charged monopole of the other protein. Thus, the *c551* system has a retarding influence from the monopole/monopole and dipole/dipole interactions, but an enhancing influence from the monopole/dipole interaction. As a result, the net ionic strength dependence is slightly positive and virtually negligible.

The Smoluchowski/Debye (SD) theory (Smoluchowski, 1916; Debye, 1942) vastly overestimates the rates in all cases, obviously because it ignores the stereospecific reaction criteria. Rate constants from SD theory at $I = 0.1$ ($\sqrt{I} = 0.316$) are 1.24×10^{10}, 1.31×10^{10}, and 3.16×10^9 $M^{-1}s^{-1}$ for *cytc(r)/cyp*, *cytc(o)/cytb5*, and *c551(o)/c551(r)*, respectively. This theory also predicts an I-dependence which, although it is in the correct direction in the cases studied here, is far too weak. The computed interaction between two central monopoles inside macroparticles quite underestimates the strength of electrostatic interaction operating between real proteins, which interact through specific surface charge interactions. Furthermore, the SD theory would not predict the more complicated I-dependence when competing electrostatic effects are present.

The DSM with the *reactive patch* model qualitatively reproduces the experimental behavior in the sense of predicting the correct direction of the ionic strength effect in all cases and predicts a much steeper *I*-dependence than the SD theory. However, when the actual computed values of the dipole position are used in the DSM simulation, i.e., finite γ values which reflect that the dipoles are not perfectly correlated with the reaction surface, the DSM theory underestimates the steepness of the *I*-dependence in both the *cytc(r)/cyp* and *cytc(o)/cytb5* cases. Furthermore, the dipolar charges in their structurally computed positions do not provide enough steering to reproduce the large rates observed in the experiments. Unrealistically large sizes for the reactive patches ($\sim 20°$, 20°) must be chosen to reproduce experimental results at the low ionic strengths, overestimating the rate at high ionic strength. An alternate model was constructed in which $\gamma = 0°$ for *cytc* and $\gamma = 180°$ for the negative reaction partners *cyp* and *cytb5*. That is, the dipole vectors were artificially realigned to coincide with the likely region of ET, apart from structural considerations, to give optimum correlation of the dipole axis with the reactive zone. As shown in Figures 16.2 and 16.3, the *I*-dependence with aligned dipoles does correspond to experiment somewhat better in both the *cytc(r)/cyp* and *cytc(o)/cytb5* reactions, and smaller, more realistic patch sizes ($\sim 5°$, 5°) provide correlation with experiment. Overall, however, the dipolar sphere model with reactive patches fails to adequately reproduce the ionic strength dependence of the experimental values quantitatively.

On the other hand, the results generated by the DSM using the *exponential reactivity* criterion produce substantially better agreement with experiment, as evidenced in Figure 16.3. The ionic strength dependence of the experiment is reproduced throughout the entire ionic-strength range, except for the lowest ionic strength, where the theory underestimates the rate constant. In the exponential model, the electron transfer distance parameter β takes the place of the patch angle θ as a measure of the extent of the reactive region. The larger the β value, the smaller the extent of reactivity. McLendon (1988) and co-workers and Northrup and Herbert (1990) have both estimated values of $\beta = 1.2$ Å^{-1} for these types of systems, latter for self-exchange in *c551* self-exchange. In Figure 16.3, for the *cytc(o)/cytb5* system the exponential model predicts an ionic strength dependence that is virtually identical with the experiment using the aligned dipole model and value $\beta = 1.2$ Å^{-1}. The success of the exponential model of reactivity and failure of the patch model is due to the fact that the latter treats the reaction as an instantaneous event on the time scale of diffusion, which appears not to be the case in the *cytc(o)/cytb5* reaction. The exponential model, on the other hand, embodies a true dynamical coupling between diffusion and reaction. In other words, the rate will depend not only on the probability of attaining ET complexes, but also on the *lifetimes* of these complexes.

C. Estimate of Electron-Transfer Distance Dependence in *c551* Self-Exchange

In the *c551* self-exchange reaction, the DSM theory was capable of reproducing the weak ionic strength dependence of the experiment (Timkovich *et al.*, 1988) with an appropriate choice of reactive patch size (10°, 10°). In this simple reaction there is only a modest electrostatic effect, the electromotive driving force is zero, and the structures are well characterized, making it a good choice for studying the ET event

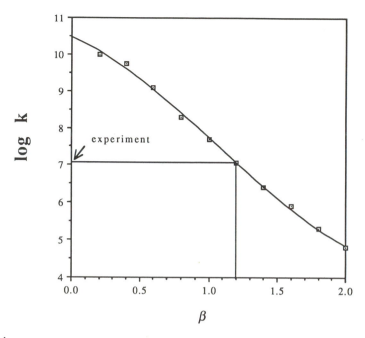

Figure 16.4

Dependence of the *c551* self-exchange second-order rate constant on the ET distance factor β (units Å^{-1}) computed by the DSM model. A value of $\beta = 1.2\ \text{Å}^{-1}$ is obtained by fitting to experimental results (Timkovich *et al.*, 1988).

more carefully using the exponential distance-dependent model to obtain a calibration of the distance factor β. Employing the intrinsic k_{et} in the simulation, with an adjustable distance factor β, an estimate of this quantity was obtainable by comparison with experiment. DSM simulations were performed at $I = 0.1$. Figure 16.4 shows the variation of the self-exchange rate constant vs. β. By interpolating on the graph with the experimental value of $\log_{10} k = 7.08$, an estimate of $\beta = 1.2\ \text{Å}^{-1}$ was obtained. This agrees with $\beta = 1.1\ \text{Å}^{-1}$ estimated for cytochrome ET by McLendon (1988), and the $1.2\ \text{Å}^{-1}$ value estimated for disordered polymers, glasses, and highly viscous solutions (Guarr *et al.*, 1983). Further studies of this system have been conducted using the rigorous BD model that we now discuss.

IV. Irregular-Body Brownian Dynamics

A. Development of Rigorous Models

More rigorous treatments of the electrostatic fields and the irregular shapes of the proteins have been accomplished (Ganti *et al.*, 1985; Allison *et al.*, 1988; Northrup *et al.*, 1987a, 1987b, 1988; Sharp *et al.*, 1987; Reynolds *et al.*, 1990). Treatment of the excluded volume effect which is cognizant of the rough surface topography of the proteins is done by developing and storing for later use in the BD simulation program a cubic spatial *exclusion grid* of 1.0 Å resolution centered on Protein I (usually the larger) which is used to define its excluded volume. A probe atom is placed at grid points, and the distance from the probe to each Protein I atom is determined. If the probe falls within 2.7 Å of any atom, that grid cell is defined to be an excluded region. The probe represents any atom of Protein II, the shape of which is treated by selecting a limited number of surface atoms which approximately represent its shape and which will diffuse by rigid body motion and be tested for penetration with the Protein I exclusion grid. The number of such atoms is limited for computational efficiency. Initial studies of cytochrome reactions are currently being enhanced to allow surface side chains to dynamically adjust their conformations, which should enhance the "fit" of the two proteins [e.g., allowing the hemes to attain a slightly closer encounter distance, as observed in molecular dynamics simulations (Wendoloski *et al.*, 1987)]. This feature is discussed in Section IV.G.

Electrostatic treatment has employed the numerical solution of the linearized Poisson–Boltzmann (LPB) equation on a cubic lattice. Details of this procedure are more adequately described in Chapter 9 by Gunner and Honig, this volume. The use of the LPB equation rather than the simple Coulomb/Debye treatment allows us to account for (i) the atomic-scale rough topography of the protein surfaces, (ii) the screening influence of diffusible ions in solvent medium which cannot penetrate the protein, and (iii) the discontinuity in the dielectric constant across protein surfaces. The electrostatic potential field surrounding Protein I is computed by iterating finite-difference solutions of the LPB equation by the method of Warwicker and Watson (1982) as adapted by Klapper *et al.*, 1986. Protein I is represented as an irregularly shaped cavity of low dielectric constant ($\varepsilon = 4$) and zero internal ionic strength and having fixed embedded charges in the crystallographic configuration. Surrounding this cavity is a continuum dielectric with $\varepsilon \sim 80$ representative of bulk water and having appropriate ionic strengths. A "focusing" method is used in which first a solution is iterated on an "outer" coarse-grained lattice of dimension 51^3 elements having a resolution $d = 4$ Å and encompassing a region of size $(200$ Å$)^3$ centered on Protein I. With such a large region, the boundary potential can be set to the Coulomb/Debye result. Having obtained a solution on an outer coarse grid, an additional calculation was performed on a refined grid of resolution $d = 1.666$ Å [for *cytc(r)/cyp*] and 1.0 Å [for *cytc(o)/cytb5* and *c551(o)/c551(r)*] of lattice size 61^3. Trial solutions and boundary values for this "inner" lattice were constructed from the outer lattice solution.

For a more accurate treatment of the interaction between two proteins, one would need to include an additional low dielectric cavity representing the Protein II interior. However, for this one would need to iterate a solution for every possible mutual configuration of the two protein cavities, a monumental computational task. A simplification is made ignoring the low dielectric in the Protein II interior, as well as allowing the electrolyte screening effect to operate throughout its interior. Thus, the direct force between the two proteins and the torque operating on Protein II are determined at each time step by placing the Protein II array of charges into the field around the Protein I cavity and, consulting the stored grid of forces, performing a summation over all Protein II charges.

The foregoing procedure allows for treatment of rotation of Protein II in a field of torque generated by Protein I. The rotation of Protein I in the field of Protein II may be included by generating a similar inner and outer force lattice around Protein II. These lattices rotate in rigid body rotation with Protein II. A limited set (e.g., dipolar pair) of charges are included on Protein I to serve as test charges which interact with the field around Protein II and are used to compute the approximate torque on Protein I in the field of Protein II. This feature of two rotating proteins is essential in treating the protein pairs which are comparably sized.

Reaction criteria in the rigorous studies are implemented by monitoring the heme-edge distance d_{edge}, the minimum distance between porphyrin atoms on the two proteins. This value is compared to a series of possible cutoff distances which define whether reaction has occurred. This cutoff choice varies from system to system and remains a somewhat arbitrary parameter of the model. An additional parameter of the reaction criterion is the angle ψ, which is the angle between heme plane normal vectors. This is included to determine whether hemes must be coparallel for efficient ET.

B. Bimolecular Rate Constants for *cytc/cyp*, *cytc/cytb5*, and *c551/c551*

Let us now consider the results for rate constants when the irregular surface topography of the proteins is being treated, with electrostatic potentials being computed by the LPB equation iterated on a grid. In Figure 16.2 we note that for the *cytc(r)/cyp* reaction (Northrup and Herbert, 1990), the rigorous treatment gives a reproduction of the experimental ionic strength dependence comparable to the dipolar sphere approximation with the aligned dipole. The experimental ionic strength dependence is steeper than any of the theoretical models. In Figure 16.3 we see that for the *cytc(o)/cytb5* reaction, the rigorous BD theory gives an excellent representation of the ionic strength dependence, and somewhat better than the best dipolar sphere model, particularly at low ionic strength. For the *c551* self-exchange reaction, a very weak and positive ionic-strength dependence is predicted, as also observed in experiment (see Table 16.2).

Rate constants were obtained for various choices of heme plane orientation ψ, which is an undetermined parameter of the distance cutoff reaction criterion. For similar choices of ψ [30° in the *cytc(o)/cytb5* reaction, and 40° in the *cytc(r)/cyp* and *c551(o)/c551(r)* reaction], the actual magnitudes of the experimental values are well reproduced in addition to the slope of the ionic-strength plot. Ionic-strength dependence agreement is especially excellent in the *cytc(o)/cytb5* reaction.

In all three reactions discussed in the preceding paragraph, when using the distance cutoff criterion, the best comparison with experiment is always obtained by choosing a d_{edge} which is about 2 Å larger than the minimum possible heme edge distance the proteins can attain. Of course, the available volume of orientation space rapidly vanishes as the heme edge distance reaches the smallest physically attainable value. By allowing a 2 Å tolerance window, one is still assured of having proteins which are essentially at contact. Here one expects there to be strong hydrophobic forces and highly specific complementary ionic contacts which take over and lock interfaces into a more intimate contact than is afforded by the rigid-body **BD** treatment.

C. Rate Constants and Reorganization Energy of Self-Exchange Reactions

We also studied the self-exchange reactions of *cytc* and *cytb5* with the rigorous BD method employing the exponential reaction criterion, which is our most advanced method of treating the reaction event (Andrew *et al.*, 1993). Table 16.3 and Figure 16.5 show the theoretical and experimental ionic-strength dependence of the bimolecular rate constant for ET using the combination of parameters β and k_{et}° giving the best fit of BD theory to experiment. Note that the rate increases with ionic strength, as the dielectric screening increasingly shields the electrostatic repulsion between like-charged reactants. The ionic strength dependence for both the *cytb5* and *cytc* self-exchanges is optimally reproduced by the rigorous BD model using values of $k_{et}^{\circ} = 3.43 \times 10^8$ s^{-1} and $\beta = 0.9$ Å^{-1} (*cytb5*) and $k_{et}^{\circ} = 1.18 \times 10^{10}$ s^{-1} and $\beta = 0.9$ Å^{-1} (*cytc*). The experimental data for *cytc* were of higher quality than for *cytb5*, enabling a more unambiguous determination of the best fit value of β. For *cytb5*, fitting was obtained using β values ranging all the way from 0.8 Å^{-1} to

Table 16.3
Experimental (Dixon *et al.*, 1989, 1990) and rigorous BD-simulated bimolecular rate constants k (M^{-1}s^{-1}) for electron self-exchange in *cytb5* and *cytc* as a function of ionic strength I.[a,b]

I (M)	cytb5		cytc	
	Experiment	Theory	Experiment	Theory
0.1	2.6×10^3	$1.46(\pm0.25) \times 10^3$		
0.12			5.4×10^3	$4.32(\pm1.06) \times 10^3$
0.3	4.6×10^3	$6.62(\pm0.77) \times 10^3$	1.6×10^4	$2.02(\pm0.41) \times 10^4$
0.5			2.8×10^4	$3.12(\pm0.44) \times 10^4$
0.6	1.6×10^4	$1.93(\pm0.36) \times 10^4$		
0.8			4.2×10^4	$4.23(\pm1.06) \times 10^4$
1.0	2.8×10^4	$3.00(\pm0.36) \times 10^4$	5.0×10^4	$4.68(\pm1.06) \times 10^4$
1.5	4.5×10^4	$4.12(\pm0.52) \times 10^4$	5.9×10^4	$5.80(\pm0.68) \times 10^4$

[a] $T = 25$ °C, pH = 7.0.
[b] The theoretical values are those using the best-fit value of the adjustable parameters $k_{et}^{\circ} = 3.43 \times 10^8$ s^{-1} (*cytb5*) and 1.18×10^{10} s^{-1} (*cytc*), and $\beta = 0.9$ Å^{-1} (both systems).

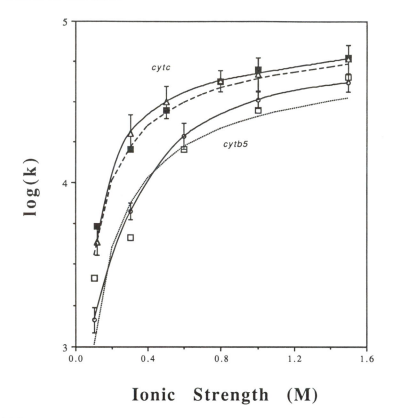

Ionic   Strength   (M)

Figure 16.5

Ionic strength dependence of the bimolecular rate constants k for self-exchange reactions of *cytb5* and *cytc*. Experimental values (Dixon *et al.*, 1989, 1990) for *cytb5* (□) and *cytc* (■) are compared to rigorous BD-simulated values for *cytb5* (○) and *cytc* (△) and van Leeuwen theory for *cytb5* ($\cdots$) and *cytc* (———). Best-fit parameters for the BD and van Leeuwen theories are given in text and in Dixon *et al.* (1989, 1990), respectively.

1.1 Å^{-1}, depending upon the inclusion or exclusion of suspicious-looking low-ionic-strength data values. Using the Marcus equation and our optimum value of $k_{et}°$, we extracted an estimate of the reorganization energy $\lambda = 1.06$ eV for *cytb5* and 0.69 eV for *cytc*. These values correspond with the estimates 1.2 and 0.7 eV made by Dixon *et al.* (1989, 1990), but are obtained with fewer estimated parameters, since the diffusion–collision part of the problem is treated explicitly by our theory and no estimate need be made of steric factors, the association constant and heme exposure. Also shown in Figure 16.5 is the best-fit theoretical curve given by the van Leeuwen (1983) equation, with parameters specified in Dixon *et al* (1989, 1990), including the net monopole charges of the proteins, the components of the dipole moments throughout the exposed heme edge, the sum of spherical radii of the two partners, and the rate constant at infinite ionic strength. Although the ionic-strength dependence it predicts is comparable, the BD theory appears to give a slightly better fit at higher ionic strengths, and with fewer parameters.

D. Docking Dynamics and Profiles

Docking dynamics and profiles have been extensively studied for the *cytc(r)/cyp* reaction (Northrup *et al.*, 1987b, 1988). *Cytc* and *cyp* are able to successfully achieve realistic geometric criteria for ET ($d_{edge} < 20$ Å and $\psi > 30°$ or $60°$) from a large ensemble of encounter complexes that are virtually all electrostatically stable rather than from a single dominant protein–protein complex. This multitude of potentially productive ET complexes form at least two and possibly three distinct and widely separated domains on *cyp*, in a belt approximately in the *cyp* heme plane (cf. Figure 16.6) and coincident with the electrostatically attractive regions observed in that plane. Figure 16.7 is an interprotein potential energy contour map around *cyp* in its heme plane as a function of the center of mass of the incoming *cytc*. At each relative position of the centers of mass of the two proteins, a Boltzmann average was performed of the potential energy over all accessible rotational orientations of the incoming *cytc* protein by Monte Carlo sampling. An extensive electrostatic channel of a depth of 1 to 2 units of $k_B T$ spans the three docking regions. The most predominant of these regions matches the region around Asp 34 of *cyp* hypothesized (Poulos and Kraut, 1980) in molecular graphics model building. The second region is substantially removed from the first region and is centered around Asp 148. A third, less populated region, intermediate between these two dominant areas, is the area near Asp 217. This channel provides a lower-dimensional region by

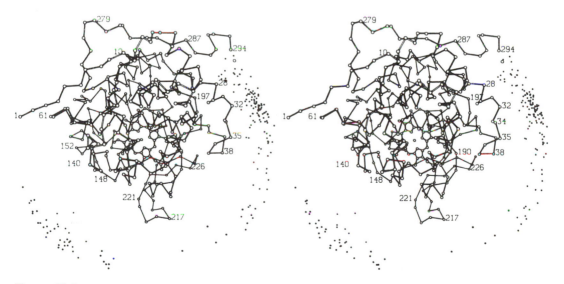

Figure 16.6

Docking profile showing the C_α skeleton of *cyp* surrounded by centers of mass of incoming *cytc(r)* at instants when heme-edge distances are sufficiently small for reaction ($d_{edge} = 20$ Å). Accordingly, ET may occur from a wide range of docking geometries rather than from a single dominant encounter complex.

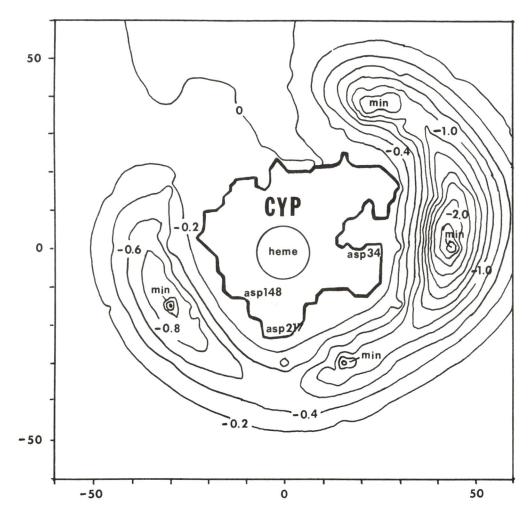

Figure 16.7

The Boltzmann-averaged total electrostatic potential energy of interaction between *cyp* and *cytc* in units of $k_B T$ as a function of the center-of-mass position of *cytc*. A cross-section of the potential energy function has been taken in the heme plane of *cyp* in the same orientational perspective as Figure 16.6. A Monte Carlo sampling method has been used to average the potential over all accessible orientations of *cytc* at each center-of-mass position.

which the incoming translating and rotating *cytc* may engage in an extended exploration of the surface of *cyp* in search of a productive ET configuration. From analysis of dynamics of trajectories, it is found that in a typical encounter of these proteins, the encounter is quite prolonged (10^2 ns); *cytc* undergoes numerous rotational reorientations (rotational relaxation time ~ 6 ns) as it explores an extensive region on the *cyp* surface, and it may even include all three major docking regions in one excursion (Northrup *et al.*, 1988).

E. Amino Acid Contacts in *Cytc*–*Cyp* Docked Complexes

In the study of the *cytc/cyp* system, the frequencies in which individual charged residues are among the top 10 closest contacts in 243 docked complexes with favorable geometric criteria for ET were tallied (Northrup *et al.*, 1987b). Lysines 13, 25, 27, 72, 79, 86, and 87 are most heavily involved in favorable electrostatic interactions with *cyp*. Lysines 7, 8, and 73 make favorable contacts, but to a much lesser extent. Negatively charged Asp 50, also located on the front of *cytc*, makes frequent contact but of a destabilizing nature, most often with negatively charged *cyp* residues Glu 35 and Glu 290. In a similar manner, negatively charged *cytc* residue Glu 90 makes frequent destabilizing contacts, most regularly with Glu 35 on *cyp*.

By analysis of lists of the triads of electrostatic contacts that occur most frequently in 243 complexes with favorable geometric criteria for ET, a frequency distribution was obtained which implies a nonspecific Coulombic binding process in which a predominantly positive surface of *cytc* interacts with several predominantly negative surfaces of *cyp*. Table 16.4 clearly shows that no dominant complex forms, but rather a surprisingly even distribution over a variety of complexes. The most

Table 16.4

Frequencies of the most frequent occurrences of triads of contacts in docked complexes of *cytc* and *cyp*.

cyp residue	*cytc* residue	Frequency
Asp34/Glu35	Lys79	11
Glu209	Lys13	
Glu290	Lys25/Lys27	
Asp34/Glu35	Lys13	8
Glu188	Lys25/Lys27	
Asp217	Lys25/Lys27	
Glu201	Lys79	8
Glu209	Lys79	
Glu290	Lys72	
Asp34/Glu35	Lys13	7
Glu201	Lys79	
Glu209	Lys79	
Asp34/Glu35	Lys79	7
Glu201	Lys25/Lys27	
Glu290	Lys25/Lys27	
Asp34/Glu35	Lys13	6
Glu201	Lys79	
Glu290	Lys72	
Asp34/Glu35	Lys13	6
Asp34/Glu35	Lys86	
Glu201	Asp79	

frequent complex which forms (but still only 11 times out of the 243 analyzed) is the triad in which *cyp* residues Asp 34/Glu 35, Glu 209, and Glu 290 contact *cytc* Lys 79, Lys 13, and Lys 25/27, respectively.

The contact analysis of complexed states revealed that their electrostatic stabilities are typically due to the interaction of two to four ion pairs with varying degrees of complementarity. This provides a possible explanation of the ionic-strength dependence of the rate constant as shown in Figure 16.2. Despite the large net monopole on these proteins, the ionic-strength dependence was much weaker and reflected the interaction of approximately 4e charges of opposite sign, in keeping with the finding of the contact study. This lends support to the hypothesis that it is the net electrostatic charge in the neighborhood of the interaction domain that primarily affects the association rate ionic-strength dependence, and not the global properties of the proteins.

F. Site-Specific Chemical Modifications of *cytc*: Reactions with *cyp*

To identify the binding domain of *cytc*, kinetics experiments have been performed (Kang *et al.*, 1978) in which various chemically modified horse *cytc*'s were reacted with yeast *cyp*. Lysines 8, 13, 25, 27, 60, 72, 73, 86, and 87 were derivatized by preparing the carboxydinitrophenyl (CDNP) derivative of the lysyl sidechain. These modified proteins have structural and redox properties which are unchanged from the native horse *cytc*. In similar fashion, the Brownian association rate has been simulated (Northrup *et al.*, 1987b) of *cyp* associating with this series of CDNP-modified *cytc* proteins for comparison purposes, including additional simulations of modified residues Lys 7, 22, 39, and 79. Additionally the −1e charge of Asp 50 has been flipped to +1e to confirm its role as a detractor of the association process, as deduced from the contact analysis. In this first series of simulations, CDNP-derivatized lysines were approximated simply by changing the charge on the lysine nitrogen from +1e to −1e without changing its location. This ignores the steric effect of the chemical modification. Simulations were performed at an ionic strength of 0.1 M, and results are tabulated in Table 16.5 in terms of both association rate constants and *relative activity*, which is defined as the ratio of the rate constant of the modified protein to that of the native protein.

In the experiment, the lowest relative activities (meaning largest detraction from the docking rate) were obtained when derivatizing Lys 72, 86, 87, 27, and 13. Only moderate effects were observed for modifications of Lys 73, 25, and 8, while no perturbation was observed for Lys 60, which is on the opposite side of *cytc* from the heme edge. In the simulations, qualitatively similar effects were observed of the chemical modification series, as shown in Table 16.5. Modification of lysines 27, 72, 86, and 13 produced the largest retardations of the rate, just as observed experimentally, though not to the same degree. Lys 87 was not found to be in this group of largest perturbations, its influence being more moderate. In the range of more moderate influences were Lys 25, 8, and 73, just as the experiment showed. However, Lys 60 also showed a slight retardation of the rate in the simulation when chemically modified, while no effect was observed in the experiment. Simulations were also performed of derivatized Lys 79, 7, 22, and 39, which were not performed in

Table 16.5

Kinetic results of chemical modification studies of *cytc* reaction with *cyp* at 0.1 M ionic strength.[a]

Modification	BD relative activity	Experimental relative activity
Lys27 → −1e	0.38	0.24
Lys79 → −1e	0.41	—
Lys72 → −1e	0.52	0.13
Lys86 → −1e	0.54	0.16
Lys13 → −1e	0.59	0.24
Lys25 → −1e	0.64	0.57
Lys87 → −1e	0.70	0.19
Lys8 → −1e	0.78	0.63
Lys73 → −1e	0.81	0.33
Lys60 → −1e	0.82	1.00
Lys7 → −1e	0.91	—
Native horse *cytc*	1	1
Lys22 → −1e	1.01	—
Lys39 → −1e	1.11	—
Asp50 → +1e	2.00	—
cyp heme propionates charged	1.27	—

[a] Comparison is made to experimental results of Kang *et al.*, 1978. BD results are shown for a heme plane criterion of $\psi = 60°$.

the experiment. Derivatization of lysine 79 showed one of the largest retardations of the rate, just as one would expect from its significance in the contact analysis study of native protein complexation. No effects were observed by modification of Lys 7, 22, and 39. This is also to be expected, as these are located on the "sides" of *cytc* and away from the exposed heme edge region. Since Asp 50 is a prominent negative residue in the region of the heme edge and is anticipated to have a negative influence on complexation as seen in the contact study, an additional simulation was performed in which the charge of Asp 50 on *cytc* was changed from −1e to +1e, while the charge location was not changed. Just as expected, the removal of a −1e charge from the predominantly positive heme edge surface of *cytc* and replacement with a +1e charge resulted in an improvement of the docking rate. What is surprising is that the rate actually *doubled* from the native simulations. This suggests a strong incentive to explore new site modifications to enhance a protein's effectiveness to form complexes with other proteins when electrostatic forces are involved.

G. Site-Directed Mutagenesis of *ycytc*: Reactions with *cytb5*

The reduction of wild type yeast iso-1-ferricytochrome *c* (*ycytc*) and several of its mutants by trypsin-solubilized bovine liver ferrocytochrome b_5 (*cytb5*) has been studied under conditions in which the electron-transfer reaction is bimolecular (Northrup et al., 1993). The effect of electrostatic charge modifications and steric changes on the kinetics has been determined by experimental and theoretical observations of the electron transfer rates of *ycytc* mutants **K79A** (Lys79 → Ala79), **K′72A** (Tml72 → Ala72) **K79A/K′72A** (Lys79, Tml72 → Ala79, Ala72), and **R38A** (Arg38 → Ala38). The rigorous Brownian dynamics method simulating diffusional docking and electron transfer was employed to predict the mutation effect on the rate constants. The electron transfer event was treated by the exponential distance dependent reactivity model, as previously described. Not only did the rigorous BD method quantitatively predict rate constants over a considerable range of ionic strength, but also quantitative prediction of mutant rates was made possible by taking into account the perturbing influence of the mutations on the redox potentials as well as on the electrostatics of docking (Table 16.6). Using these variations in $E°$ and the Marcus equation assuming a fixed reorganization energy of 0.7 eV, we esti-

Table 16.6
Net charge, dipole moments, and redox potential information of *cytb5* and various *ycytc* mutants at pH = 7.0 and $I = 0.19$ M.

Species	Net charge[a] (e)	Dipole mag (Debye)	γ[b]	$\Delta(\Delta G°_{el})$[c] (kcal/mol)	$E°$[d] (mV)
cytb5	−8.15	597	144°		
native *ycytc*	7.31	485	78°	0	290
ycytc **K79A**	6.35	506	86°	−0.046	292
ycytc **K′72A**	6.32	467	85°	+0.138	284
ycytc **K79A/K′72A**	5.36	497	93°	+0.046	288
ycytc **R38A**	6.39	524	80°	+1.176	239

[a] In units of electron charge magnitude.

[b] γ = angle of dipole moment relative to Fe position vector in center of mass frame.

[c] $\Delta(\Delta G°_{el})$ = change in free energy of reduction in kcal/mol relative to native *ycytc* derived from observed reduction potentials shown in the last column.

[d] $E°$ = potential of the reduction half-reaction of wild type or mutant *ycytc* determined from potentiometric titrations (A. G. Mauk, private communication).

Table 16.7

Theoretical and experimental bimolecular rate constants for the oxidation of *cytb5* by *ycytc* and four mutants.[a]

Species	Experimental	BD, distance cutoff model[b]	BD,[c] expo model, fixed $k_{et}°$	BD,[d] expo model driving-force corrected $k_{et}°$	$k_{et}°$ (ps^{-1})
Native *ycytc*	136 ± 3	172 ± 63	136 ± 17	136 ± 17	0.130
ycytc **K79A**	67.82 ± 3.02	62 ± 40	110 ± 10	112 ± 9	0.133
ycytc **K'72A**	37.21 ± 3.11	55 ± 15	63 ± 3	59 ± 6	0.121
ycytc **K79A/K'72A**	30.14 ± 1.71	15 ± 23	47 ± 6	46 ± 6	0.127

[a] pH = 7.0; T = 25 °C; ionic strength = 0.19 M.

[b] Reaction criterion = distance cutoff model with $d_{edge} = 12$ Å, $\psi = 30°$.

[c] Reaction criterion = exponential reactivity model with $\beta = 1.0$ Å^{-1}, fixed $k_{et}° = 1.3 \times 10^{11}$ s^{-1} (i.e., mutant perturbation of redox potential not included).

[d] Reaction criterion = exponential ractivity model with $\beta = 1.0$ Å^{-1}, $k_{et}°$ varying according to the Marcus equation (i.e., mutant perturbation of redox potential included).

mated the variation in the intrinsic electron transfer rate constant factor $k_{et}°$. BD simulations were performed with these varying values of $k_{et}°$ (see Table 16.7) for the different mutants. Both the experimentally observed rate constants and those predicted by BD descend in the order as follows: native *ycytc* > **K79A** > **K'72A** > **K79A/K'72A**. The proteins dock through essentially a single domain, with a distance of closest approach of the two heme groups in rigid-body docking typically around 12 Å. Two predominant classes of complexes were observed, the most frequent involving the quartet of *cytb5/ycytc* interactions Glu48–Lys13, Glu56–Lys87, Asp60–Lys86, and heme–Tml72, having an average electrostatic energy of -13.0 kcal/mol. The second most important type of complexes were of the type previously postulated (Salemme, 1976; Rodgers *et al.*, 1988) with interactions Glu44–Lys27, Glu48–Lys13, Asp60–Tml72, and heme–Lys79, and having an energy of -6.4 kcal/mol.

H. Inclusion of Limited Surface Side-Chain Flexibility

Brownian simulations of the diffusional association of the *cytc/cyp* reaction pair have also been performed (Northrup and Herbert, 1990) with the incorporation of limited side-chain flexibility rather than treating proteins as rotating rigid bodies. To include a limited amount of side-chain mobility in BD, each molecule is partitioned into a rigid and dynamic portion. The dynamic portion is a set of charged surface side chains on each molecule expected to have a high degree of mobility most likely to influence docking, typically including a number of the lysine, arginine, glutamate, and aspartate side chains. The rigid portion of each molecule is treated as a rigidly rotating array according to our usual definitions, their excluded volume interactions being treated by a spatial exclusion look-up table, and their potential fields being computed by the LPB equation on a lattice and stored in a look-up table. The total electrostatic potential is the sum of rigid and dynamic potential, the rigid portion being taken from the LPB look-up table, and the dynamic portion being computed by pairwise Coulomb–Debye summation over all dynamical charge sites. The Coulomb–Debye field is empirically modified to approximately reflect dielectric inhomogeneities. The dynamic side chains on each molecule are subject to forces from both the rigid and dynamic portions of both molecules. Atoms of protein II are tested for exclusion both relative to the exclusion grid around protein I and also by pairwise distance testing with the dynamic side-chain atoms of I.

Motional freedom of the side-chain atoms is treated in the following fashion. First, it is expected that time scales of torsional motion of side chains are fast (a few picoseconds per torsional state change) on the time scale of the relative diffusional motion of two bulky proteins (10 or more picoseconds per BD step). It would be computationally very demanding to treat the motional freedom of the side chains according to a BD algorithm, because the time step for properly integrating such motions accurately on the torsional potential surface would be subpicosecond. Instead, the time scale separation between torsional motion and relative diffusional motion is exploited. At each time step (~ 10 ps), Monte Carlo is used to choose a new torsional conformation of mobile side chains from an equilibrium distribution of conformers subject to the instantaneous electrostatic potential field at the side-chain charge site and a sinusoidal torsion potential energy term. Every torsion angle of each side chain is given the opportunity to take a random step, subject to excluded volume interactions, and when the new trial conformation is thus generated, the standard Monte Carlo energy criterion is used to accept or reject the step.

In these preliminary tests, rate constants with flexibility were essentially unchanged compared to the rigid study on this system (Northrup et al., 1987a,b), indicating that the more computationally convenient rigid body model gives reasonable results, at least for association rate constants in this system. The similar results with and without surface flexibility are in part due to the relative sparsity of surface atoms used to define the excluded volume shape of the incoming protein. With a more robust set of surface atoms, the inclusion of flexibility will have a more dramatic effect on rates. Even when little effect is registered in the rate constant, the inclusion of flexibility of charged surface groups is expected to modify the interaction potentials and distribution of the important ionic contacts in a meaningful way. More intimate encounters are likely to occur between proteins because of the increased opportunity for interdigitation of surface groups. This is now under investigation.

V. Conclusions

Brownian dynamics simulation is an important theoretical tool in understanding the role of electrostatics, solvent mediation, and other dynamics of proteins in their biological function. Through its use, several essential principles have been discovered and analyzed which account for the behavior of cytochromes relative to ET. The negative ionic-strength dependence in the bimolecular rates of the $cytc(o)/cytb5$

and $cytc(r)/cyp$ reactions confirms that the electrostatic forces strongly enhance the encounter of these proteins in favorable geometries for reaction. Rigorous Poisson–Boltzmann electrostatics treatment using a robust charge distribution is found to be essential for quantitative prediction of the correct ionic-strength dependence. Treatment of these systems as spherical molecules with monopoles and dipoles is inadequate to provide more than a qualitative understanding of electrostatic influences.

Studies based on a heme–heme distance cutoff model of ET reactivity provide evidence that at least partial heme-plane alignment is necessary for ET to occur. The feature of optimal alignment for heme–heme ET is presently under intense speculation. Since improvement of the method is likely to introduce features which *increase* the simulated rate under all of the geometric cases considered, the implication is that the electron transfer mechanism will involve at least partial alignment of the heme planes, probably with a ψ even smaller than the 30° to 40° range predicted here. A particularly unsatisfying feature of the BD simulation work on the cytochrome ET reactions has always been the somewhat indeterminate nature of the reaction parameters d_{edge} and ψ. However, comparison of three different systems evidences a similarity, in which values of ψ between 30° and 40° appear to provide reasonable results, with d_{edge} required to be approximately 2 Å larger than the minimum attainable distance for each pair. Association rate constants simulated here are at the upper limit in which ET is instantaneous upon achieving very liberal geometric criteria, and are still in agreement with actual experimental rate constants for ET. This implies that these reactions are near the diffusion-controlled limit.

The rigorous BD theory was used in conjuction with an exponential distance-dependent model of ET to study self-exchange in *c551* and *cytb5*. The model successfully reproduced the ionic strength dependence of the reaction. By fitting the BD-generated rate constants to the experimental curve, we were able to extract reorganization energy estimates which we believe to be a more rigorous estimate than that of Dixon *et al.* (1989, 1990). Furthermore we obtained a best-fit value of the distance decay parameter, which turned out to be $\beta = 0.9$ Å^{-1} for both proteins, in agreement with the value $\beta = 0.91$ Å^{-1} determined by Gray and Malmström (1989) by fitting ET distances and rates for five ruthenated heme metalloproteins. Since BD provides a more detailed description of the collision stage of the process, determined by the actual atomic-scale irregularity of the proteins (steric factors) and the mutual electrostatic interactions, no estimates of the association constant and heme exposure are required, and so we have fewer parameters to contend with than the model employed by Dixon *et al.* to make estimates of the reorganization energy.

We have also demonstrated that the rigorous BD method is capable of predicting the effects of site-directed mutations on the rate of docking and ET. We determined the effects of selective charge site perturbations by mutating residues Tml72 and Lys79 on the heme-exposed face of *ycytc* into neutral alanine residues. The trend in rates of the various mutants of *ycytc* was well-reproduced by the simulations for the Tml72, Lys79, and double mutation. The experimentally observed rate constants descend in the order as follows (at $I = 0.16$ M): native *ycytc* > **K79A** > **K′72A** > **K79A/K′72A**. Thus, all mutants react more slowly than the wild-type protein. Using the finite-difference linearized Poisson–Boltzmann treatment of the electrostatic potentials to calculate the influence of charged amino acid mutations on the reduction potential of *ycytc*, and using Marcus theory and an estimated reorganization energy, we estimated the perturbation in the intrinsic electron transfer rate constant caused by the charge mutations. Inclusion of this effect in the BD simulations did not change the predicted trend, but brought the exponential reactivity model into substantially better agreement for the prediction of mutant behavior.

In addition to computing rate constants, BD provides a wealth of information on the docking dynamics and profiles of these reactions. The *cytc(o)/cytb5* and *cytc*, *cytb5*, and *c551* self-exchange reaction criterion is met through a single surface region of each protein. However, for the *cytc(r)/cyp* reaction there exist two and perhaps three distinct, separated domains on *cyp* in a belt approximately in the *cyp* heme plane which meet the geometric criteria. These interestingly coincide with the electrostatically attractive regions. This shows strong correlation between electrostatic details and ET geometry restrictions in the structure of these proteins.

The existence of a unique dominant stable reactive complex is found in none of these reactions. A whole ensemble of complexes with slightly different orientations are indicated. The widely distributed frequency of ionic contacts (with three or four interprotein salt bridges stabilizing encounter complexes) shows that there is not a strict charge complementarity in operation that locks the proteins into a single ET arrangement, but that association is more nonspecific. The charge mutation study provides additional evidence that global dipole–dipole and monopole–dipole "steering" interactions between these proteins are of much less consequence relative to the local ionic contacts between exposed heme regions. The electrostatic forces provide the appropriate range of attractive potential energy to allow proteins to explore a wide variety of mutual orientations in a single encounter, rather than being deterministically steered by Coulomb forces over long distances into one selective docking arrangement. Recent experimental evidence in support of this hypothesis is provided by studies of ET reactions in electrostatic and covalently cross-linked complexes of plastocyanin and *cytc* (Peerey and Kostic, 1989). The covalently linked complex is entirely unreactive, even though it is thought to be docked in the same fashion as the electrostatically stable complex. The tight cross-links seem to prevent *cytc* and plastocyanin from the necessary exploration of a spectrum of nearby stable but less than energy-optimal docking geometries. Although significant perturbations of the redox properties due to cross-linking have been ruled out, further comparative studies are needed to fully elucidate the effects of cross-linking on the intracomplex electron-transfer reaction. Another quite interesting experimental study (Nocek *et al.*, 1990) of photoinitiated long-range ET within the complex between zinc-substituted *cyp* and ferric *cytc* has been used to probe the dynamics of intracomplex docking rearrangements. There it is found that the complex appears to undergo a docking conformational rearrangement to a low-temperature ET-inactive form at a very narrow range of temperature which is independent of solvent composition. Thus, the complex, when trapped in low-lying energetic states, is unable to sample other less stable docking geometries which are ET-active.

The utility of the BD method for simulation of ET rates between metalloproteins will increase as more detailed modeling of the intrinsic electron-transfer step is incorporated. For instance, our most advanced reactivity model, the simple exponential distance decay model, is probably inadequate. The knowledge of the variable reactivity of the protein surfaces is being made available by bonding pathways analysis (Beratan *et al.*, 1992), and these theories could be included in our modeling by special reactivity functions.

VI. References

Adam, G., & Delbruck, M. (1968) in *Structural Chemistry and Molecular Biology* (Rich, A., & Davidson, N., eds.), pp. 198–215, Freeman, San Francisco.

Allison, S. A., & McCammon, J. A. (1985) *J. Phys. Chem.* **89**, 1072.

Allison, S. A., Ganti, G., & McCammon, J. A. (1985a) *Biopolymers* **24**, 1323.

Allison, S. A., Northrup, S. H., & McCammon, J. A. (1985b) *J. Chem. Phys.* **83**, 2894.

Allison, S. A., McCammon, J. A., & Northrup, S. H. (1986) *ACS Symp. Ser.* **302**, 216.

Allison, S. A., Bacquet, R. J., & McCammon, J. A. (1988) *Biopolymers* **27**, 251.

Andrew, S. M., Thomasson, K. A., & Northrup, S. H. (1993) *J. Am. Chem. Soc.* **115**, 5516.

Beratan, D. N., Betts, J. N., & Onuchic, J. N. (1992) *J. Phys. Chem.* **96**, 2852.

Berg, O. G., & von Hippel, P. H. (1985) *Ann. Rev. Biophys. Biophys. Chem.* **14**, 131.

Debye, P. (1942) *Trans. Electrochem. Soc.* **82**, 265.

Dickinson, E. (1985) *Chem. Soc. Rev.* **14**, 421.

Dixon, D. W., Hong, X., & Woehler, S. E. (1989) *Biophys. J.* **56**, 339.

Dixon, D. W., Hong, X., Woehler, S. E., Mauk, A. G., & Sishta, B. P. (1990) *J. Am. Chem. Soc.* **112**, 1082.

Eltis, L. D., Herbert, R. G., Barker, P. D., Mauk, A. G., & Northrup, S. H. (1991) *Biochemistry* **30**, 3663.

Ermak, D. L., & McCammon, J. A. (1978) *J. Chem. Phys.* **69**, 1352.

Ganti, G., McCammon, J. A., & Allison, S. A. (1985) *J. Phys. Chem.* **89**, 3899.

Gray, H. B., & Malmström, B. G. (1989) *Biochemistry* **28**, 7499.

Guarr, T., McGuire, M., Strauch, S., & McLendon, G. J. (1983) *J. Amer Chem. Soc.* **105**, 616.

Herbert, R. G., & Northrup, S. H. (1989) *J. Mol. Liq.* **41**, 207.

Kang, C. H., Brautigan, D. L, Osheroff, N., & Margoliash, E. (1978) *J. Biol. Chem.* **253**, 6502.

Klapper, I., Hagstrom, R., Fine, R., Sharp, K., & Honig, B. (1986) *Proteins* **1**, 47.

Koppenol, W. H., & Margoliash, E. (1982) *J. Biol. Chem.* **257**, 4426.

Lamm, G. (1984) *J. Chem. Phys.* **80**, 2845.

Lamm, G., & Schulten, K. (1983) *J. Chem. Phys.* **78**, 2713.

Margoliash, E., & Bosshard, H. R. (1983) *Trends Biochem. Sci.* **8**, 316.

Matthew, J. B. (1985) *Ann. Rev. Biophys. Biophys. Chem.* **14**, 387.

Matthew, J. B., Weber, P. C., Salemme, F. R., & Richards, F. M. (1983) *Nature* **301**, 169.

Mauk, M. R., Mauk, A. G., Weber, P. C., & Matthew, J. B. (1986) *Biochemistry* **25**, 7085.

McLendon, G. (1988) *Acc. Chem. Res.* **21**, 160.

Nocek, J. M., Liang, N., Wallin, S. A., Mauk, A. G., & Hoffman, B. M. (1990) *J. Am. Chem. Soc.* **112**, 1623.

Northrup, S. H., & Herbert, R. G. (1990) *Intl. J. Quant. Chem.* **17**, 55.

Northrup, S. H., Allison, S. A., & McCammon, J. A. (1984) *J. Chem. Phys.* **80**, 1517.

Northrup, S. H., Curvin, M., Allison, S. A., & McCammon, J. A. (1986a) *J. Chem. Phys.* **84**, 2196.

Northrup, S. H., Smith, J. D., Boles, J. O., & Reynolds, J. C. L. (1986b) *J. Chem. Phys.* **84**, 5536.

Northrup, S. H., Reynolds, J. C. L., Miller, C. M., Forrest, K. J., & Boles, J. O. (1986c) *J. Amer. Chem. Soc.* **108**, 8162.

Northrup, S. H., Boles, J. O., & Reynolds, J. C. L. (1987a) *J. Phys. Chem.* **91**, 5991.

Northrup, S. H., Luton, J. A., Boles, J. O., & Reynolds, J. C. L. (1987b) *J. Computer-Aided Molec. Des.* **1**, 291.

Northrup, S. H., Boles, J. O., & Reynolds, J. C. L. (1988) *Science* **241**, 67.

Northrup *et al.* (1992) p. 6.

Northrup, S. H., Thomasson, K. A., Miller, C. M., Barker, P. D., Eltis, L. D., Guillemette, J. G., Inglis, S. C., & Mauk, A. G. (1993), *Biochemistry* **32**, 6613.

Peerey, L. M., & Kostic, N. M. (1989) *Biochemistry* **28**, 1861.

Poulos, T. L., & Kraut, J. (1980) *J. Biol. Chem.* **255**, 10322.

Reynolds, J. C. L., Cooke, K. F., & Northrup, S. H. (1990) *J. Phys Chem.* **94**, 985.

Rodgers, K. K., Pochapsky, T. C., & Sligar, S. G. (1988) *Science* **240**, 1657.

Salemme, F. R. (1976) *J. Mol. Biol.* **102**, 563.

Sharp, K., Fine, R., & Honig, B. (1987) *Science* **236**, 1460.

Shire, S. J., Hanania, G. I. H., & Gurd, F. R. N. (1974) *Biochemistry* **13**, 2967.

Siders, P., Cave, R. J., & Marcus, R. A. (1984) *J. Chem. Phys.* **81**, 5613.

Simondsen, R. P., Weber, P. C., Salemme, F. R., and Tollin, G. (1982) *Biochemistry* **21**, 169.

Smoluchowski, M. V. (1916) *Phys. Z.* **17**, 557.

Tanford, C., & Kirkwood, J. G. (1957) *J. Amer. Chem. Soc.* **79**, 5333.

Tanford, C., & Roxby, R. (1972) *Biochemistry* **11**, 2192.

Timkovich, R., Cai, M. L., & Dixon, D. W. (1988) *Biochem. Biophys. Res. Comm.* **150**, 1044.

van Leeuwen, J. W. (1983) *Biochim. Biophys. Acta* **743**, 408.

Warwicker, J., & Watson, H. C. (1982) *J. Mol. Biol.* **157**, 671.

Weber, P. C., & Tollin, G. (1985) *J. Biol. Chem.* **260**, 5568.

Wendoloski, J. J., Matthew, J. B., Weber, P. C., & Salemme, F. R. (1987) *Science* **238**, 794.

PART
VI

Modification of Cytochrome *c*

17

Chemical Modification of Surface Residues on Cytochrome *c*

FRANCIS MILLETT
BILL DURHAM

Department of Chemistry and Biochemistry
University of Arkansas

I. Introduction

Since its discovery by Keilin in 1925, cytochrome *c* has become the most thoroughly characterized member of the mitochondrial respiratory chain. However, the mechanism by which it transports electrons from the cytochrome bc_1 complex to cytochrome oxidase remains poorly understood, in part because of the complexity of its membrane-bound partners. A central question posed in the early 1970s was whether cytochrome *c* has separate binding sites for the cytochrome bc_1 complex and cytochrome oxidase, and forms an "electron bridge" between them, or instead has a single binding site for both proteins, and functions as a mobile electron shuttle. In an early paper, Takano *et al.* (1973) suggested that the cytochrome bc_1 complex might bind to the "left channel" of cytochrome *c*, defined by residues 55 to 75 that form a loop surrounding hydrophobic residues that lead from the surface of the protein to the heme. The eight lysine residues at the periphery of the channel could bind the cytochrome bc_1 complex electrostatically, while the aromatic residues Tyr 74 and Tyr 67 could be involved in electron transfer. It was suggested that cytochrome oxidase might bind to the "right channel" bounded by residues 1 to 20 and 89 to 101. This channel is also surrounded by lysine residues and contains the aromatic residues Phe 10 and Tyr 97, which could transfer electrons from the heme to cytochrome oxidase. This hypothesis was supported by several early antibody inhibition and chemical modification studies, which revealed differential effects on the reactions of cytochrome c with the cytochrome bc_1 complex and cytochrome oxidase (Smith et al., 1973; Margoliash et al., 1973). In contrast, Salemme et al. (1973) suggested that both the cytochrome bc_1 complex and cytochrome oxidase bind to the front of cytochrome c, and that reduction and oxidation occur by direct electron transfer through the exposed edge of the heme. The highly conserved lysines surrounding the heme crevice of cytochrome c would be involved in electrostatic interactions with the cytochrome bc_1 complex and cytochrome oxidase.

Beginning in 1973, several different laboratories addressed this question in an extensive series of chemical modification studies to map the interaction domain on cytochrome c for the cytochrome bc_1 complex and cytochrome oxidase. Because of the known importance of lysine residues in binding, these studies focused on preparing derivatives in which single lysine amino groups were modified to form uncharged or negatively charged groups. Margoliash and co-workers prepared 12 different derivatives modified at single lysine residues with the negatively charged 4-carboxy-2,5-dinitrophenyl (CDNP) group (Brautigan et al., 1978a,b; Osheroff et al., 1980a), while Millett and co-workers prepared 17 derivatives modified with a neutral trifluoracetyl (TFA) or trifluoromethylphenylcarbamoyl (TFC) group (Staudenmayer et al., 1976, 1977; Smith et al., 1977; Smith et al., 1980). Enzyme kinetic studies with these derivatives revealed that modification of any of the lysines immediately surrounding the heme crevice inhibited the interaction with the cytochrome bc_1 complex and cytochrome oxidase to nearly the same extent (Staudenmayer et al., 1976, 1977; Smith et al., 1977; Ferguson-Miller et al., 1978; Ahmed et al., 1978; Speck et al., 1979; Smith et al., 1980; Osheroff et al., 1980a). In an entirely different experimental approach, Rieder and Bosshard (1980) found that the same group of lysine residues were protected from acetylation by complex formation with either the cytochrome bc_1 complex or cytochrome oxidase (see Chapter 10 by Bosshard, this volume). Taken together, these results indicated that the interaction domain for both proteins is located at the heme crevice, and that the lysines immediately surrounding the heme crevice are involved in electrostatic interactions with the redox partners. Since the interaction domains for the cytochrome bc_1 complex and cytochrome oxidase are essentially the same, cytochrome c must function as a mobile shuttle during electron transport.

As with any experimental technique, the chemical modification procedures just discussed have a number of limitations and caveats. An important requirement for any structure–function study is to design and fully characterize the assay that is used to test the effect of the modification on the function of the protein. This especially applies in the case of cytochrome c, where a given lysine modification can often increase the reaction rate at low ionic strength, decrease it at high ionic strength, and have no effect at intermediate ionic strength. A second problem is that the modification might cause a long-range global change in protein conformation which could indirectly cause the observed effect on function. This possibility needs to be ruled out before the results can be interpreted as a localized electrostatic or steric effect. The major limitation of the differential modification method is that it can only be applied to stable product complexes (e.g., between oxidized cytochrome c and oxidized cytochrome oxidase) and provides no information on the activated substrate complex that is involved in the electron transfer reaction. The possibility always exists that the binding domain for the activated substrate complex might be different than that for the product complex, as appears to be the case for the reaction between *R. rubrum* cytochrome c_2 and photosynthetic reaction centers (see Section III.E). It is especially fortunate that the same conclusions were reached using the single-lysine modification method and the differential modification method to characterize the interaction of cytochrome c with the cytochrome bc_1 complex and cytochrome oxidase. Since these techniques are complementary and have entirely different limitations, there is greater confidence in the conclusions reached about the interaction of cytochrome c with its redox partners.

II. Preparation and Characterization of Cytochrome c Derivatives

A. Chemical Modification of Cytochrome c Lysine Amino Groups

The number of chemical reagents available for the specific modification of lysine amino groups is without doubt larger than for any other type of amino acid side chain, and many of these have been utilized to modify cytochrome c lysines, as shown in Table 17.1. Since most of these modification reactions involve nucleophilic attack of the amino group on the reagent, the reactions are generally carried out at high pH. When the goal is to prepare singly modified derivatives, the reaction conditions are adjusted to yield a crude reaction mixture that is labeled with one equivalent of reagent. To prepare the CDNP derivatives, Brautigan *et al.* (1978a) treated 1 mM horse ferricytochrome c with 5 mM CDNB in 0.2 M sodium bicarbonate, pH 9, for several hours at 23 °C until an average of 1 mol CDNP was incorporated per mole of cytochrome c. Altered selectivity was achieved by substituting borate for carbonate buffer (Osheroff *et al.*, 1980a). Carbonate apparently binds on the left side of the molecule, shielding lysines 72, 73, 86, and 87 from reaction. Different selectivity can also be achieved by running the reaction at lower pH, where a smaller fraction of the lysines are deprotonated. The TFA derivatives were prepared by treating 10 mM cytochrome c with 44 mM ethylthioltrifluoroacetate for 1 hour at pH 8.0 (Staudenmayer *et al.*, 1976), while the TFC derivatives were prepared by treating 10 mM cytochrome c with 40 mM trifluoromethylphenylisocyanate for 1 hour at pH 7.5 (Smith *et al.*, 1977). The protein is separated from excess reagent at the end of the reaction by passage through Sephadex G-10 or BioGel P-2 columns. A number of reagents yield a product retaining the positive charge on the modified lysine amino group. These include guanidination with *O*-methylisourea (Wilgus and Stellwagen, 1974), acetimidylation (Wallace, 1984), and reductive dimethylation (Wallace and Corthésy, 1987). It has not been possible to separate singly modified derivatives of this type, and reaction conditions are generally utilized to quantitatively modify all 19 lysine amino groups.

Pan *et al.* (1988) have developed a novel two-step procedure to prepare cytochrome c derivatives labeled at single lysine amino groups with ruthenium (bipyridine)$_2$(dicarboxybipyridine) [Ru^{II}(bpy)$_2$(dcbpy)]. In the first step, 10 mM cytochrome c was treated with the mono-*N*-hydroxysuccinimide ester of 4,4'-

Table 17.1
Modification of lysine amino groups.

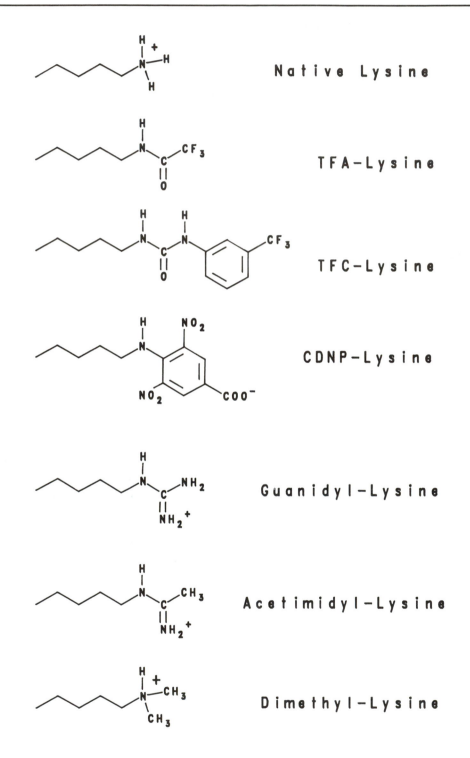

Native Lysine

TFA-Lysine

TFC-Lysine

CDNP-Lysine

Guanidyl-Lysine

Acetimidyl-Lysine

Dimethyl-Lysine

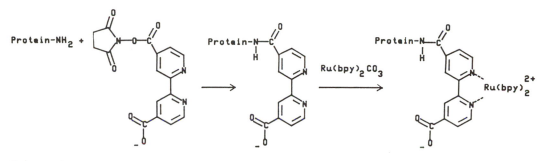

Scheme I
Preparation of Ru(bpy)$_2$(dcbpy)–cytochrome *c* derivatives (Pan *et al.*, 1988).

dicarboxy-2,2'-bipyridine (dcbpy) to form dcbpy–lysine cytochrome *c* derivatives (Scheme I). The dcbpy–lysine group has a negative charge, allowing singly labeled dcbpy–cytochrome *c* derivatives to be purified by ion-exchange chromatography, as described later. In the second step, each purified dcbpy–cytochrome *c* derivative was treated with Ru^{II}(bpy)$_2$CO$_3$ to form singly labeled Ru^{II}(bpy)$_2$(dcbpy)–cytochrome *c* derivatives. The final Ru^{II}(bpy)$_2$(dcbpy)–cytochrome *c* derivatives have the same net charge as native cytochrome *c*, allowing them to bind to cytochrome oxidase and other electron transfer proteins.

B. Purification of Singly Labeled Cytochrome *c* Derivatives

The ability to separate different singly labeled lysine derivatives of cytochrome *c* is quite remarkable, because they all have the same net charge, either 1 or 2 units less positive than that of native cytochrome *c*. The interaction of cytochrome *c* with cation-exchange resins is highly specific, allowing cytochrome *c* derivatives modified at different lysines to be resolved. In general, derivatives modified at lysine residues located in clusters of positive charge, such as those surrounding the heme crevice, interact much less strongly with the resin than derivatives modified at backside lysines. The preparation of pure derivatives requires several different chromatographic steps using different resins or elution conditions. The crude reaction mixture of CDNP–cytochrome *c* derivatives was first separated on a Whatman CM-32 carboxymethylcellulose column eluted with a sodium phosphate, pH 7.8 gradient (Brautigan *et al.*, 1978a). This separated the mono-CDNP–cytochrome *c* derivatives into four different fractions. Each fraction was then repurified on one or more columns using different elution conditions. For example, the derivatives modified at lysines 8, 27, and 87 were separated by elution with cacodylate buffer at pH 6.5, while the lysine 7 and 25 derivatives were resolved by elution with sodium borate, pH 8.6, containing 2.5 M ethanol (Osheroff *et al.*, 1980a). The TFA- and TFC–cytochrome *c* derivatives were first separated on a Bio-Rex 70 cation-exchange column eluted with ammonium phosphate, pH 7.2, and then each fraction was repurified on CM-32 eluted with sodium phosphate, pH 6.0.

578

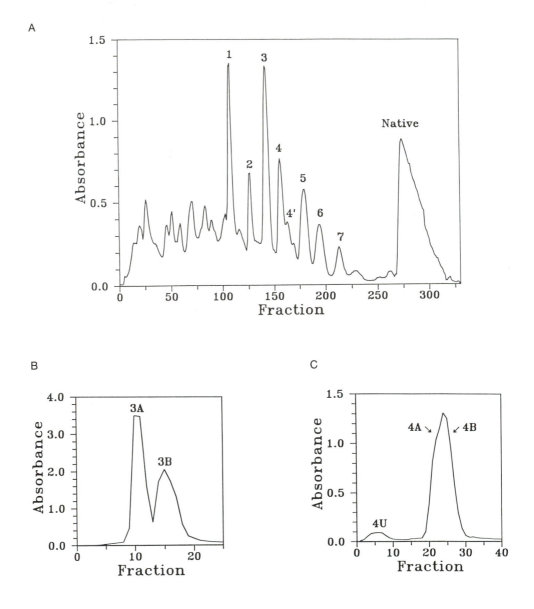

Figure 17.1

Purification of Ru(bpy)$_2$(dcbpy)–cytochrome c derivatives (Durham $et\ al.$, 1989). A: The crude reaction mixture of dcbpy–cytochrome c (500 mg) was chromatographed on a 2.5 × 70 cm Bio-Rex 70 column using an exponential gradient from 50 mM ammonium phosphate, pH 7.2, to 160 mM ammonium phosphate, pH 7.2. The flow rate was 25 mL/h and the fraction size was 3.8 mL. The absorbance was measured at 542 nm. B: Fraction 3 from A was rechromatographed on a 1.5 × 25 cm column of Whatman sulfopropyl SE-53 using an exponential gradient from 20 to 250 mM sodium phosphate, pH 6.0. The fraction size was 1 mL, and the absorbance was measured at 542 nm. C. Repurified fraction 4 was treated with Ru^{II}(bpy)$_2$CO$_3$ and chromatographed on a 0.6 × 45 cm Whatman CM-32 column using a gradient from 20 to 400 mM sodium phosphate, pH 6.0. The fraction size was 1 mL, and the absorbance was measured at 542 nm. The fraction marked 4U contained unmodified dcbpy–cytochrome c.

The separation of the dcbpy–cytochrome *c* derivatives provides a typical example and will be described here. The crude reaction mixture of dcbpy–cytochrome *c* derivatives was separated into eight different fractions on a Bio-Rex–70 column eluted with ammonium phosphate, pH 7.2, as described in Figure 17.1a. Each of these fractions was then rechromatographed on a Whatman sulfopropyl SE-53 column eluted with sodium phosphate, pH 6.0. (Figure, 17.1b). Each of the purified dcbpy–cytochrome *c* fractions was then incubated with $Ru^{II}(bpy)_2CO_3$ at pH 4 to form the $Ru(bpy)_2(dcbpy)$–cytochrome *c* derivatives. A final purification on a CM-32 column eluted with sodium phosphate, pH 6.0 (Figure 17.1c) then resulted in eight pure derivatives, and two fractions each containing two derivatives.

C. Identification of Singly Labeled Cytochrome *c* Derivatives

The specific lysine labeled in the purified cytochrome *c* derivative is identified by hydrolyzing the derivative with trypsin or chymotrypsin, and then separating the resulting peptides by an appropriate chromatographic technique. The CDNP–cytochrome *c* derivatives were originally identified using a sensitive thin-layer technique that indicated each derivative was >99% pure (Brautigan *et al.*, 1978b; Osheroff *et al.*, 1980a). The TFC and TFA derivatives were identified by an ion-exchange peptide mapping technique (Staudenmayer *et al.*, 1976), while more recent studies have utilized reversed-phase HPLC to carry out the peptide mapping. The HPLC peptide mapping studies of the $Ru(bpy)_2(dcbpy)$–cytochrome *c* derivatives will be described as an example of this technique (Pan *et al.*, 1988). The derivative was dialyzed into 0.1 M Bicine, pH 8.0, at a concentration of 1 $\mu g/\mu L$ and digested with 50 ng/μL trypsin for 15 hours at 37 °C. The tryptic digest was then eluted on a Dynamax 300 Å C-8 reversed-phase column using a linear gradient from 0.01 % aqueous trifluoroacetic acid to 100% methanol. The eluent was monitored at 210 nm and 290 nm to detect both the peptide and the ruthenium complex, respectively. In the HPLC chromatogram of fraction 4A shown in Figure 17.2, the only change relative to the chromatogram of native cytochrome was the presence of a new ruthenium-labeled peptide, which was shown by amino acid analysis to contain the sequence 23–27. Fraction 4A therefore contained a single $Ru(bpy)_2(dcbpy)$ group on lysine 25. The other derivatives were identified in a similar fashion to be: Fraction 1A: Lys 86; 1B: Lys 87; 2B: Lys 13; 3A: Lys 72; 3B: Lys 8; 4A: Lys 25; 4B: Lys 27; 4': Lys 7; 5: Lys 39 and Lys 60; 6: Lys 99 and Lys 100.

D. Characterization of Cytochrome *c* Derivatives

As mentioned in the introduction, it would not be possible to interpret the effect of a specific chemical modification on function if it resulted in a long-range change in the conformation of the protein. Therefore, it is important to fully characterize the structural integrity of the cytochrome *c* derivatives using a number of different techniques. Each of the CDNP, TFA, TFC, and $Ru(bpy)_2(dcbpy)$ derivatives was found to have the same visible spectrum as native cytochrome *c*, except for the absorbance due to the chromophore. The conformationally sensitive 695-nm band

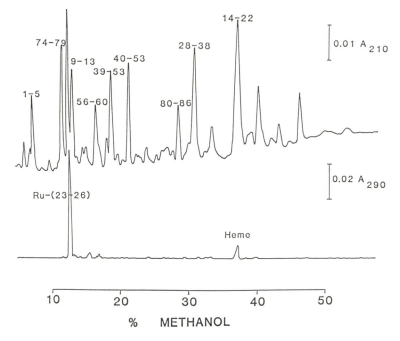

Figure 17.2
HPLC chromatogram of the tryptic digest of Ru(bpy)₂(dcbpy)–cytochrome *c* fraction 4A (Pan *et al.*, 1988). The tryptic digest (50 μg) was eluted on a Dynamax 300 Å column using a linear gradient from 0.01% aqueous trifluoroacetic acid to 100% methanol.

was unaffected in each derivative, which indicates the methionine-80 ligand to the heme iron was not disturbed. This band is very sensitive to the integrity of the heme crevice of cytochrome *c* and is lost as a result of modification of tyrosine and methionine residues (Myer *et al.*, 1980; Feinberg *et al.*, 1986) . Several of the derivatives did, however, affect the alkaline isomerization of cytochrome *c*, in which the Met 80 ligand to the heme iron is replaced by another ligand, possibly lysine 79, with an effective pK of 9.1. The pK for this isomerization was decreased to 8.1 for the CDNP- and TNP–Lys 13 derivatives, suggesting that in the native protein a salt bridge between lysine 13 and the carboxyl group of Glu 90 might stabilize the native conformation of the heme crevice (Osheroff *et al.*, 1980b). However, the TFA- and TFC–Lys 13 derivatives had the same pK as native cytochrome *c* (Smith, H. T. and Millett, 1980). The pK was actually increased to 9.3 and 9.6 for the TFC- and TFA–Lys 72 derivatives, respectively, suggesting that the native conformation was stabilized in these derivatives. The redox potentials of the derivatives were all in the range of 250 to 270 mV vs. NHE, essentially the same as that of native cytochrome *c*, 260 ± 10 mV (Brautigan *et al.*, 1978b; Staudenmayer *et al.*, 1976, 1977; Smith *et al.*, 1980). The redox potential is highly sensitive to protein conformational changes in the heme crevice region resulting from chemical modification or mutation (Pearce *et al.*, 1989; Louie *et al.*, 1988). ^{1}H NMR studies have also shown that the CDNP and TFA derivatives have essentially the same heme crevice conformation as native cytochrome *c* (Brautigan *et al.*, 1978b; Staudenmayer *et al.*, 1976). However, small differences in the hyperfine-shifted resonances were observed for the CDNP–Lys 13 and CDNP–Lys 72 derivatives, which were interpreted as a consequence of small movements of the Met 80 –S–CH_3 group relative to the heme group (Falk *et al.*, 1981). The observed shifts suggested that the CDNP–Lys 13 and CDNP–Lys 72 derivatives stabilized two different conformations of the protein that are normally in rapid equilibrium in native cytochrome *c* (Falk *et al.*, 1981). Taken as a whole, these results indicate that all of the CDNP, TFA, TFC, and Ru(bpy)$_2$(dcbpy) derivatives retain the overall conformation of native cytochrome *c*, with relatively minor alterations. This is a consequence of the external location of the modified lysine residues and is in marked contrast to the major changes in conformation which accompany modification of internal residues (Myer *et al.*, 1980; Feinberg *et al.*, 1986). It is, of course, likely that the local surface conformation in the immediate vicinity of the modified lysine is altered, and this possibility should be further explored with 2-D NMR and x-ray crystallographic studies.

III. Reactions of Cytochrome *c* Derivatives with Other Proteins

A. Reactions of Cytochrome *c* Derivatives with Cytochrome Oxidase

The steady-state reaction between cytochrome *c* and cytochrome oxidase is normally measured by following either the oxidation of cytochrome *c* spectrophotometrically, or the consumption of oxygen polarographically. Both assays display biphasic kinetics at low ionic strength, as discussed in Chapter 13 by Millett, this volume, but they are affected by specific lysine modifications quite differently. The Michaelis constant for the high-affinity phase of the polarographic assay, K_m, was found to have the same value as the equilibrium dissociation constant of the 1 : 1 complex between ferricytochrome *c* and cytochrome oxidase, K_d, suggesting that K_m is a measure of binding strength (Ferguson-Miller *et al.*, 1978; Osheroff *et al.*, 1980a). Modification of lysine residues surrounding the heme crevice with CDNP, TFA, or TFC increased the value of K_m significantly, while modification of lysines in other regions of the molecule had a much smaller effect (Table 17.2). The K_d values for most of the CDNP derivatives were in good agreement with the K_m values, providing additional evidence that the major effect of these modifications was on binding strength. The low-affinity phase of the polarographic assay was also affected by lysine modifications, but the molecular interpretation of this phase is more complex (Garber and Margoliash, 1990). The effect of specific lysine modifications on the high-affinity phase of the spectrophotometric assay is highly dependent on pH and ionic strength (Ferguson-Miller *et al.*, 1978; Smith *et al.*, 1982). At low ionic strength, many of the derivatives actually have higher rates than native cytochrome *c*. For example, in 25 mM Tris-cacodylate, pH 7.8, the V_{max} for the high-

Table 17.2

Effect of specific lysine modifications on the reactions of cytochrome c with cytochrome oxidase and the cytochrome bc_1 complex.[a]

Derivative:	Cytochrome oxidase			Cytochrome bc_1		
	CDNP[b]	TFA[c]	TFC[d]	CDNP[e]	TFA[f]	TFC[f]
Lysine						
5						
7	1.7			2.0		
8	13.3		2.5	2.8		1.7
13	66.7	4.9	9.0	10.0	4.2	3.8
22		1.1			1.0	
25	2.3	2.7		2.8	1.0	
27	10.0		2.1	7.1		2.1
39	1.3			1.7		
53						
55		1.0			1.0	
60	1.3			1.0		
72	33.3	4.3	3.0	9.1	2.1	2.1
73	6.7			2.3		
79		2.1	2.6		2.5	2.3
86	26.7			14.3		
87	23.3	3.0		14.3	2.0	
88		1.4			1.5	
99	1.1	1.0		1.1	1.0	
100			1.0			1.0

[a] The relative Michaelis constants $(K_m)_{\text{derivative}}/(K_m)_{\text{native}}$ are given for the high-affinity phase of the reaction with cytochrome oxidase measured by the polarographic assay at low ionic strength, and the reaction with the cytochrome bc_1 complex measured spectrophotometrically at high ionic strength.
[b] Measured in 25 mM Tris acetate, pH 7.8 (Osheroff et al., 1980a; Ferguson-Miller et al., 1978).
[c] Measured in 50 mM potassium Mops, pH 7.5 (Staudenmeyer et al., 1976, 1977; Smith et al., 1980).
[d] Measured in 50 mM potassium Mops, pH 7.5 (Smith et al., 1977).
[e] Measured in 100 mM Tris acetate, pH 7.5 (Speck et al., 1979).
[f] Measured in 200 mM Tris HCl, pH 7.5 (Ahmed et al., 1978; Smith et al., 1980).

affinity phase of the TFC–Lys 13 derivative was twofold greater than that of native cytochrome c, while the K_m value was about 1.5-fold greater (Smith et al., 1982). These results are consistent with the interpretation that at low ionic strength the rate-limiting step of the reaction is the dissociation of the ferricytochrome c–cytochrome oxidase complex. Thus, a lysine modification that decreases the strength of this complex can actually increase the observed rate. In contrast, at ionic strengths above 120 mM, the TFC–Lys 13 derivative reacted much more slowly than native cytochrome c, with a K_m value sevenfold greater than that of native cytochrome c, and with the same V_{max} value. At high ionic strength, only a single phase is observed in the spectrophotometric assay, and the effect of specific lysine modifications on the K_m value for this phase is essentially the same as observed for the high-affinity phase of the polarographic assay. The second-order rate constants for CDNP derivatives measured by stopped-flow at 133 mM ionic strength were also found to have the same pattern of inhibition (Veerman *et al.*, 1983). Thus, the same heme crevice reaction domain on cytochrome c is utilized in the high-affinity phase of the polarographic assay at low ionic strength, the equilibrium binding of ferricytochrome c, and the pre–steady-state and steady-state spectrophotometric assays at high ionic strength. The same heme crevice domain also appears to be involved in the low-affinity reaction of cytochrome c, but the location of this binding site on cytochrome oxidase is less well defined (Veerman *et al.*, 1983).

The decrease in binding strength due to specific lysine modification could be a result of both electrostatic effects and steric effects. A comparison of the effect of modifying a given lysine with TFA, TFC, and CDNP could help to distinguish among these factors, since the TFA and TFC groups are neutral, while the CDNP group is negatively charged. An approximation to the contribution a positively charged lysine amino group on lysine i makes to the total electrostatic free energy of binding is given by

$$V_i = \Delta G^\circ_{native} - \Delta G^\circ_{derivative} = -RT \ln Y_i,$$

where Y_i is the ratio of the K_m value of a derivative modified at lysine i with TFA or TFC to that of native cytochrome c (Smith *et al.*, 1981). The corresponding formula for the CDNP derivatives is $V_i = -0.5RT \ln Y_i$, since the change in charge is twice as large. Figure 17.3a shows that the V_i values for the three different types of modifications at a given lysine are in reasonable agreement, suggesting that the major effect of modification is in fact electrostatic. Any steric effect would be expected to be larger for the bulky TFC and CDNP groups than for the TFA group. Smith *et al.* (1981) have developed a semi-empirical relationship for the electrostatic interaction between cytochrome c and cytochrome oxidase, assuming that it can be represented by specific complementary charge-pair interactions between lysine amino groups on cytochrome c and carboxylate groups on cytochrome oxidase. This relationship is in quantitative agreement with the ionic-strength dependence of the reaction of native cytochrome c (Figure 17.4) when the parameters are based on the V_i values shown in Figure 17.3. The interaction is dominated by eight specific charge-pair interactions, with interaction energies ranging from -1.2 kcal/mol to -0.3 kcal/mol (Figure 17.5). The total electrostatic energy of binding was estimated to be -6.0 kcal/mol at 50 mM ionic strength. In a different approach, Rush and Koppenol (1988) assumed that the electrostatic interaction could be represented by the interaction between the dipole moment on cytochrome c and the net charge on cytochrome oxidase. They observed a linear relationship between the differences in the binding energies of the CDNP cytochrome c derivatives and their relative dipole moments. The entire 50-fold range of binding energies for the different derivatives could be accounted for by the dipolar interaction if it was assumed that the oxidase had a net charge of -50. However, Sinjorgo *et al.* (1986) found little difference in the ionic-strength dependence of K_m at pH 7.8, where cytochrome oxidase has a negative net charge, and pH 6.2, where it has a positive net charge.

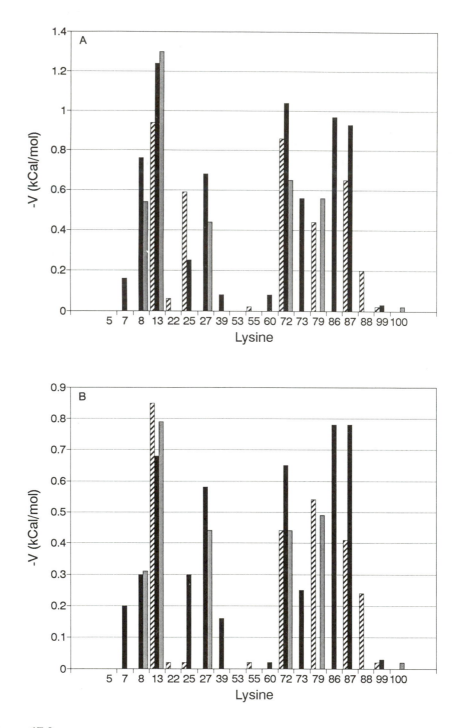

Figure 17.3
A: Effect of specific lysine modifications on the reaction between cytochrome c and cytochrome oxidase. The V_i values were calculated from the relative K_m values given in Table 17.2 as described in the text. B: Effect of specific lysine modifications on the reaction of cytochrome c with the cytochrome bc_1 complex. The V_i values were calculated from the relative K_m values given in Table 17.2 as described in the text. Stripes: TFA; black: CDNP; gray: TFC.

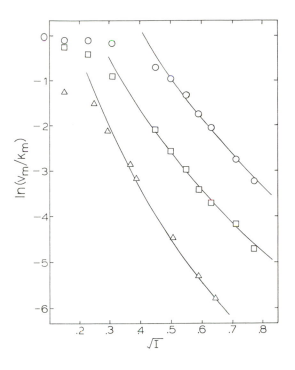

Figure 17.4
Effect of ionic strength on the reaction of cytochrome c with cytochrome oxidase and the cytochrome bc_1 complex (Smith *et al.*, 1981). The kinetics were measured spectrophotometrically in 50 mM Tris-HCl containing 0–0.6 M NaCl. o: reaction of native cytochrome c with succinate–cytochrome c reductase. □: reaction of TFA–Lys 13 cytochrome c with succinate–cytochrome c reductase. Δ: reaction of native cytochrome c with cytochrome oxidase. The lines give the semi-empirical relationship developed by Smith *et al.* (1981).

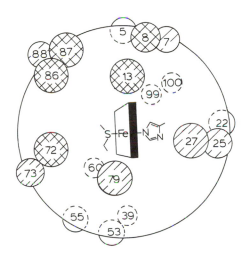

Figure 17.5
Diagram of horse cytochrome c viewed from the front of the heme crevice. The approximate positions of the β-carbon atoms of the lysine residues are indicated by closed and dashed circles for residues located toward the front and back of cytochrome c, respectively. The electrostatic contribution of lysine i to the interaction with cytochrome oxidase, V_i, is indicated by the number of diagonal hatch marks in the circle, with -0.1 kcal/mol/hatch mark.

The cytochrome c modification studies discussed earlier lead one to predict that the interaction site on cytochrome oxidase will consist of a cluster of about eight carboxylate residues. A number of cross-linking studies have indicated that the high-affinity binding site is located primarily on subunit II (Bisson *et al.*, 1980, 1982; Millett *et al.*, 1982). Millett *et al.* (1983) found that the water-soluble carbodiimide EDC specifically modified five carboxylate groups on subunit II of cytochrome oxidase and inhibited the reaction with cytochrome c. Binding one molecule of cytochrome c to cytochrome oxidase protected Asp 112, Glu 114, Asp 115, and Glu 198 from modification by EDC and prevented the loss in electron-transfer activity. These studies indicate that the negatively charged carboxylate groups on these residues are involved in cytochrome c binding. The involvement of Glu 198 is particularly interesting, since it is located between the conserved cysteines 196 and 200 that have been shown to ligand Cu_A (Hall *et al.*, 1988). It is therefore possible that Cu_A is the initial electron acceptor in cytochrome oxidase. Asp 112, Glu 114, and Asp 115 are located in a highly conserved sequence (104–115) containing alternating acidic and aromatic residues. These aromatic residues could serve as an electron-transfer pathway from cytochrome c to cytochrome a or Cu_A.

B. Reactions of Cytochrome c Derivatives with the Cytochrome bc_1 Complex

Although the spectrophotometric assay for the reaction between ferricytochrome c and the cytochrome bc_1 complex displays only a single kinetic phase, the effects of ionic strength and pH on the K_m and V_{max} values for this phase are extremely complex (see Chapter 13 by Millett, this volume). The V_{max} and K_m values are relatively unaffected by ionic strength over the range of 0.025 to 0.1 M , but then the K_m value increases rapidly with further increases in ionic strength, causing a large decrease in the V_{max}/K_m parameter (Ahmed *et al.*, 1978; Speck *et al.*, 1979) (Figure 17.4). Specific lysine modifications also have little or no effect on the kinetics at low ionic strength, but cause large increases in K_m at ionic strengths above 0.1 M. For example, the V_{max}/K_m parameter of the TFC–Lys 13 derivative was essentially the same as that of native cytochrome c at low ionic strength, but was fourfold smaller at ionic strengths above 0.2 M (Ahmed *et al.*, 1978) (Figure 17.4). Speck and Margoliash (1984) found that at low ionic strength, K_m was much larger than K_d, the equilibrium dissociation constant for ferricytochrome c, while at ionic strengths above 0.1 M, the two parameters were nearly the same. They suggested that the kinetics at low ionic strength reflect the reaction of solution cytochrome c with an ES complex already containing a bound cytochrome c, thus accounting for the insensitivity of the kinetics to specific lysine modifications and changes in ionic strength. At ionic strengths above 0.1 M, the agreement between K_m and K_d indicates that the kinetics reflect the reaction of cytochrome c with the free enzyme. Modification of lysine amino groups immediately surrounding the heme crevice with TFA, TFC, or CDNP caused a significant increase in K_m, while modification of lysines on other surface domains of cytochrome c had a relatively small effect (Table 17.2). Similar results were obtained for the CDNP derivatives using both steady-state kinetics and pre–steady-state kinetics techniques at high ionic strength

(Konig *et al.*, 1980). Comparison of the V_i values for the three different types of modifications at a given lysine indicate that the major effect was electrostatic (Figure 17.3b). The contribution of an individual lysine amino group to the electrostatic interaction ranges from about 0.8 kcal/mol for lysines 13, 72, 86, and 87 immediately surrounding the heme crevice, down to about 0.3 kcal/mol for the lysines at the periphery of the interaction domain. The semi-empirical relationship developed by Smith *et al.* (1981) was in quantitative agreement with the ionic-strength dependence of native cytochrome c (Figure 17.4), when the parameters were based on the V_i values shown in Figure 17.3b. The total electrostatic energy of binding was estimated to be -5.0 kcal/mol at 0.2 M ionic strength. The relative binding energies of the CDNP derivatives could be accounted for by the dipole moment treatment of Rush and Koppenol (1988), assuming a net charge on cytochrome c_1 of -70. The interaction domains on cytochrome c for cytochrome c_1 and cytochrome oxidase are nearly identical (Figures 17.3a and 17.3b), suggesting that cytochrome c must function as a diffusional carrier during electron transport.

Stonehuerner *et al.* (1985) have used the water-soluble carbodiimide EDC to identify the binding domain on the cytochrome bc_1 complex for cytochrome c. EDC was found to selectively label carboxylate groups in the sequence 63–81, and cytochrome c binding protected these residues from labeling. This highly acidic sequence contains seven Asp and Glu residues that appear to be involved in binding cytochrome c. Broger *et al.* (1983) have shown that arylazido-lysine 13 cytochrome c is cross-linked to cytochrome c_1 somewhere in the highly acidic sequence 165–174, suggesting that carboxylate groups in this region are also involved in cytochrome c binding. Taken together, these studies indicate that at least two different regions of the cytochrome c_1 sequence are folded together to form the cytochrome c binding site. The highly acidic "hinge" protein in the cytochrome bc_1 complex may also form part of the cytochrome c binding site (Kim and King, 1983).

C. Reaction of Cytochrome c Derivatives with Cytochrome c Peroxidase

The steady-state reaction between cytochrome c and yeast cytochrome c peroxidase is biphasic under most conditions and has a complex dependence on ionic strength, pH, and specific anions (Kang *et al.* 1978). The V_{max} for the high-affinity phase of the reaction is maximal at an ionic strength of 0.05 M, decreasing at both higher and lower ionic strengths. At very low ionic strength (10 mM), the CDNP derivatives modified at heme crevice lysines had larger V_{max} values than native cytochrome c, but nearly the same K_m values. This suggests that the rate is limited by product dissociation, and decreasing the binding strength increases the rate. In contrast, at high ionic strength (50 mM), these same derivatives had smaller V_{max} values than native cytochrome c. The TFA derivatives behaved in a similar fashion, except that the changes in V_{max} were smaller (Smith, M. B. and Millett, 1980). Although it is not clear why specific lysine modifications affect V_{max} rather than K_m, it appears that the V_{max} for the high-affinity phase of the reaction is sensitive to the interaction between the two proteins. By this criterion, the interaction domain for cytochrome c peroxidase is nearly the same as that for cytochrome oxidase and the cytochrome bc_1

complex. Intracomplex electron transfer studies have been carried out on the reaction between cytochrome *c* peroxidase and yeast cytochrome *c* mutants produced by site-directed mutagenesis (Hazzard *et al.*, 1988). Substitution of Lys 13 with Ile led to a fourfold increase in the rate of intracomplex electron transfer at low ionic strength, suggesting that lysine 13 in native cytochrome *c* might stabilize a complex at low ionic strength that is not optimal for electron transfer. This effect is comparable to the increase in the V_{max} of the high-affinity phase of the steady-state reaction at low ionic strength due to modification of lysine 13 with CDNP, although of course the kinetic parameters are quite different. The interaction domain on cytochrome *c* peroxidase has been studied by differential modification and cross-linking studies using the water-soluble carbodiimide EDC (Bechtold and Bosshard, 1985; Waldmeyer and Bosshard, 1985). The results indicated that carboxylate groups on the acidic residues 33, 34, 35, 37, 221, and 224 are involved in binding cytochrome *c*. These results are in partial, but not complete, agreement with the X-ray crystal structure for the 1:1 complex between yeast cytochrome *c* and cytochrome *c* peroxidase recently determined by Pelletier and Kraut (1992). The crystal structure indicates that the acidic residues 33, 34, 35, and 290, but not residues 221 or 224, contribute to the electrostatic interaction with cytochrome *c*.

D. Reaction of Cytochrome *c* Derivatives with Cytochrome b_5 and Sulfite Oxidase

Cytochrome b_5 is a small heme protein found in the endoplasmic reticulum, the outer mitochondrial membrane, and erythrocytes. Although the reaction between cytochrome b_5 and cytochrome *c* appears to have rather limited physiological significance (Matlib and O'Brien, 1976), it provides a good model system since both proteins are small, well-characterized, and have known x-ray crystal structures. In the first molecular modeling study of an electron transfer complex, Salemme (1976) proposed that the two proteins form a 1 : 1 complex stabilized by complementary charge-pair interactions between negatively charged carboxylate groups surrounding the heme crevice of cytochrome b_5 and lysine amino groups surrounding the heme crevice of cytochrome *c*. In the hypothetical complex, the heme groups of the two proteins are nearly coplanar, with their edges separated by 8.4 Å. Ng *et al.* (1977) and Smith *et al.* (1980) found that modification of lysines 13, 25, 27, 72, or 79 surrounding the heme crevice of cytochrome *c* with TFA or TFC significantly increased the K_m value for the steady-state reaction with cytochrome b_5, while modification of lysines in other regions of the molecule had very little effect. Rodgers *et al.* (1988) found that substitution of Glu 48, Asp 60, Glu 43, or Glu 44 on cytochrome b_5 with glutamine or asparagine residues each decreased the binding strength for the complex with cytochrome *c* by a factor of about 2. Esterification of the heme propionate group had a similar effect. These results provide experimental support for Salemme's model and suggest that lysines 13, 25, 27, 72, and 79 on cytochrome *c* form charge-pair interactions with the cytochrome b_5 carboxylate groups on Asp 48, Glu 43, Glu 44, Asp 60, and the most exposed heme propionate group, respectively. Stonehuerner *et al.* (1979) developed a semi-empirical relationship based on this model that was in quantitative agreement with the ionic-strength dependence of K_m for the reaction.

Sulfite oxidase from rat liver is a dimer consisting of two identical polypeptide chains, each containing molybdenum and heme prosthetic groups that are present in separate domains (Southerland *et al.*, 1978). The heme domain of sulfite oxidase is similar to the heme domains of cytochrome b_5 and flavocytochrome b_2, but there are significant differences in the amino acid sequences (Ozols *et al.*, 1976). All three cytochromes are highly acidic and react rapidly with cytochrome *c*. Modification of the lysine amino groups surrounding the heme crevice of cytochrome *c* with TFA, TFC, or CDNP significantly increased the K_m value for the steady-state reaction (Webb *et al.*, 1980; Speck *et al.*, 1981). The interaction domain on cytochrome *c* for sulfite oxidase thus appears to be very similar to that for cytochrome b_5. Similar results have been obtained for the reaction of cytochrome *c* with flavocytochrome c_2 from yeast (Matsushima *et al.*, 1986) and flavocytochrome *c*-552 from *Chromatium vinosum* (Bosshard *et al.*, 1986).

E. Reaction of Cytochrome c_2 Derivatives with the Cytochrome bc_1 Complex and Photosynthetic Reaction Centers from Bacteria

Hall *et al.* (1989) have carried out CDNP modification studies to map the interaction domain on *Rb. sphaeroides* cytochrome c_2 for the *Rb. sphaeroides* cytochrome bc_1 complex. Six different singly labeled CDNP cytochrome c_2 derivatives were purified for these studies. The K_m values for the derivatives modified at lysines 10, 55, 95, 97, 99, and 106 surrounding the heme crevice were significantly increased relative to the K_m value for native cytochrome c_2. The lysines located in the sequence 97–106 on the left side of the heme crevice have the greatest involvement in binding the cytochrome bc_1 complex. The involvement of lysine 97 is especially significant because it is located in an extra loop comprising residues 89–98 that is not present in eukaryotic cytochrome *c*. Kinetic studies with these derivatives have shown that the same heme crevice domain on cytochrome c_2 is used to bind to the *Rb. sphaeroides* photosynthetic reaction center (Long *et al.*, 1989), indicating that cytochrome c_2 functions as a diffusional carrier during electron transport in *Rb. sphaeroides*. CDNP modification studies have also shown that lysines surrounding the heme crevice of *R. rubrum* cytochrome c_2 are involved in binding both the cytochrome bc_1 complex and photosynthetic reaction centers from *R. rubrum* (Hall *et al.*, 1987a,b). This is in marked contrast to the differential modification studies carried out by Rieder *et al.* (1985), which showed that the formation of a complex between *R. rubrum* reaction centers and cytochrome c_2 protected only three residues on the "backside" of the cytochrome from modification with acetic anhydride. In this case it appears that the domain used to bind the product, ferricytochrome *c*, to the reaction center is quite different from the domain on cytochrome c_2 used for formation of the active ES complex that is involved in electron transfer.

IV. Acknowledgment

This work was supported in part by NIH grant GM 20488.

V. References

Ahmed, A. J., Smith, H. T., Smith, M. B., & Millett, F. (1978) *Biochemistry* 17, 2479.

Bechtold, R., & Bosshard, H. R. (1985) *J. Biol. Chem.* 260, 5191–5200.

Bisson, R., Jacobs, B., & Capaldi, R. A. (1980) *Biochemistry* 19, 4173.

Bisson, R., Steffens, G. M., Capaldi, R. A., & Buse, G. (1982) *FEBS Lett.* 144, 359.

Bosshard, H. R., Davidson, M. W., Knaff, D. B., & Millett, F. (1986) *J. Biol. Chem.* 261, 190–193.

Brautigan, D. L., Ferguson-Miller, S., & Margoliash, E. (1978a) *J. Biol. Chem.* 253, 130–139.

Brautigan, D. L., Ferguson-Miller, S., Tarr, G. E., & Margoliash, E. (1978b) *J. Biol. Chem.* 253, 140–148.

Broger, C., Salardi, S., & Azzi, A. (1983) *Eur. J. Biochem.* 131, 349–352.

Durham, B., Pan, L. P., Long, J. E., & Millett, F. (1989) *Biochemistry* 28, 8659–8665.

Falk, K., Jovall, P. A., & Angstrom, J. (1981) *Biochem. J.* 193, 1021–1024.

Feinberg, B. A., Bedore, J. E., & Ferguson-Miller, S. (1986) *Biochim. Biophys. Acta* 851, 157–165.

Ferguson-Miller, S., Brautigan, P. L., & Margoliash, E. (1978) *J. Biol. Chem.* 253, 149–159.

Garber, E. A. E., & Margoliash, E. (1990) *Biochim. Biophys. Acta* 1015, 279–287.

Hall, J., Ayres, M., Zha, X., O'Brien, P., Durham, B., Knaff, D., & Millett, F. (1987a) *J. Biol. Chem.* 262, 1046–1051.

Hall, J., Kriaucionas, A., Knaff, D., & Millett, F. (1987b) *J. Biol. Chem.* 262, 14005–14009.

Hall, J., Moubarak, A., O'Brien, P., Pan, L. P., Cho, I., & Millett, F. (1988) *J. Biol. Chem.* 263, 8142–8149.

Hall, J., Zha, X., Yu, L., Yu, C.-A., & Millett, F. (1989) *Biochemistry* 28, 2568–2571.

Hazzard, J. T., McLendon, G., Cusanovich, M. A., Das, G., Sherman, F., & Tollin, G. (1988) *Biochemistry* 27, 4445–4451.

Kang, C. H., Brautigan, D. L., Osheroff, N., & Margoliash, E. (1978) *J. Biol. Chem.* 253, 6502–6510.

Keilin, D. (1925) *Proc. Roy. Soc. Series B* 98, 312–339.

Kim, C. H., & King, R. E. (1983) *J. Biol. Chem.* 258, 13543–13551.

Konig, B. W., Osheroff, N., Wilms, J., Muijsers, A. O., Dekker, H. L., & Margoliash, E. (1980) *FEBS Lett.* 111, 395–398.

Long, J., Durham, B., Okamura, M., & Millett, F. (1989) *Biochemistry* 28, 6970–6974.

Louie, G. V., Pielak, G. J., Smith, M., & Brayer, G. D. (1988) *Biochemistry* 27, 7870–7876.

Margoliash, E., Ferguson-Miller, S., Tulloss, J., Kang, H. C., Feinberg, B. A., Brautigan, D. L., & Morrison, M. (1973) *Proc. Nat. Acad. Sci. USA* 70, 5234.

Matlib, M. A., & O'Brien, P. J. (1976) *Arch. Biochem. Biophys.* 173, 27.

Matsushima, A., Yoshimura, T., & Aki, K. (1986) *J. Biochem.* 100, 543–551.

Millett, F., Darley-Usmar, V., & Capaldi, R. A., (1982) *Biochemistry* 21, 3857.

Millett, F., deJong, C., Paulson, L., & Capaldi, R. A. (1983) *Biochemistry* 22, 546.

Myer, Y. P., Thallum, K. K., Pande, J., & Verma, B. C. (1980) *Biochem. Biophys. Res. Comm.* 94, 1106–1112.

Ng, S., Smith, M. B., & Millett, F. (1977) *Biochemistry* 16, 4975.

Osheroff, N., Brautigan, D. L., & Margoliash, E. (1980a) *J. Biol. Chem.* 255, 8245–8251.

Osheroff, N., Borden, D., Koppenol, W. H., & Margoliash, E. (1980b) *J. Biol. Chem.* 255, 1689–1697.

Ozols, J., Gerard, C., & Nobrega, I. G. (1976) *J. Biol. Chem.* 251, 6767–6774.

Pan, L. P., Durham, B., Wolinska J., & Millett, F. (1988) *Biochemistry* 27, 7180–7184.

Pearce, L. L., Gartner, A. L., Smith, M., & Mauk, A. G. (1989) *Biochemistry* 28, 3152–3156.

Pelletier, H., & Kraut, J. (1992) *Science* 258, 1748–1755.

Poulos, T. L., & Kraut, J. (1980) *J. Biol. Chem.* 255, 10322–10330.

Rieder, R., & Bosshard, H. R. (1980) *J. Biol. Chem.* 255, 4732–4741.

Rieder, R., Wienken, V., Bachofen, R., & Bosshard, H. R. (1985) *Biochem. Biophys. Res. Commun.* 128, 120–126.

Rodgers, K. K., Pochapsky, T. C., & Sligar, S. G. (1988) *Science* 240, 1657–1659.

Rush, J. D., & Koppenol, W. H. (1988) *Biochim. Biophys. Acta* 936, 187–198.

Salemme, F. R. (1976) *J. Mol. Biol.* 102, 563–569.

Salemme, F. R., Kraut, J., & Kamen, M. D. (1973) *J. Biol. Chem.* 248, 7701.

Sinjorgo, K. M. C., Steinebach, O. M., Dekker, H. L., & Muijsers, A. O. (1986) *Biochim. Biophys. Acta* 850, 108–115.

Smith, H . T., & Millett, F. (1980) *Biochemistry* **19**, 1117–1120.

Smith, H. T., Staudenmeyer, N., & Millett, F. (1977) *Biochemistry* **16**, 4971.

Smith, H. T., Ahmed, A., & Millett, F. (1981) *J. Biol. Chem.* **256**, 4984.

Smith, L., Davies, H. C., Reichlin, M., & Margoliash, E. (1973) *J. Biol. Chem.* **248**, 273.

Smith, L., Davies, H. C., Nava, M. E., Smith, H. T., & Millett, F. (1982) *Biochim. Biophys. Acta* **700**, 184–191.

Smith, M. B., & Millett, F. (1980) *Biochim. Biophys. Acta* **626**, 64–71.

Smith, M. T., Stonehuerner, G., Ahmed, A. J., Staudenmeyer, N., & Millett, F. (1980) *Biochim. Biophys . Acta* **592**, 303.

Southerland, W. M., Winge, D. R., & Rajagopalan, K. V. (1978) *J. Biol. Chem.* **353**, 8747–8752.

Speck, S. H., & Margoliash, E. (1984) *J. Biol. Chem.* **259**, 1064–1072.

Speck, S. H., Ferguson-Miller, S., Osheroff, N., & Margoliash, E. (1979) *Proc. Natl. Acad. Sci. USA* **76**, 155–160.

Speck, S. H., Koppenol, W. H., Dethmers, J. K., Osheroff, N., Margoliash, E., & Rajagopalan, K. B. (1981) *J. Biol. Chem.* **256**, 7394–7400.

Staudenmeyer, N., Smith, M. B., Smith, H. T., Spies, F. K., & Millett, F. (1976) *Biochemistry* **15**, 3198.

Staudenmeyer, N., Smith, M. B., Ng, S., & Millett, F. (1977) *Biochemistry* **16**, 600.

Stonehuerner, J., Williams, J. B., & Millett, F. (1979) *Biochemistry* **18**, 5422–5429.

Stonehuerner, J., O'Brien, P., Geren, L., Millett, F., Steidl, J., Yu, L., & Yu, C.-A. (1985) *J. Biol. Chem.* **260**, 5392–5398.

Takano, T., Kallai, O., Swanson, R., & Dickerson, R. E. (1973) *J. Biol. Chem.* **248**, 5234.

Veerman, E. C. I., Wilms, J., Dekker, H. L., Muijsers, A. O., van Buuren, K. J. H., van Gelder, B. F., Osheroff, N., Speck, S. H., & Margoliash, E. (1983) *J. Biol. Chem.* **258**, 5739–5745.

Wallace, C. J. A. (1984) *Biochem. J.* **217**, 595–599.

Wallace, C. J. A., & Corthésy, B. E. (1987) *Eur. J. Biochem.* **170**, 293–298.

Webb, M., Stonehuerner, J., & Millett, F. (1980) *Biochim. Biophys. Acta* **543**, 290–298.

Waldmeyer, B., & Bosshard, H. R. (1985) *J. Biol. Chem.* **260**, 5184–5190.

Wilgus, H., & Stellwagen, E. (1974) *Proc. Nat. Acad. Sci. USA* **71**, 2892.

18

Covalent Modification, Cross-Linking, and Cleavage of Cytochrome c with Metal Complexes

NENAD M. KOSTIĆ

Department of Chemistry
Iowa State University

I. Metal Complexes as Bioconjugation Reagents

Covalent modification of amino-acid side chains has proved useful in structural, spectroscopic, and mechanistic studies of proteins (Lundblad and Noyes, 1984; Means and Feeney, 1971; Glazer, 1977). The various chromophores, fluorophores, spin labels, and radioactive labels developed so far are mostly organic compounds. Except as heavy-atom scatterers for x-ray crystallography (Blundell and Johnson, 1976; Petsko, 1985), metal complexes have not been widely used for covalent modification of proteins. Because substitution reactions of transition-metal complexes can be controlled precisely and because these complexes have various spectroscopic and electrochemical properties, they are well suited to many biochemical and biophysical applications. They can serve as absorption chromophores, emission fluorophores, paramagnetic spin labels, NMR-active tags, NMR relaxation agents, redox agents, and radioactive markers. Selectivity in their binding to proteins (and to other biological macromolecules) can be changed in predictable ways by controlling the oxidation state, hardness or softness, and coordination number of the metal

593

and the identity and number of the ligands.[1] Well-chosen inorganic reagents can, in principle, meet all the requirements of a good labeling reagent. They can react under mild conditions; be selective toward particular side chains or surface sites; be easily detected and quantitated in the modified protein; be stable yet removable under mild conditions, so that the native protein can be restored; and be noninvasive, so that they do not alter the structure and function of the protein. All of these advantages have been achieved with cytochrome c.

A metal cation can be attached to a protein in several ways. Because reactions are usually done in aqueous solution, a "bare" cation, M^{n+}, actually is an aqua complex, $[M(H_2O)_x]^{n+}$. (1) The metal can be attached to amino-acid side chains; in this case the ambiguities of composition and charge can be avoided by using a pre-formed complex of known properties, rather than an aqua complex (Werber and Lanir, 1981; Pecoraro et al., 1984; Recchia et al., 1982). (2) The metal (without exogenous ligands) can also be attached to a prosthetic group, often as a replacement for the native metal ion. (3) Finally, the metal can be attached to designed binding sites—chemically modified side chains (Legg, 1978), chelating ligands attached to side chains (Meares and Wensel, 1984), or residues introduced by site-directed mutagenesis. This review deals only with the first method.

II. Binding Sites and Principles of Selectivity

The potential ligands in proteins are shown in Table 18.1. Their reactivity toward metal complexes depends on their accessibility from solution, i.e., their exposure on the protein surface; on pH; and on their intrinsic affinity toward the particular metal. The exposures of side chains in crystalline cytochrome c, determined by x-ray crystallography, also are relevant to the dissolved protein. Such results for the tuna protein (Takano et al., 1973) may also be useful in the study of cytochromes c that have not been crystallographically analyzed. But crystal structures should be viewed skeptically because side chains, especially those on the surface of the protein in solution, can be quite mobile. For a discussion of NMR studies of topography of cytochromes c in solution, see Chapter 5 by Pielak et al., this volume.

The kinetic and thermodynamic factors that govern metal–ligand affinity in general apply also to metal–protein complexes (Pullman and Goldblum, 1977). The principle of hard and soft acids (metal ions) and bases (donor atoms in the side chains) is a useful criterion of thermodynamic stability of adducts (complexes). Small ("light") metal ions and those in high oxidation states prefer small, electronegative donors such as oxygen and nitrogen, while large ("heavy") metal ions and those in low oxidation states prefer large, polarizable donors such as sulfur. Because

---

[1] From the inorganic point of view, a ligand is something that is attached to a metal; from the biochemical point of view, a ligand is something that is attached to a biomolecule. In a metal–protein complex either the protein or the metal can be considered the ligand; in this review the inorganic convention is used, and the protein (or a particular functional group in it) is considered the ligand.

Table 18.1
Heteroatom-containing side chains in proteins.

Potential ligand, L	Residue	Typical pK_a of conjugate acid[a]
	Asp, Glu	3.0–4.7
	Asn, Gln	
	His	5.6–7.0
—S⁻	Cys	8.3–10.0
	Cys–Cys	
	Met	
	Tyr	9.8–10.4
—NH₂	Lys	9.4–10.8
	peptide bond	10.2–10.8
	Trp	
	Arg	11.6–12.6
—O⁻	Thr, Ser	*ca.* 13.0

[a] Clark (1964); Stryer (1988); Zubay (1988).

kinetic and thermodynamic stability need not go together, one should consider the possibility of migration of the metal complex from the site of initial attachment to a site of stronger attachment, and the possibility of an equilibrium between the metal tag on the protein and the metal reagent in solution. Selectivity of labeling may be influenced not only by the metal affinity for an individual side chain (a donor atom), but also by the properties of the polypeptide backbone and the surrounding side chains. One such property is charge or electrostatic potential, especially when the metal reagent bears a high charge.

III. Labeling with [PtCl$_4$]$^{2-}$

The square-planar complex [PtCl$_4$]$^{2-}$ is routinely used by protein crystallographers for labeling of methionine residues. The discovery of this reagent exemplifies early research into metal–protein interactions. In the first study (Dickerson *et al.*, 1967), 71 compounds were tried in 132 experiments with horse ferricytochrome *c*, and signs of binding were sought by x-ray crystallography. Only [PtCl$_4$]$^{2-}$ and the organo-mercurial mersalyl, HOHgCH$_2$CH(OCH$_3$)CH$_2$NHC(O)(*o*-C$_6$H$_4$)OCH$_2$COONa, did bind. The major binding site for [PtCl$_4$]$^{2-}$ is Met 65, and a minor one is His 33 (Dickerson *et al.*, 1969). Initial notions about the reaction were later corrected (Petsko *et al.*, 1978). It is simply a displacement of one chloride ligand by the sulfur atom of methionine, to yield the square-planar complex of platinum(II), [PtCl$_3$(Met–Cyt)]$^-$. The complex [PtCl$_3$(His–Cyt)]$^-$ should have similar structure. In the general formula in Figure 18.1, L stands for an amino-acid side chain, and

Figure 18.1
[PtCl$_4$]$^{2-}$ and the amino-acid complex formed from it.

the complex charge is not shown. Reactions of [PtCl$_4$]$^{2-}$ and of *cis*-[Pt(NH$_3$)$_2$Cl$_2$] with Met 65 in tuna cytochrome *c* were studied by ^{1}H NMR spectroscopy (Boswell *et al.*, 1981, 1982). Although the chemical shifts of the SCH$_3$ group were not explained and other shifted resonances were not assigned, it was concluded that platination, unlike carboxymethylation, causes only slight, local perturbations of the protein structure. Reactions of [PtCl$_4$]$^{2-}$ with sulfur-containing amino acids, peptides, and similar ligands have been studied by NMR spectroscopy of ^{195}Pt and other nuclei (Gummin *et al.*, 1986; Galbraith *et al.*, 1987; Burgeson and Kostić, 1991). Binding of platinum(II) complexes to biomolecules in solution has been reviewed from the standpoint of inorganic chemistry (Howe-Grant and Lippard, 1980).

IV. Labeling with [Pt(trpy)Cl]⁺

Chloro(2,2′ : 6′,2″-terpyridine)platinum(II), [Pt(trpy)Cl]⁺ (Figure 18.2) is especially well suited for covalent modification of proteins. Because only the chloride ion can be displaced by amino-acid side groups L under ordinary conditions, only one-to-

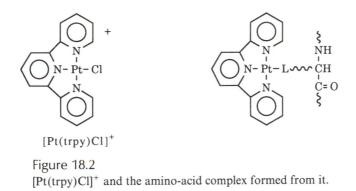

[Pt(trpy)Cl]⁺

Figure 18.2
[Pt(trpy)Cl]⁺ and the amino-acid complex formed from it.

one adducts can form. The charge of an adduct depends on the charge of L (see Table 18.1). Because the coordinated side group L can be displaced by an added ligand N, the tag can be removed from the protein as the complex [Pt(trpy)N], whose charge likewise depends on the charge of N.

A. Cysteine and Histidine

All the amino acids in Table 18.1, some other amino acids, and some peptides containing them were incubated at room temperature with equimolar amounts of [Pt(trpy)Cl]⁺ complex under the conditions described later for protein modification. The following amino acids caused no change in the UV-vis spectrum: lysine, tryptophan, aspartic acid, asparagine, glutamic acid, glutamine, proline, threonine, serine, tyrosine, cystine, and methionine (Ratilla *et al.*, 1987). Arginine reacted upon standing or heating, as will be explained later. Cysteine, homocysteine, and reduced glutathione (GSH) reacted rapidly because of the high nucleophilicity of the thiolate group toward platinum(II). Imidazole (Im), histidine, and the tripeptide Gly–His–Gly reacted more slowly because of the lesser nucleophilicity of these nitrogen ligands toward platinum(II). The rate constant for the entry of thiolate-containing ligands is approximately 300 times greater than that for the entry of imidazole-containing ligands (Brothers and Kostić, 1988). Because of this large kinetic difference, the most reactive amino acid in proteins should be free cysteine, and the next one in reactivity should be histidine. Unreactivity of methionine, cystine (Cys–Cys), and oxidized glutathione (GSSG) was surpising in view of the high affinity of the soft platinum(II) atom toward the soft thioether ligand. Persistent attempts with methionine and its various derivatives and peptides proved their complete inertness toward [Pt(trpy)Cl]$^+$ even under forcing conditions (Ratilla *et al.*, 1987). Molecular models of the pentacoordinate [Pt(trpy)ClL]$^+$ molecules, representing the intermediates or the transition states in the nonoccurring hypothetical reactions with the thioether and disulfide ligands L, showed no prohibitive crowding. But models of the expected product, [Pt(trpy)L]$^{2+}$, showed crowding. Nonparameterized molecular orbital calculations on [Pt(trpy)SMe$_2$]$^{2+}$, which mimics the expected complex with methionine, revealed strong repulsion between methyl groups in SMe$_2$ and *ortho*-H atoms in the cis pyridine rings. This interaction could not be avoided by rotation of the pyramidal thioether ligand about the Pt–S bond. Whereas the complex [Pt(trpy)SMe$_2$]$^{2+}$ seemed inherently unstable, the complex [Pt(trpy)Im]$^{2+}$ seemed devoid of steric repulsions and stable (Ratilla *et al.*, 1987). This contrast explains the unexpectedly different behavior of methionine and histidine toward [Pt(trpy)Cl]$^+$.

Labeling of each cytochrome c was accomplished simply, under mild conditions (Ratilla *et al.*, 1987; Brothers and Kostić, 1988). The protein was incubated with an equimolar amount of [Pt(trpy)Cl]Cl, in phosphate buffer at pH 5.0 or 7.0, for about one day. The reaction mixture was then chromatographed efficiently on CM 52 cation exchanger, with phosphate buffer as an eluent. The greater the number of the cationic Pt(trpy)$^{2+}$ tags on the protein, the slower its movement down the column. Singly, doubly, and triply labeled derivatives—[Pt(trpy)cyt], [{Pt(trpy)}$_2$cyt], and [{Pt(trpy)}$_3$cyt], respectively—were eluted in this order. (The overall charge of protein-containing complexes is not specified because it is variable.) Although the derivatives that are singly labeled at differently located residues of the same kind have the same overall charge, they were separated owing to their different distributions of charge. The Pt(trpy)$^{2+}$ tags remained after cation-exchange chromatography because the ligands in the protein formed inert complexes with the platinum(II) atoms.

The inorganic chromophores were clearly evident in the UV-visible spectra of the tagged protein because their bands were largely unobscured by those of the protein itself. Systematic subtractions of the spectra of the native and modified proteins from one another produced spectra of the Pt(trpy)$^{2+}$ tags. The nonspecific absorptions of the aromatic terpyridine ligand in the region 240–280 nm (ε = 19,000–31,000 M^{-1} cm^{-1}) proved useful because of their intensity. The absorption patterns in the region 320–350 nm were weaker (ε = 9,000–16,000 M^{-1} cm^{-1}), but characteristic of the ligand L in [Pt(trpy)L]$^{n+}$. Matching of the difference spectra with the spectra of model complexes containing amino acids and peptides as L allowed both counting of the tags and identification of the types of the tagged amino acids in the protein derivatives. Not only do the band positions depend on the nature of the ligand L (i.e., histidine vs. cysteine), their relative intensities depend on the environment of this ligand. For example, [Pt(trpy)His]$^{2+}$ tags at His 33, in the hydrophilic region, and at His 26, in the hydrophobic region, in horse cytochrome c both exhibited the characteristic bands at 342 and 328 nm, but the relative intensities of these bands were different. The [Pt(trpy)Cys]$^+$ tag in baker's yeast cytochrome c exhibited an additional band, at 311 nm (Brothers and Kostić, 1988).

Various methods were used to determine the binding sites in cytochromes c: peptide mapping; comparisons among the derivatives obtained from the proteins

Table 18.2
Residues[a] in cytochrome *c* that are labeled in reactions with platinum(II) chloro complexes.[b]

Complex	Organism	His 26	His 33	His 39	Met 65	Arg 91	Cys 102
$[PtCl_4]^{2-}$	horse	—	√		√	—	
	tuna	?			√	—	
$[Pt(trpy)Cl]^+$	horse	√	√		—	√	
	tuna	√			—	√	
	baker's yeast (*iso*-1)	√	√	√	—	—	√
	Candida krusei	√	√	√	—	—	
$[Pt(sbpaphy)Cl]^-$	horse	—	√		—	—	
$[Pt(dipic)Cl]^-$	horse	—	—		√	—	
$[Pt(dpes)Cl]^+$	horse	√	√		√	—	
	tuna	√			√	—	

  [a] Check, labeled; dash, not labeled; blank, unreactive amino-acid residue in that position in the sequence; question mark, uncertain.
  [b] For literature references, see the main text.

with slightly different amino-acid sequences; selective blocking, prior to labeling with $[Pt(trpy)Cl]^+$, of certain residues by protonation or by covalent modification; and measurements of the rates at which the proteins took up the inorganic chromophore. The labeled amino-acid residues are listed in Table 18.2 Yields of the platinated derivatives are approximately proportional to exposure of the respective residues on the protein surface, as determined by x-ray crystallography (Takano *et al.*, 1973).

The inorganic labels do not perturb the protein conformation, its redox potential, and the electronic and geometric structure of the active site. This general conclusion emerged from systematic comparisons of the singly and multiply labeled protein derivatives with one another and with the native proteins. The following methods were used: UV-vis spectrophotometry, cyclic voltammetry, differential-pulse voltammetry, EPR spectroscopy, and ^1H NMR spectroscopy (Ratilla *et al.*, 1987; Brothers and Kostić, 1988).

The relatively low reactivity of Cys 102 in baker's yeast cytochrome *c*—the yield of the singly labeled derivative was only *ca.* 10%—was surprising because this residue can be modified with organic reagents (Bryant *et al.*, 1985; Ramdas *et al.*, 1986) and because the protein can dimerize readily through a disulfide bond (Motonaga *et al.*, 1965). The histidine residues 33 and 39 proved considerably more reactive—the corresponding yields were *ca.* 15 and 20%—even though the thiolate ligands are far more nucleophilic than imidazole ligands toward $[Pt(trpy)Cl]^+$. The 300-fold difference between the rate constants for the two types of ligands, discussed earlier, permitted a kinetic proof of the binding group. The baker's yeast cytochrome c reacted with $[Pt(trpy)Cl]^+$ at the same rate as did its horse congener (which lacks free Cys), the tripeptide Gly–His–Gly, and other imidazole-containing ligands. This unexpected outcome of the protein labeling indicated that, contrary to the common assumption, Cys 102 is not exposed at the protein surface. Modification of this residue with various organic reagents and dimerization of the protein must be accompanied by a conformational change that makes Cys 102 in the baker's yeast cytochrome c accessible to the reagents and to another molecule of this protein. These predictions from the labeling study (Brothers and Kostić, 1988) were subsequently confirmed by the crystallographic analysis of this cytochrome c; indeed, Cys 102 points toward the protein interior (Louie et al., 1988).

Verification of the binding sites in the baker's yeast protein by peptide mapping required special precautions to ensure that the $Pt(trpy)^{2+}$ tags remained at the points of initial attachment. Experiments with model amino-acid complexes showed that the tag can migrate from histidine to free cysteine, but not to mercurated cysteine. Although Cys 102 is almost inaccessible in the native protein, it becomes fully exposed in the protein digest. A standard procedure—enzymatic cleavage of the protein tagged at histidine residues—would have been misleading; the tag would have readily migrated and would have been detected on the peptide containing Cys 102 rather than on the peptides containing His 33 and His 39. Prior to tryptic digestion, therefore, the free Cys 102 in the $Pt(trpy)^{2+}$-modified protein was blocked with $HOHg(p\text{-}C_6H_4)SO_3^-$, and the peptide mapping became reliable (Brothers and Kostić, 1988).

The complex $[Pt(trpy)Cl]^+$ differs from some of the standard labeling reagents by its noninvasiveness: the ease of its attachment to a ligand of a certain type reflects this ligand's exposure on the protein surface. This property may well render the new inorganic reagent useful as a probe of the protein topography and applicable even to proteins whose structures are unknown or only partly known. This complex was advertised (*Aldrichimica Acta*, 1988) and is sold by Aldrich Chemical Co. as a selective reagent for labeling of proteins.

B. Arginine

The recognized functions of this residue in proteins—binding of cofactors and anions—involve electrostatic attraction to the guanidinium cation. Because metal–guanidyl complexes are barely known (Mehrota, 1987), covalent binding of transition metals to arginine side chains in proteins had not even been proposed before it was actually achieved with cytochrome c (Ratilla and Kostić, 1988).

When cytochromes c from horse and tuna were incubated with $[Pt(trpy)Cl]^+$ at pH 7.0 (rather than the usual 5.0) and chromatographed as before, additional $Pt(trpy)^{2+}$ derivatives were obtained (Ratilla and Kostić, 1988). Their number and their UV-visible difference spectra indicated that the increase in pH had created a single new binding site, which is the same in all the new derivatives and is not a histidine. This site was identified indirectly, but conclusively, by monitoring of reactions between $[Pt(trpy)Cl]^+$ and amino acids, and by considerations of the cytochrome c structures. The new ligand must be accessible in both horse and tuna proteins, and there must be few such ligands because only one reacts. Only three residues—Tyr 74, Tyr 97, and Arg 91—satisfy both requirements. Thorough tests, with heating, confirmed the unreactivity of tyrosine and reactivity of arginine. *N*-Acetylarginine (AcArg) and such simple ligands as methylguanidine (MeGua) and guanidine (Gua) yielded homologous complexes, shown in Figure 18.3. Their UV-

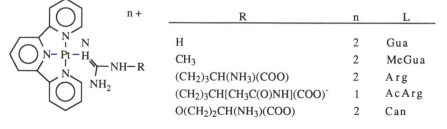

R	n	L
H	2	Gua
CH₃	2	MeGua
(CH₂)₃CH(NH₃)(COO)	2	Arg
(CH₂)₃CH[CH₃C(O)NH](COO)⁻	1	AcArg
O(CH₂)₂CH(NH₃)(COO)	2	Can

Figure 18.3
[Pt(trpy)Cl]⁺ and guanidine complexes formed from it.

visible spectra were very similar to the difference spectra of those labeled proteins that correspond to the Pt(trpy)²⁺ tag at the new site. Evidently, this site is Arg 91.

Why the side chain of Arg 91 reacts at pH 7.0 despite the guanidine basicity—its normal pK_a value is *ca.* 12.5—became evident from the structure of cytochrome *c*. Barely exposed on the surface (hence its low labeling yield of *ca.* 10%), Arg 91 abuts the N-terminus of the α-helical segment 92–102. The electric macrodipole of this segment and the hydrophobic environment together lessen the basicity of the guanidine group (Patthy and Thesz, 1980), so that it reacts with [Pt(trpy)Cl]⁺ even in neutral solution. This explanation was confirmed in experiments with canavanine (Can), a close homolog of arginine whose guanidine group has a pK_a value of 7.0. Canavanine formed the [Pt(trpy)Can]²⁺ complex readily under the conditions of the cytochrome *c* labeling (Ratilla *et al.*, 1990).

V. Labeling with Other [Pt(tridentate)Cl] Complexes

The unreactivity of [Pt(trpy)Cl]⁺ toward methionine and its reactivity toward histidine are caused by an interplay between steric and electronic properties of the terpyridine ligand. The *ortho*-H atoms repel the thioether, otherwise a very nucleophilic ligand. The strain of the terpyridine chelate in combination with aromatic character, however, labilizes the chloride ligand (Mureinik and Bidani, 1978) and facilitates its displacement by imidazole, otherwise a ligand less nucleophilic toward platinum(II). The decisive effect of the chelate ligand on the selectivity is evident from the comparison between [Pt(trpy)Cl]⁺ and [PtCl₄]²⁻. Although both of them are platinum(II) chloro complexes, the second one exhibits a large preference for methionine over histidine in proteins. Indeed, molecular orbital calculations (Ratilla *et al.*, 1987) revealed no steric crowding in [PtCl₃(SMe₂)]⁻ and in [PtCl₃Im]⁻ because the chloride ligands are sufficiently small. In the absence of steric effects, displacement reactivity is governed by nucleophilicity of the entering ligand.

The three platinum(II) complexes shown in Figure 18.4 all have the same displaceable chloride ligand, but differ from one another in the ancillary tridentate ligands and in charge. They were synthesized in the author's laboratory. Although

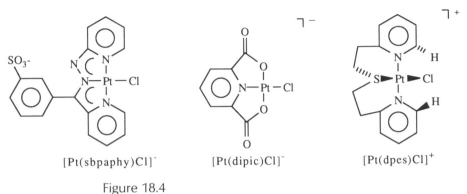

[Pt(sbpaphy)Cl]⁻   [Pt(dipic)Cl]⁻   [Pt(dpes)Cl]⁺

Figure 18.4
$[Pt(sbpaphy)Cl]^-$, $[Pt(dipic)Cl]^-$, and $[Pt(dpes)Cl]^+$.

the studies of their reactivity toward cytochromes *c* were not as thorough as studies involving $[Pt(trpy)Cl]^-$, interesting differences emerged.

The ligand 2-(3-sulfobenzoyl)pyridine-2-aldehyde 2′-pyridylhydrazone, designated sbpaphy, is similar to terpyridine, for it contains three coplanar trigonal nitrogen atoms, but it coordinates to metals as a dianion. The difference in charge between $[Pt(sbpaphy)Cl]^+$ and $[Pt(trpy)Cl]^+$ did not alter the binding selectivity—the former complex still labeled His 33 in the horse cytochrome *c*—but did change the mobility on the cation-exchange column. Although the Pt(sbpaphy) tag as a whole is neutral, the modified protein eluted before the native one, perhaps because of the electrostatic repulsion between the sulfonate group in the tag and the carboxylate groups of the CM 52 cation exchanger (Shah *et al.*, 1988).

The carboxylate "arms" of dipicolinate dianion (dipic) in $[Pt(dipic)Cl]^-$ are sterically undemanding in comparison with the terminal pyridine rings of terpyridine and sbpaphy (Zhou and Kostić, 1988). Because the chloride ligand is now unshielded, the new complex (as a K^+ salt) reacts faster with methionine than with histidine, as expected on the basis of the relative nucleophilicity of their side groups. Incubation with horse cytochrome *c* yielded a single derivative, labeled at Met 65. Binding to Met 80, an axial ligand to the iron atom, is ruled out because the spectroscopic and redox properties of the heme were unperturbed (Zhou and Kostić, 1990).

The complex $[Pt(dpes)Cl]^+$, which contains the ligand di(2-pyridyl-β-ethyl)-sulfide, is intermediate between $[Pt(trpy)Cl]^+$ and $[Pt(dipic)Cl]^-$ in its steric properties. Molecular models show that puckering of the chelate rings formed by the ligand dpes tilts the S—Pt—Cl axis away from the pyridine planes, so that the Cl^- ligand is incompletely shielded by the *ortho*-H atoms of the pyridine rings. Moreover, the chloride ligand is labilized owing to the trans effect of the thioether ligand.

The complex $[Pt(dpes)Cl]^+$ likewise is intermediate between $[Pt(trpy)Cl]^+$ and $[Pt(dipic)Cl]^-$ in its reactivity—it reacts with both Met 65 and His 33 in cytochrome *c*. But the chloride ligand must still be considerably shielded, because the yield of the His 33 derivative is greater than the yield of the Met 65 derivative (Zhou and Kostić, 1990). These studies showed that principles of inorganic stereochemistry can be used in the design of reagents for selective covalent modification of proteins (Kostić, 1988).

VI. Labeling with Metals Other Than Platinum

The classic study (Kowalsky, 1969) of the reaction between ferricytochrome *c* and chromium(II) ion, which is actually the complex $[Cr(H_2O)_6]^{2+}$, nicely demonstrated the inner-sphere mechanism for electron-transfer reactions between transition metals. Chromium(II) complexes are labile, that is, they undergo ligand substitution reactions relatively fast, whereas chromium(III) complexes are inert, that is, they undergo these reactions very slowly. As the complex $[Cr(H_2O)_6]^{2+}$ reduces iron(III) to iron(II) in cytochrome *c*, the resulting chromium(III) complex (presumably with several of the aqua ligands still coordinated to it) remains firmly bound to the amino-acid side chains at or near the site on the protein surface from which chromium(II) donated an electron to the heme. The inertness of the chromium(III) product allowed its location by peptide mapping; it was found locked between Asn 52 and Tyr 67 in the interior of cytochrome *c*, next to the tyrosine ring and the heme group (Grimes *et al.*, 1974).

Noncovalent interactions of cytochrome *c* with heteropolytungstates were studied by absorption, circular dichroism, and EPR spectroscopic methods (Chottard *et al.*, 1987). According to this study, the heme of the bound protein can have two different electron configurations. One is a low-spin configuration that is different from the configuration in the dissolved protein, and the other is a high-spin configuration. These findings were confirmed in a study of cytochrome *c* and $As_4W_{40}O_{140}^{27-}$ by resonance Raman spectroscopy (Hildebrandt and Stockburger, 1989).

The complex $[Cu(H_2O)_6]^{2+}$ binds to various groups in horse and tuna ferricytochromes *c* in the pH range 2–9. The major group in the horse protein in the pH range 4–7 is His 33; evidence came from equilibrium dialysis and visible, EPR, and ^1H NMR spectroscopy. The association constant is 1.8×10^4 M^{-1} at 25 °C and an ionic strength of 0.10 M. Minor groups have association constants not greater than *ca.* 1×10^3 M^{-1} (Augustin and Yandell, 1981). Evidently, when the ligands (in this case, H_2O) do not restrict the metal's reactivity, binding is nonselective.

Affinity of His 33 for copper(II) was used for separation of horse cytochrome *c* (which contains this residue), the tuna protein (which has the unreactive Trp 33 instead), and the *Candida krusei* protein (which has His 39 in addition to His 33) from one another by immobilized-metal affinity chromatography. A column of a gel with pendant copper(II) iminodiacetate chelates retains the *Candida krusei* protein most strongly, the horse protein less strongly, and the tuna protein not at all (Hemdan *et al.*, 1989).

VII. Cross-Linking

Cross-linking of proteins is commonly done with bifunctional organic reagents. The advantages of inorganic reagents, listed in the introduction, can be exploited in this application, too. Because only the chloride ion can be displaced easily from the tridentate complexes [Pt(chelate)Cl], only one protein molecule can bind to the platinum atom. More than one chloride ion can, however, be displaced from the unidentate complex $[PtCl_4]^{2-}$. Although this complex is routinely used as a heavy-atom tag for proteins (Dickerson *et al.*, 1969; Petsko *et al.*, 1978), cross-linking has not been observed because tagging has invariably been done with crystals, in which the protein molecules are immobile. Research in the author's laboratory has shown for the first time that proteins, exemplified by cytochrome *c*, can be cross-linked in solution selectively, under mild conditions, with metal complexes. First to be used were $[PtCl_4]^{2-}$ and its derivatives *trans*-$[PtCl_2L_2]$ (Peerey and Kostić, 1987).

Incubation of horse cytochrome *c* with $K_2[PtCl_4]$ yielded a stable diprotein complex, *trans*-$[PtCl_2(cyt)_2]$, which was separated from the native and $PtCl_3$-tagged proteins by size-exclusion (gel-filtration) chromatography on Sephadex 75–50 (Peerey and Kostić, 1987). The molecular mass of 30.7 ± 0.8 kDa, determined by this method, confirmed the composition of the diprotein complex. This value is slightly greater than expected because the mobility of a macromolecule down the Sephadex column depends not only on the mass of the macromolecule, but also on its shape. The complex, which consists of two protein molecules bridged by a $PtCl_2$ group, appeared to the Sephadex gel slightly larger than a single spheroidal molecule of the same mass.

The binding group in each of the two proteins is Met 65, as shown schematically in Figure 18.5. The diprotein complex is stable owing to the affinity of the

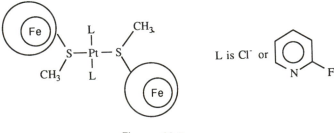

Figure 18.5
trans-$[PtCl_2(cyt)_2]$.

thioether group for platinum(II). The complex was cleaved, however, by incubation with highly nucleophilic ligands, such as thiourea, which displaced the proteins and extruded the platinum link. This combination of stability under ordinary conditions and easy removability under conditions harmless to the protein is a particularly useful feature of the $PtCl_2$ link.

This link proved detectable by infrared spectroscopy. A single band at 343 cm^{-1}, characteristic of the Pt(II)–Cl bond, indicated a trans configuration of the chloride ligands, and thus also of the protein molecules. This configuration was confirmed in experiments involving cis-[Pt(bpy)(DMF)$_2$]$^{2+}$, a complex containing two labile N,N-dimethylformamide ligands in cis positions to each other. One, but not both, of these ligands proved displaceable with cytochrome c, evidence that two protein molecules cannot occupy cis positions in a mononuclear complex. Even a cursory examination of molecular models confirmed this impossibility. The shortest distance between the two proteins in $trans$-[PtCl$_2$(cyt)$_2$] is approximately 4.5–5.0 Å, comparable with the separation between the nearest points of two cytochrome c molecules in the crystalline unit cell. Cross-linking through a $trans$-PtCl$_2$ unit apparently does not force the protein molecules too close to each other. Spectroscopic and electrochemical measurements showed no significant perturbation of the geometric and electronic structure of the cytochrome c upon cross-linking.

Since only two chloride ligands in [PtCl$_4$]$^{2-}$ are displaceable by protein molecules, the other two ligands can be changed in a purposeful way. Such an altered linking reagent is $trans$-[Pt(2-Fpy)$_2$(DMF)$_2$]$^{2+}$, whose incubation with the horse protein yielded $trans$-[Pt(2-Fpy)$_2$(cyt)$_2$]. The two 2-fluoropyridine ligands in the diprotein complex were observed by ^{19}F NMR spectroscopy; this method of detection of the bridge is more direct than infrared spectroscopy (Peerey and Kostić, 1987). Variability of the ancillary ligands (those not directly involved in the substitution reaction) is yet another advantage of the new inorganic reagents for cross-linking of proteins.

The formation of the diprotein complexes illustrates a classical concept in coordination chemistry. The first protein ligand, bonded to the platinum atom through the sulfur atom of Met 65, facilitates the entrance of the second one. This is an example of the kinetic trans effect, a well-known property of thioether ligands.

Bimetallic complexes are particularly fit for cross-linking of proteins. The metal–metal bonds or bridging ligands, or both, enrich the chemistry of these complexes and make them potentially versatile as reagents. Dirhodium(II) μ-tetracarboxylates, [Rh$_2$(RCOO)$_4$], are suitable because the two accessible vacant coordination sites allow formation of various adducts [Rh$_2$(RCOO)$_4$L$_2$] (Boyar and Robinson, 1983). The strong and short Rh(II)–Rh(II) bond and the entire "lantern" remain intact in reactions with ligands L. The changeability of the group R permits purposeful adjustment of the hydrophilicity, lipophilicity, and Lewis acidity of the complex. The feasibility of protein cross-linking through dirhodium complexes was demonstrated with [Rh$_2$(OAc)$_4$] because it is the best-studied member of the series and because the strong ^{1}H NMR signal of the four equivalent methyl groups in the acetate bridges promised (and later allowed) direct detection of the link despite the complexity of the protein ^{1}H NMR spectrum (Chen and Kostić, 1988).

A survey of amino acids showed that only cysteine, methionine, and histidine react with [Rh$_2$(OAc)$_4$]. The adducts with the last two amino acids and with imidazole were prepared as models for the possible diprotein complexes. The reaction with cysteine was not studied in detail because horse cytochrome c lacks a free cysteine residue. Incubation of the protein with [Rh$_2$(OAc)$_4$] yielded [Rh$_2$(OAc)$_4$(cyt)$_2$], shown in Figure 18.6. It was purified by size-exclusion chromatography, as before. The apparent molecular weight of the diprotein complex, deter-

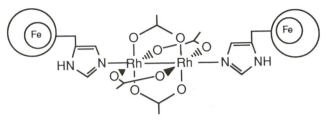

Figure 18.6
$[Rh_2(OAc)_4(cyt)_2]$.

mined by this chromatography, is 34 ± 1 kDa, or *ca.* 30% greater than the actual value. Since the $Rh_2(OAc)_4$ link keeps the protein ligands *ca.* 2.5 Å farther apart then the $PtCl_2$ link does, the bimetallic diprotein complex deviated more than the monometallic complex from the idealized spheroidal shape, and the error in its apparent molecular mass was somewhat larger (Chen and Kostić, 1988).

Several pieces of evidence pointed at His 33 as the binding site. Whereas the ^1H NMR signals of the acetate bridges and of the many detectable amino-acid residues were virtually unperturbed, the signal of C^2-H in the imidazole ring of His 33 was shifted downfield upon cross-linking. The UV-visible difference spectrum $[Rh_2(OAc)_4(cyt)_2]$ minus (2 cyt) nearly matched that of the model complex $[Rh_2(OAc)_4(Im)_2]$; small discrepancies were explicable in terms of the differences between the chromophore environments in the diprotein complex and in the model complex. The diprotein complex was formed at pH 7.0 but not at pH 5.0, because the pK_a value of His 33 is 6.4. Finally, the tuna cytochrome *c*, which has the unreactive tryptophan instead of histidine in position 33, failed to yield a diprotein complex with $[Rh_2(OAc)_4]$ (Chen and Kostić, 1988).

The complex $[Rh_2(OAc)_4(cyt)_2]$ is stable in solution, but it was cleaved, and the native protein recovered, by incubation with 2-mercaptoethanol, a highly nucleophilic ligand. The diprotein complex proved more stable (in the thermodynamic sense) than expected on the basis of the equilibrium studies of $[Rh_2(OAc)_4L_2]$ model complexes with histidine and histidine-containing dipeptides as L (Dennis *et al.*, 1982). Even the nucleophiles such as cyanide and azide, present in excess, only partly displaced the protein molecules. The inertness of the diprotein complex may have two causes. First, the protein molecules may shield the $Rh_2(OAc)_4$ core from attack by other potential ligands. Second, the oxygen atoms constituting the RhO_4 faces may form hydrogen bonds to the protein. Simulation by molecular graphics revealed the possibility of such a bond to the side chain of Asn 22 (Chen and Kostić, 1988).

The enhanced stability of the diprotein complex shows how metal complexes with small ligands, such as amino acids, may be unrealistic models for the complexes with biological macromolecules, such as proteins. A greater structural complexity of the macromolecules and a variety of functional groups in them permit secondary interactions with the inorganic label or link that are impossible in amino-acid complexes.

The two types of reagents discussed here have already shown the benefits of the new, inorganic approach to cross-linking of proteins. The $[PtL_2Cl_2]$ reagents are selective toward methionine (Peerey and Kostić, 1987), while the $[Rh_2(RCOO)_4]$ reagents are selective toward histidine (Chen and Kostić, 1988). This useful contrast is a consequence of the well-known affinities of platinum for sulfur and of rhodium for nitrogen. Those organic reagents that are selective—many are not—usually bind to amino or thiol groups, i.e., to lysine or cysteine residues. The different selectivity of the new inorganic reagents may prove advantageous in biochemical work.

Reaction of chloromercuryferrocene, FcHgCl, with baker's-yeast *iso*-1-cytochrome *c* yielded unexpected products (Lukes and Kostić, 1992). This electrophilic organomercurial did not simply bind to the nucleophilic thiolate group of Cys 102. Instead, a diprotein complex with the Fe : Hg ratio of 2 : 1 was formed; the evidence came from determination of molecular mass by gel-filtration chromatography and from metal assays by atomic-emission spectroscopy. The overall reaction, a disproportionation, is

$$2 \text{ FcHgCl} + 2 \text{ cyt–SH} \rightarrow \text{cyt–S–Hg–S–cyt} + \text{FcHgFc} + 2 \text{ HCl.} \qquad (18.1)$$

Cysteine reacts with FcHgCl in a completely analogous way (Zhu *et al.*, 1988).

Several palladium(II) aqua complexes, among them $[Pd(H_2O)_4]^{2+}$ and *cis*-$[Pd(en)(H_2O)_2]^{2+}$, selectively promote hydrolysis of the amide bond His 18–Thr 19 in horse-heart cytochrome *c* (Zhu *et al.*, 1994). This is the first example of selective hydrolytic cleavage of a protein effected by a metal complex directly attached to the protein.

IX. Conclusions and Prospects

Studies with cytochromes *c* showed advantages of transition-metal complexes over conventional organic reagents for covalent modification and cross-linking of proteins. Complexes for specific applications and those with particular selectivity can be designed on principles of inorganic reactivity and stereochemistry (Kostić, 1988). Almost all the transition-metal reagents studied so far are coordination complexes, and the potential of organometallic complexes has yet to be recognized.

Because transition-metal ions are Lewis acids, they promote hydrolytic cleavage of peptide bonds (Burgeson and Kostić, 1991; Zhu and Kostić, 1992, 1993, 1994; Parac and Kostić, in press) and even catalyze this important reaction. Because many of these ions are redox-active, they promote oxidative or other cleavage of proteins (Rana and Meares, 1990, 1991; Hoyer *et al.*, 1990; Schepartz and Cuenoud, 1990; Kim *et al.*, 1985). Because many transition metals have NMR-active isotopes and because the chemical shifts depend greatly on the ligands and on the more distant environment, these metals may prove useful as NMR tags for proteins. For instance, NMR receptivity of ^{195}Pt is 19 times greater than that of ^{13}C, and the studies of amino-acid (Gummin *et al.*, 1986; Galbraith *et al.*, 1987) and peptide (Burgeson and Kostić, 1991) complexes of platinum(II) bode well for success with cytochrome *c*.

X. Acknowledgments

Research on this subject in the author's laboratory has been funded by Iowa State University; by the U.S. Department of Energy, Division of Chemical Sciences, Office of Basic Energy Sciences, under contract W-7405-Eng-82; and by the National Science Foundation under a Presidential Young Investigator Award, CHE-8858387, and grant CHE-9404971. The work has ably been done by Herb M. Brothers II, Ingrid E. Burgeson, Jian Chen, John A. Galbraith, David D. Gummin, Joni S. Johnson, Amy J. Lukes, Kent A. Menzel, Michael S. Moxness, Tatyana N. Parac, Linda M. Peerey, Eva M. A. Ratilla, Brian K. Scott, Sonya C. Shah, Xia-Ying Zhou, and Longgen Zhu.

XI. References

Aldrichimica Acta (1988) *21*, 55.

Augustin, M. A., & Yandell, J. K. (1981) *Aust. J. Chem.* *34*, 91.

Blundell, T. L., & Johnson, L. N. (1976) in *Protein Crystallography*, Academic Press, New York, Chapter 8.

Boswell, A. P., Moore, G. R., Williams, R. J. P., Wallace, C. J. A., Boon, P. J., Nivard, R. J. F., & Tesser, G. I. (1981) *Biochem. J.* *193*, 493.

Boswell, A. P., Moore, G. R., & Williams, R. J. P. (1982) *Biochem. J.* *201*, 523.

Boyar, E. B., & Robinson, S. D. (1983) *Coord. Chem. Rev.* *50*, 109.

Brothers, H. M. II, & Kostić, N. M. (1988) *Inorg. Chem.* *27*, 1761.

Bryant, C., Strottmann, J. M., & Stellwagen, E. (1985) *Biochemistry* *24*, 3459.

Burgeson, I. E., & Kostić, N. M. (1991) *Inorg. Chem.* *30*, 4299.

Chen, J., & Kostić, N. M. (1988) *Inorg. Chem.* *27*, 2682.

Chottard, G., Michelon, M., Hervé, M., & Hervé, G. (1987) *Biochim. Biophys. Acta* *916*, 402.

Clark, J. M. (1964) *Experimental Biochemistry*, W. H. Freeman & Co., New York.

Dennis, A. M., Howard, R. A., & Bear, J. L. (1982) *Inorg. Chim. Acta* *66*, L31.

Dickerson, R. E., Kopka, M. L., Borders, C. L., Jr., Varnum, J., & Weinzierl, J. E. (1967) *J. Mol. Biol.* *29*, 77.

Dickerson, R. E.. Eisenberg, D.. Varnum, J., & Kopka, M. L. (1969) *J. Mol. Biol.* *45*, 77.

Galbraith, J. A., Menzel, K. A., Ratilla, E. M. A., & Kostić, N. M. (1987) *Inorg. Chem.* *26*, 2073.

Glazer, A. N. (1977) in *The Proteins* 3rd ed., Vol. 2 (Neurath, H., ed.), Academic Press, New York, Chapter 1 and references cited therein.

Grimes, C. J., Piszkiewicz, D., & Fleischer, E. B. (1974) *Proc. Natl. Acad. Sci. USA* *71*, 1408.

Gummin, D. D., Ratilla, E. M. A., & Kostić, N. M. (1986) *Inorg. Chem.* *25*, 2429.

Hemdan, E. S., Zhao, Y.-J., Sulkowski, E., & Porath, J. (1989) *Biochemistry* *86*, 1811.

Hildebrandt, P., & Stockburger, M. (1989) *Biochemistry* *28*, 6722.

Howe-Grant, M. E., & Lippard, S. J. (1980) *Metal Ions Biol. Syst.* *11*, 63.

Hoyer, D., Cho, H., & Schultz, P. G. (1990) *J. Am. Chem. Soc.* *112*, 3249.

Kim, K., Rhee, S. G., & Stadtman, E. R. (1985) *J. Biol. Chem.* *260*, 15394.

Kostić, N. M. (1988) *Comments Inorg. Chem.* *8*, 137.

Kowalsky, A. (1969) *J. Biol. Chem.* *244*, 6619.

Legg, J. I. (1978) *Coord. Chem. Rev.* *25*, 103.

Louie, G. V., Hutcheon, W. L. B., & Brayer, G. D. (1988) *J. Mol. Biol.* *199*, 295.

Lukes, A. J., & Kostić, N. M. (1992) *J. Inorg. Biochem.* *46*, 77.

Lundblad, R. L., & Noyes, C. M. (1984) *Chemical Reagents for Protein Modification*, Vols. I and II, CRC Press, Boca Raton, Florida.

Means, G. E., & Feeney, R. E. (1971) *Chemical Modification of Proteins*, Holden-Day, San Francisco.

Meares, C. F., & Wensel, T. G. (1984) *Acc. Chem. Res.* *17*, 202.

Mehrota, R. (1987) in *Comprehensive Coordination Chemistry* (Wilkinson, G., Gillard, R. D., & McCleverty, J. A., eds.), p. 269, Pergamon Press, New York.

Motonaga, K., Misaka, E., Nakajima, E., Veda, S., & Nakanishi, K. (1965) *J. Biochem.* **57**, 22.

Mureinik, R. J., & Bidani, M. (1978) *Inorg. Chim. Acta* **29**, 37.

Parac, T. N., & Kostić, N. M. *J. Am. Chem. Soc.*, in press.

Patthy, L., & Thesz, J. (1980) *Eur. J. Biochem.* **105**, 387.

Pecoraro, V. L., Rawlings, J., & Cleland, W. W. (1984) *Biochemistry* **23**, 153 and references cited therein.

Peerey, L. M., & Kostić, N. M. (1987) *Inorg. Chem.* **26**, 2079.

Petsko, G. A. (1985) *Methods Enzymol.* **114**, 147.

Petsko, G. A., Phillips, D. C., Williams, R. J. P., & Wilson, I. A. (1978) *J. Mol. Biol.* **120**, 345.

Pullman, B., & Goldblum, N., eds. (1977) *Metal–Ligand Interaction in Organic Chemistry and Biochemistry*, D. Reidel, Boston.

Ramdas, L., Sherman, F., & Nall, B. T. (1986) *Biochemistry* **25**, 6952.

Rana, T. M., & Meares, C. F. (1990) *J. Am. Chem. Soc.* **112**, 2457.

Rana, T. M., & Meares, C. F. (1991) *J. Am. Chem. Soc.* **113**, 1859.

Ratilla, E. M. A., & Kostić, N. M. (1988) *J. Am. Chem. Soc.* **110**, 4427.

Ratilla, E. M. A., Brothers, H. M. II, & Kostić, N. M. (1987) *J. Am. Chem. Soc.* **109**, 4592.

Ratilla, E. M. A., Scott, B. K., Moxness, M. S., & Kostić, N. M. (1990) *Inorg. Chem.* **29**, 918.

Recchia, J., Matthews, C. R., Rhee, M.-J., & Horrocks, W. DeW. (1982) *Biochim. Biophys. Acta* **702**, 105.

Schepartz, A., & Cuenoud, B. (1990) *J. Am. Chem. Soc.* **112**, 3247.

Shah, S. C., Johnson, J. S., Brothers, H. M. II, & Kostić, N. M. (1988) *J. Serb. Chem. Soc.* **53**, 139.

Stryer, L. (1988) *Biochemistry*, 3rd ed., W. H. Freeman & Co, San Francisco.

Takano, T., Kalai, O. B., Swanson, R., & Dickerson, R. E. (1973) *J. Biol Chem.* **248**, 5234.

Werber, M. M., & Lanir, A. (1981) *Isr. J. Chem.* **21**, 34 and references cited therein.

Zhou, X.-Y., & Kostić, N. M. (1988) *Inorg. Chem.* **27**, 4402.

Zhou, X.-Y., & Kostić, N. M. (1990) *Polyhedron* **9**, 1975.

Zhu, L., Daniels, L. M., Peerey, L. M., & Kostić, N. M. (1988) *Acta Cryst., Sect. C* **44**, 1727.

Zhu, L., & Kostić, N. M. (1992) *Inorg. Chem.* **31**, 3994.

Zhu, L., & Kostić, N. M. (1993) *J. Am. Chem. Soc.* **115**, 4566.

Zhu, L., & Kostić, N. M. (1994) *Inorg. Chim. Acta* **217**, 21.

Zhu, L., Qin, L., Parac, T. N., & Kostić, N. M. (1994) *J. Am. Chem. Soc.* **116**, 5218.

Zubay, G. (1988) *Biochemistry*, 2nd ed., Macmillan, New York.

19

The Alkaline Transition in Ferricytochrome *c*

MICHAEL T. WILSON

Department of Chemistry and Biological Chemistry
University of Essex

COLIN GREENWOOD

School of Biological Sciences
University of East Anglia

I. Introduction

Some 50 years ago Theorell and Åkesson (1941) investigated the effects of pH on ferricytochrome *c* and ferrocytochrome *c*. They showed that, over the pH range 0–14, the oxidized protein displays five pH-dependent spectral states linked by protonation events corresponding to four distinct pK values. The reduced protein has three states linked by two deprotonations.

These findings can be summarized as follows:

$$\text{Ferricytochrome } c \quad \text{I} \rightarrow \text{II} \rightarrow \text{III} \rightarrow \text{IV} \rightarrow \text{V} \quad \text{State}$$
$$0.4 \quad 2.5 \quad 9.35 \quad 12.8 \qquad \text{p}K$$

$$\text{Ferrocytochrome } c \quad \text{I} \rightarrow \text{II} \rightarrow \text{III} \quad \text{State}$$
$$<4 \quad >12 \qquad \text{p}K$$

The transition from state III to IV, the subject of this review and termed the *alkaline transition*, was noted by Theorell and Åkesson (1941) to be associated with the bleaching of an absorption band centered at 695 nm having a low extinction coefficient. This band and the associated alkaline transition have been universally found in both prokaryote and eukaryote ferricytochromes *c*. Figure 19.1, adapted from Greenwood and Wilson (1971), illustrates the alkaline transition by showing the variation of the 695-nm absorption as a function of pH at high ionic strength for horse cytochrome *c*. These experimental data have been simulated with a model derived from a set of simple titrations:

$$AH_3 \overset{K_1}{\rightleftharpoons} AH_2^- \overset{K_2}{\rightleftharpoons} AH^{2-} \overset{K_3}{\rightleftharpoons} A^{3-},$$

where A represents ferricytochrome *c* and K_1, K_2 and K_3 are the equilibrium constants governing the distribution of ferricytochrome *c* between the ionic forms. The pK value of the third ionization, that of the alkaline transition, calculated from Figure 19.1, was determined to be 8.9 ± 0.06. The value of the pK for the alkaline transition in horse cytochrome *c* has been determined many times with results ranging from 8.9 to 9.5 depending on solvent conditions. Each eukaryote cytochrome *c* may be characterized by the value of this pK (Osheroff *et al.*, 1980),

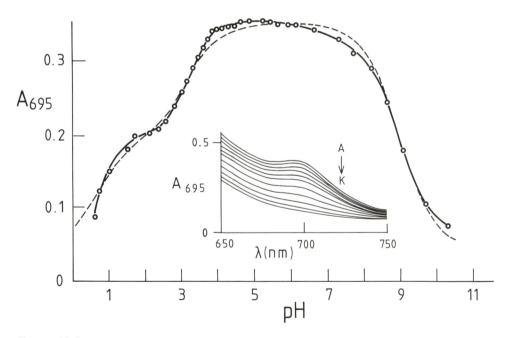

Figure 19.1

The pH dependence of the absorbance at 695 nm for horse ferricytochrome *c*. The experimental points are denoted (o). A simulation using pK values 0.75, 3.18, and 8.9 is also shown (———). The inset shows a set of spectra recorded between pH 6.5 and pH 10.5 (A → K). The temperature was 25 °C and ionic strength 0.3 M (NaCl) (adapted from Greenwood and Wilson, 1971).

for example, yeast iso-1-cytochrome c has a pK ranging from 8.4 to 8.6 again depending on conditions (see Pettigrew and Moore, 1987; Moore and Pettigrew, 1990; and Pearce *et al.*, 1989). The alkaline transition occurs with much higher pK values in bacterial cytochromes c which are structurally distinct from the eukaryote proteins, e.g., in *P. aeruginosa* cytochrome c_{551} the midpoint of the transition is around pH 11. In addition to elevated pH, the bleaching of the 695-nm absorption band, and the associated molecular reorganization, can be induced by high temperature and exposure of the protein at neutral pH to the action of protein denaturants and other chaotropic agents (Schejter and George, 1964; Greenwood and Wilson, 1971).

In 1971 the first structure of a mitochondrial cytochrome c (horse) became available (Dickerson *et al.*, 1971), confirming the suggestion (Schechter and Saludjian, 1967; Aviram and Schejter, 1969) that methionine occupies the sixth coordination position of the central iron atom. Proposals for the loss of the 695-nm band, framed in terms of the distortion or disruption of the bond between the iron atom and the sulfur of methionine thus became widely accepted (see, for example, Stellwagen and Cass, 1974). As we see later, spectroscopic techniques have now confirmed that the loss of the 695-nm band in the state III → IV transition does in fact coincide with the displacement of the methionine ligand. This conclusion together with the observation that all cytochromes having the 695-nm absorbance band contain methionine as the sixth ligand has led to the view that the 695-nm band is diagnostic of methionine coordination (see Pettigrew and Moore, 1987; Moore and Pettigrew, 1990, for extensive reviews).

There are a number of possibilities for the origin of this absorption band. The most widely accepted view is that the band arises from a porphyrin → iron(III) transition (Makinen and Churg, 1983; Smith and Williams, 1970). The band was found to be z-polarized by Eaton and Hochstrasser (1967) who thus proposed that it arose from a porphyrin π → iron d_{z^2} transition. In contrast, and more recently, however, Makinen and Churg (1983) have argued that the transition is from a porphyrin π → iron d_{xz} and d_{yz} orbitals, the energy of which is determined by the interaction of the heme with the methionine ligand.

In eukaryote cytochromes c, the 695-nm absorption band has a well-defined associated circular dichroic spectrum which is abolished at high pH values (Greenwood and Wilson, 1971). Senn *et al.* (1985) and Senn and Wüthrich (1985) have shown that the magnitude and sign of this dichroism differs among cytochromes c from different species and have correlated this, through ^{1}H NMR spectroscopy, to different orientations in which methionine may bind to the central iron atom (see Figure 19.2). The structures in Figure 19.2 depict the mode of methionine binding in tuna and *Pseudomonas aeruginosa* cytochrome c_{551} and indicate that a lone-pair orbital of the sulfur atom is directed towards different pyrrole rings in the two structures. The relative energies of the molecular orbitals involving the iron d_{xz} and d_{yz} orbitals are thus expected to differ.

The nature of the alkaline transition, that is the displacement of the intrinsic methionine ligand and rearrangement of the molecule, has been the focus of considerable interest since it was first recognized by Theorell and Åkesson (1941). This interest stems from the fact that although it is superficially simple and occurs in a

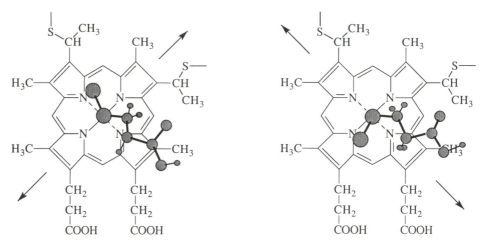

Figure 19.2
Relationship between the heme group and the intrinsic methionine ligand to the central iron atom in a eukaryote and prokaryote cytochrome *c*. Perspective drawing of tuna cytochrome *c* (left) and *Pseudomonas* cytochrome c_{551} (right). The arrow denotes the major direction of the unpaired electron density (adapted from Senn and Wüthrich, 1985).

relatively small protein, the structure of which is known in considerable detail (see Pettigrew and Moore, 1987, and Mathews, 1985, for extensive reviews), its precise mechanism remains obscure. Thus, it has provided both an illuminating example of a triggered, limited, conformational change in a heme protein and also a useful problem on which to apply the newer spectroscopic techniques now available and the technique of site-directed mutagenesis. The transition is coupled to a large redox potential change and thus, in eukaryote cytochromes *c* at least, could in principle have a functional role in the control of the electron transfer pathway and hence in energy transduction. This possibility has also helped maintain a lively interest in this area of cytochrome *c* biochemistry.

In this review we concentrate on specific features of the mechanism of the alkaline transition, such as the nature of the sixth ligand at high pH and of the ionizing species which "triggers" the transition. The wealth of informative data which has come from the comparative biochemistry we only touch lightly upon and is covered elsewhere in this volume.

II. Spectroscopic Fingerprint of the Alkaline Transition

Ferricytochrome *c* is a low-spin hexacoordinate species at both neutral and alkaline pH. The forms of the protein found under these pH conditions have been characterized by a variety of spectroscopic techniques which provide insight into the immediate environment of the iron in the two species and the nature of the transition which connects them.

From experiments in which the pH dependence of the EPR spectrum of ferri-cytochrome *c* was examined (Lambeth *et al.*, 1973), it became clear that, although the low-spin character of ferricytochrome *c* is maintained at high pH, the nature of the low-spin species is clearly different in the two states (see Table 19.1). The signal at $g = 3.07$, characteristic of ferricytochrome *c* at neutral pH values, is replaced at high pH, a majority species appearing with $g = 3.42$. The loss of the $g = 3.07$ species on increasing pH follows exactly the bleaching of the 695-nm band observed under identical conditions (see Figure 19.3). MCD spectra recorded in the near infrared spectral region (Gadsby *et al.*, 1987) similarly show pH dependence, a band at 1725 nm seen at pH 7 being replaced with another at 1465 nm, which is maxi-mally formed at pH 11.0. The pK of this transition is identical to that observed by either UV/visible or EPR spectroscopy (see Figure 19.3). These data are consistent with the view that the iron–methionine bond is broken at high pH. Proof that this ligand dissociation does indeed occur has come from two magnetic spectroscopy techniques. Firstly, Wooten *et al.* (1981), studying cytochrome *c* in which Met 80 of horse cytochrome *c* was enriched with CD_3 or $^{13}CH_3$, showed conclusively that at high pH values methionine was displaced from ligation with the paramagnetic iron center. This transition exhibited a pK_a of approximately 9.0. Secondly, near-infrared MCD has proved to be a powerful diagnostic tool for assigning heme ligands, and Thomson and his collaborators have established spectroscopic signatures for a variety of heme axial ligand sets and have further correlated these to their EPR

Table 19.1
Spectroscopic signals from the native and alkaline forms of cytochrome *c*.

Sample and conditions	Spectroscopic characteristics				Comments
	Optical	E.P.R.	M.C.D.	Coordination	
Horse heart *c* at neutral pH	695 +ve	3.07 2.23 1.21[a] 3.06 2.24 1.24[b] 3.05 2.25 1.26[c]	1725 nm	His–Met	Single low-spin species
Horse heart *c* pH 10.8	695 −ve	3.42 2.13 1.9[c]	1465 nm	His–amine	Major alkaline low-spin form, other low-spin species present
Horse heart *c* <pH 11.0	695 −ve	3.6[a,c]			Low-spin species
Leghemoglobin +*n*-butylamine pH 10.4	695 −ve	3.38 2.05[c]	1550 nm	His–amine	

[a] Lambeth *et al.* (1973).
[b] Salmeen and Palmer (1968).
[c] Gadsby *et al.* (1987).

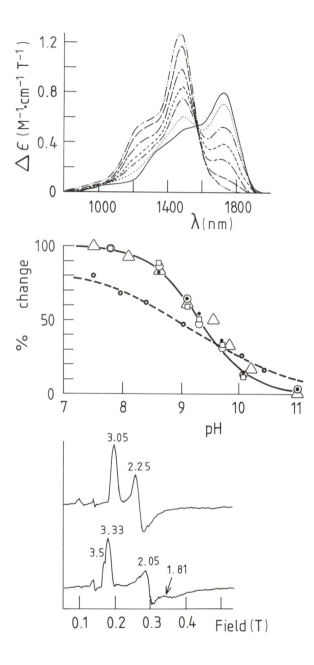

Figure 19.3

The alkaline transition in horse ferricytochrome *c* monitored by visible, MCD, EPR and NMR spectroscopies. *Top panel*: Near IR MCD spectra of horse ferricytochrome *c* as a function of pH and at 20 °C. The pH was varied from p^2H 7.8 (——) to p^2H 11.0 (——··——). Data taken from Gadsby *et al.* (1987). *Middle panel*: pH dependence of spectra signatures; EPR (△); MCD at 1465 nm (●), 1725 nm (○) (Gadsby *et al.*, 1987); optical, 695 nm (□) (Wooten *et al.*, 1981); ^{13}C (○) (Wooten *et al.*, 1981). The solid line is a single (*n* = 1) titration with p*K* = 9.3. *Bottom panel*: EPR spectra taken at p^2H 6.6 (50 mM phosphate), upper, and at p^2H 11.0 (50 mM Caps), lower (Gadsby *et al.*, 1987).

signatures. The species populated at neutral pH, having a band at 1725 nm, can be assigned unambiguously to histidine–methionine ligation. This is clearly consistent with the x-ray crystallographic structure. The loss of the 1725-nm band at high pH constitutes unequivocal evidence of the dissociation of methionine 80 from the central iron atom (Gadsby *et al.*, 1987).

In view of the fact that cytochrome *c* retains its low-spin character at high pH, one must necessarily conclude that the lost methionine is replaced by a new, strong-field ligand. The bulk of the spectroscopic evidence suggests that this incoming ligand is a lysine amino group. The new MCD band at 1425 nm observed at high pH values is within the wavelength envelope expected for histidine–amine ligation— for example, the *n*-butylamine complex of leghemoglobin exhibits a band centered at 1550 nm (Gadsby *et al.*, 1987). The EPR data for the alkaline form, shown in Table 19.1, is similarly consistent with histidine–amine ligation.

In a more recent study, Hong and Dixon (1989), using 2-D NMR to investigate the nature of the axial ligand of the alkaline form, report chemical shifts which are consistent with there being a lysine residue coordinated to the ferric iron. These authors also consider the possibility of ligation of an arginine residue, but believe this to be unlikely.

Although the transition from methionine to lysine seems to follow a simple titration curve (see Figure 19.3) the situation is more complex than it would seem at first sight since both EPR and NMR reveal that the alkaline form is in fact not a single species. The presence in the EPR spectra of components with *g* values of 3.17 and 3.6 in addition to the major species having a *g* value of 3.42 confirm the mixed nature of the alkaline form (see Table 19.1). Similarly, 2-D NMR (Hong and Dixon, 1989) has also revealed at least two coexistent basic forms at pH values above pH 9.5. Further complex pH-dependent behavior in the NMR experiment is suggested by the rather broad titration reported by Wooten *et al.* (1981), which seems to deviate from the simple ($n = 1$) protonation–deprotonation transition which is observed through the optical, magneto-optical, and EPR spectroscopies reported in Figure 19.3.

III. What Is the Likely Replacement Ligand?

Examination of the three-dimensional structure of tuna ferricytochrome *c* (Takano and Dickerson, 1981a,b) indicates a number of possible candidates for the residue which replaces Met 80 at high pH. Since it is known that the conversion of the protein from the "neutral" to the "alkaline" form does not appear to involve any gross structural reorganization (Greenwood and Wilson, 1971), one is forced to look within a restricted sector of the ferricytochrome *c* molecules for putative ligands. The top left-hand third of the structure shown in Figure 19.4 would seem to be the most likely part of the polypeptide chain to provide the incoming ligand. Within this region one can identify a number of possibilities: lysine residues 72, 73, 79, 86, 87, and 88, and arginine 91.

In spite of much work from many laboratories, the exact identity of the sixth ligand at high pH remains obscure. For example, Bosshard (1981), on the basis

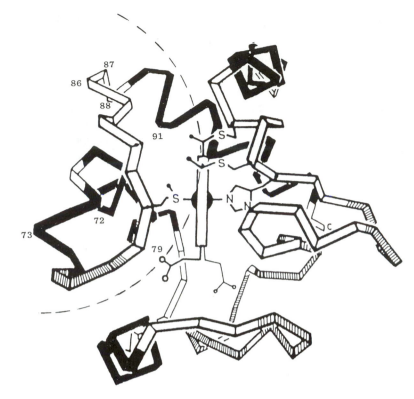

Figure 19.4
Schematic representation of the structure of tuna cytochrome c. The broken circle
encompasses that part of the molecule considered most likely to donate the replacement
ligand for Met 80 at high pH. Arg 91 is numbered, as are the lysine residues (adapted from
Takano and Dickerson, 1981a).

of an argument involving rates of acetylation of lysine on the surface of ferricytochrome c concluded that lysine 79 does not replace Met 80; indeed, this author
propounds the idea that *none* of the available lysines can be the ligand. Similar
conclusions have been drawn by Wallace and Corthésy (1987) using alkylamine derivatives of cytochrome c. This position is, however, difficult to reconcile with available MCD data and recent NMR results and has been countered by Wallace (1984),
who reports evidence, obtained using acetimidylated cytochrome c, that probably
lysine 72 or 79 can act as replacement ligands.

The problem of ligand assignment is probably made difficult because the alkaline form is not a single species, but comprises a set of proteins (at least two) in
conformational equilibrium that may each have a different lysine coordinated to the
central iron atom. Experiments to determine if a given lysine residue is the unique
ligand in the alkaline form will tend to return a negative answer, as modification of
such a lysine residue, rendering it incapable of binding to iron, will perturb the

conformational equilibrium in favor of other forms which, however, also have lysines as ligands.

IV. Initiation of the Alkaline Transition: The "Trigger"

A. The Trigger Hypothesis

While there appears to be a general consensus that lysine(s) replaces methionine at high pH, producing the alkaline forms (see NMR and MCD evidence discussed earlier), it is not thought that deprotonation of lysine itself is the primary event leading to methionine dissociation. The simple view that the amino group of a deprotonated lysine, having a high affinity for ferric iron, competitively displaces the intrinsic methionine ligand may be rejected on the basis of experiments which show that the alkaline transition takes place in protein species whose entire complement of lysine residues have been chemically modified by trifluoroacetylation or guanidination. In these modifications, unlike the native protein, ferricytochrome *c* is rendered high-spin at alkaline pH, but the observed pK_a values, monitored via the 695-nm band, are 9.9 and 8.8, respectively (Stellwagen *et al.*, 1975). The closeness of these pK_a values to that of the native species suggests that the trigger remains intact, even though the "incoming" lysine is modified such that it cannot act as a strong-field ligand for ferric iron and thus compete with the methionine.

A detailed analysis of the kinetics of the alkaline transition, which we address in a later section, has also provided evidence that an ionizable group acts as a trigger for the transition and, once deprotonated, leads to the displacement of the intrinsic ligand methionine 80 from coordination to the heme iron. Furthermore, the pK_a of this deprotonation is close to pH 11.0 (see later discussion). Any proposal concerning the identity of this "trigger" must necessarily conform with this pK value. A number of candidates have been suggested and a range of experimental evidence advanced in support of each.

B. The Nature of the Trigger

The possible candidates are tyrosine, arginine, the proximal ligand histidine 18, heme propionate HP6, or a buried water molecule.

C. Tyrosine or Arginine as the Trigger

The suggestion that tyrosine 67 is possibly the ionizing group which controls the alkaline isomerization may be discounted since replacement of this residue by phenylalanine, while increasing the pK by 1.1 pH units, does not abolish the transition (Luntz *et al.*, 1989). Furthermore, in rat cytochrome *c*, the double mutant in which tyrosine 67 is replaced by phenylalanine and proline 30 by alanine, the pK of the alkaline transition reverts to 9.5, almost that of the wild-type protein ($pK = 9.6$) (Schejter *et al.*, 1992). The backbone carbonyl of proline 30 is hydrogen-bonded to the imino nitrogen of histidine 18, the axial heme ligand. Thus, the two substitutions are on opposite sides of the heme and distant from each other. Although the mechanisms through which these substitutions exert their effects are open to discussion (Schejter *et al.*, 1992), these results seem to exclude tyrosine 67 as the trigger, but rather suggest that it is other elements of the transition (see later discussion) which are modified. Frauenhoff and Scott (1992) have argued that Tyr 67 may form a hydrogen bond with the lysine residue which fills the sixth coordination position of the iron at high pH, thus stabilizing the alkaline form of the protein. This hydrogen bond cannot form when Phe is substituted at position 67, hence destabilizing the alkaline form and raising the pK of the transition. For this argument to be sustained, however, it must also be concluded that the Tyr 67–Met 80 hydrogen bond (Figure 19.5) does not stabilize the native heme pocket structure. Tyr 67 is a highly conserved residue, and it is possible that it is important in limiting access into the heme pocket of a second water molecule and thus maintaining the high redox potential of cytochrome c. Substitution of Tyr 67 by Phe lowers the redox potential by 50 mV (Frauenhoff and Scott, 1992). Evidence to eliminate tyrosine as the ionizing group has also come from recent NMR experiments (Hong and Dixon, 1989).

Arginine appears to be excluded on the grounds of its very high pK value, but in the absence of firm experimental data this conclusion would seem to beg the question. A more telling argument is that modification of Arg 91 leaves the pK_a of the transition monitored at 695 nm unchanged.

D. Histidine as the Trigger

An interesting proposal for the identity of the trigger was made in 1987 by Gadsby *et al.*, who examined the nature of the alkaline transition by combined MCD and EPR methods. From the pH dependence of the EPR spectrum of the 1-methylimidazole complex of horse heart cytochrome c, the pK for the titration of the proximal histidine to histidinate was found to be 11.6, i.e., close to the kinetically determined pK of 11.0 for the trigger. This hypothesis has been incorporated into Scheme I, which accounts for the spectral characteristics of the alkaline form (see earlier section) and the single deprotonation event leading to the transition. In this scheme the deprotonation of the proximal histidine residue, yielding a histidinate group coordinated to the iron atom, initiates the dissociation of the *trans* ligand Met 80. This exposes the ferric iron to ligation by an incoming lysine residue, the pK of which is lowered by binding to the iron. The proton which thus dissociates from lysine returns the histidinate ligand to histidine. This sequence of events results in the substitution of the sulfur of methionine 80 by an amino group and the bleaching of the 695-nm absorption with an observed pK of 9.3.

E. Heme Propionate as the Trigger

A further alternative involving heme propionate 6 (HP6) has been advocated by Moore and his collaborators (Hartshorn and Moore, 1989; Tonge *et al.*, 1989). Experiments in which denaturation-induced proton uptake was monitored by ^{1}H NMR and through the fluorescence of tryptophan 59 have, under the assumption that the pK's of surface groups are unperturbed by denaturation, been interpreted by Hartshorn and Moore (1989) to indicate that one of the heme propionic

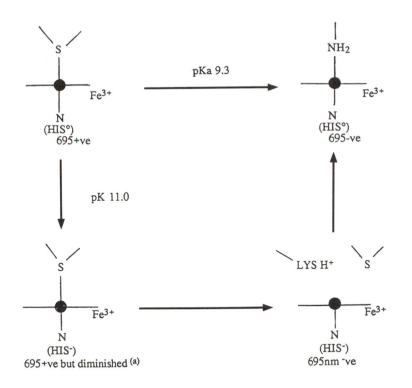

Scheme I
Adapted from Gadsby *et al.* (1987). (a) Low-spin intermediate possibly identified by Kihara *et al.* (1976).

acid substituents of ferricytochrome *c* has a pK_a of <4.5, whereas the other has a pK_a of >9.0. Using Fourier transform infrared (FTIR) spectroscopy, Tonge *et al.* (1989) identified an asymmetric carbonyl stretch at 1571 cm^{-1} indicative of carboxylate anion formation at alkaline pH. In cytochrome c_{551} this feature titrated with a pK of 6.14, in close agreement with the value of 6.2 found for HP7 by Leitch *et al.* (1984). In mitochondrial cytochrome *c* (tuna), a comparable spectral feature (1565 cm^{-1}) titrated with a pK of 9.35, i.e., close to that of the alkaline isomerizaion. In trifluoroacetylated tuna ferricytochrome *c*, in which the amino groups of all the lysines are modified, the pH-induced transition from low-spin to high-spin species exhibited a pK of 10.2, while the carboxylate formation identified by FTIR had a similar pK of 10.36. As a consequence of these similarities, Tonge *et al.* (1989) argued that the group which triggers the alkaline transition (i.e., methionine dissociation) is the deprotonation of a heme propionate. In the native ferricytochrome *c* molecule, this deprotonation should have a pK close to 11.0, if indeed it is the trigger. This propionate has tentatively been identified as HP6. In mitochondrial cytochromes *c*, this propionate is buried, probably in a hydrophobic environment, accounting for the abnormal pK value. As yet there is no confirmatory evidence to support the postulated abnormally high pK_a value of this propionate, and examination of the respective environments of HP6 and HP7 gives little insight as to why these two groups might possess such widely differing pK_a values. This is especially so since significant changes in the environment of HP7, induced by site-directed mutagenesis, e.g., Arg 38 to Ala (Cutler *et al.*, 1989) perturb the observed pK_a value of this propionate only marginally. However, if the peptide bond either before or after Arg 38 is broken, thus disrupting the structure in the vicinity of the propionate groups, then this lowers the pK of the alkaline transition markedly (e.g., cyt *c* 1–38 : 39–104, $pK = 7.1$, Wallace and Proudfoot, 1987; see also Chapter 21 by Wallace, this volume). Whether this is due to alteration in the pK of a propionate caused by increased exposure to solvent, or by a more general change in structure, is open to debate. We believe at present, therefore, that the status of HP6 as the trigger, while an intriguing possibility, remains unproven.

F. A Buried Water Molecule as the Trigger

Examination of the x-ray crystal structure of cytochrome *c* shows that, within the interior and relatively close to the iron–methionine bond, there is a water molecule. The site (see Figure 19.5) offers some possibility for hydrogen bonding, and there is thus the potential for structural changes on deprotonation of this water molecule (see also Chapter 3 by Brayer, this volume). Takano and Dickerson (1981) have suggested that it is the rupture of the hydrogen bond to threonine 78 which triggers the alkaline isomerization. Luntz *et al.* (1989) have supported the involvement of this buried water, which is stabilized by hydrogen bonds with tyrosine 67, threonine 78, and asparagine 52, since a mutant protein in which tyrosine 67 has been replaced by phenylalanine showed a raised pK of the alkaline isomerization. Further support comes from the fact that, where the water molecule is absent, as in *Pseudomonas* cytochrome c_{551}, the pK of the alkaline isomerization is raised to 11.0.

On reduction to ferrocytochrome *c*, the buried water moves approximately half an angstrom further from the iron and methionine sulfur bond (see Mathews, 1985, for review; also see Chapter 3 by Brayer, this volume). This movement may decrease the destabilizing effect on the S–Fe bond, resulting in a raised pK for the dissociation of methionine 80. However, as pointed out by Schejter and Plotkin (1988), the methionine sulfur is a particularly strong ligand for ferrous iron because of the soft π-acid nature of the methionine and the orbital overlaps. Thus, it is more likely to be for this reason that the ferrous form is more stable at alkaline pH.

G. The Trigger: Tentative Conclusions

In summary, we feel that the nature of the trigger remains open to question. However, the weight of evidence suggests that neither tyrosine 67 nor a lysine amine are candidates. Lack of firm experimental evidence makes it difficult to assess the role of arginine 91; however, its high pK_a value does appear to rule it out. The evidence that propionate HP6 triggers the transition rests on the observation that a carbonyl group in ferricytochrome *c* has an unusually high pK. In the absence of confirmatory experimental data and a plausible mechanism through which deprotonation

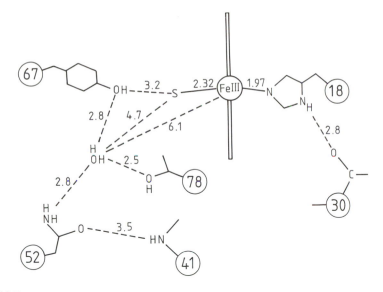

Figure 19.5
Structure of the heme pocket surrounding the buried water molecule in tuna ferricytochrome
c. The distances between the water and various neighboring atoms are given. Hydrogen
bonds are formed to Asn 52, Tyr 67, and Thr 78 (adapted from Takano and Dickerson,
1981b).

of a remote group could lead to the dissociation of methionine 80, this idea must
remain a hypothesis. Both deprotonation of buried water and the proximal histidine
have the merit of proximity to the susceptible bond. Of these two candidates, it may
seem superficially easier to propose a mechanism involving the histidine–histidinate
transition, as this is *trans* to the iron–methionine bond, and *trans* effects are well
known (e.g., binding of NO weakens the trans ferrous–histidine bond) (see, for
example, Traylor and Sharma, 1992).

However, there seems to be no straightforward way in which His⁻ would weaken
the *trans* met–ferric iron bond. If anything, coordination of His⁻ to ferric iron may
give the iron atom more ferrous character and thus bind methionine more tightly
(Schejter and Plotkin, 1988). Also, if there is a *trans* effect, why should this not also
occur in the same pH range in mitochondrial cytochrome *c* and in *P. aeruginosa*
cytochrome c_{551}, the pK for the alkaline transition of which is about 11? Finally,
Pearce *et al.* (1989) have argued that the proposal that the proximal histidine residue is the trigger does not fit comfortably with their results on Phe-82 mutants of
cytochrome *c* in which the pK of the trigger is depressed by two pH units (see later
discussion). This is considered to be an unlikely result if the trigger is the His →
His⁻ transition, mutation being expected to elevate the aberrant pK_a of the wild
type to a more normal value.

Therefore we conclude that, at present, the identity of the trigger is still to be
established, but the role of buried water seems to be attractive. In this context it is

worth noting that removal of water from the structure of cytochrome *c* by freeze-drying (Aviram and Schejter, 1972) leads to the loss of the 695-nm band, and thus presumably to the displacement of methionine from coordination. Rehydration of the protein returns it to its native state. Apart from the anomalous pK_a required of the water, possibly accounted for by the hydrophobic environment within the protein, it does appear to offer possibilities for a mechanism for the alkaline transition which are consistent with the data available. However, the details of any such mechanism have yet to be elaborated.

V. Thermodynamics of the Transition

From the spectroscopic evidence reviewed earlier, it is evident that the alkaline transition involves a ligand replacement of sulfur by nitrogen and as such may be represented in its simplest form as a two-state transition, linked to a single de-protonation event, as follows:

$$H^+ \cdot (S-Fe^{III}-N) \underset{K}{\rightleftharpoons} N-Fe^{III}-N + H^+. \tag{19.1}$$

These two species are the major forms identified in Scheme I, and while the mechanism of ligand exchange may, for example, involve low- and high-spin histidinate ion complexes, these are populated at extremely low concentrations at equilibrium.

The two species in Equation (19.1) have distinct spectroscopic signatures (see Table 19.1) which permit the thermodynamics of the transition to be investigated by monitoring the value of K as a function of temperature. An early example of such a study is that reported by Schejter and George (1964), in which they investigated the bleaching of the 695-nm band by protein denaturants and temperature. They found that the temperature-induced transition between a "cold" and a "hot" form was consistent with the simple formulation in Equation (19.1). Further investigation using kinetic and calorimetric methods confirmed the simple nature of the transition at equilibrium, and the standard thermodynamic parameters governing the events are reported in Table 19.2 for horse cytochrome *c*. All these methods give remarkably consistent values for these parameters. The somewhat unfavorable enthalpy term is consistent with the transition involving a deprotonation process, while the favorable entropy suggests a limited conformational change although the solvent contribution cannot easily be assessed. Osherhoff *et al.* (1980) have determined the values of the thermodynamic parameters for other eukaryote species, finding very similar results. These authors have monitored the temperature stability of a number of ferricytochromes *c* through the 695-nm absorbance band. These studies suggest that the heme crevice is stabilized against opening, and thus dissociation of methionine and loss of the 695-nm band, by a salt bridge and hydrogen bond at the top (Lys 13–Glu 90) and bottom (Lys 79 and backbone carbonyl of Thr or Ser 47) of the crevice, respectively.

The apparent simplicity indicated by Equation (19.1) and supported by optical and calorimetric evidence is disturbed by studies of the transition using NMR methods. We have already noted in Figure 19.3 the complexity of the pH dependence of the transitions as monitored by NMR, and this is reflected in the derived

Table 19.2
Thermodynamic parameters governing the alkaline transition in horse heart ferricytochrome *c*.

Sample	Experimental method	Thermodynamic parameter			Temp.	Ref.[a]
		ΔH_0	ΔS_0	ΔG_0		
Horse heart	NMR	6.7 ± 0.6 kcal mol^{-1}	21.9 ± 1 e.u.	0.14 kcal mol^{-1}	25°	(a)
Horse heart	Optical spec.	14.6 kcal mol^{-1}	43 e.u.	1.7 kcal mol^{-1}	25°	(b)
Horse heart	Optical spec. Kinetics (pH 9)	13 kcal mol^{-1}	44.7 e.u.	-0.3 kcal mol^{-1}	25°	(c)
Horse heart	Optical spec.	15.4 kcal mol^{-1}	47.7 e.u.	1.2 kcal mol^{-1}	25°	(d)
Horse heart	Kinetic spec.	13 kcal mol^{-1}	37 e.u.	1.97 kcal mol^{-1}	21°	(e)
Horse heart	Flow calorimetry	16 kcal mol^{-1}	43 e.u.	3.1 kcal mol^{-1}	25°	(f)

[a] (a) Hong and Dixon (1989); (b) Schejter and George (1964); (c) Wilson and Greenwood (1971); (d) Osheroff *et al.* (1980); (e) Davis *et al.* (1974); (f) Watt and Sturtevant (1969).

thermodynamic parameters (see Table 19.2 and Hong and Dixon, 1989). A possible reason for the discrepancy between NMR and optical methods may be that these do not visualize exactly the same species (Hong and Dixon, 1989). For example, the multiple alkaline forms of ferricytochrome *c* are indistinguishable optically, whereas they have distinct NMR spectra. Hong and Dixon (1989) have shown that at high pH values and high temperature (>45°), at least two alkaline forms exist, the ratio of the concentrations of which are temperature-dependent. At still higher temperatures, further alkaline forms were detected. Similar conclusions have been reached by Davies (1989), who noted the lack of agreement between the pK_a values for the alkaline transition seen in a range of mutant ferricytochromes *c* monitored optically and by NMR spectroscopy.

VI. Kinetics of the Transition

A. pH-Jump Methods

The simple transition depicted by Equation (19.1) indicates a deprotonation associated with ligand exchange, and thus the pH-jump method offers a route to investigate the kinetics of the transition. This was initially exploited by Davis *et al.* (1974), who followed the changes in absorbance at 695 nm on rapidly changing the pH of the solution. The observed rates were found to depend strongly on pH increasing rapidly above the pK of the alkaline transition. Careful analysis of these results allowed these authors to state unambiguously that a deprotonation step precedes the observed spectral changes. This conclusion is entirely compatible with, and supports, the view that deprotonation of a group triggers Met 80 dissociation and hence ligand exchange. The pH-jump data allow the simple two-state transition represented in Equation (19.1) to be elaborated in the following way:

$$
\text{H}^+ \cdot (\text{S--Fe}^{\text{III}}\text{--N}) \underset{K_\text{H}}{\overset{-\text{H}^+}{\rightleftharpoons}} \text{S--Fe}^{\text{III}}\text{--N} \underset{k_\text{b}}{\overset{k_\text{f}}{\rightleftharpoons}} \text{N--Fe}^{\text{III}}\text{--N}, \tag{19.2}
$$

$$
\text{695 nm +ve} \qquad\qquad \text{695 nm +ve} \quad K_\text{C} \quad \text{695 nm -ve}
$$

where K_H and K_C denote the equilibrium constants for the two steps. At 25 °C the value of pK_H was found to be 11.0 ± 0.1 and the conformational equilibrium constant K_C is 125 ± 36. Thus, the overall pK (given by the log of the reciprocal of the product K_H and K_C) is close to the observed pK value for the alkaline transition. The temperature dependence of K_c has allowed Davis *et al.* (1974) to determine the thermodynamic parameters of the conformational change, and these are reported in Table 19.2 where they are seen to agree closely with those found by other methods.

Analysis of the scheme in Equation (19.2) shows that the rate constant (k_obs) for the approach to equilibrium following the pH jump may be written as

$$
k_\text{obs} = k_\text{b} + k_\text{f} \frac{K_\text{H}}{K_\text{H} + [\text{H}^+]}. \tag{19.3}
$$

Values of k_b and k_f at 25 °C were found to be 0.049 s^{-1} and 6.1 ± 1.8 s^{-1}, respectively. These values are in agreement with those determined from similar pH-jump methods (Kihara *et al.*, 1976), from redox experiments (Wilson and Greenwood, 1971) and from ^{1}H NMR (Hong and Dixon, 1989) and ligand-replacement methods (Alayash and Wilson, 1979). The associated activation energies were found to be $E_\text{af} = 28$ kcal mol^{-1} and $E_\text{ab} = 16$ kcal mol^{-1} for the forward and back reactions (Davis *et al.*, 1974), respectively. Further insight into the mechanism and kinetics of the alkaline transition has emerged from pH-jump experiments using yeast iso-1-cytochromes *c* in which mutations have been engineered at position Phe 82 (Pearce *et al.*, 1989). Table 19.3 summarizes these kinetic parameters.

It is evident from this table that the overall pK falls on substituting Phe 82, and this drop is, in all cases, partly caused by a drop in pK_H of approximately 1.5 pH units. As pK_H is presumably the pK of the ionizing group that triggers the ligand exchange, Pearce *et al.* (1989) conclude that this group is unlikely to be the proximal histidine favored by Gadsby *et al.* (1987), arguing that mutation at position 82 is unlikely to lower further the abnormal pK of this group. Table 19.3 also indicates that where Phe 82 is substituted by another hydrophobic residue K_C remains largely unaltered, whereas substitution by glycine or serine lowers K_C by an order of magnitude. Caution is, however, required in making detailed interpretation of such data, since it is seen that making the substitution of Cys 102 → Thr 102 at a site remote from that part of the molecule which undergoes the conformational rearrangement nevertheless increases K_C by a factor of ~2.

Table 19.3

Kinetic and equilibrium constants for the alkaline transition of ferricytochromes *c* (see Scheme II).

Cytochrome *c* type	$k_b \times 10^2$ s^{-1}	k_f s^{-1}	K_C	pK_H	pK_H + pK_C	Ref.[a]
Horse	4.9	6.1 (1.8)	124 (36)	11 (1)	8.9 (4.3)	(a)
Horse	5.0	6.7	134	11.4	9.3	(b)
Horse	4.5	8	178		9.1	(c, d)
Horse	(pH-dependent)	4			9.5	(e)
Yeast Cys 102 wt	3.5 (1)	8.5 (3)	244 (11)	11.0 (1)	8.6 (2)	(f)
Thr 102	2.9 (4)	13.4 (7)	460 (40)	10.9 (1)	8.3 (3)	(f)
Gly 82/Thr 102	4.9 (4)	1.1 (2)	22 (3)	9.1 (3)	7.8 (6)	(f)
Ser 82/Thr 102	5.5 (5)	0.9 (2)	16 (5)	9.2 (7)	8 (1)	(f)
Ile 82/Thr 102	2.2 (2)	4 (1)	200 (40)	9.4 (6)	7 (1)	(f)
Leu 82/Thr 102	3.3 (4)	7 (2)	210 (90)	9.6 (7)	7 (1)	(f)

[a] (a) Davis *et al.* (1974); (b) Kihara *et al.* (1976); (c) Wilson and Greenwood (1971); (d) Alayash and Wilson (1979); (e) Hong and Dixon (1989); (f) Pearce *et al.* (1989).

Similar pH-jump studies on proteins in which the p*K* of the alkaline transition has been lowered by cleavage of the Gly 37 to Arg 38 bond or the Arg 38 to Lys 39 bond (Wallace and Proudfoot, 1987) would also be illuminating, as in these proteins we do not know if the effect of bond breaking is on the p*K* of the trigger or the conformational equilibrium constant, or both.

Although Davis *et al.* (1974) and Pearce *et al.* (1989) have analyzed their data assuming a single alkaline species, it is now established (see earlier) that there are at least two forms present at high pH which may have different lysines coordinated to the central iron atom. The elaboration of the scheme represented in Equation (19.2) required to take account of these two forms is not readily amenable to analytical solution, but computer simulation (Wilson and Hogg, unpublished result) leaves the essential conclusions of Davis *et al.* (1974) unchanged, i.e., that the overall p*K* of the transition comprises the p*K* of the triggering group coupled to a conformational transition. However, one may expect computed values of K_C (now comprising at least two conformational equilibrium constants) to be modified in this model. For a full description, the model presented in Scheme II may also require account to be taken of the transient intermediate with low extinction at 695 nm identified by Kihara *et al.* (1976). This intermediate is tentatively identified in Scheme I and may correspond to the hexacoordinate intermediate that precedes

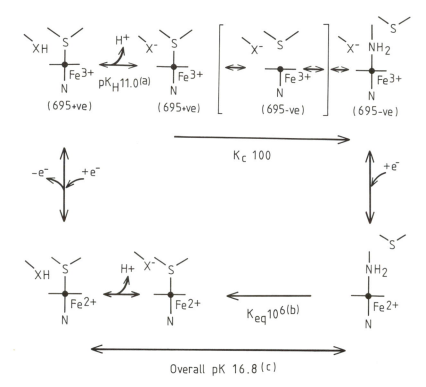

Scheme II

(a) Davis *et al.* (1974). (b) Schejter and Plotkin (1988). (c) Barker and Mauk (1992).

dissociation of methionine in Scheme II. Uno *et al.* (1984), using stopped-flow resonance Raman spectroscopy, similarly noted the presence of high- and low-spin intermediates in pH-jump experiments.

Pettigrew *et al.* (1975) studied two atypical mitochondrial cytochromes c, both characterized by having only a single thioether linkage between the heme group and the protein. Cytochrome c_{557} from *Crithidia oncopelti* conforms closely to the model proposed in Equation (19.2) with $K_C = 119$, $K_H = 2.45 \times 10^{-11}$ M, and the overall pK for the alkaline isomerization of 8.5. Cytochrome c_{558} from *Euglena gracilis*, while exhibiting an alkaline transition (p$K = 9.8$), does not obey the simple kinetic model and seems to require two deprotonation events prior to ligand exchange.

B. Oxido-Reduction Kinetics

We can summarize the pH dependence of ferricytochrome c by the upper half of Scheme II, in which deprotonation of a group designated X, e.g., the proximal histidine, a buried H_2O molecule, or HP6, leads to the dissociation of Met 80 to yield a high-spin, five-coordinate form lacking the 695-nm band. This intermediate accepts a deprotonated amino group as its sixth ligand to yield a low-spin species lacking the 695-nm band. We further conjecture that the incoming amino group may be provided by at least two different lysine residues, thus generating multiple forms at high pH.

The values of the equilibrium constants governing the steps ensure that at any given pH, ferricytochrome c comprises a mixture of the two low-spin species, i.e., the methionine-coordinated form with X protonated, and the amine-coordinated species. As the proportions of these two forms are pH-dependent, the mechanism of reduction of ferricytochrome c will similarly reflect this dependence. The kinetics of reduction of ferricytochrome c at alkaline pH values was first studied by Greenwood and Palmer (1965) and Wilson and Greenwood (1971). Figure 19.6a shows time courses for reduction of ferricytochrome c by ascorbate at two pH values and two temperatures. The biphasic character of the time course in which an ascorbate concentration–dependent rapid phase gives way to a slower, ascorbate-independent process has been interpreted in terms of Scheme II. Ascorbate reduces only the His–Met ligated form, and the slow step thus represents the reversion of the His–amine ligated species through the equilibria to the His–Met form, which is then rapidly reduced. In the scheme, the proportions of the progress curve, i.e., the fast or slow phases, accurately reflect the distribution of the major species. This conclusion is supported by the coincidence of the change of these proportions with pH and the change in the absorbance at 695 nm shown in Figure 19.6b. At pH 9.0 the rate constant for the slow phase (0.1 s^{-1}) is very close to that reported for k_{obs} by Davis et al. (1974) using pH-jump methods. The temperature dependence of the proportions yields the thermodynamic parameters reported in Table 19.2. These observations clearly established that the neutral and alkaline forms of ferricytochrome c are functionally distinct. This distinction probably reports the much lower redox potential of the amine–histidine ligated heme. With a powerful reductant such as dithionite, both species are readily and rapidly reduced, and experiments of this type have been carried out by Lambeth et al. (1973). Working at alkaline pH values, these authors detected a transient species that they identified as the ferrous alkaline form, which decays back to the His–Met ferrous form (lower half of Scheme II). The driving force for this decay is the exceedingly favorable equilibrium constant for the displacement of the amino group from c-type ferrous hemes by methionine (Schejter and Plotkin, 1988). The overall pK for the alkaline transition in the ferrous state is close to 17. The results are in general agreement with those obtained from experiments using the hydrated electron as the reducing agent (Land and Swallow, 1971; Pecht and Faraggi, 1972). Oxidation of ferrocytochrome c at pH 10.5, where ferrocytochrome c exists as the His–Met ligated form, rapidly yielded the corresponding ferric species having 695-nm absorbance, which subsequently decayed to the expected alkaline ferric form with a rate constant of 0.8 s^{-1}, consistent with the pH-jump experiments of Davis et al. (1974).

C. The Complete Redox/Conformational Cycle

The complete redox and conformational cycle represented in Scheme II has recently been investigated in yeast iso-1–cytochrome c and mutants by Barker and Mauk (1992) using cyclic voltammetry. These methods have the advantage over equilibrium techniques in that they permit the measurement of the true thermodynamic potential of individual species in systems involving interconverting species. At pH 7.1, independent of sweep rate, the protein behaved in a simple reversible redox manner with a midpoint potential of $+275$ mV. At pH 10.3, where the alkaline form predominates in the ferric state, the shape of the cyclic voltammogram is sweep-rate–dependent. At low sweep rates, no reduction of the native form was observed, but a reductive wave was seen at much lower potential (-230 mV). On reoxidation, no wave was seen at this low potential, but a wave was observed close to that found for the reoxidation of the native form of the protein. At a sweep rate of 2 V-s^{-1}, the response was essentially that of a single reversible wave with a mid-point potential of -230 mV. This behavior is consistent with the kinetic results obtained by rapid mixing experiments (Scheme II), although there are some minor differences in the rates measured by the different techniques. As expected from Scheme II, the amplitude of the wave centered at -230 mV, that associated with the alkaline form, increased with pH following a simple pH transition with a pK very close to that of the alkaline isomerization. This behavior was paralleled by the decrease in amplitude of the $+275$-mV wave. The midpoint potential of -230 mV for the alkaline form(s) of cytochrome *c* explains the functional differences revealed by ascorbate-mixing experiments (Figure 19.6). This very low redox potential is rather unusual, the bis-His coordinated form of horse cytochrome *c*, made by semi-

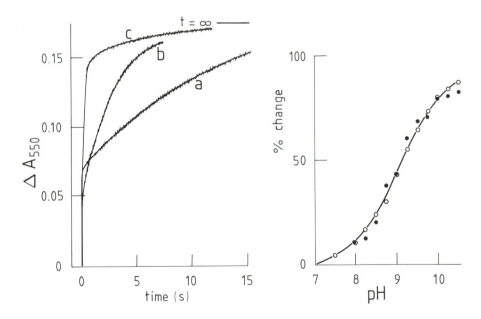

Figure 19.6

Kinetics of the reduction of horse ferricytochrome *c* by sodium ascorbate. *Left panel*: Progress curves for reduction of ferricytochrome *c* (4.3 μM) by 10 mM sodium ascorbate. (a) pH 9.3, 20 °C, (b) pH 9.3, 35 °C, (c) pH 8, 20 °C. *Right panel*: Comparison of the change in the proportion of the progress curve occurring in the slow phase (●) with the change in the absorbance at 695 nm (○).

synthetic methods, having a midpoint potential of 41 mV (Raphael and Gray, 1989); cytochrome *f* with His–amino ligation has a potential of ~360 mV (see Tanaka *et al.*, 1978; Gray, 1978). The low potential found for cytochrome *c* must therefore indicate stabilization of the ferric His–Lys species compared to the ferrous form. Barker and Mauk (1992) have suggested that this stabilization of the ferric state may occur because of the greatly increased exposure of the heme to solvent in the alkaline conformer.

VII. Some Modifications of Cytochrome *c* Which Render the Met–Iron Bond Labile: Models for the Alkaline Form of Ferricytochrome *c*

A. Cytochrome *c* Polymers

There are numerous chemical modifications of cytochrome *c* that perturb the integrity of the intrinsic Met–iron bond, some of these possibly in a fashion similar to that encountered in the alkaline form of the protein. For example, ferricytochrome *c* may, under mild denaturing conditions, form stable dimers and higher polymers (Margoliash and Lustgarten, 1962). Such dimers lack the 695-nm band and presumably the Met(80)–iron bond while remaining low-spin with the iron coordinated (probably) to a lysine. Unfortunately, we have little spectroscopic and structural information concerning these dimers which would be of interest, as this may throw light upon the nature of monomeric ferricytochrome *c* at alkaline pH values. Unlike the native protein, however, the dimers apparently do not revert to a form in which Met(80) binds to the iron when this becomes reduced. This is most simply revealed through the carbon monoxide binding kinetics of these ferrocytochrome *c* oligomers (Dupré *et al.*, 1974). While native ferrocytochrome *c* does not bind CO, the dimers and higher polymers do. The kinetics are complex, relatively slow ($k < 10^2$ $M^{-1}s^{-1}$), and heterogeneous, exhibiting a very slow CO-independent process, indicating that CO must replace a ligand, probably a Lys, before it can bind. The heterogeneity also indicates that the dimer probably comprises a family of related structures.

B. Carboxymethyl Cytochrome *c*

Carboxymethylation of methionine 80 (Schejter and George, 1965) removes this residue from coordination to the iron and abolishes the 695-nm band in the ferric protein, which may thus also act as a useful model for the alkaline form of the native protein. For example, at high pH values carboxymethyl cytochrome *c* (Cm cyt *c*) and the alkaline conformer of the native protein exhibit rather similar affinities for extrinsic ligands such as cyanide ($K_{ass} = 1.8 \times 10^4$ M^{-1}, 5.5×10^4 M^{-1}, respectively; Saleem and Wilson, 1988). At pH values around neutrality, the native protein has a lower affinity for extrinsic ligands, reflecting competition of the intrinsic ligand Met(80). In its reduced form, Cm cyt *c* undergoes a pH-dependent spin-state transition (pK = 7.2) from a pentacoordinate high-spin complex at pH 6 to a low-spin species at pH 9 (Brunori *et al.*, 1972). The high-spin form is similar in many respects to myoglobin, forming a stable oxygen adduct and combining with CO with a combination constant close to 10^6 $M^{-1}s^{-1}$ depending on conditions (Wilson *et al.*, 1973). The transition to the high-pH form is accomplished in a rapid process coupled to a deprotonation event, the pK of which is approximately 10, and has been assigned to the rapid binding of a lysine residue to the pentacoordinate iron. The process is extremely facile and is characterized by an intrinsic "on" rate of 15,200 s^{-1} and "off" rate of 25 s^{-1} (Brunori *et al.*, 1972). It is possible that this lysine is one of those that replaces methionine in the alkaline transition; as such, data relating to its identity would be valuable.

C. Nitration of Cytochrome c

Chemical modifications of tyrosine 67 are also of interest, as these lead to perturbation of the 695-nm band and conversion of the protein to an ascorbate irreducible form. Skov *et al.* (1969) have shown that nitration of Tyr 67 reduces the pK of the alkaline transition by approximately 3 pH units. This is an interesting observation and may be related to the disruption of the hydrogen-bonding network involving the buried water molecule, although other explanations involving the steric effects of the nitro group are equally convincing.

VIII. Functional Consequences of the Alkaline Transition

The structure of the left-hand side of ferricytochrome c, viewed as in Figure 19.3, is compact, places the sulfur of methionine 80 at the correct position for ligation to the central iron atom, and brings into contact with the heme a large set of amino acid residues (Figure 19.3). The interactions between the iron and the methionine ligand and the hydrophobic association of the residues with the heme and with each other stabilize the heme crevice and the whole structure. It is remarkable, therefore, that a simple deprotonation event can initiate a mechanism which replaces the methionine ligand by another, probably lysine, in a process that must require considerable restructuring and repacking of approximately one-third of the molecule (see also Chapter 4 by Nall, this volume). That this rearrangement is characterized by unexceptional values of the thermodynamic parameters governing the transition, and that the kinetics are simple and relatively rapid, is even more remarkable and lends credence to the view that this isomerization may not be a chance byproduct of the molecular architecture but is a design feature selected through evolution. However, it is not easy to ascribe an *in vivo* function to this transition.

Although the exact solution conditions between the inner and outer mitochondrial membrane may be debatable, there is little reason to suppose that *in vivo* the pH in this region is in the neighborhood of the pK of the alkaline transition, and thus the alkaline form would be present at insignificant concentration ($<5\%$, say) to play a functional role. Goldkorn and Schejter (1977) have suggested that two forms of cytochrome c, distinct in their redox behavior, exist in equilibrium at pH 7.2 at very low ionic strength. However, such conditions are also unlikely to occur in the intermembrane space.

A. Interactions with Cytochrome *c* Oxidase

An interesting proposal has been advanced by Weber *et al.* (1987), who noted that the spectrum of cytochrome *c* was altered when this protein formed a complex with its partner cytochrome *c* oxidase (see also Alleyne and Wilson, 1987, and Chapter 13 by Millett, this volume). The spectroscopic changes were tentatively interpreted to reflect the relative stabilization on binding to its oxidase of a form of cytochrome *c* in which the heme environment in the region of the Met 80 ligand was perturbed. While not equivalent to the alkaline form of the protein, this modification seemed to have features in common with it. However, subsequent experiments by Michel *et al.* (1989) have now indicated that cytochrome *c* oxidase enforces a transition from the more "open" heme crevice of the alkaline form to a state somewhat similar to the neutral conformational state. This, together with experiments in which the pK of the alkaline transition of cytochrome *c*, monitored through the 695-nm absorption band, is found to be relatively unchanged in complexes with cytochrome *c* oxidase (C. Blenkinsop and M. T. Wilson, unpublished results), reduces the functional importance of the alkaline transition *in vivo*.

From the functional view, therefore, the full alkaline transition must at present remain an intriguing curiosity. However, the puzzle of the alkaline isomerization will no doubt continue to attract attention. There are still the problems of the nature of the replacement ligand and of the "trigger" to be solved, and the conformational transition provides such a convenient system on which to test ideas of protein engineering. Finally, now that attention is being focused on the possibilities of bio-electronics, such a precise redox switch as the alkaline transition in ferricytochromes *c* will certainly contain lessons for us to learn.

IX. References

Alayash, A. I., & Wilson, M. T. (1979) *Biochem. J. 177*, 641–648.

Alleyne, T., & Wilson, M. T. (1987) in *Cytochrome Systems: Molecular Biology and Bioenergetics* (Papa, S., Chance, B., & Ernster, L., eds.), Plenum Press, New York and London, pp. 713–720.

Aviram, I., & Schejter, A. (1969) *J. Biol. Chem. 244*, 3773–3778.

Aviram, I., & Schejter, A. (1972) *Biopolymers 11*, 2141–2145.

Barker, P. D., & Mauk, A. G. (1992) *J. Amer. Chem. Soc. 114*, 3169–3624.

Bosshard, H. R. (1981) *J. Mol. Biol. 153*, 1125–1149.

Brunori, M., Wilson, M. T., & Antonini, E. (1972) *J. Biol. Chem. 247*, 6075–6081.

Cutler, R. L., Davies, A. M., Creighton, S., Warshel, A., Moore, G. R., Smith, M., & Mauk, G. (1989) *J. Am. Chem. Soc. 28*, 3188–3197.

Davies, A. (1989) Ph.D. Thesis, University of East Anglia, U.K.

Davis, L. A., Schejter, A., & Hess, G. P. (1974) *J. Biol. Chem. 249*, 2524–2632.

Dickerson, R. E. (1980) *Sci. Am. 242*, 99–110.

Dickerson, R. E., Takano, T., Eisenberg, D., Kallai, O. B., Samson, L., Cooper, A., & Margoliash, E. (1971) *J. Biol. Chem. 246*, 1511–1535.

Dupré, S., Brunori, M., Wilson, M. T., & Greenwood, C..(1974) *Biochem. J. 141*, 299–304.

Eaton and Hochstrasser (1967) *J. Chem. Phys. 46*, 2533–2537.

Frauenhoff, M. M., & Scott, R. A. (1992) *Proteins: Structure, Function and Genetics 14*, 202–212.

Gadsby, P. M. A., Peterson, J., Foote, N., Greenwood, C., & Thomson, A. J. (1987) *Biochim. J. 246*, 43–54.

Goldkorn, T., & Schejter, A. (1977) *FEBS Lett. 75*, 44–46.

Gray, T. C. (1978) *Eur. J. Biochem.* *82*, 133–141.

Greenwood, C., & Palmer, G. (1965) *J. Biol. Chem.* *240*, 3660–3663.

Greenwood, C., & Wilson, M. T. (1971) *Eur. J. Biochem.* *22*, 5–10.

Hartshorn, R. T., & Moore, G. R. (1989) *Biochem. J.* *258*, 595–598.

Hong, X., & Dixon, D. (1989) *FEBS Lett.* *246*, 105–108.

Kihara, H., Saigo, S., Nakatani, H., Haromi, K., Ikeda-Saito, M., & Iizuka, T. (1976) *Biochim. Biophys. Acta* *430*, 225–243.

Lambeth, D. O., Campbell, K. L., Zand, R., & Palmer, G. (1973) *J. Biol. Chem.* *248*, 8130–8136.

Land, E. J., & Swallow, A. J. (1971) *Arch. Biochem. & Biophys.* *145*, 365–372.

Leitch, F. A., Moore, G. R., & Pettigrew, G. W. (1984) *Biochemistry* *23*, 1831–1838.

Luntz, T. L., Schejter, A., Garber, E., & Margoliash, E. (1989) *Proc. Natl. Acad. Sci. USA* *86*, 3524–3528.

Makinen, M. W., & Churg, A. K. (1983) "Structural and analytical aspects of the electronic spectra of the proteins" in *Iron Porphyrin, Part 1*, (Lever, A. B. P., & Gray, H. B., eds.), Addison Wesley, London and Amsterdam, pp. 141–235.

Margoliash, E., & Lustgarten, J. (1962) *J. Biol. Chem.* *237*, 3397–3405.

Mathews, F. S. (1985) *Prog. Biophys. Molec. Biol.* *45*, 1–56.

Michel, B., Proudfoot, A. E. I., Wallace, C. J. A., & Bosshard, H. R. (1989) *Biochemistry* *28*, 456–462.

Moore, G. R., & Pettigrew, G. W. (1990) *Cytochromes c, Evolutionary, Structural and Physicochemical Aspects*, Springer-Verlag, Berlin.

Osheroff, N., Borden, D., Koppenol, W. H., & Margoliash, E. (1980) *J. Biol. Chem.* *255*, 1689–1697.

Pearce, L., Gärtner, A. L., & Mauk, A. G. (1989) *Biochem.* *28*, 3152—3156.

Pecht, I., & Faraggi, M. (1972) *Proc. Natl. Acad. Sci. USA* *69*, 902–906.

Pettigrew, G. W., & Moore, G. R. (1987) *Cytochromes c, Biological Aspects*, Springer-Verlag, Berlin.

Pettigrew, G. W., Aviram, I., & Schejter, A. (1975) *Biochem. J.* *149*, 155–167.

Raphael, A. L., & Gray, H. B. (1989) *Proteins: Struc., Funct., Genet.* *6*, 338–340.

Saleem, M. M. M., & Wilson, M. T. (1988) *Inorg. Chim. Acta* *153*, 93–98.

Salmeen, I., & Palmer, G. (1968) *J. Chem. Phys.* *48*, 2049–2052.

Schechter, E., & Saludjian, P. C. (1967) *Biopolymers* *5*, 788–790.

Schejter, A., & George, P. (1964) *Biochemistry* *3*, 1045–1049.

Schejter, A., & George, P. (1965) *Nature* *206*, 1150–1151.

Schejter, A., & Plotkin, B. (1988) *Biochem. J.* *255*, 353—356.

Schejter, A., Luntz, T. L., Koshy, T. I., & Margoliash, E. (1992) *Biochemistry* *31*, 8336–8343.

Senn, H., & Wüthrich, K. (1985) *Quart. Rev. Biophys.* *18*, 111–134.

Senn, H., Keller, R. M., & Wüthrich, K. (1985) *Biochem. Biophys. Res. Comm.* *92*, 1362–1369.

Skov, K., Hofmann, T., & Williams, G. R. (1969) *Canad. J. Biochem.* *47*, 750–752.

Smith, D. W., & Williams, R. J. P. (1970) *Structure and Bonding* *7*, 1–45.

Stellwagen, E., & Cass, R. (1974) *Biochem. Biophys. Res. Comm.* *60*, 371–375.

Stellwagen, E., Babul, J., & Wilgus, H. (1975) *Biochem. Biophys. Acta* *405*, 115–121.

Tanaka, K., Takahashi, M., & Asada, K. (1978) *J. Biol. Chem.* *253*, 7397–7403.

Takano, T., & Dickerson, R. E. (1981a) *J. Mol. Biol.* *153*, 79–94.

Takano, T., & Dickerson, R. E. (1981b) *J. Mol. Biol.* *153*, 95–115.

Theorell, H., & Åkesson, A. (1941) *J. Amer. Chem. Soc.* *63*, 1812–1818.

Tonge, P., Moore, G. R., & Wharton, C. W. (1989) *Biochem. J.* *258*, 599–605.

Traylor, T. G., & Sharma, V. S. (1992) *Biochemistry* *31*, 2847–2849.

Uno, T., Nishimura, Y., & Tsuboi, M. (1984) *Biochem.* *23*, 6802–6808.

Wallace, C. (1984) *Biochem. J.* *217*, 601–604.

Wallace, C., & Corthésy, B. E. (1987) *Eur. J. Biochem.* *170*, 293–298.

Wallace, C., & Proudfoot, A. E. I. (1987) *Biochem. J.* *245*, 773–779.

Watt, G. D., & Sturtevant, J. N. (1969) *Biochemistry* *8*, 4567–4571.

Weber, C., Michel, B., & Bosshard, H. R. (1987) *Proc. Natl. Acad. Sci.* *84*, 6687–6691.

Wilson, M., & Greenwood, C. (1971) *Eur. J. Biochem.* *22*, 11–18.

Wilson, M.,T., Brunori, M., Rotilio, G. C., & Antonini, E. (1973) *J. Biol. Chem.* *248*, 8162–8169.

Wooten, J. B., Cohen, J. S., Vig, I., & Schejter, A. (1981) *Biochemistry* *20*, 5394–5402.

20

The Hemepeptides from Cytochrome *c*: Preparation, Physical and Chemical Properties, and Their Use as Model Compounds for the Hemoproteins

PAUL A. ADAMS

MRC Biomembrane Research Unit
Department of Chemical Pathology
Medical School
University of Cape Town

DAVID A. BALDWIN

Materials Sciences Division
Council for Scientific and Industrial Research
Pretoria, South Africa

HELDER M. MARQUES

Centre for Molecular Design
Department of Chemistry
University of the Witwatersrand

I. Introduction

Hydrolysis of the cytochromes *c* by proteolytic enzymes or by chemical means was historically undertaken to help in the elucidation of the structure and amino acid sequence of the parent protein. Tsou (1949; 1951a,b) first reported treating cytochrome *c* with proteolytic enzymes such as pepsin, papain, trypsin, and chymotrypsin, and isolating a peptide with a molecular weight of about 2,500; this was subsequently shown (Tuppy and Paleus, 1955; Ehrenberg and Theorell, 1955) to be the heme-containing undecapeptide now generally known as microperoxidase-11 (MP-11) which retains residues 11–21 of the parent cytochrome *c* (Figure 20.1). A nonapeptide (residues 14–22) (Tuppy and Bodo, 1954), a decapeptide (residues 13–22) (Tuppy and Bodo, 1954) and, by tryptic digestion of MP-11, an octapeptide (residues 14–21) (Harbury and Loach, 1960a) were subsequently isolated. In all of these compounds, the thioether linkages between the side chains of Cys residues 14 and 17 and the porphyrin ring [as established by the early work of Theorell (1938, 1939) and Paul (1951)] are retained, as is the invariant proximal ligand, His-18 (Theorell and Åkeson, 1941; Margoliash, 1955).

In this review we shall concentrate on the small heme peptides from cytochrome *c*, i.e., those which retain ≤11 amino acid residues. The formation of noncovalent complexes between these and other larger hemepeptides, and nonhemepeptides from cytochrome *c*, are reviewed elsewhere in this book (see Chapter 21 by Wallace).

The small hemepeptides on which we shall focus are collectively termed the microperoxidases (abbreviated MP-) because some have been shown to exhibit peroxidase activity (see later discussion). Much of the current interest in these species centers on their use as structural probes for the porphyrin-binding pocket of hemoproteins and as chemical model compounds for the hemoproteins in general, in particular the hemoglobins (Hb's), myoglobins (Mb's), some of the peroxidases, and cytochrome a_3. Since these hemoproteins contain a single imidazole ring of a His residue as an axial ligand of the heme iron, the attraction of the microperoxidases as models for these hemoproteins is obvious (Figure 20.1).

Addition of imidazoles to iron-porphyrins generally results in the formation of the bis-imidazole species, usually without a detectable concentration of the mono-imidazole intermediate because of the driving force of the change in spin state of the ferric ion (Buchler, 1975; Mashiko *et al.*, 1978; Marques *et al.*, 1992).

Also, *de novo* synthesis of porphyrins with a single axial aromatic nitrogen ligand designed to model the preceding hemoproteins is onerous (Morgan and Dolphin, 1987). Many synthetic iron porphyrins are soluble only in organic solvents, and iron porphyrins often show a frustrating tendency to aggregate in solutions that are predominantly aqueous. However, conditions can usually be found with at least one of the MPs whereby one is able to circumvent some, if not all, these problems.

Here we examine the preparation and properties of the MPs, and indicate the type of information they offer as model systems and probes for structure–function relationships in the hemoproteins.

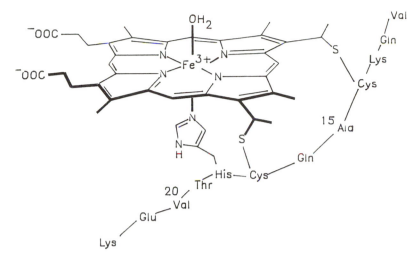

Figure 20.1
The structure of the microperoxidases, with the iron porphyrin covalently attached to the peptide chain via thioether linkages to two cysteine residues. The amino acid numbering is that of the parent protein, cytochome *c*. MP-11 contains residues 11–21; MP-9 contains 14–22; MP-8 contains 14–20; and MP-6 contains 14–19.

II. The Preparation and Purification of Hemepeptides

Short and medium-length hemepeptides derived from cytochrome *c* (cyt *c*) are normally prepared by digestion of Fe^{3+} cyt *c* or N-ε protected Fe^{3+} cyt *c* by appropriate proteolytic enzymes (Plattner *et al.*, 1975, 1977; Parr *et al.*, 1978; Schroeder and Yeager, 1978; Peterson *et al.*, 1980, 1983; Aron *et al.*, 1986; Baldwin *et al.*, 1987; Chersi *et al.*, 1987; Wallace and Proudfoot, 1987; Adams *et al.*, 1988, 1989a). However, the longer hemepeptides (≥ 65 amino acids) are obtained following cyanogen bromide cleavage of the Fe^{3+} protein (Bryant and Stellwagen, 1985; Corradin and Harbury, 1970; Wilgus and Stellwagen, 1979). Subsequent to the digestion stage, separation of the heme peptide(s) from non–heme-containing polypeptide contaminants is usually achieved by standard gel filtration/ion exchange chromatographic techniques, which, in the case of the longer peptides, provides a preliminary estimate of fragment identity using as calibration the empirical relationship which exists between elution volume of denatured peptides and their molecular weights (Wilgus and Stellwagen, 1979). Other simple procedures which may be useful for the rapid partial purification of specific heme-peptide fragments include ammonium sulfate precipitation (Peterson *et al.*, 1983) and chromatography on a bovine serum albumin–Sepharose affinity column. In contrast to the original report of the affinity chromatographic procedure (Wilchek, 1972), we have found the human serum albumin–sepharose column procedure to be markedly less efficient than that using bovine albumin (Adams *et al.*, 1989a), apparently because of the far slower reaction of heme peptides with human serum albumin.

The identity of a purified heme peptide can be conveniently confirmed by amino acid analysis using the known amino acid sequence of the appropriate cytochrome c. It is worth noting here that during the amino acid analysis procedure, cysteine decomposes to either alanine or glycine or both; this invariably results in a small glycine peak and elevated alanine relative to histidine (see Figure 20.2). Carraway et al. (1993) have recently reported the use of liquid secondary ion mass spectrometry to determine proteolytic hydrolysis sites for peptides derived from c-type cytochromes containing covalently attached heme c. Fragmentation occurs to break the disulphide bonds giving an intense peak—$M/Z = 617.2$—corresponding to protonated iron protoporphyrin IX. It is proposed that observation of this ion fragment can be used to unambiguously identify the presence of authentic ferriheme-c in peptides or peptide mixtures at the sub-nanomolar level. Salient features of the preparative protocols for short, medium, and long hemepeptides are briefly summarized in Table 20.1.

Table 20.1
Hemepeptide fragments prepared by direct enzymatic or chemical cleavage of Fe^{3+} cyt c or N-ε protected Fe^{3+} cyt c.

Cytochrome c	Hemepeptide fragment	Digestion procedure	References[a]
Horse heart	(14–19)H: MP-6	Nagarse (Sigma Chemical Co.)	
"	(13–19)H: MP-7	"	
"	(12–19)H: MP-8_1	"	
"	(14–21)H: MP-8_2	Pepsin/trypsin	i–x
"	(14–22)H: MP-9	Trypsin	
"	(11–21)H: MP-11	Pepsin	
Saccharomyces	(11–26)H: MP-16	Chymotrypsin	
Horse heart	(1–38)H	Tryptic digestion of trifluoracetylated or acetimidylated cyt c	xi, xii
Horse heart	(1–50)H	Partial acid hydrolysis of Fe^{3+} cyt c in 0.015 M HCl	xii
Human heart	(1–65)H	CNBr digestion of Fe^{3+} cyt c	xiii–xv
Horse heart	(1–80)H		
Human heart	(13–80)H	CNBr digestion of Fe^{3+} cyt c	
"	(13–104)H		
"	(13–65)H		xiii

[a] i, ii—Peterson et al., 1980, 1983; iii—Aron et al., 1986; iv, v—Adams et al., 1988, 1989; vi—Baldwin et al., 1987; vii—Chersi et al., 1987; viii—Schroeder and Yeager, 1978; ix, x—Plattner et al., 1975, 1977; xi—Parr et al., 1978; xii—Wallace and Proudfoot, 1987; xiii—Bryant and Stellwagen, 1985; xiv—Corradin and Harbury, 1970; xv—Wilgus and Stellwagen, 1979.

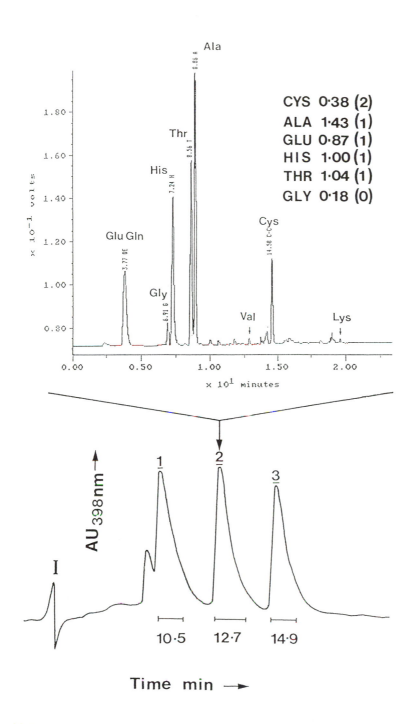

Figure 20.2
Analytical HPLC trace of Nagarse digestion of horse heart Fe^{3+} cyt c. Fractions of the major peaks 1–3 were taken as indicated and after lyophilization subjected to amino acid analysis. The analysis for peak 2 is shown; the absence of Lys and Val indicates that the heme peptide is the hexapeptide (MP-6) with amino acid residues 14–19. This is confirmed by the quantitative analysis shown on the figure—values in parentheses are the theoretical ratios. "I" is an injection artifact.

In addition to hemepeptide fragments obtained directly by digestion of the native or protected native protein, a large number of semisynthetic hemepeptide fragments have been obtained by further enzymic digestion of the hemepeptide fragments themselves—or of complexes formed between the hemepeptide fragments and non–heme-containing cytochrome *c*–derived polypeptides. These procedures are discussed elsewhere in this text (see Chapter 21 by Wallace).

The preparative regimen using classical column chromatography that was outlined earlier suffers from a number of drawbacks, the principal of these being that peptide separations are time-consuming and inefficient. Consequently, it is difficult to reproducibly obtain a sample of pure hemepeptide rapidly and in good yield by these procedures.

A rapid and efficient procedure for the monitoring of hemepeptide formation, and for separation/purification of the individual hemepeptides, utilizes high performance liquid chromatography (HPLC). Analytical HPLC is particularly suited to the rapid preparation and purification of the short hemepeptides (microperoxidases), which are of particular interest as chemical models of heme-containing enzymes. In the case of the microperoxidases, kinetic studies are usually carried out at the sub-micromolar concentration level; thus, HPLC fractions of sub-milligram quantities of these compounds can often be used in such studies directly on elution from the HPLC column (and after appropriate dilution with buffer) without recourse to lyophilization. This procedure has recently been used to optimize the preparation of the heme undecapeptide (MP-11) and heme octapeptide (MP-8) formed by sequential pepsin and trypsin digestion of horse heart cytochrome *c* (Adams *et al.*, 1989a). The feasibility of preparing MP-8 at a purity of >99% and yield of ~90% theoretical in a period of ~4 h—as opposed to a time of six to eight days at 50% yield and ~95% purity using standard chromatographic techniques—was demonstrated.[1] Analytical HPLC also provides a convenient procedure for the rapid separation and identification of individual hemepeptide components in mixtures of these species formed by the action of nonspecific proteases on cytochrome *c*. This is demonstrated in Figure 20.2, where the analytical HPLC trace obtained after a six-hour digestion of Fe^{3+} horse heart cytochrome *c* by the nonspecific protease, Nagarse, is shown (times given refer to peak maxima).

Fractions of the three major components were sampled as indicated, and amino acid analysis carried out after lyophilization. The amino acid analysis trace for HPLC peak 2 is shown as an example, and comparison with Figure 20.1 clearly identifies this peak as MP-6 (residues 14–19), since both Lys and Val are absent. Peak 1 hemepeptide contains Lys but no Val, and thus conforms to MP-7 (residues 13–19), while peak 3 hemepeptide contains Val but no Lys and is thus either MP-8 (residues 14–21) or MP-7 (residues 14–20).

---

[1] It should be pointed out, however, that proteolysis times depend upon the activity of the proteolytic enzymes being used. Hence, continual monitoring of the reaction by HPLC is recommended.

III. Physical Studies

A. The Structure of the Peptides

In the absence of an x-ray crystal structure of any of the MPs, there is no compelling evidence concerning the secondary structure of the polypeptide fragment. In the parent protein, residues 12–17 are present as part of an α-helix (Dickerson and Timkovich, 1975); furthermore, MP-11 has a high Chou–Fasman parameter (a measure of the tendency of a peptide to form an α-helix) of 1.16 (Jehlani *et al.*, 1976). A study using molecular models has shown that formation of the two thioether linkages and the bond between the imidazole ring of His-18 and Fe is feasible if the peptide in MP-11 possesses an α-helical configuration (Ehrenberg and Theorell, 1955). A ^{1}H-NMR investigation of cyanoMP-11 at 350 K, pH 10.5, indicated that all residues except for the N-terminal Val–Gln–Lys sequence lay within a shift cone centered at the Fe atom (Kimura *et al.*, 1981); this would be consistent with the peptide retaining essentially the structure of the parent protein in this region, and as was originally suggested by Harbury and Loach (1959), shielding about a quarter of the proximal face of the heme from the solvent. It has also been suggested that in lyophilized preparations of MP-11, the protein is unraveled, and both His-18 and the ε-NH$_2$ of the side chain of Lys-13 act as intramolecular ligands to the metal ion (Paleus *et al.*, 1955). This suggestion was advanced to account for the low-spin character of Fe(III) in aqueous MP-11 solutions; this spin state, however, almost certainly results from intermolecular bonding between MP-11 molecules, as discussed later.

B. Isoelectric Points

The isoelectric point of MP-11 has been determined by isoelectric focusing; values of pI = 4.75 (Jehlani *et al.*, 1976; Wilson *et al.*, 1977) and 4.85–4.90 (Peterson *et al.*, 1980) have been reported. In all cases, a small contaminant with pI = 4.46–4.65 is present, and is probably due to a species in which the side chain of Gln has been hydrolyzed during proteolysis of the parent protein under acid conditions. Furthermore, extended electrophoresis of the hemepeptide causes decomposition (Peterson *et al.*, 1980). MP-9 has pI = 4.95 (Plattner *et al.*, 1977), and the pI's of other small hemepeptides appear not to have been reported; they are, however, likely to have similar values, i.e., in the range 4.75–5.0.

C. Acid Dissociation Constants and the Coordination Sphere of Fe(III)

A compilation of the best available information concerning the behavior of MP-11, MP-9 and MP-8 in aqueous or predominantly aqueous solution leads to the schemes shown in Figure 20.3. A review of the experimental results that lead to this view is presented in the remainder of this section.

A

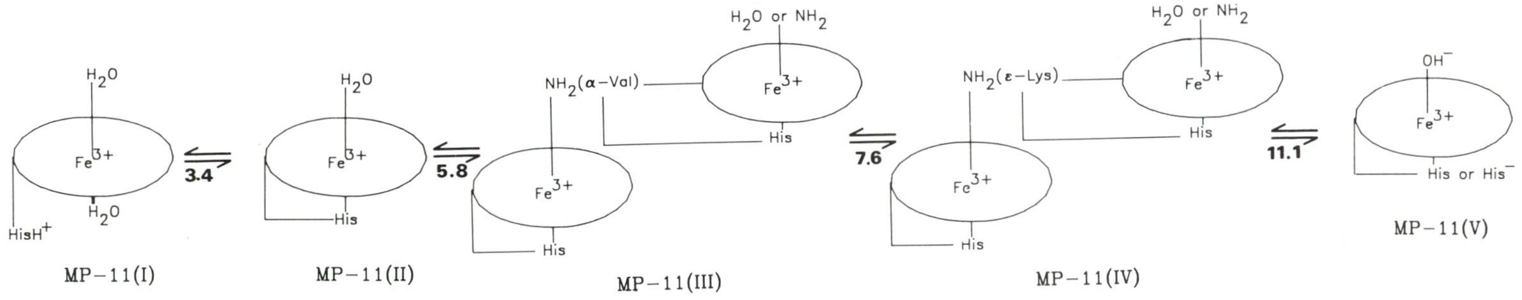

Structural scheme (left → right):

MP-11(I): HisH⁺ — Fe^{3+} with H_2O, H_2O
$\rightleftharpoons$ **3.4**
MP-11(II): Fe^{3+} with H_2O, His
$\rightleftharpoons$ **5.8**
MP-11(III): Fe^{3+} with $NH_2(\alpha\text{-Val})$, H_2O or NH_2, His
$\rightleftharpoons$ **7.6**
MP-11(IV): Fe^{3+} with $NH_2(\varepsilon\text{-Lys})$, H_2O or NH_2, His
$\rightleftharpoons$ **11.1**
MP-11(V): Fe^{3+} with OH^-, His or His^-

	MP-11(I)	MP-11(III)	MP-11(IV)	MP-11(V)
Magnetic moment $/\mu_B$	4.8	2.3		<2.3(?)
Mössbauer parameters	IS = 0.46 mm s^{-1} QS = 0.32 mm s^{-1}	IS = 0.46 and 0.40 mm s^{-1} QS = 0.32 and 2.04 mm s^{-1}		
Resonance Raman		Bands due to mixture of HS and LS Fe(III)		
ESR	g = 5.70; 2.00 (HS Fe(III))	g = 3.18; 2.19; 1.46 (LS Fe(III)) and HS lines	g = 3.25; 2.10; 1.46 predominant	
Electronic spectra /nm	495 (Q_V); 620 (CT) Q_0 unresolved		530 (Q_V); 560 (Q_0)	As for MP-11(III) but LS bands predominate

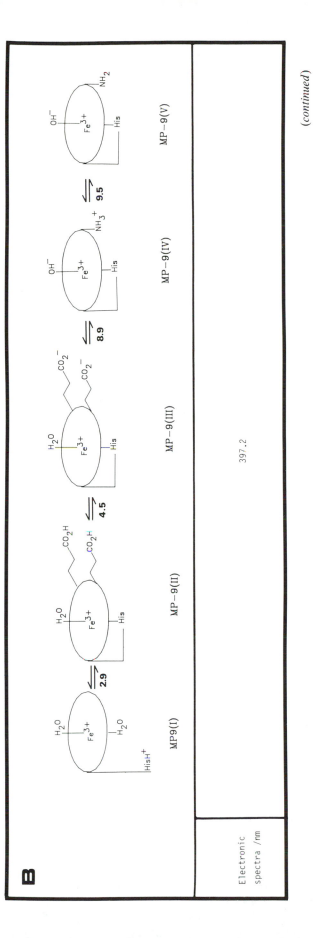

B

Electronic spectra / nm		397.2		
	MP9(I)	MP–9(II)	MP–9(III)	MP–9(IV)	MP–9(V)

Figure 20.3

The solution chemistry of the microperoxidases as deduced from spectroscopic and potentiometric studies of their solutions and selected spectroscopic parameters (see text for details). pK_a values are shown below each equilibrium.

(continued)

643

C

	MP–8(I)	MP–8(II)	MP–8(III)	MP–8(IV)	MP–8(V)
Magnetic moment /μ_B		3.72 5.42 (N-AcMP-8)	4.24 (A-AcMP-8)		2.6 4.24 (N-AcMP-8)
Resonance Raman		MP-8: Mixture of HS and LS Fe(III) N-AcMP-8: Spin equilibrium ($S = {}^5/_2 \rightleftharpoons {}^3/_2$)			
ESR		g = 6.06; 2.01 (HS) and 3.17; 2.13; 1.83 (LS) g = 5.27; 2.06 (50% ethylene glycol) N-AcMP-8: g = 6.03; 2.01			g = 2.95; 2.28; 1.93
Electronic spectra /nm	394.0 (Soret); 494 (Q_v); 619 (CT); Q_0 unresolved	397.2 (Soret); 495 (Q_v); 565 (Q_0); 622 (CT)			401.0 (Soret); 535 (Q_v); 565 (Q_0)

Figure 20.3 (*continued*)
The solution chemistry of the microperoxidases as deduced from spectroscopic and potentiometric studies of their solutions and selected spectroscopic parameters (see text for details). pK_a values are shown below each equilibrium.

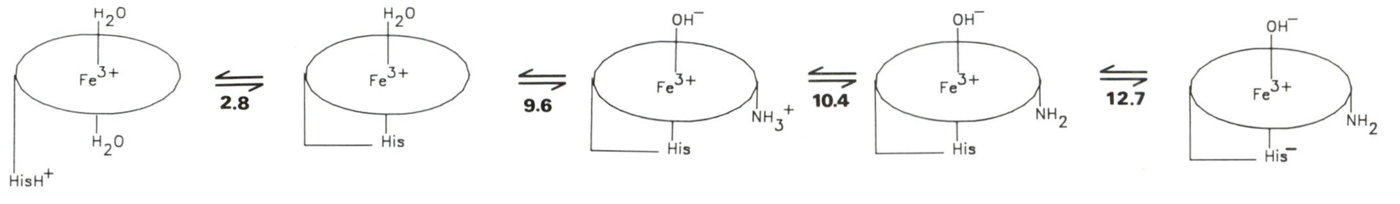

The hemepeptides for which the solution chemistry has been investigated all exhibit pH-dependent electronic spectra. Determination of pK_as associated with these spectral changes are complicated, in the absence of suitable experimental precautions, by artifacts resulting from aggregation and polymerization effects. MP-11 exhibits spectroscopically observable pK_as at 3.4, 5.8, and 7.6 (Wilson *et al.*, 1977) at 20 °C, $\mu = 0.5$ M. The first pK_a—in exact agreement with the result from a previous determination (Paleus *et al.*, 1955)—was found to be independent of peptide concentration; the second two were not, suggesting that they are associated with the formation of polymeric species. It is not surprising, therefore, that a range of values have been quoted for these pK_as [5.8 (Paleus *et al.*, 1955); 7.6 and 9.5 (Margoliash *et al.*, 1959); and 6.6 and 10.5 (Harbury and Loach, 1960a, 1959)], since the apparent value will depend upon the conditions of the experiment. In 50% SDS solutions, MP-11 has an apparent pK of 7.2 (Mazumdar *et al.*, 1991). Despite claims to the contrary by the authors, a comparison of the UV-visible spectrum obtained with that of monomeric MP-8 clearly indicates that even in detergent micelles, MP-11 is severely aggregated. The first pK_a is almost certainly associated with deprotonation and coordination of the imidazole ring of His-18 by Fe(III) (see later discussion). The other two are probably due to the intermolecular coordination of, firstly, the α-NH$_2$ group of Val-11 and, secondly, the ε-NH$_2$ group of Lys-13. Furthermore, at high pH, OH$^-$ can be expected to compete for Fe(III), and the imino group of His-18 will ionize to produce a coordinated imidazolate. Although the pK_a for formation of an imidazolate from free imidazole is high [14.44 (Hanania *et al.*, 1966)], coordination by a metal ion leads to polarization of the electron density of the heterocyclic ring, with a consequent decrease in basicity of the imino group. This occurs with $pK_a = 12.7$ in the bis-imidazole complex of ferriprotoporphyrin IX (Baldwin *et al.*, 1986a), between 9.8 and 10.4 in complexes of various imidazoles with aquocobalamin (Marques *et al.*, 1990), and as low as 6.5–7.0 in imidazoleleghemoglobin (Sievers *et al.*, 1983). Harbury and Loach (1959, 1960a) report a pK_a at 11.1 for MP-11, and Wilson *et al.* (1977) showed that a decrease in the Soret intensity of MP-11 occurred above pH 10. These effects are probably due to formation of the imidazolate of coordinated His-18, although replacement of one of the axial ligands by OH$^-$ cannot be excluded.

By monitoring the change in absorbance at 590 nm (the α-band), Harbury and Loach (1960b) demonstrated that MP-8 has an apparent pK_a of 9.9. The value was subsequently shown by potentiometry to decrease to 8.9 when the α-NH$_2$ group of Val was converted to a hydroxyl group on deamination (Myer and Harbury, 1973); this is a rather large effect if the terminal –NH$_2$ group is not involved in coordination. An examination of the reported spectroscopic titration data (Harbury and Loach, 1960b) suggests that the observed perturbation of the spectrum may be due to two overlapping pK_as; furthermore, it is known that at the concentrations of MP-8 used, the major portion of the species is present as a dimer (Aron *et al.*, 1986). Spectrophotometric titration in 20% MeOH, under conditions where MP-8 is strictly monomeric, has demonstrated at least three pK_as, at 4.43, 8.90, and 10.48 (Baldwin *et al.*, 1986b). The first—which may be a composite pK_a influenced by the ionization of remote groups such as the heme propionates—was shown by competition experiments with free imidazole to involve deprotonation and coordination of His-18. The second was attributed to ionization of bound H_2O, and the third to ionization of His-18 to form an imidazolate. The value for the second pK_a is in precise agreement with that reported for deaminated MP-8 (Myer and Harbury, 1973). A factor not taken into account was the presence of the N-terminal amino group of Cys-14, ionization of which may well perturb the electronic spectrum. A recent examination of the pK_as of monomeric N-terminal acetyl MP-8 (Marques and Munro, unpublished observations) in aqueous solution showed pK_as at 9.60 and 12.72; this suggests that the previously reported value of 8.90 is too low because of overlap with the pK_a at 10.48, which can be attributed to ionization of the N-terminal amino group. The value is in reasonable agreement with a value of 10.1 deduced from cyclic voltammetry of imidazole MP-8 (Marques, 1990).

Benzoylation of MP-8 gives a species in which water molecules occupy the two axial coordination positions of Fe(III) (Myer and Harbury, 1973); the pK_a of the coordinated H_2O decreases to 7.2, and thus demonstrates the *trans* effect of the proximal ligand on the basicity of coordinated H_2O, i.e., His > H_2O.

Using solutions of monomeric MP-9, spectroscopically observable pK_as were found to occur at 2.9 and 4.45, with overlapping pK_as at *ca.* 8.9 and 9.5 (Baldwin *et al.*, 1987a). Again, by analogy with *N*-acetyl MP-8 in acid solution [pK_a = 2.88 and 4.86 (Marques and Munro, unpublished observations)], the first pK_a is attributed to deprotonation and coordination of His-18, and the second to ionization of the heme propionates. However, the overlapping of the pK_as in alkaline solution—due to ionization of bound H_2O and the N-terminal amino group—made their reliable determination difficult.

D. Aggregation Effects

Sedimentation studies (Ehrenberg and Theorell, 1955) showed that at pH 2.3, MP-11 exists as a monomer in solution with MW $\approx$ 1800, i.e., as species MP-11(I) of Figure 20.3. At pH 8.7 the hemepeptide is extensively aggregated with an apparent MW of 10,800—i.e., MP-11(IV) in Figure 20.3 is either a pentamer or a hexamer. This aggregation is dispersed by addition of exogenous ligands such as histidine (Ehrenberg and Theorell, 1955), ammonia, or imidazole (Harbury and Loach, 1959), which compete for the axial coordination site of the metal ion. Further evidence for the aggregation of MP-11 was obtained by Harbury and Loach (1959), who noted that dilution of a concentrated solution of MP-11 with neutral buffers produced spectroscopic changes up to 400 minutes after mixing. Complete acetylation of the two primary amino groups of MP-11 eliminated this behavior, thereby implicating the N-terminal NH_2 of Val-11 and/or the ε-NH_2 group of Lys-13 in the formation of aggregates.

Aggregation of MP-11 was studied by ORD, CD, and UV-visible difference spectroscopy (Urrey, 1967) and manifests as a shift in the Soret band to longer wavelength accompanied by an increase in absorbance. This was interpreted as arising from a head-to-tail alignment of the heme planes in the aggregated species, with a heme–heme distance of about 10 Å. With MP-8, aggregation results in a decrease in the Soret band intensity without appreciable shift in band position (Urrey and Pettigrew, 1967) and was considered to be due to stacking of the hemes. CD data showed the presence of at least two types of aggregate, in each of which the heme–heme distance was interpreted to be about 7 Å. The difference in intrinsic behavior between MP-11 and MP-8 in solution argues for the participation of the ε-NH$_2$ of Lys-13 in the aggregation behavior of the former species. Studies with models have been used to show that both primary amino groups in MP-11 could coordinate intermolecularly to Fe(III) (Wilson *et al.*, 1977), but apparently there is considerable steric crowding to intermolecular coordination of the terminal NH2 group of MP-8 (Jehlani *et al.*, 1976).

A quantitative study of the aggregation of MP-8 in aqueous and aqueous methanol solutions at pH 7.0 and 12.0 has been reported (Aron *et al.*, 1986). Deviations from Beer's law could be accounted for by a simple equilibrium between monomers and dimers,

$$2 \text{ monomer} \rightleftharpoons \text{dimer} \quad (K_D),$$

with $K_D = 117 \times 10^3$, 12.1×10^3, and 2.16×10^3 M^{-1} in aqueous solution, 20% MeOH, and 50% MeOH, respectively, at 25 °C, $\mu = 0.1$ M. Acetylation of the terminal amino group of MP-8 eliminates its aggregation in neutral aqueous solution (Wang and Van Wart, 1989; Marques and Munro, unpublished observations), clearly implicating this group in the dimerization process, despite suggestions to the contrary (see earlier discussion). We have recently noted, however, that at [Ac-MP-8] > 30 μM, there is deviation from Beer's law, probably due to noncovalent aggregation of heme units (Marques and Munro, unpublished observations). At pH 12 in aqueous solution, there was no evidence for dimerization up to 32.5 μM, i.e., MP-8(IV) (Figure 20.3) is monomeric. The possibility of dimerization of Ac-MP-8, at concentrations greater than 10 μM, whereby a His[18] residue from a second Ac-MP-8 molecule coordinates to the sixth site of another to give a bis-His species has, however, been suggested by Wang *et al.* (1992).

MP-9 shows similar behavior to MP-8 (Baldwin *et al.*, 1987a), with $K_D = 150 \times 10^3$ M^{-1} in aqueous solution at pH 7.0, and 15.0×10^3 M^{-1} in 20% MeOH. In 50% MeOH there was no apparent deviation from Beer's law up to 35 μM. The aggregation of MP-9 is due to bonding of Lys-22 with the Fe center of a neighboring porphyrin, so that the dimer contains both a high-spin and a low-spin center (Arutyunyan *et al.*, 1986).

E. Further Spectroscopic Studies

The results just discussed are all consistent with the MPs existing in solution as summarized in Figure 20.3. Further evidence is available from a number of spectroscopic studies conducted on these species, which will be briefly reviewed here.

1. Infrared spectroscopy

This technique is not very useful for the characterization of the various heme peptides but is of some use in assessing the presence of contaminants such as ammonium salts which arise from the preparation and isolation of these species (Peterson *et al.*, 1983).

2. Magnetic susceptibility

The magnetic susceptibility of solutions of the MPs is most commonly measured using Evans' NMR method. MP-11 has $\mu = 4.8\mu_B$ in the pH range 1–3 and $2.3\mu_B$ in the pH range 7–11 (Jehlani *et al.*, 1976). The reported variation in μ with pH appears to suggest a further decrease in μ at pH > 11, but this may be due merely to uncertainties in the measurements. If MP-11(I) and MP-11(IV) (Figure 20.3) are high-spin and low-spin d^5 Fe(III) species, respectively, the expected spin-only values would be $5.9\mu_B$ and $1.7\mu_B$. It was suggested that the value of $2.3\mu_B$ at high pH is increased from the spin-only value by an orbital contribution, and that at low pH, MP-11 exists as dihydroxo-bridged ferric dimers (Jehlani *et al.*, 1976). If the further decrease in μ at pH > 11 noted earlier is real, this may be due to formation of MP-11(V).

At pH 7.7, 25 °C, Huang and Kassner (1981a) report $\mu = 3.72\mu_B$ for aquo MP-8; this decreases to $2.6\mu_B$ at pH 12.0. Based on the variation of the magnetic moment with temperature at both pH values, these workers concluded that MP-8 exists as an equilibrium mixture of high-spin and low-spin species. The value at pH 7.4 is compromised by the probable presence of the complex as a dimeric species [48% dimer using the equilibrium constants reported by Aron *et al.* (1986)], which, if the dimer is formed by intermolecular coordination with the terminal NH$_2$ group, would mean a significantly smaller magnetic moment than if the complex were present exclusively as the monomer. This complication is avoided in the determination of Yang and Sauer (1982), who, using N-acetylated MP-8, obtained a value of $5.42\mu_B$.

There is no dimerization of MP-8 at pH 12 (see earlier discussion), but the magnetic moment determined (Owens *et al.*, 1988) is then not due only to hydroxo MP-8 as assumed [i.e., MP-8(IV) in Figure 20.3], but has some contribution from the hydroxo-imidazolate complex [MP-8(V)]. A significantly higher value of $4.28\mu_B$ has been reported for the hydroxo complex of N-AcMP-8 (Yang and Sauer, 1982), but failure to quote the pH at which the measurement was made makes it difficult to deduce whether the measurement refers to MP-8(IV) or to MP-8(V), or to a mixture of the two.

The magnetic susceptibility of lyophilized MP-8 measured directly by susceptometry is $2.24\mu_B$ at 4 K (Owens *et al.*, 1988), strongly suggesting that the species is present largely as a low-spin complex, presumably due to intermolecular coordination of the terminal NH$_2$ group.

3. Mössbauer spectroscopy

Mössbauer studies on frozen solutions of MP-11 (pH not stated) gave isomer shifts (IS = 0.44 mm/s at 240 K) and quadrupolar splittings (QS = 2.07 mm/s) characteristic of a low-spin species (Baumgartner *et al.*, 1974). From the temperature variation of QS, the spin–orbit coupling constant was estimated to be 420 cm^{-1}, and the tetragonal and rhombic distortions 750 and 420 cm^{-1}, respectively. A more detailed study by the same workers (Nassif *et al.*, 1976) reported the pH variation in the Mössbauer parameters for MP-11. At pH 1.5, IS = 0.46 mm/s, and QS = 0.32

mm/s, characteristic of a high-spin species [i.e., MP-11(I)]; on increasing the pH to 7, two sets of lines were observed, those due to the high-spin species and a new set with IS = 0.40 mm/s and QS = 2.04 mm/s clearly due to a low-spin species [i.e., MP-11(III)]. At high pH, the Mössbauer spectrum was essentially the same as at neutral pH, but the lines due to the low-spin species were more prominent. These results were essentially confirmed by Peterson and co-workers (1983), although they found substantially less of the high-spin species in their preparations and suggest that this might have been due to an impurity in the samples in the original work.

From Mössbauer studies, MP-9 appears to exist essentially as a low-spin species in lyophilized samples, with a small amount of a high-spin species also present (Peterson *et al.*, 1983). Lyophilized MP-6 is a mixture of high-spin and low-spin Fe(III), and some Fe(II) which appears to arise from the method of preparation. In frozen solutions at pH 5.9, both Fe(III) high-spin and low-spin sites are present, with the amount of the former increasing on increasing pH to 8.5. Although these workers suggest that the terminal amino group is sterically prevented from coordinating to the metal ion, its coordination in MP-8 has been clearly demonstrated (see earlier discussion) and is probably the source of the low-spin species in their samples.

We have recently shown (Munro *et al.* 1994) from the temperature-dependence of the Mössbauer spectrum that *N*-acetyl MP-8 lyophilized from pure water contains two species. The first, which contains coordinated H_2O, approaches an $S = \frac{5}{2}$ ground state at low temperature, while the other, containing coordinated OH^-, approaches an $S = \frac{1}{2}$ ground state. With an increase in temperature, the high-spin aqua complex thermally equilibrates at a significantly higher rate to the low-spin complex than the low-spin hydroxo complex equilibrates with the high-spin complex. A similar situation was observed for the 80 K Mössbauer spectrum of hemepeptide HP1–50 (cyt *c* sequence numbering), where lyophilization was carried out from 0.1% v/v CF_3COO^-/aq. In contrast, the Mössbauer spectrum of the noncovalent complex formed between HP1–50 and the nonheme peptide (NHP 51–104) derived from cytochrome *c* reflected only one kind of iron site. The asymmetric Mössbauer spectrum found for this site was interpreted as being the envelope of slightly different spectra each originating from structurally similar six-coordinate low-spin Fe(III) heme sites reflecting varying degrees of heme-site solvation in the complex (Adams *et al.*, 1994).

4. Resonance Raman

The first report of the resonance Raman (RR) spectra of a hemepeptide was apparently that of Loehr and Loehr (1973), who found that the RR bands of Fe(III) and Fe(II) MP-11 at pH 7.3 were characteristic of low-spin complexes. A more detailed study by Wang and Van Wart appeared recently (1989). These workers compared the RR spectra of MP-8 and N-AcMP-8 and showed that at pH 7.0, MP-8 exists as a mixture of high-spin and low-spin species, with the ratio of the former to the latter increasing on decreasing the concentration. Acetylation of the terminal amino group caused the bands due to the low-spin species to be eliminated, and the temperature-dependence of the bands showed that N-AcMP-8 consists of an equilibrium mixture of high-spin ($S = \frac{5}{2}$) and intermediate-spin ($S = \frac{3}{2}$) species, with the intermediate-spin species presumably arising as a result of the weak axial ligand field of the metal. This suggests that either (i) MP-8 is a mixture of six- and five-coordinate species [but the electronic spectra are strongly suggestive of a six-coordinate species (Aron et al., 1986)] or (ii) the Fe–N bond is relatively weak, possibly because of the steric constraints imposed by the polypeptide (Jehlani et al., 1976). The existence of a high-spin/intermediate-spin equilibrium would account for the lower-than-expected magnetic moments just mentioned. Othman et al. (1993) reported RR spectra of Fe(III) MP-8 in aqueous solution pH 7–14 and demonstrate a change from aggregated (low-spin, six-coordinate) complexes to high-spin, five-coordinate monomeric structures on addition of detergent or alcohol. Furthermore, solvent-dependent (pH, detergent, alcohol) changes of the octapeptide conformation attributed to changes in the H-bonding interaction of the axial His were also demonstrated from changes in the low-frequency (150–250 cm^{-1}) region of the RR spectrum.

5. Electron spin resonance

The ESR spectrum of a frozen solution of MP-11 at pH 1.5 is typical of a high-spin d^5 Fe(III) ion with g values at 5.70 and 2.00 (Nassif et al., 1976). At pH 7, lines due to a low-spin complex are predominant ($g = 3.18$; 2.19; 1.46), and at pH 10 the low-spin Fe(III) center predominates ($g = 3.25$; 2.10; 1.46).

The aggregation of MP-8 at neutral pH is also demonstrated by ESR spectroscopy (Yang and Sauer, 1982), which shows a mixture of high-spin ($g = 6.06$; 2.01) and low-spin centers ($g = 3.17$; 2.13; 1.83) whereas N-AcMP-8 shows predominantly the high-spin center ($g = 6.03$; 2.01). This last point has recently been confirmed (Marques and Munro, unpublished). At pH 11.5 [i.e., 91% MP-8(IV) and 9% MP-8(III)], the predominant species is low-spin ($g = 2.95$; 2.28; 1.93). In 50% ethylene glycol at pH 7.1, in which MP-8 is monomeric (Blumenthal and Kassner, 1979), the ESR spectrum ($g = 5.27$; 2.06) is consistent with a high-spin species (Hamilton et al., 1984). Lyophilized MP-8, as might be expected, shows both high-spin and low-spin Fe(III) centers (Owens et al., 1988).

6. UV-visible spectroscopy

In common with other aromatic complexes of transition metal ions, the bands in the electronic spectrum of the MPs are broad. Unlike the cobalamins, the positions of the main bands are not particularly sensitive to the identity of the axial ligands, but are moderately sensitive to the spin and oxidation states of Fe. The band positions of representative porphyrin complexes of high-spin and low-spin Fe(II) and Fe(III) are listed in Table 20.2. The information is drawn largely from Makinen and Churg (1983).

The visible spectrum of MP-11 at pH 1.5 shows a band at 495 nm (Q_v) with a shoulder on the long-wavelength side (Q_0), and the L $\rightarrow$ M CT band at 620 nm (Nassif et al., 1976). At pH 10, the CT band is totally absent and the Q_v and Q_0

Table 20.2
Electronic spectra of representative iron–porphyrin complexes.

Hemoprotein	Wavelength/nm	Assignment[a]
High-spin Fe (III)		
aquometmyoglobin	357	N band (a'_{2u}, $b_{2u}(P) \rightarrow e_g(\pi^*)(P)$)
	386	B(1, 0) (vibrational overtone)
	408	B band (a_{1u}, $a_{2u}(P) \rightarrow e_g(\pi^*)(P)$)
	467	$a_{2u}(P) \rightarrow a_{1g}(d_z2\ M)$
	505	Q_v (vibrational overtone)
	545	Q_O (a_{1u}, $a_{2u}(P) \rightarrow e_g(\pi^*)(P)$)
	581	$a_{1u}(P) \rightarrow e_g(d_{xz}, d_{yz}\ M)$
	633	$a_{2u}(P) \rightarrow e_g(d_{xz}, d_{yz}\ M)$
	1010	Admixture of Q_O with a_{1u}, $a_{2u}(P) \rightarrow e_g(M)$ and $a'_{2u}(P) \rightarrow a_{1g}(d_z2\ M)$
Low-spin Fe (III)		
cyanometMb	357	N
	400	B(1, 0)
	421	B
	487	$a_{2u}(P) \rightarrow a_{1g}(d_z2\ M)$
	545	Q_v
	581	Q_O
High-spin Fe (II)		
deoxyMb	406	B(1, 0)
	433	B
	513	Q(2, 0)
	556	Q_v
	607	Q_O
	626	$a_{1g}(d_z2\ M) \rightarrow e_g(\pi^*)(P)$
	757	$b_{1g}(d_x2-_y2\ M) \rightarrow e_g(\pi^*)(P)$
	930	$e_g(d_{xy}, d_{yz}\ M) \rightarrow e_g(\pi^*)(P)$
Low-spin Fe (II)		
oxyMb	348	N
	398	B(1, 0)
	418	B
	476	$a_{2u}(P) \rightarrow a_{1g}(d_z2\ M)$

(*continued*)

Table 20.2 (*continued*)
Electronic spectra of representative iron–porphyrin complexes.

Hemoprotein	Wavelength/nm	Assignment[a]
	542	Q_v
	580	Q_0
	645	$e_g(d_{xy}, d_{yz} M) \rightarrow e_g(\pi^*)(P)$
	925	$a_{1u}, a_{2u}(P) \rightarrow O_2(\pi_g)$

[a] P = porphyrin; M = metal.

bands are at 530 and 560 nm, respectively. The lyophilized powder spectrum (obtained as a mull) shows a mixture of the two sets of bands (Nassif *et al.*, 1976). Comparison of these results with those of Table 20.2 clearly shows the spectrum in acidic solution to be consistent with a predominantly high-spin species, that in basic solution to be predominantly low-spin, and the lyophilized powder to be a mixture of the two.

Plattner *et al.* (1977) reported MP-9 at pH 7 to have a Soret band at 397.5 nm with $\varepsilon = 1.11 \times 10^5$ M^{-1} cm^{-1}; however, the dimerization of MP-9 was not taken into account in this study. Under strictly monomeric conditions (Baldwin *et al.*, 1987a), $\varepsilon = 1.59 \times 10^5$ M^{-1} cm^{-1} with $\lambda_{Soret} = 397.2$ nm. At pH 1.75 and 7.00, the spectra are characteristic of high-spin Fe(III), and at pH 12.2, of low-spin Fe(III) (see Figure 20.4).

Two groups have provided reliable UV-visible data for MP-8 (Aron *et al.*, 1986; Owens *et al.*, 1988). Monomeric MP-8 in acid solution (pH 1.5) has λ_{Soret} at 394.0 (1.77×10^5 M^{-1} cm^{-1}), Q_v at 494 nm, an unresolved Q_0 band, and the CT band at 619 nm (Marques, 1986). In neutral solution, the bands are at 397.2 ($\varepsilon = 1.57 \times 10^5$ M^{-1} cm^{-1}), 495, 565, and 622 nm, respectively. In alkaline solution, the Soret band is at 401.0 nm, Q_v and Q_0 are at 535 and 565 nm, respectively, and the CT band is absent. These data are consistent with the $H_2O–Fe–H_2O$ species [i.e., MP-8(I)] and the $H_2O–Fe–His$ [i.e., MP-8(II)] species being predominantly high-spin, and the $OH–Fe–His^-$ species [i.e., MP-8(V)] being low-spin, in agreement with the anticipated field strength of the axial ligands.

7. NMR spectroscopy

Yet further evidence for the dimerization of MP-8 and MP-11 is found in the NMR work of Smith and McLendon (1981) and Satterlee (1983), who demonstrated the presence of broad ^{1}H resonances due to monomeric high-spin species and significantly sharper resonances due to aggregated low-spin species.

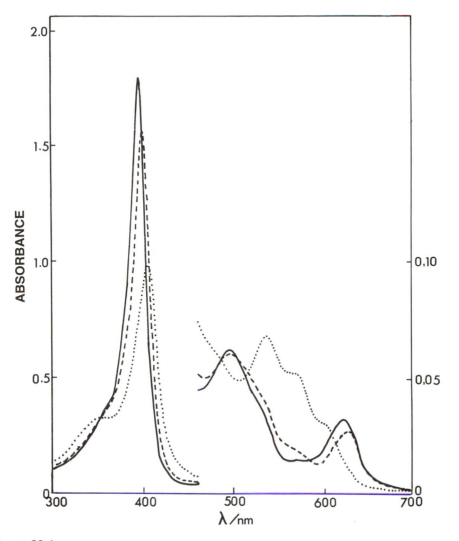

Figure 20.4

The spectra of monomeric solutions of MP-9 at (——) pH 1.75; (– – –) pH 7.00; and (· · ·) pH 12.2. From Baldwin *et al.* (1987a), with permission.

IV. Electrochemistry of the Microperoxidases

The MPs are electrochemically active species, and the use of mediators is unnecessary; indeed, a recent report (Brunori *et al.*, 1989) has shown that MP-11 may be trapped within a cellulose triacetate membrane fixed to a suitable electrode to act as a "solid state" promoter for studying the electrochemistry of soluble metalloproteins such as cytochrome *c* and azurin. MP-9 immobilized on a SnO_2 electrode responds to micromolar concentrations of H_2O_2 (Tatsuma and Watanabe, 1991) and, because of inhibition, the system can be used as a sensor for species which coordinate MP-9 such as imidazole and its derivatives (Tatsuma and Watanabe, 1992).

A. Redox-Linked Ionizations

Harbury and Loach (1959) first demonstrated that the midpoint potentials, $E_{1/2}$, of MP-11 and MP-9 and their complexes with ligands such as imidazole and pyridine are pH dependent and referred to this effect as being due to "oxidation-linked proton functions." Such an effect requires the uptake of a proton by a basic group in a molecule to be coupled to the uptake of an electron at the metal center (see Baldwin *et al.*, 1981, 1984, 1986a; Marques, 1990), as explained in Figure 20.5.

The behavior of the midpoint potential, $E_{1/2}$, in the presence of m oxidation-linked proton functions, is illustrated in Figure 20.6 and summarized by Equation (20.1) where K^{II} and K^{III} are the pK_as of the ionizable functional groups in complexes of the reduced and oxidized states of the metal, respectively:

$$E_{1/2} = E' + \frac{RT}{nF} \ln \left[\frac{\sum\limits_{j=0}^{m} \prod\limits_{i=1}^{m-j} K_i^{II} \cdot [H^+]^j}{\sum\limits_{j=0}^{m} \prod\limits_{i=1}^{m-j} K_i^{III} \cdot [H^+]^j} \right]. \tag{20.1}$$

$E_{1/2}$ for MP-8 as determined by potentiometric titration is pH-dependent (Harbury and Loach, 1960b), although the authors erroneously assumed that $E_{1/2}$ increased

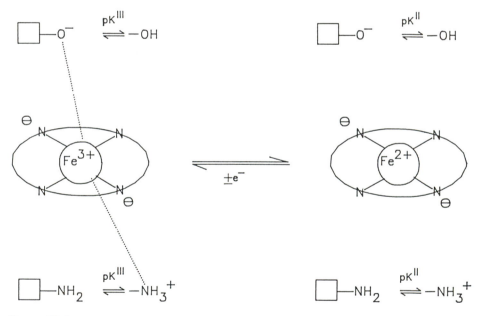

Figure 20.5
The coupling of uptake of a proton at an ionizable functional group (ROH and RNH$_2$, for example) to uptake of an electron at Fe(III) in a heme-containing complex. Electrostatic repulsion between the residual $+1$ charge at the metal center (i.e., $+3$ from Fe, -2 from pyrrole N's) and the conjugate acid RNH$_3^+$, and electrostatic attraction between the metal and the conjugate acid RO$^-$, causes a decrease in pK^{III}, the pK_a in the oxidized complex. On reduction, the residual positive charge at the metal center is neutralized, and the pK_as of the ionizable functional groups (pK^{II}) increase.

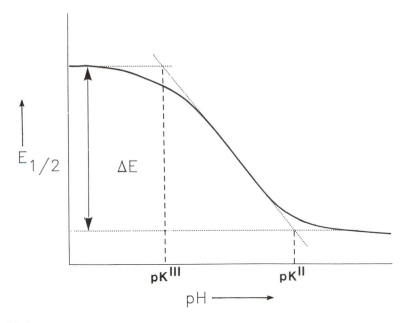

Figure 20.6

The dependence of the midpoint potential, $E_{1/2}$, on pK^{III} and pK^{II} (see Figure 20.5). If pK^{III} and pK^{II} are both above or below the operating pH range, $E_{1/2}$ is independent of pH. If they are within the operating pH range, $E_{1/2}$ decreases with increasing pH. The magnitude of $\Delta E_{1/2}$ depends upon the magnitude of the separation between pK^{III} and pK^{II}.

linearly with decreasing pH at pH values below their operating pH range (pH 5.5–12.5). A refit of their data to Equation (20.1) by standard nonlinear least-squares methods shows the presence of multiple redox-state sensitive ionizable functional groups in the molecule; $pK^{III} \approx 5$ and 9.9 ± 0.1, and $pK^{II} = 6.6 + 0.1$ and *ca.* 13.3. Since it is now known that in aqueous solution MP-8 consists of a mixture of monomeric high-spin, intermediate-spin, and polymeric low-spin complexes [in which the N-terminal amino group of Cys-14 binds intermolecularly to Fe(III)] (see earlier discussion), an assignment for the ionizing functional groups is difficult. It is worth noting that on deamination of Cys-14, pK^{III} is reported as 8.9 and pK^{II} as 6.8 (Marques, 1990); pK^{III} then agrees with the pK_a determined spectroscopically for ionization of bound H_2O in monomeric MP-8 (see earlier discussion), and pK^{II} is probably due to ionization of a heme carboxylate in the ferrous complex (see later discussion).

The data reported by Harbury and co-workers for the imidazole complexes of MP-11 and MP-8 (ImMP-X) are probably more reliable, since it has been shown subsequently (Marques, 1990) that ImMP-8, up to concentrations of at least 1.2 mM, is monomeric in aqueous solution. Multiple pK_as are seen in both oxidized and reduced forms of ImMP-11 [$pK^{III} = 6.6, 10.5, 11.1$; $pK^{II} = 4.6, 5.6, 7.2, 9.8$ (Myer and Harbury, 1973)], but for ImMP-8 the situation is simpler [$pK^{III} = 4.7$,

11.1; $pK^{II} = 6.2$ (Harbury and Loach, 1960a)], which suggests involvement of the ε-NH$_2$ group of Lys-13 in the coordination chemistry. Recently, the ImMP-8 system was reinvestigated by cyclic voltammetry at a glassy carbon electrode (Marques, 1990) (see Figure 20.7) and enabled an assignment of the groups responsible for the multiple pK_as to be identified. Three ionizations were found for the ferric complex (5.6, 10.1, 12.9) and two for the ferrous complex (6.8, 11.1). An examination of the molecular structure of ImMP-8 by molecular mechanics techniques (Marques and Hancock—unpublished observations) shows that the heme propionates and the amino group of Cys-14 are all within about 10 Å of the metal center and hence are likely candidates for the ionizing functionalities (pK_as in the ferric and ferrous complexes, respectively: heme propionate, 5.6 and 6.8; NH$_2$ of Cys-14, 10.1 and 11.5). Ionization of bound Im (and/or His-18) to form an imidazolate is responsible for the pK_a at 12.9, a process which can be followed spectroscopically. The pK_a for bound imidazole in the ferrous complex is thus clearly >14. The values are therefore in reasonable agreement with those of Harbury and co-workers, although it is suspected that the pK^{III} at 11.1 which they report is a composite pK_a due to ionization of Cys-14 and bound imidazole.

The demonstrated coupling of proton uptake at a remote ionizable functional group to electron uptake at the metal center in these and other complexes of iron porphyrins may help to explain the mechanism of proton pumping by various cytochromes of electron transport chains (Baldwin *et al.*, 1981, 1984, 1986a; Marques, 1990).

B. Direct Electrochemistry

The direct electrochemistry of MP-11 at a glassy carbon electrode and a silver electrode (Razumas *et al.*, 1989; Santucci *et al.*, 1988) has been reported. Although Santucci *et al.* found no gross effects with increasing MP-11 concentration, it is likely that in the concentration range used (50–500 μM), MP-11 exists in aqueous solution in aggregated form (see earlier discussion). Both the reported diffusion coefficient to the working electrode (2×10^{-6} cm^2 s^{-1}) and the rate constant for heterogeneous electron transfer (3×10^{-3} cm s^{-1}) are smaller than the values for ImMP-8 (4.5×10^{-5} cm^2 s^{-1} and 4×10^{-2} cm s^{-1}, respectively), which may be a consequence of aggregation in the MP-11 system, although differences in rate constants could simply reflect differences in the quality of the glassy carbon electrode surfaces used in the two studies.

The same group have also investigated the spectroelectrochemistry of ImMP-11 at bare and gold-plated reticulated vitreous carbon (RVC) electrodes (Zamponi *et al.*, 1990). As expected (see earlier discussion), $E_{1/2}$ is pH dependent, decreasing from -190.5 mV (vs. SHE) at pH 7 to -279.0 mV at pH 12. Unlike ImMP-8, however (Marques 1990), ImMP-11 only shows reversible redox behavior in a very limited pH range (7 to 8). In acidic (pH < 6) and alkaline solutions (pH > 10), considerable irreversibility is observed. The reasons for this have not been established unambiguously; the authors attributed the irreversible behavior to structural changes brought about by protonation of the heme carboxylate groups, and deprotonation of Lys-13 and ionization of coordinated imidazole.

Razumas *et al.* (1992) have extended their earlier studies (see earlier discussion) on MP-11 to the electrochemical and electroanalytical properties of the three heme peptides HP-8, -9, and -11, with regard to the use of different electrode materials, investigating the absorption and structure of the peptides by ellipsometry and sur-

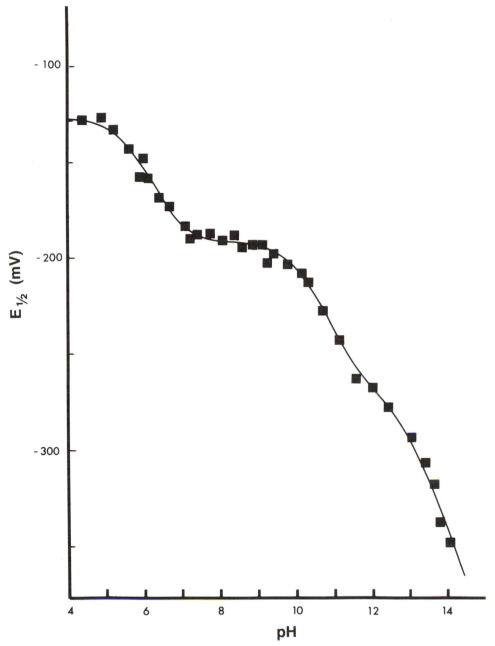

Figure 20.7
The pH dependence of the redox potential (versus standard hydrogen electrode) of ImMP-8. From Marques (1990), with permission. See text for details.

face-enhanced Raman spectroscopy. The microperoxidases were found to be useful models for investigating the interfacial behavior of heme proteins.

Hamilton *et al.* (1984) studied the variation in redox potentials of MP-8 complexes of a number of N and O donor ligands by cyclic voltammetry at a platinum electrode. They report asymmetric CV traces when using sterically crowded ligands, indicative of loss of ligand from the coordination sphere on reduction. Because of the rather broad range of ligands used, no definitive conclusions concerning the effect of structure on redox potential could be deduced.

V. Reactions of Microperoxidases: Ligand Binding Studies

The microperoxidases comprise a class of compounds in which the metal ion is coordinated in its equatorial plane by a macrocycle, and in its axial positions by His and a readily replaceable H_2O molecule. In their general architecture they are therefore reminiscent of vitamin B_{12a} (aquocobalamin), the coordination chemistry of which has been extensively studied over the last 25 years (Pratt, 1972; Dolphin, 1982). It is to be anticipated that the basic inorganic chemistry of the microperoxidases will attract increasing attention in the future. Until now, however, attention has been focused on their use as models for the hemoproteins in order to discern the effect of the protein on the fundamental chemistry of the prosthetic group.

The study of the thermodynamics and kinetics of ligand substitution processes in the MPs is not without its potential pitfalls, however. The most obvious, and that which has unfortunately so often been overlooked and has compromised a significant fraction of the results published to date, is the tendency of the MPs to aggregate in solution (see earlier discussion). It is therefore crucial that the solution chemistry of an MP under investigation be fully delineated for the information concerning ligand substitution processes to be reliable and to unambiguously refer to reactions such as (20.2) (where L is an incoming ligand; only the axial ligands are shown, and the charges are omitted for convenience):

$$His-Fe-H_2O + L \rightleftharpoons His-Fe-L + H_2O$$

$$K = [His-Fe-L]/[His-Fe-H_2O][L]$$

(20.2)

In this section we shall briefly review available information concerning (a) the thermodynamics of coordination of ligands by ferric and ferrous MPs; (b) the kinetics of ligand substitution reactions; and (c) reduction and self-exchange reactions of the MPs.

A. Thermodynamics of Ligand Coordination

In a monomeric ferric hemepeptide in which the bond between His-18 and Fe is preserved, the Fe-bound H_2O molecule is readily replaced by a wide range of ligands from solution, as symbolized in Equation (20.2). The ground state of a ferrous hemepeptide has not yet been adequately established, and the species may be either six-coordinate with bound H_2O, or five-coordinate with a vacant coordination site. Ferrous hemepeptides also coordinate a variety of ligands from solution. This property of the MPs was appreciated at an early stage; hence Tsou (1949, 1951b) reported that spectroscopic changes occurred on adding CN^- to ferric MP-11, and CN^-, CO, NO, nitrosobenzene, and methyl isonitrile to ferrous MP-11; and Harbury and Loach (1960b) attempted to determine equilibrium constants for coordination of imidazole, pyridine, and glycine by both ferric and ferrous MP-11. A selection of quantitative information concerning these processes for ferric hemepeptides is presented in Table 20.3.

As outlined in Section III, aquo-MP-X exists as an equilibrium mixture of high-spin ($S = \frac{5}{2}$) and intermediate-spin ($S = \frac{3}{2}$) species. On replacement of H_2O by a ligand L, the resulting complex is either low-spin ($S = \frac{1}{2}$), high-spin, or in spin equilibrium ($S = \frac{5}{2} \leftrightarrow S = \frac{1}{2}$), depending on the ligand field strength.

The electronic and ESR spectra of cyano-MP-8 have the characteristics of the spectra of a low-spin Fe(III) iron porphyrin (Owens *et al.*, 1988; Marques *et al.*, 1987), as does the Mössbauer spectrum of cyano-MP-11 (Baumgartner *et al.*, 1974). The magnetic moment of cyano-N-Ac-MP-8 [2.83 μ_B (Yang and Sauer, 1982)] serves as further confirmation. Other ligands which produce low-spin Fe(III) are imidazole and its derivatives (Baldwin *et al.*, 1986b; Owens *et al.*, 1988; Baumgartner *et al.*, 1974; Marques, 1986; Yang and Sauer, 1982; Harbury *et al.*, 1965), ammonia and other primary amines (Marques *et al.*—unpublished results), thiolates (Sadeque *et al.*, 1987a; Baldwin, D. A., *et al.*—unpublished results) and thioethers (Baumgartner *et al.*, 1974; Marques, 1986; Yang and Sauer, 1982). Weaker field ligands such as N_3^- (Owens *et al.*, 1988; Blumenthal and Kassner, 1979; Hamilton *et al.*, 1984; Yang and Sauer, 1982; Baldwin, D. A., *et al.*—unpublished results) and sulfite (Marques, 1986) produce a complex which is in spin equilibrium. Anions such as CNO^- (Jain and Kassner, 1984), SCN^- (Baldwin, D. A., *et al.*—unpublished results), and F^- (Baumgartner *et al.*, 1974; Hamilton *et al.*, 1984; Yang and Sauer, 1982) form high-spin compounds.

The determination of meaningful formation constants, K, for a ligand substitution process requires correction of observed formation constants, K_{obs}, to take into account acid-base equilibria present in solution as shown in Scheme I; thus Equation (20.3) applies:

$$K = K_{obs}(1 + [H^+]/K_a)(1 + K_{Fe}/[H^+]). \qquad (20.3)$$

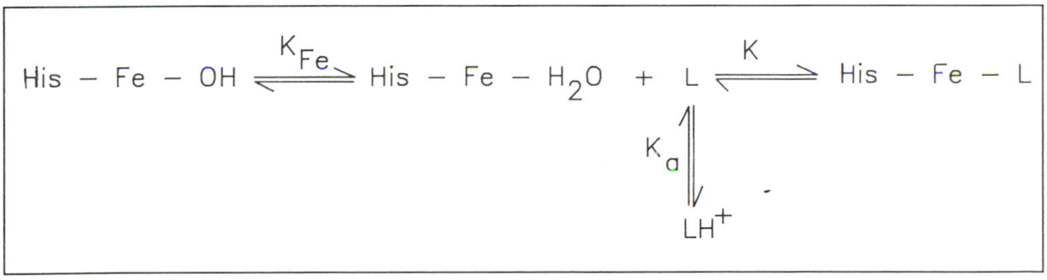

Scheme I

Table 20.3

Selected data for complexes of the microperoxidases.

Ligand	Hemepeptide	Electronic spectra[a]	ESR[a]	Mössbauer[a,b]	μ_{eff}/μ_B	log K	Thermodynamic parameters[g]	Comments	Ref.[o]
A. Small anionic ligands									
CN⁻	MP-11	537[c]							i
	MP-11	410[c]				7.57 (pH 5.34) 7.30 (pH 6.00) 7.10 (pH 7.20)		Aq solution; log K values corrected for pK_a of HCN	ii
	MP-11			QS = 1.20[d]; IS = 0.33 Δ = 270[e]; V = 160[e] Δ = 300[f]; V = 160[f]				T = 210 K	iii
	MP-9	352; 411.8*; 535; 560 (sh) *ε = 1.13 × 10⁵ M⁻¹ cm⁻¹				7.67		20% MeOH pH 7.00; corrected for pK_as of HCN and MP-9	iv
	MP-8	412.5[c]				7.54	$\Delta H°$ = −79 $\Delta S°$ = −126	50% ethylene glycol pH 7.4, corrected for pK_a of HCN	v
	MP-8	533[c]				6.18		Aq solution pH 7.0; no well-defined isosbestic points	vi
	MP-8	352; 410.5*; 532; 555 (sh) *ε = 1.09 × 10⁵ M⁻¹ cm⁻¹				7.58		20% MeOH pH-independent log K between pH 6 and 8; corrected for pK_as of HCN and MP-8	vii
	MP-8	353; 412*; 533 *ε = 9.98 × 10⁴ M⁻¹ cm⁻¹	g = 3.34; 1.80; 1.01. Δ = 1575; V = 388					50% ethylene glycol	viii
	N-AcMP-8				2.83			T = 25 °C	ix

Compound	Spectral data	g / other	Value	ΔH° / ΔS°	Conditions	Ref.
MP-6			6.16	$\Delta H^\circ = -10$; $\Delta S^\circ = 52$	pH 7.5; corrected for pK_a HCN	x, xi
MP-9	410.5		1.39		20% MeOH pH 7.00	iv
MP-8	410.5^c; $\varepsilon = 1.3 \times 10^5$ M^{-1} cm^{-1}		1.48	$\Delta H^\circ = -26.1$; $\Delta S^\circ = -61$	50% ethylene glycol; corrected for pK_a of MP-8	xii
MP-8	410.2*; 530; 565; 620; *$\varepsilon = 1.26 \times 10^5$ M^{-1} cm^{-1}		1.45		20% MeOH; corrected for pK_a of MP-8	xiii
MP-8			1.44		Aq solution, pH 7.0	vi
MP-8		$g = 2.81$; 2.14; 1.71			$T = 77$ K; at RT lines due to HS complex present	xiv
MP-8	351; 410*; 533; 567; 622; *$\varepsilon = 1.13 \times 10^5$ M^{-1} cm^{-1}	$g = 2.78$; 2.14; 1.70; $\Delta = 2198$; V = 931			50% ethylene glycol	viii
N-AcMP-8			5.10		25 °C	ix
MP-6			1.23	$\Delta H^\circ = -3.5$; $\Delta S^\circ = 7.1$	Aq solution pH 6.2: thermodynamic parameters at pH 7.5	x, xi
MP-8	411.4; 490; 516; 570; 620		0.79		20% MeOH pH 7	xv
MP-8	401; 496; 534; 567; 622^c		0.083	$\Delta H^\circ = -1.3$; $\Delta S^\circ = -3$	50% ethylene glycol pH 7.4	xvi
MP-8	400.2*; 494; 530; 568; 620; *$\varepsilon = 1.20 \times 10^5$ M^{-1} cm^{-1}		0.072		20% MeOH, pH 7.0	xiii

(continued)

Table 20.3 (*continued*)
Selected data for complexes of the microperoxidases.

Ligand	Hemepeptide	Electronic spectra	ESR[a]	μ_{eff}/μ_B	Mössbauer[a,b]	log K	Thermodynamic parameters[g]	Comments	Ref.[o]
SCN⁻	MP-8	399.8*; 492; 525; 568; 621 *ε = 1.17 × 10^5 M⁻¹ cm⁻¹				−0.052		20% MeOH, pH 7.0	xv
	MP-6					0.34		Aq solution pH 6.8	x
F⁻	MP-11				QS = 0.63; IS = 0.35				iii
	MP-8					−0.022[h]		50% ethylene glycol	xvii
	MP-8		g = 5.52; 2.17						xiv
	A-AcMP-8			5.48				25 °C	ix
	MP-6					0.46		Aq solution pH 6.8	x
B. Nitrogen donors									
Imidazole	MP-11[i]					3.92 (4.27)[j]		pH 9, aq solution	xviii
	MP-11				QS = 2.09[d]; IS = 0.40 Δ = 750; V = 410				iii
	MP-11					4.34 (pH 4) 4.11 (pH 5) 3.98 (pH 6.5)		Aq solution; corrected for pK_a of imidazole	ii
	MP-9					4.34		20% MeOH pH 7.0	iv
	MP-8	525; 560[c]							xix
	MP-8	350; 404.1*; 527; 555 *ε = 1.23 × 10^5 M⁻¹ cm⁻¹				4.45		20% MeOH; pH-independent between pH 5.50 and 7.00	xx

Ligand	Complex	Spectral data	EPR parameters	Value	Conditions	Ref
	MP-8	348; 406.5*; 528; 557 *$\varepsilon = 11.16 \times 10^5$ M^{-1} cm^{-1}	$g = 2.84; 1.51$ $\Delta = 1512; v = 768$		50% ethylene glycol	viii
	N-Ac-MP-8			2.58	25 °C	ix
	MP-6			4.05	Aq solution, pH 8.5	x
2-Methyl imidazole	MP-8	350; 403.4; 528; 555		2.61	20% MeOH pH 7.0	xv
Benzimi-dazole	MP-8	352; 401.8; 526; 555		2.71	20% MeOH pH 7.0	xv
Pyridine	MP-11^i			2.49 (2.85)^j	Aq solution, pH 9.0	xviii
	MP-8	406		2.7	20% MeOH	xxi
	MP-8			2.0	pH 7.5	xxv
	MP-8			0.92^k	pH 12	xxii
Aniline	MP-8	406		2.7	20% MeOH	xxi
	MP-8			1.34^k	pH 12	xxii
N-Methyl pyridinium	MP-8			2.16^k	pH 12	xxii
P-Fluoro aniline^l	MP-8			4.65	50% ethylene glycol pH 7.2	xxiii
NH$_3$	MP-8	350; 403.5*; 528; 555 (sh) *$\varepsilon = 1.1 \times 10^5$ M^{-1} cm^{-1}		3.34	20% MeOH, pH 7–10 Corrected for pK_as of NH$_3$ and MP-8	xxi
Glycine	MP-11^i			3.03 (3.39)^j	Aq solution pH 9	xviii
	MP-8	403		3.46	20% MeOH	xxvi

(continued)

663

Table 20.3 (*continued*)
Selected data for complexes of the microperoxidases.

Ligand	Hemepeptide	ESR[a]	Electronic spectra	Mössbauer[a,b]	μ_{eff}/μ_B	log K	Thermodynamic parameters[g]	Comments	Ref.[o]
Alanine	MP-8		403			2.89		20% MeOH	xxvi
Valine	MP-8		403			3.73		20% MeOH	xxvi
Leu	MP-8		403.5			3.99		20% MeOH	xxvi
Phe	MP-8		406			4.76		20% MeOH	xxvi
Trp	MP-8		406			5.64		20% MeOH	xxvi
NH$_3$	MP-8		403.5			3.34		20% MeOH	xxvii
C. Sulfur donors									
RSH	MP-8					~3		R = Me; Et; iPr see text	xxiii
Cysteine	MP-8		415.2 $\varepsilon = 9.1 \times 10^4\ \mathrm{M^{-1}\ cm^{-1}}$			5.4		pH-independent log K at 5 °C; see text	xiii
Thioglycolate	MP-8		414.8 $\varepsilon = 8.9 \times 10^4\ \mathrm{M^{-1}\ cm^{-1}}$			4.5		m,n	xiii
Cysteamine	MP-8		415.8 $\varepsilon = 8.8 \times 10^4\ \mathrm{M^{-1}\ cm^{-1}}$			5.8		m,n	xiii
Dithiothreitol	MP-8		415.5 $\varepsilon = 9.3 \times 10^4\ \mathrm{M^{-1}\ cm^{-1}}$			6.6		m,n	xiii
N-Acetyl methionine	MP-8		530; 560; 625						xix
	MP-8		weak band @ 695			-0.3			xxiv

Ligand	Compound	Spectroscopic data	Value	Conditions	Ref.
	MP-8	399.6; 522; 555; 620; 695	0.58	20% MeOH pH7.0	xv
	N-Ac-MP-8	g = 2.92; 2.31; 1.51; Δ = 1200; V = 780	4.02 (T-dependent)		ix
Methionine	MP-11	QS = 2.17[d]; IS = 0.45; Δ = 800; V = 370			iii
Thiodiglycol	MP-8	399.7; 490; 528; 624; 685	−0.17	20% MeOH pH 7.0	xv

[a] Δ = tetragonal splitting (cm^{-1}); V = rhombic splitting (cm^{-1}).

[b] QS = quadrupolar splitting (mm s^{-1}); IS = isomer shift (mm s^{-1}).

[c] Estimated from published spectra.

[d] Varies with temperature, which is characteristic of a low-spin complex.

[e] Frozen solutuion.

[f] Lyophilized powder.

[g] $\Delta H°$ in kJ mol^{-1}; $\Delta S°$ in JK^{-1} mol^{-1}.

[h] By NMR.

[i] MP-11 deaminated with HNO$_2$.

[j] Corrected log K assuming pK_a MP-11 = 8.9 (cf. MP-8).

[k] Noncovalent complex formation at high pH.

[l] Both o- and m-fluoroanilines have log K < 1.2. It is unclear whether coordination has occurred.

[m] Corrected for the pK_a of RSH and MP-8.

[n] Under N$_2$ and 5 °C to prevent radical destruction of porphyrin—see text.

[o] i—Tsou, 1951; ii—Wilson et al., 1977; iii—Baumgartner et al., 1974; iv—Baldwin et al., 1987b; v—Blumenthal and Kassner, 1980; vi—Smith and McLendon, 1980; vii—Marques et al., 1987; viii—Owens et al., 1988; ix—Yang and Sauer, 1982; x—Saleem and Wilson, 1988a; xi—Blumenthal and Kassner, 1979; xii—Blumenthal and Kassner, 1979; xiii—Baldwin, D. A., et al.—unpublished results; xiv—Hamilton et al., 1984; xv—Marques, 1986; xvi—Jain and Kassner, 1984; xvii—Huang and Kassner, 1981; xviii—Harbury and Loach, 1960b; xix—Harbury et al., 1965; xx—Baldwin et al., 1986b; xxi—Marques et al.—unpublished results; xxii—Adams et al., 1988; xxiii—Sadeque et al., 1987; xxiv—Smith and McLendon, 1981; xxv—Shimuzu et al., 1982; xxvi—Byfield and Pratt, 1992; xxvii—Marques, Byfield, and Pratt, unpublished results.

Of the ligands that form low-spin Fe(III) MPs, CN^- has the largest formation constant (log $K \approx 7.6$, Table 20.3), as demonstrated for monomeric MP-8 (Marques *et al.*, 1987; Blumenthal and Kassner, 1980) and monomeric MP-9 (Baldwin *et al.*, 1987a). The significantly lower binding constant reported for this process in aqueous solution (log $K = 6.2$) (Smith and McLendon, 1980) is undoubtedly due to aggregation effects, as is testified to by the absence of well-defined isosbestic points in the reported electronic spectrum. The pH dependence of the equilibrium constants for CN^- binding to MP-11 (Wilson *et al.*, 1977), even when one corrects the reported values for the pK_a of HCN, suggests the presence of intermolecular competition for the axial binding site by amino groups of the peptide chain.

Imidazole is bound by monomeric MP-8 with log $K = 4.45$ (Baldwin *et al.*, 1986b); the presence of a substituent on the C atom bound to the coordinating N, as in 2-methyl imidazole and benzimidazole (Marques, 1986), decreases the stability of the complex to log $K = 2.6-2.7$; this is undoubtedly due to steric interaction between the ligand and the porphyrin ring. Both pyridine [log $K = 2.0-2.9$ (Harbury and Loach, 1960b; Shimizu *et al.*, 1982)] and primary amines [log $K = 2.7-3$ (Harbury and Loach, 1960b; Marques *et al.*—unpublished results)] are coordinated less strongly than imidazole, although the affinity for *p*-fluoroaniline (log $K = 4.65$) is about the same (Sadeque *et al.*, 1987b). A further complication which has to be borne in mind is the formation of noncovalent (i.e., $\pi-\pi$) complexes with aromatic ligands such as pyridine, *N*-methyl pyridinium cation, and aniline (Adams *et al.*, 1988).

Byfield and Pratt (1992) have determined the binding constant for a number of amino acids to monomeric MP-8 in aqueous methanol solution. They found that the presence of an aromatic side chain significantly increases log K [e.g., Gly(3.5) < Phe(4.8) < Trp(5.6)] and causes the Soret band to shift by *ca.* 3 nm to longer wavelength. These effects were ascribed to donor–acceptor interaction between the aromatic side chain of the coordinated amino acid and the porphyrin π electron system, suggesting a role for the invariant phenylalanine of the peroxidase enzymes, whereby the residue stabilizes the highly oxidized Fe(IV) and porphyrin π cation species and furthermore facilitates charge movement in the low dielectric environment of the protein interior. Further studies on the interaction of aliphatic/aromatic amines and heterocyclic bases with MP-8 have also been reported (Marques *et al.*, 1993; Byfield *et al.*, 1993; Hamza and Pratt, 1993) and suggest that ammonia, aniline, and pyridine interact with the iron porphyrin predominantly as ligands of the iron as opposed to $\pi-\pi$ conjugates. However, recent work (Munro and Marques, unpublished) has shown that coordination of an aromatic ligand such as an imidazole or a pyridine derivative to monomeric *N*-acetyl MP-8 in aqueous solution proceeds through formation of an initial complex ascribed to a $\pi-\pi$ complex between the macrocycle and the incoming ligand.

Thiols are coordinated by MP-8 to form low-spin complexes; however, reduction of Fe(III) to Fe(II) is competitive, and the resultant thiyl radicals rapidly attack the porphyrin ring leading to irreversible bleaching of the spectrum (Baldwin, D. A., *et al.*—unpublished results). This may be seen, for example, in the spectra reported by Sadeque *et al.* (1987a, 1987b), rendering their reported formation constants highly suspect. This process is retarded on exclusion of O_2 from solution, but the rate of destruction of the chromophore is still sufficiently rapid to interfere significantly with the spectroscopic titration. We have found that at 5 °C under N_2 the rate of destruction of the porphyrin is sufficiently slow (i.e., no significant change in Soret intensity over 2 h) to allow for determination of equilibrium constants. The pH dependence of coordination of cysteine indicates that the thiolate anion is the coordinating species.

There are at least three independent determinations for the binding constant of N_3^- by MP-8 (Blumenthal and Kassner, 1979; Baldwin, D. A., *et al.*—unpublished results; Smith and McLendon, 1980); if the conditions of Equation (20.3) are applied, all three studies show log $K \approx 1.45$, and there is good spectroscopic evidence to demonstrate the existence of the complex as an equilibrium mixture of high-spin and low-spin species.

A number of thioethers have been shown to bind to ferric MPs (Baumgartner *et al.*, 1974; Marques, 1986; Smith and McLendon, 1981; Yang and Sauer, 1982); binding constants are low, emphasizing the inherently weak affinity of Fe(III) for these ligands; this explains why a variety of ligands such as CN^- will bind to cytochrome *c* itself, displacing Met-80 from the coordination sphere.

The coordination of small anionic ligands such as F^-, SO_3^{2-}, SCN^-, and CNO^- has been investigated (Table 20.3). In all cases, the resultant complex appears to be predominantly high-spin, and formation constants are small. The determination of quantitative values, however, is problematic. We have recently found (Marques and Munro—unpublished observations) that on increasing the ionic strength of a solution of *N*-acetylMP-8 with $NaClO_4$, the Soret intensity decreases because of the formation of noncovalently bonded dimers in solution. Since formation constants for binding of small anions are low, high concentrations of ligand in solution are required to effect significant complex formation. The formation of a complex is usually accompanied by a decrease in the Soret band intensity and a shift to longer wavelengths; hence, it is likely that the measured spectroscopic changes on adding increasing amounts of such a ligand to solution are a composite effect of aggregation and coordination. For this reason, equilibrium constant values which have been determined by monitoring absorbance changes in the electronic spectrum for coordination of these small anionic ligands are probably only approximate values.

Although the coordination of methyl isonitrile, nitrosobenzene (Tsou, 1951b), CO (Tsou, 1951b; Marques, 1986; Sharma *et al.*, 1975; Hasinoff, 1981a; Sharma *et al.*, 1976; Eisenstein, 1977; Alberding *et al.*, 1978; Hasinoff, 1981b), NO (Tsou, 1951b; Sharma *et al.*, 1983; Marques, 1986), imidazole (Harbury and Loach, 1960b; Myer and Harbury, 1973; Harbury *et al.*, 1965), *N*-acetyl methionine and a variety of thioethers and thiols (Harbury and Loach, 1960b), and O_2 (Marques *et al.*, 1988) by ferrous MPs has been reported, the only quantitative determination appears to be the early report by Harbury and Loach (1960b) who found log $K = 3.7$ for imidazole, 4.1 for pyridine, and 2.7 for glycine coordination by deaminated ferrous MP-11.

As mentioned, the main interest in determining equilibrium constants for binding of various ligands by MPs has been to allow a comparison of the results with analogous studies with various hemoproteins to elucidate the effect of the protein on the fundamental chemistry of the prosthetic group. This approach allowed Harbury and co-workers, for example, to suggest that the spectroscopic properties of cytochrome *c* could be accounted for by the presence of His and Met in the coordination sphere, rather than His and His as was the general view at the time (Harbury *et al.*, 1965).

Although the MPs are extremely useful models for the prosthetic group of hemoproteins such as Hb, Mb, and the peroxidases, all of which contain a single His as axial ligand, there is a difference in the porphyrin type; the hemoproteins just mentioned above all contain protoporphyrin IX, whereas the MPs contain porphyrin *c*. The difference in peripheral substituents is not expected to have a great effect, however. The influence of heme type on equilibrium constants for substitution of water by other ligands in reconstituted metMbs has been reported (Sono and Asakura, 1976; Makino and Yamazaki, 1973). The *cis* effect is small; for example, for formation of the azido complex, $\log K = 4.42$ for protoporphyrin, 4.52 for deuteroporphyrin, and 4.29 for mesoporphyrin (Sono and Asakura, 1976) and for the cyano complex the values are 4.50, 4.48, and 4.42, respectively (at pH 7, 20 °C) (Makino and Yamazaki, 1973). The electron-withdrawing effect of the heme *c* substituents are expected to be intermediate between those for proto-and deuteroporphyrin; hence, the difference in heme type makes an insignificant contribution to the observed differences in $\log K$ values between the MPs and the hemoproteins. *Trans* effects are likely to be much more important (Morishima and Inubushi, 1978). The retention of His-18, therefore, is a major positive factor in favor of the MPs as models for the hemoproteins.

The MPs bind ligands such as N_3^- (Blumenthal and Kassner, 1979; Baldwin, D. A., *et al.*—unpublished results), CN^- (Marques *et al.*, 1987), and thiols (Baldwin, D. A., *et al.*—unpublished results) as anions, as do metHb and metMb (Antonini and Brunori, 1971). This is in marked contrast to the hydroperoxidases, where these ligands are apparently bound as the undissociated conjugate acids (Millar *et al.*, 1981; Pratt, 1975). The MP results emphasize the unusual behavior of the hydroperoxidases in this respect, a feature of their chemistry which is related to the mechanism of activation of H_2O_2 at physiological pH (Baldwin *et al.*, 1985, 1987b).

A comparison of equilibrium constants for ligand coordination by the MPs and the hemoproteins enable evaluations of the effect of protein structure on the coordination chemistry of the prosthetic group to be made. The distal environment of the cofactor, in particular, is likely to have a profound effect, although other factors will play a role. Baldwin and Chothia (1979) compared the atomic coordinates of human deoxy-, horse met-, and human carbonmonoxyHb and showed that the F-helix moves across the face of the heme by *ca.* 1 Å on coordination of a ligand. This will significantly affect the nature of the bonding to the proximal His, and *trans* effects are expected to be crucial in ligand binding.

Binding constants of ligands to five-coordinate hemoproteins and to MPs are not strictly comparable. The differences are expected to be small, however. For example, $\log K$ values for binding of F^- to the five-coordinate metMb from the mollusc *Aplysia limacina* and the six-coordinate metMb from sperm whale are very similar—1.7 (Giacometti *et al.*, 1981) and 1.9 (Antonini and Brunori, 1971), respectively.

Factors such as ligand field stabilization energy, inherent complex instability due to electronic effects, and steric effects between the ligand and the porphyrin will be similar in the MPs and the hemoproteins, and differences in log K values can then be attributed to the effect of the protein itself. To facilitate comparisons, values of K_{rel}, the ratio of the equilibrium constant for ligand binding to the hemoprotein and to MP-8 respectively, are listed in Table 20.4.

The distal environment of most Hbs and Mbs consists of the side chains of a Phe and Val, and the distal His. From neutron diffraction data, the distal His has been shown to form a hydrogen bond to the uncoordinated oxygen atom of bound dioxygen in MbO_2 (Phillips and Schoenborn, 1981) and may hydrogen-bond to other ligands coordinated to the metal. The ability of a ligand to coordinate to Fe(III) can therefore be expected to be destabilized by steric interactions with the distal amino acid residues and stabilized by hydrogen-bonding to the distal His.

Log K values for binding of F^- to various metMbs and metHbs are similar. It is noteworthy, however, that the value for *Aplysia* metMb, in which the distal His is replaced by Val (Tentori *et al.*, 1971), is not significantly different. The distal His therefore imparts no special stability to F^- complexes. The affinity of the hemoproteins for F^- is some 70 times greater than that of the MPs. This may be attributed to the more hydrophobic environment of the protein, which decreases the dielectric constant and hence increases the coulombic interaction between the anion and the residual positive charge on the metal center.

For SCN^-, log K for coordination by *Aplysia* metMb $K_{(rel)}$ is of the same order as for F^- and is significantly smaller than for coordination by other hemoproteins; in this case the distal His appears to significantly stabilize the complex. No data are available for coordination of CNO^- to *Aplysia* metMb; assuming a $K_{(rel)}$ of *ca.* 50, log K is predicted to be *ca.* 1.2.

Coordination of N_3^- to sperm whale metMb involves no major changes in the structure of the protein (the anion coordinates in a bent fashion and, fitting between the distal residues, is oriented over one of the methine carbon atoms of the porphyrin ring) (Stryer *et al.*, 1964). The absence of the distal His both in *Aplysia* metMb and in *D. dendriticum* metHb (Smit and Winterhalter, 1981) results in a marked decrease in K_{rel} for binding of N_3^-; the average value, 67, for these two hemoproteins is in the range expected for the coordination of an anion without additional stabilization by interaction with distal residues.

For metHb M (Hyde Park) and metHb M (Iwate), where the distal His is replaced by Tyr, K_{rel} is of the same order as that for hemoproteins which contain the distal His. Tyr has a pK_a of 9.6–10 in proteins (Cantor and Schimmel, 1980); on this basis one might expect the distal Tyr in these mutant Hbs to be protonated at physiological pH and the evident interaction with bound ligands to be due to hydrogen bonding. However, in close proximity to a charged metal ion, this pK_a may well be <7.

Hydrogen atoms are poor diffractors of x-rays, and the existence of a hydrogen bond between the distal His and a bound ligand can only be inferred from the distance between them in structure determined by x-ray diffraction methods. McCoy and Caughey (1970) argued that the interaction between the distal His and a ligand arises not from hydrogen bonding, but from donation of electron density from a N

Table 20.4
Summary of equilibrium data for various ligands with ferric MP-8 and various ferric hemoproteins.

Ligand (L)	Hemoprotein/ hemepeptide	log K[a]	K_{rel}[b]	λ_{Scoret} nm	$\Delta\lambda$[c] nm	pH	T °C	Ref.[aa]
F⁻	MP-8	−.022[e,g]				7.6	25	i
	Sperm whale Mb⁺[f]	1.85	74	406	−3.5	7.5	20	ii
	Horse heart Mb⁺	1.84	73	406[h]	−2.0[i]	7.0	20	iii
	Aplysia Mb⁺[j]	1.7	53			7.0	20	iv
	Hb⁺ A[k]	1.77	62			7.5	20	ii
	Hb⁺ C[l]	1.89	82			7.5	20	ii
	Hb⁺ A, α-chain	1.85	64			7.5	7	ii
	Horse Hb⁺			403	−2.0[m]			
SCN⁻	MP-8	−.52[g]		399.8	2.6	7.0	25	d
	Horse heart Mb⁺	2.42	870	411.5[n]	3.5	7.0	20	iii
		2.35	740			7.1	21–23	ii
	Aplysia Mb⁺	1.05	37			7.0	20	iv
	Hb⁺ A, α-chain	2.6	1300			7.2	10	v
	Hb⁺ A	2.6	1300			7.0	20	ii
CNO⁻	MP-8	.083[g]		401	3.8	7.4	25	g
	Horse heart Mb⁺	2.58	310	426	18	7.0	20	iii
		2.38	200			7.1	21–23	ii
	Sperm whale Mb⁺	3.3	1600			7.0	25	ii
	Hb⁺ A, α-chain	3.2	1300			7.2	10	v
	Hb⁺ A	3.19	1300			7.4	25	vi
N₃⁻	MP-8	1.45[g]		410.2	13	7.0	25	d
	Horse heart Mb⁺	4.46	1000	419[n]	11	7.0	20	iii
	Sperm whale Mb⁺	4.98	3400	422	12.5	7.0	20	vii
		4.7	1800			6.2	20	viii
	Sperm whale Mb⁺ modified with BrCN	+[o]						
	D. coriacea Mb⁺[p]	5.0	3500	418[q]	9	6.2	20	viii
	Aplysia Mb	3.4	89	~418[r]	16[s]	7.0	20	iv
	Hb⁺ A	5.41	9100			7.5	20	ii
	Hb⁺ C	5.38	8500			7.5	20	ii
	Hb⁺ S[t]	5.41	9100			7.5	20	ii
	Hb⁺ A, α-chain	4.91	2900			7.5	7	ii
	D. dendriticum Hb⁺[u]	3.1	45			7.0	25	ii
	Hb⁺ M$_{Hyde Park}$[v]	5.17	5200			6.8	25	ii
	Hb⁺ M$_{Iwate}$[v]	5.27	6600			6.8	25	ii
Imidazole	MP-8	4.45[g]		404.1	6.9	7.0	25	d
	Horse heart Mb⁺	1.6	0.0015			8.2		ix
	Sperm whale Mb⁺	2.1	0.0045	416[w]	6.5	7.5	20	x
		1.6	0.0014			7.0	25	ii
	Aplysia Mb⁺	2.3	0.0071	416[w]	14	7.5	20	x
	Hb⁺ A, α-chain	2.6	0.014			7.2	10	v
		2.35	0.0079			7.2	24	xi
		2.3	0.0071			7.2		xii
	Hb⁺ A, β-chain	1.6	0.0014			7.2		xii
	Hb⁺ horse			411	6			ii
2-Me imidazole	MP	2.61[g]		403.4	6.2	7.0	25	d
	Horse heart Mb⁺	1.3	0.049			8.2		ix

Table 20.4 (*continued*)

Ligand (L)	Hemoprotein/ hemepeptide	log K^a	$K_{rel}{}^b$	λ_{Soret} nm	$\Delta\lambda^c$ nm	pH	T °C	Ref.^{aa}
Benz-	MP-8	2.71^g		401.8	4.6	7.0	25	d
imidazole	Sperm whale Mb$^+$	1.8^x	0.12			7.0	25	ii
CN$^-$	MP-8	7.58^g		410.5	13.3	7.0	25	d
	Sperm whale Mb$^+$	6.81^y	0.17	424	14.5	7.0	20	vii
		8.4	6.6	422	12.5	7.0	20	ii
	Sperm whale Mb$^+$ modified with BrCN	$-^z$						
	Hb$^+$ A	8.74	14			7.0	20	xiii
		8.9	21			7.0	20	ii
	Horse Hb$^+$			419	11			

^{a} K refers to the reaction His–Fe(III)–H_2O + L $\rightleftharpoons$ His–Fe(III)–L + H_2O, where L = ligand of column 1.

^{b} $K_{rel} = K_{hemoprotein}/K_{MP-8}$.

^{c} $\Delta\lambda = \lambda_{Soret}$ (complex) $- \lambda_{Soret}$ (aquocomplex).

^{d} In 20% MeOH; for other values see Table 20.3.

^{e} In 50% ethylene glycol solution.

^{f} Mb$^+$ = metMb; Hb$^+$ = metHb.

^{g} See Table 20.3.

^{h} pH 6.

^{i} Horse Mb$^+$(H_2O), λ_{Soret} = 408 nm^{ii}.

^{j} *Aplysia limacina* is a mollusk; the distal His is replaced by Val^{xiv}.

^{k} Human adult Hb.

^{l} Hb which causes mild hemolytic anaemia; β(A3)Glu Lys^{ii}.

^{m} Horse Hb^i(H_2O), λ_{Soret} = 405 nm^{ii}.

^{n} pH 6.

^{o} Reported to bind, but no quantiative data given.

^{p} *Dermochelys coriacea* is a marine turtle; the heme environment is virtually identical to that of sperm whale Mb.

^{q} pH 5.3^{viii}.

^{r} Estimate from Fig. 3 of Ref. iv.

^{s} *Aplysia* Mb$^+$(H_2O), $\lambda_{Soret} \sim$ 420 nm (estimated from Fig. 2 of Ref. iv).

^{t} Sickle cell Hb; β(6) Glu Val.

^{u} *Dicrocoelium dendriticum* is a small liver fluke; the distal His is replaced by Gly^{xv}.

^{v} Mutant Hb's which autoxidise readily. In Hb M$_{Hyde\,Park}$ the distal His of β chains is replaced by Tyr; in Hb M$_{Iwate}$, the same replacement occurs in the α chains.

^{w} Estimated from Fig. 1 of Ref. x.

^{x} From kinetic measurements.

^{y} Assuming only CN$^-$ coordinates, the reported value (log K = 4.50) was corrected for pK_{HCN}.

^{z} No complex formed.

^{aa} i—Huang and Kassner, 1981; ii—Antonini and Brunori, 1971; iii—Sono and Asakura, 1976; iv—Giacometti *et al.*, 1981; v—Uchida *et al.*, 1971b; vi—Scheler and Jung, 1954; vii—Makino and Yamazaki, 1973; viii—Ascenzi *et al.*, 1984; ix—Eaton and Wilkins, 1978; x—Bolognesi *et al.*, 1982; xi—Uchida *et al.*, 1971a; xii—Brittain, 1981; xiii—Anusiem *et al.*, 1966; xiv—Tentori *et al.*, 1971; xv—Smit and Winterhalter, 1981.

atom of His to the ligand. This interpretation is also favored by Kassner and co-workers (Huang and Kassner, 1981b), who suggest that the low value of K_{rel} for F^- is a consequence of the inability of this ligand to act as an electron acceptor.

With the data available, one cannot distinguish between stabilization involving hydrogen bonding and electron density donation. As far as the low value of K_{rel} for F^- is concerned, it could be argued that this small ligand does not protrude far enough to enable hydrogen bonding between it and the distal His to occur. Whatever the nature of the interaction with the distal residue (His or Tyr), it markedly stabilizes the azide complexes of metMb and metHb.

K_{rel} values for coordination of CN^- are quite small. This could be due either to the absence of the stabilizing effects of the distal His or to the fact that, unlike N_3^-, CN^- prefers linear coordination. The crystal structure of cyanometHb (Deatherage et al., 1976) shows that, because of interaction with the distal His, coordination of CN^- results in major changes both to the protein structure and to the heme itself. The distortion of the heme group and the movement of (principally) the E, F, and G helices allows for linear coordination of the ligand. Although the distal His is still oriented favorably for H-bonding, the major rearrangements required for coordination of CN^- cause log K values to decrease.

Bulky nitrogenous bases, imidazole, 2-methylimidazole, and benzimidazole have $K_{rel} < 1$. Crystal structures show that coordination of imidazole causes the distal His to be expelled from the heme cavity both in sperm whale metMb (Bolognesi et al., 1982) and in leghemoglobin (Kuranova et al., 1982). Since log K for coordination of imidazole by *Aplysia* metMb and other metMbs is similar, the steric gate set up by Val E11 and Phe CD1 is the dominant factor in affecting the stability of the complexes with this ligand.

B. Kinetics of Ligand Substitution Processes

Few quantitative studies on the kinetics of the substitution of H_2O in ferric MPs have been reported. Available data are reported in Table 20.5.

The second-order rate constants for replacement of H_2O in MP-11, MP-8, and MP-6 have been reported (Wilson et al., 1977; Smith and McLendon, 1980; Saleem and Wilson, 1988b); in all these studies it was assumed that only CN^- acted as incoming ligand, and in all cases the studies were complicated by deaggregation effects subsequent to ligand coordination. These results can therefore only be taken as approximately correct, since they do not strictly refer to substitution of H_2O in a monomeric MP, and the extent of aggregation in solution is likely to have a significant effect on the measured rate constants.

Marques et al. (1987) studied the kinetics of the reaction of monomeric MP-8 in 20% MeOH, 25 °C, as a function of pH between pH 5.5 and 12 and demonstrated that cyano-MP-8 is formed via three routes [Equations (20.4)–(20.6)]:

$$His-Fe-H_2O + CN^- \xrightarrow{k_1} His-Fe-CN + H_2O, \qquad (20.4)$$

$$His-Fe-H_2O + HCN \underset{k_{-2}}{\overset{k_2}{\rightleftharpoons}} His-Fe-NCH + H_2O, \qquad (20.5a)$$

$$His-Fe-NCH + OH^- \xrightarrow{k_3} His-Fe-NC + H_2O, \qquad (20.5b)$$

Table 20.5

Kinetics of selected reactions of microperoxidases.

(A) Ligand substitution reactions on Fe(III)

Ligand	Hemepeptide	Reaction	Rate constant $M^{-1} s^{-1}$ ($T = 25 °C$)	Comments	Ref.[a]
CN^-	MP-11	$(His–Fe–X)_n + CN^- \rightarrow His–Fe–CN$	3.4×10^6	Only CN^- considered as entering nucleophile. Deaggregation effects seen subsequent to ligand binding	i
CN^-	MP-8	$(His–Fe–X)_n + CN^- \rightarrow His–Fe–CN$	1.4×10^6 $\Delta H^{\ddagger} = 39$ kJ mol^{-1} $\Delta S^{\ddagger} = 92$ JK^{-1} mol^{-1}	Only CN^- considered as entering nucleophile; CN^--independent changes seen after coordination, probably due to deaggregation	ii
CN^-	MP-8	$His–Fe–H_2O + CN^- \rightarrow His–Fe–CN + H_2O$	6.0×10^6	In 20% aqueous MeOH	iii
		$His–Fe–H_2O + HCN \rightarrow His–Fe–CN + H^+ + H_2O$	4.8×10^3	In 20% aqueous MeOH	iii
		$^-His–Fe–OH + CN^- + H^+ \rightarrow His–Fe–CN + OH^-$	1.8×10^5	In 20% aqueous MeOH	iii
CN^-	MP-6	$(His–Fe–X)_n + CN^- \rightarrow His–Fe–CN$	c. 1.8×10^6	Reported rate constant corrected for pK_a of HCN and of MP-6 (assumed to be ca. 9)	iv

(continued)

673

Table 20.5 (*continued*)
Kinetics of selected reactions of microperoxidases.

(B) Ligand substitution reactions on Fe(II)

Ligand	Hemepeptide	Reaction	Rate constant $M^{-1} s^{-1}$	Comments	Ref.[a]
CO	MP-11	$(His-Fe(H_2O))_x + CO$ $\rightarrow His-Fe-CO$	2×10^7	$T = 20\ °C$; complications due to heme aggregation noted	v
CO	MP-11	$His-Fe(H_2O) + CO$ $\rightarrow His-Fe-CO$	2×10^7	$T = 25\ °C$; mixed solvents used to minimize heme aggregation effects	vi
NO	MP-11	$His-Fe(H_2O) + NO$ $\rightarrow His-Fe-NO$	1.1×10^6	$T = 20\ °C$; there may have been aggregation effects in this study	vii

(C) Reduction of Fe(III) to Fe(II)

Reductant	Hemepeptide	Rate constant $M^{-1} s^{-1}$	Comments	Ref.[a]
$S_2O_4^{2-}$	MP-9	1.02×10^8 4.1×10^7	$T = 10\ °C$; pH 7.0 $T = 10\ °C$; pH 8.95; in both cases SO_2^- is the reductant	viii
$S_2O_4^{2-}$	MP-9, MP-11	1.3×10^8 to 3.0×10^8	$T = 20\ °C$; independent of nature of axial ligand	ix

* i—Wilson *et al.*, 1977; ii—Smith and McLendon, 1980; iii—Marques *et al.*, 1987; iv—Saleem and Wilson, 1988b; v—Sharma *et al.*, 1975; vi—Hasinoff, 1981a; vii—Sharma *et al.*, 1983; viii—Kazmi *et al.*, 1985; ix—Peterson and Wilson, 1987.

$$His-Fe-NC \underset{k_{-4}}{\overset{k_4}{\rightleftharpoons}} His-Fe-CN, \qquad (20.5c)$$

$$^-His-Fe-OH + CN^- \overset{k_5}{\rightarrow} {}^-His-Fe-CN, \qquad (20.6)$$

and then, fast,

$$^-His-Fe-CN + H^+ \rightarrow His-Fe-CN.$$

Hence (i) both CN^- and HCN act as nucleophiles towards Fe(III); (ii) OH^- *trans* to neutral His is inert; but (iii) OH^- *trans* to anionic His is labile. The values for the rate constants k_1, k_2, and k_5 are listed in Table 20.5. Comparison of the results from this study with those from other studies indicate that aggregation effects in solution decrease the measured rate constants for attack by CN^- from 6×10^6 M^{-1} s^{-1} to $2-3 \times 10^6$ M^{-1} s^{-1}.

The results of the preceding study provide a basis for evaluating the effect of the protein on the kinetics of ligand substitution reactions of iron porphyrins. As with equilibrium constants, the *cis* effect is small, with the rate of formation of the cyanide complexes of meso-, deutero-, and protohemin reported as 3.7, 7.4, and 3.3×10^3 M^{-1} s^{-1} (Wang *et al.*, 1975). MetMb's and metHb's from various sources react insignificantly slowly with HCN and react with CN^- with rate constants of $2-5 \times 10^2$ M^{-1} s^{-1} (Awad and Badro, 1967; Goss *et al.*, 1982, and ref's therein; Seybert *et al.*, 1976), i.e., 10^4 times more slowly than with MP-8. The retardation effect of the protein may be ascribed to (i) the absence of a well-defined ligand channel to the metal site (Ringe *et al.*, 1984) and/or (ii) the considerable steric interaction between the linearly coordinated CN^- ligand and the distal environment of the protein (see earlier discussion). The heme site of the soybean leghemoglobins is much more accessible to an incoming ligand; the rate constant for reaction with CN^- is 9×10^6 M^{-1} s^{-1} (Job *et al.*, 1980), provided that a group (probably a heme propionate) with a $pK_a = 4.7$ is protonated. Deprotonation of this group sets up an electrostatic barrier to the incoming ligand and the rate constant decreases to 2×10^4 M^{-1} s^{-1}. Multiple ionizations also control the rate of reaction of the peroxidases with CN^- (Araiso and Dunford, 1981; Job and Ricard, 1975); nevertheless, the rate constants ($2-9 \times 10^4$ M^{-1} s^{-1}) are too fast for HCN to be the incoming nucleophile. Although these hemoproteins apparently bind HCN and not CN^- (see earlier discussion), the kinetics suggest the existence of a distinct proton uptake site. This is in accordance with the proposed mechanism of activation of H_2O_2 by these enzymes, where deprotonation of the substrate is envisaged to occur prior to coordination of HO_2^- (Baldwin *et al.*, 1985; 1987b). A similar mechanism may be operative in the case of cyanide inhibition of half-reduced cytochrome oxidase (rate constant of the order of 10^6 M^{-1} s^{-1} (Jones *et al.*, 1984)), in which cyanide is apparently bound as HCN.

A few studies with CO and NO as incoming ligand to ferrous MP-11 have been reported. Both Sharma and co-workers (1975) and Hasinoff (1981a) report rate constants of 2×10^7 M^{-1} s^{-1} at 20 °C and 25 °C, respectively, for binding of CO. In the first study, aggregation effects were noted; in the second, mixed solvents were used to minimize these effects. The rate constant for reaction of MP-11 with NO is reported as 1.1×10^6 M^{-1} s^{-1} at 20 °C (Sharma *et al.*, 1983); aggregation effects cannot be excluded, however.

Ultrafast flash photolysis has been used to study the rebinding of CO to MP-8 in glycerol-water glasses at 100 K (Postlewaite *et al.*, 1989). Photolysis leads to a short-lived intermediate, [Fe- - -CO], which subsequently relaxes to the deligated heme, [Fe + CO]. Recombination can be observed from both [Fe + CO] and (ultrafast) from [Fe- - -CO].

C. Reduction and Self-Exchange Reactions

Kazmi *et al.* (1985) and Peterson and Wilson (1987) have investigated the kinetics of reduction of various MPs with dithionite anion; both studies show that the reducing species is the SO_2^- radical. Since the rate of reduction (*ca.* 2×10^8 M^{-1} s^{-1}) is virtually independent of the nature of the axial ligand (CN^-, H_2O, or OH^-) and the state of aggregation of the hemepeptides, it was concluded that the reduction occurs through an outer-sphere electron transfer process. Both studies point out that the inherent reactivity of the MPs towards the reductant is considerably greater than that of cytochrome c itself [*ca.* 4×10^7 M^{-1} s^{-1} (Lambeth and Palmer, 1973)], probably because of the greater degree of exposure of the heme edge in the polypeptides, although there could be mechanistic differences in the reaction of the reductant with the cofactor in the native protein.

Kimura *et al.* (1981) related ^{1}H-NMR line widths obtained during a redox titration of cyano-MP-11 to the percentage of each oxidation state in solution and obtained a self-exchange rate constant of 1.3×10^7 M^{-1} s^{-1} at 330 K. This is in agreement with a value of $k_{11} > 10^6$ M^{-1} s^{-1} reported subsequently (McLendon and Smith, 1982) and obtained by studying the photoinduced electron transfer from $Ru(bpy)_3^{2+}$ to MP-8. The self-exchange rate constants are therefore some three to four orders of magnitude larger than for cytochrome c itself, again illustrating the retarding effect which the protein has in these reactions.

VI. Reactions of Microperoxidases: Oxygen Utilization and Peroxidase Activity

Reaction with oxygen and the catalysis of redox processes are among the most intensively studied of hemeprotein functions. Paralleling the studies carried out using the intact hemeprotein, a prolific literature has developed reporting chemical modeling of these reactions using both simple and more complex derivatized iron porphyrins specifically designed to mimic some aspect of hemeprotein architecture around the catalytically active metalloporphyrin moiety. The reaction of oxygen with synthetic metalloporphyrins, as well as the peroxidasic activity of these compounds, has been extensively reviewed (Jones *et al.*, 1979; Collman, 1977; Traylor, 1981; Sykes, 1982; Adams, 1990b).

A. $Fe^{2+}O_2MP-8$: Preparation and Reactivity

Reduction of deoxygenated solutions (50:50 (v/v) DMF/DMSO) of $Fe^{3+}MP-8$ by stoichiometric amounts of an aqueous dithionite solution gave a product with a well-defined UV/visible absorption spectrum characteristic of an Fe^{2+} porphyrin, *cf.* Table 20.6. This compound reacted with O_2 at -15 °C to give a new complex with absorption bands (Soret, α and β) similar to those observed for the O_2 complexes of synthetic Fe^{2+} porphyrins—Table 20.6—and assumed therefore to be the $Fe^{2+}O_2$ derivative of MP-8 (Marques *et al.*, 1988).

Reduction of $Fe^{3+}MP-8$ using dithionite is problematical since the reductant can attack the porphyrin ring system even under the most carefully controlled anaerobic conditions, leading to irreversible heme degradation. This problem was circumvented by a procedure, used by Valentine and others (Valentine, 1979; McCandlish *et al.*, 1980), for the preparation of $Fe^{2+}O_2$ derivatives of simple metalloporphyrins, whereby the oxygenated complex is formed on reaction of deoxygenated $Fe^{3+}MP-8$ in DMF-DMSO with a 15–20-fold molar excess of KO_2 (in DMSO solubilized with 18-crown-6) at -15 °C (Marques *et al.*, 1988). The

Table 20.6
UV-Visible spectra of reduced hemepeptide/hemepeptide adducts.

Hemepeptide/adduct	λ_{max} nm (ε mM^{-1} cm^{-1})			Comments	Ref.[c]
	Soret	α	β		
$Fe^{2+}MP-8$	419.9 (208 $\pm$ 12)	549	519	In 50:50 DMF-DMSO	i
$Fe^{2+}MP-11$	417(189)[a]	530(31)[a]	520(15)[a]	0.1 M tetraborate buffer pH 9.5	ii
$Fe^{2+}MP-8$	419.5(191)	—	—	20% aq MeOH pH_{app} 7.5–8	iii
$Fe^{2+}MP-8$	420.5(160)	—	—	pH 12; 20% aq MeOH	iii
$Fe^{2+}O_2MP-8$	413.5(101)[b]	568(10.5)[b]	538(15.3)[b]	50:50 DMF-DMSO -15 °C	i
$Fe^{2+}MP-8-An$	415(125)	—	—	pH 12; 20% MeOH prep. by addition of aniline to $Fe^{2+}MP-8$	iii
$Fe^{2+}MP-8-An$	414(128)	—	—	pH 12; 20% MeOH prep. by NADH reduction of $Fe^{3+}MP-8$ An	iii

[a] Estimated from Fig. 2 of Ref. ii.
[b] Estimated from Fig. 1 of ref. i.
[c] i—Marques *et al.*, 1988; ii—Peterson and Wilson, 1987; iii—Adams and Adams, 1988.

$Fe^{2+}O_2$MP-8 product was relatively stable ($t_{1/2} \approx 6$ h at -15 °C) and exhibited none of the degradation products observed when it was prepared via the dithionite route. CO and NO complexes of Fe^{2+}MP-8 were formed on flushing solutions of the $Fe^{2+}O_2$ complex with CO and NO gas, respectively.

Reaction of $Fe^{2+}O_2$MP-8 with organic reducing agents (HX) was used to model the reaction of HbO_2 with these compounds (Marques *et al.*, 1988). The latter reaction involves an unusual H-atom transfer to leave a free radical X, which then undergoes further reaction. $Fe^{2+}O_2$MP-8 reacted with HX in the order X = $-SH$ > hydroquinone (QH_2) > $PhNHNH_2 \gg$ others (models for amino acid side chains). The reaction with $PhNHNH_2$—which does not readily undergo electron transfer reactions—and the order QH_2 > phenol(ate) indicated that the H-transfer reaction is a characteristic of $Fe^{2+}O_2$ porphyrin complexes and does not appear to involve a protein-dependent functional moiety.

B. MP-Mediated O_2 Activation/Insertion Reactions

Adams and co-workers, extending earlier studies made using ferriprotoporphyrin IX as catalyst (Adams *et al.*, 1986, and references therein) investigated the *ab initio* MP-8 catalysis of O_2 activation (to HO_2^-) which occurs at high pH in the obligatory presence of substrate (aniline-An) and reductant (NADH), prior to regiospecific *para* hydroxylation of the aniline (Adams *et al.*, 1988; Adams and Adams, 1988). It was suggested that the first reduction by NADH to Fe^{2+}MP-8, prior to O_2 binding and metabolism, could only take place by reaction with the $Fe^{3+} \cdot$ MP-8-aniline complex. The requirement for aniline was rationalized in terms of electron transfer from NADH to Fe^{3+} occurring via the π electron system of a planar complex formed between the aromatic ring of aniline and the tetrapyrrole ring system of Fe^{3+}MP-8 in a manner similar to that previously envisaged by Bonnett (1981).

Lipid peroxidation in rat brain and liver microsomes is inhibited by MP-9 (Vodnyanszky *et al.*, 1985, 1986; Venekei *et al.*, 1986). This was explained in terms of MP-9 acting as an electron shunt for the microsomal NADPH cytochrome P-450 reductases [for which the heme peptide is a good substrate (Vegh *et al.*, 1986)], that is, inhibition occurring prior to the cytochrome P-450 catalytic cycle, or alternatively, that MP-9 bound and metabolized active oxygen intermediates formed during the cyt P-450 redox cycle. The latter possibility, combined with the observation that both O_2^- and H_2O_2 are formed during oxidation of reduced MP-9 (Meszaros *et al.*, 1984), led Rosvai and co-workers to investigate the MP-9 catalyzed *para*-hydroxylation of aniline at neutral pH, both NADPH cytochrome P-450 reductase–mediated and oxy-radical–mediated processes being studied (Rosvai *et al.*, 1988). *p*-Aminophenol formation rates were enhanced by superoxide dismutase (SOD) and inhibited by catalase in the reductase-mediated process, suggesting that the hydroxylation cycle proceeds via an $Fe^{2+}O_2$MP-9 complex—which could dissociate to give free O_2^-—followed by a second-electron reduction of the $Fe^{2+}O_2$MP-9 complex to give H_2O_2.

While in substantial agreement, the oxygen activation studies utilizing MP-8 and MP-9 just described differ in one important respect. In the high-pH MP-8-mediated *ab initio* system, SOD inhibits the observed induction phase of the reaction and is without effect on the steady-state catalytic cycling rate, as found for the hemin-mediated process (Adams—unpublished observations; Adams and Adams, 1989), while at neutral pH SOD enhances the rate of *p*-aminophenol formation (Rosvai *et al.*, 1988). The high-pH MP-8 effect parallels the Cu–Tyrosine (SOD analog) inhibition observed for microsomal cyt P-450–mediated hydroxylation (Werringloer *et al.*, 1979), i.e., inhibition without H_2O_2 formation, and was explained in terms of nonspecific inhibition at the level of electron transfer from NADH to Fe^{3+} in the initial induction region of the reaction. The lack of SOD effect during the steady-state cycling of the MP-8 system at high pH is most probably due to the well-documented (Hodgson and Fridovich, 1975), very rapid inactivation of SOD by H_2O_2 (relatively high concentrations of which are free in solution during the steady-state phase) which occurs at high pH, but which would not occur in the neutral MP-9 system.

A range of MP-11–catalyzed H_2O_2-supported reactions—sulfide oxidation, amine-*N*-demethylation, and olefin oxidation—paralleling the corresponding Cyt P-450-catalyzed reactions has recently been reported (Mashino *et al.*, 1990). These observations complement those of Adams and Adams (1986) made using alkaline hemin at high pH to catalyze *ab initio* a range of oxidative bond cleavage functions—oxidative *N*-demethylation, *O*-dealkoxylation, etc.

C. Hemepeptides as Peroxidase Models

The fact that the hemepeptides from cytochrome *c* exhibit peroxidase activity—i.e. catalytically reduce H_2O_2 to water in the presence of an appropriate reducing substrate (RH) as shown in Equations (20.7)—has long been appreciated, and the early reports of this "microperoxidase" activity, in which particular emphasis was placed on the inhibitory/stimulatory effects of nitrogenous ligands of the hemepeptides, have recently been reviewed (Adams, 1990b).

$$Fe^{3+} + HO_2^-(H_2O_2) \rightleftharpoons Fe^{3+}HO_2^-(H_2O_2),$$
$$\text{(peroxidase)}$$

$$Fe^{3+}HO_2^-(H_2O_2) \rightarrow \text{Cpd I} + OH^-(H_2O),$$
$$\text{oxo-iron(IV)}$$
$$\text{porphyrin } \pi \text{ cation}$$

$$\text{Cpd I} + RH \rightarrow \text{Cpd II} + R^\cdot + H^+,$$
$$\text{oxo-iron(IV)}$$
$$\text{porphyrin}$$

$$\text{Cpd II} + RH \rightarrow Fe^{3+} + H_2O + R^\cdot.$$

(20.7)

Several points of major relevance to the mechanism and function of peroxidase enzymes can be investigated using hemepeptides as models for protein-free monomeric hemin, with an axial histidine mimicking the invariant proximal histidine of the peroxidase enzymes. These concern (a) the active form of the substrate, i.e., HO_2^- or H_2O_2; (b) the mechanism of O–O bond cleavage in the $Fe^{3+}HO_2^-/H_2O_2$

complex, i.e., heterolytic or homolytic—this further indicates (Bruice *et al.*, 1986) whether the catalytic cycle proceeds via Cpd I and Cpd II analogs in analogy to the peroxidase cycle, or directly via a Cpd II analog; and (c) the protective effect of protein in preventing irreversible porphyrin ring system degradation in redox catalysis.

Buffer catalysis and solvent isotope effects have been used to probe the nature of –O–O– bond cleavage in the reaction of MP-11 [and other Fe(III)porphyrins] with H_2O_2, *t*-BuOOH and peracids (Traylor and Xu (1990)). Comparison with the corresponding reaction between hydroperoxides and dialkyl sulfides and those of peracids with hemin (both reactions proceeding by a well-established heterolytic cleavage mechanism) led to the conclusion that cleavage was heterolytic in the case of both the buffer- and solvent-catalyzed pathways.

Other current investigations have concentrated more specifically on modeling mechanistic aspects of the peroxidase catalytic cycle using the octapeptide MP-8, which is essentially monomeric in neutral aqueous solution at catalytically effective concentration ($< 10^{-6}$ M).

Baldwin *et al.* (1987b) investigated the MP-8/H_2O_2 peroxidasic reaction using guaiacol (*o*-methoxyphenol) as reducing substrate under conditions where the formation of a Compound I analog was rate-limiting. From the pH dependence of the rate constant for reaction of H_2O_2 with MP-8 to form Cpd I, it was inferred that the active form of the substrate was the hydroperoxide anion HO_2^-, and on this basis the pH-independent rate constant for reaction of aquo MP-8 with HO_2^- was estimated to be $3.7 \pm 0.4 \times 10^8$ M^{-1} s^{-1}, in good agreement with the value of 1.3×10^8 M^{-1} s^{-1} reported for reaction of H_2O_2 with cytochrome *c* peroxidase (Yonetani and Schleyer, 1967). Protonated guanidinium ion (GuaH$^+$) accelerated the peroxidasic activity of MP-8 more than 10-fold without coordinating to the metal. This acceleration was ascribed (by comparison with the crystalline urea-H_2O_2 and GuaH$^+$HCO$_3^-$ adducts) to formation of a GuaH$^+ \cdot HO_2^-$ ion pair via double H-bonding, each from one NH of GuaH$^+$ to one O of HO_2^-. This suggests that the role of the invariant distal arginine of the peroxidase enzymes is to facilitate a proton-coupled reaction enabling the enzyme to bind H_2O_2 as HO_2^- and H$^+$ in a pH-independent process, thus enabling the enzyme to function in a catalytically efficient manner at physiologically relevant pH.

The mechanism of the same reaction has been studied from a different standpoint using 2,2'-azinobis(3-ethylbenzthiazoline-6-sulfonate) (ABTS) as reducing substrate (Adams, 1990a). Oxidation of ABTS by hypervalent oxo-iron species results in formation of the metastable emerald green ABTS$^+$ cation radical, allowing kinetics to be monitored spectrophotometrically at 660 nm. A detailed investigation revealed three parallel processes, which were identified as the following:

(a) a normal peroxidasic catalytic cycle with free MP-8 as the catalyst;
(b) a parallel catalytic cycle to (a) in which the catalyst was an MP-8-ABTS complex; and
(c) a degradative process involving direct attack of H_2O_2 on the MP-8 porphyrin ring.

Using the criterion of Rush and Koppenol (1988), whereby the effect of Br^- on the efficiency of the reaction is measured, it was shown that $OH^{\cdot}$ radicals were not formed in significant amounts during catalysis.[2] This implied that O–O bond cleavage in the $Fe^{3+} \cdot HO_2^-$ complex is heterolytic, resulting in formation of an oxo-iron(IV) porphyrin π cation radical intermediate, a peroxidase compound I analog. This conclusion was supported by Wang *et al.* (1991), who found that the reaction of N_{α}-acetyl MP-8 with H_2O_2 using ABTS as reducing substrate resulted in formation of a compound I analog via heterolytic cleavage of the –O–O– bond. Strong evidence for formation of a compound II analog was also found in the system. Low-temperature stopped-flow studies demonstrated the formation of a transient 340-nm absorption band characteristic of the so-called cpd O, a new intermediate of unknown structure which precedes cpd I in the horseradish peroxidase catalytic cycle (Baek and Van Wart, 1989).

The relative catalytic efficiency for the MP-8-catalyzed oxidation of a range of phenols, anilines, and naphthols (as reducing substrate) by H_2O_2 has been found to be dependent on the nature, but not the concentration, of substrate (Cunningham *et al.*, 1991). This was interpreted by the authors as providing kinetic evidence for a "substrate-involved" MP-8 decomposition pathway, presumably via formation of an MP-8–substrate complex as postulated in the case of ABTS for process (b) above. The existence of such complexes for horseradish peroxidase (e.g., Childs and Bardsley, 1975) would pose interesting questions as to how the enzyme protects itself from this type of degradation during catalysis of H_2O_2 reduction.

Parallel studies have been carried out using ABTS with myoglobin as catalyst (Adams—unpublished observations). This hemeprotein catalyzes reduction of H_2O_2 in a peroxidase-like reaction; however, –O–O– bond cleavage occurs in a mixed homolytic/heterolytic manner, the former mechanism giving an oxo-iron(IV) species, a peroxidase Cpd II analog, and a protein radical. The latter is probably located on a tyrosine residue (Ortiz de Montellano, 1987) and is formed by reaction with $OH^{\cdot}$ produced on homolysis of the –O–O– bond of the obligatory $Fe^{3+}Mb \cdot H_2O_2$ Michaelis complex. In the presence of Br^-, it was found that the efficiency of ABTS oxidation in the myoglobin system is increased ~ 10–15% [*cf.* MP-8 catalysis, where Br^- was without effect (Adams, 1990a)]. Addition of other anions (Cl^-, PO_4^{3-}) in the same concentration range led to a 3–5% drop in efficiency. This observation thus appears to validate the Rush–Koppenol criterion for distinguishing heterolytic/homolytic O–O cleavage in iron porphyrin peroxidasic-type systems.

The Mb-catalyzed process could, unlike the MP-8-catalyzed reaction, be repeated several times by addition of further H_2O_2/ABTS to the same catalytic system, with little if any change in efficiency of $ABTS^+$ formation. This directly demonstrates the protective function of the protein in enzymic redox reactions, whereby degradative attack of H_2O_2/HO_2^- on the porphyrin ring system is prevented. It is

---

[2] $OH^{\cdot}$ radicals oxidize ABTS with an efficiency of *ca.* 58%, whereas bromine radicals—formed quantitatively on reaction of $OH^{\cdot}$ with Br^--oxidize ABTS with an efficiency of *ca.* 90%. A significant increase in efficiency of $ABTS^+$ formation in the presence of Br^- thus implies that a significant proportion of the –O–O– bond cleavage is proceeding via a homolytic pathway.

noted here that Bodaness (1983) has observed that in the presence of H_2O_2, MP-11 converts NADPH to the catalytically active $NADP^+$ species, with concomitant protection of heme against oxidative attack by the peroxide. He therefore proposed the further novel possibility that reduced nucleotides may constitute a final distal defense whereby intracellular heme is protected from degradative effects of H_2O_2.

The peroxidase-like activity of MP-11 has found application in the food industry for the colorimetric quantitation of lipid hydroperoxides in oil and fat samples at hydroperoxide levels as low as 0.5–0.05 μM (Akaza and Aota, 1990).

VII. Reactions of Microperoxidases: Binding to Apo Proteins

The reaction of ferriprotoporphyrin IX(hemin) with heme-binding proteins is important because of its relevance to the *in vivo* formation of heme proteins with catalytic and transport function.

Attempts to elucidate the kinetic mechanism of the heme–protein binding interaction have been obscured by the extensive aggregation of hemin in aqueous solution at neutral pH even at the sub-micromolar concentration level. Despite attempts at circumventing this problem—e.g., by using Fe^{2+} carbonmonoxy heme in place of Fe^{3+} hemin (Rose and Olson, 1983), or the use of a pH-jump technique to effect the rapid dissociation–reassociation of Fe^{3+} myoglobin (Adams, 1976)—details of the mechanism of the reaction of hemin with heme-binding proteins are still not well understood.

Characterization of the aggregation thermodynamics of the heme octapeptide MP-8 has revealed that the peptide is essentially monomeric in aqueous solution at neutral pH at peptide concentrations $< 10^{-6}$ M (Aron *et al.*, 1986); this suggested that MP-8 could well provide a useful model of monomeric hemin with which to study details of the heme/apoprotein binding process in a system where kinetic artifacts due to ligand aggregation would be absent. Furthermore, although the noncovalent interaction of medium and long hemepeptides with non-heme peptide fragments from cytochrome *c* has been extensively studied to provide insight into structure–function relationships in the cytochromes *c* (see Chapter 21 by Wallace, this volume), no investigation of the interactions of short hemepeptides with heme-binding proteins appears to have been reported prior to 1989.

A. Interaction of Hemepeptides with Human Serum Albumin and Apo Myoglobin

Adams and co-workers studied the interaction of MP-8(-11) with members of three classes of heme-binding protein. The first two of these—apo myoglobin (apo-Mb) and human serum albumin (HSA)—are representative of proteins in which the iron porphyrin exhibits O_2 transport/catalytic activity (Mb) and *in vivo* heme transport activity (HSA). The latter (HSA) has recently been reported to exist as two non-converting—or very slowly interconverting—conformers, which exhibit different affinities towards ferriprotoporphyrin IX (Moehring *et al.*, 1983; Pasternak *et al.*,

1985); these are here referred to as HSA I and II. Furthermore, since the reactions of both apo Mb and HSA with hemin and other iron porphyrins have been extensively investigated, they constitute useful reference proteins against which the relevance of other MP-/protein interactions can be assessed.

Titration of hemepeptide with HSA and apo Mb (Adams *et al.*, 1989b,c) gave spectral changes consistent with formation of a 1:1 complex in each case. Results are summarized in Table 20.7. Of interest is the observation that careful reduction of anaerobic Fe^{3+} MP-8-apo Mb solution with dithionite gives a product with a split Soret peak which, in combination with kinetic observations (see later discussion) was taken to indicate the possibility that the complex was an equilibrium mixture of mono and bis-histidyl ligated species, one histidine originating from the hemepeptide and the other being the proximal histidine of the apo-Mb. Reduction of the Fe^{3+} MP-8(11)–HSA complex gives a typical Fe^{2+} low-spin hemeprotein spectrum with no reductant-mediated heme destruction observed—again demonstrating the protective effect exercised by the protein. CN^- showed apparent reaction with both

Table 20.7
Selected spectral and thermodynamic data for hemepeptide–protein complexes.[e]

Complex	Electronic spectra-λ/nm	K_a, M^{-1}	Hill coefficient (n)	Ref.[f]
Human serum albumin–MP-8				
Fe^{3+}	406, 526, 560 (sh) [a]$\varepsilon = 1.71 \times 10^5\ M^{-1}\ cm^{-1}$	—	—	i
Fe^{2+}	414, 522, 553 [a]$\varepsilon = 2.26 \times 10^5\ M^{-1}\ cm^{-1}$ $\varepsilon_\beta = 2.07 \times 10^4\ M^{-1}\ cm^{-1}$ $\varepsilon_\alpha = 2.72 \times 10^4\ M^{-1}\ cm^{-1}$	—	—	i
Human serum albumin–MP-11				
Fe^{3+}	406, 525, 558 [a]$\varepsilon = 1.68 \times 10^5\ M^{-1}\ cm^{-1}$	[b]$5(\pm 1) \times 10^4\ M^{-1}$	[b]0.99	i
Apo Mb-MP-8				
Fe^{3+}	407, 540, 581 (sh) [a]$\varepsilon \simeq 1.23 \times 10^5\ M^{-1}\ cm^{-1}$	[c]$6.4(\pm 0.4) \times 10^5\ M^{-1}$ [d]$3.4(\pm 0.3) \times 10^5\ M^{-1}$	[c]1.04 [d]0.99	ii
Fe^{2+}	432 (418), 522, 556	—	—	

[a] Soret.

[b] Spectrophotometric titration aq buffer pH 7.0; $\mu = 0.1$; 15 °C.

[c] Spectrophotometric titration aq buffer pH 7.0; 18 °C, $\mu = 0.1$.

[d] As for footnote (c), but 20% MeOH/aq buffer.

[e] MP-8 exhibited no detectable Soret shift on binding to GST π. However, ε decreased $\sim 10\%$, indicating hydrophobic interactions as the driving force in complex formation.

[f] i—Adams *et al.*, 1989b; ii—Adams *et al.*, 1989a.

Fe^{3+}MP-11-HSA and with methemalbumin; however, in both cases the pseudo–first-order rate constant was [CN$^-$]-independent and approximately equal to the limiting "off" rate-constant from the complex. This suggested that CN$^-$ was ligating, not the complex, but the iron porphyrin on diffusion from the complex, a conclusion further supported by the lack of reaction of CN$^-$ with the final HSA–MP-8 complex for which the "off" rate constant (k_{-1}) is close to zero (Table 20.8). Thus, the heme iron in the MP-HSA and methemalbumin complexes does not appear to be accessible to small anionic ligands, in agreement with the concept that HSA transports heme in an inaccessible "internalized" form, minimizing the possibility of nonspecific and biologically undesirable catalysis of redox processes by the iron porphyrin.

The kinetic mechanisms inferred from the binding kinetics for the interaction of MP-8 and MP-11 with apo-Mb and HSA are summarized in Table 20.8. Of relevance to the correctness of these mechanisms is the required agreement between equilibrium constants evaluated using derived microscopic rate constants and values estimated either thermodynamically (i.e., by spectrophotometric titration) or as parameters in the fitting of the relevant pseudo–first-order rate constant vs. concentration function. Values of k_1—the binding rate constant for the MP-8/MP-11 HSA interaction—demonstrate clearly a steric hindrance exercised on the interaction by the peptide moiety in comparison with the monomeric hemin/HSA interaction process (Adams and Berman, 1980) ($k_1 \approx 440$ M^{-1} s^{-1} and 1.7×10^5 M^{-1} s^{-1}, respectively). This suggests, in agreement with the CN$^-$ binding data, that access to the primary heme binding site in HSA is sterically controlled.

Decreasing the bulk solvent dielectric constant, by carrying out the HSA/MP-11 reaction in MeOH/buffer solutions up to 20% v/v in MeOH, does not change the simple kinetic mechanism shown in Table 20.8 (Adams and Thumser 1993). k_1 is essentially invariant as ε decreases from 82.6 to 72.6; however, k_{-1} increases smoothly by a factor of ~500 over the same range. Interpretation of the results in terms of a simple electrostatic model (Kirkwood, 1934) for solubility of zwitterionic species in aqueous/nonaqueous solvent systems suggested that the observed increase in k_{-1} resulted from increased solvational stabilization of exposed heme peptide with decreasing ε in both bulk solvent and in the transition state. It was further suggested that this system is a particularly simple model for demonstrating the effect of localized hydrophobic changes on the *in vivo* release of a transported ligand to a receptor molecule or target site. The binding of a range of substituted hemins including MP-8 and MP-11 to both human and bovine serum albumins has been reported (Casella *et al.*, 1993). In all cases, coordination of the iron involved a protein-derived histidine with the heme in HSA held more rigidly than in BSA, in agreement with the fact that heme peptides react far more slowly with HSA than with BSA (Adams *et al.* 1989a), implying a more sterically restricted binding site in the former.

B. Reaction of Hemepeptides with Gultathione S-Transferases

The third class of heme-binding protein studied—human placental glutathione S-transferase π (GST π)—is a homodimeric enzyme which binds hemin (and other hydrophobic ligands—e.g., bilirubin), but does not utilize the bound heme in the execution of its catalytic function: namely, the conjugation of the tripeptide gluta-thione with electrophilic xenobiotic compounds. Kinetic studies (Adams and Goold, 1990a,b) indicated that GST π binds MP-8 at two sites per subunit, one of which— a kinetically "slow" site—was at or spatially close to the bilirubin binding site, while the other—a kinetically "fast" site—was at or close to the binding site for the electrophilic co-substrate 1-chloro-2,4-dinitrobenzene (CDNB). The kinetic mecha-nism for interaction of MP-8 with GST is complicated by concomitant "solva-tional" elimination of one binding site (identified as that being at or spatially close to the CDNB site); however, both the direct MP-8 binding kinetics and steady-state kinetics of the MP-8–mediated inhibition of the enzymic conjugation process are consistent with the mechanism shown in Table 20.8.

In an almost simultaneous publication to those just described, Federici and co-workers report the inhibition of GST π-catalyzed conjugation of GSH and CDNB on incubation with hemin (Caccuri *et al.*, 1990). Despite clearly apparent effects of ligand aggregation on their results, the conclusion reached—namely, that hemin binds to the GST π subunit at two discrete sites, one high- and one low-affinity—is in full agreement with that of Adams and Goold (1990b,c). Furthermore, associa-tion constants reported for the sites, 2×10^{-8} M (high) and 4×10^{-7} M (low), re-spectively, are in remarkably good agreement with the values evaluated from micro-scopic rate constants for the kinetically "fast" and "slow" MP-8 binding sites given in Table 20.8. This would appear to further validate the rationale that MP-8 provides a useful model for monomeric hemin in the study of iron porphyrin/ protein interaction mechanisms.

The human erythrocyte glutathione S-transferase, an enzyme closely related to GST π binds MP-8, MP-9, and MP-11 at a single site on each subunit that appears to be spatially close to or identical with the CDNB binding site (Thumser, 1991; Thumser and Adams, 1994). The binding and inhibition kinetics exhibited by the system show clear evidence of steric effects (which, it is postulated, prevent interac-tion with the bilirubin binding site). Such effects have also been found on compari-son of the binding of MP-11 and protohemin with an NADPH-dependent methe-moglobin reductase found in erythrocytes and liver (Xu *et al.*, 1992). It appears, therefore, that the heme peptides may have application in assessing small differences in the accessibility of heme-binding sites of otherwise closely related or identical proteins.

VIII. Microperoxidases as Ultrastructure Probes in Cell Biology

The small size, relatively low cytotoxicity, and high peroxidasic activity (relative to cyt *c* and soluble hemin analogs) of the microperoxidases has resulted in their use as histochemical tracers in ultrastructural studies to demonstrate tissue permeability pathways, and, in their derivatized forms, to probe the distribution of ionic binding sites on cell surfaces and as covalent tags for binding to antibody fragments. These applications have recently been reviewed (Adams, 1990b).

Table 20.8
Kinetics of reactions of microperoxidases with heme-binding proteins.

Protein	Hemepeptide	Kinetic mechanism	Rate constants (pH 7.00, $T = 15$ °C)	Comments	Ref.[a]
Human serum albumin (HSA)	MP-8	HSA I + MP-8 $\overset{k_1}{\underset{k_{-1}}{\rightleftharpoons}}$ Int[b] Int $\overset{k_2}{\underset{k_{-2}}{\rightleftharpoons}}$ Complex	$k_1 = 4.3(\pm 0.4) \times 10^2$ M^{-1} s^{-1} $k_{-1} = 0.03(\pm 0.003)$ s^{-1} $k_2 = 0.023(\pm 0.004)$ s^{-1} $k_{-2} = 0.002(\pm 0.0004)$ s^{-1}	—	i
		HSA II + MP-8 $\overset{k_1}{\underset{k_{-1}}{\rightleftharpoons}}$ Int	$k_1 = 79(\pm 3)$ M^{-1} s^{-1} $k_{-1} < 2 \times 10^{-5}$ s^{-1}		
		HSA I + MP-11 $\overset{k_1}{\underset{k_{-1}}{\rightleftharpoons}}$ Int	$k_1 = 4.4(\pm 0.6) \times 10^2$ M^{-1} s^{-1} $k_{-1} = 0.011(\pm 0.001)$ s^{-1}	$K_a = k_1/k_{-1} = 4.1(\pm 0.5) \times 10^4$ M^{-1} cf. value obtained by spectrophotometric titration quoted in Table 20.7.	i

Protein	Hemepeptide	Kinetic mechanism	Rate constants (pH 7.00, $T = 18$ °C) 20% MeOH/aq buffer	Comments	Ref.[a]
Apo myoglobin (apo-Mb)	MP-8	apo Mb + MP-8 $\overset{k_1}{\underset{k_{-1}}{\rightleftharpoons}}$ Int Int $\overset{k_2}{\underset{k_{-2}}{\rightleftharpoons}}$ Complex	$k_1 = 1.6(\pm 0.2) \times 10^6$ M^{-1} s^{-1} $k_{-1} = 12(\pm 1.5)$ s^{-1} $k_2 = 7.3(\pm 0.7)$ s^{-1} $k_{-2} = 1.9(\pm 0.4)$ s^{-1}	Overall association constant calculated from k_1, etc. = $5.1(\pm 1.2) \times 10^5$ M^{-1}. cf. value of K_a in aq and aq MeOH soln given in Table 20.7.	ii

Protein	Hemepeptide	Kinetic mechanism	Rate constants (pH 7.00, $T = 22.5\ °C$)	Comments	Ref.[a]
Human placental glutathione S-transferase π	MP-8		$k_1 = 5.5(\pm 0.23) \times 10^4\ M^{-1}\ s^{-1}$ $k_{-1(max)} = 4 \times 10^{-4}\ s^{-1}$ $k_2 = 6.3(\pm .25) \times 10^2\ s^{-1}$ $k_{-2} = 3.7(\pm 0.8) \times 10^{-4}\ s^{-1}$ $k_s = 2.2(\pm 0.4) \times 10^{-3}\ s^{-1}$	"f" and "s" refer to kinetically fast and slow binding sites, respectively. "S" is the solvationally inactivated GST π in which "f" is eliminated. $K_{a(f)max} = 7.2 \times 10^{-8}\ M$, $K_{a(s)} = 5.9 \times 10^{-7}\ M$.	iii

[a] i, ii—Adams et al., 1989a,b; iii—Adams and Goold, 1990a.
[b] Int ≡ Intermediate species.

687

IX. Acknowledgment

The authors are indebted to Lucille Odes for her expert preparation of the manuscript.

X. References

Adams, P. A. (1976) *Biochem. J. 159*, 371–376.
Adams, P. A. (1990a) *J. Chem. Soc. Perkin Trans. 2*, 1407–1414.
Adams, P. A. (1990b) in *Peroxidases in Chemistry and Biology*, Vol. 2, Chapter 7 (Everse, J. & Grisham, M. B., eds.), pp. 171–200, CRC Press, Boca Raton, Florida.
Adams, C., & Adams, P. A. (1986) *S. Afr. J. Sci. 82*, 113 –114.
Adams, P. A., & Adams, C. (1988) *J. Inorg. Biochem. 34*, 177–187.
Adams, C., & Adams, P. A. (1989) *J. Inorg. Biochem. 37*, 29–34.
Adams, P. A., & Berman, M. C. (1980) *Biochem. J. 191*, 95–102.
Adams, P, A., & Goold, R. D. (1990a) *J. Chem. Soc. Chem. Comms. 2*, 97–98.
Adams, P. A., & Goold, R. D. (1990b) *J. Chem. Soc. Faraday Trans. 1, 86*, 1797–1801.
Adams, P. A., & Goold, R. D. (1990c) *J. Chem. Soc. Faraday Trans. 1, 86*, 1803–1806.
Adams, P. A., & Thumser, A. E. (1993) *J. Inorg. Biochem. 50*, 1–7.
Adams, P. A., Adams, C., & Baldwin, D. A. (1986) *J. Inorg. Biochem. 28*, 441–454.
Adams, P. A., Byfield, M. P., Milton, R. C. deL., & Pratt, J. M. (1988) *J. Inorg. Biochem. 34*, 167–175.
Adams, P. A., Byfield, M. P., Goold, R. D., & Thumser, A. E. (1989a) *J. Inorg. Biochem. 37*, 55–59.
Adams, P. A., Goold, R. D., & Thumser, A. E. (1989b) *J. Inorg. Biochem. 37*, 91–103.
Adams, P. A., Goold, R. D., & Thumser, A. E. (1989c) *J. Chem. Soc. (Lond.) Faraday Trans. 1, 85*, 3845–3852.
Adams, P. A., Milton, R. C. de L. and Silver, J. (1994) *Biometals 7*, 217–220.
Akaza, I., & Aota, N, (1990) *Talanta 37*, 925–929.
Alberding, N., Austin, R. H., Chan, S. S., Eisenstein, L., Frauenfelder, H., Good, D., Kaufmann, K., Marden, M., Norlund, T. M., Reinisch, L., Reynolds, A. H., Sorenson, L. B., Wagner, G. C., & Yue, K. T. (1978) *Biophys. J. 24*, 319–334.
Antonini, E., & Brunori, M. (1971) *Hemoglobin and Myoglobin in Their Reactions with Ligands*, Elsevier, Amsterdam.
Anusiem, A. C., Beetlestone, J. G., & Irvine, D. H. (1966) *J. Chem. Soc. A*, 357–363.
Araiso, T., & Dunford, H. B. (1981) *J. Biol. Chem. 256*, 10099–10104.
Aron, J., Baldwin, D. A., Marques, H. M., Pratt, J. M., & Adams, P. A. (1986) *J. Inorg. Biochem. 27*, 227–243.
Arutyunyan, A. M., Surkov, S. A., Luchkov, V. A., Lysko, A. I., Mironov, A. F., Rumyantseva, V. D., & Evstigneeva, R. P. (1986) *Mol. Biol. (Moscow) 20*, 502–513.
Ascenzi, P., Condo, S. G., Bellelli, A., Barra, D., Bannister, W. H., Giardina, B., & Brunori, M. (1984) *Biochim. Biophys. Acta. 788*, 281–289.
Awad, E. S., & Badro, R. G. (1967) *Biochemistry 6*, 1785–1791.
Baek, H. K., & Van Wart, H. E. (1989) *Biochemistry 28*, 5714–5719.
Baldwin, D. A., Campbell, V. M., Carleo, L. A., Marques, H. M., & Pratt, J. M. (1981) *J. Am. Chem. Soc. 103*, 186–188.
Baldwin, D. A., Campbell, V. M., Marques, H. M., & Pratt, J. M. (1984) *FEBS Letts. 167*, 339–342.
Baldwin, D. A., Marques, H. M., & Pratt, J. M. (1985) *FEBS Lett. 183*, 309–312.
Baldwin, D. A., Marques, H. M., & Pratt, J. M. (1986a) *S. Afr. J. Chem. 39*, 189–196.
Baldwin, D. A., Marques, H. M., & Pratt, J. M. (1986b) *J. Inorg. Biochem. 27*, 245–254.
Baldwin, D. A., Mabuya, M. B., & Marques, H. M. (1987a) *S. Afr. J. Chem. 40*, 103–110.
Baldwin, D. A., Marques, H. M., & Pratt, J. M. (1987b) *J. Inorg. Biochem. 30*, 203–217.
Baldwin, J., & Chothia, C. (1979) *J. Mol. Biol. 129*, 175–220.
Baumgartner, C. P., Sellers, M., Nassif, R., & May, L. (1974) *Eur. J. Biochem. 46*, 625–629.
Blumenthal, D. C., & Kassner, R. J. (1979) *J. Biol. Chem. 254*, 9617–9620.

Blumenthal, D. C., & Kassner, R. J. (1980) *J. Biol. Chem.* **255**, 5859–5863.

Bodaness, R. S. (1983) *Biochem. Biophys. Res. Comm.* **113**, 710–716.

Bolognesi, M., Cannillo, E., Ascenzi, P., Giacometti, G. M., Merli, A., & Brunori, M. (1982) *J. Mol. Biol.* **158**, 305–315.

Bonnett, R. (1981) *Essays in Biochemistry* **17**, 1–51.

Brittain, T. (1981) *J. Inorg. Biochem.* **15**, 243–252.

Bruice, T. C., Zipplies, M. F., & Lee, W. A. (1986) *Proc. Natl. Acad. Sci. USA* **83**, 4646–4649.

Brunori, M., Santucci, R., Campanella, L., & Tranchida, G. (1989) *Biochem. J.* **264**, 301–304.

Bryant, C., & Stellwagen, E. (1985) *J. Biol. Chem.* **260**, 332–336.

Buchler, J. W. (1975) in *Porphyrins and Metalloporphyrins* (Smith, K. M., ed.), Chapter 5, Elsevier, Amsterdam.

Byfield, M. P., & Pratt, J. M. (1992) *J. Chem. Soc. Chem. Commun.*, 214–215.

Byfield, M. P., Hamza, M. S. A., & Pratt, J. M. (1993) *J. Chem. Soc. Dalton Trans.*, 1641–1645.

Caccuri, A. M., Aceto, A., Piemonte, F., Diilio, C., Rosato, N., & Federici, G. (1990) *Eur. J. Biochem.* **189**, 493–497.

Cantor, C. R., & Schimmel, P. R. (1980) *Biophysical Chemistry*, Part 1, p. 49, W.H. Freeman, San Francisco.

Carraway, A. D., Burkhalter, R. S., Timkovich, R., & Peterson, J. (1993) *J. Inorg. Biochem.* **52**, 201–207.

Casella, L., Gulotti, M., Poli, S., & De Gioia, L., (1993) *Gazz. Chim. Ital.* **123**, 149–156.

Chersi, A., Trinca, M. L., & Muratti, E. (1987) *J. Chromatogr.* **410**, 463–465.

Childs, R. E., & Bardsley, W. G. (1975) *Biochem. J.* **145**, 93.

Collman, J. P. (1977) *Accounts of Chemical Research* **10**, 265–272.

Corradin, G., & Harbury, H. A. (1970) *Biochim. Biophys. Acta* **221**, 489–496.

Cunningham, I. D., Bachelor, J. L., & Pratt, J. M. (1991) *J. Chem. Soc., Perkin Trans. 2*, 1839–1843.

Deatherage, J. F., Loe, R. S., Anderson, C. M., & Moffat, K. (1976) *J. Mol. Biol.* **104**, 687–706.

Dickerson, R. E., & Timkovich, R. (1975) in *The Enzymes* (Boyer, P. D., ed.), pp. 397–547, Vol. 11, Academic Press, London.

Dolphin, D., ed. (1982) B_{12} (2 Vols.), Wiley, New York.

Eaton, D. R., & Wilkins, R. G. (1978) *J. Biol. Chem.* **253**, 908–915.

Ehrenberg, A., & Theorell, H. (1955) *Acta Chem. Scand.* **9**, 1193–1205.

Eisenstein, L. (1977) *Int. J. Quantum Chem., Quantum Biol. Symp.* **4**, 363–374.

Giacometti, G. M., Ascenzi, P., Bolognesi, M., & Brunori, M. (1981) *J. Mol. Biol.* **146**, 363–374.

Goss, D. J., LaGow, J. B., & Parkhurst, L. J. (1982) *Comp. Biochem. Physiol.* **71B**, 229–233, and references therein.

Hamilton, G. H., Owens, J. W., O'Connor, C. J., & Kassner, R. J. (1984) *Inorg. Chim. Acta* **93**, 55–60.

Hamza, M. S. A., & Pratt, J. M. (1993) *J. Chem. Soc. Dalton Trans.*, 1647–1650.

Hanania, G. I. H., Irvine, G. H., & Irvine, M. V. (1966) *J. Chem. Soc. A*, 296–299.

Harbury, H. A., & Loach, P. A. (1959) *Proc. Natl. Acad. Sci. USA* **45**, 1344–1359.

Harbury, H. A., & Loach, P. A. (1960a) *J. Biol. Chem.* **235**, 3640–3645.

Harbury, H. A., & Loach, P. A. (1960b) *J. Biol. Chem.* **235**, 3646–3653.

Harbury, H. A., Cronin, J. R., Fanger, M. W., Hettinger, T. P., Murphy, A. J., Myer, Y. P., & Vinogradov, S. N. (1965) *Proc. Natl. Acad. Sci. USA* **54**, 1658–1664.

Hasinoff, B. B. (1981a) *Arch. Biochem. Biophys.* **211**, 396–402.

Hasinoff, B. B. (1981b) *J. Phys. Chem.* **85**, 526–531.

Hodgson, E. K., & Fridovich, I. (1975) *Biochemistry* **14**, 5294–5299.

Huang, Y., & Kassner, R. J. (1981a) *J. Am. Chem. Soc.* **103**, 4927–4932.

Huang, Y., & Kassner, R. J. (1981b) *J. Biol. Chem.* **256**, 5327–5331.

Jain, A., & Kassner, R. J. (1984) *J. Biol. Chem.* **259**, 10309–10314.

Jehlani, A. M. T., Stotter, D. A., & Wilson, M. T. (1976) *Eur. J. Biochem.* **71**, 613–616.

Job, D., & Ricard, J. (1975) *Arch. Biochem. Biophys.* **170**, 427–437.

Job, D., Zleba, B., Puppo, A., & Rigaud, J. (1980) *Eur. J. Biochem.* **107**, 491–500.

Jones, M. G., Bicker, D., Wilson, M. T., Brunori, M., Colosimo, A., & Sarti, P. (1984) *Biochem. J.* **220**, 57–60.

Jones, R. D., Basolo, F., & Summerville, D. A. (1979) *Chemical Reviews* **79**, 139–179.

Kazmi, S. A., Mills, M. A., Pitluk, Z. W., & Scott, R. A. (1985) *J. Inorg. Biochem.* **24**, 9–12.

Kimura, K., Peterson, J., Wilson, M., Cookson, D. J., & Williams, R. J. P. (1981) *J. Inorg. Biochem.* **15**, 11–25.

Kirkwood, J. G. (1934) *J. Chem. Phys.* **2**, 351–364.

Kuranova, I. P., Teplyakov, A. V., Obmolova, G. V., Voronova, A. A., Popov, A. N., Kheiker, D. M., & Arutyunyan, E. G. (1982) *Bioorg. Khim.* **8**, 1625–1636.

Lambeth, D. O., & Palmer, G. (1973) *J. Biol. Chem.* **248**, 6095–6103.

Loehr, T. M., & Loehr, J. S. (1973) *Biochem. Biophys. Res. Commun.* **55**, 218–223.

Makinen, M. W., & Churg, A. K. (1983) in *Iron Porphyrins, Part 1, Physical Bioinorganic Chemistry Series* (Lever, A. B. P., & Gray, H. B., eds.), pp. 141–235, Addison-Wesley, Reading, Massachusetts.

Makino, R., & Yamazaki, I. (1973) *Arch. Biochem. Biophys.* **157**, 356–368.

Margoliash, E. (1955) *Nature* **175**, 293–295.

Margoliash, E., Frohwirt, N., & Wiener, E. (1959) *Biochem. J.* **71**, 559–570.

Marques, H. M. (1986) Ph.D. Thesis, University of the Witwatersrand, Johannesburg, South Africa.

Marques, H. M. (1990) *Inorg. Chem.* **29**, 1597–1599.

Marques, H. M., Baldwin, D. A., & Pratt, J. M. (1987) *J. Inorg. Biochem.* **29**, 77–91.

Marques, H. M., Baldwin, D. A., & Pratt, J. M. (1988) *S. Afr. J. Chem.* **41**, 68–70.

Marques, H. M., Marsh, J. H., Mellor, J. R., & Munro, O. Q. (1990) *Inorg. Chim. Acta* **170**, 259–269.

Marques, H. M., Munro, O. Q., & Crawcour, M. (1992) *Inorg. Chim. Acta* **196**, 221–229.

Marques, H. M., Byfield, M. P., & Pratt, J. M. (1993) *J. Chem. Soc. Dalton Trans.*, 1633–1639.

Mashiko, T., Kastner, M. E., Spartalian, K., Scheidt, W. R., & Reed, C. A. (1978) *J. Am. Chem. Soc.* **100**, 6354l–6362.

Mashino, T., Nakamura, S., Hirobe, M., & Masaaki, H. (1990) *Tetrahedron Lett.* **31**, 3163–3166.

Mazumdar, S., Medhi, O. K., & Mitra, S. (1991) *Inorg. Chem.* **30**, 700–705.

McCandlish, E., Miksztal, A. R., Nappa, M., Sprenger, A. Q., Valentine, J. S., Strong, J. D., & Spiro, T. G. (1980) *J. Am. Chem. Soc.* **102**, 4268–4271.

McLendon, G., & Smith, M. (1982) *Inorg. Chem.* **21**, 847–850.

McCoy, S., & Caughey, W. S. (1970) *Biochemistry* **9**, 2387–2393.

Meszaros, L., Vegh, M., & Horvath, I. (1984) *Oxidation Communications* **6**, 47–54.

Millar, F., Wrigglesworth, J. M., & Nicholls, P. (1981) *Eur. J. Biochem.* **117**, 13–17.

Moehring, G. A., Chu, A. H., Kurlansik, L., & Williams, T. J. (1983) *Biochemistry* **22**, 3381–3386.

Morgan, B., & Dolphin, D. (1987) *Structure and Bonding* **64**, 115.

Morishima, I., & Inubushi, T. (1978) *J. Am. Chem. Soc.* **100**, 3568–3574.

Munro, O. Q., Marques, H. M., Pollak, H., & Malas, N., (1993) *Nucl. Instr. and Meth. Phys. Res.*, 315–317.

Myer, Y. P., & Harbury, H. A. (1973) *Ann. N. Y. Acad. Sci.* **206**, 685–700.

Nassif, R., Baumgartner, C. P., Sellers, M., & May, L. (1976) *z. Naturforsch.* **C31**, 232.

Ortiz de Montellano, P. R. (1987) *Acc. Chem. Res.* **20**, 289–294.

Othman, S., Le Lirzin, A., & Desbois, A., (1993) *Biochemistry* **32**, 9781–9791.

Owens, J. W., O'Connor, C. J., & Kassner, R. J. (1988) *Inorg. Chim. Acta* **151**, 107–116.

Paleus, S., Ehrenberg, A., & Tuppy, H. (1955) *Acta Chem. Scand.* **9**, 365–374.

Parr, G. R., Hantgan, R. R., & Taniuchi, H. (1978) *J. Biol. Chem.* **253**, 5381–5388.

Pasternak, R. F., Gibbs, E. J., Mauk, G. A., Reid, L. S., Wong, N. M., Kurokama, Ko., Hashim, M., & Muller-Eberhard, V. (1985) *Biochemistry* **24**, 5443–5448.

Paul, K. G. (1951) *Acta Chem. Scand.* **5**, 389–405.

Peterson, J., & Wilson, M. T. (1987) *Inorg. Chim. Acta* **135**, 101–107.

Peterson, J., Silver, J., Wilson, M. T., & Morrison, I. E. G. (1980) *J. Inorg. Biochem.* **13**, 75–82.

Peterson, J., Saleem, M. M. M., Silver, J., Wilson, M. T., & Morrison, I. E. G. (1983) *J. Inorg. Biochem.* **19**, 165–178.

Phillips, S. E., & Schoenborn, B. P. (1981) *Nature* **292**, 81–82.

Plattner, H., Wolfram, D., Bachmann, L. & Wachter, E. (1975) *Histochemistry* **45**, 1–21.

Plattner, H., Wachter, E., & Gröbner, P. (1977) *Histochemistry* **53**, 223–242.

Postlewaite, J. C., Miers, J. B., & Dlott, D. D. (1989) *J. Am. Chem. Soc.* **111**, 1248–1255.

Pratt, J. M. (1972) *Inorganic Chemistry of Vitamin B_{12}*, Academic Press, London.

Pratt, J. M. (1975) in *Techniques and Topics in Bioinorganic Chemistry* (McAuliffe, C. A., ed.), pp. 107–204, Macmillan, London.

Proudfoot, A. E. I., & Wallace, C. J. E. (1987) *Biochem. J.* **248**, 965–967.

Razumas, V. J., Gudavicus, A. V., Kazlauskaite, J. D., & Kulys, J. J. (1989) *J. Electroanal. Chem.* **271**, 155–160.

Razumas, V., Kazlauskaite, J. Ruzgas, T., & Kulys, J. (1992) *Bioelectrochemistry and Bioenergetics* **28**, 159–176.

Ringe, D., Petsko, G. A., Kerr, D. E., & Ortiz de Montellano, P. R. (1984) *Biochemistry* **23**, 2–4.

Rose, M. Y., & Olson, J. S. (1983) *J. Biol. Chem.* **258**, 4298–4303.

Rosvai, E., Vegh, M., Kramer, M., & Horvath, I. (1988) *Biochem. Pharm.* **37**, 4574–4577.

Rush, J. D., & Koppenol, W. H. (1988) *J. Am. Chem. Soc.* **110**, 4957–4966.

Sadeque, A. J. M., Shimizu, T., & Hatano, M. (1987a) *Inorg. Chim. Acta* **135**, 109–113.

Sadeque, A. J. M., Shimizu, T., Ogoma, Y., & Hatano, M. (1987b) *Inorg. Chim. Acta* **135**, 203–205.

Saleem, M. M. M., & Wilson, M. T. (1988a) *Inorg. Chim. Acta* **153**, 93–98.

Saleem, M. M. M., & Wilson, M. T. (1988b) *Inorg. Chim. Acta* **153**, 99–104.

Saleem, M. M. M., & Wilson, M. T. (1988c) *Inorg. Chim. Acta* **153**, 105–113.

Santucci, R., Reinhard, H., & Brunori, M. (1988) *J. Am. Chem. Soc.* **110**, 8536–8537.

Satterlee, J. M. (1983) *Inorg. Chim. Acta* **79**, 195–196.

Scheler, W., & Jung, F. (1954) *Biochem. Z.* **325**, 515–524.

Schroeder, H. R., & Yeager, F. M. (1978) *Anal. Chem.* **50**, 1114–1120.

Seybert, D. W., Moffat, K., & Gibson, Q. H. (1976) *J. Biol. Chem.* **251**, 45–52.

Sharma, V., Ranney, H. M., Geibel, J. F., & Traylor, T. G. (1975) *Biochem. Biophys. Res. Commun.* **66**, 1301–1306.

Sharma, V. S., Schmidt, M. R., & Ranney, H. M. (1976) *J. Biol. Chem.* **251**, 4267–4272.

Sharma, V. S., Isaacson, R. A., John, M. E., Waterman, M. R., & Chevion, M. (1983) *Biochemistry* **22**, 3897–3902.

Shimizu, T., Nozawa, T., & Hatano, M. (1982) *J. Biochem. (Tokyo)* **91**, 1951–1958.

Sievers, G., Gadsby, P. M. A., Peterson. J., & Thomson, A. J. (1983) *Biochim. Biophys. Acta* **742**, 637–647.

Smit, J. D. G., & Winterhalter, K. H. (1981) *J. Mol. Biol.* **146**, 641–647.

Smith, M. C., & McLendon, G. (1980) *J. Amer. Chem. Soc.* **102**, 5666–5670.

Smith, M., & McLendon, G. (1981) *J. Am. Chem. Soc.* **103**, 4912–4921.

Sono, M., & Asakura, T. (1976) *J. Biol. Chem.* **251**, 2664–2670.

Stryer, L., Kendrew, J. C., & Watson, H. C. (1964) *J. Mol. Biol.* **8**, 96–104.

Sykes, A. G. (1982) *Advances in Inorganic and Bioinorganic Mechanisms* **1**, 121.

Tatsuma, T., & Watanabe, T. (1991) *Anal. Chem.* **63**, 1580–1585.

Tatsuma, T., & Watanabe, T. (1992) *Anal. Chem.* **64**, 143–147.

Tentori, L., Vivaldi, G., Carta, S., Marinucci, M., Massa, A., Antonini, E., & Brunori, M. (1971) *FEBS Lett.* **12**, 181–185.

Theorell, H. (1938) *Biochem. Z.* **298**, 242–267.

Theorell, H. (1939) *Enzymologia* **6**, 88.

Theorell, H., & Åkeson, A. (1941) *J. Am. Chem. Soc.* **63**, 1804–1811.

Thumser, A. E. (1991) Ph.D. Thesis, University of Cape Town, South Africa.

Thumser, A. E., & Adams, P. A. (1994) *J. Inorg. Biochem.* **53**, 157–168.

Traylor, T. G. (1981) *Accounts of Chemical Research* **14**, 102–109.

Traylor, T. G., & Xu, F. (1990) *J. Am. Chem. Soc.* **112**, 178–186.

Tsou, C. L. (1949) *Nature* **164**, 1134–1135.

Tsou, C. L. (1951a) *Biochem. J.* **49**, 362–367.

Tsou, C. L. (1951b) *Biochem. J.* **49**, 367–374.

Tuppy, H., & Bodo, G. (1954) *Monatshefte Chem.* **85**, 1182–1186.

Tuppy, H., & Paleus, S. (1955) *Acta Chem. Scand.* **9**, 353–364.

Uchida, H., Heystek, J., & Klapper, M. H. (1971a) *J. Biol. Chem.* **246**, 2031–2034.

Uchida, H., Heystek, J., & Klapper, M. H. (1971b) *J. Biol. Chem.* **246**, 6843–6848.

Urrey, D. W. (1967) *J. Am. Chem. Soc.* **89**, 4190–4196.

Urrey, D. W., & Pettigrew, J. W. (1967) *J. Am. Chem. Soc.* **89**, 5276–5283.

Valentine, J. S. (1979) in *Biochemical and Clinical Aspects of Oxygen* (Caughy, W. ed.), pp. 659–677, Academic Press, New York.

Vegh, M., Kramer, M., & Horvath, I. (1986) *Biochim. Biophys. Acta* **882**, 6–11.

Venekei, I., Kittel, A., & Horvath, I. (1986) *Acta Biochim. Biophys. Hung.* **21**, 13–22.

Vodnyanszky, L., Marton, A., Venekei, I., Vegh, M., Blazovits, A., Kittel, A., & Horvath, I. (1985) *Biochim. Biophys. Acta* **835**, 411–414.

Vodnyanszky, L., Marton, A., Vegh, M., Blazovits, A., Auth, F., Vertes, A., & Horvath, I. (1986) *Acta Biochim. Biophys. Hung.* **21**, 3–11.

Wallace, C. J. E., & Proudfoot, A. E. I. (1987) *Biochem. J.* **245**, 773–779.

Wang, J.-S., & Van Wart, H. E. (1989) *J. Phys. Chem.* **93**, 7925–7931.

Wang, J.-S., Baek, H. K., & Van Wart, H. E. (1991) *Biochem. Biophys. Res. Comm.* **179**, 1320–1324.

Wang, J.-S., Tsai, A.-L., Heldt, J., Palmer, G., & Van Wart, H. E. (1992) *J. Biol. Chem.* **267**, 15310–15318.

Wang, J. T., Yeh, H. J. C., & Johnson, D. F. (1975) *J. Am. Chem. Soc.* **97**, 1968–1971.

Werringloer, J., Kawano, S., Chanos, N., & Estabrook, R. W. (1979) *J. Biol. Chem.* **254**, 11839–11846.

Wilchek, M. (1972) *Anal. Biochem.* **49**, 572–575.

Wilgus, H., & Stellwagen, E. (1979) *Anal. Biochem.* **94**, 228–230.

Wilson, M. T., Ranson, R. J., Masiakowski, P., Czarnecka, E., & Brunori, M. (1977) *Eur. J. Biochem.* **77**, 193–199.

Xu, F., Quandt, K. S., & Hulquist, D. E. (1992) *Proc. Natl. Acad. Sci. USA* **89**, 2130–2134.

Yang, E. K., & Sauer, K. (1982) in *Electron Transport and Oxygen Utilization* (Chien, H., ed.), pp. 89–94, Elsevier, Amsterdam.

Yonetani, T., & Schleyer, H. (1967) *J. Biol. Chem.* **242**, 1974–1981.

Zamponi, S., Santucci, R., Brunori, M., & Marassi, R. (1990) *Biochim. Biophys. Acta* **1034**, 294–297.

21

Semisynthesis of Cytochrome *c* Derivatives and Noncovalent Complexes

CARMICHAEL J. A. WALLACE

Department of Biochemistry
Dalhousie University
and
Département de Biochimie Médicale
Université de Genève

I. Introduction

Rational protein engineering for structure–function studies requires knowledge of the three-dimensional structure of the protein and of the evolutionarily conserved residues within it. By the early 1970s, sequence comparisons and high-resolution x-ray crystallography were so well developed that cytochrome *c* featured as a *Scientific American* centerfold (Dickerson, 1972). This structural knowledge had inevitably led to some fairly detailed mechanistic proposals (Dickerson *et al.*, 1972) which, although incorrect (Dickerson and Timkovich, 1975), stimulated attempts to verify the hypotheses by the study of site-specific structural analogs.

Obviously, at that time only protein chemical techniques were available for the generation of such derivatives, by synthesis or chemical modification. Specific chemical reactions were already much used in studies of cytochrome *c* but suffered from a number of limitations; nonetheless, some extremely valuable studies have been performed, especially with singly modified lysine derivatives. However, the only synthetic methods that provide the possibility of replacing any residue with a wide range of alternatives are total chemical synthesis or semisynthesis. Total synthesis can be attempted in one of two ways. The stepwise approach uses an insoluble support upon which the peptide chain is built by the repetitive addition of amino acids. Failure to achieve quantitative reaction in each cycle, and the lack of intermediate

693

purification steps, often results in a heterogeneous product. The fragment condensation method calls for the construction of short peptides by solution chemistry, and their subsequent coupling in the correct order to yield the desired sequence. Purification can be effected at each stage in the procedure, which is inevitably extremely laborious.

Both strategies have been essayed for the preparation of cytochrome *c* analogs. A synthesis of the tuna sequence by the solid-phase method was reported by Sano (1972). Given the state of development of the technology, this was an extremely courageous attempt, and the results point out the difficulties he faced. These included the sensitivity to side reactions of some amino acids, especially Met and Trp, the lack of suitable side-chain protecting groups (for His), the harsh conditions for covalent heme reinsertion, and the physical breakdown of the solid support during an extended synthesis. The project was abandoned, but in the subsequent 20 years the techniques of solid-phase peptide synthesis have been so much improved (Kent, 1988) that it now has a major role in the production of specific protein analogs including cytochrome *c* (see Section V.B.3).

A strategy for the synthesis of the entire sequence of yeast apocytochrome by fragment condensation was developed by Scoffone's group, and substantial development work was performed over a number of years (e.g., Moroder *et al.*, 1973, 1975), but final reassembly was not achieved. Nevertheless, the strategy, and the intermediates produced, have also proved of ultimate use in the goal of producing informative analogs via the intermediacy of semisynthesis. The limitations and technical difficulty of the other chemical approaches to specific analog preparation have meant that the main source of informative cytochrome *c* derivatives has been protein semisynthesis.

The advent of site-directed mutagenesis of cloned genes has resulted in a new and highly efficient method for protein engineering. Cytochrome *c* was adopted by Michael Smith's group as a model for their pioneering work in the area, and a great deal has already been learnt from mutants of yeast cytochrome *c* produced in this way. However, in the foreseeable future, semisynthesis will remain the only possible, or only economically viable, means of inserting noncoded amino acids or labeled amino acids at specific sites in a sequence. Sometimes, for reasons yet unexplained, a cloned gene, mutant or not, is simply not expressed by a heterologous host. As well as remaining an alternative, semisynthesis can be a true partner of oligonucleotide-directed mutagenesis, since the latter technology can be used to generate large amounts of an otherwise scarce protein for the semisynthetic purposes just described, or to build into a protein specific cleavage and resynthesis sites to facilitate the semisynthetic process.

II. Rationale and Techniques of Semisynthesis

A. Overview

Semisynthesis is a strategy for the synthesis of a protein analog in which the bulk of the novel structure is composed of a fragment or fragments derived from the naturally occurring protein (or, these days, of the native sequence over-expressed in a heterologous host). The logic of this approach is that for interpretable structure–function results, only one or a few residues should be modified in any analog, and thus most of the sequence will be identical to that of the native protein and can, for simplicity, be derived from it. Semisyntheses have been attempted of a wide range of well-known proteins, and the field has been well reviewed by Offord (1980, 1987). Like total chemical synthesis of proteins, semisynthesis has its stepwise and fragment-condensation forms, but thus far derivatives of cytochrome *c* have been obtained exclusively by fragment-condensation routes.

Having selected a candidate residue for substitution, the essential steps in any semisynthesis are fragmentation of the protein into a number of peptides, and separation of those fragments from one another; modification in some way of the peptide containing the residue of interest, to introduce a new structure at that position; then, reassembly of modified and native fragments in the correct order. Additionally, it may be necessary to introduce prior protection of side-chain functional groups, either because of their sensitivity to the conditions of manipulation, or to distinguish between α- and ω- COOH and NH groups. If so, a final deprotection step may also be required.

B. Side-Chain Protection

Total chemical synthesis of peptides and proteins tends to employ maximal side-chain protection to avoid damage to sensitive residues in lengthy syntheses. The philosophy in semisynthesis is to avoid protection wherever possible. The reasoning behind this minimal protection strategy is that (a) simplicity is desirable, and any unnecessary step is to be avoided, and (b) side-chain protection tends to drastically reduce the solubility of the intermediates in aqueous solution, rendering manipulations and purification very much more difficult. In semisynthesis, especially those in which conformational catalysis (see Section III) is employed in fragment condensation, even distinguishing between side-chain and terminal –COOH and –NH$_2$ groups may be unnecessary.

Nonetheless, there are instances where at least some protection is required. In these circumstances, groups that preserve the natural high solubility of cytochrome *c* and its fragments are favored, as are those for which the final deprotection conditions are not damaging to the protein. A particular application of side-chain protection is to limit the extent of fragmentation by proteolysis (see next subsection).

C. Protein Fragmentation

Two conflicting objectives govern the tactics of protein fragmentation in semisynthesis. Firstly, the fewer the number of fragments generated, the simpler will be the reassembly process. However, for the reasons to be considered here, the residue of interest should ideally reside in a small fragment. Much then depends on the choice of cleavage agent—or rather, cleavage agents, since the two objectives can be reconciled by, if necessary, subfragmentation of a larger fragment containing that residue by a second agent. A number of chemical cleavage procedures exist that are often specific to infrequently encountered residues. Fontana's group has worked extensively in this area in the context of cytochrome *c* (Fontana and Gross, 1986). Although proteases are rarely highly specific and generally give more extensive cleavage than desired, their action can often be limited by short exposures to substrate, or by masking of a class of specificity sites through chemical modification.

D. Fragment Modification

A number of tactics to effect a substitution of the candidate residue have evolved. If the residue is close to either terminus of the fragment, techniques of sequential degradation and resynthesis adapted from protein sequencing and chemical synthesis can be used to replace it. For more inaccessible residues, replacement of the entire peptide by a synthetic product is the simplest way, although if the residue is susceptible to chemical modification, that approach provides an easy alternative. Substitution of the relevant peptide by its homolog from another species can provide an informative chimeric protein.

E. Fragment Reassembly

This stage of a semisynthesis is likely to be the most demanding. Although many methods have been developed for the formation of peptide bonds between fragments, many are unsuited to naturally obtained, minimally protected fragments. Those that are useful are often inefficient, requiring large excesses of one component over another, when of course natural fragments will be obtained in roughly equimolar quantities. For this reason, semisyntheses from many fragments can assume heroic proportions (e.g., Rees and Offord, 1976). In most semisyntheses, however, the number of fragments has been kept to a strict minimum, and practitioners have looked for ways to make religation more efficient. Since many fragments are generated by proteolysis and any enzyme-catalyzed reaction is in principle reversible, could not proteases be used for peptide bond synthesis? Normally the equilibrium lies heavily in the direction of cleavage, so that exceptional circumstances would be required to observe appreciable synthesis. The work of Laskowski's group revealed what those circumstances might be (Homandberg *et al.*, 1978): firstly, that the fragments be physically held together, either by noncovalent forces or disulphide bridges (entropic factors); and secondly, that a large proportion of the water be replaced by organic cosolvent, to reduce the heat of proton transfer between substrate amino and carboxyl groups (enthalpic factors). In these circumstances up to 50% of resynthesis from equimolar mixtures is observed. Unfortunately, the number of cases where protease treatment yields two-fragment complexes so that reverse proteolysis can be applied to semisynthesis has proved to be limited. Nonetheless, the principles revealed by studies in this area have led to the development, with cytochrome *c* as model, of a novel and probably general method of fragment condensation (Section IV). Finally, if fragments do form a stable noncovalent complex that is functional, then the fragment-condensation step may be irrelevant to structure–function studies of semisynthetic analogs.

III. Noncovalent Complexes

A noncovalent complex, as the name implies, is the association of fragments representing more or less the entire primary sequence of the native protein to form a structure approximating the conformation of the parent protein. "More or less" is used advisedly, since viable noncovalent complexes may be formed that lack some fraction of the sequence, or indeed they may be composed of overlapping fragments. In any case, the normal noncovalent forces that define and maintain tertiary structure operate, in the absence of a peptide or disulfide bond linking the pieces, to hold the fragments together, and these constructs generally exhibit normal or partial enzymatic activity. They are stable enough to survive chromatography by a variety of methods, and so can be obtained in highly purified form.

Such complexes are useful for protein structure–function studies not only because the incorporation of modified fragments can yield informative analogs more simply than complete semisynthesis, but also because they in themselves can be usefully compared to the native, unclipped protein. Furthermore, the kinetics and thermodynamics of complex formation can reveal much about the normal processes of protein folding.

A. Preparation

Noncovalent complexes are formed when two or more protein fragments associate strongly. Typical values of K_D are $10^{-6}-10^{-8}$ M. It is assumed that if the complex functions like the parent protein, then its structure closely mimics the native conformation. The first example of a functional complex to be discovered was the well-known ribonuclease S system (Richards, 1958), which was later shown by crystallography to have a similar structure to ribonuclease A. This report was followed by others covering a wide range of proteins, including cytochrome *c* (Wallace and Proudfoot, 1987). The techniques described earlier for limiting cleavage to yield small numbers of fragments for semisynthesis are also likely to give rise to complexing sets of fragments, and this has proved particularly so for the cytochrome (Wallace and Proudfoot, 1987). The question is raised of whether cytochrome *c* is particularly suited to complex formation, and it may be that the heme group does act as a nucleation center for the cooperative refolding of the fragments. However, the range of other protein types in which the phenomenon has been noted, and the number of differing cleavage sites within that range, suggests that the apparent propensity of cytochrome *c* may be simply a consequence of the intensive study given it (as in so many other areas) by a number of groups.

The first reported complex resulted from limited cyanogen bromide cleavage of the protein. Of the two methionine residues in the protein, that at position 65 is not essential, so that purified fragments (1–65)H and (66–104) were seen to complex strongly (Corradin and Harbury, 1971). Furthermore, on standing this initial complex underwent a change that restored full biological activity. Only later was it realized that this change was in fact the spontaneous reformation of the missing peptide bond 65–66 (Corradin and Harbury, 1974b). This phenomenon, of conformational catalysis of religation, is analogous to that seen in reverse proteolysis systems, when the proximity of the two termini in the complex helps overcome the thermodynamic barrier to synthesis. It has proved of great value in making easier the semisynthesis of very many cytochrome *c* analogs (Section IV).

Shortly after this first example, it was shown that overlapping complexes, as well as contiguous ones, could be stable (Fisher *et al.*, 1973). (1–65)H derived from the CNBr cleavage was combined with the apoprotein produced when heme is cleaved from the holoprotein. The spectrum of the product was like that of native cytochrome *c* and not the (1–65)H fragment.

A novel, stable contiguous complex was derived after lysine residues of horse cytochrome *c* were protected by acetimidylation (Wallace and Harris, 1984). The action of trypsin is thus limited to arginyl residues, of which there are two in the protein. In fact, even after denaturation, only one, Arg^{38}, proves susceptible (Harris and Offord, 1977). The product complex, (1–38)H:(39–104), showed spectral properties identical to those of the parent protein, but had only limited biological activity in cytochrome *c*–depleted mitochondria. Similar conclusions were reached by other workers (Westerhuis *et al.*, 1979), using a different ε-amino group protection. They further showed that while activity with reductase is diminished, the complex is more efficient at electron transfer to cytochrome oxidase.

Stellwagen's group undertook a thorough investigation of the complementarity of both contiguous and overlapping fragments, by examining absorbance and CD spectra, redox potentials, and biological activity. They conclude that in the systems they examined, extensive overlap seemed necessary for functionality (Wilgus *et al.*, 1978). Although obviously unaware of the results with (1–38)H:(39–104) (Harris and Offord, 1977), they cited the work of Taniuchi's group (Hantgan and Taniuchi, 1977) showing that a new [(1–53)H:(54–104)] contiguous complex could have substantial biological activity.

Taniuchi and co-workers had made the important observation that overlapping complexes could be trimmed by a strictly limited exposure to trypsin to remove all or much of the redundant portions and yield fully or near-contiguous complexes. Thus, the mixture of (1–65)H and apo (1–104) gave (1–53)H:(39–104), (1–53)H:(54–104), and (1–53)H:(56–104). They rightly surmised that if the cleavage point were in the 38–57 region (the so-called bottom loop), then contiguous complexes would retain high functionality. This surmise has received considerable support from later work (Wallace and Proudfoot, 1987).

Shortly thereafter, Taniuchi's group showed that breakpoints in the 20s loop were also allowable (Parr *et al.*, 1980) in a study where 1–38 and apo 1–104 were combined and trypsin-treated. Two alternative products, (1–25)H:(23–104) and (1–38)H:(56–104), were recovered; the viability of the latter complex implied that the bottom loop was in fact irrelevant to at least some of the functions of the protein. Taniuchi's group (Juillerat *et al.*, 1980) were the first to report a three-fragment complex exhibiting some biological activity. They prepared this complex by tryptic trimming of the redundant portions of an overlap complex, yielding (1–25)H:(28–38):(56–104), which was reduced by yeast lactate dehydrogenase at 16% of the rate for native cytochrome *c*. (1–25)H:(56–104) was not reduced. Wallace and Proudfoot (1987) also used the trimming of overlap complexes to prepare (1–55)H:(56–

104) from (1–65)H:(56–104) and showed that partial acid hydrolysis would yield yet another novel contiguous complex, (1–50)H:(51–104).

Confirmation that breakpoints outside the bottom and 20s loops will be much less well tolerated than those within were recently obtained. A new overlap complex, (1–65)H:(60–104), was proteolytically trimmed with chymotrypsin, giving a complex, (1–59)H:(60–104), that exhibited no activity in depleted mitochondria (Wallace and Campbell, 1990). The known two-fragment contiguous or near-contiguous complexes of cytochrome *c* ("primary complexes") are collated in Figure 21.1. These primary complexes have been the precursors of both secondary complexes in which the native sequence is maintained while the breakpoint is shifted, and structural analogs of both primary and secondary complexes (Section V.B.5), but they have also proved valuable in their own right for studies of the thermodynamics and kinetics of protein folding and of electron transfer.

B. Functional Studies

Taniuchi's group have used the (1–53)H:(54–104) complex to investigate the conformational dynamics of cytochrome *c* (Hantgan and Taniuchi, 1978). By labeling fragment (54–104), fragment exchange kinetics could be examined over a range of temperatures. Thus, thermodynamic parameters for the unfolding (and refolding) process could be determined for both ferric and ferrous forms of the complex. The results implied that despite the apparent similarity of static conformations, the dynamic conformation of the protein differed greatly between oxidation states.

This system and others were used in their kinetic studies of complex formation (Parr and Taniuchi, 1979, 1980, 1982) to reveal second-order, biphasic kinetics and a very high degree of sensitivity of these rates to the sequence of the heme-containing fragment. Kinetic studies on the three-fragment complex (Juillerat *et al.*, 1980) revealed the nature of the folding pathway, and a conformational dynamics study (Juillerat and Taniuchi, 1982) revealed a thermal transition at 30° in the mechanism of unfolding, and presumably folding, of the three-fragment complex. Fisher and Taniuchi (1992) have recently examined a set of hybrid complexes composed of vertebrate and yeast components to define the "core domains" for the folding process, and to establish a folding order.

Such studies provide insights into the folding and unfolding that occur in native proteins, while those of Wallace and co-workers (Wallace and Proudfoot, 1987; Proudfoot *et al.*, 1986) also examined the relationship between the exact position of the missing peptide bond in the complex and biological activity. Taking the (1–38)H:(39–104) complex as a starting point, semisynthesis was used to create a secondary complex with shifted breakpoint (1–37)H:(38–104). It had already been noted that although differing complexes contained identical sequences, some variation in functionality could be observed (Hantgan and Taniuchi, 1977). We reasoned that comparative studies of related complexes could provide useful information on the functional role of the residues at the breakpoints, and of the secondary structural elements that contained them. In fact, the seemingly minor change from (1–38)H:(39–104) to (1–37)H:(38–104) produced striking consequences. Biological activity increased from 25% to 75% of that of the parent protein, and the redox

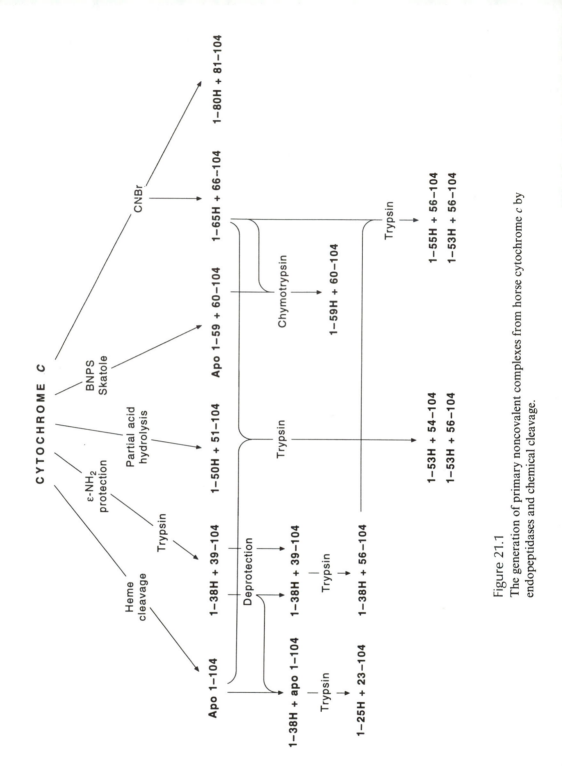

Figure 21.1
The generation of primary noncovalent complexes from horse cytochrome c by endopeptidases and chemical cleavage.

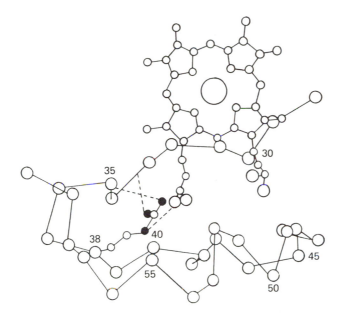

Figure 21.2
The conventional bottom loop of cytochrome c (residues 36–59) and its relationship to the heme group. The nitrogen atoms of the guanido group of arginine 38 are in black and are linked by hydrogen bonds and a salt bridge to the polypeptide backbone on the right side of the molecule and the heme propionate, thus integrating these structural elements. From Proudfoot *et al.* (1986), with permission.

potential increased from 150 mV to 220 mV. It was also noted that while the ferrous complex (1–38)H:(39–104) was completely converted to (1–38)H:(56–104) upon trypsin treatment, the nick-shifted product was to a large extent resistant to the enzyme.

The latter result indicated that the bottom loop (Figure 21.2) is much more tightly integrated into the overall protein structure in the derivative and that this stabilization must be due to the presence in the loop structure of the residue arginine 38. In the crystal structure this residue is seen to make extensive hydrogen bonds, and a salt bridge, with surrounding residues. The rare occurrence of a buried arginine residue performing such a role led us to suggest that this was the primary function of the residue, important enough to ensure its evolutionary invariance. This conclusion is confirmed by differences noted in the pKs for both acid and alkaline conformational transitions in site-directed mutants of position 38. For example, [Gly^{38}] human cytochrome c has a pK_{695} of 8.05, more than 1 pH unit lower than that of the native protein (Wallace and Tanaka, 1994).

The concerted changes noted above in activity in the depleted mitochondria assay and in redox potential had been apparent with other cytochrome c analogues (Wallace and Corthésy, 1986; Wallace *et al.*, 1986), so a systematic study of this

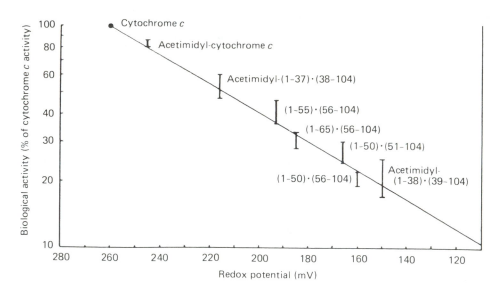

Figure 21.3

The relationship between relative biological activity in the cytochrome c–depleted mitochondria reductase assay and oxidation–reduction potential, at pH 7, for a set of two-fragment complexes and their parent cytochromes. From Wallace and Proudfoot (1987), with permission.

phenomenon was undertaken with a wide range of two-fragment complexes (Wallace and Proudfoot, 1987). The results confirmed that there was indeed a relationship between these two parameters, and that the logarithm of activity varied directly with redox potential (Figure 21.3). In any rate equation for electron transfer, there should be a driving force term, expressed as the difference between redox potentials of donor and acceptor (Marcus and Sutin, 1985). In the respiratory chain, for steps not coupled to proton translocation, such values are normally just slightly negative. A more positive value means a lower driving force, and the values of the difference between redox potentials of these complexes and reductase steadily increase with decreasing electron transfer rates. The implication is that the reductase–cytochrome c step is rate-limiting in this system, a view that is supported by rate measurements of the isolated components (Antalis and Palmer, 1982). If this reasoning is correct, then these analogs should have values of potential difference from oxidase that are more negative than normal, and a greater driving force. As a result, electron transfer rates to oxidase should increase. Measurements of activities of the complex (1–38)H:(39–104) with isolated components (Westerhuis et al., 1979) show lower activity with reductase and higher activity with oxidase, compared with the native protein, clearly supporting the foregoing interpretation. Noncovalent complexes have also been used in studies of the conformational factors underlying cytochrome c antigenicity (Collawn et al., 1988; and see Chapter 11 by Paterson, this volume) and of the anion-binding properties of the protein (Corthésy and Wallace, 1986). Currently a set of 16 different complexes is being studied in peroxide binding and reduction assays (see previous chapter) to examine the effect of position and size of polypeptide chain rupture on the accessibility of the heme iron to small molecule ligands (Adams, P. A., and Wallace, C. J. A., unpublished data).

C. Complex Formation and the Catalysis of Fragment Condensation

The realization that the $(1-65)$H:$(66-104)$ complex of cytochrome *c* spontaneously undergoes conversion to the covalent form [Hse^{65}] cytochrome *c* (Corradin and Harbury, 1974b) was a consequence of the discovery of a similar phenomenon in pancreatic trypsin inhibitor (Dykes *et al.*, 1974). Peptide bond formation results from aminolysis of the internal ester, homoserine lactone, generated at the C-terminus of fragments by the CNBr cleavage reaction. In free solution, using model compounds, this reaction proceeds very slowly and is not competitive with hydrolysis (Wallace, 1976). Hence, the high efficiency of peptide bond synthesis (60–90%) must be a consequence of catalysis by the structure of the complex itself. Harbury (1978) suggested that the conformational requirements for this kind of resynthesis are stringent, based on the observation that reaction only proceeds in the ferrous, and not the ferric, complex. This view has been amply confirmed since. A one-residue gap or overlap at the breakpoint (Wilgus *et al.*, 1978; Wallace *et al.*, 1986; Harbury, 1978) or even a one-residue shift in the breakpoint (Wallace *et al.*, 1986) is enough to eliminate religation. While other complex system have been discovered in which spontaneous peptide bond formation between fragments will occur (Galpin and Hoyland, 1985), others [e.g., lactalbumin and superoxide dismutase (C. J. A. Wallace, unpublished results)] show no evidence of recombination. Attempts to induce the phenomenon artificially have also proved largely unrewarding. R. C. Sheppard (personal communication) built homoserine lactone into the C-terminus of S peptide, but observed little reformation of ribonuclease A from the resulting ribonuclease S analog. We have incorporated the lactone into the cytochrome *c* $(1-38)$H:$(39-104)$ complex system (Proudfoot *et al.*, 1989), but very little religation of the fragments resulted.

As discussed earlier, complex formation is a necessary condition for fragment condensation by reverse proteolysis, so it is assumed that an analogous autocatalysis operates in these circumstances. Indeed, the conformational requirements might be less stringent, given the wider range of examples of the phenomenon (Kullman, 1987). A protease-mediated resynthesis of the $(1-38)$H:$(39-104)$ complex was reported using the Arg-specific clostripain (Juillerat and Homandberg, 1981). Others were unable to show peptide bond formation between 38 and 39, although the two fragments were held together by acrolein (a contaminant of some glycerol)-mediated cross-links (Proudfoot *et al.*, 1984).

Despite these disappointments, the basic principle revealed by these studies has been the inspiration of a general method that uses the autocatalytic properties of noncovalent complexes to promote fragment condensation. The method is discussed in Section IV.D.

IV. Practical Considerations in the Semisynthesis of Cytochrome *c*

A. The Heme Group

Cytochromes *c* are unusual in containing covalently bound heme, and careful consideration must be given to this sometimes chemically sensitive group in planning a semisynthesis. In practice this has meant that manipulations of the sequence of heme-containing fragments have not been attempted, and efforts to achieve classical chemical fragment condensation of heme fragments to the remainder of the molecule have been abandoned (Harris, 1977).

An alternative strategy would be to reinsert heme into a semisynthetic apoprotein in the final step. This can be achieved chemically; in trial experiments with apoproteins derived from native bovine cytochrome *c*, protoporphyrinogen was reacted with the protein, and the product porphyrin apoprotein *c* was combined with ferrous acetate under anaerobic conditions (Sano and Tanaka, 1986). The yield of crude product was 7–10%. The purified material exhibited 85% of native biological activity. Alternatively, there exists in yeast mitochondria an enzyme catalyzing the specific insertion of heme into yeast apoprotein once it has crossed the outer membrane (Basile *et al.*, 1980). The enzyme, which can be solubilized (Taniuchi *et al.*, 1983), is also capable of reincorporating heme into the horse apoprotein (Veloso *et al.*, 1981), so the suggestion was made that the determinants of specificity were simple and located near the attachment points, cysteines 14 and 17 (Veloso *et al.*, 1984). If so, then the enzyme would be capable of inserting heme into semisynthetic apoproteins or smaller apofragments destined for inclusion in noncovalent complexes. Both approaches have been shown to be feasible (Veloso *et al.*, 1984; Vita *et al.*, 1991) although at present the low activity of the mitochondrial cytochrome *c* synthetase preparations inevitably means that product yields are low.

B. Side-Chain Protection

Preliminary ε-amino protection of cytochrome *c* has proved useful in many semisyntheses in three roles. Restriction of tryptic cleavage to arginine residues allows strictly limited fragmentation (Harris and Offord, 1977). In the remodeling of fragments by sequential degradation and resynthesis, the Edman reaction has been frequently employed (Wallace and Corthésy, 1986; Wallace, 1979), requiring that the ε-NH$_2$ groups be blocked to prevent permanent derivatization. Finally, in chemical fragment condensations (e.g., Wallace, 1979), α- and ε-amino groups must be differentiated. A number of blocking groups have been exploited for these objectives, the first of which requires retention of water solubility. These are the citraconyl group (Juillerat *et al.*, 1980) which because of acid lability is only suitable for the first role; the methylsulfonylethyloxycarbonyl (Msc) group (Boon and Tesser, 1988), the acetimidyl (Acim) group (Wallace and Harris, 1984), and the trifluoroacetyl (TFA) group. These latter three are all in principle base-labile, but experience had shown incomplete removal of TFA groups (Ledden *et al.*, 1979). The other two groups have been extensively used, Msc by Tesser and coworkers, and Acim by ourselves. Both have characteristics to recommend them. The Msc group is very rapidly deprotected in strong base, but since it gives an uncharged derivative, it tends to reduce the water solubility of the protein and fragments. The Acim group is only slowly removed, at high pH, but if care is taken, cytochrome *c* is undamaged (Wallace and Harris, 1984). Because the Acim group preserves the positive charge of the amino group, solubility is maintained, and deprotection is not strictly necessary. This measure of convenience has been questioned (Shaw, 1987), but by very many criteria the acetimidylated protein is so little different from the native protein (Wallace, 1984a) that it constitutes a valid point of comparison for structure–function studies of analogs.

One group has experimented with post-cleavage amino group protection (Ledden *et al.*, 1979) employing the conventional blocking agents of total chemical synthesis for temporary use in those steps where water solubility is not a priority. They found conditions in which a significant differentiation of α- and ε-amino groups could be achieved. The same tactic has also been used where temporary side-chain carboxyl protection is required, as in the carboxyl component of a chemical fragment condensation. Where this fragment is prepared by solid phase peptide synthesis, specific ω-COOH protection can be built in (Nix and Warme, 1979). If the peptide is from a natural source, then CNBr fragments provide an elegant solution to the problem (Offord, 1972), for the C-terminal homoserine is in the lactone form, which may be simply hydrolyzed to the free carboxyl form after protection of the side-chain groups. For this purpose use has been made of the *p*-methoxyphenyl ester (Wallace, 1979; Wallace and Offord, 1979) or the methyl ester (Wallace and Offord, 1979). Tesser's group's ingenious approach avoided the need for side-chain carboxyl protection in fragment condensation altogether (Boon *et al.*, 1978). In a solution synthesis of (66–79), they introduced the C-terminal amino acid as a protected hydrazide; at the completion of the synthesis, the peptide hydrazide could be converted directly into the peptide azide, an activated species that will combine with the α-NH$_2$ of fragment (80–104).

C. Accessibility of Residues

Figure 21.1 demonstrated that the variety of limited cleavage techniques available to protein chemists could give, even considering two-fragment systems alone, breakpoints well distributed throughout the sequence. Obviously, when we consider methods that may yield three or more fragments, accessibility to most regions of the sequence is further increased. The importance of this factor lies in the limitations of the sequential degradation and resynthesis techniques, which means that an amino acid intended for substitution should lie, at most, three or four residues from a breakpoint. If no convenient cleavage site can be found, then a residue remains inaccessible. Fortunately, the recent improvements in solid-phase peptide synthesis mean that a synthetic fragment could be deployed with confidence in these circumstances, and the use of site-directed mutagenesis, described in the following subsection, to create cleavage and religation sites at any desired sequence position should allow easy access to almost any residue.

D. Fragment Reassembly

The difficulty and inefficiency of the chemical coupling of large peptide fragments has already been touched upon, as has the major drawback of the cytochrome *c* system, the sensitivity of heme to the coupling conditions. The fortuitous discovery of the spontaneous resynthesis of CNBr fragments (Corradin and Harbury, 1974b) provided one way to avoid the problems, so that until 1987, all fragment condensation semisyntheses of cytochrome *c* depended on this phenomenon for the final coupling step.

An automatic limitation to residues in the 66–104 region ensues. Although it proved possible to prepare semisynthetic noncovalent complexes with substitutions outside this area, and useful insights were gained, there has always been a prejudice towards analogs of the full covalent sequence of a protein. As related earlier, the first attempts to artificially induce the conformationally assisted spontaneous religation at other sites in proteins met with little success. We reasoned that perhaps homoserine lactone was insufficiently activating of the carbonyl carbon for all but a very few special cases, and that a more activating group might promote religation generally.

The problem of introducing the ideal activating agent was solved by Rose and collaborators (Rose *et al.*, 1987, 1988). Using trypsin, a set of aminoacyl esters was added by reverse proteolysis to the C-terminus of fragment (1–38). The esters employed varied in activating power. Some, such as the widely-used *N*-hydroxysuccinimide ester, proved unstable under reverse proteolysis conditions. Of the others, the dichlorophenyl ester proved to combine sufficient stability with adequate activating power for the next step. In this step, the now (1–39) dichlorophenyl ester was mixed with equimolar (40–104) in dilute neutral aqueous solution. Raising the pH led to rapid aminolysis of the ester to give about 40% of product cytochrome *c*. Thus, a conformationally assisted religation had been achieved, demonstrating that where a looser complex structure meant homoserine lactone was an ineffective activating agent, the principle of conformational assistance could nonetheless be harnessed by use of stronger but still mild reagents. It could also be shown that the system would operate when noncontiguous complexes were prepared, as long as the termini were in reasonable proximity in the tertiary structure (Figure 21.4); thus, (1–39)H and (56–104) were combined in 60% yield to give the bottom loop-deleted analog of cytochrome *c* (Wallace, 1987).

These promising results were followed up by an extensive study aimed at optimizing the system (Proudfoot *et al.*, 1989). The parameters examined for the activation step (protease-catalyzed addition of amino acid esters) included pH, solvent, nucleophile concentration, enzyme: substrate ratio, and most importantly, the nature of the added amino acid. A wide range of substrates were found to give almost uniformly high yields for the addition. In the coupling step, the influence of the C-terminal residue and the location of the breakpoint were considered. In the great majority of the tested systems, coupling yields in the 30–60% range were noted. These latter two observations hinted that the approach that had been developed could prove to be truly universal: that as long as the fragments complexed, any breakpoint in any protein could prove a suitable site for conformationally catalyzed peptide bond synthesis, now called autocatalytic fragment religation (AFR).

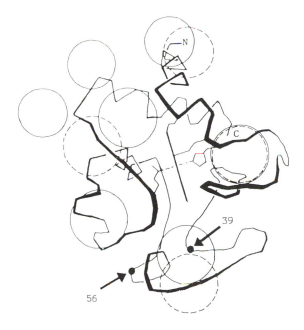

Figure 21.4

A sketch based on the published three-dimensional structure of mitochondrial cytochrome *c* showing (arrow) the reactive termini of the loop-deleted complex, (1–39)H:(56–104). From Wallace (1987), with permission.

To sustain a claim for true generality as a tool for protein engineering, three conditions should be met. (1) It should be possible to introduce breakpoints at any point in a sequence. Figure 21.1 showed the range of primary complexes that could be prepared by the use of just a few limited cleavage methods. Figure 21.5 shows how a primary complex can be manipulated by sequential degradation and resynthesis techniques to give a family of secondary complexes. (2) Meeting the first condition means that any residue might form the C-terminus of a fragment, and it is therefore necessary to be able to activate any and all such residues. To satisfy this condition, other enzymes than trypsin should be competent for reverse proteolysis. We have been examining this point (Wallace and Campbell, 1990; Wallace, 1991) and find that serine proteases are generally suitable. Within this class, an extremely wide range of substrate specificities exists. (3) Activated fragments should always combine in high yield with contiguous partners of a noncovalent association. In almost every case so far examined, fragments activated by dichlorophenyl esters religate with high efficiency. The two exceptions that have been noted occur in complexes with breakpoints at tight bends, so that it is possible that in such structures, the termini are held apart, rather than together. If this proves general, it will be necessary to avoid this class of complex for resynthesis.

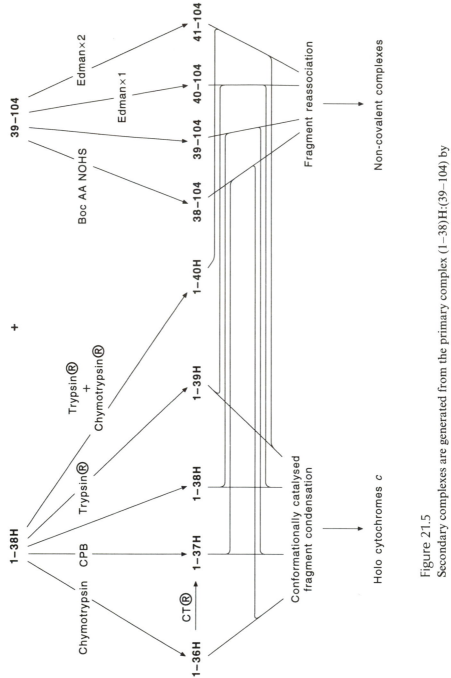

Figure 21.5

Secondary complexes are generated from the primary complex (1–38)H:(39–104) by enzymatic and chemical means for use in functional studies or as intermediates in a conformationally catalyzed fragment condensation. The constituent fragments may have the native, or a modified, sequence. R denotes the use of the enzyme in reverse proteolysis (synthetic mode).

The fact that only a few of the two-fragment systems in which homoserine lactone is found at the C-terminus will undergo AFR has been noted earlier. One such case is that of the cytochrome *c* from *Saccharomyces cerevisiae*, where methionine is found at position 64 in the vertebrate numerotation, but not at 65. This protein can be cleaved to give a two-fragment complex, but no religation occurs. We had in the past speculated on why that should be (Wallace *et al.*, 1986).

We realized the potential synergy of site-directed mutagenesis and semisynthesis in that the former could be used to introduce AFR sites at any desired position in a sequence to facilitate a final religation. The yeast system seemed an ideal test case for this premise, and so Met^{64} was changed to Leu, and Ser^{65} to Met (Wallace *et al.*, 1991, 1992). Subsequent cleavage and efficient religation have proved the point that the two technologies can operate synergistically. We have since inserted methionines at six different positions distributed throughout the sequence and seen very mixed results for religation efficiency (C. J. A. Wallace and J. G. Guillemette, unpublished results), emphasising that this parameter is very dependent on local conformational factors. In the case of the difference between positions 64 and 65, it appears to matter very much whether the breakpoint is in the hydrophobic or hydrophilic face of an amphipathic helix (Wallace *et al.*, 1991, 1992). However, the synergistic approach can clearly also be exploited to introduce sites of cleavage and activation by proteases, so that the generalized AFR strategy that we have developed can be further enhanced.

E. The State of the Art

Three recent developments mean that semisynthesis is better able than ever to furnish unusual protein analogs for academic, clinical or industrial purposes. The breakthrough in fragment condensation technique described earlier greatly simplifies the process for any protein, and in the context of cytochrome *c* this approach is yielding informative analogs modified in regions not previously accessible (Proudfoot *et al.*, 1989). The improvements in solid-phase peptide synthesis mean that now large fragments can be prepared with confidence to replace crucial stretches of the sequence. Finally, the advent of genetic methods for the manipulation of protein sequence means that a structure can to some extent be redesigned to facilitate the application to it of semisynthesis.

V. Semisynthetic Cytochrome *c* Analogs for Structure–Function studies

A. Introduction

The primary goal of protein semisynthesis has always been the generation of defined analogs for structure–function studies, although both clinical studies with (Shoelson *et al.*, 1984) and industrial production of (Markussen and Schoenburg, 1983) semisynthetic insulins have been reported. Cytochrome *c* was chosen as a target for semisynthetic studies for its intrinsic interest as a model both for general questions concerning protein folding and structural stabilization and for the specific study of biological electron transfer reactions. The availability of high-resolution structures of both oxidation states was an essential prerequisite for these goals, but clearly the choice was also influenced by the relative abundance and rugged nature of the protein. In addition to providing information about known functions, such studies may also reveal the existence of novel ones.

B. Approaches Employed for Analog Generation

1. Chimeric sequences

The first true semisynthetic analogs of the protein followed rapidly upon the discovery of the spontaneous religation of CNBr fragments (Corradin and Harbury, 1974b). Taking homologous fragments from different species, the same workers were able to show that a high degree of recombination could also occur in heterologous pairs (Corradin and Harbury, 1974a; Harbury, 1978) to give fully active cytochromes. The disadvantage of the approach is that the substitutions achievable are only those that are well tolerated in the course of evolutionary history. An important point made by these chimeras and the original religated material is that the product [Hse^{65}] cytochrome c is functionally equivalent to the native Met^{65} form (a point confirmed by many authors), and so conclusions can be confidently drawn from any analog prepared by use of this reaction.

2. Fragment-specific chemical modification

A completely or partially specific modification of a protein sequence can be achieved by a combination of semisynthesis and traditional chemical modification techniques. By confining the reaction to a single fragment, only one or a few defined residues will be altered in the reassembled protein. With cytochrome c, this approach has been taken, not unexpectedly, with the (1–65)H+ (66–104) pair, derivatizing both common and rare residues. Only two arginine residues are encountered in the horse sequence, at positions 38 and 91, so that this tactic allows independent derivatization of each (Wallace and Rose, 1983). Similarly, fragment (1–65)H contains only one of the four tyrosine residues of the horse sequence and has been modified by iodination or acetylation prior to recombination with (66–104) (Wallace, 1986). Clearly the success of this method also depends on the availability of highly specific reagents, but a broader specificity may be enhanced by selective protection. For example, acetylation of (1–65)H by acetylimidazole will modify both tyrosine and lysine residues. Prior protection by acetimidylation confines the modification to Tyr^{48}. The acetimidylation reaction itself has been used to produce specific partial modification of the protein (Wallace, 1984) following an earlier study employing the guanidinyl group (Wilgus and Stellwagen, 1974).

3. Total synthesis of fragment (66–104)

The most direct route to sequence modifications of this peptide was first attempted in 1977 (Barlow *et al.*, 1977) when synthesis of a 39-residue peptide was still an ambitious undertaking with solid-phase techniques, so that only 1–2% of the crude product would couple with (1–65)H. A later attempt using alternative resins and reagents was more successful (Atherton *et al.*, 1980), but side reactions due to the unavailability of suitable arginine protection led to a heterogeneous product.

The attempt by the Scoffone group, described in Section 1, to synthesize the whole molecule by solution methods (Moroder *et al.*, 1973, 1975) was eventually to become a semisynthesis project. Borin adapted the fragment condensation strategy to yield the yeast homolog of (66–104), (71–108), and an analog of the yeast CNBr fragment (70–108) (Borin *et al.*, 1987). The former was combined in good yield with horse (1–65)H to give a horse-yeast chimeric sequence (Wallace *et al.*, 1986), but the (70–108) fragment of yeast would not recombine with the corresponding yeast heme fragment (1–69)H, even though the breakpoint is shifted by just one residue.

Solution total synthesis was also adopted by Tesser's group (Ten Kortenaar *et al.*, 1983a, b, 1985; Tesser *et al.*, 1985) for the preparation of (66–104) incorporating substitutions in the 81–104 segment (see also Section IV.B.4). Low yields of final product were recorded, but it should be remembered that the nature of the introduced replacement amino acid, as much as any difficulties with the synthesis, can have a profound effect on the spontaneous religation yield.

Most recently, interest has refocussed on solid-phase methods for (66–104) synthesis using automated systems. In one study, the native sequence was prepared in pure form at 24% overall yield, which recombined with (1–65)H as efficiently as native (66–104) (Di Bello *et al.*, 1988, 1990). Both this and the Met(SO$_2$)^{80} analog sequence were also prepared by manual methods in somewhat lower yields. The analog would not religate with (1–65)H.

We have used automated methods to make the native sequence and five analogs incorporating change at three conserved amino acids [67, 78, 83] in the 66–104 sequence (Wallace *et al.*, 1989a, b) and to prepare a set of nine different substitutions at the crucial Met^{80} position (Wallace and Clark-Lewis, 1992). In these syntheses, the protocols developed by the Caltech group (Kent, 1988) produced the target peptide as 80% of the crude product, which permitted purification in good yield. Religation yields varied from very low to (in many cases) as high as the natural fragment. Replacement of Met^{80} by His, Cys, and Leu, and of Tyr^{67} by Phe and p-F-Phe, have also been reported by others (Raphael and Gray, 1989, 1991; Frauenhoff and Scott, 1992). Recently syntheses of peptides incorporating changes at residues 68, 70, 71, 74, 91, and 97 have been undertaken and the semisynthetic proteins derived from them are under active study (C. J. A. Wallace and I. Clark-Lewis, unpublished results).

4. Semisynthesis of fragment (66–104)

Modifications could also be made by manipulating the sequence of natural (66–104) or by total synthesis of a part of the fragment and its recombination with a naturally derived peptide. The partial synthesis route has been favored by the groups of Tesser and Warme. For both the strategy has been to prepare synthetic (66–79) by, respectively, solution or solid-phase methods, followed by coupling to

(80–104) derived from the natural CNBr fragment (81–104). The first report of a successful synthesis of this type (Boon *et al.*, 1978; Tesser *et al.*, 1979) and reconstitution of the native sequence of cytochrome *c* (Boon *et al.*, 1979a) came from Tesser's laboratory. The strategy was then exploited to make analogs of conserved residues in the 66–79 region: lysines 72, 73, and 79, tyrosine 74 (Boon *et al.*, 1979b), and the invariant tyrosine 67 (Boon *et al.*, 1981). They have also reported other analogs at position 67, a substitution at Thr[78] (Ten Kortenaar *et al.*, 1985), and the sequential degradation and resynthesis of natural (81–104) to prepare modifications to residue 81 (Boots and Tesser, 1987).

The alternative, partial solid-phase strategy was demonstrated first with the native sequence (Nix and Warme, 1979) and then exploited to swap phenylalanine and *p*-fluorophenylalanine for tyrosine at 67 and leucine for tyrosine at 74 (Koul *et al.*, 1979). Unfortunately, the very small yields of final products did not permit their complete purification for biological tests.

We chose to attempt a semisynthesis of the protein from the three natural fragments formed by CNBr cleavage. Reconstruction of the native sequence required restoring the essential Met[80] (Wallace and Offord, 1979; Wallace, 1978). It was reported that an [*o*-fluoroPhe[82]] analog of (66–104) prepared at the same time would not religate with (1–65)H (Wallace, 1978), but an improved route (Wallace 1979) yielded this and an [ε-Cbz Lys[81]] substitution. The same study reported the shifting of the essential methionine from position 80 to 81. We have also used the sequential degradation and resynthesis techniques to replace [Glu[66]] in the natural fragment (66–104) by Gln, Lys, and norvaline (Wallace and Corthésy, 1986).

5. Solid-phase synthesis of (1–65)

Gozzini and co-workers have suggested an elegant way to avoid the problems posed to such a synthesis by the need to insert the heme group before spontaneous religation of (1–65)H and (66–104) will occur. They observed that in the presence of (1–25)H, the apofragments (1–65) or (23–65) would combine with natural (66–104) in up to 80% yield (Gozzini *et al.*, 1991; Vita *et al.*, 1992). Thus substitutions could be built into the 1–65 segment, and the apoprotein would then, one hoped, be a substrate for the enzyme of yeast mitochondria, cytochrome *c*–heme lyase (Basile *et al.*, 1980; Taniuchi *et al.*, 1983; Veloso *et al.*, 1981). Alternatively, for substitutions in 23–65, the semisynthetic complex (1–25)H: (23–104) could be studied. Successful incorporation of heme has been reported (Vita *et al.*, 1991) and thus a novel route to analog production is open.

6. Semisynthetic modifications of noncovalent complexes

The discovery of the first functional contiguous complex of cytochrome *c*, (1–38)H: (39–104), was followed by its use as the starting point for semisynthesis (Harris and Offord, 1977; Harris, 1978). Residue 39 was removed and replaced, and residue 38 was removed, to make a set of complexes with substitutions at residue 39, or in which one or two of residues 38 and 39 were absent. Subsequent work led to the reincorporation of Arg[38] at the N-terminus of (39–104), to give the first nick-shifted or secondary complex (Figure 21.4) (Proudfoot *et al.*, 1986; Harris, 1979). We later extended this work to incorporate substitute residues into the complex at position 38 (Proudfoot and Wallace, 1987), and made hybrid complexes of lysine-modified and unmodified fragments (Wallace, 1984a). It was noted (Westerhuis *et al.*, 1982) that the stable truncated complex (1–38)H:(60–104), unlike the previously reported (1–38)H:(56–104) (Parr *et al.*, 1980), was quite inactive. The suggestion that tryptophan 59 was essential to a functional complex (Myer *et al.*, 1979) in ordering the correct iron ligation through its hydrophobic interaction with the heme was supported by the quenching by heme of the tryptophan fluorescence seen in such complexes (Hantgan and Taniuchi, 1978). The proposition was tested and supported in the truncated complex system by the addition of N-terminal tryptophan to (60–104), giving the active complex (1–38)H:(59–104) (Westerhuis *et al.*, 1982). Semisynthetic analogs of the three-fragment complex (1–25)H:(28–38):(39–104) incorporating synthetic fragments (28–38) prepared by solid-phase methods have been used to test the roles of leucines 32 and 35 and threonine 28 (Juillerat and Taniuchi, 1986), and of proline 30 and glycine 34 (Poerio *et al.*, 1986).

7. Modifications to fragment (39–104)

The development of the conformationally assisted coupling of protease-activated fragments opened up new regions of the sequence for substitution by covalent semisynthesis. Obvious targets for modification by this technique were residues 39 and 40 at the N-terminus of (39–104). Having been removed by the Edman degradation, these residues were reinstated by reverse proteolysis at the C-terminus of (1–38)H, prior to the religation reaction (Proudfoot *et al.*, 1989).

C. Structure–Function Studies of Semisynthetic Cytochrome *c* Analogs

1. Experimental systems

Semisynthesis has provided point mutations of 30 of the 104 residues in the horse sequence, including 13 of the 21 invariant residues (Table 21.1). In most cases, more than one replacement amino acid has been introduced. If one includes noncovalent complexes and chimeric proteins, more than 100 analogs have been studied, and it comes as no surprise that a very wide range of experimental techniques have been applied to them.

To avoid a detailed description of each technique, the reader is referred to a representative reference that gives an illuminating example of its use, and some description of the methods. It should be emphasized that in many cases, far more than one reference is available for this role. Further details can be sought in the relevant chapters of this volume.

Absorbance spectroscopy has, of course, been widely applied in the study of derivatives; characteristic bands in the visible region can reveal much about the coordination and spin states of the heme iron, and hence the conformational or electronic changes consequent on modification. In particular, perturbations in spectra

Table 21.1
A compilation of amino acid substitutions introduced by semisynthesis into horse cytochrome c.[a]

Residue	Evolutionary conservatism	Mutation	Functional effects	Suggested cause	Reference[b]
Thr 28	Variable	T → V, deletion	No effect		a
Pro 30	Invariant	P → G	Decreased bioactivity	Lower E'_m: Fe^{2+} destabilized more than Fe^{3+}	b
Leu 32	Invariant	L → V,F,I, NVal	Decreased stability	Hydrophobic core destabilized. Side-chain stereochemistry crucial	a
Gly 34	Invariant	G → A,S	Internal destabilization		b
Leu 35	Highly conserved	L → I	No effect	Increased bulk causes more fluctuation	a
		L → K,T	Some decrease in stability	Hydrophobic core destabilized Stereochemistry not crucial	a
Gly 37	Highly conserved	G → A	Not tested		c
Arg 38	Invariant	R → Q,K,G, etc	Decreased E'_m	Ω-loop destabilized	d, e, f
Lys 39	Conserved	K → A,F,G,E, etc	Decreased E'_m	Ω-loop destabilized	g, h, f
Thr 40	Highly conserved	T → K,V,F	Decreased E'_m	H-bond to 157 eliminated	g
Tyr 48	Invariant	Y → I Tyr,O-Ac-Tyr	Increased E'_m	Increased hydrophobicity at core	i
Lys 53	Conserved	K → Acim Lys	Decreased E'_m	Conformational change (by NMR)	j
Trp 59	Invariant	W → deletion	Inactive	Trp packs heme	k
Lys 60	Variable	K → A	Reduced activity	Shifts dipole axis?	l
Met 65	Variable	M → Homoserine	No effect		m, n
Glu 66	Highly conserved	E → K,Q	Increased E'_m	General electrostatic effect	o
			ATP affinity reduced	Part of binding site	
		E → NVal	Inactive	60s helix disrupted	
Tyr 67	Highly conserved	Y → L,F,pFPhe	Decreased E'_m	Crevice H_2O accessible, polarity change	p, q, r, s
Asn 70	Highly conserved	N → Homoserine	Mild destabilization	Together, these two residues form tight bend initiating 70s loop?	l
Pro 71	Invariant	P → NVal	Nonfunctional		l
Lys 72	Invariant	K → Acetyl Lys	Increased K_m with oxidase	+ve charge for oxidase binding	t
Lys 73	Invariant	K → Acetyl Lys	Increased K_m with oxidase	+ve charge for oxidase binding	t
Tyr 74	Highly conserved	Y → L	Decreased thermostability	Loss of structural rigidity?	t, s
Ile 75	Highly conserved	I → K	Decreased E'_m, 695 band lost	Severe destabilization	u
Thr 78	Highly conserved	T → N,Aba,V	Decreased E'_m	Structural destabilization	p, q
Lys 79	Invariant	K → Acetyl Lys	Increased K_m with oxidase	+ve charge for oxidase binding	t

Met 80	Invariant	M → Very many	Varied	Met sets E'_m, assists electron transfer	v, w
Ile 81	Highly conserved	I → CbzK	Low activity	Increased polarity/bulk in electron port	x
		I → A,L,V	Little effect		u
Phe 82	Invariant	F → o-FPhe	Diminished activity	Polarity change at electron port	x
		F → L	Increased K_m with oxidase	Aromatic ring aids binding?	p
Ala 83	Variable	A → P	Normal E'_m, low e^- transfer	Loss of flexibility in electron port	q
Arg 91	Invariant	R → Dmp Orn etc	High-affinity ATP binding lost	Essential part of site	e
Tyr 97	Highly conserved	Y → L	Increased E'_m	Influences F10 packing to heme?	p
		Y → BrY	Diminished activity	Increased polarity in interior?	l

[a] 30 residues of 104, including 13 of 21 invariant side chains, have been the object of semisynthetic modification. Suggested causes are those proffered in the original reference, except where indicated by a question mark. Abbreviations are NVal, norvaline; I Tyr, iodotyrosine; O-Ac-Tyr, O-acetyltyrosine; Acim Lys, ε-acetimidyl lysine; p-FPhe, p-fluorophenylalamine; Aba, α-aminobutyric acid; CbzK, ε-carbobenzoxylysine; o-FPhe, 4-fluorophenylalanine; BrY, 3-bromotyrosine; Dmp Orn, dimethylpyrimidylornithine. Conserved means unchanged within a kingdom; highly conserved means complete or very nearly complete functional group conservatism.

[b] a, Juillerat and Taniuchi (1986); b, Poerio et al. (1986); c, Wallace and Campbell (1990); d, Proudfoot et al. (1986); e, Wallace and Rose (1983); f, Proudfoot and Wallace (1987); g, Proudfoot et al. (1989); h, Harris (1978); i, Wallace (1986); j, Wallace (1984); k, Westerhuis et al. (1982); l, C. J. A. Wallace, unpublished data; m, Wallace and Offord (1979); n, Tesser et al. (1979); o, Wallace and Corthésy (1986); p, Ten Kortenaar et al. (1985); q, Wallace et al. (1989); r, Boon et al. (1981); s, Koul et al. (1979); t, Boon et al. (1979); u, Boots and Tesser (1987); v, Wallace and Clark-Lewis (1992); w, Raphael and Gray (1989); x, Wallace (1979).

715

of analogs induced by changing environmental, especially chaotropic, conditions are revealing about the forces that stabilize protein structure and the residues that provide them (Wallace *et al.*, 1989). More detailed information can be obtained from nuclear magnetic resonance spectroscopy, which has been applied to several analogs (Boon *et al.*, 1981). The development of two-dimensional NMR techniques will allow detailed analysis of subtle conformational change. Equally sensitive to small changes at the surface of the molecule are monoclonal antibodies. Using panels of antibodies, different changes in affinity can be used to map mutationally induced conformational change (Collawn *et al.*, 1988). Presently the ultimate test of conformational state is x-ray crystallography. As yet no semisynthetic analogs have been analyzed in this way.

The kinetics of fragment reassociation in noncovalent complexes have been studied by stopped-flow techniques that monitor both absorbance and fluorescence changes (Parr and Taniuchi, 1979). Thermodynamic studies of complex stability have used both fragment exchange (with labeled and cold fragments) (Hantgan and Taniuchi, 1978) and equilibrium gel filtration and dialysis (Juillerat and Taniuchi, 1986).

The interaction of semisynthetic cytochrome c with physiological partners has also been subject to scrutiny. K_ms for the oxidase reaction give a sense of the relative affinities (Boon *et al.*, 1979), but a more direct measure of K_d and stoichiometry with oxidase can be had from the study of spectral changes accompanying binding (Michel *et al.*, 1989). Circular dichroism (CD) and magnetic circular dichroism (MCD) spectra can be informative about conformational changes due to the interaction (Michel *et al.*, 1989).

A principal attribute of cytochrome c is its redox potential. E'_m of semisynthetic cytochromes has generally been measured using the uncomplicated method of mixtures, in which the absorbance difference at 550 nm is titrated with the ferricyanide–ferrocyanide couple (Wallace *et al.*, 1986). This technique is suitable for the determination of E'_m values up to about 100 mV less than that of the native protein (Figure 21.6). Other related spectrophotometric methods have used the Fe–EDTA couple (Corthésy and Wallace, 1988). An alternative to redox titrations is the use of differential pulse polarograpy (Raphael and Gray, 1989).

The prime role of the protein is electron transfer between complexes III and IV of the respiratory chain, so measures of the electron transfer efficiency of derivatives are essential to an understanding of structure–function relationships in cytochrome c. Generally, kinetic studies have employed either cytochrome oxidase, yeast lactate dehydrogenase (Hantgan and Taniuchi, 1977), or the succinate oxidase activity of cytochrome c–depleted whole mitochondria (Wallace and Proudfoot, 1987) (Figure 21.7). The oxidase reaction can be followed either spectrophotometrically (Michel *et al.*, 1989) or polarographically (Ten Kortenaar *et al.*, 1975).

The relationship between activity in the depleted mitochondria (reductase) assay and redox potential of analogs has been discussed in Section III.B. It was noted that exceptions to that relationship did exist (Wallace and Proudfoot, 1987), but that they all involved residues that were part of the electron port of cytochrome c, the surface region bounded by a ring of positive charge and within which the exposed heme edge emerged. Thus, if correlation of activity and E'_m data show devia-

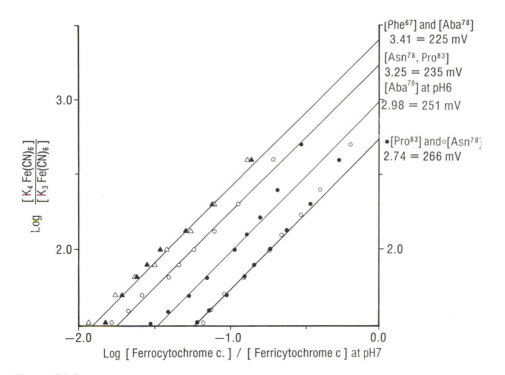

Figure 21.6
Determining the redox potential of cytochrome *c* analogs by the method of mixtures. The ferro/ferricyanide ratio is varied and the resulting proportions of reduced and oxidized cytochrome calculated from the 550 nm absorbance of cytochrome *c* and analogs. From Wallace *et al.* (1989a), with permission.

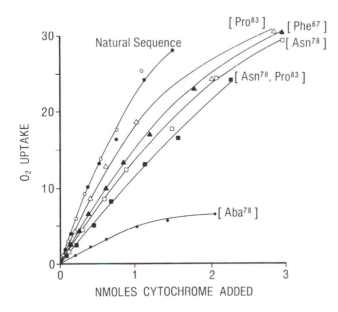

Figure 21.7
Examples of activity curves of cytochrome *c* and analogs obtained with cytochrome *c*-depleted mitochondria. O_2 uptake is measured polarographically in a Clark-type electrode. Succinate is substrate. Since the mitochondrial membrane is intact, electron transport is still coupled to phosphorylation, and therefore the ATP concentration is maintained at a low level by glucose/hexokinase. Relative activities are calculated by determining the tangents to initial slopes. From Wallace *et al.* (1989a), with permission.

tion from the established relationship, such deviation is a useful diagnostic of a direct role in electron transfer for the modified residue.

We have been using semisynthetic analogs of the protein to examine another functional property of cytochrome *c* that we consider biologically significant, the strong affinity for specific anions. Such studies require measures of the number of anions bound under varying conditions. We found that equilibrium gel filtration was most suitable under conditions of low ionic strength and with limited quantities of analog (Corthésy and Wallace, 1986), while for higher ionic-strength conditions, equilibrium dialysis has proved most valuable (Craig and Wallace, 1991). Another useful means of determining relative affinities has been competition experiments using affinity gels with an ATP ligand, to which cytochrome *c* binds most strongly (Craig and Wallace, 1991).

2. Direction and stabilization of the Cytochrome *c* fold

Studies of semisynthetic analogs, as well as those of the noncovalent complexes described earlier, have helped to identify the crucial primary structural elements responsible for establishing the folding pattern. It has long been clear that the heme group has a major role in directing folding, since holo- and apo-protein structures are radically different, so that the final conformation is not established until after the heme group is enzymatically added. Although no analog structures have been determined, biological activity provides a very good indicator of whether the correct fold has been assumed, since the association with physiological partners is crucially dependent on electron port conformation. Thus, the radically different behavior of (1–38)H:(60–104) (totally inactive) from that of semisynthetic (1–38)H:(59–104) (50% activity) confirmed a functional role for the evolutionary conserved Trp[59] that was attributed by the authors to its ability to stack with the heme, as in the native structure, and thus define a productive active site structure (Westerhuis *et al.*, 1982).

The 695-nm charge transfer band of the ferric iron–thioether sulfur coordination bond is a very sensitive indicator of the perturbation of conformation in the heme crevice, and its absence or weakening in an analog will suggest that the substituted residue had a structural role to play. Thus, unsurprisingly, the placement of a norvaline residue in the hydrophilic face of the perfect amphipathic helix that forms the conventional left side of the molecule leads to complete loss of the band (Wallace and Corthésy, 1986). In the L-α-aminobutyric acid (Aba) 78 derivative (Wallace *et al.*, 1989), the pK for the loss of this band (the alkaline transition) is shifted from 9.3 to <7 (Figure 21.8). This phenomenon may explain the apparent total lack of a 695-nm band noted by Ten Kortenaar *et al.* for the Val[78] analog (Ten Kortenaar *et al.*, 1985). The crystal structure of the cytochrome shows that the almost completely conserved Thr[78] residue is internal and is part of a hydrogen bond network found in the heme crevice (Takano and Dickerson, 1981). The bond provided by threonine, at least, is important in inhibiting the methionine displacement by an alternative ligand. The only natural substitution known is Asn[78], in *Chlamydomonas* cytochrome *c* (Amati *et al.*, 1988). We made this change in the horse sequence (Wallace *et al.*, 1989) and found reduced structural stability. Appar-

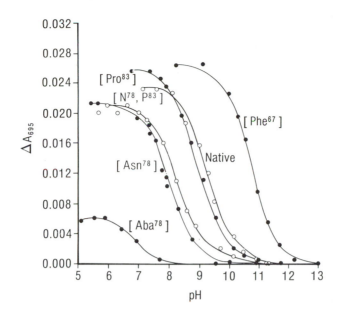

Figure 21.8
Titrations of the 695-nm absorbance band of native horse and some semisynthetic cytochromes c, showlng how pK for the transition and, therefore, the stability of the heme cleft to ligand exchange can vary with the nature of the substitution. From Wallace *et al.* (1989a), with permission.

ently in the alga either tolerance of the change is much greater than in any other species, or there is a compensatory change in sequence. In sharp contrast, the replacement in the same network of Tyr[67] by Phe actually increases stability to alkaline and thermal denaturation. Thermodynamic calculations suggested that the entropic cost of burying the hydroxyl group in the heart of the molecule outweighed the enthalpic advantage of H-bonding (Wallace *et al.*, 1989), so that the role of tyrosine is not a structural one.

The severe destabilization of the Aba[78] analog is reflected in the lower yield experienced for the (1–65)H:(66–104) religation reaction used in its preparation; the same is true for the [Nva[66]] protein. Such low yields are, therefore, likely to be diagnostic of mutations that affect conformation. Conversely, the ease with which active chimeric molecules may be formed, even from sequences as distant as horse and yeast (Wallace *et al.*, 1986), demonstrates how well-conserved the stabilizing interactions have been during the last billion years of evolution. The pair of residues Asn[70] and Pro[71] are absolutely conserved and form a sharp right-angle bend between two α-helices. This "Asx-turn" and its role in defining the cytochrome fold have been probed by independent change at the two residues. The substitution of Pro[71] by norvaline gives a quite inactive protein, as a consequence of a shift in the pK of alkaline transition (*vide infra*) from 9.3 to 5.2, which indicates the crucial role played by this residue (C. J. A. Wallace and I. Clark-Lewis, unpublished results).

Semisynthetic derivatives have also revealed residues that are unnecessary to the assumption or maintenance of the cytochrome *c* fold. In particular, any change in the bottom loop (Proudfoot *et al.*, 1989) or elimination of the loop altogether (Wallace, 1987) does not significantly alter the conformation of the remainder of the molecule. This result was foreshadowed by the observation that the crystal structure of some bacterial cytochromes, which among many other sequence changes lack this very loop, exhibit the characteristic fold. The apparent tolerance by the folding pattern of elimination of the loop even when the rest of the molecule is unchanged provides strong support for the proposal that Ω-loops, of which the bottom loop of cytochrome *c* is one, are independent units of folding (Wallace, 1987). As such, they will rely on intraloop interactions for their stability: We have demonstrated the importance of residues 39 and 40, at the neck of the Ω-loop, in this role (Proudfoot, *et al.*, 1989). The integration of the independent units into the whole, however, requires external interactions. In this context Arg[38] can be viewed as a screw fixing the loop to the rest of the box-like structure enclosing the heme (Proudfoot *et al.*, 1986; Proudfoot and Wallace, 1987).

3. The Alkaline Transition

Ferricytochrome *c* shows five distinct electronic spectra with changing pH (Theorell and Åkesson, 1941; also see Chapter 19 by Wilson and Greenwood, this volume). The change from the neutral form (state III) to state IV is known as the alkaline transition and in the horse protein occurs with a pK of 8.9–9.4, dependent on ionic strength. The spectroscopic changes, which include loss of the 695-nm S-Fe charge-transfer band, are believed to be due to displacement of methionine as sixth coordinating ligand. Since the iron remains low-spin, the replacement is presumed to be a strong-field ligand. The transition has been analyzed as a single-proton ionization of p$K \approx 11$, on which is superimposed a conformational transition.

The origins of this phenomenon, which may have biological significance, have concerned workers since it was first observed, and semisynthesis has been used to address the question. If there is indeed displacement of the sixth ligand methionine, then by what and how far?

The most popular view in the past, that the displacing agent is the side chain of lysine 79, and the evidence for it, is set out in Dickerson's reviews (Dickerson and Timkovich, 1975; Takano and Dickerson, 1981). The results of the first study using semisynthetic analogs were interpreted as supporting this view, in that lysine modifications caused significant change in the transition (Wallace, 1984b), even though Bosshard (1981) had been unable to detect any difference in susceptibility to chemical modification of the most likely lysine residues between neutral and alkaline states. The results of specific chemical modification of lysines (Smith and Millett, 1980) had in fact supported the view that lysine 79 could not be the substitute, but did not exclude Lys[72]. However, a normal alkaline transition for the plant and fungal cytochromes, where residue 72 is the permanently charged trimethyl lysine (Wallace and Boulter, 1988), implies that this residue cannot be the substitute either. NMR evidence (Boswell *et al.*, 1983) that only minor conformational change occurs during the transition suggests that just those lysine side chains located near-by can be considered as candidates for the substitution reaction. Apart from 72 and 79, there are other lysines that could conceivably approach the iron atom with modest structural perturbation, including lysine 73. The recent semisynthesis of an ornithine[80] analog of the horse protein (Wallace and Clark-Lewis, 1992) obviously has bearing on this issue. We found that the product was low-spin and had identical electrostatic properties to the native protein, indicating that an uncharged amino group can successfully coordinate the heme iron at, or even below, neutral pH. The UV-visible spectrum had properties that are similar, but not identical, to those of state IV of the horse protein. However, the redox potential (-40 mV) is substantially higher than that measured for state IV. The [Nva[71]] analog of the horse protein described earlier has no detectable 695-nm band at pH 7, and an extrapolated pK value for the "alkaline" transition of 5.2. The UV-visible spectrum at pH 7 is almost identical to that of native state IV and, like it, is resistant to ascorbate reduction. Increased flexibility in the 70s loop is a probable, but yet to be confirmed, consequence of the substitution. Therefore, the 4 pH-unit change in pK_{695} (which is likely to be primarily due to a change in the ease of the conformational transition, rather than the pK of proton ionization) suggests that the structures involved in the conformational transition are in, or in contact with, the 70s loop.

Alternatives to lysine side chains have been considered. Histidines 26 or 33 are believed to provide the sixth ligand in low-spin heme peptides such as (1–65)H, but have been excluded by studies of chemically modified or natural variants (Wallace and Corthésy, 1987). The possibility that methionine remained ligated, but in an alternative conformation with a shifted (and thus obscured) charge-transfer band, or that OH– is the alternate, have been raised (Bosshard, 1981; Pettigrew *et al.*, 1976). The behavior of some new semisynthetic analogs (Wallace and Clark-Lewis, 1992) argues against these suggestions. When methionine is replaced by either alanine or norleucine, neither of which can ligate, the resulting ferricytochromes remain low-spin at neutral pH. Thus, some alternative ligand must be provided by the protein, in all probability the same that replaces methionine in the alkaline transition, since their spectra closely resemble those of state IV. Another possibility to have been aired is Tyr[67] (Theorell and Åkesson, 1941; Salemme *et al.*, 1973). Its –OH group is so close to the iron atom that little conformational change would occur during the transition, in accordance with the NMR evidence (Boswell *et al.*, 1983). Modifications of this residue that affect the pK_a of the phenolic hydroxyl do cause shifts in pK_{695} (Skov and Williams, 1971). Finally, elimination of the hydroxyl group in [Phe[67]] cytochrome *c* (Wallace *et al.*, 1989) leads to a marked increase in the pK of the loss of methionine coordination. However, if the tyrosine–OH were the coordinating group, then Phe67 cytochrome *c* could never exist in the alkaline state IV and would presumably, at that higher pH, transform directly to state V. In fact, the visible spectrum of the [Phe–67] analog at pH 11.6 closely resembles that of state IV, not state V (C. J. A. Wallace, unpublished data). The most recent evidence supports the candidacy of Lys[79], but somewhat confuses the issue by suggesting that there is more than one isomer of the alkaline conformer. It is to be hoped that structural studies of the [Nva[71]] analog, which is in state IV at pH 7, will aid in resolving the issue.

4. Binding to physiological partners

The strength and specificity of interaction with reductase, oxidase, and ancillary partners in the mitochondrial membranes and intermembrane space have been largely revealed by chemical modification studies (see Chapter 17 by Millett and Durham, this volume). Nonetheless, some semisynthetic derivatives have been made to address the issue. The analogs developed by Tesser's group have invariably been tested with purified oxidase, so that K_m values for this reaction are known in each case. Modifications of lysine residues in the ring of positive charge that encloses the electron port on the molecule's surface increase K_m, most strikingly when two or more are simultaneously changed (Boon et al., 1979), providing confirmation of the chemical modification studies. However, they have been able to show that while changing internal hydrophobic residues leaves K_m of the oxidase reaction unaffected (Boon et al., 1981), residues in the hydrophobic stretch that parallels the exposed heme edge of the electron port do contribute to the strength of binding (inasmuch as it is signified by K_m). Thus, mutations at isoleucine 81 (Boots and Tesser, 1987) and phenylalanine 82 (Ten Kortenaar et al., 1985) produce a noticeable effect on K_m.

A novel use to which semisynthetic analogs are being put is in studies of the orientation of the protein relative to the membrane surfaces with which it interacts. X-ray standing waves can be used for this purpose, if the protein contains a detectable heavy atom. Initial studies with SeMet[80]-labeled protein proved successful (Wang et al., 1994), confirming the "face-down" postulate of cytochrome c-phospholipid bilayer interaction (Salemme, 1977). We have also introduced bromine as the label, in 3-bromotyrosine 97 cytochrome c, and made the double-labeled SeM[80], BrY[97] protein, to establish the absolute topography.

5. Setting the redox potential

All known eukaryotic cytochromes c have E'_m values that fall in the range 260 ± 20 mV, and the vast majority are near the middle of this range. Since the protein forms part of an interdependent chain of electron carriers, it is unsurprising that such evolutionary conservatism occurs, and a principal function of the protein coat that enfolds the heme must be to maintain a stable and appropriate redox potential.

How it does so, and at a level strongly divergent from that of free heme, has been the subject of intense speculation. The following protein factors have been suggested to play a major role in modulating heme redox potential:

(a) axial ligation—the nature of the fifth and sixth ligands (Moore and Williams, 1977);
(b) the polarity of the residues packing the heme (Kassner, 1973);
(c) the specific interaction of heme and Trp 59 (Myer et al., 1979);
(d) the degree of heme exposure to solvent (Stellwagen, 1978);
(e) surface charge and internal dielectric constant (Rees, 1980);
(f) special electrostatic interactions (Moore, 1983).

The evaluation of the relative importance of these factors in eukaryotic cytochrome *c* has been hampered by the simple fact that they all possess approximately the same midpoint potential, so that analysis has often involved comparison with prokaryotic cytochromes and the consequent simultaneous variation of several different factors.

Protein engineering provides a unique opportunity to vary potential within a common structural framework, and semisynthetic analogs have proved valuable in quantitating the effects of many of these parameters. Those modifications encompassing the bottom loop have demonstrated that its primary role is probably to close off the bottom of the heme from solvent and maintain a high potential (Wallace and Proudfoot, 1987; Wallace, 1987). Its elimination causes a 120 mV drop in E'_m (Wallace, 1987). Many analogs have caused changes in surface charge, and often the result is a modified redox potential. Replacing Glu[66] by glutamine leads to an increase of 16 mV; replacement with lysine increases the potential by 24 mV (Wallace and Corthésy, 1986). Elimination of the negative charge of position 69 raises E'_m by 8 mV (Wallace and Corthésy, 1986). In general the changes induced by charge change are small [reversing the charge on all 19 lysines by citraconylation only leads to a 100-mV drop (Wallace and Corthésy, 1987)], although Lys[39] → Ala or Phe makes a 50 mV difference (Proudfoot *et al.*, 1989). This change probably more reflects a weakening of the structure of the bottom loop (Proudfoot *et al.*, 1989) than it does a specially potent electrostatic interaction due to low dielectric constant of the protein matrix intervening between residue 39 and the iron atom (Proudfoot and Wallace, 1987).

A special electrostatic interaction proposed for Arg[38] (Moore, 1983) has been tested by substitutions in a noncovalent complex by lysine and glutamine. Since [Lys[38]] cytochrome *c* (1–37H):(38–104) has a potential much lower than that of the Arg[38] parent, and only a little higher than that of the [Glu[38]] analog, it was concluded that the primary role of Arg[38] is to stabilize the bottom loop of the protein by H-bonding, and not to provide a counterion to the heme propionate (Proudfoot and Wallace, 1987). Thus, the evidence for special electrostatic interactions as controlling influences is sketchy, and it is clear that electrostatic effects in general are of minor importance.

As yet, few semisynthesis experiments have been directed to modulating the polarity of the heme environment. Tyr[48] has been made the slightly more hydrophobic iodotyrosine or *O*-acetyl tyrosine, with a consequent E'_m rise of 10 mV in both cases (Wallace, 1986), but the results of modifying Tyr[67] are puzzling. The drop in E'_m on introducing *p*-fluorophenylalanine might be reconciled with the slightly greater polarity of this side chain (Ten Kortenaar *et al.*, 1985). However, the clearly less polar phenylalanine (Wallace *et al.*, 1989) and leucine (Boon *et al.*, 1979) show the same degree of change. We have proposed that this influence is a consequence of another modulating factor, the electron distribution within the heme ring itself, which is likely to be controlled by a closely proximate polar group (Wallace *et al.*, 1989). This view is supported by the shifts in absorbance band wavelengths shown by the Phe[67] derivative and by the observation that another influence on electron distribution, the nature of the substituent groups on the heme, also modifies redox potential. *Crithidia* cytochrome *c*, with one vinyl group replacing a thioether, has a high E'_m (Moore *et al.*, 1984).

Most recently, the quantitative influence of axial ligation has been addressed (Wallace and Clark-Lewis, 1992; Raphael and Gray, 1989, 1991). The results give general support for the view that this is a prime modulator of potential, and that substitute ligands with particular affinity for ferric iron will cause a substantial drop in potential by stabilizing that state vs. the ferrous one (Moore and Williams, 1977). Thus, $Met^{80} \rightarrow$ His or Orn causes a 200–300 mV drop (Wallace and Clark-Lewis, 1992; Raphael and Gray, 1989), while $Met^{80} \rightarrow$ Cys reduces E_m by 300 or 600 mV (Wallace and Clark-Lewis, 1992; Raphael and Gray, 1991). Alternative thioether ligands, which, like Met, ligate Fe^{2+} well, maintained a high potential, although the seemingly subtle replacement of sulfur by selenium in the selenomethionine 80 analog led to a 50-mV potential drop. The introduction of nonligating residues also gives relatively high E_ms, although the issue is confused by the spin-state change that occurs on reduction of $[Ala^{80}]$ cytochrome c. As well as displaying a myoglobin-like E'_m and absorbance spectrum, this analog binds O_2 in the reduced state (Wallace and Clark-Lewis, 1992).

It is apparent that all of the suggested modulating factors are in fact operative within the cytochrome c molecule, and that their qualitative influence, with the exception of electrostatic effects, can be very substantial.

6. The catalysis of electron transfer

Electrons do not readily move from an orbital of a redox center to one in another, distant center, and so it seems likely that the protein coat adopts a catalytic role to ensure a high rate of electron transfer. Some of the proposed mechanisms (applicable to transfers to and from any physiological partner) (see Chapter 15 by Scott in this volume) invoke the participation of residues within the electron port, and these have been the subject of semisynthetic modification for kinetic analysis. As pointed out earlier, rates are also dependent on thermodynamic driving force, represented by $\Delta E'_m$, so discrepancies between observed rates and E_ms can pinpoint mechanistically important residues (Wallace and Proudfoot, 1987).

Because of its absolute conservation and its central position in the active site, Phe^{82} has always been a prime target for replacement. Early experiments showed that substitution by o-fluorophenylalanine would greatly reduce activity with reductase, but not enough of the analog was available for E'_m determinations (Wallace, 1979). The $Phe^{82} \rightarrow$ Leu substitution leaves both E'_m and V_{max} for the oxidase reaction unchanged, so that the observed difference, an increased K_m, suggests that the primary role of this residue may be in binding rather than catalysis (Ten Kortenaar et al., 1985).

Ile^{81} is a functionally conserved residue (Ala or Val are evolutionarily acceptable alternatives), and a drastic modification to ε-carbobenzoxylysine diminishes electron transfer rates (Wallace, 1979). More subtle changes, to Ala, Val, or Leu, leave E'_m unaffected (Boots and Tesser, 1987). The Ala^{81} and Val^{81} proteins have unchanged V_{max} (although K_m is increased), but the electron transfer rate with oxidase is lower for $[Leu^{81}]$ cytochrome c, hinting that conformation in this region may in fact be important to electron transfer efficiency.

Residue 83 is Ala (or Val) in all animals, Gly in fungi, and Pro in higher plants. We noticed that the plant cytochrome c electron transfer rates in rat mitochondria were anomalously low, although the yeast cytochromes behave normally, and we suspected the cause to be this change (Wallace and Boulter, 1988). We introduced the same substitution into the horse protein (Wallace *et al.*, 1989), and indeed found unchanged redox potential and a significantly lower rate of electron transfer with reductase. If this is a consequence of reduced conformational flexibility in the electron port, then the implication is that some movement occurs during the electron transfer act, consistent with one set of mechanistic proposals (Poulos and Kraut, 1980) and the results of molecular dynamics simulations (Wendoloski *et al.*, 1987).

The replacement of threonine at position 78 by other residues also provoked a disproportionately large change in activity (Wallace *et al.*, 1989). This finding implied either a direct mechanistic role for the residue, or an indirect one via the methionine 80 to which it is H-bonded. This latter possibility was tested by the series of substitutions we recently introduced at this site (Wallace and Clark-Lewis, 1992). Some deviations from the redox potential–reductase activity relationship were noted. Diminished (selenomethionine, alanine, norleucine), normal (S-methyl cysteine, thienylalanine), and enhanced (histidine) activities support the view that the ligating residue also make a significant contribution to the act of electron transfer.

7. Regulation of the respiratory chain

We reported the unusual observation that the gross modification of an absolutely conserved residue (Arg^{91}) did not affect the major functional indicators of cytochrome *c* (Wallace and Rose, 1983). This necessitated the proposal that the residue was conserved for some less obvious function, and we subsequently used semisynthetic analogs of Arg^{91} and other residues to show that cytochrome *c* was capable of binding ATP in a strong and specific manner at a site that incorporated this residue (Corthésy and Wallace, 1986, 1988; Craig and Wallace, 1991). If this binding had a functional role, as was indeed suggested by the conservation of the residue, then an obvious possibility was to mediate feedback inhibition. The effects of site occupancy on the interaction of cytochrome *c* with its physiological partners have been studied in the absence of free ATP (which could affect other system components) by preparing affinity-labeled protein. The adducts have been shown to be labeled at the Arg^{91}-containing site, which has little effect on the physiochemical properties of the protein (Craig and Wallace, 1993). However, reactivity with the respiratory chain, in particular the oxidase, is profoundly inhibited. We have also observed that blocking ([DHCH–Arg^{91}] cytochrome *c*) or eliminating ([Nle^{91}] cytochrome *c*) the guanidino group changes the sensitivity to ATP of redox reactions of cytochrome *c* with its partners (D. B. Craig, I. Clark-Lewis, and C. J. A. Wallace, unpublished results).

VI. Acknowledgments

My contribution to the work described here was made possible by the financial support of the Natural Sciences and Engineering Research Council of Canada, the Swiss National Science Foundation, and the Medical Research Council of the United Kingdom; the many collaborators whose names appear on the relevant publications, especially Dr. Amanda Proudfoot; and the support of my mentor, Professor Robin Offord. I am very grateful to them all and to Charles Bradshaw, Monique Rychner, Barbara Battistolo, and Angela Brigley for valuable technical assistance.

VII. References

Amati, B. B., Goldschmidt-Clermont, M., Wallace, C. J. A., & Rochaix, J.-D. (1988) *J. Mol. Evol.* **28**, 151–160

Antalis, T. M., & Palmer, E. (1982) *J. Biol. Chem.* **257**, 6194–6206.

Atherton, E., Wooley, V., & Sheppard, R. C. (1980) *J. Chem. Soc., Chem. Commun.*, 970–971.

Barstow, L. E., Young, R. S., Yakali, E., Sharp, J. J., O'Brien, J. C., Berman, P. W., & Harbury, H. A. (1977) *Proc. Natl. Acad. Sci. USA* **74**, 4248–4250.

Basile, G., Di Bello, C., & Taniuchi, H. (1980) *J. Biol. Chem.* **255**, 7181–7191.

Boon, P. J., & Tesser, G. I. (1988) *Int. J. Pept. Prot. Res.* **25**, 510–516.

Boon, P. J., Tesser, G. I., & Nivard, R. J. F. (1978) in *Semisynthetic Peptides and Proteins* (Offord, R. E., & Di Bello, C., eds.), pp. 115–126, Academic Press, London.

Boon, P. J., Tesser, G. I., & Nivard, R. J. F. (1979a) *Proc. Natl. Acad Sci. USA* **76**, 61–65.

Boon, P. J., van Raay, A. J. M., Tesser, G. I., & Nivard, R. J. F. (1979b) *FEBS Lett.* **108**, 131–135.

Boon, P. J., Tesser, G. I., Brinkhof, H. H. K., & Nivard, R. J. F. (1981) *Proc. Eur. Pept. Symp.* **16**, 301–305.

Boots, H. A., & Tesser, G. I. (1987) *Proc. Eur. Pept. Symp.* **19**, 211–214.

Borin, G., Corradin, G., Calderan, A., Marchiori, F., & Wallace, C. J. A. (1987) *Biopolymers* **25**, 2269–2279.

Bosshard, H. R. (1981) *J. Mol. Biol.* **153**, 1125–1149.

Boswell, A. P., Moore, G. R., Williams, R. J. P., Harris, D. E., Wallace, C. J. A., Bocieck, S., & Welti, D. (1983) *Biochem. J.* **213**, 679–686.

Collawn, J. F., Wallace, C. J. A., Proudfoot, A. E. I., & Paterson, Y. (1988) *J. Biol. Chem.* **263**, 8625–8634.

Corradin, G., & Harbury, H. A. (1971) *Proc. Natl. Acad. Sci. USA* **68**, 3036–3039.

Corradin, G., & Harbury, H. A. (1974a) *Fed. Proc.* **33**, 1302.

Corradin, G., & Harbury, H. A. (1974b) *Biochem. Biophys. Res. Commun.* **61**, 4100–4106.

Corthésy, B. E., & Wallace, C. J. A. (1986) *Biochem. J.* **236**, 359–364.

Corthésy, B. E., & Wallace, C. J. A. (1988) *Biochem. J.* **252**, 349–355.

Craig, D. B., & Wallace, C. J. A. (1991) *Biochem. J.* **279**, 781–786.

Craig, D. B., & Wallace, C. J. A. (1993) *Protein Science* **2**, 966–976.

Di Bello, C., Tonellato, M., Lucciari, A., Buso, O., & Gozzini, L. (1988) in *Peptide Chemistry 1987* (Shiba, T., & Sakakibara, S., eds.), pp. 409–412, Protein Res. Found., Osaka.

Di Bello, C., Tonellato, M., Lucciari, A., Buso, O., Gozzini, L., & Vita, C. (1990) *Int. J. Pept. Prot. Res.* **35**, 336–345.

Dickerson, R. E. (1972) *Sci. Amer.*, April, 58–72.

Dickerson, R. E., & Timkovich, R. (1975) *The Enzymes* **11**, 397–547.

Dickerson, R. E., Takano, T., Kallai, O. B., & Samson, L. (1972) in *Structure and Function of Oxidation–Reduction Enzymes* (Akeson, A. & Ehrenberg, A., eds.), pp. 69–83, Pergamon, Oxford.

Dykes, D. F., Creighton, T., & Sheppard, R. C. (1974) *Nature* **247**, 202–204.

Fisher, A., & Taniuchi, H. (1992) *Arch. Biochem. Biophys.* **296**, 1–16.

Fisher, W. R., Taniuchi, H., & Anfinsen, C. B. (1973) *J. Biol. Chem.* **248**, 3188–3195.

Fontana, A., & Gross, E. (1986) in *Practical Protein Chemistry* (Darbe, A., ed.), pp. 67–120, Wiley & Sons, New York.

Frauenhoff, M. M., & Scott, R. A. (1992) *Proteins: Struct., Funct., Genet.* **14**, 202–212.

Gadsby, P. M. A., Peterson, J., Foote, N., Greenwood, C., & Thomson, A. J. (1987) *Biochem. J.* **246**, 43–54.

Galpin, I. J., & Hoyland, D. A. (1985) *Tetrahedron* **41**, 907–910.

Gozzini, L., Taniuchi, H., & Di Bello, C. (1991) *Int. J. Pept. Prot. Res.* **37**, 293–298.

Hantgan, R. R., & Taniuchi, H. (1977) *J. Biol. Chem.* **252**, 1367–1374.

Hantgan, R. R., & Taniuchi, H. (1978) *J. Biol. Chem.* **253**, 5373–5380.

Harbury, H. A. (1978) in *Semisynthetic Peptides and Proteins* (Offord, R. E., & Di Bello, C., eds.), pp. 73–89, Academic Press, London.

Harris, D. E. (1977) D. Phil. Thesis, Oxford University.

Harris, D. E. (1978) in *Semisynthetic Peptides and Proteins* (Offord, R. E., & Di Bello, C., eds.), pp. 127–138, Academic Press, London.

Harris, D. E. (1979) *Proc. Amer. Pept. Symp.* **6**, 613–616.

Harris, D. E., & Offord, R. E. (1977) *Biochem. J.* **161**, 21–24.

Homandberg, E. A., Mattis, J. A., & M. Laskowski, J. (1978) *Biochemistry* **17**, 5220–5227.

Juillerat, M., & Homandberg, G. A. (1981) *Int. J. Pept. Prot. Res.* **18**, 335–342.

Juillerat, M., & Taniuchi, H. (1982) *Proc. Natl. Acad. Sci. USA* **79**, 1825–1829.

Juillerat, M. A., & Taniuchi, H. (1986) *J. Biol. Chem.* **261**, 2697–2701.

Juillerat, M., Parr, G. R., & Taniuchi, H. (1980) *J. Biol. Chem.* **255**, 845–853.

Kassner, R. J. (1973) *J. Am. Chem. Soc.* **95**, 2674–2677.

Kent, S. B. H. (1988) *Annu. Rev. Biochem.* **57**, 957–989.

Koul, A. K., Wasserman, G. F., & Warme, P. K. (1979) *Biochem. Biophys. Res. Commun.* **89**, 1253–1259.

Kullman, W. (1987) *Enzymatic Peptide Synthesis*, CRC Press, Boca Raton, Florida.

Ledden, D. J., Nix, P. T., & Warme, P. K. (1979) *Biochim. Biophys. Acta* **578**, 401–412.

Marcus, R. A., & Sutin, N. (1985) *Biochim. Biophys. Acta* **811**, 265–322.

Markussen, J., & Schoenburg, K. (1983) *Proc. Eur. Pept. Symp.* **17**, 387–390.

Michel, B., Proudfoot, A. E. I., Wallace, C. J. A., & Bosshard, H. R. (1989) *Biochemistry* **28**, 456–462.

Moore, G. R. (1983) *FEBS Lett.* **161**, 171–175.

Moore, G. R., & Williams, R. J. P. (1977) *FEBS Lett.* **79**, 229–232.

Moore, G. R., Harris, D. E., Leitch, F. A., & Pettigrew, G. W. (1984) *Biochim. Biophys. Acta* **764**, 331–342.

Moroder, L., Borin, G., Marchiori, F., & Scoffone, E. (1973) *Biopolymers* **12**, 477–492.

Moroder, L., Filippi, B., Borin, G., & Marchiori, F. (1975) *Biopolymers* **14**, 2061–2074.

Myer, Y. P., Saturno, A. F., Verma, B. C., & Pande, A. (1979) *J. Biol. Chem.* **254**, 11202–11206.

Nix, P. T., & Warme, P. K. (1979) *Biochim. Biophys. Acta* **578**, 413–427.

Offord, R. E. (1972) *Biochem. J.* **129**, 499–501.

Offord, R. E. (1980) *Semisynthetic Proteins*, Wiley & Sons, Chichester.

Offord, R. E. (1987) *Protein Engineering* **1**, 151–157.

Parr, G. R., & Taniuchi, H. (1979) *J. Biol. Chem.* **254**, 4836–4842.

Parr, G. R., & Taniuchi, H. (1980) *J. Biol. Chem.* **255**, 8914–8918.

Parr, G. R., & Taniuchi, H. (1982) *J. Biol. Chem.* **257**, 10103–10111.

Parr, G. R., Hantgan, R. R., & Taniuchi, H. (1980) *J. Biol. Chem.* **255**, 5381–5388.

Pettigrew, G. W., Aviram, I., & Schejter, A. (1976) *Biochem. Biophys. Res. Commun.* **68**, 807–813.

Poerio, G., Parr, G. R., and Taniuchi, H. (1986) *J. Biol. Chem.* **261**, 10976–10989.

Poulos, T. L., & Kraut, J. (1980) *J. Biol. Chem.* **255**, 10322–10330.

Proudfoot, A. E. I., & Wallace, C. J. A. (1987) *Biochem. J.* **248**, 965–967.

Proudfoot, A. E. I., Offord, R. E., Schmidt, M., & Wallace, C. J. A. (1984) *Biochem. J.* **221**, 325–331.

Proudfoot, A. E. I., Wallace, C. J. A., Harris, D. E., & Offord, R. E. (1986) *Biochem. J.* **239**, 333–337.

Proudfoot, A. E. I., Rose, K., & Wallace, C. J. A. (1989) *J. Biol. Chem.* **254**, 8764–8770.

Raphael, A. L., & Gray, H. B. (1989) *Proteins: Struct., Funct., Genet.* **6**, 338–340.

Raphael, A. L., & Gray, H. B. (1991) *J. Am. Chem. Soc.* **113**, 1038–1040.

Rees, A. R., & Offord, R. E. (1976a) *Biochem. J.* **159**, 487–493.

Rees, A. R., & Offord, R. E. (1976b) *Biochem. J.* **159**, 467–486.

Rees, D. C. (1980) *J. Mol. Biol.* **141**, 323–326.

Richards, F. M. (1958) *Proc. Natl. Acad. Sci. USA* **44**, 162–166.

Rose, K., Herrero, C., Proudfoot, A. E. I., Offord, R. E., & Wallace, C. J. A. (1988) *Biochem. J.* **249**, 83–88.

Rose, K., Herrero, C., Proudfoot, A. E. I., Wallace, C. J. A., & Offord, R. E. (1987) *Proc. Eur. Pept. Symp.* **19**, 219–222.

Salemme, F. R. (1977) *Annu. Rev. Biochem.* **46**, 299–329.

Salemme, F. R., Kraut, J., & Kamen, M. D. (1973) *J. Biol. Chem.* **248**, 7701–7716.

Sano, S. (1972) in *Structure and Function of Oxidation–Reduction Enzymes* (Akeson, A., & Ehrenberg, A., eds.), pp. 35–45, Pergamon, Oxford.

Sano, S., & Tanaka, K. (1964) *J Biol. Chem.* **239**, 3109–3110.

Shaw, W. V. (1987) *Biochem. J.* **246**, 1–17.

Shoelson, S. E., Polonsky, K. S., Zeidler, A., Rubenstein, A. H., & Tager, A. S. (1984) *J. Clin. Invest.* **73**, 1351–1358.

Skov, K., & Williams, G. R. (1971) *Can. J. Biochem.* **49**, 441–447.

Smith, H. T., & Millett, F. (1980) *Biochemistry* **19**, 1117–1120.

Stellwagen, E. (1978) *Nature* **275**, 73–74.

Takano, T., & Dickerson, R. E. (1981) *J. Mol. Biol.* **153**, 95–115.

Taniuchi, H., Basile, G., Taniuchi, M., & Veloso, D. (1983) *J. Biol. Chem.* **258**, 10963–10966.

Ten Kortenaar, P. B. W., Tesser, G. I., & Nivard, R. J. F. (1983a) *Proc. Eur. Pept. Symp.* **17**, 349–352.

Ten Kortenaar, P. B. W., Tesser, G. I., & Nivard, R. J. F. (1983b) *Proc. Amer. Pept. Symp.* **7**, 211–214.

Ten Kortenaar, P. B. W., Adams, P. J. H. M., & Tesser, G. I. (1985) *Proc. Natl. Acad. Sci. USA* **82**, 8279–8282.

Tesser, G. I., Boon, P. J., & Nivard, R. J. F. (1979) *Proc. Eur. Pept. Symp.* **15**, 671–675.

Tesser, G. I., Ten Kortenaar, P. B. W., & Boots, H. A. (1985) *Proc. Eur. Pept. Symp.* **18**, 221–224.

Theorell, H., & Åkesson, A. (1941) *J. Am. Chem. Soc.* **63**, 1804–1820.

Veloso, D., Basile, G., & Taniuchi, H. (1981) *J. Biol. Chem.* **256**, 8646–8651.

Veloso, D., Juillerat, M., & Taniuchi, H. (1984) *J. Biol. Chem.* **259**, 6067–6073.

Vita, C., Gozzini, L., & Di Bello, C. (1991) *Proc. Eur. Pept. Symp.* **21**, 255–256.

Vita, C., Gozzini, L., & Di Bello, C. (1992) *Eur. J. Biochem.* **204**, 631–640.

Wallace, C. J. A. (1976) D. Phil. Thesis, University of Oxford.

Wallace, C. J. A. (1978) in *Semisynthetic Peptides and Proteins* (Offord, R. E., & Di Bello, C., eds.), pp. 101–114, Academic Press, London.

Wallace, C. J. A. (1979) *Proc. Amer. Pept. Symp.* **6**, 609–612.

Wallace, C. J. A. (1984a) *Biochem. J.* **217**, 595–599.

Wallace, C. J. A. (1984b) *Biochem. J.* **217**, 601–604.

Wallace, C. J. A. (1986) in *Proceedings of the First Forum on Peptides* (Castro, B., & Martinez, J., eds.), pp. 434–438, Centre de Pharmacologie-Endocrinologie, Montpellier.

Wallace, C. J. A. (1987) *J. Biol. Chem.* **262**, 16767–16770.

Wallace, C. J. A. (1991) *Proc. Eur. Pept. Symp.* **21**, 260–263.

Wallace, C. J. A., & Boulter, D. (1988) *Phytochemistry* **27**, 1947–1950.

Wallace, C. J. A., & Campbell, L. A. (1990) *Proc. Am. Pept. Symp.* **11**, 1043–1045.

Wallace, C. J. A., & Clark-Lewis, I. (1992) *J. Biol. Chem.* **267**, 3852–3861.

Wallace, C. J. A., & Corthésy, B. E. (1986) *Protein Engineering* **1**, 23–27.

Wallace, C. J. A., & Corthésy, B. E. (1987) *Eur. J. Biochem.* **170**, 293–297.

Wallace, C. J. A., & Harris, D. E. (1984) *Biochem. J.* **217**, 589–594.

Wallace, C. J. A., & Offord, R. E. (1979) *Biochem. J.* **179**, 169–182.

Wallace, C. J. A., & Proudfoot, A. E. I. (1987) *Biochem. J.* **245**, 773–779.

Wallace, C. J. A., & Rose, K. (1983) *Biochem. J.* **215**, 651–656.

Wallace, C. J. A., Corradin, G., Marchiori, F., & Borin, G. (1986) *Biopolymers* **24**, 2121–2132.

Wallace, C. J. A., Mascagni, P., Chait., B. T., Collawn, J. F., Paterson, Y., Proudfoot, A. E. I., & Kent, S. B. H. (1989a) *J. Biol. Chem.* **264**, 15199–15209.

Wallace, C. J. A., Proudfoot, A. E. I., Mascagni, P., & Kent, S. B. H. (1989b) *Proc. Eur. Pept. Symp.* **20**, 283–285.

Wallace, C. J. A., Guillemette, J. G., Hibiya, Y., & Smith, M. (1991) *J. Biol. Chem.* **266**, 21355–21357.

Wallace, C. J. A., Guillemette, J. G., Smith, M., & Hibiya, Y. (1992) in *Techniques in Protein Chemistry III* (Angeletti, R. H., ed.), pp. 209–217, Academic Press, New York.

Wang, J., Wallace, C. J. A., Clark-Lewis, I., and Caffrey, M. (1994) *J. Mol. Biol.* **237**, 1–4.

Wendeloski, J. J., Mathew, J. B., Weber, P. C., & Salemme, F. R. (1987) *Science* **238**, 794–797.

Westerhuis, L. W., Tesser, G. I., & Nivard, R. J. F. (1979) *Rec. Trav. Chim. Pays-Bas* **98**, 109–112.

Westerhuis, L. W., Tesser, G. I., & Nivard, R. J. F. (1982) *Int. J. Pept. Prot. Res.* **19**, 290–299.

Wilgus, H., & Stellwagen, E. (1974) *Proc. Natl. Acad. Sci. USA* **71**, 2892–2894.

Wilgus, H., Ranweiler, J. S., Wilson, G. S., & Stellwagen, E. (1978) *J. Biol. Chem.* **253**, 3265–3272.

Index

A

Absorption of cytochrome *c*, 695-nm
  circular dichroism, 613
  origin of, 613
  pH dependence of, 612
  p*K* of bleaching of, 611, 612

Acetic anhydride, in differential protection technique, 376–381

Ag electrode, cytochrome *c* adsorbed on, 300–306

Alcohol dehydrogenase-associated cytochrome *c*, 60

Algal cytochrome c_{550}, 54

Algal cytochrome c_{553}, 53

Alkaline transition
  conformational equilibrium, 618, 629
  kinetics
    activation energy, 626
    rate constants, 627
  NMR studies of, 235–236, 270
  in semisynthetic analogs, 720
  temperature dependence, 624, 630
  thermodynamic parameters, 624, 625, 626, 632
  trigger
    arginine, 619
    heme propionate, 620
    histidine, 620
    tyrosine, 619
    water, 622, 623

Alkylation of Met-80 of cytochrome *c*, 340, 342

Amide protons, NMR of
  model for exchange of, 240–241, 245

Amino acid sequence of cytochrome *c*
  chemical modification, 15
  electrostatic charge of the protein, 16
  evolutionary studies, 14, 15
  mendelian inheritance, 15

Antibody
  antigen interaction with
    biochemical techniques by, 408–411
    conformational changes in, 405–407

Antibody, antigen interaction with (*continued*)
  conformational requirements for, 418–421
  H-D exchange and 2-D NMR by, 411–416
  limitations in, 416–418
  x-ray crystallography of, 404–407

binding to cytochrome *c*, NMR studies of, 265–266

complementary determining regions, 406

elbow bend, 407

monoclonal, 402
  to cytochrome *c*, 403
    from cow, 403
    differential protection of, 390
    from horse, 403
    from human, 403
    from *Paracoccus denitrificans*, 403
    from rat, 403
    from yeast (iso-1), 403
  to lysozyme, 402, 403, 405, 406, 407
  to neuraminidase, 405
  produced by splenic fragments, 403
    to horse cytochrome *c*, 404
    to rat cytochrome *c*, 404
  to staphylococcal nuclease, 403

serum, 399, 401
  to guanaco cytochrome *c*, 401
  to horse cytochrome *c*, 401
  to human cytochrome *c*, 401
  to mouse cytochrome *c*, 401, 404
  to rabbit cytochrome *c*, 401, 402, 404

Antigen
  cytochrome *c* as, 378
    from chicken, 400, 441
    from cow (bovine), 400, 425, 441
    from guanaco, 401, 441
    from horse, 400, 401, 404, 408–416, 441
    apo-, 425
    from human, 400, 401

Antigen, cytochrome *c* as (*continued*)
  from moth, 400, 425–430
  from mouse, 400, 401, 441
  from pigeon, 400, 425–430
  from rat, 400, 404
  evolutionary variability of 401, 403
  insulin, 398
  lysozyme, 398, 402, 403, 405, 406, 407, 424, 425
  myoglobin, 398, 424, 425
  presenting cells, 399, 421
  processing, 399, 424, 439, 440–441
    endogenous pathway of, 441
    exogenous pathway of, 441
  protein-engineered, 420
  surface mobility of, 402, 414, 415
  thermodynamics and, 416
    measured by isothermal titration calorimetry, 416

Antigenicity
  definition, 398
  surface mobility and, 402

apocytochrome *c*, NMR of interaction with sodium dodecyl sulfate, 268
  p*K* of histidines, 231–233, 268

"autoreduction" of ferricytochrome *c*, 23

Azotobacter vinelandii flavodoxin, 265

B

Bacillus cytochrome *c*, 61

Bacillus oxidase-associated cytochrome *c*, 61

Bacterial cytochrome *c* peroxidase, 72

B-cell
  activation, 399, 406–407
  tolerance, 400, 402

Bioconjugation, 593 (*see also* Chemical modification)
  attachment methods, 594
  with inorganic chromophores, 598
  selectivity, 594
  side-chains, heteroatom-containing, 595
  pK_a of, 595